Advances in Intelligent Systems and Computing

Volume 211

Series editor

Janusz Kacprzyk, Polish Academy of Sciences, Warsaw, Poland
e-mail: kacprzyk@ibspan.waw.pl

For further volumes:
http://www.springer.com/series/11156

About this Series

The series "Advances in Intelligent Systems and Computing" contains publications on theory, applications, and design methods of Intelligent Systems and Intelligent Computing. Virtually all disciplines such as engineering, natural sciences, computer and information science, ICT, economics, business, e-commerce, environment, healthcare, life science are covered. The list of topics spans all the areas of modern intelligent systems and computing.

The publications within "Advances in Intelligent Systems and Computing" are primarily textbooks and proceedings of important conferences, symposia and congresses. They cover significant recent developments in the field, both of a foundational and applicable character. An important characteristic feature of the series is the short publication time and world-wide distribution. This permits a rapid and broad dissemination of research results.

Bing-Yuan Cao · Hadi Nasseri
Editors

Fuzzy Information & Engineering and Operations Research & Management

Editors
Bing-Yuan Cao
School of Mathematics and Information Science, Guangzhou Higher Education Mega Center
Guangzhou University
Guangzhou
People's Republic of China

Hadi Nasseri
Department of Mathematics and Computer Sciences
Mazandaran University
Babolsar
Iran

ISSN 2194-5357 ISSN 2194-5365 (electronic)
ISBN 978-3-642-38666-4 ISBN 978-3-642-38667-1 (eBook)
DOI 10.1007/978-3-642-38667-1
Springer Heidelberg New York Dordrecht London

Library of Congress Control Number: 2013949431

Printed on acid-free paper

Springer is part of Springer Science+Business Media (www.springer.com)

Editorial Committee

Preface

This book is a monograph from submissions from the 6th International Conference on Fuzzy Information and Engineering (ICFIE2012) on October 25–26, 2012 in Babolsar, Iran and from the 6th academic conference from Fuzzy Information & Engineering Branch of Operation Research Society of China (FIEBORSC2012) during December 18–24, 2012 in Shenzhen, China. The monograph is published by *Advances in Intelligent Systems and Computing* (AISC), Springer, ISSN: 2194-5357.

This year, we have received more than 300 submissions. Each paper has undergone a rigorous review process. Only high-quality papers are included in it. The book, containing papers, is divided into Seven main parts:

Part I—themes on "Programming and Optimization".
Part II—subjects on "Lattice and Measures".
Part III—topics are discussed on "Algebras and Equation" appearance.
Part IV—ideas circle around "Forecasting, Clustering and Recognition".
Part V—focuses on "Systems and Algorithm".
Part VI—thesis on "Graph and Network".
Part VII—dissertations on "Others".

We appreciate the organizations sponsored by Mazandaran University, Iran; Fuzzy Information and Engineering Branch of China Operation Research Society; China Guangdong, Hong Kong and Macao Operations Research Society, and Guangdong Province Operations Research Society.

We are showing gratitude to the Iranian Fuzzy Systems Society; Iranian Operation Research Society; Fuzzy Information and Engineering Branch of International Institute of General Systems Studies in China (IIGSS-GB), and Guangzhou University for Co-sponsorships.

We wish to express our heart-felt appreciation to the Editorial Committee, reviewers, and our students. In particular, we are thankful to Doctoral: Ren-jie Hu, Hong Mai; Master: Zhi-ping Zhu, who have contributed a lot to the development of this issue. We appreciate all the authors and participants for their great contributions that made these conferences possible and all the hard work worthwhile. Meanwhile, we are thankful to China Education and Research Foundation; China Science and Education Publishing House, and China Charity Press Publishing and its president Mr. Dong-cai Lai for sponsoring.

Finally, we thank the publisher, Springer, for publishing the AISC (Notes: Our series of conference proceedings by Springer, like *Advances in Soft Computing* (ASC), AISC, (ASC 40, ASC 54, AISC 62, AISC 78, AISC 82 and AISC 147, have been included into EI and all are indexed by Thomson Reuters Conference Proceedings Citation Index (ISTP)), and thank the supports coming from international magazine *Fuzzy Information and Engineering* by Springer.

March, 2013

Bing-yuan Cao
Hadi Nasseri

Contents

Part II Lattice and Measures

Part III Algebras and Equation

Part VI Graph and Network

Part VII Others

Part I
Programming and Optimization

Fuzzy Modeling of Optimal Initial Drug Prescription

Mostafa Karimpour, Ali Vahidian Kamyad and Mohsen Foroughipour

Abstract This paper focused on a fuzzy approach in migraineurs drug prescription. There is no denying that medicine data records are mixed with uncertainty and probability so all methods concerning migraine drug prescription should be considered in fuzzy environment and designed according to the knowledge of an expert specialist. Overall it seems logical to propose fuzzy approach which could cover all uncertainties. According to fuzzy rule base concepts which are obtained by co-operation of an expert, fuzzy control has been used to model drug prescription system. Finally clinical experiences are used to confirm efficacy of drug prescription model. It should be considered that in most cases this disease may cause health social problems as well as financial obstacle for the companies and in upper stage for governments as a result of reduce work time among the employees. So prescribing optimal initial drug to make the disease stable is obligatory. With the corporation of experts 25 rule bases are introduced and tested on 50 different patient records, result shows that the accuracy of model is 94 % ± 5.5 with r-square equal to 0.9148.

Keywords Fuzzy control · Mamdani's inference approach · Migraine drug prescription

M. Karimpour · A. V. Kamyad
Department of Electrical Engineering, Ferdowsi University of Mashhad, Mashhad, Iran
e-mail: m_karimpour@yahoo.com

A. V. Kamyad
e-mail: avkamyad@yahoo.com

M. Foroughipour (✉)
Department of Neurology, Mashhad University of Medical sciences, Mashhad, Iran
e-mail: ForoughipourM@mums.ac.ir

B.-Y. Cao and H. Nasseri (eds.), *Fuzzy Information & Engineering and Operations Research & Management*, Advances in Intelligent Systems and Computing 211,
DOI: 10.1007/978-3-642-38667-1_1, © Springer-Verlag Berlin Heidelberg 2014

1 Introduction

Migraine is a French term derived from Hemi crania, a Greek word that means one side of the head [1]. But in practical migraine is a common type of headache usually felt as a severe unilateral throbbing headache that lasts 4–72 h and sometimes accompanied with nausea, vomiting and other symptoms [2]. Approximately 13 billion dollars in USA is lost as a result of reduced work productivity in migrainures and nearly 27 billion euros in European community yearly [3]. Research shows that only 56 % of people suffering migraine know that they have migraine [4].

International classification of headache disorders 2nd (ICHD) presented all types of headaches and migraines with different criteria for diagnosing each one [5]. Medical diagnosing could be done in a variety of classification methods to name a few one could refer to the following: artificial neural networks, fuzzy control, support vector machines, etc. Previously some research is done in migraine diagnosing and migraine state to state rate of transition with drug. Mendes et al. [6] used artificial neural networks to classify headache into four different classes as tension type headache, medication-overuse, migraine with aura and migraine without aura. Simić et al. [3] Classified different types of migraine by employing rule-based fuzzy logic, which is suitable for knowledge-based decision making.

Maas et al. [7] proposed markov models as a powerful way of nonlinear system modeling in migraine drug response modeling. To get more insight into the concentration-effect markov model is proposed to describe the course of headache response to either placebo or sumatriptan. Figure 1 shows the structure of the markov response model.

As it could be seen from Fig. 4 The model is consist of two layers: first, state layer (hidden layer) which represents three different states of migraine (no relief, relief, and pain free). Second score layer (open layer) which shows the headache score (no pain, mild pain, moderate pain and severe pain).

The mentioned model is able to predict headache states and headache recurrence in migraine patients who receive placebo or oral sumatriptan. Scientist showed that sumtriptans are very effective in reducing migraine attacks. Oral doses of 50 or

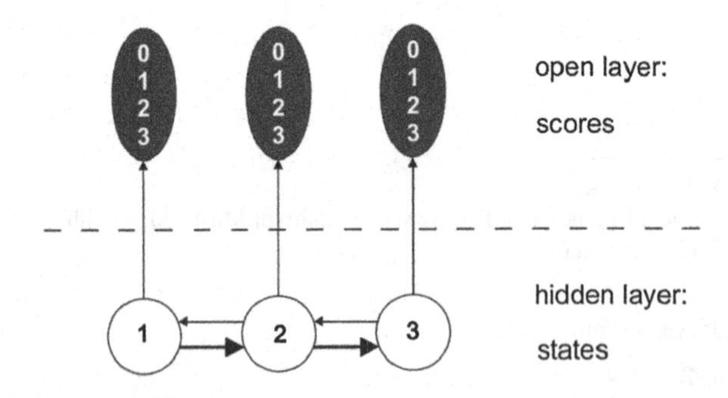

Fig. 1 Structure of a markov model

100 mg are usually prescribed to treat migraine attacks. The impression of sumatriptan on forward transition rates was expected to follow (1).

$$Rate(t)_{x,y} = Rate(0)_{x,y} \cdot exp\left(\frac{Emax_{x,y} \cdot C(t)}{EC50_{x,y} + C(t)}\right) \quad (1)$$

In Eq. 3 $Rate(t)_{x,y}$ shows the rate of drug-induced in forward transitions, $Rate(0)_{x,y}$ is the transmission rate from state x to y in the absence of sumtriptan. $C(t)$ is the sumtriptan plasma concentration. $Emax_{x,y}$ reveals the maximum effect of sumatriptan on forward transition rate. $EC50_{x,y}$ is the sumatriptan concentration related to half of the maximum effect. Anisimov et al. [8] showed that if x>y

$$Rate(t)_{x,y} = Rate(0)_{x,y} \quad (2)$$

Equations (1) and (2) were chosen based on the fact that the increase dose of sumtriptan drug leads to the headache pain decrease. In this theory this observation can be modeled as (a) Increasing forward transition rate is a function of drug concentration. (b) Reducing backward transition rate is a function of drug concentration.

In migraine, headache severity depends on age so the effectiveness of medication may also depend on age. Maas et al. [9] parameterized the interaction between age and drug exposure. The proposed model also reveals the clinical observations such as: (a) The rate which changes state pain relief to pain-free relates to age inversely. (b) In placebo-treatment, the mean transition time from 'no relief' to 'relief' is 3 h for young adolescents and 6 h for patients older than 30 years old. (c) Sumatriptan reduces the transition time to 2 h, without considering age. In this case by considering age Eq. (1) could be improved to Eq. (3).

$$Rate(t)_{x,y} = \mathrm{Rate}_{\min} + \frac{Rate_{\max} - Rate_{\min}}{1 + f(age, C(t))}$$
$$f(age, C(t)) = E_0 \cdot \exp\left(\frac{E_{\max_{age}} \cdot \exp(age))}{\exp(E50_{age} + \exp(age)} - \frac{E_{\max_{C(t)}} \cdot \exp(C(t))}{\exp(E50_{C(t)}) + \exp(C(t))}\right) \quad (3)$$

$\mathrm{Rate}_{\min}$ and $\mathrm{Rate}_{\max}$ represents the minimum and maximum values of transition rate. f is the function that describes the association between patient age and sumatriptan concentration (c(t)) to the transition rate. At t=0 $\exp(C(t))$ must be replaced with zero.

Gradual titration of drugs used for treatment of migraine (like sodium valproate) takes a long time to reach the best dosage and also starting with high range dosages increases the side effect of the drug so choosing the best initial dosage plays an important role in migraine treatment. In this paper a novel fuzzy modelling method is proposed in which the best dosage of initial drug could be detected.

2 Methods and Materials

Sodium valproate is the sodium salt of valproic acid. Sodium valproate can be used to control migraine. In migraine sodium valproate drug prescription depends on different factors but the most important factors are 1- pain intensity, 2- pain frequency and 3- weight. So in this paper fuzzy rule bases are proposed through the help of an expert specialist to find optimal initial sodium valproate drug prescription in migraineurs to cure disease in the minimum time by considering drug side effects as model constraints. A fuzzy questionnaire should be complete with these parameters through the patients. This study is done in neurology ward of Ghaem hospital in mashhad from Mar 2011 to Aug 2012.

The related questions are presented in Table 1.

There are some basic hypothesis used on this research which should be considered in sodium valproate prescription system, presented as follows:

1. Patient should be diagnosed and detected as migrainures.
2. There are some restrictions such as low blood pressure, depression, drugs side effects which is better for the patient to use sodium valproate.

Membership function of pain intensity and pain frequency are proposed as follows:

1. Pain intensity which is measured through using virtual analogue scaling (VAS), can be described well with the following membership function shown in Fig. 2.

Table 1 Questioner

1- The intensity of pain
The answer referring to this question could be in the form of an integer number between 1 (low pain) to 10 (The maximum imaginable pain) for parameters
2- Frequency of pain during one month. An integer number between 0 and 10 (upper than 10 is considered as 10)
3- Patient's weight (kg)

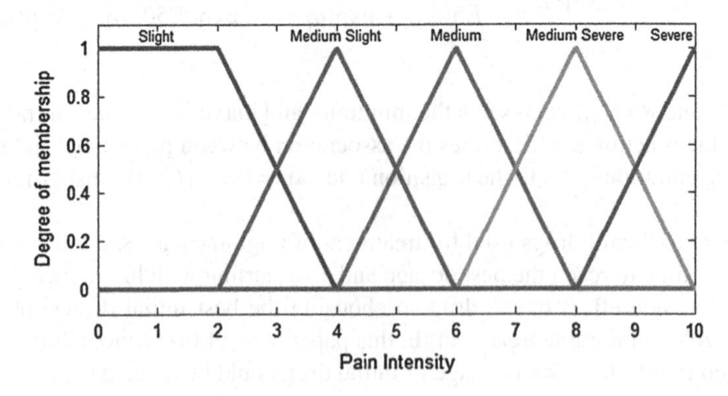

Fig. 2 Membership function of pain intensity

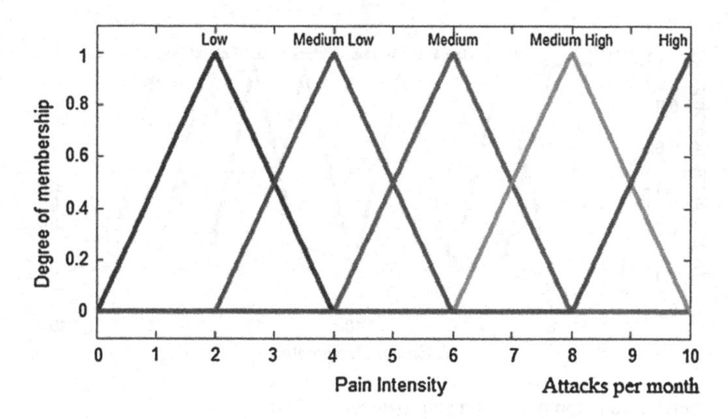

Fig. 3 Membership function of attack frequency

At the first part as the pain intensity is not detectable correctly and declared with the patient's feeling, so it is considered as trapezoidal membership function for slight in our approach, and a triangular membership function for the rest. It is divided into five parts 'Slight, Medium Slight, Medium, Medium Severe, Severe' to get the optimal decision making.

2. The frequency of migraine attacks which is presented in the form of number of attack per month is well presented with the chart provided in Fig. 3.

Pain frequency is completely measurable so all the membership functions are considered as triangular. Pain frequency is also divided into five classes 'Low, Medium Low, Medium, Medium High, High' which is really helpful in final decision making. It should be considered that increasing these classes increase mathematical computation and curse of dimensionality. Attack frequency more than 10 times per month, is considered as 'High'.

The consequent part of each rule is the membership function of sodium valproate dosage, the membership function is considered as Fig. 4.

Mamdani's fuzzy inference system is employed in sodium valproate optimal dosage calculation. As there are 2 different inputs with 5 different membership functions assumed, 25 different rule bases are introduced. Fuzzy rule bases are proposes through the consultation of expert specialists. The experimental results at the final part of this paper proof that the fuzzy control result with the comparison to reality (based on patient's records) is acceptable.

Fuzzy rule bases are presented in Table 2.

In real systems as the partition of uncertainty increases the efficacy of fuzzy logic rule bases increases. For complex systems usually knowledge exists and just few numerical data is available.

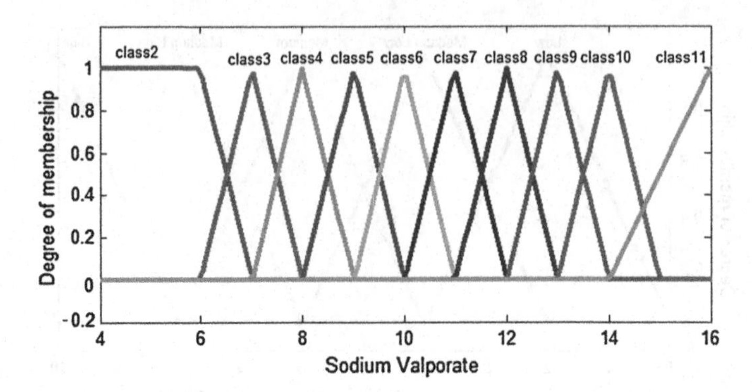

Fig. 4 Membership function of sodium valproate

Table 2 Rules developed for two inputs

Premises		Consequent
Paint intensity	Pain frequency	Sodium valproate
Severe	High	Class 11
Severe	Medium high	Class 10
Severe	Medium	Class 09
Severe	Medium low	Class 08
Severe	Low	Class 01
Medium severe	High	Class 11
Medium severe	Medium high	Class 10
Medium severe	Medium	Class 09
Medium severe	Medium low	Class 04
Medium severe	Low	Class 01
Medium	High	Class 10
Medium	Medium high	Class 08
Medium	Medium	Class 07
Medium	Medium low	Class 04
Medium	Low	Class 01
Medium slight	High	Class 04
Medium slight	Medium high	Class 03
Medium slight	Medium	Class 03
Medium slight	Medium low	Class 02
Medium slight	Low	Class 01
Slight	High	Class 01
Slight	Medium high	Class 01
Slight	Medium	Class 01
Slight	Medium low	Class 01
Slight	Low	Class 01

3 Results

Mamdani introduced a fuzzy inference system and developed a strategy which is usually referred to as max-min method. Mamdani's fuzzy inference system is a way of linking linguistic inputs to the linguistic outputs with just using min and max functions and allows gaining approximate reasoning [10]. Figure 5 represents model flow chart of our approach.

50 different patients who recursed to Ghaem hospital of Mashhad were inserted to the system. Data were gathered from Mar 2011 to Aug 2012 and were tested according to the rule bases presented in part 2. The experimental results are as presented in Table 3.

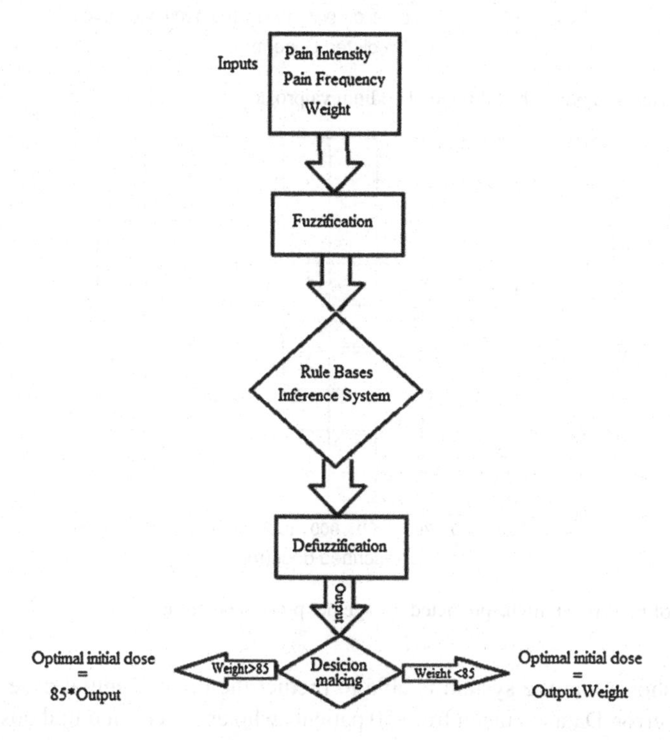

Fig. 5 Model block diagram

Table 3 Experimental results

Parameters	Test data
APE	5.9916
R square	0.9148
MSE	5229
RMSE	72.3561
Residual mean	44.6517

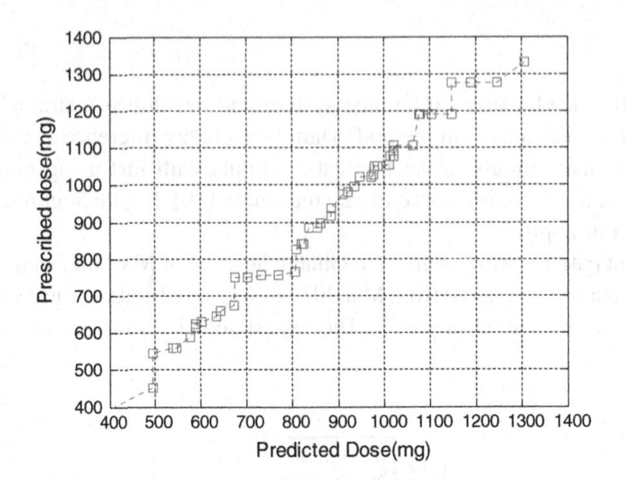

Fig. 6 Predicted and prescribed dosage of sodium valproate

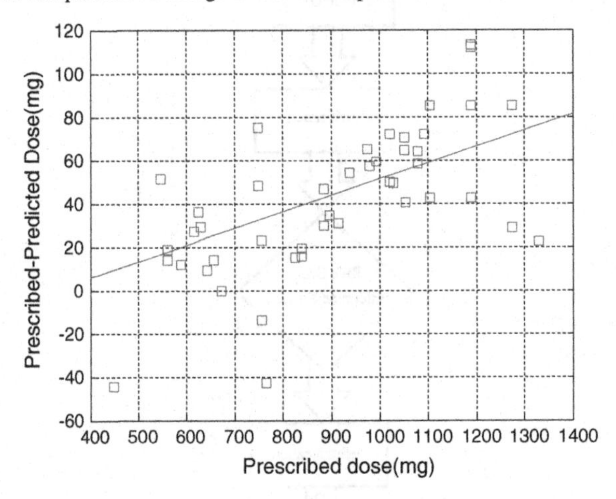

Fig. 7 Plot of prescribed minus predicted dosage per prescribed dosage

Result shows that the system is able to predict the optimal initial dose with an acceptable error. Data are taken from 50 patients who used certain initial dosage and pain frequency is reduced and intensity of their pain is relieved.

Figure 6 shows predicted and prescribed dosage of sodium valproate. Figure 7 presents plot of prescribed dosage minus predicted dosage per prescribed dosage of sodium valproate.

4 Conclusion

Defining the patterns of migraine acute treatment in the population is an important step in evaluating migraine treatment in relation to guiding principles, and improving health care [11].

Research shows that migraine headaches are not often treated effectively in many people who suffer from them. Preventive treatments are effective in about 38 % of people suffering migraine. Less than one-third of those people currently use these types of treatments. Fortunately, in many people the frequency and severity of migraine attacks which are the important factors of this study can be reduced with preventive treatment. In fact, some studies show that migraine attacks may be reduced more than half. Some epilepsy drugs are useful in preventing migraine. Strong evidence shows divalproex sodium, sodium valproate, and topiramate are helpful in preventing migraine [12].

Research shows that extended-release divalproex sodium 500–1,000 mg per day had an average reduction in 1 month migraine headache prevalence with a rate (−1.2 attacks per week) from 4.4 per week (baseline) to 3.2 per week. In most headache trials, patients taking divalproex sodium or sodium valproate reported no adverse events [13]. According to what has been mentioned one of the most effective and most widely used drugs in treating migraine is sodium valproate, but there are two factors that force the optimal usage of this drug: the side effect and the cost [14] but in Iran side effect is much more important because of the insurance cover on drug price.

Gradual titration of sodium valproate is one way of avoiding side effects but it takes a long time to reach the best dose. Recommended daily dosage ranges from 800 to 1,500 mg [15]. In Iran the start point could be 400 mg or lower [16].

In this paper for the first time a new method in optimal migraine drug prescription system is proposed. Mamdani fuzzy inference system is developed in predicting optimal sodium valproate initial dosage. Gradual titration of sodium valproate is one way of avoiding side effects but it takes a long time to reach the best dosage. So a mathematical model is proposed through the knowledge-driven from expert specialists. Simulation result indicates that model is able to predict sodium valproate optimal initial dosage with the accuracy of 94 % ±5.5.

Acknowledgments Center of Excellence in Modeling of Linear and Non-linear Systems (CRMCS) at Ferdowsi university of mashhad is appreciated for its partially financial support of this research.

References

1. Chawal, J., Blanda, M., Braswell.R., Lutsep.H.: Migrain Headache. E medscape (2011)
2. Wilson, JF.: In the clinic: Migraine. Ann. Int.. Med. **147**(9),11–16 (2007)
3. Simić.S., Simić.D., Slankamenac, P., and Simić-ivkov, M.: Rule-Based Fuzzy Logic System for, pp. 383–388, Springer, NY (2008)
4. Evans, R.: Migraine: A Question and Answer Review. Elsevier **93**(2), 245–262 (2009)

5. Society, H.: The International Classification Of Headache Disorders. Cephalagia, **24**, 1–160 (2004)
6. Mendes, K.B., Fiuza, R.M., Teresinha, M., Steiner, A.: Diagnosis of Headache using Artificial Neural Networks. J. Comput. Sci. **10**(7), 172–178 (2010)
7. Maas, H.J., Danhof, M.: Della Pasqua OE. Prediction of headache response after model. Cephalalgia **26**, 416–422 (2006)
8. Anisimov, V., Maas, H.: J., Danhof M, and Della Pasqua OE. Analysis of responses in migraine modelling using hidden Markov models. Statistics in medicine **26**(22), 4163–78 (2007)
9. Maas H.J., Danhof M., Della Pasqua O.E.: Model-based quantification of the relationship between age and anti-migraine therapy. Blackwell Publishing Ltd, Cephalalgia (2009)
10. Mlynek D., Leblebici Y.: Design of VLSI systems. Chapter9 (1998)
11. Bigal M.E., Borucho S., Serrano, D., Lipton R.B.: The acute treatment of episodic and chronic migraine in the USA. Cephalalgia (2009)
12. AAN Summary of Evidence-Based Guideline for Patients and their Families.: Prescription drug treatment for migraine prevention in adults. American Academy of Neurology (2012)
13. Silberstein S.D., Holland, S., Freitag, F., Dodick, D.W., Argoff C, Ashman, E.: Evidence-based guideline update: Pharmacologic treatment for episodic migraine prevention in adults. Nuerology **78**:1337–1345 (2012)
14. National Institute for Health and Clinical Excellence.: Drugs used in the management of bipolar disorder (mental health). Nice Clinical Guideline **38** (2006)
15. Mirjana, S., Miroslava, Z., Stevo, L.: Prophylactic treatment of migraine by valporate. Medicine and Biology. **10**(3), 106–110 (2003)
16. Afshari, D., Rafizadeh, S., Rezaei, M.: A Comparative Study of the Effects of Low-Dose Topiramate Versus Sodium Valproate in Migraine Prophylaxis. Int. J. Neurosci. **122**(2), 60–68 (2012)

Decision Parameter Optimization of Beam Pumping Unit Based on BP Networks Model

Xiao-hua Gu, Zhi-qiang Liao, Sheng Hu, Jun Yi and Tai-fu Li

Abstract Beam pumping unit is the most popular oil recovery equipment. One of the most common problems of beam pumping unit is its high energy consumption due to its low system efficiency. The main objective of this study is modeling and optimization a beam pumping unit using Artificial Neural Network (ANN). Among the various networks and architectures, multilayer feed-forward neural network with Back Propagation (BP) training algorithm was found as the best model for the plant. In the next step of study, optimization is performed to identify the sets of optimum operating parameters by Strength Pareto Evolutionary Algorithm-2 (SPEA2) strategy to maximize the oil yield as well as minimize the electric power consumption. Forty-nine sets of optimum conditions are found in our experiments.

Keywords Modeling · Artificial neural network · Optimization · Strength pareto evolutionary algorithm-2 · Beam pumping unit

1 Introduction

Beam pumping unit (BPU) is one of most important oil recovery equipments, the occupancy of which reaches to 70 % [1]. However, because of the negative torque, long gear train, poor working conditions and other reasons, the system efficiency of

X. Gu (✉) · T. Li · J. Yi
Department of Electrical and Information Engineering, Chongqing University of Science and Technology, Chongqing 401331, People's Republic of China
e-mail: xhgu@cqu.edu.cn

Z. Liao
College of Electronic Engineering, Xi'an Shiyou University, Xi'an 710065, People's Republic of China

S. Hu
School of Mathematics and Statistics, Chongqing University of Technology, Chongqing 400054, People's Republic of China

B.-Y. Cao and H. Nasseri (eds.), *Fuzzy Information & Engineering and Operations Research & Management*, Advances in Intelligent Systems and Computing 211, DOI: 10.1007/978-3-642-38667-1_2, © Springer-Verlag Berlin Heidelberg 2014

BPU is very low. So the research for energy saving of BPU is very important and necessary [1, 2].

Currently, methods of BPU energy saving are mainly mechanical ones and electrical ones [3]. In the first aspect, researchers focus on changing the structure of the pumping unit, adjusting the balance of the pumping unit and so on [4]. Whereas, electrical methods attempt to improve the motor control technology [5, 6]. In recent years, intelligent algorithm used to BPU energy saving is a new trend [7]. In this paper, we aim at building up an ANN for simulating the BPU, and then searching the optimum operation parameters based on the trained model.

The process of BPU is very complicated in nature typically due to unknown dynamic behaviors, nonlinear relations and numerous involved variables. Developing an accurate mathematical model are very hard even impossible. ANN modeling is a new choice to manage the complexities mentioned since it only requires the input-output data as opposed to a detailed knowledge of a system [8]. To identify optimum operating parameters, optimization based on ANN model is also performed. In this case, we consider to minimizing the electric power consumption and maximizing the oil yield, which leads to a multi-objective problem (MOP). SPEA2 [9], which has characteristics of stability, global search capability, nearest neighbor density estimation and new truncation method, is an ideal choice. Consequently, in this paper, BP networks is employed to model the BPU and SPEA2 is applied to the trained BP network model to determine the parameter values to energy saving and yield increasing.

2 Method Study

2.1 BP Networks

Multilayer feed-forward neural network with Back Propagation (BP) training algorithm is proposed by Rumelhart and Mcclalland [10] to solve the multi-layer network learning algorithm problem in 1986. It consists of input layer, hidden layer, and output layer. In BP networks training stage, the error between the experimental data and the corresponding predicted data is counted and back-propagated through the network. The algorithm adjusts the weights in each successive layer to reduce the error. This procedure is circulated until the error between the experimental data and the corresponding predicted data satiety certain error criterion. BPN models compute the output value as a sum of non-linear transformations of linear combinations of the inputs. The data predicted by BPN models were plotted against the corresponding experimental data to visualize the modeling abilities of the BPN models.

2.2 SPEA2

SPEA2 which proposed by Zitzler [9] in 2001 is a multi-objective evolutionary algorithm characterized by the concepts of strength and density. SPEA2 is an extension of SPEA, it has an improved fitness assignment strategy and hence can search the global optimum of all objective functions. Consequently, SPEA2 becomes one of the most popular optimization techniques in multi-objective combinatorial optimization problems, as well as nonlinear ones.

3 BPU Modeling and Optimization Based on BPN-SPEA2

3.1 BPU

Beam pumping unit production system includes two parts: the ground part and the underground part. The mainly ground devices include the dynamic force and the balance of beam-pumping unit, while the underground part includes the oil pump and the corresponding valve. With the in-depth study of the process, number of punching as the decision parameter is proposed, which updates the optimum number of punching that could response the change of status. Running in optimum number of punching always saves more energy. But just number of punching as input parameter to model of beam pumping unit production system is hardly approximation real model as the amount information is lacking. In order to solve this problem, the environment parameters are supplied as the input parameter to model. So the input parameters include number of punching(NP), maximum load(MAXL), minimum load(MINL), effective stroke(ES) and computational pump efficiency(CPE), while the output parameters are electric power consumption(EPC) and oil yield(OY). All the inputs and outputs of BP networks are show in Table 1.

3.2 BPU Modeling Based on BPN

BP networks algorithm is developed to model BPU. As shown in Fig. 1, the network is three-layer. The input neurons node in the network is five and output neurons node is two. Hidden layer's neuron number t is determined by an empirical formula $t = \sqrt{n+m} + \alpha$, where n is the number of input layer neurons, m is the number of output layer neurons, and α is a constant between 1 and 10.

The training was executed systematically with different number of nodes in hidden layer. Tan-sigmoid transfer function was used as the activation function for the hidden layers, and linear transfer function was used for the output layer. The values of the test data were normalized to within the range from −1 to 1.

Table 1 Input/output parameters of BP neural networks

Inputs		Outputs
Decision parameter	Number of punching (time/min)	Electric power consumption (kw/h)
	Maximum load (kN)	
Environment parameters	Minimum load (kN)	
	Effective stroke (m)	Oil yield (t/d)
	Computational pump efficiency (%)	

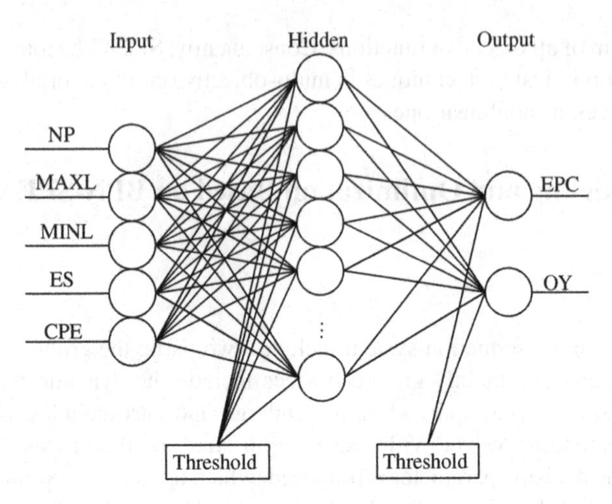

Fig. 1 BPN model of beam pumping unit production system

3.3 SPEA2 Optimization of Decision Parameter of BPU

Once the beam pumping unit production system process model based on BP networks is developed, it can be used to determine its fitness value for optimization to obtain the optimal values of the input variables that minimizing electric power consumption and maximizing oil yield. Considering that SPEA2 always searches the minimize fitness, hence the maximum objective is converted to minimum one by taking its negative. Finally, the beam pumping system energy saving multi-objective problem with 5 variables, 2 objectives is described as follow:

$$\hat{y} = \min F(\hat{x}) = \min(f_1(\hat{x}), f_2(\hat{x})) \tag{1}$$

In which, $\hat{x} = (x_1, x_2, \cdots, x_n) \in X$, $X = \{(x_1, x_2, \cdots, x_n) | l_i \leq x_i \leq u_i, i = 1, 2, \cdots 5\}$, $L = (l_1, l_2, \cdots, l_5)$, $\hat{y} = (y_1, y_2, \cdots, y_n) \in Y$, $U = (u_1, u_2, \cdots, u_5)$, where L is the lower boundary of optimal parameters and U is the upper boundary of optimal parameters.

Obviously, in above model, the electric power consumption and the oil yield are the two objectives of SPEA2, which means that the decision parameters are optimization to response the change status.

4 Experimental Results and Analysis

4.1 The Training and Simulation of BP Networks

To build up a BP network for simulation BPU, necessary data is provided. Among the available 3,234 data from DaGang oil field, three examples of the samples are

Table 2 Examples of samples

NO.	NP (time/min)	MAXL (kN)	MINL (kN)	ES (m)	CPE (%)	EPC (kw/h)	OY (t/d)
1	3.01	97.2	44.7	3.074646	69.26006	11.97	31.05339
2	2.99	98.4	43.5	3.131836	71.01529	12.44	31.48944
...	...	...	...	...	...	...	...
3,234	2.99	98.7	43.4	3.027187	69.71369	12.23	30.47277

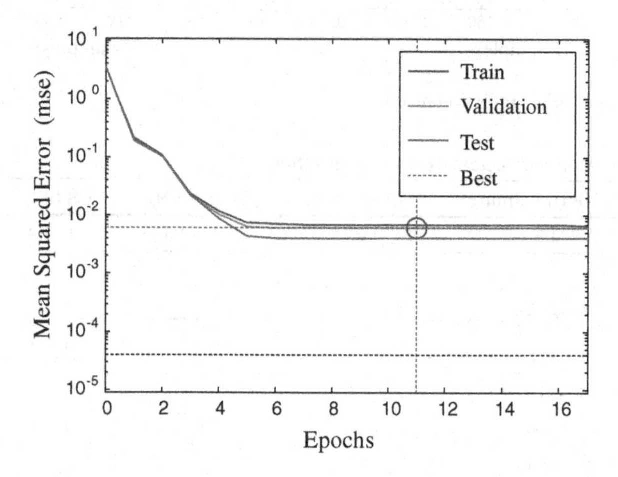

Fig. 2 The BP networks training process

shown in Table 2. The 3,234 samples are divided into two parts, 3,000 samples data are used to build up the model of BPU, while the other 234 are applied to test the generalization ability of trained model.

BP networks training was a supervisory learning process with cross validation. The Levenberg-Marquardt (LM) algorithm is repeatedly applied until the error threshold or the stop criterion is reached. Among many converged cases, the configuration 5-9-2 appeared the most optimal. Training started after setting BP network structure and parameters. When the BPN training precision meet the error limit expectation, as showed in Fig. 2, the training finished.

To evaluate the precision of the obtained model, the performance of the obtained model on the test set is tested. The comparisons of the predicted objectives and the real objectives are given in Fig. 3.

From the simulation results, we find that the predicted values and the real ones are very close. In other words the simulated values match well with the measured ones. It achieves a high prediction, and completely meets with the actual production needs. This confirms that the BPN based beam pumping process model is stable and reliable and could be regarded as a knowledge source for follow-up process parameters optimization.

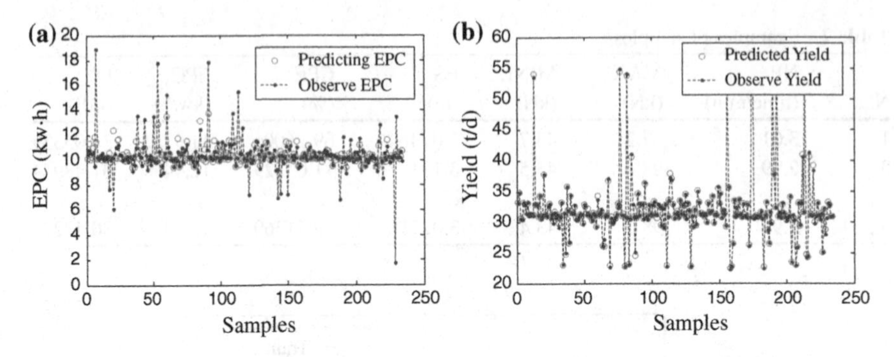

Fig. 3 The BP networks prediction results

Table 3 The limits of parameters used in optimization

Parameters	NP (time/min)	MAXL (kN)	MINL (kN)	ES (m)	CEP (%)
Limits	2.5–3.5	93.3–93.5	42.3–42.5	3.1–3.2	71–72

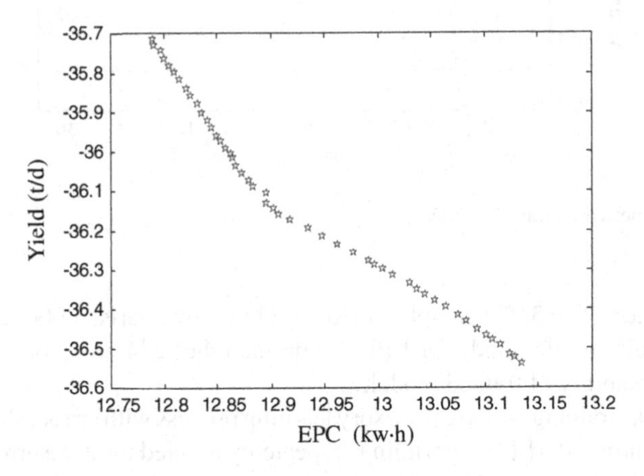

Fig. 4 The Pareto optimal front of SPEA2

4.2 Decision Parameter Optimization by SPEA2

Using the designed network, the effect of the operating parameters of BPU, like the number of punching, maximum load, minimum load, effective stroke, computational pump efficiency, on electric power consumption and oil yield are studied. Table 3 shows domain of change for input parameters.

In the SPEA2 algorithm, initial population, evolutionary generation, offspring individuals and parent individual are set to 80, 500, 40, and 40.The results of beam pumping unit Pareto frontier are shown in Fig. 4. Three instances of solutions of Pareto optimal set is shown in Table 4.

Table 4 The Pareto optimal front of SPEA2

NO.	NP	MAXL	MINL	ES	CPE	EPC	OY
1	3.5	93.4979	42.4996	3.10132	71.9995	12.90683	36.159
2	3.49999	93.3	42.4999	3.10004	71.7451	12.85295	35.973
...	...	...	...	...	...	...	...
49	3.49999	93.3104	42.4998	3.10047	71.7582	12.82161	35.740

As shown in Table 4, different value of number of punching from Pareto optimal set can be set according to corresponding status, the result shows that electric power consumption is decreased above 5 % and oil yield is increased above 6 %.

5 Conclusion

In this paper, a hybrid BPN-SPEA2 strategy is proposed to achieve the optimum decision parameters in the beam pumping unit production system for the aim of saving energy. BPN is applied to build up the model of beam pumping unit, and decision parameters are optimization by SPEA2 based on the trained BPN model. The experiments are conducted on 3,234 real samples from DaGang oil field. The results show that the system performance using the optimum parameters is improved significantly. Specifically the electric power consumption decreases more than 5 % and oil yield increases more than 6 %. It provided the proposed method is a alternative effective solution for energy saving of oil field.

Acknowledgments The authors would like to acknowledge DaGang oil field for providing industrial data. Thanks to the support by National Natural Science Foundation of China (No. 51075418 and No. 61174015), Chongqing Natural Science Foundation (cstc2013jcyA40044) and Project Foundation of Chongqing Municipal Education Committee (KJ121402).

References

1. Bai, L.P., Ma, W.Z., Yang, Y.: The Discussing of Energy-saving of Motor for Beam balanced Pump. Oil Machinery **27**(3), 41–44 (1999)
2. Su, D.S., Liu, X.G., Lu, W.X.: Overview of Energy-saving Mechanism of Beam-balanced Pump. Oil Machinery **19**(5), 49–63 (2001)
3. Xu, J., Li, J., Chen, J., Han, M., Han, X.: Research on Power Saving Positive Torque and Constant Power Pumping Unit and Tracking Technique System. Procedia Engineering **29**, 1034–1041 (2012)
4. Zhu, J., Ruan, J.Q., Sun, H.F.: Comparative Test Study of Energy-saving Pumping and Their Effect. Oil Field Equipment **35**(3), 60–62 (2006)
5. Gu, Y.H., Xiao, W.S., Zhou, X.X., Zhang, S.C., Jin, Y.H.: Full Scale Test of ZXCY-Series Linear Motor Pumping Units. Petroleum Exploration and Development **35**(3), 366–372 (2008)
6. Li W., Yin Q., Cao J., Li L.: The Optimization Calculation and Analysis of Energy-saving Motor used in Beam Pumping Unit based on Continuous Quantum Particle Swarm Optimization. Bio-Inspired Computing: Theories an Applications: 1–8 (2010)

7. Qi W., Zhu X., Zhang Y.: Study of Fuzzy Neural Network Control of Energy-saving of Oil Pump. Proceedings of the CSEE: 137–140 (2004)
8. Zahedi, G., Parvizian, F., Rahimi, M.R.: An Expert Model for Estimation of Distillation Sieve Tray Efficiency based on Artificial Neural Network Approach. Journal of Applied Science **10**(12), 1076–1082 (2010)
9. Zitzler E., Laumanns M., Thiele L.: SPEA2: Improving the strength Pareto evolutionary algorithm. (2001)
10. Rumelhart D. E., McClelland J. L.: Parallel Distributed Processing: Psychological and Biological Models vol. 2: The MIT Press (1986)

Optimization Strategy Based on Immune Mechanism for Controller Parameters

Xian-kun Tan, Chao Xiao and Ren-ming Deng

Abstract Control quality of the controller depends on correct tuning of control parameter, and it is directly related to the control effect of whole control system. Aimed at the puzzle that the parameter of controller has been difficult to tune, the paper proposed a sort of optimization model based on immune mechanism for tuning of controller parameters. Firstly it defined the antibody, antigen and affinity of tuning parameter, and secondly explored the process of parameter tuning based on immune mechanism in detail, then explained the tuning method by means of optimizing parameters of PID controller as well as the seven parameters of HSIC controller. Finally it took a high-order process control as the example, and made the simulation to a large time delay process, and to a highly non-minimum phase process as well as HSIC controller. The simulation experiment results demonstrated that it is better in comparison with some other tuning methods for dynamic and steady performance. The research result shows that the proposed method is more effective for controller parameter tuning.

Keywords Immune mechanism · Genetic algorithm · Clone selection · Mutation · Parameter tuning

X. Tan (✉)
School of Polytechnic, Chongqing Jiaotong University, Chongqing 400074, China
e-mail: sngeet@163.com

C. Xiao · R. Deng
College of Automation, Chongqing University, Chongqing 400030, China
e-mail: txkcx11@163.com

R. Deng
e-mail: dengrenming65106683@126.com

B.-Y. Cao and H. Nasseri (eds.), *Fuzzy Information & Engineering and Operations Research & Management*, Advances in Intelligent Systems and Computing 211,
DOI: 10.1007/978-3-642-38667-1_3,

1 Introduction

For recently several decade years,although lots of advanced control algorithms have been continually presented, but until nowadays, the conventional control algorithm of PID is always applied in process industrial control field widely, and also it has still taken the most part in the actual engineering applications. The key puzzle in actual engineering of industrial control fields is very difficult to tune the parameters of PID controller so as to influence the control quality being not so good. Therefore it is necessary to research the method of tuning parameter for PID controller. According to the partition of development stage, it can be divided as the routine and intelligent tuning method of PID parameters [1]. In terms of the partition of controlled object number, it is generally divided as single-variable and multi-variable tuning method of PID parameters. By means of combination form of control quantity, also it can be divided as linear and non-linear parameter tuning method of PID controller, the former is used to the classical PID controller, and the latter is used to the nonlinear PID controller created by combination mode of non-linear track-differentiator and non-linear combination. With the rapid development of intelligent control technology, in order to improve the performance of PID controller, it has been come forth to lots of intelligent PID controller. Those controllers fused the multi-aspect technologies such as fuzzy control [2], NN [3], ant algorithm [4], genetic arithmetic [5, 6] and so on. It made PID controller own the self-learning ability and therefore extended the application bound of PID controller. Based on the enlightening of biology immunity system, the paper proposed a sort of parameter tuning model of controller based on the immune mechanism.

2 Process of Parameter Tuning for Controller

Due to the classical control algorithm of PID being conventional tuning manner, in order to improve the system performance, for convenience, here it takes the PID parameter tuning as an example, and makes the anatomy explain why it is the puzzle. Therefore it is necessary to research the influence on control system characteristic for each parameter in the control algorithm of PID. The following would respectively explain the influence on dynamic and steady performance of the whole system for each node parameter of PID.

1) The action of proportional node is to reduce the deviation of system. If the proportional coefficient KP is enlarged then it would be expedited to the response speed, and reduced to the steady error and enhanced to the control precision for the whole system. But too large KP always results in overshoot, and leads to the system being instable. And if KP is too small then the overshoot of the whole system would be reduced so as to enlarge the steady-state abundant extent. But the system precision would be reduced, and it would make the system precision be reduced so as to result in time delay of transition process.

2) The action of integral node is to eliminate steady-state deviation, and to reduce the system error. But it would make that the system response speed becomes slow, the system overshoot gets large, and the system results in bringing oscillation. But it is propitious to reduce system error for increasing integral coefficient KI, however too strong integral action is able to make the system bring overshoot so as to bring system oscillation. Although reducing integral coefficient KI makes the system improve the stability, and avoids the system bringing oscillation, and reduces the system overshoot, but it is adverse for eliminating steady deviation.
3) The function of differential node can reflect the change trend of deviation signal. It is able to introduce a early modifying signal that it redounds to reduce system overshoot before deviation signal value going into too large. It redounds to reduce the system error, to overcome the oscillation, to make the system going to stable state fleetly, to enhance the system response speed, and to reduce the adjusting time, therefore the dynamic characteristic of system is improved. The disadvantages are poor in anti-jamming ability, and great in influence on process response. If the differential coefficient KD gets large then it is propitious to expedite the system response, to reduce the system overshoot, and to increase the system stability. But it would result in disturbance sensitive and weaken the restraining disturbance ability, and if the KD gets too large then it would result in response process advancing to apply the brake, and delay the adjusting time. And reversely, if the differential coefficient KD gets too small then the speed-down of system adjusting process would be delayed, and the system overshoot would be increased. It makes system response speed slow-down, and finally it would result in system stability being bad.

The above analysis shows that the tuning of PID parameter is very complicated, and even it is very difficult to make adjusting and controlling in the complicated system. It is the most difficult to select the three control parameters, KP, KI and KD correctly. Therefore it is necessary to seek the new method of parameter tuning so as to obtain the satisfactory control quality for whole system.

3 Description of Optimization for Control Parameter Tuning

Now consider a closed-loop negative feedback control system shown as in Fig. 1, in which, Gp(s) is the controlled object model, Gc(s) is the controller, r is the input of set value, e is the error signal, u is the controller output, y is the system output.

The aim of controller parameter tuning is to select a set of control parameter of PID controller for given controlled object by means of a sort of tuning method. Aimed at the given input, the tuning result of the system constituted by controlled object and controller is to make the one or several performance indexes obtain optimization under the condition of certain criteria. Therefore there are two main problems needed solving in optimization design of the controller, they are the selection of performance indexes and the choice of tuning method for seeking optimization parameters.

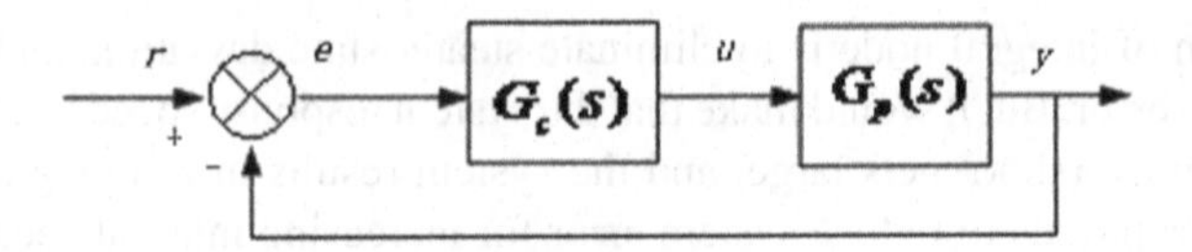

Fig. 1 Control system with closed-loop negative feedback

There are three aspects of indexes to weigh the control system performance. They are respectively the stability, precision and speediness. For example, the rising time reflects the system speediness, the shorter the rising time is, the sooner the control process response is, and the better the system control quality is.

If it only hankers for dynamic characteristic of system simply then it is very possible to make the gained parameter be too large for control signal. And therefore it would result in system being instable because of connatural saturation character in the system for actual engineering application. In order to get better control effect, it proposes that the control quantity, system error and rising time should have certain constraint condition. Because the accommodating function is related to the objective function, after determining the objective function, it could directly be as accommodating function to make the parameter seek the optimization value. The optimal parameter is the control parameter that is corresponding to x under the condition of satisfying constraints, and it makes the function f(x) to reach the extremum.

In order to obtain the satisfactory dynamic character of transition process, it adopted the performance index of error absolute time integral to be as selection parameter for least objective function. To avoid the control energy to be too large, it added a square item of control input in the objective function. The formula (1) is chosen as the optimal index of parameter selection.

$$J = \int_0^{\infty} (\omega_1 |e(t)| + \omega_2 u^2(t))dt + \omega_3 \times t_u \tag{1}$$

In which, e(t) is the system error, u(t) is the control output, t_u is the rising time, ω_1, ω_2 and ω_3 is respectively the different weighting value.

To avoid the system bringing overshoot, the castigation function is adopted. Namely, once the overshoot happens, the overshoot would be added an item of optimal performance indexes, here the optimal performance index would be expressed as formula (2).

$$\textit{If } e(t) < 0, \quad J = \int_0^{\infty} (\omega_1 |e(t)| + \omega_2 u^2(t) + \omega_4 |e(t)|)dt + \omega_3 \times t_u \tag{2}$$

In which, ω_1, ω_2 and ω_3 and ω_4 is respectively the weight value, and $\omega_4 \gg \omega_1$.

Table 1 Mapping between human-body immune and immune based tuning model

Human-body immune system	Model of parameter tuning
Antigen	Optimal solution
Antibody	Feasible solution
Cell clone	Antibody copy
Binding of antibody and antigen	Value of antibody replacing the antigen
Cell B, Cell T	Vector
Increase of antibody density	Increase of approximate feasible solution

4 Design on Model Algorithm

In this paper, the proposed self-tuning algorithm of control parameter based on immune mechanism is similar to the working process of human body immunity system. The relationship between the human body immunity system and the proposed method of parameter tuning model is shown as in Table 1.

In the candidate solution generated by random, through the affinity computing of antigen and antibody, the superiority antibody is voted in, they hold the superiority gene, and make the mutation in the clone process and therefore lots of new antibodies are produced. Then it can reappraise for new antibody set and the new brought antibody also renews the antibody set. The updating mechanism of antibody is that the newly brought antibody, owned higher affinity, washes out the antibody that the affinity is low in the antibody. The antibody in the set is ranked according to the sort ascending of affinity. Through the specified evolution generation, the optimal antibody can be distilled. Therefore the optimal solution is obtained.

4.1 Basic Conception

In order to solve the problems, it is necessary to find the optimal solution of the problem needed by solving. The optimal solution is abstracted as the antigen, and the feasible solution of the problem is abstracted as an antibody that represents a candidate solution of the problem. For convenience to discuss the problem, here some definitions would be given as the following.

Definition 1. The antigen is specified as the optimized objective function.

Definition 2. The antibody is specified as the candidate solution of objective function. In the real number encoding, usually the antibody is a multi-dimension vector, $X = vX1, X2, ...XN)$, each antibody is represented by a point in n-dimension space.

Definition 3. The affinity of antibody-antigen is the value after computing antibody to replace antigen (optimized objective function).

4.2 Model of Clone Selection

In the human body immune system, when the antigen is in inbreak for human body, the immune system of human body would bring lots of antibody to match the antigen. Meanwhile, the antibody density, which affinity with antigen is large, would get higher, and it is propitious to eliminate the antigen. When the antigen dies out, this kind of antibody reproduction would be restrained. And at the same time, the antibody density would be reduced so as to make the immune system keep the balance all the time. In the antibody set, the superiority antibody, which is large in affinity with antigen, is activated. In order to eliminate antigen carrying through large number of clone, the encoding of clone selection process is designed as the following.

```
Procedure CloneSelect ( )
Begin
Assume antigen ag; /* s.t. min(J) */
Antibody set of random initialization;
While (evolution generation < m) / *m is the evolution generation * /
Begin
Computing f affinity (ag, ab) for each antibody until condition end,
Carrying through sort ascending array according to the affinity value
Selecting the front θ antibody brings new antibody ab new according to
the f num (ab(i))
Carrying through mutation for new brought antibody ab new
Joining N new antibody of random bringing into ab new
While (ab new non-empty)
Begin
Selecting the least affinity cell from the antibody set;
If (the affinity of selected cell > f affinity ( ag, ab new )
Then make replacing using new antibody;
    End
  End
End
Output the optimal solution from antibody set;
End
```

To make the new producing cell (come from clone selection process [7]) join to antibody set, the antibody density would be increased. It shows that the amount of approximate solution is increasing. But if this kind of antibody is excessively centralized then it is very difficult to keep antibody diversity. And the antibody owned better evolution potential would be lost, therefore it is able to go into the local optimization [8].

To avoid getting in local optimization, it is restricted to the amount of cloned memory cell in the clone selection process, shown as in formula (3).

$$f_{num} = \sum_{i=1}^{q} \lceil \frac{\beta \cdot \theta}{i} \rceil \tag{3}$$

In which, f num is the total clone number, the *ith* item represents the clone number of *ith* cell, β is a parameter factor of pre-enactment, θ is the amount of superiority cell. It can know from formula (3), the large the cell affinity is, also the more the clone amount is, and contrarily it would be less. For instance, β= 2, θ= 100, because the affinity magnitude of cloned memory cell is to arrange according to the sequence, therefore the cloned amount of maximum antibody cell of the affinity is 200, the cloned number is 100, the rest may be deduced by analogy method. Meanwhile in order to prevent getting in local optimization, it introduces certain new antibody produced by random in each generation evolution process.

The antibody cell updates dynamically in the evolutionary process. Each antibody cell always selects optimal antibody from the current antibody and new cloned antibody. Accordingly the dynamic update of antibody cell set is realized, and the antibody scale keeps the steady-state balance.

4.3 Selection for Mutation

The objective of mutation is to make change for encoding of filial generation antibody so as to obtain the better solution than the father generation. Because the antibody adopts real number encoding, so the mode of Gaussian mutation is used in the algorithm. And also the mutation does not be acted on barbarism species. In order to centralize search around high affinity antibody, and to guarantee the antibody diversity, it is introduced to a sort of self-adaptive mutation, namely it acts on individual component for each mutation operator shown as in formula (4).

$$X_i = |x_i + N_m * N(0,1) * x_i| \tag{4}$$

In which, N(0,1) is a random number subjected to the standard Gaussian distribution, $|\cdot|$ is to find absolute value because of control parameter being not able to be negative, Nm is the mutation rate corresponding to the antibody, it is determined by formula (5).

$$Nm_i = \rho \frac{f(x_i)}{max(f(x_i))} \tag{5}$$

Obviously the mutation rate of the antibody is inversely proportional to its affinity, the higher the affinity is, and the smaller the mutation rate is. The antibody adjusts adaptively the mutation step-length in terms of affinity magnitude for each iterative process. It makes search to be centralized around the high affinity antibody so as to enhance the convergent speed and also it keeps the species diversity. The is the mutation constant used as adjusting the mutation intensity, it is related to the search space size and species scale.

5 Simulation and Its Analysis

For the above algorithm, here we can take the parameter tuning of human simulated intelligent controller (HSIC) as an examples to validate its correctness. For validating the performance of algorithm proposed in this paper, the controlled object [6] is selected as high order process, its transfer function is expressed by formula (7).

$$G_1(s) = 1/(1+s)8 \tag{6}$$

By means of the proposed method as mentioned above, it can be used as the tuning of HSIC algorithm [9]. The control algorithm is expressed as the formula (7).

$$\mu_{u_n} = \begin{cases} sgn(e_n)U_{max} & (e,\dot{e})\epsilon\Phi_1 \\ K_p * e_n & (e,\dot{e})\epsilon\Phi_2 \\ K_p * \dot{e}_n + K_d * \dot{e}_n & (e,\dot{e})\epsilon\Phi_3 \cup \Phi_6 \\ -K_p^{'} * \dot{e}_n + K_{d'} * \dot{e}_n & (e,\dot{e})\epsilon\Phi_4 \\ K_p * \dot{e}_n + K_i * \int e & (e,\dot{e})\epsilon\Phi_5 \end{cases} \tag{7}$$

where:
u_n : The *N*th output of the controller,
e_n : The *N*th deviation,
$\dot{e}_n$: The *N*th deviation change rate,
$K_p, K_{p'}$: Proportional coefficient,
$K_d, K_{d'}$: Differential coefficient,
K_i : Integral coefficient,
U_{max} : The maximum output of the controller,
e_1, e_2 : The threshold of the deviation,
$\dot{e}_1$: The threshold of the deviation change rate

From the formula (7), it can be seen that there are seven parameters needed to be tuned, they are respectively, K_p, K_p', K_d, K_d', K_i and e_1, e_2. It is very difficult to tune the seven parameters of HSIC. In this paper, it selects a high order process as controlled plant [10] its transfer function is given in formula (6).

Firstly, the parameter needed to be optimized should be encoded, in the paper, real value encoding is adopted. Encoding K_p, K_p', K_d, K_d', K_i and $e_1, e_2, \dot{e}_1$ as an antibody, the antibody evolution generation is 100, population size is 50 the initial range of parameter, K_p and K_p' are over interval [0, 30], K_d, K_d' and K are over interval [0, 5], $e_1, e_2, \dot{e}_1$ are over interval [0, 1]. The others are respectively $\omega_1 = 0.999$, $\omega_2 = 0.001$, $\omega_4 = 100$, $\omega_3 = 2.0$, $\theta = 20$, $\beta = 5$, To increase global search ability, the five antibodies are added in each generation.

The paper realized the simulation of optimization control in MATLAB by means of optimization strategy based on immune mechanism, and we compared it with Chien-Hrones-Reswick (CHR) , Refined Ziegler Nichols (RZN) and Genetic Programming with ZN(GP) to tune controllers their control parameters come from Ref. [6].

Table 2 Parameters and objective values obtained of G1(s)

Parameters and variables	IHSIC	GP	RZN	CHR
KP		0.68	0.35	1.48
Ti	*	4.67	4.53	9.06
Td		1.47	1.14	2.02
ess (%)	0.00	0.00	0.00	0.00
mp (%)	0.00	4.40	0.00	48.54
ts (%)	13.59	10.02	30.04	23.22

* 0.43, 27.56, 1.59, 11.27, 0.01, 0.27, 0.25, 0.50

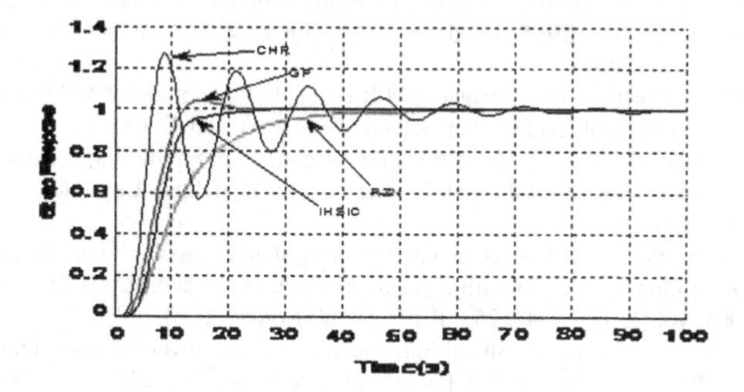

Fig. 2 Comparison curve for different algorithm

Table 2 is corresponding to the formula (7), for convenience contrast to the PID, the parameters in the Table 2 is still to use KP, Ti, Td to express, but the parameters of HSIC is expressed as "*", the actual value is under Table 2, they are respectively the above seven parameters. In the Fig. 2 and Table 2, where ess represents steady state error, mp represents overshoot, and ts is the settling time. From Fig. 2, it can be concluded that IHSIC adjusting time is much better than CHR and RZN tuning PID control method, and the overshoot is obvious less than CHR. Compare with GP tuning PID, although the adjusting time is less than GP, but IHSIC has no overshoot.

6 Conclusion

Based on the clone selection mechanism of biology immune system, the paper proposed a sort of immune based optimization method for controller parameter tuning, and established the algorithm model that could solve the parameter tuning problems of control system. And in this paper, it is realized to the parameter tuning of PID and HSIC by the proposed immune model. The simulation experiment result shows that it owns the generality and effectiveness in method, and also it is suitable for parameter tuning of other algorithm such as optimal control etc.

References

1. Wang, Wei, Zhang, Jingtao, Chai, Tianyou: A survey of advanced PID parameter tuning methods. Acta Automatica Sinica **26**(3), 347–355 (2000)
2. Khan, A.A., Rapal, N.: Fuzzy PID controller: design, tuning and comparison with conventional PID controller. In: IEEE Proceedings of the ICEIS2006, vol. 4, pp. 1–6. IEEE, Islamabad, Pakistan (2006)
3. Zhang, M.G., Wang, Z.G., Wang, P.: Adaptive PID decoupling control based on RBF neural network and its application. In: Proceedings of ICWAPR2007, vol. 11, pp. 727–731, Beijing, China (2007)
4. Pillay, N., Govender, P.: A particle swarm optimization approach for model independent tuning of PID control loops. In: IEEE Proceedings of the AFRICON2007, Windhoek, vol. 9, pp. 1–7. IEEE, South Africa (2007)
5. Arruda, L.V.R., Swiech, M.C.S., Delgado, M.R.B., et al.: PID control of MIMO process based on rank niching genetic algorithm. Appl. Intell. **29**(3), 290–305 (2008)
6. de Almeida, G.M., Re Silva, V.V.: Application of genetic programming for fine tuning PID controller parameters designed through Ziegler-Nichols technique. Lect. Notes Comput. Sci. **3612**, 313–322 (2005)
7. Kim, J., Bentley, P.J.: Towards an artificial immune system for network intrusion detection: an investigation of clonal selection with a negative selection operator. In: IEEE Proceedings of the ICEC2001, vol. 2, pp. 1244–1252. IEEE, Seoul, Korea (2006)
8. de Castro, L.N., Timmis, J.I.: Artificial immune systems as a novel soft computing paradigm. Soft Comput. **7**(8), 526–544 (2003). A Fusion of Foundations, Methodologies and Applications
9. Li, Zhusu, Yaqing, Tu: Human Simulated Intelligent Controller. National Defence Industry Press, Beijing (2003)
10. de Almeida, G.M., e Silva, VVR et al.: Application of genetic programming for fine tuning PID controller parameters designed through Ziegler-Nichols Technique. LNCS(Aug), 313–322 (2005)

Preinvex Fuzzy-valued Function and Its Application in Fuzzy Optimization

Zeng-tai Gong, Yu-juan Bai and Wen-qing Pan

Abstract Based on the ordering of fuzzy numbers proposed by Goetschel and Voxman, in this paper, the representations and characterizations of semi-E-preinvex fuzzy-valued function are defined and obtained. As an application, the conditions of strictly local optimal solution and global optimal solution in the mathematical programming problem are discussed.

Keywords Fuzzy numbers · semi-E-preinvexity · fuzzy optimization

1 Introduction

The concept of fuzzy set was introduced by Zadeh in [11]. Since then, many applications of fuzzy set have been widely developed. Just as many systems with parameter uncertainty, the optimization theory with parameter uncertainty such as in objective function, constraints, or both of objective function and constraints, is often dealt. It is well known that the classical theory of convex analysis and mathematical programming are closely linked each other. Some authors have discussed the convexity, quasi-convexity and $B-$convex of fuzzy mappings [6, 8]. In 1994, Noor [5] introduced the concept of preinvex fuzzy-valued functions over the field of real numbers R, and obtained some properties of preinvex fuzzy-valued functions. After that, the properties of preinvex fuzzy-valued functions have been developed and generalized by many authors [7–10] and applied in fuzzy optimization problem [3]. The essence

Z. Gong (✉)
College of Mathematics and Statistics, Northwest Normal University, 730070 Lanzhou, China
e-mail: zt-gong@163.com

Y. Bai
Department of Mathematics, Longdong University, 745000 Qingyang, China

W. Pan
School of Economics and Management, Tsinghua University, 100084 Beijing, China

B.-Y. Cao and H. Nasseri (eds.), *Fuzzy Information & Engineering and Operations Research & Management*, Advances in Intelligent Systems and Computing 211,
DOI: 10.1007/978-3-642-38667-1_4,

of preinvex fuzzy-valued functions is investigated and some judge theorems and a characterization of preinvex fuzzy-valued functions are obtained by using the upper (lower) semi-continuity [4]. In this paper, some representations and characterizations of semi-E-preinvex fuzzy-valued functions are obtained. As an application, the conditions of strictly local optimal solution and global optimal solution in the mathematical programming problem are discussed.

2 Preliminaries and Definitions

A fuzzy number is a mapping u: $R \to [0, 1]$, with the following properties:

1. u is normal, i.e., there exists $x_0 \in R$ with $u(x_0) = 1$;
2. u is convex fuzzy set;
3. u is semicontinuous on R and
4. $\overline{x \in R : u(x) > 0}$ is compact.

Let $\mathcal{F}$ be the set of all fuzzy numbers on R. For $u \in \mathcal{F}$, we write

$$[u]^{\alpha} = [u^{-}(\alpha), u^{+}(\alpha)],$$

then the following conditions are satisfied:

1. $u^{-}(\alpha)$ is abounded left continuous non-decreasing function on $(0, 1]$;
2. $u^{+}(\alpha)$ is abounded left continuous non-increasing function on $(0, 1]$;
3. $u^{-}(\alpha)$ and $u^{+}(\alpha)$ are right continuous at $\alpha = 0$ and left continuous at $\alpha = 1$.
4. $u^{-}(1) \leq u^{+}(1)$.

Conversely, if the pair of functions $u^{-}(\alpha)$ and $u^{+}(\alpha)$ satisfy conditions (1) – (4), then there exists a unique $u \in \mathcal{F}$ such that

$$[u]^{\alpha} = [u^{-}(\alpha), u^{+}(\alpha)]$$

for each $\alpha \in [0, 1]$ (see, e.g.[1])

For brevity, we write
$V = \{(u^{-}(\alpha), u^{+}(\alpha), \alpha) | 0 \leq \alpha \leq 1, u^{-} : I \to R, u^{+} : I \to R$ are bounded functions$\}$; $\hat{V} = \{(u^{-}(\alpha), u^{+}(\alpha), \alpha) | 0 \leq \alpha \leq 1, u^{-}(\alpha), u^{+}(\alpha)$ are Lebesgue integrable$\}$; $\mathcal{F} = \{(u^{-}(\alpha), u^{+}(\alpha), \alpha) | 0 \leq \alpha \leq 1, u^{-}(\alpha)$ is left continuous non-decreasing function, $u^{+}(\alpha)$ is left continuous non-increasing function and they are right continuous at $\alpha = 0\}$.

The addition and scalar multiplication in V are defined as follows:
$(u^{-}(\alpha), u^{+}(\alpha), \alpha) + (v^{-}(\alpha), v^{+}(\alpha), \alpha) = (u^{-}(\alpha) + v^{-}(\alpha), u^{+}(\alpha) + v^{+}(\alpha), \alpha)$,
$k(u^{-}(\alpha), u^{+}(\alpha), \alpha) = (ku^{-}(\alpha), ku^{+}(\alpha), \alpha)$.

Let $x = (x_1, x_2, \cdots x_n) \in R^n$, $u = (u_1, u_2, \cdots u_n) \in \hat{V}^n$, then scalar product of x and u is defined by

$$\langle x, u\rangle = x_1u_1 + x_2u_2 + \cdots + x_nu_n.$$

Assume that $u, v \in \hat{V}$,

$$u = \{(u^-(\alpha), u^+(\alpha), \alpha)|0 \le \alpha \le 1\},$$

$$v = \{(v^-(\alpha), v^+(\alpha), \alpha)\ \|0 \le \alpha \le 1\}$$

are members of $\hat{V}$, then u precedes v $(u \le v)$ if

$$\int_0^1 f(\alpha)(u^-(\alpha) + u^+(\alpha))d\alpha \le \int_0^1 f(\alpha)(v^-(\alpha) + v^+(\alpha))d\alpha.$$

For a fuzzy-valued function

$$F(x) = \{(F^-(\alpha, x), F^+(\alpha, x), \alpha)|0 \le \alpha \le 1\},$$

we define

$$T_{F(x)} = \int_0^1 f(\alpha)[F^-(\alpha, x) + F^+(\alpha, x)]d\alpha,$$

where f is monotone non-decreasing function and $f(0) = 1$, $\int_0^1 f(\alpha)d\alpha = \frac{1}{2}$. f can be interpreted as weighting function, its nondecreasing ensure a closer level of nuclear cuts and determine the relationship between the greater role. Especially, if $f(\alpha) = \alpha$, then the ordering relation between two $u, v \in \hat{V}$ could be found in [2] which defined by Goetschel and Voxman.

Definition 2.1 [5] *Let $S \subset R^n$ be an open set, $F : S \to \mathcal{F}$ be a fuzzy-valued function. If $\frac{\partial}{\partial x_i}F^-(\alpha, x)$ and $\frac{\partial}{\partial x_i}F^+(\alpha, x)$ $(i = 1, 2, \ldots, n)$ are continuous, then $F(x)$ is said to be differentiable on S, and*

$$\nabla F(x) = \left[\frac{\partial}{\partial x_1}F(x), \frac{\partial}{\partial x_2}F(x), \ldots, \frac{\partial}{\partial x_n}F(x)\right]$$

is called the gradient of fuzzy-valued function $F(x)$.

Let $y \in (S \subset R^n)$, we say S is invex at y with respect to $\eta : S \times S \to R^n$, if for each $x \in S, \lambda \in [0, 1]$, $y + \lambda\eta(x, y) \in S$.

S is said to be an invex set with respect to η if S is invex at each $y \in S$.

In particular, when $\eta(x, y) = x - y$, invex set degrades a general convex set.

$\eta : S \times S \to R^n$ is called a skew mapping if

$$\eta(x, y) + \eta(y, x) = 0, x, y \in S,\ \lambda \in [0, 1].$$

Definition 2.2 *A fuzzy-valued function $F : S \to \mathcal{F}$ is said to be preinvex on invex set S with respect to $\eta : S \times S \to R^n$, if for any $x, y \in K, \lambda \in [0, 1]$,*

$$T_{F(y+\lambda\eta(x,y))} \leq \lambda T_{F(x)} + (1-\lambda)T_{F(y)}.$$

Definition 2.3 *A set $S(\subset R^n)$ is said to be E-invex set with respect to $\eta : S\times S \rightarrow R^n$ on S, if there is a mapping $E : R^n \rightarrow R^n$ such that*

$$E(y) + \lambda\eta(E(x), E(y)) \in S,$$

for $0 \leq \lambda \leq 1$.

Definition 2.4 *A fuzzy-valued function $F : S \rightarrow \mathcal{F}$ is said to be semi-E-preinvex respect to $\eta : S \times S \rightarrow R^n$ on S, if there is a mapping $E : R^n \rightarrow R^n$ such that*

$$E(y) + \lambda\eta(E(x), E(y)) \in S$$

and

$$T_{F(E(y)+\lambda\eta(E(x),E(y)))} \leq \lambda T_{F(x)} + (1-\lambda)T_{F(y)}$$

for each $x, y \in S$ and $0 \leq \lambda \leq 1$.

3 The Judgement and Characterization of Semi-E-preinvex Fuzzy-valued Functions

Theorem 3.1 *Let $F : S \rightarrow \mathcal{F}$ be a semi-E-preinvex fuzzy-valued function on E-invex set S, then $T_{F(E(y))} \leq T_{F(y)}$ for each $x \in S$.*

Proof. Let $F : S \rightarrow \mathcal{F}$ be a semi-E-preinvex fuzzy-valued function on E-invex set S and $\lambda \in [0, 1]$. Then

$$T_{F(E(y)+\lambda\eta(E(x),E(y)))} \leq \lambda T_{F(x)} + (1-\lambda)T_{F(y)}.$$

Thus for $\lambda = 0$, then $T_{F(E(y))} \leq T_{F(y)}$ for each $x \in S$.

Theorem 3.2 *Let $F : S \rightarrow \mathcal{F}$ be a semi-E-preinvex fuzzy-valued function on S iff*

$$T_{F(E(y)+\lambda\eta(E(x),E(y)))} \leq \lambda T_u + (1-\lambda)T_v$$

for $u, v \in \mathcal{F}$ satisfying $T_{F(x)} \leq T_u$, $T_{F(y)} \leq T_v$, and $x, y \in S$, where $0 \leq \lambda \leq 1$.

Proof. Note that $T_{F(x)} < T_u$, $T_{F(y)} < T_v$, and F is semi-E-preinvex on S, then

$$T_{F(E(y)+\lambda\eta(E(x),E(y)))} \leq \lambda T_{F(x)} + (1-\lambda)T_{F(y)} < \lambda T_u + (1-\lambda)T_v.$$

Conversely, write

$$F(x) = (F^-(\alpha, x), F^+(\alpha, x), \alpha),\ F(y) = (F^-(\alpha, y), F^+(\alpha, y), \alpha)$$

for $x, y \in S, \lambda \in (0, 1)$. $\forall \varepsilon > 0$, let

$$u = \{(F^-(\alpha, x) + \frac{\varepsilon}{2}, F^+(\alpha, x) + \frac{\varepsilon}{2}, \alpha) |\ 0 \le \alpha \le 1\},$$

$$v = \{(F^-(\alpha, y) + \frac{\varepsilon}{2}, F^+(\alpha, y) + \frac{\varepsilon}{2}, \alpha) |\ 0 \le \alpha \le 1\},$$

then

$$T_{F(x)} = \int_0^1 \alpha[F^-(\alpha, x) + F^+(\alpha, x)]d\alpha < \int_0^1 \alpha[F^-(\alpha, x) + F^+(\alpha, x) + \varepsilon]d\alpha = T_u,$$

$$T_{F(y)} = \int_0^1 \alpha[F^-(\alpha, y) + F^+(\alpha, y)]d\alpha < \int_0^1 \alpha[F^-(\alpha, y) + F^+(\alpha, y) + \varepsilon]d\alpha = T_v.$$

That is,
$T_{F(E(y)+\lambda\eta(E(x),E(y)))}$

$$< \lambda \int_0^1 \alpha[F^-(\alpha, x) + F^+(\alpha, x) + \varepsilon]d\alpha + (1-\lambda) \int_0^1 \alpha[F^-(\alpha, y) + F^+(\alpha, y) + \varepsilon]d\alpha.$$

It follows that

$$T_{F(E(y)+\lambda\eta(E(x),E(y)))} \le \lambda T_{F(x)} + (1 - \lambda)T_{F(y)}$$

when $\varepsilon \to 0$.

Theorem 3.3 *Let $F : S \to \mathcal{F}$ be a semi-E-preinvex fuzzy-valued function on E-invex set S. Then*

$$K_u(F) = \{x | x \in S,\ T_{F(x)} \le T_u\} \quad \forall u \in \mathcal{F}$$

is E-invex set.

Proof. For any $x, y \in K_u(F)$, we have $T_{F(x)} \le T_u$, $T_{F(y)} \le T_u$. Since S is E-invex set, i.e.

$$E(y) + \lambda\eta(E(x), E(y)) \in S,$$

and $F : S \to \mathcal{F}$ is semi-E-preinvex on E-invex set S, i.e.

$$T_{F(E(y)+\lambda\eta(E(x),E(y)))} \le \lambda T_{F(x)} + (1 - \lambda)T_{F(y)} \le T_u,$$

we have

$$E(y) + \lambda\eta(E(x), E(y)) \in K_u(F).$$

Hence $K_u(F)$ is a E-invex set.

Given a mapping $E : R^n \to R^n$. Let the mapping

$$E \times I : R^n \times \mathcal{F} \to R^n \times \mathcal{F}$$

to be

$$E \times I(x, u) = (E(x), u), \ \ \forall (x, u) \in R^n \times \mathcal{F}.$$

Definition 3.1 *Let $S \subset R^n \times \mathcal{F}$. S ($\subset R^n \times \mathcal{F}$) is said to be a $E \times I$-invex set if there exists $E : R^n \to R^n$ such that*

$$\begin{aligned} &E \times I(y, v) + \lambda\eta[E \times I(x, u), E \times I(y, v)] \\ &= [E(y) + \lambda\eta(E(x), E(y)), \lambda u + (1 - \lambda)v] \in S, \end{aligned}$$

for $(x, u), (y, v) \in S$ ($x, y \in R^n$, $u, v \in \mathcal{F}$) and $\lambda \in [0, 1]$.

Theorem 3.4 *Let $\{S_i\}_{i \in J}$ ($S_i \subset R^n \times \mathcal{F}$) be $E \times I$-invex set. Then $\bigcap_{i \in J} S_i (\subset R^n \times \mathcal{F})$ is a $E \times I$-invex set.*

Proof. Let $(x, u), (y, v) \in \bigcap_{i \in J} S_i$, $\lambda \in [0, 1]$, then $\forall i \in J$, we have (x, u), $(y, v) \in S_i$. Since each $S_i (\subset R^n \times \mathcal{F})$ is $E \times I$-invex, i.e. there exists a mapping $E : R^n \to R^n$ such that for $(x, u), (y, v) \in S_i$ and $\lambda \in [0, 1]$, we have

$$\begin{aligned} &E \times I(y, v) + \lambda\eta[E \times I(x, u), E \times I(y, v)] \\ &= [E(y) + \lambda\eta(E(x), E(y)), \lambda u + (1 - \lambda)v] \in S_i \end{aligned}$$

for $i \in J$. It follows that

$$[E(y) + \lambda\eta(E(x), E(y)), \lambda u + (1 - \lambda)v] \in \bigcap_{i \in J} S_i.$$

That is, $\bigcap_{i \in J} S_i (\subset R^n \times \mathcal{F})$ isa $E \times I$-invex set.

Theorem 3.5 *Let S be a $E-$invex set. Then F is a semi-E-preinvex fuzzy-valued function on S iff*

$$\{(x, u) | x \in S, u \in \mathcal{F}, \ T_{F(x)} < T_u\}$$

is a $E \times I-$ invex set.

Proof. Let $S(F) = \{(x, u) | x \in S, u \in \mathcal{F}, \ T_{F(x)} < T_u\}$. Since

$$\begin{aligned} &E \times I(y, v) + \lambda\eta[E \times I(x, u), E \times I(y, v)] \\ &= [E(y) + \lambda\eta(E(x), E(y)), \lambda u + (1 - \lambda)v] \in S(F), \end{aligned}$$

i.e. for any $x, y \in S$, $\lambda \in [0, 1]$ and $u, v \in \mathcal{F}$ satisfying $T_{F(x)} < T_u$, $T_{F(y)} < T_v$, thus

$$T_{F(E(y)+\lambda\eta(E(x),E(y)))} < \lambda T_u + (1-\lambda)T_v.$$

According to Theorem 3.2, F is semi-E-preinvex on S iff $S(F)$ is a $E \times I-$ invex set.

We define an epigraph of F as follows:

$$epi(F) = \{(x,u)|x \in S\ , u \in \mathcal{F},\ T_{F(x)} \leq T_u\}.$$

Theorem 3.6 *Let S be an $E-$invex set. Then $F : S \to \mathcal{F}$ is a semi-E-preinvex fuzzy-valued function on S iff*

$$epi(F) = \{(x,u)|x \in S\ , u \in \mathcal{F},\ T_{F(x)} \leq T_u\}$$

is an $E \times I-$ invex set.

Proof. Let F be a semi-E-preinvex fuzzy-valued function on S. Then for any (x,u), $(y,v) \in epi(F)$ and $\lambda \in [0,1]$, we have

$$E(y) + \lambda\eta(E(x), E(y)) \in S$$

and

$$T_{F(E(y)+\lambda\eta(E(x),E(y)))} \leq \lambda T_{F(x)} + (1-\lambda)T_{F(y)} \leq \lambda T_u + (1-\lambda)T_v.$$

Hence

$$\begin{aligned} &E \times I(y,v) + \lambda\eta[E \times I(x,u), E \times I(y,v)] \\ &\quad = [E(y) + \lambda\eta(E(x),E(y)), \lambda u + (1-\lambda)v] \in epi(F). \end{aligned}$$

i.e. $epi(F)$ is an $E \times I-$ invex set.

Conversely, $\forall (x,y) \in S$, $\lambda \in [0,1]$, we have $(x, F(x)) \in epi(F)$, $(y, F(y)) \in epi(F)$. Since $epi(F)$ is an $E \times I-$ invex set, we have

$$\begin{aligned} &E \times I(y, F(y)) + \lambda\eta[E \times I(x, F(x)), E \times I(y, F(y))] \\ &\quad = [E(y) + \lambda\eta(E(x),E(y)), \lambda F(x) + (1-\lambda)F(y)] \in epi(F). \end{aligned}$$

Hence

$$T_{F(E(y)+\lambda\eta(E(x),E(y)))} \leq \lambda T_{F(x)} + (1-\lambda)T_{F(y)}.$$

Therefore, F is semi-E-preinvex on S.

Theorem 3.7 *Let $\{F_i|i \in J\}$ be a collection of semi-E-preinvex function on S, if for any $x \in S$, $\sup\{F_i(x)|i \in J\}$ exists, then $F(x) = \sup\{F_i(x)|i \in J\}$ is semi-E-preinvex on S.*

Proof. $\forall i \in J$, $\{F_i\}$ is semi-E-preinvex on S, then

$$epi(F_i) = \{(x,u)|x \in S\,, u \in \mathcal{F},\ T_{F_i(x)} \leq T_u\}$$

is an $E \times I-$ invex set, Furthermore,

$$\bigcap_{i \in J} epi(F_i) = \{(x,u)|x \in S\,, u \in \mathcal{F},\ T_{F_i(x)} \leq T_u,\ i \in J\}$$

is an $E \times I-$ invex set. Note that

$$\begin{aligned}
&(x,u) \in \bigcap_{i \in J} epi(F_i) \\
&\Leftrightarrow x \in S\,, u \in \mathcal{F},\ T_{F_i(x)} \leq T_u, i \in J \\
&\Leftrightarrow x \in S\,, u \in \mathcal{F},\ T_{F(x)} \leq T_u \\
&\Leftrightarrow (x,u) \in epi(F).
\end{aligned}$$

Hence $\bigcap_{i \in J} epi(F_i) = epi(F)$ is an $E \times I-$ invex set. According Theorem 3.6, F is semi-E-preinvex on S.

Theorem 3.8 *Let $F_i : S \to \mathcal{F}$ $(i = 1, 2, \cdots, k)$ be a semi-E-preinvex fuzzy-valued function on S with a mapping $E : R^n \to R^n$. Then*

$$h(x) = \Sigma_{i=1}^{k} a_i F_i(x)\ (a_i \geq 0,\ i = 1, 2, \cdots, k)$$

is semi-E-preinvex on S.

Proof. Since $F_i : S \to \mathcal{F}$ $(i = 1, 2, \cdots, k)$ is semi-E-preinvex on S with a mapping $E : R^n \to R^n$, i.e. for any $x, y \in S$ and $\lambda \in [0, 1]$, then

$$T_{F_i(E(y)+\lambda\eta(E(x),E(y)))} \leq \lambda T_{F_i(x)} + (1-\lambda)T_{F_i(y)},\ \ i = 1, 2, \cdots, k.$$

Hence

$$T_{\Sigma_{i=1}^{k} F_i(E(y)+\lambda\eta(E(x),E(y)))} \leq \lambda T_{\Sigma_{i=1}^{k} F_i(x)} + (1-\lambda)T_{\Sigma_{i=1}^{k} F_i(y)}.$$

Therefore

$$T_{h(E(y)+\lambda\eta(E(x),E(y)))} \leq \lambda T_{h(x)} + (1-\lambda)T_{h(y)}.$$

i.e. h is semi-E-preinvex on S.

Theorem 3.9 *Let F be semi-E-preinvex on S. Then the following states are true:*

1. If $\phi : \mathcal{F} \to \mathcal{F}$ *is a nondecreasing convex function, then the composite function* $\phi \circ F : S \to \mathcal{F}$ *is semi-E-preinvex on S;*
2. If $\phi : \mathcal{F} \to \mathcal{F}$ *is positively homogeneous nondecreasing additive function, then the composite function* $\phi \circ F : S \to \mathcal{F}$ *is semi-E-preinvex on S.*

Proof. For any $x, y \in S,\ \lambda \in [0, 1]$, we have $E(y) + \lambda\eta(E(x), E(y)) \in S$ and

$$T_{F(E(y)+\lambda\eta(E(x),E(y)))} \leq \lambda T_{F(x)} + (1-\lambda)T_{F(y)},$$

1. When $\phi : \mathcal{F} \to \mathcal{F}$ is a nondecreasing convex function,

$$\begin{aligned}
&T_{\phi\circ F(E(y)+\lambda\eta(E(x),E(y)))}\\
&= T_{\phi(F(E(y)+\lambda\eta(E(x),E(y))))}\\
&\leq T_{\phi(\lambda F(x)+(1-\lambda)F(y))}\\
&\leq T_{\lambda\phi F(x)+(1-\lambda)\phi F(y)}\\
&= T_{\lambda\phi\circ F(x)+(1-\lambda)\phi\circ F(y)}.
\end{aligned}$$

   i.e. $\phi \circ F : S \to \mathcal{F}$ is semi-E-preinvex on S.
2. When $\phi : \mathcal{F} \to \mathcal{F}$ be positively homogeneous nondecreasing additive function,

$$\begin{aligned}
&T_{\phi\circ F(E(y)+\lambda\eta(E(x),E(y)))}\\
&= T_{\phi(F(E(y)+\lambda\eta(E(x),E(y))))}\\
&\leq T_{\phi(\lambda F(x)+(1-\lambda)F(y))}\\
&\leq T_{\phi(\lambda F(x))+\phi((1-\lambda)F(y))}\\
&\leq T_{\lambda\phi F(x)+(1-\lambda)\phi F(y)}\\
&= T_{\lambda\phi\circ F(x)+(1-\lambda)\phi\circ F(y)}.
\end{aligned}$$

   i.e. $\phi \circ F : S \to \mathcal{F}$ is semi-E-preinvex on S.

4 The Optimization of Preinvex Fuzzy-valued Function

Let $F : S \to \mathcal{F}$ be a fuzzy-valued function on S. We consider the following fuzzy optimization problem.

$$(P)\ \ \min\ F(x),\ s.t.\ x \in S = \{x \in R^n | G_i(x) \leq \tilde{0}, i = 1, 2, \ldots, m\},$$

where $F : R^n \to \mathcal{F}$ and $G_i : R^n \to \mathcal{F}$ are semi-E-preinvex on S.

Theorem 4.1 *Let $F : R^n \to \mathcal{F}$ and $G_i : R^n \to \mathcal{F}$ be semi-E-preinvex fuzzy-valued functions on R^n. Then S is $E-$invex.*

Proof. Let $G_i : R^n \to \mathcal{F}\ (i = 1, 2, \ldots, m)$ be semi-E-preinvex fuzzy-valued function on R^n,

$$S_i = \{x \in R^n | T_{G_i(x)} \leq 0\}\ (i = 1, 2, \ldots, m).$$

Then

$$T_{G_i(E(y)+\lambda\eta(E(x),E(y)))} \leq \lambda T_{G_i(x)} + (1-\lambda)T_{G_i(y)} \in S_i.$$

Hence

$$S = \bigcap_{i=1}^{m} S_i = \{x \in R^n | T_{G_i(x)} \leq 0, i = 1, 2, \ldots, m\}$$

is $E-$invex.

Similarly to the proof of Theorem 4.1, we have the following theorem.

Theorem 4.2 *Let $F : R^n \to \mathcal{F}$ and $G_i : R^n \to \mathcal{F}$ be semi-E-preinvex fuzzy-valued function on R^n. Then $F : R^n \to \mathcal{F}$ is semi-E-preinvex on S.*

Theorem 4.3 *Let $F : R^n \to \mathcal{F}$ be semi-E-preinvex on S and $\bar{x}$ a solution of the following problem:*

$$(P_E) \ \min(F \circ E)(x), \qquad x \in S.$$

Then $E(\bar{x})$ is a solution of the problem P.

Proof. Let $E(\bar{x})$ be a nonsolution of the problem P. Then there exists $y \in S$ such that

$$T_{F(y)} < T_{F(E(\bar{x}))}.$$

Then according 4.1,we have $T_{F(E(y))} \leq T_{F(y)}$. Hence

$$T_{F(E(y))} < T_{F(E(\bar{x}))},$$

which contradicts the optimality of $\bar{x}$ for the problem (P_E). Therefore $E(\bar{x})$ be a solution of problem P.

Theorem 4.4 *Let $F : R^n \to \mathcal{F}$ be semi-E-preinvex fuzzy-valued function on S and $\bar{x} = E(\bar{x}) \in S$ a local solution of the problem of (P). Then $\bar{x}$ be a global solution of problem P.*

Proof. Let $\bar{x} = E(\bar{x}) \in S$ be a local solution of the problem of (P). Then there exists $\delta > 0$, such that $\forall x \in U(\bar{x}, \delta) \bigcap S$, we have

$$T_{F(\bar{x})} < T_{F(x)}.$$

Suppose $\bar{x}$ is a nonsolution of problem of (P), then there exists $y \in S$ such that

$$T_{F(y)} < T_{F(\bar{x})} = T_{F(E(\bar{x}))}.$$

$\forall \lambda \in (0, 1)$, we have

$$\lambda T_{F(y)} + (1-\lambda)T_{F(\bar{x})} < T_{F(\bar{x})}.$$

Since F is semi-E-preinvex on S and $\bar{x} = E(\bar{x}) \in S$, we have

$$T_{F(E(\bar{x})+\lambda\eta(E(y),E(\bar{x})))} < \lambda T_{F(y)} + (1-\lambda)T_{F(\bar{x})}.$$

Hence

$$T_{F(E(\bar{x})+\lambda\eta(E(y),E(\bar{x})))} < T_{F(\bar{x})}.$$

Since λ may be arbitrarily small, we have

$$E(\bar{x}) + \lambda\eta(E(y), E(\bar{x})) \in U(\bar{x}, \delta) \bigcap S.$$

which contradicts (1), therefore $\bar{x}$ is a global solution of problem P.

Theorem 4.5 *Let $F : R^n \to \mathcal{F}$ be semi-E-preinvex fuzzy-valued function on S and $\bar{x} \in S$ satisfy*

$$T_{F(E(\bar{x}))} = T_{\min_{x\in S} F(E(x))},$$

if $T_u = T_{\min_{x\in S} F(E(x))}$. Then $\Omega = \{x \in S | T_{F(x)} = T_u\}$ is $E-$invex.

Proof. $\forall x, y \in \Omega, 0 \le \lambda \le 1$, then

$$x, y \in S, \quad T_{F(x)} = T_u, \quad T_{F(y)} = T_u.$$

Since $F : R^n \to \mathcal{F}$ is semi-E-preinvex on S, we have

$$E(y) + \lambda\eta(E(x), E(y)) \in S,$$

and

$$T_u \le T_{F(E(y)+\lambda\eta(E(x),E(y)))} \le \lambda T_{F(x)} + (1-\lambda)T_{F(y)} = T_u.$$

Hence

$$T_{F(E(y)+\lambda\eta(E(x),E(y)))} = T_u.$$

i.e.

$$E(y) + \lambda\eta(E(x), E(y)) \in \Omega,$$

therefore Ω is $E-$invex.

5 Conclusions

In the real world there are many linear programming problems where all decision parameters are fuzzy numbers. Many authors have considered various types of fuzzy linear programming problems and proposed several approaches for solving these problems. One of the approaches for solving fuzzy linear programming problems is based on the concept of comparison of fuzzy numbers by use of ranking functions. In this paper, we used the ordering of fuzzy numbers proposed by Goetschel and Voxman, obtained representations and characterizations of semi-E-preinvex fuzzy-valued function. As an application, the conditions of strictly local optimal solution and global optimal solution in the mathematical programming problem have been discussed.

Acknowledgments Thanks to the support by National Natural Science Foundations of China (71061013, 61262022) and Scientific Research Project of Northwest Normal University (NWNU-KJCXGC-03-61).

References

1. Diamond, P., Kloeden, P.E.: Metric Spaces of Fuzzy Sets: Theory Applications. World Scientific, Singapore (1994)
2. Goetschel, R., Voxman, J.W.: Elementary fuzzy calculus. Fuzzy Sets Syst. **18**, 31–43 (1986)
3. Gong, Z.T., Li, H.X.: Saddle point optimality conditions in fuzzy optimization problems. In: Kacprzyk, J. (ed.) Advances in Soft Computing 54: Fuzzy Information and Engineering, pp. 7–14. Springer, BerLin (1998)
4. Gong, Z.T., Bai, Y.J.: Preinvex fuzzy- valued functions and its application. J. Northwest Normal Univ. **46**(1), 1–5 (2010). (in Chineses)
5. Noor, M.A.: Fuzzy preinvex functions. Fuzzy Sets Syst. **64**, 95–104 (1994)
6. Panigrahi, M., Panda, G., Nanda, S.: Convex fuzzy mapping with differentiability and its application in fuzzy optimization. Eur. J. Oper. Res. **185**, 47–62 (2007)
7. Syau, Y.R.: Generalization of preinvex and B-vex fuzzy mappings. Fuzzy Sets Syst. **120**, 533–542 (2001)
8. Syau, Y.R., Lee, E.S.: Preinvexity and Φ_1-convexity of fuzzy mappings through a linear ordering. Comput. Math. Appl. **51**, 405–418 (2006)
9. Wu, Z.Z., Xu, J.P.: Generalized convex fuzzy mappings and fuzzy variational-like inequality. Fuzzy Sets Syst. **160**, 1590–1619 (2009)
10. Yang, X.M., Zhu, D.L.: On properties of preinvex functions. J. Math. Anal. Appl. **256**, 229–241 (2001)
11. Zadeh, L.A.: Fuzzy sets. Inf. Control **8**, 338–356 (1965)

Posynomial Geometric Programming with Fuzzy Coefficients

Ren-jie Hu, Bing-yuan Cao and Guang-yu Zhang

Abstract In practice, there are many problems in which all decision parameters are fuzzy numbers, and such problems are usually solved by either possibilistic programming or multiobjective programming methods. Unfortunately, all these methods have shortcomings. In this note, using the concept of comparison of fuzzy numbers, we introduce a very effective method for solving these problems. Then we propose a new method for solving posynomial geometric programming problems with fuzzy coefficients.

Keywords Fuzzy number · Posynomial geometric programming · Fuzzy posynomial geometric programming.

1 Introduction

Fuzzy geometric programming was first proposed by Cao Bingyuan in 1987 in Tokyo, in the second session of the International Fuzzy Systems Association (IFSA) conference held in Japan. The direct algorithm and dual algorithm of fuzzy geometric programming were studied in references [1, 2]. By mainly utilizing original algorithm and duality algorithm to the multi-objective fuzzy geometric programming, the multi-objective fuzzy geometric programming has been solved in [3–5]. The refer-

R. Hu · G. Zhang (✉)
School of Management, Guangdong University of Technology, 510520 Guangdong, China
e-mail: renjiehu2005@163.com

G. Zhang
e-mail: guangyu@gdut.edu.cn

B. Cao
School of Mathematics and Information Science,Guangzhou University,
510006 Guangdong, China
e-mail: caobingy@163.com

B.-Y. Cao and H. Nasseri (eds.), *Fuzzy Information & Engineering and Operations Research & Management*, Advances in Intelligent Systems and Computing 211,
DOI: 10.1007/978-3-642-38667-1_5,

ences [6] studied the method for solving fuzzy posynomial geometric programming based on comparison method of fuzzy numbers developed by Roubens. The references [7] studied the method for solving fuzzy posynomial geometric programming based on comparison method of fuzzy numbers developed by Renjie Hu. In this note, we discuss the problem on how to solve the fuzzy posynomial geometric programming with fuzzy coefficients. For solving this problem, we proposed a new method on comparison of fuzzy numbers. By utilizing the concept of comparison of fuzzy numbers; posynomial geometric programming with fuzzy coefficients is reduced to a posynomial geometric programming with crisp coefficient. In other words, we get a new method different from the method mentioned in [1, 2, 6, 7]. And the method is testified to be effective by numerical examples.

2 Preliminaries

Definition 2.1 *Let* $\widetilde{A} \in F(X)$, $\forall \alpha \in [0,1]$, *written down as* $A_\alpha = \{x \in X \,|\, \mu_{\widetilde{A}}(x) \geq \alpha\}$, A_α *is said to be* $\alpha - cut$ *set of a fuzzy set* $\widetilde{A}$.

Definition 2.2 *Let* $\widetilde{A}$ *be a fuzzy number, i.e. a convex normalized fuzzy subset of the real line in the sense that:*

(a) $\exists x_0 \in R$ *and* $\mu_{\widetilde{A}}(x_0) = 1$, *where* $\mu_{\widetilde{A}}(x)$ *is the membership function specifying to what degree* x *belongs to* $\widetilde{A}$.
(b) $\mu_{\widetilde{A}}$ *is a piecewise continuous function.*

According definition 2.1 and definition 2.2, the $\alpha - cut$ of $\widetilde{A}$ is $A_\alpha = [A_\alpha^L, A_\alpha^R]$, $A_\alpha^L = \inf\{x \in X \,|\, \mu_{\widetilde{A}}(x) \geq \alpha\}$, $A_\alpha^R = \sup\{x \in X \,|\, \mu_{\widetilde{A}}(x) \geq \alpha\}$, where $\alpha \in [0,1]$.

Definition 2.3 *Fuzzy number* $\widetilde{A}$ *is said to be an trapezoidal number,* $\tilde{A} = (a_1, a_2, a_3, a_4)$, *if its membership function has the following form:*

$$\mu_{\tilde{A}}(x) = \begin{cases} \frac{x-a_1}{a_2-a_1}, a_1 \leq x \leq a_2, \\ 1, a_2 \leq x \leq a_3, \\ \frac{x-a_4}{a_3-a_4}, a_3 \leq x \leq a_4, \\ 0, others. \end{cases}$$

Definition 2.4 *A fuzzy number* $\widetilde{A}$ *is said to be a triangular fuzzy number,* $\widetilde{A} = (a_1, a_2, a_3)$, *if its membership function has the following form*

$$\mu_{\tilde{A}}(x) = \begin{cases} \frac{x-a_1}{a_2-a_1}, a_1 \leq x \leq a_2, \\ \frac{x-a_3}{a_2-a_3}, a_2 \leq x \leq a_3,. \\ 0, others. \end{cases}$$

Definition 2.5 *A fuzzy number $\widetilde{A}$ is said to be a symmetry triangular fuzzy number, $\tilde{A} = (m, a)$, if its membership function has the following form,*

$$\mu_{\tilde{A}}(x) = \begin{cases} \frac{1}{a}x - \frac{m-a}{a}, m - a \leq x \leq m, \\ -\frac{1}{a}x + \frac{m+a}{a}, m \leq x \leq m + a, \\ 0, others. \end{cases}$$

3 Fuzzy Posynomial Geometric Programming

Definition 3.1 *$\bar{X}(A_\alpha)$ is said to be an α level mean of $\widetilde{A}$, if it has the following form,*

$$\bar{X}(A_\alpha) = \frac{2}{\frac{1}{a} + \frac{1}{b}},$$

where, $a = \inf\left\{x \in X \left|\mu_{\tilde{A}}(x) \geq \alpha\right.\right\}, b = \sup\left\{x \in X \left|\mu_{\tilde{A}}(x) \geq \alpha\right.\right\}$.

Definition 3.2 *Let $F(A) \triangleq \int_0^1 \bar{X}(A_\alpha)d\alpha$, $F(A)$ is called the mean of $\widetilde{A}$, then $F(A) \geq F(B) \Leftrightarrow \tilde{A} \geq \tilde{B}$.*

Lemma 3.1 *Let a symmetry triangular fuzzy number $\tilde{A} = (m, a)$, then $F(\widetilde{A}) = -\frac{a^2}{3m} + m$, that is $\forall \tilde{A} = (m, a)$, $\exists$ a real number $A = -\frac{a^2}{3m} + m$ corresponding to it.*

Proof. *$\forall \tilde{A} = (m, a)$, according to the definition 2.5 and definition 2.6,*

$$F(\widetilde{A}) = \int_0^1 \bar{X}(A_\alpha)d\alpha$$

$$= \int_0^1 \frac{2}{\frac{1}{ax+m-a} + \frac{1}{-ax+m+a}} d\alpha$$

$$= -\frac{a^2}{3m} + m.$$

Lemma 3.2 *Let symmetry triangular fuzzy number $\tilde{A} = (m, a)$, $\widetilde{B} = (n, b)$ then,*

$$\widetilde{A} \geq \widetilde{B} \Leftrightarrow -\frac{a^2}{3m} + m \geq -\frac{b^2}{3n} + n.$$

Proof. *According to the Lemma 3.2,*

$$F(\widetilde{A}) = -\frac{a^2}{3m} + m, F(\widetilde{B}) = -\frac{b^2}{3n} + n.$$

According to the comparison of fuzzy numbers based on the sense of Harmonic mean,

$$F(\widetilde{A}) \geq F(\widetilde{B}) \Leftrightarrow \widetilde{A} \geq \widetilde{B}.$$

Hence,

$$\widetilde{A} \geq \widetilde{B} \Leftrightarrow -\frac{a^2}{3m} + m \geq -\frac{b^2}{3n} + n.$$

We consider the following fuzzy posynomial geometric programming problem:

$$\begin{aligned} &\min \tilde{g}_0(x) \\ &s.t. \tilde{g}_i(x) \leq \tilde{b}_i (1 \leq i \leq p), \\ &x > 0. \end{aligned} \tag{1}$$

Objectives and constraint function is: $\tilde{g}_i(x)(0 \leq i \leq p)$

$$\tilde{g}_i(x) = \sum_{k=1}^{J_i} \tilde{c}_{ik} \prod_{k=1}^{m} x_l^{r_{ikl}} (0 \leq i \leq p),$$

where $\tilde{c}_{ik}$ *and* $\tilde{b}_i$ *are positive symmetry triangular fuzzy number,* r_{ikl} *is a real number.*

Theorem 3.1 *Problem (1) is equivalent to the following posynomial geometric programming problem:*

$$\begin{aligned} &\min g_0(x) \\ &s.t. \tfrac{1}{b_i} g_i(x) \leq 1 (1 \leq i \leq p), \\ &x > 0. \end{aligned} \tag{2}$$

Objectives and constraint function is: $g_i(x)(0 \leq i \leq p)$*:*

$$g_i(x) = \sum_{k=1}^{J_i} c_{ik} \prod_{k=1}^{m} x_l^{r_{ikl}} (0 \leq i \leq p),$$

where, c_{ik} *and* b_i *are real number correspond to* $\tilde{c}_{ik}$ *and* $\tilde{b}_i$*, and* $c_{ik} > 0, b_i > 0, r_{ikl}$ *is a real number.*

Proof. *Let* Q_1 *and* Q_2 *be the feasible solution sets of (1) and (2), respectively. Firstly, we prove* $Q_1 = Q_2$*, where* $1 \leq i \leq p$*.*

Then $x \in Q_1$ *if and only if:*

$$\sum_{k=1}^{J_i} \tilde{c}_{ik} \prod_{l=1}^{m} x_l^{r_{ikl}} \leq \tilde{b}_i, i = 1, \ldots, p,$$

if and only if:

$$\sum_{k=1}^{J_i} c_{ik} \prod_{l=1}^{m} x_l^{r_{ikl}} \leq b_i, i = 1, \ldots, p,$$

if and only if:

$$x \in Q_2.$$

Hence $Q_1 = Q_2$.

Now suppose that x^0 is optimal solution for (1), then for all $x \in Q_1$ we have: $\widetilde{g}_0(x) \geq \widetilde{g}_0(x^0)$,

if and only if:

$$\sum_{k=1}^{J_0} \tilde{c}_{0k} \prod_{l=1}^{m} x_l^{r_{0kl}} \geq \sum_{k=1}^{J_0} \tilde{c}_{0k} \prod_{l=1}^{m} (x_l^0)^{\gamma_{0kl}},$$

if and only if:

$$g_0(x) \geq g_0(x^0).$$

We conclude that x^0 is an optimal solution for (2).

According to the duality theory of geometric programming, the duality of problem (2) is:

$$\max d(\beta) = \prod_{i=0}^{p} \prod_{k=1}^{J_i} (\frac{1}{b_i})^{\delta_i \beta_{ik}} (\frac{c_{ik}}{\beta_{ik}})^{\beta_{ik}} \prod_{i=1}^{p} (\beta_{i0})^{\beta_{i0}}$$

$$s.t. \begin{cases} \sum_{k=1}^{J_0} \beta_{0k} = 1, \\ \sum_{i=0}^{p} \sum_{k=1}^{J_i} r_{ikl} \beta_{ik} = 0, 1 \leq l \leq m,, \\ \beta \geq 0, \end{cases} \quad (3)$$

where, $\beta = (\beta_{01}, \ldots, \beta_{0J_0}, \beta_{11} \ldots, \beta_{1J_1}, \ldots \beta_{p1}, \ldots \beta_{pJ_p},)^T, \beta_{i0} = \sum_{k=1}^{J_i} \beta_{ik}$, when $\beta_{ik} = 0$, $(\frac{1}{\beta_{ik}})^{\beta_{ik}} = 1$, when $i = 0$, $\delta_i = 0$; when $i \neq 0$, $\delta_i = 1$.

Theorem 3.2 If posynomial geometric programming (2) are super-compatible [8], then,

(1) There exist maxima of the duality problem (3).
(2) The minimum of problem (2) is equal to the maxima of problem (3). That is $g_0(X^*) = d(\beta^*)$, where $X^*$is the optimal solution to the problem (2), and β^* is an optimal solution to the problem (3).
(3) The relation between original variable X^* and dual variable β^* is:

$$\beta_{ik}^* = \begin{cases} \frac{U_{0k}(X^*)}{g_0(X^*)}, (i = 0, 1 \leq k \leq J_0), \\ \frac{\mu_i^* U_{ik}(X^*)}{g_0(X^*)}, (i \neq 0, 1 \leq i \leq p, 1 \leq k \leq J_i), \end{cases} \quad (4)$$

where the Lagrange multiplier $\mu^*=(\mu_1, \mu_2, \cdots, \mu_p)^T$ satisfying $\mu_i^*(g(X^*)-1)=0, (1 \leq i \leq p)$.

Proof. *See Reference [8].* Numerical example

$$\min \tilde{g}_0(x) = \tilde{c}_{01}x_1^{-1}x_2^{-1}x_3^{-1} + \tilde{c}_{02}x_1x_2$$

$$s.t. \begin{cases} \tilde{g}_1(x) = \tilde{c}_{11}x_1x_3 + \tilde{c}_{12}x_1x_2 \leq \tilde{b}_1, \\ x_1, x_2, x_3 > 0, \end{cases} \tag{5}$$

where, $\tilde{c}_{01} = (50, 10\sqrt{15}), \tilde{c}_{02} = (60, 60), \tilde{c}_{11} = (1, \frac{\sqrt{6}}{2}), \tilde{c}_{12} = (1, 1.5), \tilde{b}_1 = (1, \sqrt{6})$.

Solution. Consider the corresponding posynomial geometric programming to problem (5):

$$\min g_0(x) = 40x_1^{-1}x_2^{-1}x_3^{-1} + 40x_1x_2$$

$$s.t. \begin{cases} g_1(x) = \frac{1}{2}x_1x_3 + \frac{1}{4}x_1x_2 \leq 1, \\ x_1, x_2, x_3 > 0. \end{cases} \tag{6}$$

The dual programming of problem (6) is:

$$\max d(\beta) = (\frac{40}{\beta_{01}})^{\beta_{01}}(\frac{40}{\beta_{02}})^{\beta_{02}}(\frac{40}{2\beta_{11}})^{\beta_{11}}(\frac{1}{4\beta_{12}})^{\beta_{12}}(\beta_{11}+\beta_{12})^{\beta_{11}+\beta_{12}} \tag{7}$$

$$s.t. \begin{cases} \beta_{01} + \beta_{02} = 1, \\ -\beta_{01} + \beta_{11} + \beta_{12} = 0, \\ -\beta_{01} + \beta_{02} + \beta_{12} = 0, \\ \beta_{01} + \beta_{02} - \beta_{11} = 0, \\ \beta \geq 0. \end{cases}$$

Solve problem (7) and we obtain: $\beta_{01}^* = \frac{2}{3}, \beta_{02}^* = \frac{1}{3}, \beta_{11}^* = \frac{1}{3}, \beta_{12}^* = \frac{1}{3}$ $d(\beta^*) = 60$.

To acquire the optimal solution x^*, we try to solve the following equation set.

$$\begin{cases} 40x_1^{-1}x_2^{-1}x_3^{-1} = (\frac{2}{3}) \cdot 60, \\ 40x_2x_3 = (\frac{1}{3}) \cdot 60, \\ \frac{1}{2}x_1x_3 = \frac{1}{3}/\frac{2}{3}, \\ \frac{1}{4}x_1x_2 = \frac{1}{3}/\frac{2}{3}, \end{cases} \tag{8}$$

We obtain $x_1^* = 2, x_2^* = 1, x_3^* = \frac{1}{2}, x_1^* = 2$.

Hence, $\tilde{g}_0(x^*) = g_0(x^*) = d(\beta^*) = 60$

4 Conclusion

According to the comparison of fuzzy numbers, this paper obtains the corresponding relationship between fuzzy numbers and real numbers. Then we transform the posynomial geometric programming with fuzzy coefficients into a normal posynomial geometric programming and obtain two solutions of the posynomial geometric programming with fuzzy coefficients. And the method is testified to be effective by numerical examples

Acknowledgments Thanks to the support by National Natural Science Foundation of China (No. 70771030, No. 70271047 and No. 71173051).

References

1. Cao B.Y.: Solution and theory of question for a kind of fuzzy positive geometric program. Proceedings of IFSA Congress, vol. 1, pp. 205–208. Toyo, (1987). <error l="70" c="Undefined command " />.
2. Cao, B.Y.: Fuzzy geometric programming (I). Int. Fuzzy Sets Syst. **53**, 135–153 (1993)
3. Verma, R.K.: Fuzzy geometric programming with several objective function. Int. J. Fuzzy Sets Syst. **35**, 115–120 (1990)
4. Biswal, M.P.: Fuzzy programming technique to solve multi objective geometric programming Problems. Int. J. Fuzzy Sets Syst. **51**, 67–71 (1992)
5. Cao, B.Y.: Some type of multi-objective geometric programming. Hunan Annals Math. **1**, 99–106 (1995)
6. Zhou, X.G., Li, P.L.: A method for solving posynomial geometric programming with fuzzy number. Fuzzy Syst. Math. **23**(2), 144–148 (2009)
7. Hu, R.J.: A method for solving posynomial geometric programming. J. Guangzhou Univ. **9**(3), 23–25 (2010)
8. Wu F., Yun Y.Y.: Geometric programming Practice and understanding of mathematics, the periodical 1 to periodical 4, (1982)

Global Optimization of Linear Multiplicative Programming Using Univariate Search

Xue-gang Zhou

Abstract We show that, by using suitable transformations and introducing auxiliary variables, linear multiplicative program can be converted into an equivalent parametric convex programming problem, parametric concave minimization problem or parametric D.C. programming. Then potential and known methods for globally solving linear multiplicative program become available.

Keywords Linear multiplicative programming · Convex programming · Univariate search.

1 Introduction

Consider linear multiplicative programming problems as follows:

$$\text{(LMP)} \quad \begin{cases} \max \sum_{i=1}^{p} (a_i^T x + b_i)(c_i^T x + d_i), \\ \text{s.t.} \quad x \in X, \end{cases} \tag{1}$$

where $p \geqslant 2$, $X \subseteq R^n$ is a nonempty, compact convex set, for each $i \in I$, a_i, c_i are n-dimension column vector. In general, the problem (LMP) is a special case of non-convex programming problem, which is known to be NP-hard even when $p = 1$ [1].

Since from 1990s, there has been a resurgence of interest in problem (1). This interest arises from two factors. First, multiplicative programming problems arise in a variety of practical applications [2–6]. Second, encouraged by rapid advances in high-speed computing, researchers in recent years have been developing and testing

X. Zhou (✉)
Department of Applied Mathematics, Guangdong University of Finance, Guangzhou, 510521 Guangzhou, China
e-mail: zhouxg@aliyun.com

B.-Y. Cao and H. Nasseri (eds.), *Fuzzy Information & Engineering and Operations Research & Management*, Advances in Intelligent Systems and Computing 211,
DOI: 10.1007/978-3-642-38667-1_6,

new methods for solving global optimization problems in practice, including problem (1). Hence, it is very necessary to present good algorithm for (LMP) problem. In the past 20 years, many solution algorithms have been proposed for globally solving the problem (LMP). The methods can be classified as parameterization based methods [7], outer-approximation and branch-and-bound methods [8, 9], vertex enumeration methods [10], a method based on image space analysis [11], an outcome-space cutting plane method [12], heuristic methods [13]. The article intends to show that, by using suitable transformations, there is a potential to globally solve problem (LMP) by techniques that are well know techniques.

The organization and content of this article can be summarized as follows. In Sect. 2, we demonstrate the potential in at least cases to globally solve problem (LMP) by standard parametric convex programming techniques.

2 Preliminary Result and Potential Approach

For convenience, for each $x \in X$ and $i \in I$, let

$$f_i(x) = (a_i^T x + b_i)(c_i^T x + d_i),$$

and for each $x \in X$, let

$$f(x) = \sum_{i=1}^{p} f_i(x).$$

The potential approach and the test problem construction methos to be presented rely upon the following result.

Theorem 1. *Let $i \in I$ and let*

$$Z = \{x \in X | \alpha(a_i^T x + b_i) + \beta(c_i^T x + d_i) = K\},$$

where α, β, K are scalars and α, β are not both zero.

(a) If $\alpha\beta > 0$, then f_i is a concave function on Z.
(b) If $\alpha\beta < 0$, then f_i is a convex function on Z.
(c) If $\alpha\beta = 0$, then f_i is a linear function on Z.

Proof. First, notice that, since X is a convex set, Z is a convex set. To prove parts (a) and (b), we will derive a key inequality. Let λ be a scalar such that $0 \leqslant \lambda \leqslant 1$ and suppose that $x_1, x_2 \in X$. It is not difficult to show

$$\begin{aligned}&[\lambda(c_i^T x_1 + d_i) + (1-\lambda)(c_i^T x_2 + d_i)]^2 - [\lambda(c_i^T x_1 + d_i)^2 + (1-\lambda)(c_i^T x_2 + d_i)^2]\\&= -\lambda(1-\lambda)(c_i^T x_1 - c_i^T x_2)^2 \leqslant 0.\end{aligned}$$

Then, for any $\lambda \in [0, 1]$, $x_1, x_2 \in X$,

$$[\lambda(c_i^T x_1+d_i)+(1-\lambda)(c_i^T x_2+d_i)]^2 \leqslant [\lambda(c_i^T x_1+d_i)^2+(1-\lambda)(c_i^T x_2+d_i)^2]. \quad (2)$$

(a) Suppose that $\lambda \in R$ satisfies $0 \leqslant \lambda \leqslant 1$, and that $x_1, x_2 \in Z$. Since $\alpha\beta > 0$ and $x_1, x_2 \in Z$, it follows that, for each $j = 1, 2$,

$$a_i^T x_j + b_i = \frac{K}{\alpha} - \frac{\beta}{\alpha}(c_i^T x_j + d_i). \quad (3)$$

We have that

$$\begin{aligned}
&f_i[\lambda x_1 + (1-\lambda)x_2] = [a_i^T(\lambda x_1 + (1-\lambda)x_2) + b_i][c_i^T(\lambda x_1 + (1-\lambda)x_2) + d_i] \\
&= [\lambda(a_i^T x_1 + b_i) + (1-\lambda)(a_i^T x_2 + b_i)][\lambda(c_i^T x_1 + d_i) + (1-\lambda)(c_i^T x_2 + d_i)] \\
&= \{\frac{K}{\alpha} - \frac{\beta}{\alpha}[\lambda(c_i^T x_1 + d_i) + (1-\lambda)(c_i^T x_2 + d_i)]\}[\lambda(c_i^T x_1 + d_i) + (1-\lambda)(c_i^T x_2 + d_i)] \\
&= \frac{K}{\alpha}[\lambda(c_i^T x_1 + d_i) + (1-\lambda)(c_i^T x_2 + d_i)] - \frac{\beta}{\alpha}[\lambda(c_i^T x_1 + d_i) + (1-\lambda)(c_i^T x_2 + d_i)]^2 \\
&\geqslant \frac{K\lambda}{\alpha}(c_i^T x_1 + d_i) + \frac{K(1-\lambda)}{\alpha}(c_i^T x_2 + d_i) - \frac{\beta}{\alpha}[\lambda(c_i^T x_1 + d_i)^2 + (1-\lambda)(c_i^T x_2 + d_i)^2] \\
&= \lambda(c_i^T x_1 + d_i)\left[\frac{K}{\alpha} - \frac{\beta}{\alpha}(c_i^T x_1 + d_i)\right] + (1-\lambda)(c_i^T x_2 + d_i)\left[\frac{K}{\alpha} - \frac{\beta}{\alpha}(c_i^T x_2 + d_i)\right] \\
&= \lambda(c_i^T x_1 + d_i)(a_i^T x_1 + b_i) + (1-\lambda)(c_i^T x_2 + d_i)(a_i^T x_2 + b_i) \\
&= \lambda f_i(x_1) + (1-\lambda) f_i(x_2).
\end{aligned}$$

where the second equation follows by substituting (3) in the first equation and gathering terms, the inequality can be shown by using that $\alpha\beta > 0$ and (2). By the choices of λ, x_1, x_2, this proves (a).

(b) Assume that $\lambda \in [0, 1]$ and $x_1, x_2 \in Z$. The remainder of the proof of part (b) follows by using the same steps as in the prof of part (a), but with the directions of the inequalities reversed.

(c) Since $\alpha\beta = 0$, without loss of generality, we assume $\alpha = 0, \beta \neq 0$. Then Z is given by

$$Z = \left\{x \in X | c_i^T x + d_i = \frac{K}{\beta}\right\}. \quad (4)$$

So, part (c) easily follows from (4).

From the proof of Theorem 1, notice that the theorem holds also without the assumption that $a_i^T x + b_i$ and $c_i^T x + d_i, i \in I = \{i, 2, \cdots, p\}$, are positively-value functions on X. For $i \in I$, f_i is neither convex nor concave on X. However, Theorem 1 shows that, on subsets of appropriate hyperplanes in R^n, f_i can be convex, concave, or linear. This result may allow us in some cases to globally solve problems (LMP) by convex parametric programming techniques. To illustrate this, consider the following result and example.

For all $i \in I$, assume that there exist $\alpha_i, \beta_i \in R$ such that

$$\begin{aligned}
&\alpha_1(a_1^T x + b_1) + \beta_1(c_1^T x + d_1) \\
&= \alpha_i(a_i^T x + b_i) + \beta_i(c_i^T x + d_i) \\
&= a^T x + b = K,
\end{aligned} \quad (5)$$

where $a^T = (\alpha_1 a_1^T + \beta_1 c_1^T) \in R^n, b = (\alpha_1 b_1 + \beta_1 d_1) \in R, K \in R$. Now we construct problem (LMP_K) as follows:

$$(\text{LMP}_K)\ \max_{x\in X} \sum_{i=1}^{p} (a_i^T x + b_i)(c_i^T x + d_i),$$
$$\text{s.t.}\ \ a^T x + b = K.$$

Let $K_{\max} = \max_{x\in X}(a^T x + b)$, $K_{\min} = \min_{x\in X}(a^T x + b)$. Since X is nonempty, compact convex set, for each $K \in [K_{\min}, K_{\max}]$, $X^0 = \{x \in X | a^T x + b = K\}$ is nonempty, compact convex set.

Proposition 1. *For any $i \in I$, let α_i, β_i be not both zero.*

(a) *For all $i \in I$, if $\alpha_i\beta_i \geqslant 0$, problem ($LMP_K$) is parametric convex programming problem over convex set.*
(b) *For all $i \in I$, if $\alpha_i\beta_i < 0$, problem (LMP_K) is parametric concave minimization problem over convex set.*
(c) *Let $I' \subset I$ and $I - I' \neq \emptyset$, If $\alpha_i\beta_i \geqslant 0$ for all $i \in I'$, and $\alpha_i\beta_i < 0$ for all $x \notin I'$, problem (LMP_K) is parametric d.c. programming over convex set.*

Proof. (a) On the basis of Theorem 1, for all $i \in I$, if $\alpha_i\beta_i \geqslant 0$, $f_i(x) = (a_i^T x + b_i)(c_i^T x + d_i)$ are concave function over convex set $Z = \{x \in X | a^T x + b = K, K \in [K_{\min}, K_{\max}]\}$. Then $f(x) = \sum_{i=1}^{p} f_i(x)$ is concave function over Z. Since $\max f(x) = \min -f(x)$, problem (LMP_K) is parametric convex programming over convex set.

(b) According to Theorem 1, for all $i \in I$, if $\alpha_i\beta_i < 0$, $f_i(x) = (a_i^T x + b_i)(c_i^T x + d_i)$ are convex function over convex set $Z = \{x \in X | a^T x + b = K, K \in [K_{\min}, K_{\max}]\}$. Then $f(x) = \sum_{i=1}^{p} f_i(x)$ is convex function over Z. So problem LMP_K is parametric concave minimization problem over convex set.

(c) Without loss of generality, let $I' = \{1, 2, \cdots, k\}, k < p$. Since $\alpha_i\beta_i \geqslant 0$ for all $i = 1, 2, \cdots, k$, $f^1(x) = \sum_{i=1}^{k} f_i(x)$ is concave function over Z, and since $\alpha_i\beta_i < 0$ for all $x \notin I'$, $f^2(x) = \sum_{i=k+1}^{p} f_i(x)$ is convex function over Z. Then $f(x) = f^1(x) + f^2(x)$ is d.c. function over Z. This completes proof.

Example 1. Consider problem (LMP) with

$$\max_{x\in X} f(x) = f_1(x) + f_2(x),$$

where $f_1(x) = (-2x_1+3x_2-6)(3x_1-5x_2+3)$, $f_2(x) = (4x_1-5x_2-7)(-3x_1+3x_2+4)$, $X = \{x \in R^2 | x_1 + x_2 \leqslant 1.5, x_1 - x_2 \leqslant 0, x_1, x_2 \geqslant 0\}$.

Notice that

$$\begin{aligned}&0.5(-2x_1 + 3x_2 - 6) + 0.5(3x_1 - 5x_2 + 3)\\&= 0.5(4x_1 - 5x_2 - 7) + 0.5(-3x_1 + 3x_2 + 4)\\&= 0.5x_1 - x_2 - 1.5.\end{aligned}$$

Let $K = 0.5x_1 - x_2 - 1.5$, then we obtain

$$K_{\max} = \max_{x \in X}(0.5x_1 - x_2 - 1.5) = -1.5,$$
$$K_{\min} = \min_{x \in X}(0.5x_1 - x_2 - 1.5) = -3$$

By part (a) of Theorem 1, this implies that, for any $K \in [-3, -1.5]$, the parametric programming problem

$$(\text{LMP}_K) \ \max_{x \in X} f_1(x) + f_2(x), \quad \text{s.t. } 0.5x_1 - x_2 - 1.5 = K$$

involves maximizing a concave, continuous function over a nonempty, compact polyhedron. This implies also that if, for each $K \in [-3, -1.5]$, we let $v(K)$ denote the optimal objective function value of the parametric programming problem $(\text{LMP})_K$, then the global optimal value of problem (LMP) is equal to the optimal value of the problem

$$(\text{P1}) \ \max \ v(K), \quad \text{s.t. } -3 \leqslant K \leqslant -1.5,$$

and that, if K^* is an optimal solution for problem (P1), then any optimal solution for problem $(\text{LMP})_{K^*}$ is a globally optimal solution for problem (LMP). Since the function $v(K)$ is pointwise maximum of the concave function $f_1 + f_2$ over convex set X, the function $v(K)$ is concave on $K \in [-3, -1.5]$. Thus, to globally solve problem (P1), we may use, for instance, a simple bisection search for the maximum of $v(K)$ over $K \in [-3, -1.5]$. In this search, each evaluation of $v(K)$ involves solving the convex programming $(\text{LMP})_K$. We have performed this search and found, after 13 iterations, that K^* is approximately equal to -1.87492 and that the approximate globally optimal value $v(K^*)$ of problem (LMP) is -38.87628. From the solution of the convex programming problem $(\text{LMP})_K$ with $K = K^*$, we obtained the approximate global optimal solution $(x_1^*, x_2^*) = (0.74984, 0.74984)$ for problem (LMP).

Remark 1. If problem (P) is the following form:

$$(\text{LMP}') \quad \begin{cases} \min \sum_{i=1}^{p} (a_i^T x + b_i)(c_i^T x + d_i), \\ \text{s.t.} \ \ x \in X, \end{cases}$$

On the basis of part (b) of Theorem 1, we may also convert (LMP′) to the parametric programming problem that can involve a convex, concave, or d.c. continuous function over a nonempty, compact polyhedron. Then we may use well-known techniques, such as bisection search, to globally solve problem (LMP′).

3 Conclusion

In this paper, we show that, by using suitable transformations, there is a potential to globally solve problem (LMP) by techniques that are well know techniques.

Acknowledgments Thanks to the support by National Natural Science Foundation of China (No.70771030 and No.70271047) and Project Science Foundation of Guangdong University of Finance(No.11XJ02-12 and No.2012RCYJ005).

References

1. Matsui, T.: NP-Hardness of Linear Multiplicative Programming and Related Problems. J. of Global Optimization. **9**, 113–119 (1996)
2. Maranas, C.D., Androulakis, I.P., Floudas, C.A., Berger, A.J., Mulvey, J.M.: Solving long-term financial planning problems via global optimization. Journal of Economic Dynamics and Control. **21**, 1405–1425 (1997)
3. Quesada I., Grossmann I.E.: Alternative bounding approximations for the global optimization of various engineering design problems, in: I.E. Grossmann (Ed.), Global Optimization in Engineering Design, Nonconvex Optimization and Its Applications, Vol. 9, Kluwer Academic Publishers, Norwell, MA,pp. 309–331 (1996).
4. Mulvey, J.M., Vanderbei, R.J., Zenios, S.A.: Robust optimization of large-scale systems. Operations Research. **43**, 264–281 (1995)
5. Dorneich, M.C., Sahinidis, N.V.: Global optimization algorithms for chip design and compaction. Engineering Optimization. **25**(2), 131–154 (1995)
6. Bennett, K.P., Mangasarian, O.L.: Bilinear separation of two sets in n-space. Computational Optimization and Applications. **2**, 207–227 (1994)
7. Konno, H., Kuno, T., Yajima, Y.: Global optimization of a generalized convex multiplicative function. Journal of Global Optimization. **4**, 47–62 (1994)
8. Benson, H.P.: An outcome space branch and bound-outer approximation algorithm for convex multiplicative programming. Journal of Global Optimization. **15**, 315–342 (1999)
9. Zhou, X.G., Wu, K.: A accelerating method for a class of multiplicative programming with exponent. Journal of Computational and Applied Mathematics **223**, 975–982 (2009)
10. Bennett, K.P.: Global tree optimization: a non-greedy decision tree algorithm. Computing Sciences and Statistics. **26**, 156–160 (1994)
11. Falk, J.E., Palocsa, S.W.: Image space analysis of generalized fractional programs. Journal of Global Optimization. **4**(1), 63–88 (1994)
12. Benson, H.P., Boger, G.M.: Outcome-space cutting-plane algorithm for linear multiplicative programming. Journal of Optimization Theory and Applications. **104**(2), 301–322 (2000)
13. Benson, H.P., Boger, G.M.: Multiplicative programming problems: analysis and efficient point search heuristic. Journal of Optimization Theory and Applications. **94**(2), 487–510 (1997)

Affine SIFT Based on Particle Swarm Optimization

Xue-lian Cao, Guo-Rong Cai and Shui-li Chen

Abstract As for ASIFT, ASIFT has been proven to be invariant to image scaling and rotation. Specially, ASIFT enables matching of images with severe view point change and outperforms significantly the state-of-the-art methods. It accomplished this by simulating several views of the original images. However, we found that the simulated parameters are continuous, namely, transformations acquired by ASIFT cant express the real relationship between reference and input images. Therefore, a particle swarm optimization based sample strategy is presented in this paper. The basic idea is to search the best transform in continuous parameter space. Experimental results show that the proposed PSO-ASIFT algorithm could get more matches compared with the original ASIFT and SIFT.

1 Introduction

Image matching is the process of aligning different images of the same scene acquired at different periods of time, different viewing angles or different sensors [1]. Typically, feature-based image matching processes are made of the following four steps [2]: 1) feature detection, 2) feature matching, 3) mapping function design, 4) image transformation and re-sampling.

Some feature detectors, including Harris points [3], Harris-Laplace [4] and Hessian-Laplace detectors [4],have been proven to be invariant to rotation. On the

X. Cao
College of Mathematical and Computer Science, Fuzhou University, Fuzhou 350002, China
e-mail: cao_xuelian@126.com

G-R. Cai
College of Science, Jimei University, Xiamen 361021, China
e-mail: xiser@jmu.edu.cn

S. Chen (✉)
College of Science, Jimei University, Xiamen 361021, China
e-mail: sgzx@jmu.edu.cn

B.-Y. Cao and H. Nasseri (eds.), *Fuzzy Information & Engineering and Operations Research & Management*, Advances in Intelligent Systems and Computing 211,
DOI: 10.1007/978-3-642-38667-1_7,

other hand, to achieve viewpoint invariant, Hessian-Affine [5] and MSER (maximally stable extremal region) [6] have been proposed. The most widely used feature point detection and extraction method is the Scale-Invariant Feature Transform (SIFT) [7], which has been shown to be robust to image scaling and rotation and partially invariant to viewpoint changes. As pointed out in [8], SIFT descriptor is superior to many descriptors such as the distribution-based shape context [9], derivative-based complex filters [10], and moment invariants [11]. In recent years, a number of SIFT descriptor variants and extensions have been proposed including PCA-SIFT [12], GLOH (gradient location-orientation histogram) [13] and SURF (speeded up robust features) [14].

In 2009, Yu and Morel proposed a new variant of SIFT called Affine-SIFT (ASIFT) [15]. Mathematical proof reveals that ASIFT method is fully affine invariant. Also, experimental results obtained on multi-view images show that AISFT can be a robust tool for image matching. It is worth noting that AISFT uses discrete setting for image sampling. As a result, the more samples ASIFT generated, the more good matches will get. This strategy, although efficient, is a brute force strategy to find out the best transform between images to be matched. Since the sampling parameters t and ϕ are continuous in ASIFT, optimization method can be used to find out the best combination of t and ϕ. Therefore, in this paper, particle swarm optimization (PSO) has been introduced in parameter searching. Particularly, PSO algorithm is used to optimize the sampling parameters of ASIFT by searching the best matching in a continuous space. The purpose is to extract the best transformation between reference and input images. Experimental results show that the proposed method outperforms significantly the state-of-the-art ASIFT and SIFT.

2 Affine Scale Invariant Feature Transform (ASIFT)

2.1 The Affine Camera Model

Theorem 1 *[15]: Any affine map* $A = \begin{pmatrix} a & b \\ c & d \end{pmatrix}$ *with strictly positive determinant which is not a similarity has a unique decomposition:*

$$A = H_\lambda R_1(\psi) T_t R_2(\phi) = \lambda \begin{pmatrix} cos\psi & -sin\psi \\ sin\psi & -cos\psi \end{pmatrix} \begin{pmatrix} t & 0 \\ 0 & 1 \end{pmatrix} \begin{pmatrix} cos\phi & -sin\phi \\ sin\phi & -cos\phi \end{pmatrix}, \quad (1)$$

where $\lambda > 0$ and $\lambda^2 t$ is the determinant of A, R_i(i = 1, 2) are rotations matrices, $\phi \in [0, \pi]$, T_t is a tilt, namely a diagonal matrix with first eigenvalue $t > 1$ and the second one equal to 1.

Fig. 1(a) is the geometrical interpretation of the affine decomposition Eq. (1). The angles ϕ and θ are respectively the camera optical axis longitude and latitude. The other angle ψ denotes the camera spin, and λ is a zoom parameter. The tilt t is

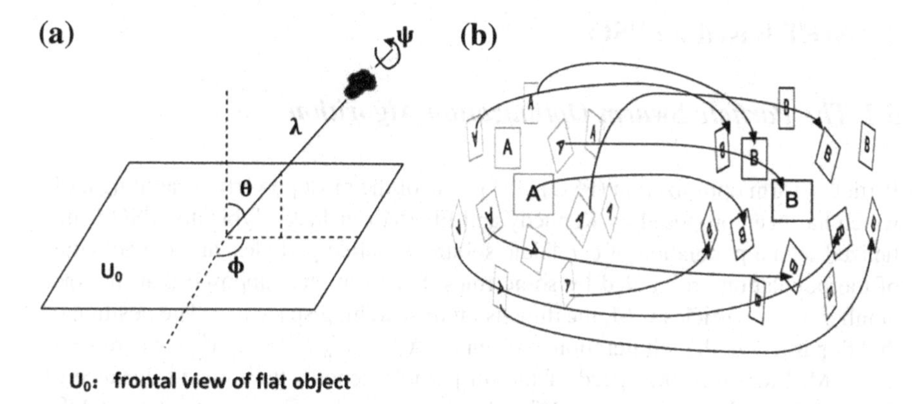

Fig. 1 Basic idea of ASIFT. **a** Geometric interpretation of the affine mapping; **b** Simulated images of reference image *A* and *B*. These images are used for feature extraction and matching

correlate with the latitude θ, namely $t = 1/cos\theta$. Since SIFT is invariant to rotation and scaling, only ϕ and t are sampled.

The parameter t in Eq. (1) is absolute tilt, which measures the tilt between the frontal view and a slanted view. The amount of tilt between two slanted images is quantified by the transition tilt τ. The tilt goes up when the transition tilt τ increases.

2.2 The Procedure of ASIFT Algorithm

In ASIFT, the author proposed a two-resolution strategy to speed up ASIFT. The reference images are sub-sampled by a 3×3 factor in the low-resolution phase. Then, the high-resolution ASIFT simulates the 5 best affine transforms from among all simulated images that yield most matches in the low-resolution process. Specially, the low resolution stage is conducted via the following two steps. 1) Each image to be matched is transformed by simulating many distortions caused by the change of camera optical axis orientation from a frontal position. The images undergo ϕ -rotations followed by tilts with parameter $t = |1/\cos\theta|$ as shown in Fig. 1(b). The square image A and B represent the compared images. The parallelograms represented the simulated images. 2) As shown in Fig. 1(b), all simulated affine images are feature extracted by the SIFT method and features are compared by the KD-Tree algorithm [16].

3 ASIFT Based on PSO

3.1 The Particle Swarm Optimization Algorithm

Particle swarm optimization (PSO) [17] is one of the most powerful searching tools which has been proposed by Kennedy and Eberhart in 1995. Typically, PSO is initialized with a population of candidate solutions, called particles, and the activities of the population are guided by some rules. For example, supposed that the total number of the particles is M, the dimension of searching space is d. The position of the ith particle in the nth iteration is given as: $X_i^n = (x_{i1}^n, x_{i2}^n, \ldots, x_{id}^n)$, where i = 1, 2, ..., M, Moreover, the speed of the ith particle denotes the moving distance of the particle and be expressed as: $V_i^n = (v_{i1}^n, v_{i2}^n, \ldots, v_{id}^n)$. Then the position and the speed of the $j(j = 1, 2, \ldots, d)$ dimension of the ith partical in the n+1 iteration are given as the following equations:

$$V_{ij}^{n+1} = \omega v_{ij}^n + c_1 r_1 (p_{ij} - x_{ij}^n) + c_2 r_2 (g_j - x_{ij}^n)), \tag{2}$$

$$X_{ij}^{n+1} = x_{ij}^n + v_{ij}^{n+1}, \tag{3}$$

where, ω denote the inertia weight, c_1 and c_2 stand for the acceleration factors, r_1 and r_2 are two random numbers in the ranges of [0, 1], the constant v_{max} is used to control the searching range. p_{ij} and g_j are respectively the best position of the current particle and the whole population.

3.2 ASIFT Algorithm Based on PSO

It is pointed out that the mutating particle swarm optimization (MPSO) [18] has been proven to be an efficient tool for parameter searching. Therefore, in this paper, MPSO is employed to optimize the sampling of the latitude and the longitude. We define a global parameter R and a random parameter r_i for each particle. During the iteration, the ith particle will mutate if its random parameter r_i is bigger than R. Otherwise the particle will update normally. Experiments show that PSO with the mutation factor is not only increased the diversity of the population but also prevented the algorithm converging untimely.

The coding strategy of each particle is given as follows:

t_{i1}	ϕ_{i1}	t_{i2}	ϕ_{i2}

where t_{i1} and t_{i2} are respectively the tilt parameters corresponding the reference and the input images of the ith particle, ϕ_{i1} and ϕ_{i2} are the camera optical axis longitude of two images, respectively.

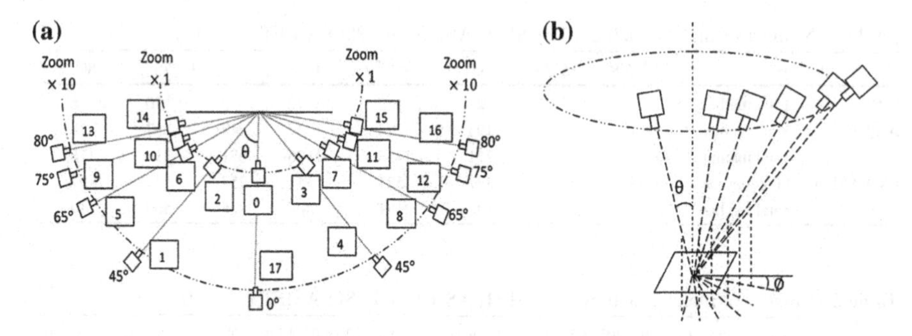

Fig. 2 Geometric interpretation of the **a** Absolute tilt tests, and **b** Transition tilt tests

With the code value of each particle, affine sample can be extract and then SIFT is used to extract feature points. Therefore, the fitness value is defined as the number of good matches achieved by affine samples.

In short, PSO-ASIFT can be summarized by the following five steps.

Step 1: Initialize the population of particles (randomly initialize M particles). Randomize the positions and velocities for entire population. Record the global best location p_g and the local best locations p_i of the ith particle, regarding to their fitness values.

Step 2: Updating the best position of each particle and the whole group according to Eq. (2) and Eq. (3).

Step 3: Check up the value of random number r_i of each particle, if r_i is bigger than the R, then turn to Step5, otherwise, turn to Step 4.

Step 4: Randomize the position for the current particle.

Step 5: If the stopping conditions are satisfied then stop iterating, otherwise, return to Step 2.

4 Experimental Results and Discussions

4.1 Description of the Experiments

In our experiments, PSO-ASIFT will be compared with the state-of-the-art method SIFT and original ASIFT. As for PSO-ASIFT, the population size is $M = 10$ and the maximal iterations are 100. To evalute the performance of PSO-ASIFT, the experimental content setting and the testing images of the literature [15] are used in this paper. Fig. 2 explain the testing set, and the resolution of these images is 600 × 450. The Fig. 2(a) illustrates the setting of the absolute tilt tests by the image be photographed with a latitude angle θ varying from 0° (frontal view) to 80°, from distances varying between 1 and 10. The Fig. 2(b) illustrates the setting of the transition

Table 1 Number of matches achieved by SIFT, ASIFT and PSO-ASIFT ($Z \times 10$)

	θ_2/t_2	$-80°/5.8$	$-75°/3.9$	$-65°/2.3$	$-45°/1.4$	$45°/1.4$	$65°/2.3$	$75°/3.9$	$80°/5.8$
SIFT	Total mathes	4	9	21	152	142	9	10	6
ASIFT	Max mathes	23	110	193	284	267	247	67	26
	Total mathes	149	354	771	1435	1506	631	313	116
PSO-ASIFT	Max mathes	39	190	337	521	541	257	145	97
	Total mathes	172	615	1137	2191	1808	826	560	219

Table 2 Number of matches achieved by SIFT, ASIFT and PSO-ASIFT ($Z \times 10$)

	ϕ_2/τ_2	$10°/1.9$	$20°/3.3$	$30°/5.3$	$40°/7.7$	$50°/10.2$	$60°/12.4$	$70°/14.3$	$80°/15.6$	$90°/16$
SIFT	Total mathes	27	23	2	1	11	2	0	0	0
ASIFT	Max mathes	118	88	52	34	29	25	20	18	13
	Final mathes	710	363	182	124	66	65	46	33	90
PSO-ASIFT	Max mathes	150	117	61	44	38	33	22	19	22
	Final mathes	1311	611	408	208	238	165	101	52	97

tilt tests by the images be photographed with a longitude angle ϕ that varied from 0° to 90°, from a fixed distance.

4.2 Results and Discussions

Tables 1, 2 summarize the performance of three algorithms in terms of the max number of good matches obtained via a single transformation, and the total number of matches.

Table 1 show the results of the experiment performed with the images taken with Zoom $\times$ 10 and the latitude angles range from 0° to 80°. The latitude angles and the absolute tilts are listed in the first row. In Table 1 we can see the performance of SIFT decays considerably when the angle goes over 75°, where ASIFT and PSO-ASIFT are sitll robust in the angle of 80°, and the performance of PSO-ASIFT exceed ASIFT.

Typical matching results are illustrated in Fig. 3 and Fig. 4, where Fig. 3 show the matches with −65° angle (tilt $t = 2.3$). The ASIFT and PSO-ASIFT find respectively 771 pairs and 1137 pairs matching. Note that SIFT only find 21 matches, this implies that SIFT isn't robust to large viewpoint change. The values of parameters of the best transformation achieved by ASIFT and PSO-ASIFT are given in Table 3. Figure 4 show the matches with tilt $t = 5.8$. The number of correspondences by ASIFT is 116 and PSO-ASIFT is 219. Note that the matching number achieved by PSO-ASIFT is about twice that of ASIFT, while SIFT completely failed. The parameters for the best transformation obtained by ASIFT and PSO-ASIFT are given in Table 3.

It is estimated the performance of the ASIFT and the PSO-ASIFT algorithms with different absolute tilts under comparison in above experiments. The perfor-

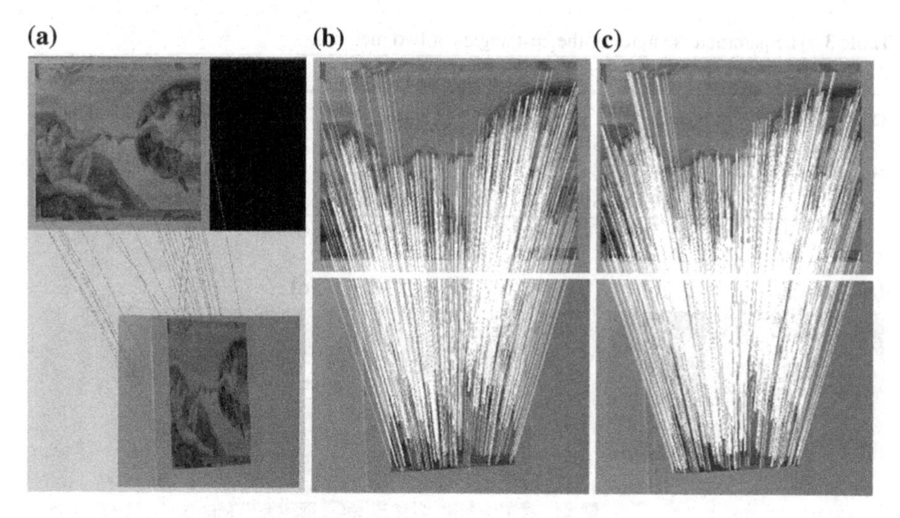

Fig. 3 Matches with *Zoom* × 10, $\theta_1 = 0°, t_1 = 1, \theta_2 = -65°, t_2 = 2.3$. PSO-ASIFT, ASIFT and SIFT find respectively **a** 1137, **b** 771 and **c** 21 matches

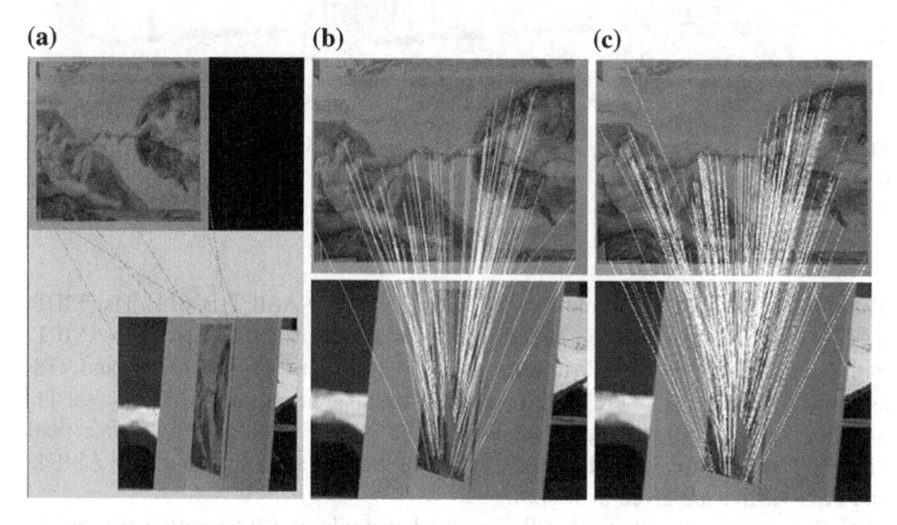

Fig. 4 Matches with *Zoom* × 10, $\theta_1 = 0°$, $t_1 = 1$, $\theta_2 = 80°$, $t_2 = 5.8$. PSO-ASIFT, ASIFT and SIFT find respectively **a** 219, **b** 116 and **c** 6 matches

mance of the two methods with different transition tilts is estimated by the following experiments.

Table 2 compares the performance of three algorithms with the images taken with the optimal *Zoom* × 4 and the absolute tilt $t = 4$ while the longitude angle ϕ growing from 0° to 90°. $\phi_1 = 0°$, ϕ_2 and the transition tilts τ are listed in the first row. Fig. 5 (a) and (b) show respectively the matching results with the longitude angle $\phi_2 = 20°$ (transition tilt $\tau = 3.3$) achieved by the SIFT, ASIFT and the PSO-ASIFT methods.

Table 3 The parameters values of the best angles of two methods

z × 10		t_1	$\phi_1(\text{o})$	t_2	$\phi_2(\text{o})$	mathes
$\theta_2 = -65°$ $t_2 = 2.3$	ASIFT	1.00	0	2.00	0	193
	PSO-ASIFT	1.93	87.09	1.00	91.67	337
$\theta_2 = 80°$ $t_2 = 5.8$	ASIFT	1.41	45.26	8.00	162.15	26
	PSO-ASIFT	5.05	89.38	1.00	99.12	97

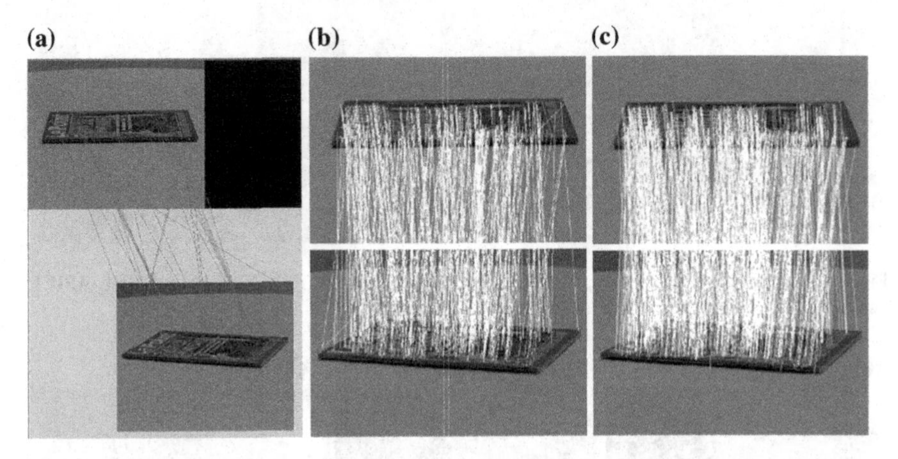

Fig. 5 Matches with $\phi_1 = 0°$, $\phi_2 = 20°$, $\tau = 3.3$. PSO-ASIFT, ASIFT and SIFT find respectively **a** 611, **b** 363 and **c** 23 matches

The number of matches found by ASIFT is 363, PSO-ASIFT is 611. The SIFT method only find 23 matches. The number of matches of ASIFT and PSO-ASIFT with latitude angle $\phi_2 = 50°$ (transition tilt $\tau = 10.2$) are respectively 66 and 238 as shown in Fig. 6 (b) and (c). Fig. 6 (a) show the matches found by SIFT, total 11 matches. The number of matches achieved by PSO-ASIFT is almost four times that of ASIFT. In addition, the parameters values of the best angles obtained by ASIFT and PSO-ASIFT are given in Table 4.

Under an absolute tilt $t = 4$, SIFT method struggle at 1.9 transition tilt, but it fail completely when the transition tilt gets bigger. ASIFT and PSO-ASIFT work perfectly up to the 16 transition tilt. The above experiments show that the maximum transition tilt, about 2 for SIFT, is by far insufficient. ASIFT outperforms significantly the SIFT method, while PSO-ASIFT superior compared with ASIFT.

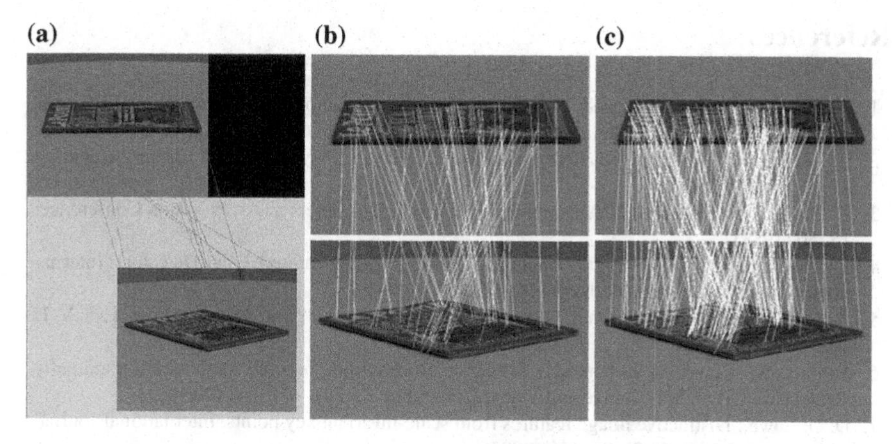

Fig. 6 Matches with $\phi_1 = 0°, \phi_2 = 50°, \tau = 10.2$. PSO-ASIFT, ASIFT and SIFT find respectively **a** 238, **b** 66 and **c** 11 matches

Table 4 The parameters values of the best angles of two methods

		t_1	$\phi_1(o)$	t_2	$\phi_2(o)$	mathes
$\phi_1 = 0°$ $\phi_2 = 20°$ $\tau = 3.3$	ASIFT	2.00	107.72	2.00	72.19	88
	PSO-ASIFT	1.00	13.75	3.53	64.74	117
$\phi_1 = 0°$ $\phi_2 = 50°$ $\tau = 10.2$	ASIFT	4.00	90.00	4.00	90.00	32
	PSO-ASIFT	4.89	91.72	3.67	91.72	38

5 Conclusion

In this paper, we presented a powerful method for image matching. The proposed PSO-SIFT algorithm employed PSO to optimize the sampling parameters of ASIFT by searching the best transformation in the continuous space. Experimental results demonstrated that the proposed PSO-ASIFT method achieved more matches compared with ASIFT. However, the time complexity of PSO-ASIFT is higher than ASIFT, since it generates more samples than ASIFT. Therefore, our future work will focus on how to reduce the complexity of the optimization-based methods, such as find out best transformation in an interpolated parameter space, rather than generate samples for feature generation and feature matching.

Acknowledgments The work was supported by Natural Science Foundation of Fujian Province of China (No.2011J01013), and Special Fund of Science, Technology of Fujian Provincial University of China (JK2010013) and Fund of Science, Technology of Xiamen (No. 3502Z20123022). Corresponding author: Professor Shui-li Chen

References

1. A. Wong, D. A. Clausi.: ARRSI: Automatic registration of remote-sensing images. IEEE Transaction on Geoscience and Remote Sensing 45(5): 1483–1493 (2007).
2. B. Zitova' and J. Flusser.: Image registration methods: a survey. Image and Vision Computing 21: 977–1000 (2003).
3. C. Harris and M. Stephens.: A combined corner and edge detector. Alvey Vision Conference: 15–50 (1988).
4. K. Mikolajczyk and C. Schmid.: Scale and Affine Invariant Interest Point Detectors. International Journal of Computer Vision 60(1): 63–86 (2004).
5. K. Mikolajczyk and C. Schmid.: An affine invariant interest point detector. Proc. ECCV 1: 128–142 (2002).
6. J. Matas, O. Chum, M. Urban, and T. Pajdla.: Robust wide-baseline stereo from maximally stable extremal regions. Image and Vision Computing 22(10): 761–767 (2004).
7. D. G. Lowe.: Distinctive image features from scale-invariant key points. International Journal of Computer Vision 60(2): 91–110 (2004).
8. K. Mikolajczyk and C. Schmid.: A performance Evaluation of Local Descriptors. IEEE Trans. PAMI 27(10): 1615–2005 (2005).
9. S. Belongie, J. Malik, and J. Puzicha.: Shape matching and object recognition using shape contexts. IEEE Transactions on Pattern Analysis and Machine Intelligence 24 (24): 509–522 (2002).
10. A. Baumberg.: Reliable feature matching across widely separated views. Proc. of the IEEE Conf. on Computer Vision and Pattern Recognition 1: 774–781 (2000).
11. L. J. V. Gool, T. Moons, and D. Ungureanu.: Affine/Photometric Invariants for Planar Intensity Patterns. Proceedings of the 4th European Conference on Computer Vision 1: 642–651 (1996).
12. Ke, Y.: R. Sukthankar.: PCA-SIFT: A more distinctive representation for local image descriptors. Proc. of the IEEE Conf. on Computer Vision and. Pattern Recognition **2**, 506–51 (2004)
13. K. Mikolajczyk and C. Schmid.: A Performance Evaluation of Local Descriptors. IEEE Transactions on Pattern Analysis and Machine Intelligence 27 (10): 1615–1630 (2005).
14. H. Bay, T. Tuytelaars, and L. Van Gool.: Surf: Speeded up robust features. European Conference on Computer Vision 1: 404–417 (2006).
15. J. M. Morel and G. Yu.: ASIFT: A New Framework for Fully Affine Invariant Image Comparison. SIAM Journal on Imaging Sciences 2(2): 438–469 (2009).
16. J. L. Bentley.: Multidimensional binary search trees used for associative searching. Communications of the ACM 18(9): 509–517 (1975).
17. Kennedy and Eberhart RC.: Particle Swarm Optimization. Proceeding of IEEE International Conference on Neural Networks. Piscataway, NJ: IEEE service center: 1942–1948 (1995).
18. B. Y. You, G. L. Chen, and W. Z. Guo.: Topology control in wireless sensor networks based on discrete particle swarm optimization. Proceeding of IEEE International Conference on Intelligent Computing and Intelligent Systems 11: 269–273 (2009).

Genetic Quantum Particle Swarm Optimization Algorithm for Solving Traveling Salesman Problems

Wei-quan Yao

Abstract This paper presents a Genetic Quantum Particle Swarm Optimization (GQPSO) algorithm to solve Traveling Salesman Problems (TSPs). This algorithm is proposed based on the concepts of swap operator and swap sequence by introducing crossover, mutation and inverse operators in Genetic Algorithm (GA). Our algorithm overcomes such drawbacks as low convergence rate and local optimum when using Particle Swarm Optimization (PSO) or Quantum Particle Swarm Optimization (QPSO) algorithm to solve TSP. The experiment result shows that GQPSO algorithm has very powerful global search ability and its convergence rate is sharply accelerated compared to that of QPSO algorithm. GQPSO algorithm will have very good application prospects in solving combinational optimization problems.

Keywords TSP · PSO algorithm · GQPSO algorithm · Inverse operator · Combinational optimization.

1 Introduction

Particle Swarm Optimization (PSO) algorithm was firstly proposed by Kennedy and Eberhart in 1995 [1]. Since PSO is simple and easy to be realized, with just fewer control parameters to tune, it caught attentions from researchers from both domestic and overseas. Theoretically, however, PSO is not an algorithm with global convergence. To overcome disadvantages of PSO, Jun Sun et al. proposed a new version of PSO, Quantum-behaved Particle Swarm Optimization (QPSO) algorithm in 2004, which is global convergent and powerful in locating optimal solutions with fewer control parameters and fast convergence rate. Since it was proposed, this algo-

W. Yao (✉)
Mathematical teaching and research section of Guangzhou City Polytechnic,
Gungzhou 510405, P.R.China
e-mail: yaowq6916@126.com

B.-Y. Cao and H. Nasseri (eds.), *Fuzzy Information & Engineering and Operations Research & Management*, Advances in Intelligent Systems and Computing 211,
DOI: 10.1007/978-3-642-38667-1_8,

rithm has been used in solving many optimization problems [3]. Compared to other intelligent algorithms, such as genetic algorithm (GA), simulated annealing (SA) algorithm, etc., QPSO is more powerful on some function optimization problems. QPSO was successfully applied in dealing with continues optimization problems, but little attention was focused on discrete problems.

The travelling salesman problem (TSP) could be described as finding out a shortest path that could visit each city once and only once. TSP is a famous and typical NP problem, belonging to combinatorial optimization problems [4]. In literature [5], a special PSO algorithm for TSP was proposed by introducing the concepts of swap operator and swap sequence. But this PSO is slow in solving TSP, QPSO is employed in [6] to overcome this shortcoming. However, there are few works that combine GA and QPSO for travelling salesman problems. In this paper, we introduced crossover and mutation operators into QPSO and proposed Genetic Quantum-behaved Particle Swarm Optimization (GQPSO) algorithm for TSP. It was shown by the experimental results that GQPSO is faster than QPSO with higher efficiency. The proposed algorithm in this paper is a new try for combinatorial optimization problems.

2 Particle Swarm Optimization Algorithm

2.1 The Standard PSO Algorithm

Just like genetic algorithm, PSO is also based on population and evolution mechanism, which is specifically described as follows: In a D-dimensional target searching space, M particles representing potential solutions form a population, denoted as $X = \{X_1; X_2; \cdots; X_M\}$. At t th generation, the position of particle i is $X_i(t) = [X_{i,1}(t), X_{i,2}(t), \cdots, X_{i,D}(t)]$ and the velocity is $V_i(t) = [V_{i,1}(t), V_{i,2}(t), \cdots, V_{i,D}(t)]$, $i = 1, 2, \cdots, M$. The personal best position of particle i is denoted as $\mathbf{pbest}_i(t) = [pbest_{i,1}(t), pbest_{i,2}(t), \cdots, pbest_{i,D}(t)]$, while the global best position is $\mathbf{gbest}_i(t) = [gbest_{i,1}(t), gbest_{i,2}(t), \cdots, gbest_{i,D}(t)]$ such that $\mathbf{gbest}(t) = pbest_g(t)$, where $g \in \{1, 2, \cdots, M\}$is the subscript at which the global best position locates among all the personal best positions. At t+1 th generation, the velocity and position of particle i are updated by the following equations [3]:

$$V_{i,j}(t+1) = \omega V_{i,j}(t) + c_1 r_{1,j}[pbest_{i,j}(t) - X_{i,j}(t)] + c_2 r_{2,j}[gbest_j(t) - X_{i,j}(t)], \tag{1}$$

$$X_{i,j}(t+1) = X_{i,j}(t) + V_{i,j}(t+1), \tag{2}$$

where subscript $i(= 1, 2, \cdots, M)$ means the i th particle in the population and $j(= 1, 2, \cdots, D)$ is the j th dimension of this particle; D is the dimension of the searching space and M is the population size. c_1 and c_2 are acceleration constants, while $r_{1,j}$ and $r_{2,j}$ are two random numbers uniformly distributed over interval [0,1]. ω called inertia weight is used to balance global and local search ability. In order to control the flying velocity within a reasonable range, $V_{i,j}$ is often restricted within interval $[-V_{\max}, V_{\max}]$, where $V_{\max}$ is a positive value. If the searching area is limited to $X_{i,j} \in [lb, ub]$, where lb and ub are the lower and upper bounds of the searching area respectively, then $V_{\max}$ could be set to $V_{\max} = k \cdot (ub - lb), 0.1 \leq k \leq 1$.

2.2 *Quantum Particle Swarm Optimization Algorithm*

Sun et al. [3] proposed a new PSO model which is based on DELTA potential well from the viewpoint of quantum mechanics. They believed particles had quantum behaviors, and put forward QPSO algorithm based on this model. In quantum space, particle could search in the whole feasible solution space, so the global convergence of QPSO is much better than that of standard PSO. In QPSO, particles only have position vector without velocity vector, and the updating equations of position are :

$$p_{i,j} = \phi_j(t) \cdot p_{i,j}(t) + [1 - \phi_j(t)] \cdot G_j(t) \quad \phi_j(t) : U(0,1), \tag{3}$$

$$X_{i,j}(t,1) = p_{i,j}(t) \pm \alpha \cdot |C_j(t) - X_{i,j}(t)| \cdot \ln[1/u_{i,j}(t)] \quad u_{i,j}(t) : U(0,1), \tag{4}$$

where

$$C(t) = (C_1(t), C_2(t), \cdots, C_N(t)) = \frac{1}{M}\sum_{i=1}^{M} P_i(t)$$
$$\left(\frac{1}{M}\sum_{i=1}^{M} P_{i,1}(t), \frac{1}{M}\sum_{i=1}^{M} P_{i,2}(t), \cdots, \frac{1}{M}\sum_{i=1}^{M} P_{i,N}(t)\right). \tag{5}$$

In formulas (3),(4) and (5), $p_i(t)$ is the local attractor of particle i, and $C(t)$ denotes the mean best position of the whole swarm defined as the mean value of all personal best positions; $\phi_j(t)$ and $u_{i,j}(t)$ are random numbers distributed on [0,1] uniformly. α is called contraction-expansion (CE) coefficient, the only control parameter to be tuned, and it is usually set to $\alpha = (1.0-0.5) * (G_{\max} - t)/G_{\max} + 0.5$. The meanings of $G_j(t)$,i, j, M and N are the same in Sect. 2.1

The pseudo code of QPSO algorithm is described as follows:

```
Initialize position vectors of particles by using certain strategy
While terminal condition is not met
for i=1:M
        if f(X_i) < f(P_i)
                P_i = X_i
        end
end for
Pg = min(P_i)
Calculate the mean best position C
```

```
for i = 1 : M
        for j = 1 : N
                φ = rand(0, 1)
                p_{i,j} = φ * P_{i,j} + (1 − φ) * G_j
                u = rand(0, 1)
                if rand(0,1) < 0.5
                        X_{i,j} = p_{i,j} + α* |C_j − X_{i,j}|* ln(1/u)
            else
                        X_{i,j} = p_{i,j} − α* |C_j − X_{i,j}|*ln(1/u)
            end
        end for
end for
end while
```

3 The Traveling Salesman Problems

Given the distances of each pair of cities among N cities, determine the shortest path that visits each city one and only once. And this could be described as follows [7]: Suppose the set of N cities is $C = \{c_1, c_2, \cdots, c_N\}$, and the distance of each pair of cities is denoted by $d(c_i, c_j) \in R^+$, where $c_i, c_j \in C(1 \le i, j \le N)$. Our aim is to find out a sequence of cities $\{c_{\pi(1)}, c_{\pi(2)}, \cdots, c_{\pi(N)}\}$ that would make the objective function

$$T_d = \sum_{i=1}^{N-1} d(c_{\pi(i)}, c_{\pi(i+1)}) + d(c_{\pi(N)}, _{\pi(1)}), \tag{6}$$

minimum, where $\pi(1), \pi(2), \cdots, \pi(N)$ is a full permutation of $1, 2, \cdots, N$. Now we give some definitions of TSP.

Definition 3.1 Suppose the solution of TSP with N cities is $X = (x_k), k = 1, 2, \cdots, N$, define swap operator $SO(i, j)$ as interchanging x_i and x_j in solution X. Then $X' = X + SO(i, j)$ is a new solution after executing operator $SO(i, j)$, and here "+" indicates the implementation of operator.

Example 3.1 Let $X = (354126)$ and $SO(3, 4)$, then $X' = X + SO(i, j) = (35412\,6) + \mathrm{SO}(3, 4) = (351426)$.

Definition 3.2 An ordered sequence made up of one or more swap operators is called a swap sequence, denoted by SS, and it is defined as $SS = (SO_1, SO_2, \cdots, SO_n)$. Difference swap sequences may result in the same new solution after executing on a same solution. The set of swap sequences that product the same results is called an equivalent set of swap sequences. In this equivalent set o, the swap sequence holds the minimum number of swap sequences is basic swap sequence, whose construction method is referred to literature [5].

4 GQPSO Algorithm for TSP

4.1 The Design of GQPSO

The updating equations of QPSO, i.e. (3),(4) and(5), are for continuous optimization problems. In order to apply QPSO algorithm in TSP, the updating equations should be reconstructed as follows [8]:

$$p_i(t) = \phi(t) \cdot P_i(t) + [1 - \phi(t)] \cdot G(t) \quad \phi(t) : U(0, 1), \tag{7}$$

$$X_i(t + 1) = p_i(t) \pm \alpha \cdot |C(t) - X_i(t)| \cdot \ln[1/u(t)] \quad u(t) : U(0, 1), \tag{8}$$

where $\phi(t)$ and $u(t)$ are random numbers uniformly distributed over [0,1]; p_i in (7) can be obtained by crossover operation as in genetic algorithm, i.e. generate two offspring with P_i and G, and randomly select one as p_i. The mean best position $C(t)$ in (8) could be got by the following method: first, calculate the mean value *faverage* of paths corresponding toP_i, then the one that is closest to *faverage* among all P_is is chosen as $C(t)$. $|C(t) - X_i(t)|$ means the basic swap sequence of $C(t)$ and $X_i(t)$, and $\alpha \cdot |C(t) - X_i(t)| \cdot \ln[1/u(t)]$ indicates that the swap operator in $|C(t) - X_i(t)|$ is kept as the probability of $\alpha \ln[1/u(t)]$. If $\alpha \ln[1/u(t)] > 1$, then it set to 1.

Based on the reconstructed equations, this paper introduces crossover, mutation and inverse operators as in genetic algorithm and proposes GQPSO algorithm. The details are described as follows [9, 10].

4.1.1 Coding Scheme and Initialization of Particles

Just like GA, chromosomes in GQPSO are also coded as the order of visited cities. This coding scheme implies the legal constraint condition of TSP, that is there is no repeated emergence of any city. This paper modifies the initialization method of particles using greedy algorithm. Specifically, first randomly select a starting city, then pick out a city that is closest to this starting city. Don't stop this operation until all cities are visited. In this paper, 20 % particles are generated by this greedy method, and the remaining are randomly produced.

4.1.2 The Design of Local Attractor P_i

The local attractor Pi of particle i is obtained by crossing P_i and G. The crossover operator is Partially Matched Crossover (PMX) that randomly selects a section in parents and uses elements in this section to define a series of swaps. By executing these swaps on each parent string independently, the offspring chromosomes will be generated. For example, suppose $P_i = [6 \;\; 3 \;\; 7\vdots 8 \;\; 5 \;\; 1 \;\; 2\vdots 4 \;\; 9 \;\; 10]$, $G = [6 \;\; 1 \;\; 7\vdots 4 \;\; 9 \;\; 5 \;\; 8\vdots 3 \;\; 10 \;\; 2]$, then the swaps in the selected section are $8 \leftrightarrow$

4|5 ↔ 9|1 ↔ 5|2 ↔ 8. Please notice that 8 and 5 are linking elements of 4,2 and 9,1 respectively, so the merged swaps are 4 ↔ 2|9 ↔ 1. By using these swaps, the generated offspring strings are $q_1 = [6\ \ 3\ \ 7\ \ 8\ \ 5\ \ 9\ \ 4\ \ 2\ \ 1\ \ 10]$, $q_2 = [6\ 9\ 7\ 2\ 1\ 5\ 8\ 3\ 10\ 4]$. P_i is randomly selected among q_1 and q_2.

4.1.3 Mutation and Inverse Operators

In this paper, the swap mutation is chosen: randomly select two points in strings and swap their values. For example, for $p = [5\ 4\ 10\ \bar{1}\ 8\ 6\ \bar{9}\ 3\ 2\ 7]$, if the two randomly selected points are 4 and 7, then we get $p = [5\ 4\ 10\ \bar{\bar{9}}\ 8\ 6\ \bar{\bar{1}}\ 3\ 2\ 7]$after swap. In order to improve the local search ability, we introduce evolutionary inverse operator. The so called inverse is randomly selecting two points and inserting word strings between the above two points in the original positions in inverted sequence [3]. For example, suppose $p = [3\ 4\ 9\ 7\ 8\ 6\ 10\ 5\ 2\ 1]$, the new string after inverse is $p' = [3\ \ 10\ \ 6\ \ 8\ \ 7\ \ 9\ \ 4\ \ 5\ \ 2\ \ 1]$ if the inverse points are 2 and 7 respectively. In this paper, we adopt a one-way and multi-time inverse operator [3].

4.2 The Flowchart of GQPSO for TSP

The flowchart of GQPSO for TSP is described as follows in pseudo-code form [11].

```
Use greedy strategy initializing particles
t = 0
  while t < Gmax
    α = (1.0 - 0.5)*(Gmax - t)/Gmax + 0.5;
      for i=1:M
                if f(Xi) < f(Pi)
                        Pi = Xi
                end
end for
Pg = min(Pi)
Calculate the mean best position C
for i=1:M
        pi = crossover(Pi , G)
        Get basic swap sequence v = |C(t) − Xi(t)|
        u = rand(0,1), w = α* ln(1/u)
        Xi = change(pi,v, w) % Keep the swap operators in v with probability w
and execute it on pi
        Xi = mutation(Xi, Pm) % Mutate Xi with probability Pm
        Xi = inverse(Xi) % Inverse Xi
end for
t = t + 1
end while
```

Table 1 Performance comparison between algorithms

	This paper	Literature [6]	Literature [5]
Particle swam size	14	14	100
Mean iterations	278	4,000	20,000
Mean searching space	14×278=3892	14×4000=56000	$100 \times 20{,}000$=2000000
Searching space/ solution space	3892/(13!/2) % = 0.000125 %	56000/(13!/2) = 0.0018 %	2000, 000/(13!/2) = 0.064 %

5 Experimental Results

We use 14-cities TSP benchmark problem to test and verify the effectiveness of our GQPSO. The coordinates of 14 cities are: CP = [16.47 96.10; 16.47 94.44; 20.09 92.54; 22.39 93.37; 25.23 97.24; 22.00 96.05; 20.47 97.02; 17.20 96.29; 16.30 97.38; 14.05 98.12; 16.53 97.38; 21.52 95.59; 19.41 97.13; 20.09 94.55]. And the real minimum value of this problem is d*=30.8785. We have the following experimental parameters: mutation probability P_m=0.4, particle swarm size N = 14, max number of iterations Gmax = 4000, the inverse length decrease from (1/5)*n to 2 progressively, where n is the length of the code. The testing environment is : Pentium(R) Dual-Core CPU T4500 @2.30GHz 2.29 GHz, 1.99GB RAM. Among 20 runs, there are 15 runs that convergent the optimal solution: 7→13→8→11→9→10→1→2→14→3→4→5→6→12. And the mean iterations are 278. From the above results we could see that GQPSO can find the known optimal solution in a very short time. Now we compare our algorithm to the algorithms in [5] and [6], and the results are listed in Table 1.

From Table 1 we can see that the mean iterations of GQPSO is 7/100 times of that of QPSO, and the mean searching space of our algorithm is 0.000125 % of the whole solution space. Therefore, GQPSO obviously overmatches QPSO in terms of efficiency.

6 Conclusion

To overcome the shortcomings of PSO and QPSO when solving TSP, this paper proposed GQPSO based on the concepts of swap operator and swap sequence by introducing crossover, mutation and inverse operators in Genetic Algorithm. One 14 cities benchmark problem was used to test GQPSO. It was shown by the experimental results that GQPSO could find the real minimum in very short time whose convergence speed was obviously faster than that of QPSO. GQPSO has wide application prospect for combinatorial optimization problems.

References

1. Eberhart R, Kennedy J.A: New optimizer using particles Swarm theory. In: Proceeding Sixth International Symposium on Micro Machine and Human Science. Nagoya, Japan: IEEE Service Center, Piscataway, 39–43 (1995)
2. Sun, J., Feng, B., Xu, W.: Particle swarm optimization with particles having quantum behavior. Evol. Comput. **1**, 325–331 (2004)
3. Sun J.: Particle swarm optimization with particles having quantum behavior. Wuxi:Jiangnan University (2009)
4. Zhou, Weiyuan: Review on particle swarm optimization algorithm for TSP. Sci. Technol. Inf. **149**(11), 207–221 (2008)
5. Huang, L., Wang, K.P., Zhou, C.G.: Particle swarm optimization for traveling salesman problems. J. Jilin Univ. Ed. Sci. **42**(4), 477–480 (2003)
6. Pan-rong, L.I., Wenbo, X.: Solve traveling salesman problems based on QPSO. Comput. Eng. Des. **28**(19), 4738–4740 (2007)
7. Liang, Y., Wu, C., Shi, X.: Swarm intelligence optimization algorithm theory and Application. Science Press, Beijing (2009)
8. Jingsong, Su, Zhou, Changle, Jiang, J.: An Improved evolutionary algorithm for traveling salesman problem based on inver-over operator. Comput. Technol. Dev. **17**(7), 94–97 (2007)
9. Zhong, Yubin: The thought of fuzzy mathematics modeling is inltratedin MSMT. Adv. Intell. Soft Comput. **2**(62), 233–242 (2009)
10. Yubin, Z.: The FHSE model of software system for synthetic valuating engerprising. J. Guangzhou Univ. **4**(4), 316–320 (2005)
11. Wang, Y., Wanliang, W.: An improved quantum particle swarm optimization for solving production planning problem. Comput. Simul. **28**(1), 234–237 (2011)

An Improved Optimization Algorithm for Solving Real Eigenvalue and Eigenvector of Matrixes

Wei-quan Yao

Abstract In this paper, aiming to papers [1], [2]'s deficiency, we propose a method which changes to use the Excel's single variable equations and regional operations with the Excel VBA programming. Finally, we get a method to solve the matrix eigenvalues and the eigenvectors by using Excel. The facts show that the method in this paper is better than the references' methods.

Keywords Excel VBA · Matrix · Eigenvalue · Eigenvector · Iteration.

1 Introduction

Due to the visible then output characteristics, it has unique advantages to sort data and management data by using Excel. But if we want to do a further deepen processing on the data, it often require complex mathematical calculations, for instance, to solve the matrix eigenvalues and eigenvectors. About in Excel how to solve the matrix eigenvalues and eigenvectors, paper [1] and paper [2] show us different methods. But those methods have shortages, for example, it is difficult to determine the initial value or unable to solve multiple linear independence eigenvectors of a same eigenvalue. They even have defect.

For convenience, in this paper we put the method of solving matrix eigenvalues and eigenvectors in paper [1], and the method of solving matrix eigenvalues in paper [2] respectively called paper [1] method and paper [2] method. Below let's first discuss their shortages.

Paper [1] method is through the given initial value of an eigenvalue, and then using the method of solving single variable equations in Excel, to obtain the eigenvalue

W. Yao (✉)
Mathematical teaching and research section of Guangzhou City Polytechnic,
Guangzhou 510405, People's Republic of China
e-mail: yaowq6916@126.com

B.-Y. Cao and H. Nasseri (eds.), *Fuzzy Information & Engineering and Operations Research & Management*, Advances in Intelligent Systems and Computing 211,
DOI: 10.1007/978-3-642-38667-1_9, © Springer-Verlag Berlin Heidelberg 2014

which is the most close to the initial value; then we take an approximate value of the eigenvalue and a nonzero initial eigenvector, iteration solving the eigenvectors by using inverse power method.

In this method, there are four shortages: (1) It didn't give a method for determining the initial value of an eigenvalue; (2) We can not find out the multiple root of a characteristic equation. (3) Are the eigenvector initial values corresponding to different eigenvalues the same? How to solve it if its corresponding eigenvalue has multiple linear independence eigenvectors? (4) This method does not conducive to the application for further deepening because it is using step-by-step "manual" to solve and the last result at the end of the iteration when seeking eigenvectors is also by the people subjective judgment [3–5].

Paper [2] method is transforming the QR algorithm which is calculating the real symmetric matrix eigenvalues into a VBA program, and then to achieve auto-complete solving the real symmetric matrix eigenvalues.

This method has three shortages: (1) For a real non-symmetric matrix, the solution obtained by this method may be wrong, for example, the solving results of $\begin{bmatrix} 3 & 1 & 0 \\ 1 & 3 & 2 \\ 1 & 1 & 2 \end{bmatrix}$ is 5, 2, 1, but 5 and 1 are not the eigenvalues of original matrix. (2) It does not give a method to solve the eigenvectors, which makes its application at a discount. (3) The algorithm has defects, for example, in the process wilkinsonQR we would use the value d = (v(1, 1) − v(2, 2)) / 2 as the denominator, if this value is zero, it would appear error interrupt, such as solving the eigenvalues of real symmetric matrix $\begin{bmatrix} 3 & 0 & 3 \\ 0 & 1 & 3 \\ 3 & 3 & 1 \end{bmatrix}$. Here aim at the shortages of paper [1] method and paper [2] method, we change to use the Excel's single variable equations and regional operations, propose an improved method and demonstrate the efficiency of this algorithm through the practical examples. Below we will introduce the theory and the application of this improved method [6–10].

2 The Design of the Improved Optimization Algorithm

2.1 Solving the Eigenvalue Initial Value

Here we modify the QR algorithm of paper [2] to calculate the eigenvalue initial values of an order phalanx A preliminarily, and make $\lambda_1 \geq \lambda_2 \geq \cdots \geq \lambda_n$.

(1) Insert a command line
If d = 0 Then d = 1E − 13
after the command line d = (v(1, 1) − v(2, 2)) / 2 in the process wilkinsonQR( ).
This makes us can solve the approximations for real matrix eigenvalue(s) even the denominator is zero.

(2) In the process QR( ), in order to sort and checkout the operational results of 1) and to finish the subsequent calculations, we add local variables and add program segments after command line Loop Until q = 1.

The VBA macro code of added program segments is as follows:

```
.Cells(1, 0).Value = "eigenvalue"
For I = 1 To n
.Cells(1, I).Value = .Cells(I, I).Value
Next I
Range(.Cells(1, 1), .Cells(1, n)).Select
Selection.Sort Key1:=.Cells(1, 1), Order1:=xlDescending, Header:=xlGuess _
, OrderCustom:=1, MatchCase:=False, Orientation:=xlLeftToRight, _
SortMethod:=xlPinYin, DataOption1:=xlSortNormal
Range(.Cells(2, 1), .Cells(n + 1, n)).Clear
For I = 1 To n
For J = 1 To n
rngI.Cells(I, J).Value = 0
Next J
rngI.Cells(I, I).Value = 1
Next I
For I = 1 To n
Cells(0,I).Formula = "=MDETERM("+rng0.Address+"-"+.Cells(1,I).Address_
+ "*" + rngI.Address + ")"
d = .Cells(0, I).Value
.Cells(0, I).Clear
If Abs(d) > 1E-6 Then
Rd1 rng0
Exit For
End If
Next I
rngI.Clear
Ru rng0
.Cells(3, 0).Value = " eigenvector "
```

2.2 To Checkout the Eigenvalue Initial Value

We checkout the eigenvalue initial value λ_i $(i = 1, 2, \cdots, n)$, if it satisfy the characteristic equation, we turn to solve the eigenvectors or else we use $\lambda_{0i} = \lambda_i$ $(i = 1, 2, \cdots, n)$ as the initial value to implement iteration solving eigenvalues.

(1) To establish the equation $|A - \lambda E| = 0$;
The VBA macro code which is calculating characteristic equation determinant corresponding to i-th eigenvalue is as follows:

.Cells(0, I).Formula = "=MDETERM(" + rng0.Address + "-" + .Cells(1, I). Address + "*" + rngI.Address + ")"

(2) To test and to verify the eigenvalue λ_{0i} $(i = 1, 2, \cdots, n)$, if all the eigenvalues satisfy equation, turn to 4, or else implement 3.

2.3 *Solving Eigenvalues Iteratively*

To initialize λ_{0i} $(i = 1, 2, \cdots, n)$ into λ'_i $(i = 1, 2, \cdots, n)$, establish single variable equations, solve the true solution λ_i $(i = 1, 2, \cdots, n)$ of eigenvalues iteratively.

(1) To establish single equation $|A - \lambda E| = 0$;
(2) Do initialization processing on real eigenvalues obtained previously. Make $\lambda_{0i} = \lambda_i$ $(i = 1, 2, \cdots, n)$, $\lambda_1 = \lambda_{01} + 2|\lambda_{01}|$ and $\lambda_n = \lambda_{0n} - 2|\lambda_{0n}|$. If $\lambda_{0i} = \lambda_{0i+1} = \cdots = \lambda_{0i+j}$ $(i = 1, 2, \cdots, n - j - 1, j > 1)$, when i>1, then $\lambda_i = \lambda_{0i} + |\lambda_{0i} - \lambda_{0i-1}|/3$ and $\lambda_{i+j} = \lambda_{0i+j} - |\lambda_{0i+j} - \lambda_{0i+j+1}|/3$. This makes the distribution of initial value more reasonable.
(3) Process the real eigenvalues initial value dynamically, to solve eigenvalues by using VBA macro code to program through the method of solving single variable equation. Make $\lambda_{0i} = \lambda_i$ $(i = 1, 2, \cdots, n)$, λ_{0i} $(1 \le i \le n)$ is the initial value, solving the equation $|A - \lambda E| = 0$, the solution denoted as λ_i $(1 \le i \le n)$. If $\lambda_{0i} \ge \lambda_{i-1}$ $(i = 2, 3, \cdots, n)$, then we make $\lambda_{0i} = \lambda_{i-1} - 2|\lambda_{i-1} - \lambda_i|/3$ as the initial value to solve the equation $|A - \lambda E| = 0$.

    The VBA code of solving single variable equations is as follows:

```
Cells(0, I).Formula = "=MDETERM(" + rng.Address + "-" + _
rngRd.Cells(1, I).Address + "*" + .Address + ")"
.Cells(0, I). Select
.Cells(0, I).GoalSeek Goal:=0, ChangingCell:=rngRd.Cells(1, I)
```

(4) Make verification and validation on the results of iteration once again and mark λ_i $(i = 1, 2, \cdots n)$ which is not satisfy$|A - \lambda_i E| = 0$, so that it does not participate in the subsequent corresponding eigenvectors calculations, and then complete the iterative operation.

In order to make the iterated results more accurate, at the beginning of this process, we set the most iterations and maximum error value through below VBA code:

```
With Application
  .MaxIterations = 1000 'The most iterations
  .MaxChange = 0.000001 ' The maximum error value
End With
ActiveWorkbook. PrecisionAsDisplayed = False
```

2.4 Solving the Eigenvectors Corresponding to Eigenvalue λ_i $(i = 1, 2, \cdots, n)$

We take eigenvector initial values $X_{0i} = (1, 1, \cdots, 1)^T$ for eigenvalues which are not equal, iterative solving eigenvectors. We take eigenvector initial values $X_{0i} = (1, 1, \cdots, 1, 10000000000, 1, \cdots, 1)^T$ $(i = 2, 3, \cdots, n)$ for eigenvalues which are equal, iterative solving eigenvectors. The process is designed as follows:

(1) For each eigenvalue $\lambda_i, i = 1, 2, \cdots, m$, perform 2) to 5);

(2) Setting the approximate value $\tilde{\lambda}_i = \lambda_i + 0.001$ of eigenvalue λ_i, make $|A - \tilde{\lambda}_i E| \neq 0$;

(3) Setting eigenvector initial value $X_{0i} = (1, 1, \cdots, 1)^T$ corresponding to λ_i, if $\lambda_i = \lambda_{i-1}$ $(i = 2, 3, \cdots, n)$, then $X_{0i} = (1, 1, \cdots, 1, 10000000000, 1, \cdots, 1)^T$ $(i = 2, 3, \cdots, n)$;

(4) Calculating $Y_k, X_k, k = 1, 2, 3, \cdots$ according to the formula $\left.\begin{array}{l}\left(A - \tilde{\lambda}_i E\right) Y_k = X_{k-1} \\ X_K = Y_K / \|Y_K\|_\infty\end{array}\right\}, k = 1, 2, 3, \cdots$ until k>1000 or $\underset{1 \le i \le n}{Max}\{x_{ki}/x_{k-1,i}\} - \underset{1 \le i \le n}{Min}\{x_{ki}/x_{k-1,i}\} < 1E-13$ (if $x_{k-1,i} = 0, x_{k-1,i-1} \neq 0$, then $x_{ki}/x_{k-1,i} = sgn(x_{ki-1}/x_{k-1,i-1})$);

(5) Record the result(s) of 4): $X_i = X_{ki}$ $(i \le m)$.

The VBA macro code of solving eigenvectors is as follows:

```
KK = 0
Do
rngX0.Value = rngX1.Value
rngY.FormulaArray = "=mmult(minverse(" + rng.Address + ")," + rngX0.Address
+ ")"
rngY1.FormulaArray = "=max(abs(" + rngY.Address + "))"
For J = 1 To n
   rngX1.Cells(J, 1).Value = rngY.Cells(J, 1).Value / rngY1.Cells(J, 1).Value
   If Abs(rngX0.Cells(J, 1).Value) > 0 Then
      rngL.Cells(J, 1).Value = rngX1.Cells(J, 1).Value / rngX0.Cells(J, 1).Value
      sgnL = Sgn(rngL.Cells(J, 1).Value)
   Else
      rngL.Cells(J, 1).Value = sgnL
   End If
Next J
Range(rngL.Cells(1, 1), rngL.Cells(n, 1)).Select
Selection.Sort Key1:=rngL.Cells(1, 1), Order1:=xlDescending, Header:=xlGuess _
   , OrderCustom:=1, MatchCase:=False, Orientation:=xlTopToBottom, _
   SortMethod:=xlPinYin, DataOption1:=xlSortNormal
   d = rngL.Cells(1, 1).Value - rngL.Cells(n, 1).Value
KK = KK + 1
```

```
Loop Until KK > 1000 Or d < 0.0000000000001 '1E-13
rngRu.Value = rngX0.Value
```

2.5 Unitize the Eigenvectors

(1) For each eigenvector X_i, $i = 1, 2, \cdots, m$, perform (2) to (4);
(2) If$|x_{ij}| < 9.9E - 6\,(j = 1, 2, \cdots, n)$, then $x_{ij} = 0\,(j = 1, 2, \cdots, n)$;
(3) Calculating the modulus $d = \sqrt{\sum_{j=1}^{n} x_{ij}^2}$ of eigenvector X_i;
(4) Use modulus d multiplied by eigenvector X_i.

3 The Simulation and Operation of Optimization Algorithm

Below we use solving matrix eigenvalues and eigenvectors as an example to make simulation and operation on the new method.

As shown in Fig. 1. input matrix in Excel worksheet cell A1 : D4, obtained the eigenvalues and eigenvectors of matrix A after performing the macro process. The simulation of its operation process is as follows:

(1) Using QR algorithm to calculate, appear $d = (v(1, 1) - v(2, 2))/2 = 0$ when calling procedure wilkinsonQR( ) in the calculation, to give d a small nonzero value $d = 1E - 13$, preliminary to figure out the eigenvalues 4, 2, 2, −2;

Microsoft Excel - 代数运算

文件(F) 编辑(E) 视图(V) 插入(I) 格式(O) 工具(T) 数据(D) 窗口(W) 帮助

A3 | 2

	A	B	C	D	E	F	G
3	2	0	0	3			
4	0	2	0	3			
5	0	0	1	3			
6	3	3	3	1			
7							
8							
9	特征值	6.562817	2	1.33062	-3.89344		
10							
11	特征向量	-0.44784	0.70711	-0.4033	0.369872		
12		-0.44784	-0.7071	-0.4033	0.369872		
13		-0.36733	0	0.81648	0.445457		
14		-0.68114	0	0.08998	-0.72661		

Fig. 1 Input matrix A and its operation results

(2) Due to 4 is not the eigenvalue of matrix A, so the initial value 4, 2, 2, −2 is adjusted for 12, 2.6667, 0.6667, −6;
(3) With adjusted value for initial value, we obtain all the eigenvalues of matrix A are 6.562817, 2, 1.33062, −3.89344 by using the single variable equation solving method;
(4) Because the four eigenvalues each are not identical, we take vector quantity $X_0 = (1\ ,\ 1\ ,\ 1\ ,\ 1)^T$ for initial vector, to solve the eigenvectors of each eigenvalue, and get below results:

$$\begin{bmatrix} -0.65749 \\ -0.65749 \\ -0.5393 \\ -1 \end{bmatrix}, \begin{bmatrix} 1 \\ -1 \\ 0 \\ 0 \end{bmatrix}, \begin{bmatrix} -0.4939 \\ -0.4939 \\ 1 \\ 0.11021 \end{bmatrix}, \begin{bmatrix} 0.509041 \\ 0.509041 \\ 0.613066 \\ -1 \end{bmatrix}.$$

(5) Make standardization on the above vectors, we obtain

$$\begin{bmatrix} -0.44784 \\ -0.44784 \\ -0.36733 \\ -0.68114 \end{bmatrix}, \begin{bmatrix} 0.70711 \\ -0.7071 \\ 0 \\ 0 \end{bmatrix}, \begin{bmatrix} -0.40328 \\ -0.40328 \\ 0.816477 \\ 0.089982 \end{bmatrix}, \begin{bmatrix} 0.369872 \\ 0.369872 \\ 0.445457 \\ -0.72661 \end{bmatrix}$$

We get the same results by using the special mathematical software MATLAB to take operations.

We also use the optimization algorithm to calculate the eigenvalues and eigenvector of matrix $B = \begin{bmatrix} 4\,0\,2\,3 \\ 0\,4\,0\,3 \\ 0\,0\,1\,3 \\ 0\,0\,3\,1 \end{bmatrix}$, and the results are shown in Fig. 2.

The results obtained by making operation on MATLAB are: the eigenvalue are $4, 4, 4, -2$, the corresponding eigenvectors are

$$(1, 0, 0, 0)^T, (0, 1, 0, 0)^T, (-8575, -0.5145, 0, 0)^T, \quad (-0.1104, -0.3313, -0.6626, 0.6626)^T$$

4 Analysis of Results

The algorithm which is optimized can solve matrix real eigenvalues and eigenvectors by using the most commonly used office software–Excel. This algorithm but also has the following advantages:

(1) It can solve all real eigenvalues of a real matrix, and it is able to figure out the multiple root of a characteristic equation;
(2) It can calculate all real eigenvalues of the corresponding eigenvectors;
(3) The results of the operation are highly accurate. It can rival professional mathematical software;

Microsoft Excel - 代数运算

文件(F) 编辑(E) 视图(V) 插入(I) 格式(O) 工具(T) 数据(D) 窗口(W) 帮助

A3 fx 4

	A	B	C	D	E	F	G
3	4	0	2	3			
4	0	4	0	3			
5	0	0	1	3			
6	0	0	3	1			
7							
8							
9	特征值	4	4	4	-2		
10							
11	特征向量	0.857493	1	0.0005	0.110432		
12		0.514496	0.0003	1	0.331295		
13		1.71E-07	1E-10	1E-10	0.662589		
14		1.71E-07	1E-10	1E-10	-0.66259		

Fig. 2 Input the matrix B and the operation results

(4) The advantage is based on Excel sorting data. It is easier for data processing;
(5) It is based on VBA programming, all the calculation done automatically without the need of more intermediate to operation, and this is convenient to application promotion.

5 Conclusion

In this paper, results the optimization algorithm in sovling the general matrix real eigenvalues and eigenvectors with specialized mathematical software MATLAB-run results fit together well. And by adjusting the relevant parameters, it can make the accuracy improved.

In programming, the choice of some parameters is the results of repeated experiments. As among the initialization vectors of the same eigenvalues iteration eigenvectors, we make i-th component displacement into10000000000, and we found that the greater the absolute value of this number, the higher the accuracy of the results. This is more consistent with the results of MATLAB running.

References

1. Yang, M., Lu, J.: In excel naming and calculating eigenvalues eigenvectors of matrix. Comput. Knowl. Technol. **14**(3), 1295–1296 (2007)
2. Rong, S.: Excel VBA for then numerical calculation of the matrix eigenvalue. Jiangnan univ. J. (Nat. Sci. Ed.) **24**(4), 418–422 (2009)

3. Golub, F.: Matrix Calculation. Science Press, Beijing (2001)
4. Lin, C.: Numerical methods (Part ii). Science Press, Beijing (2005)
5. ExcelHome. The actual combat skills essence of Excel VBA[M]. People's Posts & Telecom Press, Beijing (2008)
6. Visual Basic 6.0 Chinese version, Introduction and Enhancement. Tsinghua University Press, Beijing (1999)
7. Zhong, Y.: The optimization for location for large commoditys regional distribution center. Adv. Soft Comput. **40**, 969–979 (2007)
8. Zhong, Y.: The Design of A Controller in Fuzzy PETRI NET. Fuzzy Optim. Decis. Making **7**(4), 399–408 (2008)
9. Zhong Y.: The structures and expressions of power group. Fuzzy Inf. Eng. **2**(2):(2010)
10. Zhong, Y., Deng, H., Chen, H.: A fuzzy logic based mobile intelligent system for eeEvaluating and selecting hotels in tourism, pp. 733–737. IEEEof International Conference on Web Information Systems and Mining (2009)

Part II
Lattice and Measures

Fuzzy Diameter Approximate Fixed Point in Normed Spaces

S. A. M. Mohsenialhosseini and H. Mazaheri

Abstract We define fuzzy diameter approximate fixed point in fuzzy norm spaces. We prove existence theorems, we also consider approximate pair constructive mapping and show its relation with approximate fuzzy fixed point.

Keywords Fuzzy norm space · F^z-approximate fixed point · Fuzzy diameter approximate fixed point.

1 Introduction

Chitra and Mordeson [6] introduce a definition of norm fuzzy and thereafter the concept of fuzzy norm space has been introduced and generalized in different ways by Bag and Samanta in [1–3]. The definitions are as follows:

Definition 1.1. Let U be a linear space on $\mathbf{R}$. A function $N : U \times \mathbf{R} \to [0, 1]$ is called fuzzy norm if and only if for every $x, u \in U$ and every $c \in \mathbf{R}$ the following properties are satisfy:

(F_{N1}) $N(x, t) = 0$ for every $t \in \mathbf{R}^- \cup \{0\}$,

(F_{N2}) $N(x, t) = 1$ if and only if $x = 0$ for every $t \in \mathbf{R}^+$,

(F_{N3}) $N(cx, t) = N(x, \frac{t}{|c|})$ for every $c \neq 0$ and t $\in \mathbf{R}^+$,

(F_{N4}) $N(x + u, s + t) \geq \min\{N(x, s), N(u, t)\}$ for every $s, t \in \mathbf{R}^+$,

(F_{N5}) the function $N(x, .)$ is nondecreasing on $\mathbf{R}$, and $\lim_{t\to\infty} N(x, t) = 1$.

S. A. M. Mohsenialhosseini (✉)
Faculty of Mathematics, Vali-e-Asr University, Rafsanjan, Iran
e-mail: amah@vru.ac.ir

H. Mazaheri
Faculty of Mathematics, Yazd University, Yazd, Iran
e-mail: hmazaheri@yazduni.ac.ir

B.-Y. Cao and H. Nasseri (eds.), *Fuzzy Information & Engineering and Operations Research & Management*, Advances in Intelligent Systems and Computing 211,
DOI: 10.1007/978-3-642-38667-1_10,

A pair (U, N) is called a fuzzy norm space. Sometimes, We need two additional conditions as follows:

$$(F_{N6})\ \forall t \in \mathbf{R}^+\ N(x,t) > 0 \Rightarrow x = 0.$$

(F_{N7}) function $N(x, .)$ is continuous for every $x \neq 0$, and on subset

$$\{t : 0 < N(x,t) < 1\}$$

is strictly increasing.
Let (U, N) be a fuzzy norm space. For all $\alpha \in (0, 1)$, we define α norm on U as follows:

$$\|x\|_\alpha = \wedge\{t > 0 : N(x,t) \geq \alpha\}\ for\ every\ x \in U.$$

Then $\{\|x\|_\alpha : \alpha \in (0, 1)\}$ is an ascending family of normed on U and they are called $\alpha - norm$ on U corresponding to the fuzzy norm N on U. Some notation, lemmas and example which will be used in this paper are given below:

Lemma 1.2. *Bag and Samanta [1] Let (U, N) be a fuzzy norm space such that satisfy conditions F_{N6} and F_{N7}. Define the function $N' : U \times \boldsymbol{R} \to [0, 1]$ as follows:*

$$N'(x,t) = \begin{cases} \vee\{\alpha \in (0,1) : \|x\|_\alpha \leq t\} & (x,t) \neq (0,0) \\ 0 & (x,t) = (0,0) \end{cases}$$

Then

(a) N' is a fuzzy norm on U.
(b) $N = N'$.

Lemma 1.3. *Bag and Samanta [1] Let (U, N) be a fuzzy norm space such that satisfy conditions F_{N6} and F_{N7}. and $\{x_n\} \subseteq U$, Then $\lim_{n\to\infty} N(x_n - x, t) = 1$ if and only if*

$$\lim_{n\to\infty} \|x_n - x\|_\alpha = 0$$

for every $\alpha \in (0, 1)$.
Note that the sequence $\{x_n\} \subseteq U$ converges if there exists a $x \in U$ such that

$$\lim_{n\to\infty} N(x_n - x, t) = 1\ for\ every\ t \in \boldsymbol{R}^+.$$

In this case x is called the limit of $\{x_n\}$.

Example 1.4. Bag and Samanta [1] Let V be the Real or Complex vector space and let N define on $V \times R$ as follows:

$$N(x,t) = \begin{cases} 1 & t > |x| \\ 0 & t \leq |x| \end{cases}$$

for all $x \in V$ and $t \in R$. Then (N, V) is a fuzzy norm space and the function N satisfy conditions F_{N6} and $\|x\|_\alpha = |x|$ for every $\alpha \in (0, 1)$.

Notation 1.5. Let (U, N) be a fuzzy norm space and $\{\|.\|_\alpha : \alpha \in (0, 1)\}$ be the set of all $\alpha-$norms on U. For two subset A and B of U, we consider:

$$\delta(A, B) = \wedge\{\|x - y\|_\alpha : x \in A, y \in B, \alpha \in (0, 1)\}.$$

Definition 1.6. Mohsenalhosseini et al. [12] Let (U, N) be a fuzzy norm space satisfy condition F_{N6} and F_{N7} and $\{\|.\|_\alpha : \alpha \in (0, 1)\}$ the set of $\alpha-$norms defined on U. Suppose A and B are nonempty subsets of U and $T : A \cup B \to U$.

For some $\epsilon > 0$ and for any $x \in A \cup B$ is said to be a F^z-approximate fixed point for T if for some $\alpha \in (0, 1)$

$$\|x - Tx\|_\alpha \leq \delta(A, B) + \epsilon.$$

Proposition 1.7. *Mohsenalhosseini et al. [12] Let (U, N) be a fuzzy norm space such that satisfy conditions F_{N6} and F_{N7} and $\{\|.\|_\alpha : \alpha \in (0, 1)\}$ be the set of $\alpha-$norms defined on U. Suppose A, B are nonempty subsets of U and $T : A \cup B \to U$. If for $x \in A \cup B$ and $\alpha \in (0, 1)$*

$$Lim_{n\to\infty}\|T^n x - T^{n+1} x\|_\alpha = \delta(A, B),$$

then there exists a F^z-approximate fixed point in $A \cup B$.

Definition 1.8. Mohsenalhosseini et al. [12] let (U, N) be a fuzzy norm space such that satisfy conditions F_{N6} and F_{N7} and $\{\|.\|_\alpha : \alpha \in (0, 1)\}$ be the set of $\alpha-$norms defined on U. Suppose A and B are nonempty subsets of U and $T : A \cup B \to U$ and $S : A \cup B \to U$. A point (x, y) in $A \times B$ is said a F^z-approximate fixed point for (T, S), if for some $\alpha \in (0, 1)$ there exists an $\epsilon > 0$ such that

$$\|(Tx, Sy)\|_\alpha \leq \delta(A, B) + \epsilon.$$

In this paper we will denote the set of all F^z-approximate fixed points of T and (T, S), for a given $\epsilon > 0$, by
$F^z_T(A, B) = \{x \in A \cup B : \|x - Tx\|_\alpha \leq \delta(A, B) + \epsilon, \; for \; some \; \alpha \in (0, 1)\}$, and
$F^z_{(T,S)}(A, B) = \{(x, y) \in A \times B : \|Tx - Sy\|_\alpha \leq \delta(A, B) + \epsilon \; for \; some \; \alpha \in (0, 1)\}$,
respectively.

Proposition 1.9. *Mohsenalhosseini et al. [12] let (U, N) is a fuzzy norm space such that satisfy conditions F_{N6} and F_{N7} and $\{\|.\|_\alpha : \alpha \in (0, 1)\}$ be the set of $\alpha-$norms defined on U. Suppose A and B are nonempty subsets of U and $T : A \cup B \to U$ and $S : A \cup B \to U$. If for a $(x, y) \in A \times B$ and for some $\alpha \in (0, 1)$*

$$Lim_{n\to\infty}\|T^n x - S^n y\|_\alpha = \delta(A, B),$$

then there exists a F^z-approximate fixed point in $A \times B$.

2 Fuzzy Diameter and Fuzzy Radius Approximate in Fixed Point

In this section, we give the definitions of fuzzy diameter and fuzzy radius. Also, we will obtain the theorems and result about fuzzy diameter and fuzzy radius approximate fixed point.

Definition 2.1. Let (U, N) be a fuzzy normed linear space and $F_T^z(A, B)(\neq) \subset A \cup B$. We define fuzzy diameter of $F_T^z(A, B)$ as

$$diam(F_T^z(A, B)) = \bigvee_{\alpha \in (0,1)} [\bigwedge \{t > 0 : N(u - v, t) \geq \alpha \; \forall \, u, v \in F_T^z(A, B)\}]$$

and it is denoted by $T - \theta(F_T^z(A, B))$.

Notation 2.2. Let (U, N) be a fuzzy normed linear space satisfying F_{N6} and $F_T^z(A, B)$ $(\neq)$ $\subset A \cup B$. Then $\bigvee\{\|x - y\|_\alpha : x, y \in F_T^z(A, B)\}$ is denoted by $\alpha - \theta(F_T^z(A, B))$ ($\| \|_\alpha$ is the $\alpha-$norm of N), $0 < \alpha < 1$.

Proposition 2.3. *Let (U, N) be a fuzzy normed linear space satisfying F_{N6} and $F_T^z(A, B)(\neq) \subset A \cup B$. Then $T - \theta(F_T^z(A, B)) = \bigvee\{\alpha - \theta(F_T^z(A, B)) : \alpha \in (0, 1)\}$.*

Proof: If $F_T^z(A, B)$ is singleton then clearly $\alpha - \theta(F_T^z(A, B)) = T - \theta(F_T^z(A, B)) = 0$ for all $\alpha \in (0, 1)$. So we suppose that $F_T^z(A, B)$ is not singleton.
Now $k > T - \theta(F_T^z(A, B))$

$$\Rightarrow k > \bigvee_{\alpha \in (0,1)} [\bigwedge \{t > 0 : N(u - v, t) \geq \alpha \; \forall \, u, v \in F_T^z(A, B)\}]$$

$$\Rightarrow k > \bigwedge \{t > 0 : N(u - v, t) \geq \alpha \; \forall \, u, v \in F_T^z(A, B)\} \; \forall \alpha \in (0, 1).$$
$$\Rightarrow N(u - v, k) \geq \alpha \; \forall \, u, v \in F_T^z(A, B)\} \; \forall \alpha \in (0, 1).$$
$$\Rightarrow N(u - v, k) \geq \alpha \; \forall \, u, v \in F_T^z(A, B).$$
$$\Rightarrow \|u - v\|_\alpha \leq k \; \forall \, u, v \in F_T^z(A, B)\} \; and \; \forall \alpha \in (0, 1).$$
$$\Rightarrow \bigvee\{\|x - y\|_\alpha : x, y \in F_T^z(A, B)\} \leq k \; \forall \alpha \in (0, 1).$$
$$\Rightarrow \alpha - \theta(F_T^z(A, B)) \leq k \; \forall \alpha \in (0, 1).$$
$$\Rightarrow T - \theta(F_T^z(A, B)) \geq \alpha - \theta(F_T^z(A, B)) \; \forall \alpha \in (0, 1).$$
$$\Rightarrow T - \theta(F_T^z(A, B)) \geq \bigvee\{\alpha - \theta(F_T^z(A, B)) : \alpha \in (0, 1)\}.$$

Thus

$$T - \theta(F_T^z(A, B)) \geq \bigvee\{\alpha - \theta(F_T^z(A, B)) : \alpha \in (0, 1)\}. \quad (1)$$

Now $k < T - \theta(F_T^z(A, B))$

$$\Rightarrow \bigvee_{\alpha \in (0,1)} [\bigwedge \{t > 0 : N(u - v, t) \geq \alpha \, \forall \, u, v \in F_T^z(A, B)\}] > k$$

$$\Rightarrow \exists \alpha_0 \in (0, 1) \; such \; that \; \bigwedge \{t > 0 : N(u - v, t) \geq \alpha_0 \, \forall \, u, v \in F_T^z(A, B)\} > k$$

$$\Rightarrow \exists u_0, v_0 \in F_T^z(A, B) \; such \; that \; N(u_0 - v_0, k) < \alpha_0. \tag{2}$$

Now $\|u_0 - v_0\|_{\alpha_0} = \bigwedge \{t > 0 : N(u - v, t) \geq \alpha_0\} \geq k$ by Eq. (2).
So $\alpha_0 - \theta(F_T^z(A, B)) = \bigvee \{\|u - v\|_{\alpha_0} : \; \forall \, u, v \in F_T^z(A, B)\} \geq \|u_0 - v_0\|_{\alpha_0} \geq k$.
Thus $\bigvee \{\alpha - \theta(F_T^z(A, B)) : \alpha \in (0, 1)\} \geq \alpha_0 - \theta(F_T^z(A, B)) \geq k$, i.e. $\bigvee \{\alpha - \theta(F_T^z(A, B)) : \alpha \in (0, 1)\} \geq k$. Thus

$$\bigvee \{\alpha - \theta(F_T^z(A, B)) : \alpha \in (0, 1)\} \geq T - \theta(F_T^z(A, B)). \tag{3}$$

Now from Eqs. (1) and (3) we get $T - \theta(F_T^z(A, B)) = \bigvee \{\alpha - \theta(F_T^z(A, B)) : \alpha \in (0, 1)\}$.

Definition 2.4. Let (U, N) be a fuzzy normed linear space and $F_1^z T(A, B)$, $F_2^z T(A, B) \subset A \cup B$. We define fuzzy radius of $F_1^z T(A, B)$ as

$$T - r_{A \cup B}(F_1^z T(A, B)) = \bigvee_{\alpha \in (0,1)} [\bigwedge \{t > 0 : N(u - v, t) \geq \alpha \, \forall \, v \in F_1^z T(A, B)\}] \; (u \in F_2^z T(A, B))$$

Notation 2.5. Let (U, N) be a fuzzy normed linear space satisfying F_{N6}. For any non-empty subsets $F_1^z T(A, B)$, $F_2^z T(A, B)$ of $A \cup B$ and $\alpha \in (0, 1)$, we denote $\alpha - r_{A \cup B}(F_1^z T(A, B)) = \bigvee \{\|u - v\|_\alpha : v \in F_1^z T(A, B)\} \; (u \in F_2^z T(A, B)$.

Proposition 2.6. *Let (U, N) be a fuzzy normed linear space satisfying F_{N6}. Then For any subsets $F_1^z T(A, B)$ of $A \cup B$, $\alpha \in (0, 1)$ and $u \in A \cup B$, $T - r_u(F_1^z T(A, B)) = \bigvee \{\alpha - r_u(F_1^z T(A, B)) : \alpha \in (0, 1)\}$.*

Proof: If $F_1^z T(A, B)$ is singleton then clearly

$$\alpha - r_u(F_1^z T(A, B)) = T - r_u(F_1^z T(A, B)) = 0$$

for all $\alpha \in (0, 1)$. So we suppose that $F_1^z T(A, B)$ is not singleton.
Now $k > r_u(F_1^z T(A, B))$

$$\Rightarrow k > \bigvee_{\alpha\in(0,1)} [\bigwedge\{t > 0 : N(u - v, t) \geq \alpha \; \forall \, u, v \in F^z{}_T(A, B)\}]$$

$$\Rightarrow k > \bigwedge\{t > 0 : N(u - v, t) \geq \alpha \; \forall \, v \in F_1^z{}_T(A, B)\} \; \forall \alpha \in (0, 1).$$

$$\Rightarrow N(u - v, k) \geq \alpha \; \forall \, v \in F_1^z{}_T(A, B)\} \; \forall \alpha \in (0, 1).$$

$$\Rightarrow N(u - v, k) \geq \alpha \; \forall \, v \in F_1^z{}_T(A, B).$$

$$\Rightarrow \|u - v\|_\alpha \leq k \; \forall \, v \in F_1^z{}_T(A, B)\} \; and \; \forall \alpha \in (0, 1).$$

$$\Rightarrow k \geq \bigvee\{\|u - v\|_\alpha : \; \forall v \in F_1^z{}_T(A, B)\} = \alpha - r_u(F_1^z{}_T(A, B)) \; \forall \alpha \in (0, 1).$$

$$\Rightarrow \alpha - r_u(F_1^z{}_T(A, B)) \leq k \; \forall \alpha \in (0, 1).$$

$$\Rightarrow T - r_u(F_1^z{}_T(A, B)) \geq \alpha - r_u(F_1^z{}_T(A, B)) \; \forall \alpha \in (0, 1).$$

$$\Rightarrow T - r_u(F_1^z{}_T(A, B)) \geq \bigvee\{\alpha - r_u(F_1^z{}_T(A, B)) : \alpha \in (0, 1)\}.$$

Thus

$$T - r_u(F_1^z{}_T(A, B)) \geq \bigvee\{\alpha - r_u(F_1^z{}_T(A, B)) : \alpha \in (0, 1)\}. \tag{4}$$

Now $T - r_u(F_1^z{}_T(A, B)) > k$

$$\Rightarrow \bigvee_{\alpha\in(0,1)} [\bigwedge\{t > 0 : N(u - v, t) \geq \alpha \; \forall \, v \in F_1^z{}_T(A, B)\}] > k$$

$$\Rightarrow \exists \alpha_0 \in (0, 1) \; such \; that \; \bigwedge\{t > 0 : N(u - v, t) \geq \alpha_0 \; \forall \, v \in F_1^z{}_T(A, B)\} > k$$

$$\Rightarrow \exists v_0 \in F_1^z{}_T(A, B) \; such \; that \; N(u - v_0, k) < \alpha_0. \tag{5}$$

Now $\|u - v_0\|_{\alpha_0} = \bigwedge\{t > 0 : N(u - v_0, t) \geq \alpha_0\} \geq k$ by Eq. (8).
So $\alpha_0 - r_u(F_1^z{}_T(A, B)) = \bigvee\{\|u - v\|_{\alpha_0} : \; \forall \, v \in F^z{}_T(A, B)\} \geq \|u - v_0\|_{\alpha_0} \geq k$.
Thus $\bigvee\{\alpha - r_u(F_1^z{}_T(A, B)) : \alpha \in (0, 1)\} \geq \alpha_0 - r_u(F_1^z{}_T(A, B)) \geq k$. Therefore

$$\bigvee\{\alpha - r_u(F_1^z{}_T(A, B)) : \alpha \in (0, 1)\} \geq T - r_u(F_1^z{}_T(A, B)). \tag{6}$$

Now from Eqs. (4) and (6) we get $T - r_u(F_1^z{}_T(A, B)) = \bigvee\{\alpha - r_u(F_1^z{}_T(A, B)) : \alpha \in (0, 1)\}$.

Definition 2.7. Let (U, N) be a fuzzy normed linear space and $F^z_{(T,S)}(A, B)(\neq) \subset A \times B$. We define fuzzy diameter of $F^z{}_{(T,S)}(A, B)$ as

$$diam(F^z{}_{(T,S)}(A, B)) = \bigvee_{\alpha\in(0,1),\beta\in(0,1)} [\bigwedge\{t > 0 : N((u, v) - (u_1, v_1), t) \geq (\alpha, \beta) \; \forall \, (u, v), (u_1, v_1) \in F^z_{(T,S)}(A, B)\}]$$

and it is denoted by $(T, S) - \theta(F^z{}_{(T,S)}(A, B))$.

Notation 2.8. Let (U, N) be a fuzzy normed linear space satisfying F_{N6} and $F^z_{(T,S)}$ $(A, B)(\neq) \subset A \times B$. Then

$$\bigvee\{\|(x, y) - (x_1, y_1)\|_\alpha : (x, y), (x_1, y_1) \in F^z_{(T,S)}(A, B)\}$$

is denoted by $(\alpha, \beta) - \theta(F^z_{(T,S)}(A, B))$ ($\| \|_\alpha$ and $\| \|_\beta$ are the of N), $0 < \alpha < 1$ and $0 < \beta < 1$.

Proposition 2.9. *Let (U, N) be a fuzzy normed linear space satisfying F_{N6} and $F^z_{(T,S)}(A, B)(\neq) \subset A \times B$. Then*

$$(T, S) - \theta(F^z{}_{(T,S)}(A, B)) = \bigvee\{(\alpha, \beta) - \theta(F^z{}_{(T,S)}(A, B)) : \alpha, \beta \in (0, 1)\}.$$

Proof: If $F^z{}_{(T,S)}(A, B)$ is then clearly $(\alpha, \beta) - \theta(F^z{}_{(T,S)}(A, B)) = (T, S) - \theta(F^z{}_{(T,S)}(A, B)) = 0$ for all $\alpha, \beta \in (0, 1)$. So we suppose that $F^z_{(T,S)}(A, B)$ is not singleton.
Now $k > (T, S) - \theta(F^z{}_{(T,S)}(A, B))$

$$\Rightarrow k > \bigvee_{\alpha\in(0,1),\beta\in(0,1)} [\bigwedge\{t > 0 : N((u, v) - (u_1, v_1), t)$$
$$\geq \wedge\{\alpha, \beta\} \,\forall\, (u, v), (u_1, v_1) \in F^z{}_{(T,S)}(A, B)\}]$$
$$\Rightarrow k > \bigwedge\{t > 0 N((u, v) - (u_1, v_1), t)$$
$$\geq \wedge\{\alpha, \beta\} \,\forall\, (u, v), (u_1, v_1) \in F^z{}_{(T,S)}(A, B)\} \,\forall \alpha, \beta \in (0, 1).$$
$$\Rightarrow N((u, v) - (u_1, v_1), k) \geq \wedge\{\alpha, \beta\} \,\forall\, (u, v), (u_1, v_1) \in F^z{}_{(T,S)}(A, B) \,\forall \alpha, \beta \in (0, 1).$$
$$\Rightarrow N((u, v) - (u_1, v_1), k) \geq \wedge\{\alpha, \beta\} \,\forall\, (u, v), (u_1, v_1) \in F^z{}_{(T,S)}(A, B).$$
$$\Rightarrow \|(u, v) - (u_1, v_1)\|_{\wedge\{\alpha,\beta\}} \leq k \,\forall\, (u, v), (u_1, v_1) \in F^z{}_{(T,S)}(A, B) \textit{ and } \forall \alpha, \beta \in (0, 1).$$
$$\Rightarrow \bigvee\{\|(x, y) - (x_1, y_1)\|_{\wedge\{\alpha,\beta\}} : (x, y), (x_1, y_1) \in F^z{}_{(T,S)}(A, B)\} \leq k \,\forall \alpha, \beta \in (0, 1).$$
$$\Rightarrow \wedge\{\alpha, \beta\} - \theta(F^z{}_{(T,S)}(A, B)) \leq k \,\forall \alpha, \beta \in (0, 1).$$
$$\Rightarrow (T, S) - \theta(F^z{}_{(T,S)}(A, B)) \geq \wedge\{\alpha, \beta\} - \theta(F^z{}_{(T,S)}(A, B)) \,\forall \alpha, \beta \in (0, 1).$$
$$\Rightarrow (T, S) - \theta(F^z{}_{(T,S)}(A, B)) \geq \bigvee\{\wedge\{\alpha, \beta\} - \theta(F^z{}_{(T,S)}(A, B)) : \alpha, \beta \in (0, 1)\}.$$

Thus

$$(T, S)-\theta(F^z{}_{(T,S)}(A, B)) \geq \bigvee\{\wedge\{\alpha, \beta\}-\theta(F^z{}_{(T,S)}(A, B)) : \alpha, \beta \in (0, 1)\}. \quad (7)$$

Now $k < (T, S) - \theta(F^z{}_{(T,S)}(A, B))$

$$\Rightarrow \bigvee_{\alpha\in(0,1),\beta\in(0,1)} [\bigwedge\{t > 0 : N((u,v) - (u_1, v_1), t) \geq$$
$$\wedge\{\alpha, \beta\} \,\forall\, (u, v), (u_1, v_1) \in F^z_{(T,S)}(A, B)\}] > k$$
$$\Rightarrow \exists \alpha_0, \beta_0 \in (0, 1) \textit{ such that } \bigwedge\{t > 0 : N((u, v) - (u_1, v_1), t) \geq$$
$$\wedge\{\alpha_0, \beta_0\} \,\forall\, (u, v), (u_1, v_1) \in F^z{}_{(T,S)}(A, B)\} > k$$

$$\Rightarrow \exists (x, y), (x_1, y_1) \in F^z{}_{(T,S)}(A, B) \textit{ such that } N((x, y) - (x_1, y_1), k) < \wedge\{\alpha_0, \beta_0\}. \quad (8)$$

Now $\|(x, y) - (x_1, y_1)\|_{\wedge\{\alpha_0,\beta_0\}} = \bigwedge\{t > 0 : N(u - v, t) \geq \wedge\{\alpha_0, \beta_0\}\} \geq k$ by Eq. (5). So

$$\wedge\{\alpha_0, \beta_0\} - \theta(F^z{}_{(T,S)}(A, B)) = \bigvee\{\|(u, v) - (u_1, v_1)\|$$
$$\wedge\{\alpha_0, \beta_0\} : \,\forall\, (u, v) - (u_1, v_1) \in F^z{}_{(T,S)}(A, B)\} \geq$$
$$\|(x, y) - (x_1, y_1)\|_{\wedge\{\alpha_0,\beta_0\}} \geq k.$$
$$\wedge\{\alpha_0, \beta_0\} - \theta(F^z{}_{(T,S)}(A, B)) = \bigvee\{\|(u, v) - (u_1, v_1)\|$$
$$\wedge\{\alpha_0, \beta_0\} : \,\forall\, (u, v) - (u_1, v_1) \in F^z{}_{(T,S)}(A, B)\} \geq$$
$$\|(x, y) - (x_1, y_1)\|_{\wedge\{\alpha_0,\beta_0\}} \geq k.$$

Thus

$$\bigvee\{\wedge\{\alpha, \beta\} - \theta(F^z{}_{(T,S)}(A, B)) : \alpha \in (0, 1)\} \geq \wedge\{\alpha_0, \beta_0\} - \theta(F^z{}_{(T,S)}(A, B)) \geq k,$$

i.e. $\bigvee\{\wedge\{\alpha, \beta\} - \theta(F^z{}_{(T,S)}(A, B)) : \alpha, \beta \in (0, 1)\} \geq k$. Thus

$$\bigvee\{\wedge\{\alpha, \beta\} - \theta(F^z{}_T(A, B)) : \alpha, \beta \in (0, 1)\} \geq (T, S) - \theta(F^z{}_T(A, B)). \quad (9)$$

Now from Eqs. (4) and (6) we get $(T, S) - \theta(F^z{}_{(T,S)}(A, B)) = \bigvee\{\wedge\{\alpha, \beta\} - \theta(F^z{}_{(T,S)}(A, B)) : \alpha, \beta \in (0, 1)\}$.

3 Conclusion

The theory of fuzzy approximate fixed points is not less interesting than that of fuzzy fixed points and many results formulated in the latter can be adapted to a less restrictive framework in order to guarantee the existence of the fuzzy approximate fixed points. We proved results about fuzzy diameter and fuzzy radius approximate on fuzzy norm spaces. we think that this paper could be of interest to the researchers working in the field fuzzy functional analysis in particular, fuzzy approximate fixed point theory are used.

Acknowledgments The authors are extremely grateful to the referees for their helpful suggestions for the improvement of the paper.

References

1. Bag, T., Samanta, S.K.: Finite dimensional fuzzy normed linear spaces. J. Fuzzy Math. **11**(3), 687–705 (2003)
2. Bag, T., Samanta, S.K.: Fuzzy bounded linear operators. Fuzzy Sets syst. **151**(3), 513–547 (2005)
3. Bag, T., Samanta, S.K.: Some fixed point theorems in fuzzy normed linear spaces. Inf. Sci. **177**, 3271–3289 (2007)
4. Espinola, R.: A new approach to relatively nonexpansive mappings. Proc. Amer. Math. Soc. **136**(6), 1987–1995 (2008)
5. Browder, F.E.: Nonexpansive nonlinear operators in a Banach spaces. Proc. Natl. Acad. Sci. **54**, 1041–1044 (1965)
6. Chitra, A., Mordeson, P.V.: Fuzzy linear operators and fuzzy normed linear spaces. Bull. Cal. Math. Soc. **74**, 660–665 (1969)
7. Cadariu, L., Radu, V.: On the stability of the Cauchy functional equation: a fixed point approach. Grazer Math. Ber. **346**, 43 (2004)
8. Golet, I.: On fuzzy normed spaces. Southest Asia Bull. Math. **31**(2), 245–254 (2007)
9. Cancan, M.: Browders fixed point theorem and some interesting results in intuitionistic fuzzy normed spaces. Fixed Point Theory Appl. (2010), Article ID 642303, 11 pages doi:10.1155/(2010)/642303.
10. Grabic, M.: Fixed points in fuzzy metric spaces. Fuzzy Sets Syst. **27**(3), 385–389 (1988)
11. Marudai, M., Vijayaraju, P.: Fixed point theorems for fuzzy mapping. Fuzzy Sets Syst. **135**(3), 402–408 (2003)
12. Mohsenalhosseini, S.A.M., Mazaheri, H., Dehghan, M.A.: F^z-approximate fixed point in fuzzy normed spaces for nonlinear maps. Iranin J. Fuzzy Sys. (2012) amah@vru.ac.ir.
13. Saadati, R., Vaezpour, S.M., Cho, Y.J.: Quicksort algorithm: application of a fixed point theorem in intuitionistic fuzzy quasi-metric spaces at a domain of words. J. Comput. Appl. Math. **228**(1), 219–225 (2009)
14. Sisodia, K.S., Rathore, M.S., Singh, D., Khichi, S.S.: A Common Fixed Point Theorem in Fuzzy Metric Spaces. Int. J. Math. Anal. **5**(17), 819–826 (2011)
15. Zikic, T.: On fixed point theorems of Gregori and Sapena. Fuzzy Sets Syst. **144**(3), 421–429 (2004)

Cascade and Wreath Products of Lattice-Valued Intuitionistic Fuzzy Finite State Machines and Coverings

Li Yang and Zhi-wen Mo

Abstract The concepts of the cascade products and the wreath products of lattice-valued intuitionistic fuzzy finite state machines, homomorphisms and weak coverings are given. At the same time, the covering relations of two homomorphisms lattice-valued intuitionistic fuzzy finite state machines are studied. The covering relations among the full direct products, cascade products, wreath products are disscussed. Some transitive properties of covering relations are obtained in the product machines.Therefore,it is an important step to study lattice-valued intuitionistic fuzzy finite state machines.

Keywords Cascade product · Wreath product · Homomorphism · Covering

1 Introduction

The theory of fuzzy sets proposed by Zadeh in [1]. The mathematical formulation of a fuzzy automaton was first proposed by Wee in [2]. Afterwards,out of several higher order fuzzy sets, intuitionistic fuzzy sets introduced by Atanassov [3, 4] which have been found to be highly useful to deal with vagueness. Using the notion of intuitionistic fuzzy sets, Jun [5–7] introduced the concepts of intuitionistic fuzzy finite state machines as a generalization of fuzzy finite state machines, intuitionistic successors, intuitionistic subsystems, intuitionistic submachines, intuitionistic q-twins and so on. Intuitionistic fuzzy recognizers was introduced by Zhang and Li [8]. The theory of lattice-valued intuitionistic fuzzy sets was introduced by Atanassov [9]. Thus, on the basis of lattice-valued intuitionistic fuzzy sets, the present authors [10] introduced the concepts of lattice-valued intuitionistic fuzzy finite state machines, the full direct

L. Yang (✉) · Z. Mo
College of Mathematics and Software Science, Sichuan Normal University,
Chengdu 610066, China
e-mail: 505603431@qq.com

B.-Y. Cao and H. Nasseri (eds.), *Fuzzy Information & Engineering and Operations Research & Management*, Advances in Intelligent Systems and Computing 211,
DOI: 10.1007/978-3-642-38667-1_11, © Springer-Verlag Berlin Heidelberg 2014

products, the restricted direct products and coverings. In this thesis, the concepts of the cascade products and the wreath products of lattice-valued intuitionistic fuzzy finite state machines, homomorphisms and weak coverings are given.

2 Preliminaries

Definition 2.1. [9] Let X be a nonempty set, a complete lattice L with involutive order reversing unary operation' : $L \longrightarrow L$. A lattice-valued intuitionistic fuzzy set A in a set X is an object of the form

$$A = \{\langle x, \mu_A(x), \nu_A(x)\rangle | x \in X\},$$

where μ_A and ν_A are functions $\mu_A : X \longrightarrow L$, $\nu_A : X \longrightarrow L$, such that for all $x \in X$, $\mu_A(x) \leq (\nu_A(x))\prime$. For the sake of simplicity, we shall use the notation $A = (\mu_A, \nu_A)$ instead of $A = \{\langle x, \mu_A(x), \nu_A(x)\rangle | x \in X\}$.

Definition 2.2. [10] A lattice-valued intuitionistic fuzzy finite state machine (LIFFSM, for short) is a triple $M = (Q, X, A)$, where Q and X are finite nonempty sets, called the set of states and the set of input symbols, respectively, and $A = (\mu_A, \nu_A)$ is a lattice-valued intuitionistic fuzzy set in $Q \times X \times Q$.

Let X^* denote the set of all words of elements of X of finite length. Let Λ denote the empty word in X^* and $|x|$ denote the length of x for every $x \in X^*$.

Definition 2.3. [10] Let $M_i = (Q_i, X_i, A_i)$ be a LIFFSM, $i = 1, 2$. Then $M_1 \times M_2 = (Q_1 \times Q_2, X_1 \times X_2, A_1 \times A_2)$ is called the full direct product of M_1 and M_2,

$$\mu_{A_1 \times A_2}((q_1, q_2), (x_1, x_2), (p_1, p_2)) = \mu_{A_1}(q_1, x_1, p_1) \wedge \mu_{A_2}(q_2, x_2, p_2),$$

$$\nu_{A_1 \times A_2}((q_1, q_2), (x_1, x_2), (p_1, p_2)) = \nu_{A_1}(q_1, x_1, p_1) \vee \nu_{A_2}(q_2, x_2, p_2),$$

where $\mu_{A_1 \times A_2} : (Q_1 \times Q_2) \times (X_1 \times X_2) \times (Q_1 \times Q_2) \longrightarrow L$, $\nu_{A_1 \times A_2} : (Q_1 \times Q_2) \times (X_1 \times X_2) \times (Q_1 \times Q_2) \longrightarrow L, \forall (q_1, q_2), (p_1, p_2) \in Q_1 \times Q_2, \forall (x_1, x_2) \in X_1 \times X_2$.

Definition 2.4. [10] Let $M_i = (Q_i, X, A_i)$ be a LIFFSM, $i = 1, 2$. Then $M_1 \wedge M_2 = (Q_1 \times Q_2, X, A_1 \times A_2)$ is called the restricted direct product of M_1 and M_2,

$$\mu_{A_1 \wedge A_2}((q_1, q_2), a, (p_1, p_2)) = \mu_{A_1}(q_1, a, p_1) \wedge \mu_{A_2}(q_2, a, p_2),$$

$$\nu_{A_1 \wedge A_2}((q_1, q_2), a, (p_1, p_2)) = \nu_{A_1}(q_1, a, p_1) \vee \nu_{A_2}(q_2, a, p_2),$$

where $\mu_{A_1 \wedge A_2} : (Q_1 \times Q_2) \times X \times (Q_1 \times Q_2) \longrightarrow L$, $\nu_{A_1 \wedge A_2} : (Q_1 \times Q_2) \times X \times (Q_1 \times Q_2) \longrightarrow L, \forall (q_1, q_2), (p_1, p_2) \in Q_1 \times Q_2, \forall a \in X$.

Theorem 2.1 *[10] Let $M_i = (Q_i, X_i, A_i)$ be a LIFFSM, $i = 1, 2$. Then the following assertions hold.*

(1) $M_1 \times M_2$ be a LIFFSM,
(2) $M_1 \wedge M_2$ be a LIFFSM, where $X_1 = X_2 = X$.

Definition 2.5. [10] Let $M_i = (Q_i, X, A_i)$ be a LIFFSM, $i = 1, 2$. Let $\eta : Q_2 \longrightarrow Q_1$ be a surjective partial function and $\xi : X_1 \longrightarrow X_2$ be a function. Then the ordered pair (η, ξ) is called a covering of M_1 by M_2, written $M_1 \leq M_2$, if

$$\mu_{A_1}(\eta(p), x_1, \eta(q)) \leq \mu_{A_2}(p, \xi(x_1), q),$$

$$\nu_{A_1}(\eta(p), x_1, \eta(q)) \geq \nu_{A_2}(p, \xi(x_1), q),$$

for all $x_1 \in X_1$ and p, q belong to the domain of η.

Theorem 2.2 *[10] Let $M_i = (Q_i, X_i, A_i)$ be a LIFFSM, $i = 1, 2, 3$. If $M_1 \leq M_2$ and $M_2 \leq M_3$, then $M_1 \leq M_3$.*

Theorem 2.3 *[10] Let $M_i = (Q_i, X, A_i)$ be a LIFFSM, $i = 1, 2$. Then $M_1 \wedge M_2 \leq M_1 \times M_2$.*

Theorem 2.4 *[10] Let $M_i = (Q_i, X_i, A_i)$ be a LIFFSM, $i = 1, 2, 3$. If $M_1 \leq M_2$, then*

(1) $M_1 \times M_3 \leq M_2 \times M_3$ and $M_3 \times M_1 \leq M_3 \times M_2$,
(2) $M_1 \wedge M_3 \leq M_2 \wedge M_3$ and $M_3 \wedge M_1 \leq M_3 \wedge M_2$, where $X_1 = X_2 = X_3 = X$.

Corollary 2.1 [10] Let $M_i = (Q_i, X_i, A_i)$ be a LIFFSM, $i = 1, 2, 3$. If $M_1 \leq M_2$, then

(1) $M_1 \wedge M_3 \leq M_2 \times M_3$, where $X_1 = X_3 = X$,
(2) $M_3 \wedge M_1 \leq M_3 \times M_2$, where $X_1 = X_3 = X$.

Corollary 2.2 [10] Let $M_i = (Q_i, X_i, A_i)$ be a LIFFSM, $i = 1, 2, 3, 4$. If $M_1 \leq M_2$ and $M_3 \leq M_4$, the following assertions hold.

(1) $M_1 \times M_3 \leq M_2 \times M_4$,
(2) $M_1 \wedge M_3 \leq M_2 \wedge M_4$, where $X_1 = X_2 = X_3 = X_4 = X$,
(3) $M_1 \wedge M_3 \leq M_2 \times M_4$, where $X_1 = X_3 = X$.

3 Cascade and Wreath Products of Lattice-Valued Intuitionistic Fuzzy Finite State Machines

Definition 3.1. Let $M_i = (Q_i, X_i, A_i)$ be a LIFFSM, $i = 1, 2$. Then $M_1 \omega M_2 = (Q_1 \times Q_2, X_2, A_1 \omega A_2)$ is called the cascade product of M_1 and M_2,

$$\mu_{A_1\omega A_2}((q_1,q_2),b,(p_1,p_2)) = \mu_{A_1}(q_1,\omega(q_2,b),p_1) \wedge \mu_{A_2}(q_2,b,p_2),$$

$$\nu_{A_1\omega A_2}((q_1,q_2),b,(p_1,p_2)) = \nu_{A_1}(q_1,\omega(q_2,b),p_1) \vee \nu_{A_2}(q_2,b,p_2),$$

where $\mu_{A_1\omega A_2} : (Q_1 \times Q_2) \times X_2 \times (Q_1 \times Q_2) \longrightarrow L, \nu_{A_1\omega A_2} : (Q_1 \times Q_2) \times X_2 \times (Q_1 \times Q_2) \longrightarrow L, \omega : Q_2 \times X_2 \longrightarrow X_1$ be a function, $\forall (q_1,q_2),(p_1,p_2) \in Q_1 \times Q_2, \forall b \in X_2$.

Definition 3.2. Let $M_i = (Q_i, X_i, A_i)$ be a LIFFSM, $i = 1, 2$. Then $M_1 \circ M_2 = (Q_1 \times Q_2, X_1^{Q_2} \times X_2, A_1 \circ A_2)$ is called the wreath product of M_1 and M_2,

$$\mu_{A_1\circ A_2}((q_1,q_2),(f,b),(p_1,p_2)) = \mu_{A_1}(q_1,f(q_2),p_1) \wedge \mu_{A_2}(q_2,b,p_2),$$

$$\nu_{A_1\circ A_2}((q_1,q_2),(f,b),(p_1,p_2)) = \nu_{A_1}(q_1,f(q_2),p_1) \vee \nu_{A_2}(q_2,b,p_2),$$

where $\mu_{A_1\circ A_2} : (Q_1 \times Q_2) \times (X_1^{Q_2} \times X_2) \times (Q_1 \times Q_2) \longrightarrow L, \nu_{A_1\circ A_2} : (Q_1 \times Q_2) \times (X_1^{Q_2} \times X_2) \times (Q_1 \times Q_2) \longrightarrow L, X_1^{Q_2} = \{f | f : Q_2 \longrightarrow X_1\}, \forall ((q_1,q_2),(f,b),(p_1,p_2)) \in (Q_1 \times Q_2) \times (X_1^{Q_2} \times X_2) \times (Q_1 \times Q_2)$.

Theorem 3.1 *Let $M_i = (Q_i, X_i, A_i)$ be a LIFFSM, $i = 1, 2$. Then*

(1) $M_1 \omega M_2$ be a LIFFSM,
(2) $M_1 \circ M_2$ be a LIFFSM.

Proof. By the same method of Theorem 3.1 in Ref. [10], it is easy to proof that the result is true.

4 Coverings Properties of Products

Definition 4.1. Let $M_1 = (Q_1, X_1, A_1)$ and $M_2 = (Q_2, X_2, A_2)$ be LIFFSM. Let $\eta : Q_2 \longrightarrow Q_1$ be a surjective partial function and $\xi : X_1 \longrightarrow X_2$ be a partial function. Then the ordered pair (η, ξ) is called a weak covering of M_1 by M_2, written $M_1 \leq_w M_2$, if

$$\mu_{A_1}(\eta(p), x_1, \eta(q)) \leq \mu_{A_2}(p, \xi(x_1), q),$$

$$\nu_{A_1}(\eta(p), x_1, \eta(q)) \geq \nu_{A_2}(p, \xi(x_1), q),$$

for all x_1 in the domain of X_1 and p, q in the domain of η.

A weak covering differs from a covering only in that ξ in Definition 4.1 is a partial function, while ξ in Definition 2.5 is a function. Thus every covering is a weak covering.

Definition 4.2. Let $M_1 = (Q_1, X_1, A_1)$ and $M_2 = (Q_2, X_2, A_2)$ be LIFFSM. A pair (α, β) of mappings, $\alpha : Q_1 \longrightarrow Q_2$ and $\beta : X_1 \longrightarrow X_2$, is called a homomorphism, written $(\alpha, \beta) : M_1 \longrightarrow M_2$, if

$$\mu_{A_1}(q, a, p) \leq \mu_{A_2}(\alpha(q), \beta(a), \alpha(p)),$$

$$\nu_{A_1}(q, a, p) \geq \nu_{A_2}(\alpha(q), \beta(a), \alpha(p)),$$

$\forall q, p \in Q_1$ and $\forall a \in X_1$.

The pair (α, β) is called a strong homomorphism, if

$$\mu_{A_2}(\alpha(q), \beta(a), \alpha(p)) = \vee\{\mu_{A_1}(q, a, t) | t \in Q_1, \alpha(t) = \alpha(p)\},$$

$$\nu_{A_2}(\alpha(q), \beta(a), \alpha(p)) = \wedge\{\nu_{A_1}(q, a, t) | t \in Q_1, \alpha(t) = \alpha(p)\},$$

for all $q, p \in Q_1$ and $\forall a \in X_1$.

A homomorphism(strong homomorphism) $(\alpha, \beta) : M_1 \longrightarrow M_2$ is called an isomorphism(strong isomorphism) if α and β are both one-one and onto.

Theorem 4.1 *Let $M_1 = (Q_1, X_1, A_1)$ and $M_2 = (Q_2, X_2, A_2)$ be LIFFSM. Let $(\alpha, \beta) : M_1 \longrightarrow M_2$ be a homomorphism. If (α, β) is a strong homomorphism with α one-one, then*

$$\mu_{A_2}(\alpha(q), \beta(x_1), \alpha(p)) = \mu_{A_1}(q, x_1, p),$$

$$\nu_{A_2}(\alpha(q), \beta(x_1), \alpha(p)) = \nu_{A_1}(q, x_1, p),$$

for all $q, p \in Q_1$, $x_1 \in X_1$.

Proof. Since (α, β) is a strong homomorphism, we have

$$\mu_{A_2}(\alpha(q), \beta(x_1), \alpha(p)) = \vee\{\mu_{A_1}(q, x_1, t) | t \in Q_1, \alpha(t) = \alpha(p)\},$$

$$\nu_{A_2}(\alpha(q), \beta(x_1), \alpha(p)) = \vee\{\nu_{A_1}(q, x_1, t) | t \in Q_1, \alpha(t) = \alpha(p)\},$$

for all $q, p \in Q_1, x_1 \in X_1$. Since α is an one-one and $\alpha(t) = \alpha(p)$, we have $t = p$. Thus

$$\mu_{A_2}(\alpha(q), \beta(x_1), \alpha(p)) = \mu_{A_1}(q, x_1, p),$$

$$\nu_{A_2}(\alpha(q), \beta(x_1), \alpha(p)) = \nu_{A_1}(q, x_1, p).$$

Theorem 4.2 *Let $M_1 = (Q_1, X_1, A_1)$ and $M_2 = (Q_2, X_2, A_2)$ be LIFFSM. Let $(\alpha, \beta) : M_1 \longrightarrow M_2$ be a homomorphism, then*

(1) If this homomorphism be an onto strong homomorphism and α is an one-one, then $M_2 \leq M_1$,

(2) If α be an one-one, then $M_1 \leq M_2$.

Proof. (1) Since $(\alpha, \beta) : M_1 \longrightarrow M_2$ be an onto strong homomorphism, there exist surjective functions $\alpha : Q_1 \longrightarrow Q_2$ and $\beta : X_1 \longrightarrow X_2$. Let $\eta : Q_1 \longrightarrow Q_2$ and $\xi : X_2 \longrightarrow X_1$, $\eta = \alpha$. Since β be a surjective function. Hence there exists at least one original image $a \in X_1$ such that $\beta(a) = a'$ for $\forall a' \in X_2$. Let $\xi(a') = a$. (α, β) be a strong homomorphism with α one-one. Then

$$\mu_{A_2}(\alpha(q), \beta(a), \alpha(p)) = \mu_{A_1}(q, a, p),$$

$$\nu_{A_2}(\alpha(q), \beta(a), \alpha(p)) = \nu_{A_1}(q, a, p),$$

$\forall q, p \in Q_1, a' \in X_2$. If $\xi(a') = a$, we can view

$$\mu_{A_2}(\eta(q), a', \eta(p)) = \mu_{A_2}(\alpha(q), \beta(a), \alpha(p)) = \mu_{A_1}(q, a, p) = \mu_{A_1}(q, \xi(a'), p),$$

$$\nu_{A_2}(\eta(q), a', \eta(p)) = \nu_{A_2}(\alpha(q), \beta(a), \alpha(p)) = \nu_{A_1}(q, a, p) = \nu_{A_1}(q, \xi(a'), p).$$

Hence the ordered pair (η, ξ) is a covering of M_2 by M_1, $M_2 \leq M_1$.

(2) Since $(\alpha, \beta) : M_1 \longrightarrow M_2$ be a homomorphism, there exist functions $\alpha : Q_1 \longrightarrow Q_2$ and $\beta : X_1 \longrightarrow X_2$, such that

$$\mu_{A_1}(q_1, a_1, p_1) \leq \mu_{A_2}(\alpha(q_1), \beta(a_1), \alpha(p_1)),$$

$$\nu_{A_1}(q_1, a_1, p_1) \geq \nu_{A_2}(\alpha(q_1), \beta(a_1), \alpha(p_1)),$$

$\forall q_1, p_1 \in Q_1$ and $\forall a_1 \in X_1$. Let $\eta : Q_2 \longrightarrow Q_1$. If $\alpha(q_1) = q_2$, then $\eta(q_2) = q_1$. Since α be an one-one, we can view q_1 is determined uniquely. Thus η be a surjective partial function. Let $\xi : X_1 \longrightarrow X_2$, $\xi = \beta$, then

$$\mu_{A_1}(\eta(q_2), a_1, \eta(p_2)) \leq \mu_{A_2}(q_2, \xi(a_1), p_2),$$

$$\nu_{A_1}(\eta(q_2), a_1, \eta(p_2)) \geq \nu_{A_2}(q_2, \xi(a_1), p_2).$$

Hence the ordered pair (η, ξ) is a covering of M_1 by M_2, $M_1 \leq M_2$.

Corollary 4.1 Let $M_1 = (Q_1, X_1, A_1)$ and $M_2 = (Q_2, X_2, A_2)$ be LIFFSM. Let $(\alpha, \beta) : M_1 \longrightarrow M_2$ be a homomorphism, then

(1) If this homomorphism be an strong homomorphism and α is a both one-one and onto, then $M_2 \leq_w M_1$,
(2) If α be an one-one, then $M_1 \leq_w M_2$.

Proof. (1) The proof is similar to that of Theorem 4.2(1).

(2) we can know $M_1 \leq M_2$ by Theorem 4.2(2). Since every covering is a weak covering. Thus $M_1 \leq_w M_2$.

Theorem 4.3 *Let $M_i = (Q_i, X_i, A_i)$ be a LIFFSM, $i = 1, 2, 3$. If $M_1 \leq_w M_2$ and $M_2 \leq M_3$, then $M_1 \leq_w M_3$.*

Proof. Since $M_1 \leq_w M_2$, there exists a partial surjective function $\eta_1 : Q_2 \longrightarrow Q_1$ and a partial function $\xi_1 : X_1 \longrightarrow X_2$ such that

$$\mu_{A_1}(\eta_1(p_1), x_1, \eta_1(q_1)) \leq \mu_{A_2}(p_1, \xi_1(x_1), q_1),$$

$$\nu_{A_1}(\eta_1(p_1), x_1, \eta_1(q_1)) \geq \nu_{A_2}(p_1, \xi_1(x_1), q_1),$$

for all x_1 belong to the domain of ξ_1 and p_1, q_1 belong to the domain of η_1.

Since $M_2 \leq M_3$, there exists a surjective partial function $\eta_2 : Q_3 \longrightarrow Q_2$ and a function $\xi_2 : X_2 \longrightarrow X_3$ such that

$$\mu_{A_2}(\eta_2(p_2), x_2, \eta_2(q_2)) \leq \mu_{A_3}(p_2, \xi_2(x_2), q_2),$$

$$\nu_{A_2}(\eta_2(p_2), x_2, \eta_2(q_2)) \geq \nu_{A_3}(p_2, \xi_2(x_2), q_2),$$

for all $x_2 \in X_2$ and p_2, q_2 belong to the domain of η_2.

Let $\eta = \eta_1 \circ \eta_2 : Q_3 \longrightarrow Q_1$, $\xi = \xi_2 \circ \xi_1 : X_1 \longrightarrow X_3$. Clearly, η is a surjective partial function and ξ is a partial function. If $\forall x_1 \in domain(\xi) = domain(\xi_1)$ and $p, q \in domain(\eta) \subseteq domain(\eta_2)$, then

$$\begin{aligned}
\mu_{A_1}(\eta(p), x_1, \eta(q)) &= \mu_{A_1}(\eta_1 \circ \eta_2(p), x_1, \eta_1 \circ \eta_2(q)) \\
&= \mu_{A_1}(\eta_1(\eta_2(p)), x_1, \eta_1(\eta_2(q))) \\
&\leq \mu_{A_2}(\eta_2(p), \xi_1(x_1), \eta_2(q)) \\
&\leq \mu_{A_3}(p, \xi_2(\xi_1(x_1)), q) \\
&= \mu_{A_3}(p, \xi_2 \circ \xi_1(x_1), q) \\
&= \mu_{A_3}(p, \xi(x_1), q).
\end{aligned}$$

Similarly, we can prove $\nu_{A_1}(\eta(p), x_1, \eta(q)) \geq \nu_{A_3}(p, \xi(x_1), q)$. Clearly, (η, ξ) is a required weak covering of M_1 by M_3.

Theorem 4.4 *Let $M_i = (Q_i, X_i, A_i)$ be a LIFFSM, $i = 1, 2$. Then*

(1) $M_1 \omega M_2 \leq M_1 \circ M_2$,
(2) $M_1 \circ M_2 \leq M_1 \times M_2$,
(3) $M_1 \omega M_2 \leq M_1 \times M_2$.

Proof. (1) Let $\omega_b : Q_2 \longrightarrow X_1$ be a function defined by $\omega_b(p_2) = \omega(p_2, b)$ for all $p_2 \in Q_2$ and $b \in X_2$. Define $\xi : X_2 \longrightarrow X_1^{Q_2} \times X_2$ by $\xi(b) = (\omega_b, b)$ and let η be an identity map on $Q_1 \times Q_2$.

(2) Define $\xi : X_1^{Q_2} \times X_2 \longrightarrow X_1 \times X_2$ by $\xi(f, b) = (f(p_2), b)$ and let η be an identity map on $Q_1 \times Q_2$.

(3) Since $M_1 \omega M_2 \leq M_1 \circ M_2$ and $M_1 \circ M_2 \leq M_1 \times M_2$. We have $M_1 \omega M_2 \leq M_1 \times M_2$ by Theorem 2.2.

Theorem 4.5 *Let $M_i = (Q_i, X_i, A_i)$ be a LIFFSM, $i = 1, 2, 3$. If $M_1 \leq M_2$, then*

(1) Given $\omega_1 : Q_3 \times X_3 \longrightarrow X_1$, there exists $\omega_2 : Q_3 \times X_3 \longrightarrow X_2$ such that $M_1 \omega_1 M_3 \leq M_2 \omega_2 M_3$. If (η, ξ) is a covering of M_1 by M_2 and ξ is a surjective, then for all $\omega_1 : Q_1 \times X_1 \longrightarrow X_3$, there exists $\omega_2 : Q_2 \times X_2 \longrightarrow X_3$ such that $M_3 \omega_1 M_1 \leq M_3 \omega_2 M_2$,
(2) $M_1 \circ M_3 \leq M_2 \circ M_3$ and $M_3 \circ M_1 \leq_w M_3 \circ M_2$.

Proof. Since $M_1 \leq M_2$, there exists a surjective partial function $\eta_1 : Q_2 \longrightarrow Q_1$ and a function $\xi_1 : X_1 \longrightarrow X_2$ such that

$$\mu_{A_1}(\eta_1(p_2), x_1, \eta_1(q_2)) \leq \mu_{A_2}(p_2, \xi_1(x_1), q_2),$$

$$\nu_{A_1}(\eta_1(p_2), x_1, \eta_1(q_2)) \geq \nu_{A_2}(p_2, \xi_1(x_1), q_2),$$

for all $x_1 \in X_1$ and p_2, q_2 belong to the domain of η_1.

(1) Given $\omega_1 : Q_3 \times X_3 \longrightarrow X_1$, set $\omega_2 = \xi_1 \circ \omega_1$ and ξ_2 as an identity mapping on X_3. Define $\eta_2 : Q_2 \times Q_3 \longrightarrow Q_1 \times Q_3$ by $\eta_2((q_2, q_3)) = (\eta_1(q_2), q_3)$. Clearly, we can prove (η_2, ξ_2) is a required covering, $M_1 \omega_1 M_3 \leq M_2 \omega_2 M_3$.
Given $\omega_1 : Q_1 \times X_1 \longrightarrow X_3$, set $\omega_2 : Q_2 \times X_2 \longrightarrow X_3$ such that $\omega_2(q_2, \xi_1(x_1)) = \omega_1(\eta_1(q_2), x_1)$. Since ξ_1 is a surjective and X_1 is finite, such ω_2 exists. Clearly, ω_2 is not unique. Define $\eta_2 : Q_3 \times Q_2 \longrightarrow Q_3 \times Q_1$ by $\eta_2((q_3, q_2)) = (q_3, \eta_1(q_2))$ and set $\xi_2 = \xi_1$. Clearly, we can prove (η_2, ξ_2) is a required covering, $M_3 \omega_1 M_1 \leq M_3 \omega_2 M_2$.
(2) Define $\eta_2 : Q_2 \times Q_3 \longrightarrow Q_1 \times Q_3$ by $\eta_2((q_2, q_3)) = (\eta_1(q_2), q_3)$ and $\xi_2 : X_1^{Q_3} \times X_3 \longrightarrow X_2^{Q_3} \times X_3$ by $\xi_2(f, x_3) = (\xi_1 \circ f, x_3)$. Obviously, η_2 is a surjective partial function and ξ_2 is a function.
Define $\eta_2 : Q_3 \times Q_2 \longrightarrow Q_3 \times Q_1$ by $\eta_2((q_3, q_2)) = (q_3, \eta_1(q_2))$ and $\xi_2 : X_3^{Q_1} \times X_1 \longrightarrow X_3^{Q_2} \times X_2$ by $\xi_2(f, x_1) = (f \circ \eta_1, \xi_1(x_1))$. Obviously, η_2 is a surjective partial function and ξ_2 is a partial function.

Corollary 4.2 *Let $M_i = (Q_i, X_i, A_i)$ be a LIFFSM, $i = 1, 2, 3$. If $M_1 \leq M_2$, then*

(1) $M_1 \omega M_3 \leq M_2 \circ M_3$ and $M_3 \omega M_1 \leq_w M_3 \circ M_2$,
(2) $M_1 \circ M_3 \leq M_2 \times M_3$ and $M_3 \circ M_1 \leq M_3 \times M_2$,
(3) $M_1 \omega M_3 \leq M_2 \times M_3$ and $M_3 \omega M_1 \leq M_3 \times M_2$.

Proof. By Theorems 4.4, 4.5 and 2.2, we can prove $M_1 \omega M_3 \leq M_2 \circ M_3$. According to Theorems 4.4, 2.4 and 2.2, we can prove (2), (3) is true. Now we will prove $M_3 \omega M_1 \leq_w M_3 \circ M_2$.

$M_3 \omega M_1 = (Q_3 \times Q_1, X_1, A_3 \omega A_1)$, $M_3 \circ M_2 = (Q_3 \times Q_2, X_3^{Q_2} \times X_2, A_3 \circ A_2)$. Since $M_1 \leq M_2$, there exists a surjective partial function $\eta_1 : Q_2 \longrightarrow Q_1$ and a function $\xi_1 : X_1 \longrightarrow X_2$ such that

$$\mu_{A_1}(\eta_1(p_2), x_1, \eta_1(q_2)) \leq \mu_{A_2}(p_2, \xi_1(x_1), q_2),$$

$$\nu_{A_1}(\eta_1(p_2), x_1, \eta_1(q_2)) \geq \nu_{A_2}(p_2, \xi_1(x_1), q_2),$$

for all $x_1 \in X_1$ and p_2, q_2 belong to the domain of η_1.

Define $\eta_2 : Q_3 \times Q_2 \longrightarrow Q_3 \times Q_1$ by $\eta_2((q_3, q_2)) = (q_3, \eta_1(q_2))$ and $\xi_2 : X_1 \longrightarrow X_3^{Q_2} \times X_2$ by $\xi_2(x_1) = (f_{x_1}, \xi_1(x_1))$. Define $f_{x_1} : Q_2 \longrightarrow X_3$ by $f_{x_1}(p_2) = \omega(\eta_1(p_2), x_1)$. Clearly η_2 be a surjective partial function and ξ_2 be a partial function. $\forall (p_3, p_2), (q_3, q_2)$ in the domain of η_2 and x_1 in the domain of ξ_2,

$$\begin{aligned}
&\mu_{A_3 \omega A_1}(\eta_2((p_3, p_2)), x_1, \eta_2((q_3, q_2))) \\
&\quad = \mu_{A_3 \omega A_1}((p_3, \eta_1(p_2)), x_1, (q_3, \eta_1(q_2))) \\
&\quad = \mu_{A_3}(p_3, \omega(\eta_1(p_2), x_1), q_3) \wedge \mu_{A_1}(\eta_1(p_2), x_1, \eta_1(q_2)) \\
&\quad = \mu_{A_3}(p_3, f_{x_1}(p_2), q_3) \wedge \mu_{A_1}(\eta_1(p_2), x_1, \eta_1(q_2)) \\
&\quad \leq \mu_{A_3}(p_3, f_{x_1}(p_2), q_3) \wedge \mu_{A_2}(p_2, \xi_1(x_1), q_2) \\
&\quad = \mu_{A_3 \circ A_2}((p_3, p_2), (f_{x_1}, \xi_1(x_1)), (q_3, q_2)) \\
&\quad = \mu_{A_3 \circ A_2}((p_3, p_2), \xi_2(x_1), (q_3, q_2)).
\end{aligned}$$

Similarly, we can prove

$$\nu_{A_3 \omega A_1}(\eta_2((p_3, p_2)), x_1, \eta_2((q_3, q_2))) \geq \nu_{A_3 \circ A_2}((p_3, p_2), \xi_2(x_1), (q_3, q_2)).$$

Hence (η_2, ξ_2) is the required weak covering.

Corollary 4.3 Let $M_i = (Q_i, X_i, A_i)$ be a LIFFSM, $i = 1, 2, 3, 4$. If $M_1 \leq M_2$ and $M_3 \leq M_4$, then

(1) $M_1 \circ M_3 \leq_\omega M_2 \circ M_4$,
(2) $M_1 \omega M_3 \leq_\omega M_2 \circ M_4$,
(3) $M_1 \circ M_3 \leq M_2 \times M_4$,
(4) $M_1 \omega M_3 \leq M_2 \times M_4$.

Proof. By Theorems 4.5 and 4.3, we can prove $M_1 \circ M_3 \leq_\omega M_2 \circ M_4$. Similarly, we can prove (2), (3), (4) are true.

5 Conclusion

It is well known that product is a basic operation in automata theory. In this paper,we have investigated the products of lattice-valued intuitionistic fuzzy finite state machines and coverings.

Acknowledgments The authors are very grateful to the anonymous referee for his/her careful review and constructive suggestions. This work is supported by the National Natural Science Foundation of China (Grant No.11071178), the Research Foundation of the Education Department of Sichuan Province (Grant No. 12ZB106) and the Research Foundation of Sichuan Normal University (Grant No. 10MSL06).

References

1. Zadeh, L.A.: Fuzzy sets. Inf. Control **8**(3), 338–353 (1965)
2. Wee, W.G.: On generalizations of adaptive algorithm and application of the fuzzy sets concept to pattern classification, Purdue University (1967)
3. Atanassov, K.T.: Intuitionistic fuzzy sets. Fuzzy Sets Syst. **20**, 87–96 (1986)
4. Atanassov, K.T.: New operations defined over the intitionistic fuzzy sets. Fuzzy Sets Syst. **61**, 137–142 (1994)
5. Jun, Y.B.: Intuitionistic fuzzy finite state machines. J. Appl. Math. Comput. **17**(1–2), 109–120 (2005)
6. Jun, Y.B.: Intuitionistic fuzzy finite switchboard state machines. Appl. Math. Comput. **20**(1–2), 315–325 (2006)
7. Jun, Y.B.: Intuitionistic fuzzy transformation semigroups. Inf. Sci. **177**, 4977–4986 (2007)
8. Zhang, X., Li, Y.: Intuitionistic fuzzy recognizers and intuitionistic fuzzy finite automata. Soft Comput. **13**, 611–616 (2009)
9. Atanassov, K., Stoeva, S.: Intuitionistic L-fuzzy sets. In: Trappl, R. (ed.) Cybernetics and Systems Research 2, pp. 539–540. Elsevier, North-Holland (1984)
10. Yang, L., Mo, Z.W.: Direct product of lattice-valued intuitionistic fuzzy finite state automata and coverings. J. Sichuan Normal Univer. (Nat. Sic.)(To appear)

Attribute Reduction of Lattice-Value Information System Based on L-Dependence Spaces

Chang Shu, Zhi-wen Mo, Xiao Tang and Zhi-hua Zhang

Abstract Lattice is a wide concept. All different kinds of information systems come down to lattice-value information system. Attribute reduction of different kinds of information systems could be boiled down to that of lattice-value information systems. In this paper, L-dependence space is established on lattice-value information system. Then attribute reduction of theory and algorithm is put forward and the effectiveness and feasibility of algorithm are explained by an example. Finally, the result of attribute reduction is compared with other algorithms by computational complexity.

Keywords Lattice-value · Information system · Attribute reduction · L-dependence space

1 Introduction

One of the fundamental goals of artificial intelligence (AI) is to build artificially computer-based systems which make computer simulate, extend and expand human's intelligence and empower computers to perform tasks which are routinely performed by human beings. Due to the fact that human intelligence actions are always involved with uncertainty in information processing, one important task of AI is to study how to make the computer simulate human being to deal with uncertain information. Among major ways in which human being deal with uncertainty of information, the uncertainty reasoning becomes an essential mechanism in AI.

C. Shu (✉) · Z. Mo
College of Mathematics and Software Science, Sichuan Normal University, Chengdu 610066, Sichuan, China
e-mail: yaroer2002@tom.com

X. Tang · Z. Zhang
School of Mathematical Science, University of Electronic Science and Technology of China, Chengdu 6111731, Sichuan, China

B.-Y. Cao and H. Nasseri (eds.), *Fuzzy Information & Engineering and Operations Research & Management*, Advances in Intelligent Systems and Computing 211,
DOI: 10.1007/978-3-642-38667-1_12,

A general model of many finite structures is introduced and investigated in computer science. The model is referred as dependence space. The main feature of the model is that it enables us to deal with the indiscernibility-type incompleteness of information that a modeled structure might be burdened with. The model provides a general framework for expressing the concept of independence of sets and the concept of dependency between sets with respect to a dependence space. It is shown that these concepts are the foundation on which many applied structures rest. The theory of dependence spaces is developed aimed at providing tools for studying the problems relevant to the theory of mod reduction and algorithm is put forward.

Lattice is a widely used concept. All different kinds of information system come down to lattice-value information system. Attribute reduction of all different kinds of information system come down to that of lattice-value information system. In this paper, the concept dependence space is established on lattice-value information system. Then effectiveness and feasibility of algorithm are explained by an example. Finally, the result of attribute reduction is compared with other algorithms by computational complexity [1–3].

2 Lattice-value Information System and Properties

Lattice is a widely concept. All sorts of different information systems can attribute to lattice-value information systems, so reduction of all sorts of different information systems can attribute to one of lattice-value information systems [4–11].

Definition 1. *Data is represented as an lattice-value information system* (U, A, F)*, where* $U = \{x_{1,}x_2, \cdots x_n\}$ *is an object set,* $A = \{a_1, a_2, \cdots a_n\}$ *is an attribute set and* $F = \{f_l : l \leq m\}$ *is a set of object attribute value mappings (also is information function)* $f_l : U \rightarrow V_l \ (l \leq m)$*, where* V_l *is the domain of the attribute* a_l *and a finite lattice with the maximum element 1 and the minimum element 0. For convenience, let the same symbols* $\geq$ *explain order relation of* V_l*.*

Definition 2. *Let* (U, A, F) *be an lattice-value information system, given a binary relation* $R_B = \{(x, y) \in U \times U : f_l(x) \geq f_l(y) \, (\forall a_l \in B)\}$ *for any attribute subset* $B \subseteq A$ *, we define* $[x]_B = \{y \in U : (x, y) \in R_B\} = \{y \in U : f_l(x) \geq f_l(y) \, (\forall a_l \in B)\}, x \in U$*.*

Theorem 1. *Let* (U, A, F) *be an lattice-value information system, given the following properties:*

1. R_B *is reflexive and transitive, but it is not symmetrical. In general,* R_B *is not an equivalence relation.*
2. *When* $B_1 \subseteq B_2 \subseteq A$*, there holds* $R_{B_1} \supseteq R_{B_2} \supseteq R_A$*.*
3. *When* $B_1 \subseteq B_2 \subseteq A$*, the relation* $[x]_{B_1} \supseteq [x]_{B_2} \supseteq [x]_{R_A}$ *exists.*
4. $\Im = \{[x]_B : x \in U\}$ *is a cover of U.*
5. *When* $y \in [x]_B$*, there is* $[y]_B \subseteq [x]_B$*.*

Theorem 2. *Let (U, A, F) be an lattice-value information system and R is a relation on L-lattice. We have $R_l = \{(x, y) \in U \times U : (f_l(x), f_l(y)) \in R\}$, then there have the following properties:*

1. $R_{\{R_l\}} = R'_{\{R_l\}} = R_l$.
2. $R_B = \cap\{R_l : a_l \in B\}, R'_B = \cup\{R_l : a_l \in B\}$.
3. $R_\phi = U \times U, R'_\phi = \phi$.
4. $R_{B\cup C} = R_B \cap R_C, R'_{B\cup C} = R'_B \cup R'_C$.
5. *If* $B \subseteq C$,*then* $R_B \supseteq R_C, R'_B \subseteq R'_C$.
6. *If* $B \neq \phi$, *then* $R_B \subseteq R'_B$.
7. $\sim R_B = R'_{-B}, \sim R'_B = R_{-B}$.

Proof. It is clear to prove by definitions.

Theorem 3. *Let (U, A, F) be an lattice-value information system and R is a relation on L-lattice. If $B \subseteq A$, and $B \neq \phi$, then*

1. *If R is reflexive, then R_B and R'_B are also reflexive.*
2. *If R is symmetrical, then R_B and R'_B are also symmetrical.*
3. *If R is transitive, then R_B and R'_B are also transitive.*

Proof. It is immediate by definitions.

3 L-Dependence Spaces Based on Lattice-Value Information Systems

Definition 3. *Let (U, A, F) be an lattice-value information system, R is an equivalence relation of $P(A)$.*

1. *Let $(B_1, C_1) \in R$, $(B_2, C_2) \in R$. If $(B_1 \cup B_2, C_1 \cup C_2) \in R$, R is a consistent relation.*
2. *If R is a consistent relation, (A, R) is defined as an L-dependence space.*

Theorem 4. *Let (U, A, F) be an lattice-value information system, S is a binary relation on L. Note:*

$$R_B = \left\{(x_i, x_j) \left| \left(f_l(x_i), f_l(x_j)\right) \in S\,(a_l \in B)\right.\right\},$$

$$R = \{(B, C) \,|R_B = R_C\}.$$

Then (A, R) is an L-dependence space.

Table 1 Lattice-value information system

U	a_1	a_2	a_3
x_1	$\{22, 23, \ldots 26\}$	$\{48, 49, \ldots 54\}$	$\{154, 155, \ldots 157\}$
x_2	$\{26, 27, \ldots 33\}$	$\{73, 74, \ldots 78\}$	$\{170, 171, 175\}$
x_3	$\{24, 25, \ldots 29\}$	$\{51, 52, \ldots 58\}$	$\{159, 160, \ldots 162\}$
x_4	$\{31, 32, \ldots 37\}$	$\{75, 76, \ldots 82\}$	$\{157, 158, \ldots 165\}$

Proof. It is easy to prove that R is an equivalence relation. If it is noted $R_l = R_{\{a_l\}}$, then $R_B = \bigcap_{a_l \in B} R_l$ and $R_{B \cup C} = R_B \cap R_C$. If $(B_1, C_1) \in R$, $(B_2, C_2) \in R$, then $R_{B_1} = R_{C_1}, R_{B_2} = R_{C_2}$, and then $R_{B_1 \cup B_2} = R_{B_1} \cap R_{B_2} = R_{C_1} \cap R_{C_2} = R_{C_1 \cup C_2}$. Thus, we prove that $(B_1 \cup B_2, C_1 \cup C_2) \in R$ and R is a consistent relation. This means that (A, R) is an L-dependence space.

Theorem 5. *Let (U, A, F) be an lattice-value information system and $H \subseteq P(A)$. We have $T(H) = \{(B, C) \in P(A)^2 \mid \forall D \in H, B \subseteq D \Leftrightarrow C \subseteq D\}$, and then $(A, T(H))$ is an L-dependence space.*

Proof. Clearly, $T(H)$ is an equivalence relation.

Let $(B_1, C_1) \in T(H), (B_2, C_2) \in T(H)$.

Then the following relations hold: $\forall D \in H, B_1 \subseteq D \Leftrightarrow C_1 \subseteq D, B_2 \subseteq D \Leftrightarrow C_2 \subseteq D$, $B_1 \cup B_2 \subseteq D \Leftrightarrow B_1 \subseteq D \quad B_2 \subseteq D \Leftrightarrow C_1 \subseteq D \text{and} C_2 \subseteq D \Leftrightarrow C_1 \cup C_2 \subseteq D$. Therefore, $(B_1 \cup B_2, C_1 \cup C_2) \in T(H)$, and $T(H)$ is a consistent relation.

Definition 4. *Let (A, R) is an L-dependence space. If there exists $H \subseteq P(A)$, then $T(H) = R$ holds, therefore we call that H is dense in R.*

Theorem 6. *Let (U, A, F) be lattice-value information system and S is a binary relation on L. Note:*

$$R_B = \{(x_i, x_j) \mid (f_l(x_i), f_l(x_j)) \in S, (a_l \in B)\},$$
$$R = \{(B, C) \mid R_B = R_C\},$$
$$C_{ij} = \{a_l \in A \mid (f_l(x_i), f_l(x_j)) \in S\} \, (i, j \leq n).$$

We denote coordination matrix of R on lattice-value information system as follows: $M_R = (C_{ij} : i, j \leq n)$. Therefore M_R is dense in R and that is $T(M_R) = R$.

Proof. Clearly, R and $T(M_R)$ are all consistent relations on $P(A)$.

If $(B, C) \in R$, then $R_B = R_C$.

Hence, $B \subseteq C_{ij} \Leftrightarrow (x_i, x_j) \in R_B \Leftrightarrow (x_i, x_j) \in R_C \Leftrightarrow C \in C_{ij}$, for any $i, j \leq n$.

That is $(B, C) \in T(M_R)$. Hence $R \subseteq T(M_R)$.

In turn, if $(B, C) \in T(M_R)$, then $B \subseteq C_{ij} \Leftrightarrow C \subseteq C_{ij}$ $(i, j \leq n)$.

Furthermore, $(x_i, x_j) \in R_B \Leftrightarrow B \subseteq C_{ij} \Leftrightarrow C \subseteq C_{ij} \Leftrightarrow (x_i, x_j) \in R_C$. Therefore, $R_B = R_C$ and that is $(B, C) \in R$. Hence, $T(M_R) \subseteq R$.

Generally, $T(M_R) = R$.

Clearly, M_R has the properties as follows:

1. For any $B \subseteq A$, $(x_i, x_j) \in R_B \Leftrightarrow B \subseteq C_{ij}$.
2. For any $B \subseteq A$, $(x_i, x_j) \in R'_B \Leftrightarrow B \cap C_{ij} \neq \phi$.

4 Attribute Reduction

Definition 5. *Let (U, A, F) be an lattice-value information system and R is a relation of lattice L. If $R_B = R_A$ and $R_{B-\{b\}} \neq R_B$ $(\forall b \in B)$, then $B \subseteq A$ is defined as a reduction of A.*

Theorem 7. *Let (U, A, F) be an lattice-value information system. If (A, R) is a L-dependence space and H is dense in R , then for $B \subseteq A$, B is reduction of A if and only if B is the minimal element of $H_0 = \{D' \subseteq A : D' \cap D \neq \phi\, (\forall D \in \bar{H})\}$ where $\bar{H} = \{D \neq \phi : -D \in H\}$.*

Example 1. *Let (U, A, F) be an lattice-value information system, where $U = \{x_1, x_2, x_3, x_4\}$,*

We denote $L = P(V) = \{1, 2, \cdots, 200\}$ and define a relation on L as follows: $R = \{(E, F) \in P(V) \times P(V) : E \cap F \neq \phi\}$. Then

$$R_B = \{(x_i, x_j) \in U \times U : f_l(x_i) \cap f_l(x_j) \neq \phi\, (\forall a_l \in B)\},$$

$$R'_B = \{(x_i, x_j) \in U \times U : f_l(x_i) \cap f_l(x_j) \neq \phi\, (\exists a_l \in B)\}$$

We put forward a coordination matrix of A, as shown in Table 2.

Table 2 A coordination matrix M_R

U	x_1	x_2	x_3	x_4
x_1	A	$\{a_1\}$	$\{a_1, a_2\}$	$\{a_3\}$
x_2	$\{a_1\}$	A	$\{a_1\}$	$\{a_1, a_2\}$
x_3	$\{a_1, a_2\}$	$\{a_1\}$	A	$\{a_3\}$
x_4	$\{a_3\}$	$\{a_1, a_2\}$	$\{a_3\}$	A

Thus $H_1 = \{C_{ij} : i, j \le n\} = \{\{a_1\}, \{a_3\}, \{a_1, a_2\}, A\}$ is dense in R where $R = \{(B, C) \in P(A) \times P(A) : R_B = R_C\}$, $\bar{H}_1 = \{\{a_2, a_3\}, \{a_1, a_2\}, \{a_3\}\}$, $H_0^1 = \{\{a_1, a_3\}, \{a_2, a_3\}, A\}$.

Obviously, both $\{a_1, a_3\}$ and $\{a_2, a_3\}$ are minimal elements in H_0^1. Therefore, $\{a_1, a_3\}$ and $\{a_2, a_3\}$ are all reduction of A in relation of R.

Similarly, $H_2 = \{-C_{ij} : i, j \le n\} = \{\phi, \{a_3\}, \{a_1, a_2\}, \{a_2, a_3\}\}$ is dense in R' where

$R' = \{(B, C) \in P(A) \times P(A) : R'_B = R'_C\}$, $\bar{H}_2 = \{\{a_1\}\{a_3\}, \{a_1, a_2\}, A\}$, $H_0^2 = \{\{a_1, a_3\}, A\}$. Obviously, $\{a_1, a_3\}$ is a minimal element in H_0^2. Therefore, $\{a_1, a_3\}$ and $\{a_2, a_3\}$ are all reduction of A in relation of R'.

5 Conclusion

L-dependence space is established on lattice-value information system. Then attribute reduction of theory and algorithm is put forward and effectiveness and feasibility of algorithm are explained by an example. Finally, the result of attribute reduction is compared with other algorithms by computational complexity.

Acknowledgments Thanks to the support by National Natural Science Foundation of China (No. 11071178).

References

1. Novotny, M., Pawlak, Z.: On a problem concerning dependence spaces. Fundamenta Informaticae **16**, 275–287 (1992)
2. Novotny, M., Pawlak, Z.: Independence of attributes. Bull. Polish Acad. Sci. Math. **36**, 459–465 (1988)
3. Novotny, M.: Dependence spaces of information systems. In: Orlowska, E. (ed.) Logical and algebra-ic investigations in rough set theory (to appear)
4. Belohlavek, R.: Concept lattices and order in fuzzy logic. Ann. Pure Appl. Logic **128**, 277–298 (2004)
5. Wille, R.: Restructuring Lattice theory: an approach based on hierarchies of concepts. In: Rival, I. (ed.) Ordered Sets, pp. 445–470. Reidel, Dordrecht-Boston (1982)
6. Yao, Y.Y.: Concept lattices in rough set theory. In: Proceedings of 23rd International Meeting of the North American Fuzzy Information Processing Society, pp. 796–800 (2004)
7. Zhang, W.-X., Wei, L., Qi, J.-J.: Attribute reduction in concept lattice based on discernibilit-y matrix. In: Slezak, D. et al. (eds.): RSFDGrC 2005, LNAI 3642, pp. 157–165 (2005)
8. Zhang, W.-X., Mi, J.-S., Wu, W.-Z.: Approaches to knowledge reduction in inconsistent syst-ems. Int. J. Intell. Syst. **18**, 989–1000 (2003)
9. Zedeh, L.A.: Fuzzy sets. Inf. Control **8**, 338–353 (1965)
10. Yager, R.R.: An approach to ordinal decision making. Int. J. Approximate Reasoning **12**, 237–261 (1995)
11. Liang, J.-Y., Xu, Z.-B.: The algorithm on knowledge reduction in incomplete information systems. Int. J. Uncertainty Fuzziness Knowl. Based Syst. **10**(1), 95–103 (2002)

A New Similarity Measures on Vague Sets

Zheng-qi Cai, Ya-bin Shao, Shuang-liang Tian and Yong-chun Cao

Abstract Vague set is a valid tool for processing uncertain information. The similarity measure of two uncertain patterns is important for intelligent reasoning. It is also a key problem to measure the similarity of vague values or vague sets in vague information processing systems. In this paper, according to the theory of similarity measure between intervals, it is shown that four factors influencing vague sets and vague values should be taken into account while calculating the similarity degree. Some existing similarity measures are reviewed and compared. Some faults of existed methods are pointed out. A new method for similarity measures between vague sets (values) is put forward, and is proved to satisfy some rules. The validity and advantage of this method are illustrated by an example.

Keywords Vague sets · Fuzzy sets · Similarity measure

1 Introduction

Zadeh [1] proposed fuzzy theory in 1965. The most important feature of a fuzzy set is that a fuzzy set is a class of objects that satisfy a certain property. The membership function of fuzzy set assigns each object a number which is on interval [0, 1] as its membership degree. It not only includes the evidence that the element belongs to the set, but also includes the evidence that the element does not belong to the set, but it cannot represent both at the same time too.

Z. Cai (✉) · Y. Shao · S. Tian · Y. Cao
School of mathematics and Computer Science, Northwest University for Nationalities, 730030 Lanzhou, China
e-mail: caizhengqi@126.com

Y. Shao
e-mail: yb_shao@163.com

B.-Y. Cao and H. Nasseri (eds.), *Fuzzy Information & Engineering and Operations Research & Management*, Advances in Intelligent Systems and Computing 211,
DOI: 10.1007/978-3-642-38667-1_13,

In order to deal with this problem, Gau and Buehrer [2] proposed the concept of vague set in 1993, by replacing the value of an element in a set with a sub-interval of [0, 1]. Namely, a truth-membership function $t_v(x)$ and a false-membership function $f_v(x)$ are used to describe the boundaries of membership degree. These two boundaries form a sub-interval $[t_v(x), 1 - f_v(x)]$ of [0, 1]. The vague set theory improves description of the objective real world, becoming a promising tool to deal with inexact, uncertain or vague knowledge. Many researchers have applies this theory to many situations, such as fuzzy control, decision-making, knowledge discovery and fault diagnosis. And the tool has presented more challenging than that with fuzzy sets theory in applications.

In intelligent activities, it is often needed to compare and couple between two fuzzy concepts. That is, we need to check whether two knowledge patterns are identical or approximately same, to find out functional dependence relations between concepts in a data mining system. Many measure methods have been proposed to measure the similarity between two vague sets (values). Each of them is given from different side, having its own counterexamples.

In this chapter, existing similarity measures between vague sets are analyzed, compared and summarized. Some faults of existed methods are pointed out. According to the theory of similarity measure between intervals, it is shown that four factors influencing vague sets and vague values should be taken into account while calculating the similarity degree. A new method for similarity measures between vague sets (values) is put forward, and is proved to satisfy some rules. The validity and advantage of this method are illustrated by an example.

2 Preliminaries

In this section, we review some basic definitions and terms of vague values and vague sets from [2, 3].

Definition 1 *Let X be a space of points (objects), with a generic element of X denoted by x, A vague set V in X is characterized by a truth-membership function $t_v(x)$ and a false-membership function $f_v(x)$, $t_v(x)$ is a lower bound on grade of membership of x derived from the evidence for x, and $f_v(x)$ is a lower bound on the negation of x derived from the evidence against x, $t_v(x)$ and $f_v(x)$ both associate a real number in the interval [0, 1] with each point in X, where $0 \le t_v(x) + f_v(x) \le 1$. That is*

$$t_v : X \to [0, 1], \quad f_v : X \to [0, 1], \forall x \in X.$$

The membership degree of vague set V is denoted by $V(x) = [t_v(x), 1 - f_v(x)]$.

A vague set can be denoted by $V = \{(x, t_v(x), f_v(x)) | x \in X\}$, $[t_v(x), 1 - f_v(x)]$ is called as the vague value of point x in V. Note: $x_v = [t_v(x), 1 - f_v(x)]$.

In a voting model, the vague value $x = [t_A(x), 1 - f_A(x)] = [0.5, 0.8]$, then $t_A(x) = 0.5$, $f_A(x) = 0.2$, it can be interpreted as: The vote for a resolution is 3 in favor, 3 against and 4 abstentions.

When X is continuous, a vague set V can be written as

$$V = \int_X [t_V(x), 1 - f_V(x)]/x, \quad x \in X. \tag{1}$$

When X is discrete, a vague set V can be written as

$$V = \sum_{i=1}^{n} [t_V(x_i), 1 - f_V(x_i)]/x_i, \quad x \in X \tag{2}$$

Definition 2 *Let x and y be two vague values, where $x = [t_v(x), 1 - f_v(x)]$ and $y = [t_v(y), 1 - f_v(y)]$. If $t_v(x) = t_v(y)$ and $f_v(x) = f_v(y)$, then the vague values x and y are called equal.*

Definition 3 *A vague set A is contained in the other vague set B, $A \subseteq B$ if and only if $\forall x \in X, t_A(x) \leq t_B(x), f_A(x) \geq f_B(x)$.*

Definition 4 *Two vague sets A and B are equal, written as $A = B$, if and only if $A \subseteq B$, and $B \subseteq A$; that is $t_A = t_B$ and $1 - f_A = 1 - f_B$. $\forall x \in X, t_A(x) = t_B(x), 1 - f_A(x) = 1 - f_B(x)$.*

Definition 5 *The complement of a vague set A is denoted by $\overline{A}$ and is defined by $t_{\overline{A}}(x) = f_A(x), 1 - f_{\overline{A}}(x) = 1 - t_A(x)$*

Definition 6 *The union of two vague sets A and B with respective truth-membership and false-membership function t_A, f_A, t_B and f_B is a vague set C, written as $C = A \cup B$, whose truth-membership and false-membership functions are related to those of A and B by*

$$t_C(x) = max(t_A(x), t_B(x)), 1 - f_C(x) = max(1 - f_A(x), 1 - f_B(x))$$

Definition 7 *The intersection of two vague sets A and B with respective truth-membership and false-membership function t_A, f_A, t_B and f_B is a vague set C, written as $C = A \cap B$, whose truth-membership and false-membership functions are related to those of A and B by*

$$t_C(x) = min(t_A(x), t_B(x)), \quad 1 - f_C(x) = min(1 - f_A(x), 1 - f_B(x))$$

In the sequel, instead of writing $t_A(x)$ and $f_A(x)$ for all x in X, we sometimes write more simply t_A and f_A.

Definition 8 *Let A be a vague set on X, $x = [t_A(x), 1 - f_A(x)]$ is a vague value, this approach bounds the grade of membership of x to a subinterval $[t_A(x), 1 - f_A(x)]$*

of [0,1], $t_A(x)$ is called the left end point of x, $1 - f_A(x)$ is called the right end point of x. Let $S(x) = t_A(x) - f_A(x)$ called the score of x or degree of support of x, $\phi(x) = (t_A(x)+1-f_A(x))/2$ called the middle point of x, $\pi(x) = 1-f_A(x)-t_A(x)$ called the length(the interval length) of x. Use $\pi(x)$ represent the unknowable degree of vague set A. Correspondently, it knowable degree can be described by $K(x) = 1 - \pi(x) = f(x) + t(x)$, it can also reflect the degree of supporting. Obviously, we have the following properties:

(1) $S(x) \in [-1, 1]$;
(2) $|\Phi(x) - \Phi(y)| = |S(x) - S(y)|/2 \in$ *[0,1];*
(3) $|\pi(x) - \pi(y)| = |K(x) - K(y)| \in$ *[0,1];*
(4) $|K(x) - K(y)| = |\Phi(x) - \Phi(y)|/2 \in$ *[0,1].*

3 Some Main Factors of Vague Value Similarity Measure

In the similarity measure of vague set, people usually adopt the similarity measure method of vague value of vague set.

In generally speaking the characteristic of interval has four important parameters: the left (right) end point, the interval length and the middle point. Therefore, above factors should be considered when we measure the similarity of interval. Since a vague value is a subinterval of [0, 1], so similarity measure of vague values is equivalent to similarity measure of intervals.

From the voting model, we can see that the vague value reflected three kinds of information, namely “supporting number”, “against number” and “abstentions number”. Therefore we should consider these information and the approve tendency information when we have to measure the similarity of two vague values. In essential speaking, there are four characteristics should be considered in measuring the similarity degree of vague sets (value).

(1) Distance of the interval end points. For two vague value x and y, the distance of the left end point, right end point are respectively $|t_x - t_y|$, $|f_x - f_y|$. Intuitively, the smaller distance of interval end points, then the bigger the similarity degree of x and y. Therefore, the distance of the interval end points should be considered in similarity measure. In fact, $|t_x - t_y|$ is difference of supporting degree; $|f_x - f_y|$ is difference of opposing degree.
(2) Difference of interval length. For vague value x, it’s interval length is $1 - f_x - t_x$, namely is the unknowable degree $\pi(x)$. $\pi(x)$ is meaningful to measure the similarity between two vague values. Under the same condition, if x has smaller uncertainty degree, and y has bigger uncertainty, then the smaller the similarity degree of x and y, which is adaptive to our intuition .So, $|\pi(x) - \pi(y)|$ is a main factor for the similarity measure.
(3) Distance of the interval middle point. If x and y have smaller distance of the middle point, then the bigger the similarity degree of x and y. Therefore, the distance of the interval middle point is an important factor to similarity measure

between two vague values. For vague value x and y, the distance of the interval middle point is $|\Phi(x) - \Phi(y)|$.
In addition, the background of uncertainty information processing should also be considered. According to the voting model, besides the supporting number and against number, the abstentions number and the effect of voted results on abstentions should also to be considered. In fact, as soon as the publication of the first voting result, the voter of voting abstain will reselect the second voting. Therefore, the uncertainty degree should be considered in measuring the similarity degree of vague sets (value). So, the fourth factor as follow.

4) The effect of uncertainty on supporting and against degree. In generally, the effect of uncertainty on supporting and against degree is hard to determine. From the voting model, the larger number of against and abstentions is, the smaller effect of uncertainty on supporting and against degree should be. On the contrary, the smaller number of against and abstentions is, the larger effect of uncertainty on supporting and against degree should be. Based on the discussion, while measure the difference of supporting degree and against degree, we adopt weighting method to embody the effect of uncertainty on supporting and against degree.

4 Researches into Similarity Measure

4.1 Analysis on Existing Similarity Measure between Vague Values

In this section, comprehensive analyses of similarity measures between vague values are provided. Suppose that $x = [t(x), 1 - f(x)]$ and $y = [t(y), 1 - f(y)]$ are two vague values over the discourse university U.

Chen [4, 5] considered the middle point, gave the following similarity measure formula:

$$M_c(x, y) = 1 - |S(x) - S(y)|/2 = 1 - |\Phi(x) - \Phi(y)| \tag{3}$$

From the definition of the $M_c(x, y)$, we can see that the above similarity measure only considering the interval middle point.

Hong and Kim [6] gave the similarity measure like below:

$$M_{HK}(x, y) = 1 - \frac{|t_x - t_y| + |f_x - f_y|}{2} \tag{4}$$

Obviously, the $M_{HK}(x, y)$ similarity measure only considering the distance of the interval end points. The $M_{HK}(x, y)$ pays equal attention both to the difference of two truth-membership degrees and to the difference of two false-membership degrees, between two vague values. Pairs of vague values, which have both the same difference of truth-membership degrees and the same difference of false-membership degrees, have the same similarity. But it does not distinguish the positive difference and negative difference between true- and false-membership degrees.

Li and Xu [7] gave the formula of similarity measure based on interval end point:

$$M_{LX}(x, y) = 1 - [|S(x) + S(y)| + |t(x) - t(y)| + |f(x) - f(y)|]/4 \quad (5)$$

In fact,

$$M_{LX}(x, y) = [M_c(x, y) + M_{HK}(x, y)]/2 \quad (6)$$

$M_{LX}(x, y)$ similarity measure only considers the distance of the interval end point and the interval middle point. The $M_{LX}(x, y)$ model inherits the advantages of the $M_C(x, y)$ and $M_{HK}(x, y)$ models, paying equal attentions to the support of vague value, truth-membership degree, and false-membership degree, respectively.

Jiulun [8] colligated the formula mentioned above and gave the following similarity measure formula:

$$M_F(x, y) = 1 - \frac{|t_x - t_y| + |f_x - f_y| + |\pi_x - \pi_y|}{2} \quad (7)$$

$$M'_F(x, y) = 1 - \frac{|t_x - t_y| + |f_x - f_y| + |\pi_x - \pi_y| + |\phi_x - \phi_y|}{3} \quad (8)$$

$M_F(x, y)$ similarity measure paying equal attentions to the distance of the interval end points and the interval middle point. Based on $M_F(x, y)$, $M'_F(x, y)$ similarity measure considering the difference of interval length, but which is not consider the effect of uncertainty on support and opposite degree.

Wenbin and Yu [9] gave the formula of similarity measure based on true-membership, false-membership, score and vague degree:

$$M_Z(x, y) = 1 - \frac{1}{28}\{8|S(x) - S(y)| + 5|t_x - t_y| + 5|f_x - f_y| + 2|t_x + f_x - t_y - f_y| \quad (9)$$

$M_Z(x, y)$ similarity measure only considers interval end points, interval middle point and uncertainty degree, but not considering the effect of uncertainty on support and opposite degree.

Each similarity measure expression has its own measuring focus although they all evaluate the similarities in vague sets. We think all existing similarity measures are valuable. There exist two reasons behind this thought: First, a new similarity measure is proposed, always accompanying with explanations of overcoming counter-intuitive cases of other methods. Second, there are different selection criteria and requirements during specific application procedure of similarity measure.

4.2 Some Basic Rules of Similarity Measure Between Vague Values

Definition 9 *Let* $x = [t_x, 1 - f_x]$, $y = [t_y, 1 - f_y]$, $z = [t_z, 1 - f_z]$ *be three vague values in vague set V,* $M(x, y)$ *is said to be the degree of similarity between x and y, if* $M(x, y)$ *satisfies the properties condition (P1–P6).*

$P1: 0 \leq M(x, y) \leq 1$
$P2: M(x, y) = M(y, x)$
$P3: M(x, y) = M(1 - x, 1 - y)$
$P4: M(x, x) = 1$
$P5: M(x, y) = 0$ *if and only if* $x = [0, 0], y = [1, 1]$ *or* $x = [1, 1], y = [0, 0]$
$P6:$ *If* $x \leq y \leq z$, *then* $M(x, z) \leq min \{M(x, y), M(y, z)\}$

5 New Method for Similarity Measure

5.1 New Method for Similarity Measure Between Vague Values

Definition 10 *Let* $x = [t_x, 1 - f_x], y = [t_y, 1 - f_y]$ *be two vague values in vague set A, then the similarity degree of x and y can be evaluated by the function M(x, y):*

$$M(x, y) = 1 - \frac{(2 - t_x - t_y)|t_x - t_y| + (2 - f_x - f_y)|f_x - f_y| + |\phi_x - \phi_y| + |\pi_x - \pi_y|}{3} \tag{10}$$

Since $\pi_x + \pi_y + f_x + f_y = 2 - t_x - t_y, \pi_x + \pi_y + t_x + t_y = 2 - f_x - f_y$, Coefficient $(\pi_x + \pi_y + f_x + f_y)$ of $|t_x - t_y|$ implies that the uncertainty and against information of x and y have a very important effect on $|t_x - t_y|$. If $2 - t_x - t_y$ is larger, the impact of $|t_x - t_y|$ should be smaller, similarity should be smaller. If $2 - t_x - t_y$ is smaller, the impact of $|t_x - t_y|$ should be larger, similarity should be larger. Coefficient $(2 - f_x - f_y)$ of $|f_x - f_y|$ means that when $|f_x - f_y|$ is the same, similarity should be smaller if $|f_x - f_y|$ is smaller.

Above similarity measure considering the distance of the interval end points, the distance of the interval middle points, interval length and the effect of uncertainty on supporting and against degree.

Theorem 1 $M(x, y)$ *satisfies the properties condition P1–P6.*

Proof. (1)From the Definition 1, $0 \leq |t_x - t_y| \leq 1, 0 \leq |f_x - f_y| \leq 1, 0 \leq |\phi_x - \phi_y| \leq 1,$

$$0 \leq |\pi_x - \pi_y| \leq 1, 0 \leq 2 - t_x - t_y \leq 2, 0 \leq 2 - f_x - f_y \leq 2.$$

If $|t_x - t_y| = 0, |f_x - f_y| = 0$,then $|\phi_x - \phi_y| = 0, |\pi_x - \pi_y| = 0$,so $M(x, y) \leq 1 - 0 = 1$.

If $|t_x - t_y| = 1$, we have $t_x = 0, t_y = 1$ or t_x=1, $t_y = 0$. If $|f_x - f_y| = 1$, then $f_x = 0$, f_y=1 or f_x=1, f_y=0. Corresponding to the two cases, there are four combinatorial forms. From the definition of Vague set, it is only t_x=0,t_y=1, f_x=1, f_y =0 and t_x =1, t_y=0, f_x=0, f_y=1 holds. Hence, $|\phi_x - \phi_y| + |\pi_x - \pi_y| = 1$. So, if $|t_x - t_y| = 1$ and $|f_x - f_y| = 1$ satisfy, then $|\phi_x - \phi_y| + |\pi_x - \pi_y| = 1$, $M(x, y) \geq 1 - \frac{1+1+1}{3} = 0$. Therefore, $0 \leq M(x, y) \leq 1$ holds.

(2) P2 and P3 are obviously.

(3) If $x = y$, it is clear that $M(x, y) = 1$. On the contrary, if $M(x, y)$ 1, then $(2 - t_x - t_y)|t_x - t_y| + (2 - f_x - f_y)|f_x - f_y| + |\phi_x - \phi_y| + |\pi_x - \pi_y| = 0$, we have: $t_x = t_y$, $f_x = f_y$, namely $x = y$, therefore P4 holds.

(4) If $x = [0,0]$, $y = [1,1]$ or $x = [1,1]$, $y = [0,0]$, From above (1), we have $M(x, y) = 0$, On the contrary, if $M(x, y)$0,we can know $|t_x - t_y| = 1$ and $|f_x - f_y| = 1$, namely x=[0,0],y=[1,1]or x=[1,1],y=[0,0]. Therefore, P5 holds.

(5) If $x \leq y \leq z$, then $t_x \leq t_y \leq t_z$ and $f_z \geq f_y \geq f_x$,
Since $|\phi_x - \phi_y| = \frac{1}{2}(t_y - t_x + f_x - f_y)$, $|\pi_x - \pi_y| = |(t_y - t_x) - (f_x - f_y)|$,
Therefore, $|\phi_x - \phi_y| + |\pi_x - \pi_y| = \frac{3}{2}\max\{(t_y - t_x), (f_x - f_y)\}$.

Let $r_x = 1 - t_x$, $g_x = 1 - f_x$, from the premise condition $r_x \geq r_y \geq r_z$, $g_x \leq g_y \leq tg_z$, we have

$$M(x, y) = 1 - \frac{(r_x + r_y)(r_x - r_y) + (g_x + g_y)(g_y - g_x) + 3\max\{(r_x - r_y), (g_y - g_x)\}/2}{3}$$

$M(x, y) \geq M(x, z)$ equivalent to

$$\frac{(r_x^2 - r_y^2) + (g_y^2 - g_x^2) + 3\max\{(r_x - r_y), (g_y - g_x)\}/2}{3} \leq \frac{(r_x^2 - r_z^2) + (g_z^2 - g_x^2) + 3\max\{(r_x - r_z), (g_z - g_x)\}/2}{3}$$

Let $\mathrm{m} = \mathrm{r_x}$, $\mathrm{n} = \mathrm{r_y}$, $\mathrm{z} = \mathrm{g_y}$, $\mathrm{w} = g_x$, then $1 \geq \mathrm{m} \geq \mathrm{n} \geq 0$, $1 \geq \mathrm{z} \geq \mathrm{w} \geq 0$

Supposing $k(m, n, z, w) = \begin{cases} \frac{m^2 - n^2 + z^2 - w^2 + 3(m-n)/2}{3} & m - n \geq z - w \\ \frac{m^2 - n^2 + z^2 - w^2 + 3(z-w)/2}{3} & m - n < z - w \end{cases}$

If $m - n \geq z - w$, $\frac{\partial k}{\partial m} = \frac{1}{3}(2m + \frac{3}{2}) \geq 0$, $\frac{\partial k}{\partial n} = \frac{1}{3}(-2n - \frac{3}{2}) \leq 0$

If $m - n < z - w$, $\frac{\partial k}{\partial m} = \frac{2m}{3} \geq 0$, $\frac{\partial k}{\partial n} = \frac{1}{3}(-2n) \leq 0$

From the monotonicity, $M(x, y) \geq M(x, z)$ holds. The same procedure may be easily adapted to obtain $M(y, z) \geq M(x, z)$.

5.2 New Method for Similarity Measure Between Vague Sets

Definition 11 *Let A and B be two vague sets in* $X = \{x_1, x_2, \ldots, x_n\}$. *If* $V_A(x) = [t_A(x), 1 - f_A(x)]$, $V_B(x) = [t_B(x), 1 - f_B(x)]$ *are vague value of point x in A and B respectively. Then the similarity degree of A and B can be evaluated by the function T.*

$$T(A, B) = \frac{1}{n}\sum_{i=1}^{n} M(V_A(x_i), V_B(x_i)) \tag{11}$$

Theorem 2 *In Definition 11,* $T(A,\ B)$ *has the following properties:*

(1) $0 \le T(A, B) \le 1$
(2) $T(A, B) = T(B, A)$
(3) $T(A, B) = T(\bar{A}, \bar{B})$
(4) $T(A, B) = 0 \Leftrightarrow \{V_A(x_i), V_B(x_i)\} = \{[0, 0], [1, 1]\}$
(5) $T(A, B) = 1 \Leftrightarrow V_A(x_i) = V_B(x_i), \pi_A(x_i) = \pi_A(x_i) = 0, i = 1, 2, \ldots, n$
(6) If $A \subseteq B \subseteq C$ *,then* $T(A, C) \le min\{T(A, B), T(B, C)\}$

From the Theorem 1, it is easily to proving Theorem 2 by the same method.

6 Numerical Examples

In Table 1, six groups of vague values (x, y) are given. We compare our measure method with others

From Table 1,we can see sometimes the similarity gained by formulae of $M_C(x, y)$ and $M_Z(x, y)$ are counterintuitive. For example, for the first group data pair ([0, 1], [0.5, 0.5]), $x \neq y$, apparently we know the similarity of them is absolutely not 1, but $M_C([0, 1], [0.5, 0.5]) = 1$. For the second data pair ([0, 0], [1, 1]), they are totally

Table 1 Comparisons of various similarity measures

	1	2	3	4	5	6
x	[0,1]	[0,0]	[0,1]	[0.4,0.8]	[0.4,0.8]	[0,0.4]
y	[0.5,0.5]	[1,1]	[0,0]	[0.5,0.7]	[0.5,0.8]	[0.6,1]
$M_C(x, y)$	1	0.5	0.5	1	0.95	0.7
$M_{HK}(x, y)$	0.5	0	0.5	0.9	0.95	0.4
$M_{LX}(x, y)$	0.75	0	0.5	0.95	0.95	0.4
$M_F(x, y)$	0	0	0	0.86	0.92	0.4
$M'_F(x, y)$	0.33	0	0.17	0.8	0.9	0.4
$M_Z(x, y)$	0.75	0.07	0.46	0.95	0.94	0.63
M(x,y)	0.33	0	0.17	0.85	0.91	0.24

opposite,$M_C([0, 0], [1, 1]) = 0.5$, $M_Z([0, 0], [1, 1]) = 0.07$. $M(x, y)$ can be accordant with our intuition.

For the third group data pair ([0, 1], [0, 0]), intuitively, the support degree of them are equal to zero, so, the similarity of them should not equal to zero. But M_F([0, 1], [0, 0])=0,this is counterintuitive..

Then compare the similarity of the fourth group data pair ([0.4, 0.8], [0.5, 0.7]) with the fifth group data pair ([0.4, 0.8], [0.5, 0.8]). Intuitively, the similarity of the fourth group and the fifth group should satisfy $M([0.4, 0.8], [0.5, 0.7]) < M([0.4,0.8], [0.5,0.8])$, but from the above result we can see only $M_{HK}(x, y)$, $M_F(x, y)$, $M'_F(x, y)$ and $M(x, y)$ can satisfy the limitation. However, $M_{LX}([0.4, 0.8], [0.5, 0.7]) = M_{LX}([0.4, 0.8], [0.5, 0.8])$, thus , $M_{LX}(x, y)$ can not distinguish them.

For the sixth group data pair ([0, 0.4], [0.6, 1]), intuitively, the similarity of them should smaller, but we can see only $M(x, y)$ are more accordant with our intuition, the result of other method are bigger.

Through above example, Each similarity measure expression has its own measuring focus as showed in Table 1 although they all evaluate the similarities in vague sets.we can see M_C, M_{HK}, M_{LX}, M_F, M'_F and M_Z are existing some defect of different extent. However, compare with other methods, numerical example show that the method of this paper is more effective.

7 Conclusions

In this chapter, existing similarity measures between vague sets are analyzed, compared and summarized. These methods of similarity measure can be used to solve the problem of how to determine the similarity between two vague values in a certain extent. We think all existing similarity measures are valuable. But each of them focuses on different aspects. Some faults of existed methods are pointed out. According to the theory of similarity measure between intervals, it is shown that four factors influencing vague sets and vague values should be taken into account while calculating the similarity degree. A new method for similarity measures between vague sets (values) is put forward, and is proved to satisfy some rules. The validity and advantage of this method are illustrated by an example.

Acknowledgments This work is supported by the National Scientific Fund of China (No.11161041), Fundamental Research Funds for the Central Universities (No. zyz2012081).

References

1. Zadeh, L.A.: Fuzzy sets. Inf. Control **8**, 338–356 (1965)
2. Gau, W.L., Buehrer, D.J.: Vague sets. IEEE Trans. Syst. Man Cybernetics (Part B) **23**(2), 610–614 (1993)

3. Chengyi, Z., Pingan, D.: On measures of similarity between vague sets. Compu. Eng. Appl. **39**(17), 92–94 (2003)
4. Chen, S.M.: Measures of similarity between vague sets. Fuzzy Sets Syst. **74**(2), 217–223 (1995)
5. Chen, S.M.: Similarity measures between vague sets and between elements. IEEE Trans. Syst. Man Cybern. (Part B) **27**(1), 153–158 (1997)
6. Hong, D.H., Kim, C.: A note on similarity measures between vague sets and elements. Inf. Sci. **115**, 83–96 (1999)
7. Li, F., Xu, Z.Y.: Similarity measures between vague sets. Chinese J. Softw. **12**(6), 922–927 (2001)
8. Jiulun, F.: Similarity measures on vague values and vague sets. Syst. Eng. Theory Pract. **26**(8), 95–100 (2006)
9. Wenbin, Z., Yu, J.: New method of measurement measure for measuring the degree of similarity between vague sets. **41**(24), 53–56 (2005)

Characterizations of α-Quasi Uniformity and Theory of α-P.Q. Metrics

Xiu-Yun Wu, Li-Li Xie and Shi-Zhong Bai

Abstract In Wu Fuzzy Systems and Mathematics 3:94–99, 2012, the author introduced concepts of α-remote neighborhood mapping and α-quasi uniform, and obtained many good results in α-quasi uniform spaces. This chapter will further investigate properties of α-remote neighborhood mapping, and give some characterizations of α-quasi uniforms. Based on this, this chapter also introduces concept of α-P.Q. metric, and establishes the relations between α-quasi uniforms and α-P.Q. metrics.

Keywords α-Quasi uniform · α-Homeomorphism · α-P.Q. metric · α-Remote neighborhood mapping

1 Introduction

Theory of quasi-uniformity in completely distributive lattices was firstly introduced by Erceg [1] and Hutton [2]. Then it was developed into various forms and was extended into different topological spaces [3–9]. In [10], the author introduced the concept of α-quasi uniform in α-layer order-preserving operator spaces, and revealed the relations between α-layer topological spaces and α-quasi uniform spaces. In this chapter, firstly, we further study properties of α-remote neighborhood mappings.

X. Wu (✉)
Institute of Computational Mathematics, Department of Mathematics,
Hunan Institute of Science and Engineering, Yongzhong 425100, China
e-mail: wuxiuyun2000@126.com

L. Xie
Department of English Teaching, Hunan Institute of Science and Engineering,
Yongzhong425100, China

S. Bai
Institute of Mathematics, Wuyi university, Jiangmen 529020, China

B.-Y. Cao and H. Nasseri (eds.), *Fuzzy Information & Engineering and Operations Research & Management*, Advances in Intelligent Systems and Computing 211,
DOI: 10.1007/978-3-642-38667-1_14,

Then we discuss some characterizations of α-quasi uniformities. Secondly, we introduce the concept of α-P.Q. metrics, and establish the relations between α-quasi uniforms and α-P.Q. metrics.

2 Preliminaries

In this chapter, X, Y will always denote nonempty crisp sets, A mapping $A : X \to L$ is called an L-fuzzy set. L^X is the set of all L-fuzzy sets on X. An element $e \in L$ is called an irreducible element in L, if $p \vee q = e$ implies $p = e$ or $q = e$, where $p, q \in L$. The set of all nonzero irreducible elements in L will be denoted by $M(L)$. If $x \in X, \alpha \in M(L)$, then x_α is called a molecule in L^X. The set of all molecules in L^X is denoted by $M^*(L^X)$. If $A \in L^X, \alpha \in M(L)$, take $A_{[\alpha]} = \{x \in X \mid A(x) \geq \alpha\}$ [3] and $A^\alpha = \vee\{x_\alpha \mid x_\alpha \not\leq A\}$ [11]. It is easy to check $(A_{[\alpha]})' = A^\alpha_{[\alpha]}$.

Let (L^X, δ) be an L-fuzzy topological space, $\alpha \in M(L)$. $\forall A \in L^X$, $D_\alpha(A) = \wedge\{G \in \delta' \mid G_{[\alpha]} \supset A_{[\alpha]}\}$. Then the operator D_α is a α-closure operator of some co-topology on L^X, denoted by $D_\alpha(\delta)$. We called α-layer topology. The pair $(L^X, D_\alpha(\delta))$ is called α-layer co-topological space [11]. An α-layer topological space$(L^X, D_\alpha(\delta))$ is called an α-C_{II} space, if there is a countable base $\mathscr{B}_\alpha$ of $D_\alpha(\delta)$.

A mapping $F_\alpha : L^X \to L^Y$ is called an α-mapping, if $F_\alpha(A)_{[\alpha]} = F_\alpha(B)_{[\alpha]}$ whenever $A_{[\alpha]} = B_{[\alpha]}$, and $F_\alpha(A) = 0_X$ whenever $A_{[\alpha]} = \emptyset$. The mapping F_α^{-1} : $L^Y \to L^X$ is called the reverse mapping of F_α, if for each $B \in L^Y$, $F_\alpha^{-1}(B) = \vee\{A \in L^X \mid F_\alpha(A)_{[\alpha]} \subset B_{[\alpha]}\}$. Clearly, F_α^{-1} is also an α-mapping.

An α-mapping $F_\alpha : L^X \to L^Y$ is called an α-order-preserving homomorphism (briefly α-*oph*), iff both F_α and F_α^{-1} are α-union preserving mappings.

An α-mapping $F_\alpha : L^X \to L^Y$ is called an α-Symmetric mapping, if for every $A, B \in L^X$, we have

$$\exists C_{[\alpha]} \not\subset A^\alpha_{[\alpha]}, B_{[\alpha]} \not\subset F_\alpha(C)_{[\alpha]} \Leftrightarrow \exists D_{[\alpha]} \not\subset B^\alpha_{[\alpha]}, A_{[\alpha]} \not\subset F_\alpha(D)_{[\alpha]}.$$

An α-mapping $f_\alpha : L^X \to L^X$ is called an α-remote neighborhood mapping, if for each $A \in L^X$ with $A_{[\alpha]} \neq \emptyset$, we have $A_{[\alpha]} \not\subset f_\alpha(A)_{[\alpha]}$. The set of all α-remote neighborhood mappings is denoted by $\mathscr{F}_\alpha(L^X)$, (briefly by $\mathscr{F}_\alpha$).

For $f_\alpha, g_\alpha \in \mathscr{F}_\alpha$, let's define

(1) $f_\alpha \leq g_\alpha \Leftrightarrow \forall A \in L^X, f_\alpha(A)_{[\alpha]} \subset g_\alpha(A)_{[\alpha]}$.
(2) $(f_\alpha \vee g_\alpha)(A) = f_\alpha(A) \vee g_\alpha(A)$.
(3) $(f_\alpha \odot g_\alpha)(A) = \wedge\{f_\alpha(B) \mid \exists B \in L^X, B_{[\alpha]} \not\subset g_\alpha(A)_{[\alpha]}\}$.

An non-empty subfamily $\mathscr{D}_\alpha \subset \mathscr{F}_\alpha$ is called an α-quasi-uniform, if $\mathscr{D}_\alpha$ satisfies:

(α-U1) $\forall f_\alpha \in \mathscr{D}_\alpha, g_\alpha \in \mathscr{F}_\alpha$ with $f_\alpha \leq g_\alpha$, then $g_\alpha \in \mathscr{D}_\alpha$.
(α-U2) $\forall f_\alpha, g_\alpha \in \mathscr{D}_\alpha$ implies $f_\alpha \vee g_\alpha \in \mathscr{D}_\alpha$.
(α-U3) $\forall f_\alpha \in \mathscr{D}_\alpha$, then $\exists g_\alpha \in \mathscr{D}_\alpha$, such that $g_\alpha \odot g_\alpha \geq f_\alpha$.
$(L^X, \mathscr{D}_\alpha)$ is called an α-quasi-uniform space. A subset $\mathscr{B}_\alpha \subset \mathscr{D}_\alpha$ is called a base of $\mathscr{D}_\alpha$, if $\forall f_\alpha \in \mathscr{D}_\alpha$, there is $g_\alpha \in \mathscr{B}_\alpha$, such that $f_\alpha \leq g_\alpha$. A subset $\mathscr{A}_\alpha \subset \mathscr{D}_\alpha$ is

called a subbase of $\mathscr{D}_\alpha$, if all of finite unions of the elements in $\mathscr{A}_\alpha$ consist a base of $\mathscr{D}_\alpha$. An α-quasi-uniform $\mathscr{D}_\alpha$ is called an α-uniform, if $\mathscr{D}_\alpha$ possesses a base whose elements are α-symmetric. Usually, we call this base α-symmetric base.

In [10], the author discussed the relation between an α-quasi uniform space and an α-layer co-topological space as following:

Let $\mathscr{D}_\alpha$ be an α-quasi uniform on L^X. $\forall A \in L^X$, Let's take $c_\alpha(A) = \vee\{B \in L^X \mid \forall f_\alpha \in \mathscr{D}_\alpha, A_{[\alpha]} \not\subset f_\alpha(B)_{[\alpha]}\}$. Then c_α is an α-closure operator of some L-fuzzy co-topology, which is denoted by $\eta_\alpha(\mathscr{D}_\alpha)$. Each α-layer topological space $(L^X, D_\alpha(\delta))$ can be α-quasi uniformitale, i.e., there is an α-quasi uniform $\mathscr{D}_\alpha$, such that $D_\alpha(\delta) = \eta_\alpha(\mathscr{D}_\alpha)$.

Other definitions and notes not mentioned here can be seen in [12].

3 Properties of α-Remote Neighborhood Mappings

Theorem 3.1 *Let $f_\alpha, g_\alpha \in \mathscr{F}_\alpha$. Then*

(1) $f_\alpha \vee g_\alpha \in \mathscr{F}_\alpha$, $f_\alpha \odot g_\alpha \in \mathscr{F}_\alpha$.
(2) $f_\alpha \odot g_\alpha \leq f_\alpha$, $f_\alpha \odot g_\alpha \leq g_\alpha$.
(3) $(f_\alpha \odot g_\alpha) \vee h_\alpha = (f_\alpha \vee h_\alpha) \odot (g_\alpha \vee h_\alpha)$,

$(f_\alpha \vee g_\alpha) \odot h_\alpha = (f_\alpha \odot h_\alpha) \vee (g_\alpha \odot h_\alpha)$.

Theorem 3.2 *Let $f_\alpha \in \mathscr{F}_\alpha$. If for each $A \in L^X$,*

$$f_\alpha^{\nabla}(A) = \wedge\{f_\alpha(B) \mid A_{[\alpha]} \subset \bigcup_{G_{[\alpha]} \not\subset B^\alpha_{[\alpha]}} f_\alpha(G)^\alpha_{[\alpha]}\},$$

and

$$f_\alpha^{\diamond}(A) = \wedge\{f_\alpha(B) \mid \bigcup_{G_{[\alpha]} \not\subset B^\alpha_{[\alpha]}} f_\alpha(G)^\alpha_{[\alpha]} \not\subset f_\alpha(A)_{[\alpha]}\}.$$

Then

(1) $f_\alpha^{\nabla}, f_\alpha^{\diamond} \in \mathscr{F}_\alpha$.
(2) $f_\alpha^{\diamond} \leq f_\alpha^{\nabla} \leq f_\alpha$.
(3) $f_\alpha^{\diamond}(A) = \wedge\{f_\alpha^{\nabla}(B) \mid B_{[\alpha]} \not\subset f_\alpha(A)_{[\alpha]}\} = (f_\alpha^{\nabla} \odot f_\alpha)(A)$.
(4) f_α^{∇} *is α-Symmetric.*

Proof (1) By $A_{[\alpha]} \subset \bigcup_{G_{[\alpha]} \not\subset A^\alpha_{[\alpha]}} f_\alpha(G)^\alpha_{[\alpha]}$, we have $A_{[\alpha]} \not\subset f_\alpha^{\nabla}(A)_{[\alpha]}$. Thus $f_\alpha^{\nabla} \in \mathscr{F}_\alpha$. Again by $A_{[\alpha]} \not\subset f_\alpha(A)_{[\alpha]}$, we get $\bigcup_{G_{[\alpha]} \not\subset A^\alpha_{[\alpha]}} f_\alpha(G)^\alpha_{[\alpha]} \not\subset f_\alpha(A)_{[\alpha]}$. Therefore, $f_\alpha^{\diamond} \in \mathscr{F}_\alpha$.

(2) The proof is obvious.

(3) For each $B \in L^X$,

$$\bigcup_{G_{[\alpha]}\not\subset B^{\alpha}_{[\alpha]}} f_\alpha(G)^{\alpha}_{[\alpha]} \not\subset f_\alpha(B)_{[\alpha]} \Leftrightarrow \exists D_{[\alpha]} \subset \bigcup_{G_{[\alpha]}\not\subset A^{\alpha}_{[\alpha]}} f_\alpha(G)^{\alpha}_{[\alpha]}, D_{[\alpha]} \not\subset f_\alpha(B)_{[\alpha]}$$

$$\Leftrightarrow \exists D_{[\alpha]} \subset \bigcup_{G_{[\alpha]}\not\subset B^{\alpha}_{[\alpha]}} f_\alpha(G)^{\alpha}_{[\alpha]}, D_{[\alpha]} \not\subset f_\alpha(B)_{[\alpha]}.$$

So

$$f^{\diamond}_\alpha(A) = \wedge\{f^{\nabla}_\alpha(D) \mid D_{[\alpha]} \not\subset f_\alpha(D)_{[\alpha]}\} = (f^{\nabla}_\alpha \odot f_\alpha)(A).$$

(4) $\forall D, E \in L^X$. If there is $A_{[\alpha]} \not\subset D^{\alpha}_{[\alpha]}$, such that

$$E_{[\alpha]} \not\subset f^{\nabla}_\alpha(A)_{[\alpha]} = \cap\{f_\alpha(B)_{[\alpha]} \mid A_{[\alpha]} \subset \bigcup_{G_{[\alpha]}\not\subset B^{\alpha}_{[\alpha]}} f_\alpha(G)^{\alpha}_{[\alpha]}\}.$$

Then there is $B \in L^X$, satisfying $E_{[\alpha]} \not\subset f_\alpha(B)_{[\alpha]}$ and $A_{[\alpha]} \subset \bigcup_{G_{[\alpha]}\not\subset (B^{\alpha}_{[\alpha]}} f_\alpha(G)^{\alpha}_{[\alpha]}$.

Clearly, $\bigcup_{G_{[\alpha]}\not\subset B^{\alpha}_{[\alpha]}} f_\alpha(G)^{\alpha}_{[\alpha]} \not\subset D^{\alpha}_{[\alpha]}$, i.e., $D_{[\alpha]} \not\subset \bigcap_{G_{[\alpha]}\not\subset B^{\alpha}_{[\alpha]}} f_\alpha(G)_{[\alpha]}$. Thus, there is $C_{[\alpha]} \not\subset B^{\alpha}_{[\alpha]}$, such that $D_{[\alpha]} \not\subset f_\alpha(C)_{[\alpha]}$. Conclusively, we have

$$D_{[\alpha]} \subset f_\alpha(B)^{\alpha}_{[\alpha]} \subset \bigcup_{G_{[\alpha]}\not\subset B^{\alpha}_{[\alpha]}} f_\alpha(G)^{\alpha}_{[\alpha]}.$$

and

$$D_{[\alpha]} \not\subset \cap\{f_\alpha(C)_{[\alpha]} \mid D_{[\alpha]} \subset \bigcup_{H_{[\alpha]}\not\subset C^{\alpha}_{[\alpha]}} f_\alpha(H)^{\alpha}_{[\alpha]}\} = f^{\nabla}_\alpha(D)_{[\alpha]}.$$

Therefore, f^{∇}_α is α-Symmetric.

Theorem 3.3 *Let f_α be an α-Symmetric remote neighborhood mapping, then*

(1) $C_{[\alpha]} \not\subset f_\alpha(A)_{[\alpha]} \Rightarrow A_{[\alpha]} \subset \bigcup_{D_{[\alpha]}\not\subset C^{\alpha}_{[\alpha]}} f_\alpha(D)^{\alpha}_{[\alpha]}$.

(2) $A_{[\alpha]} \subset \bigcup_{D_{[\alpha]}\not\subset C^{\alpha}_{[\alpha]}} f_\alpha(D)^{\alpha}_{[\alpha]} \Rightarrow C_{[\alpha]} \not\subset f_\alpha(A)_{[\alpha]}$.

Proof (1) Since f_α is an α-Symmetric mapping. $\forall D_{[\alpha]} \not\subset A^{\alpha}_{[\alpha]}$, there is $B_{[\alpha]} \not\subset C^{\alpha}_{[\alpha]}$, satisfying $D_{[\alpha]} \not\subset f_\alpha(B)_{[\alpha]}$. Therefore,

$$A^{\alpha}_{[\alpha]} \supset \bigcap_{B_{[\alpha]}\not\subset C^{\alpha}_{[\alpha]}} f_\alpha(B)_{[\alpha]}, \quad i.e., A_{[\alpha]} \subset \bigcup_{D_{[\alpha]}\not\subset C^{\alpha}_{[\alpha]}} f_\alpha(D)^{\alpha}_{[\alpha]}.$$

(2) By $A_{[\alpha]} \subset \bigcup_{D_{[\alpha]}\not\subset C^{\alpha}_{[\alpha]}} f_\alpha(D)^{\alpha}_{[\alpha]}$, we have $A^{\alpha}_{[\alpha]} \supset \bigcap_{D_{[\alpha]}\not\subset C^{\alpha}_{[\alpha]}} f_\alpha(D)_{[\alpha]}$. So $\forall x \notin A^{\alpha}_{[\alpha]}$, there is $D_{[\alpha]} \not\subset C^{\alpha}_{[\alpha]}$, such that $x \notin f_\alpha(D)_{[\alpha]}$. Since f_α is α-Symmetric.

There is $B^x_{[\alpha]} \not\subset x^{\alpha}_{[\alpha]}$, such that $C_{[\alpha]} \not\subset f_\alpha(B^x)_{[\alpha]}$. Take $E = \vee\{B^x \mid x \notin A^{\alpha}_{[\alpha]}\}$, then $x \not\subset E^{\alpha}_{[\alpha]}$. This implies $E^{\alpha}_{[\alpha]} \subset A^{\alpha}_{[\alpha]}$, i.e., $A_{[\alpha]} \subset E_{[\alpha]}$. Furthermore, we can conclude $C_{[\alpha]} \not\subset f_\alpha(A)_{[\alpha]}$. Otherwise, if $C_{[\alpha]} \subset f_\alpha(A)_{[\alpha]} \subset f_\alpha(E)_{[\alpha]}$, then it contradicts with the statement: for each $B^x \le E$, $C_{[\alpha]} \not\subset f_\alpha(B^x)_{[\alpha]}$.

Theorem 3.4 *Let $f_\alpha \in \mathscr{F}_\alpha$. Then*

$$(f_\alpha^{\nabla})_\alpha^{\nabla} \le f_\alpha^{\nabla} \odot f_\alpha^{\nabla} \le f_\alpha^{\diamond} \le f_\alpha \odot f_\alpha.$$

Proof $\forall A \in L^X$,

$$\begin{aligned}(f_\alpha^{\nabla})_\alpha^{\nabla}(A) &= \wedge\{f_\alpha^{\nabla}(B) \mid A_{[\alpha]} \subset \bigcup_{G_{[\alpha]} \not\subset B^{\alpha}_{[\alpha]}} f_\alpha^{\nabla}(G)^{\alpha}_{[\alpha]}\} \\ &\le \wedge\{f_\alpha^{\nabla}(C) \mid C_{[\alpha]} \not\subset f_\alpha^{\nabla}(A)_{[\alpha]}\} \\ &= (f_\alpha^{\nabla} \odot f_\alpha^{\nabla})(A).\end{aligned}$$

By Theorem 2 (2), $f_\alpha^{\nabla} \le f_\alpha$, we have

$$\begin{aligned}(f_\alpha^{\nabla} \odot f_\alpha^{\nabla})(A) &= \wedge\{f_\alpha^{\nabla}(C) \mid C_{[\alpha]} \not\subset f_\alpha^{\nabla}(A)_{[\alpha]}\} \\ &\le \wedge\{f_\alpha^{\nabla}(C) \mid C_{[\alpha]} \not\subset f_\alpha(A)_{[\alpha]}\} \\ &= f_\alpha^{\diamond}(A).\end{aligned}$$

Therefore

$$f_\alpha^{\diamond}(A) = \wedge\{f_\alpha^{\nabla}(C) \mid C_{[\alpha]} \not\subset f_\alpha(A)_{[\alpha]}\} \le \wedge\{f_\alpha(C) \mid C_{[\alpha]} \not\subset f_\alpha(A)_{[\alpha]}\} = (f_\alpha \odot f_\alpha)(A).$$

Theorem 3.5 *Let $f_\alpha \in \mathscr{F}_\alpha$. Then*

(1) $f_\alpha^{\nabla} \le f_\alpha \odot f_\alpha$.
(2) $f_\alpha \odot f_\alpha \odot f_\alpha = f_\alpha^{\diamond}$.

Proof (1) By Theorem 2, we have

$$\begin{aligned}f_\alpha^{\nabla}(A) &= \wedge\{f_\alpha(B) \mid A_{[\alpha]} \subset \bigcup_{G_{[\alpha]} \not\subset B^{\alpha}_{[\alpha]}} f_\alpha(G)^{\alpha}_{[\alpha]}\} \\ &\le \wedge\{f_\alpha(B) \mid A_{[\alpha]} \not\subset f_\alpha(A)_{[\alpha]}\} \\ &= f_\alpha \odot f_\alpha(A).\end{aligned}$$

(2) By Theorem 3 (2), we have

$$\bigcup_{D_{[\alpha]} \not\subset C^{\alpha}_{[\alpha]}} f_\alpha(D)^{\alpha}_{[\alpha]} \not\subset f_\alpha(A)_{[\alpha]} \Rightarrow \exists B_{[\alpha]} \subset \bigcup_{D_{[\alpha]} \not\subset C^{\alpha}_{[\alpha]}} f_\alpha(D)^{\alpha}_{[\alpha]}, B_{[\alpha]} \not\subset f_\alpha(A)_{[\alpha]}$$
$$\Rightarrow \exists B_{[\alpha]} \not\subset f_\alpha(A)_{[\alpha]}, C_{[\alpha]} \not\subset f_\alpha(B)_{[\alpha]}$$
$$\Rightarrow C_{[\alpha]} \not\subset f_\alpha \odot f_\alpha(A)_{[\alpha]}.$$

This shows $f_\alpha \odot f_\alpha \odot f_\alpha \leq f_\alpha^\diamond$. On the other hand, by Theorem 2 (2) and (1) above, it is easy to find $f_\alpha^\diamond \leq f_\alpha^\nabla \odot f_\alpha \leq f_\alpha \odot f_\alpha \odot f_\alpha$. Therefore, (2) holds.

4 Characterizations of α-Quasi Uniformities

Theorem 4.1 *An non-empty subfamily $\mathscr{D}_\alpha \subset \mathscr{F}_\alpha$ is an α-uniform, iff $\mathscr{D}_\alpha$ satisfies:*

(1) $f_\alpha \in \mathscr{D}_\alpha, g_\alpha \in \mathscr{F}_\alpha$ with $f_\alpha \leq g_\alpha$, then $g_\alpha \in \mathscr{D}_\alpha$.
(2) $f_\alpha, g_\alpha \in \mathscr{D}_\alpha$ implies $f_\alpha \vee g_\alpha \in \mathscr{D}_\alpha$.
(3) $f_\alpha \in \mathscr{D}_\alpha$, then $\exists g_\alpha \in \mathscr{D}_\alpha$, such that $g_\alpha^\diamond \geq f_\alpha$.

Proof Necessity. Since $\mathscr{D}_\alpha$ is an α-uniform, (1) and (2) hold. Furthermore, if $\mathscr{B}_\alpha \subset \mathscr{D}_\alpha$ is an α-symmetric base. So for every $f_\alpha \in \mathscr{D}_\alpha$, there is $g_\alpha \in \mathscr{B}_\alpha$, such that $g_\alpha \odot g_\alpha \odot g_\alpha \odot g_\alpha \geq f_\alpha$. By Theorem 5 (2), $g_\alpha^\diamond = g_\alpha \odot g_\alpha \odot g_\alpha \geq g_\alpha \odot g_\alpha \odot g_\alpha \odot g_\alpha \geq f_\alpha$.

Sufficiency. If $\mathscr{B}_\alpha = \{g_\alpha^\nabla \mid g_\alpha \in \mathscr{D}_\alpha\}$. For every $f_\alpha \in \mathscr{D}_\alpha$, there is $g_\alpha \in \mathscr{D}_\alpha$, such that $g_\alpha^\diamond \geq f_\alpha$. By Theorem 4, $g_\alpha \odot g_\alpha \geq g_\alpha^\diamond \geq f_\alpha$. Then $\mathscr{D}_\alpha$ is an α-quasi-uniform. By Theorem 2 (2)and (4), we know $g_\alpha^\nabla \geq g_\alpha^\diamond \geq f_\alpha$. This shows $\mathscr{B}_\alpha$ is an α-symmetric base of $\mathscr{D}_\alpha$. Therefore $\mathscr{D}_\alpha$ is an α-uniform.

Theorem 4.2 *An non-empty subfamily $\mathscr{D}_\alpha \subset \mathscr{F}_\alpha$ is an α-uniform, iff $\mathscr{D}_\alpha$ satisfies:*

(1) $f_\alpha \in \mathscr{D}_\alpha, g_\alpha \in \mathscr{F}_\alpha$ with $f_\alpha \leq g_\alpha$, then $g_\alpha \in \mathscr{D}_\alpha$.
(2) $f_\alpha, g_\alpha \in \mathscr{D}_\alpha$ implies $f_\alpha \vee g_\alpha \in \mathscr{D}_\alpha$.
(3) $f_\alpha \in \mathscr{D}_\alpha$, then $\exists g_\alpha \in \mathscr{D}_\alpha$, such that $g_\alpha^\nabla \geq f_\alpha$.

Proof Necessity. Since $\mathscr{D}_\alpha$ is an α-uniform, (1) and (2) hold. By Theorem 6, For every $f_\alpha \in \mathscr{D}_\alpha$, there is $g_\alpha \in \mathscr{D}_\alpha$, such that $g_\alpha^\diamond \geq f_\alpha$. So according to Theorem 2 (2), $g_\alpha^\nabla \geq g_\alpha^\diamond \geq f_\alpha$. Thus (3) holds.

Sufficiency. If $\mathscr{B}_\alpha = \{g_\alpha^\nabla \mid g_\alpha \in \mathscr{D}_\alpha\}$. By (3), For every $f_\alpha \in \mathscr{D}_\alpha$, there is $g_\alpha \in \mathscr{D}_\alpha$, such that $g_\alpha^\nabla \geq f_\alpha$. Again, there is $h_\alpha \in \mathscr{D}_\alpha$, such that $h_\alpha^\nabla \geq g_\alpha$. By Theorem 4, we get

$$h_\alpha \odot h_\alpha \geq h_\alpha^\nabla \odot h_\alpha^\nabla \geq (h_\alpha^\nabla)_\alpha^\nabla \geq g_\alpha^\nabla \geq f_\alpha.$$

So $\mathscr{B}_\alpha$ is an α-symmetric base of $\mathscr{D}_\alpha$. Therefore, $\mathscr{D}_\alpha$ is an α-uniform.

5 α-P.Q. Metric and its Properties

A binary mapping $d_\alpha : L^X \times L^X \to [0, +\infty)$ is called an α-mapping, if $\forall (A, B), (C, D) \in L^X \times L^X$ satisfying $A_{[\alpha]} = C_{[\alpha]}$ and $B_{[\alpha]} = D_{[\alpha]}$, then $d_\alpha(A, B) = d_\alpha(A, B)$.

Definition 5.1 An α-mapping $d_\alpha : L^X \times L^X \to [0, +\infty)$ is called an α-P.Q metric on L^X, if

(α-M1) $d_\alpha(A, A) = 0$.

(α-M2) $d_\alpha(A, C) \leq d_\alpha(A, B) + d_\alpha(B, C)$.

(α-M3) $d_\alpha(A, B) = \bigwedge_{C_{[\alpha]} \subset B_{[\alpha]}} d_\alpha(A, C)$.

Theorem 5.1 *Let d_α be an α-P.Q. metrics on L^X. $\forall r \in (0, +\infty)$, a mapping $P_r : L^X \to L^X$ is defined by $\forall A \in L^X$,*

$$P_\alpha^r(A) = \vee\{B \in L^X \mid d_\alpha(A, B) \geq r\}.$$

Then

(1) P_α^r is α-symmetric mapping.
(2) $\forall A, B \in L^X$, $B_{[\alpha]} \subset P_\alpha^r(A)_{[\alpha]} \Leftrightarrow d_\alpha(A, B) \geq r$.
(3) $\forall A \in L^X, r > 0$, $A_{[\alpha]} \not\subset P_\alpha^r(A)_{[\alpha]}$.
(4) $\forall r, s \in (0, \infty)$, $P_\alpha^r \odot P_\alpha^s \geq P_\alpha^{r+s}$.
(5) $\forall A \in L^X, r > 0$, $P_\alpha^r(A)_{[\alpha]} = \bigcap_{s<r} P_\alpha^s(A)_{[\alpha]}$.
(6) $\forall A \in L^X$, $\bigcap_{r>0} P_\alpha^s(A)_{[\alpha]} = \emptyset$.

Proof Since d_α is an α-mapping, (1), (5) and (6) are easy.

(2) Clearly, $d_\alpha(A, B) \geq r \Rightarrow B_{[\alpha]} \subset P_\alpha^r(A)_{[\alpha]}$. Conversely. $\forall x \in B_{[\alpha]}$, $\exists x \in D_x \in L^X$, such that $d_\alpha(A, D_x) \geq r$. So $d_\alpha(A, \{x_\alpha\}) \geq d_\alpha(A, D_x) \geq r$. Thereby $d_\alpha(A, B) = \bigwedge_{x \in B_{[\alpha]}} d_\alpha(A, \{x_\alpha\}) \geq r$.

(3) Suppose $A_{[\alpha]} \subset P_\alpha^r(A)_{[\alpha]}$. By (2), we get $d_\alpha(A, A) \geq r$. This is a contradiction with (α-M1).

(4) $\forall A, B \in L^X$, if $B_{[\alpha]} \not\subset P_\alpha^r \odot P_\alpha^s(A)_{[\alpha]}$, then there is $D \in L^X$, such that $D_{[\alpha]} \not\subset P_\alpha^s(A)_{[\alpha]}$ and $B_{[\alpha]} \not\subset P_\alpha^r(D)_{[\alpha]}$. By (2), we have $d_\alpha(A, D) < s$, $d_\alpha(D, B) < r$. Then by (α-M2), we gain $d_\alpha(A, B) < r + s$. Therefore $B_{[\alpha]} \not\subset P_\alpha^{r+s}(A)_{[\alpha]}$. This means $P_\alpha^r \odot P_\alpha^s \geq P_\alpha^{r+s}$.

Theorem 5.2 *If a family of α-mappings $\{P_\alpha^r \mid P_r : L^X \to L^X, r > 0\}$ satisfies the conditions (2)–(5) in Theorem 8. For each $A, B \in L^X$, let's denote*

$$d_\alpha(A, B) = \wedge\{r \mid B_{[\alpha]} \not\subset P_\alpha^r(A)_{[\alpha]}\}.$$

Then

(1) $d_\alpha(A, B) < r \Leftrightarrow B_{[\alpha]} \not\subset P_\alpha^r(A)_{[\alpha]}$.
(2) d_α *is* α*-P.Q. metric on* L^X.

Proof (1) By Theorem 8 (2),(5), we have

$$d_\alpha(A, B) < r \Leftrightarrow \exists s < r, B_{[\alpha]} \not\subset P_\alpha^s(A)_{[\alpha]} \Leftrightarrow B_{[\alpha]} \not\subset P_\alpha^r(A)_{[\alpha]}.$$

(2) $\forall A, B \in L^X, r, s > 0$, if $d_\alpha(A, B) > r + s$, then

$$B_{[\alpha]} \subset P_\alpha^{r+s}(A)_{[\alpha]} \subset (P_\alpha^r \odot P_\alpha^s(A))_{[\alpha]}.$$

Thus $\forall C \in L^X$, we have $C_{[\alpha]} \not\subset P_\alpha^s(A)_{[\alpha]}$ and $B_{[\alpha]} \not\subset P_\alpha^r(C)_{[\alpha]}$. this implies $d_\alpha(A, C) > s$, and $d_\alpha(A, B) > r$. Hence $d_\alpha(A, B) + d_\alpha(A, C) > r + s$. Consequently, we obtain $d_\alpha(A, B) + d_\alpha(A, C) \geq d_\alpha(A, B)$.

Theorem 5.3 *An* α*-mapping* $d_\alpha : L^X \times L^X \to [0, +\infty)$ *satisfies* $(\alpha - M1)$*,*$(\alpha - M2)$ *and* $(\alpha - M3^*)$*, then for each* $C_{[\alpha]} \subset B_{[\alpha]}$*,* $d_\alpha(C, B) = 0$.
$(\alpha - M3^*)\forall A \in L^X, r > 0, A_{[\alpha]} \not\subset P_\alpha^r(A)_{[\alpha]}$.

Proof By (α-M2), for each $B, C \in L^X$, satisfying $C_{[\alpha]} \subset B_{[\alpha]}$, we have $d_\alpha(A, B) < d_\alpha(A, C) + d_\alpha(C, B)$. Here we can conclude $d_\alpha(C, B) = 0$. Otherwise, if $d_\alpha(C, B) = s > 0$, then $B_{[\alpha]} \subset P_\alpha^s(C)_{[\alpha]}$. it contracts with $C_{[\alpha]} \not\subset P_\alpha^s(C)_{[\alpha]}$ according to (α-M3*). Therefore $d_\alpha(C, B) = 0$.

Theorem 5.4 *An* α*-mapping* $d_\alpha : L^X \times L^X \to [0, +\infty)$ *is an* α*-P.Q metric on* L^X*, iff* d_α *satisfies (*α*-M1),(*α*-M2) and (*α*-M3*).*

Proof We only need to prove (α-M3)$\Leftrightarrow$(α-M3*). By Theorem 8 (2), it is easy to check (α-M3)$\Rightarrow$(α-M3*). If the converse result is not true, then there are $r, s > 0$, such that

$$d_\alpha(A, B) < s < r \leq \bigcap_{C_{[\alpha]} \subset B_{[\alpha]}} d_\alpha(A, C).$$

so for each $C_{[\alpha]} \subset B_{[\alpha]}$,

$$r \leq d_\alpha(A, C) \leq d_\alpha(A, B) + d_\alpha(B, C) < s + d_\alpha(B, C).$$

Thus, $0 < r - s < d_\alpha(B, C)$, which implies $C_{[\alpha]} \subset P_\alpha^{r-s}(B)_{[\alpha]}$. As a result

$$B_{[\alpha]} = (\vee\{C \mid C_{[\alpha]} \subset B_{[\alpha]}\})_{[\alpha]} \subset P_\alpha^{r-s}(B)_{[\alpha]}.$$

However, it contradicts with (α-M3*). Therefore (α-M3) holds.

Theorem 5.5 d_α *is* α*-P.Q. metric on* L^X*. Then* $\{P_\alpha^r \mid r > 0\}$ *satisfying (3)–(5) in Theorem 8 is an* α*-base of some* α*-uniform, which is called the* α*-uniform induced by* d_α*.*

Given an α-quasi-uniform $\mathscr{D}_\alpha$, if we say $\mathscr{D}_\alpha$ is metricable, we mean there is an α-P.Q. metric d_α, such that $\mathscr{D}_\alpha$ is induced by d_α.

Theorem 5.6 *An α-quasi-uniform spaces $(L^X, \mathscr{D}_\alpha)$ is α-P.Q. metricable iff it has a countable α-base.*

Proof By Theorem 12, the necessity is obvious. Let's prove the sufficiency.

Let $\mathscr{D}_\alpha$ be an α-uniformity on L^X, which has a countable α-base $\mathscr{B}_\alpha = \{P_\alpha^n \mid n \in N\}$. Let's take $g_\alpha^1 = P_\alpha^1$, then there is $g_\alpha^2 \in \mathscr{B}_\alpha$, such that $g_\alpha^2 \odot g_\alpha^2 \odot g_\alpha^2 \geq g_\alpha^1 \vee P_\alpha^2$. In addition, there is $g_\alpha^3 \in \mathscr{B}_\alpha$, such that $g_\alpha^3 \odot g_\alpha^3 \odot g_\alpha^3 \geq g_\alpha^2 \vee P_\alpha^3$. The process can be repeated again and again, then $\{g_\alpha^n \in n \in N\}$ is also an α-base. Obviously, $g_\alpha^{n+1} \odot g_\alpha^{n+1} \odot g_\alpha^{n+1} \geq g_\alpha^n$. Let's take $\varphi_\alpha : L^X \to L^X$, defined by: $\forall A \in L^X$,

$$\varphi_\alpha^r(A) = \begin{cases} g_\alpha^n(A), & \frac{1}{2^n} < r \leq \frac{1}{2^{n-1}}, \\ 0_X, & r > 1. \end{cases}$$

Clearly, $\forall r > 0$, $A_{[\alpha]} \not\subset \varphi_\alpha^r(A)_{[\alpha]}$. And $\forall \frac{1}{2^n} < r \leq \frac{1}{2^{n-1}}$, we have

$$\varphi_\alpha^r \odot \varphi_\alpha^r \odot \varphi_\alpha^r = g_\alpha^n \odot g_\alpha^n \odot g_\alpha^n \geq g_\alpha^{n-1} = \varphi_\alpha^{2r}.$$

Let's define

$$f_\alpha^r(A) = \wedge \left\{ (\varphi_\alpha^{r_1} \odot \varphi_\alpha^{r_2} \odot \cdots \odot \varphi_\alpha^{r_k}(A)) \mid \sum_{i=1}^{k} r_i = r \right\}.$$

Then $\mathscr{B}_\alpha = \{f_\alpha^r \mid r > 0\} \subset \mathscr{D}_\alpha$.

Finally, let's prove $\mathscr{B}_\alpha$ is an α-base of $\mathscr{D}_\alpha$ satisfying (3)-(6) in Theorem 8.

Step 1. For $f_\alpha \in \mathscr{D}$, there is $n \in N$, such that $g_\alpha^n \geq f_\alpha$. So if $r \in (\frac{1}{2^{n+1}}, \frac{1}{2^n}]$, then $\varphi_\alpha^{2r} = g_\alpha^n$. Besides, it is easy to check, $\varphi_\alpha^{r_1} \odot \varphi_\alpha^{r_2} \odot \cdots \odot \varphi_\alpha^{r_k} \geq \varphi_\alpha^{2r}$ whenever $\sum_{i=1}^{k} r_i = r$. Thus $f_\alpha^r \geq \varphi_\alpha^{2r} = g_\alpha^n \geq f_\alpha$. This means $\mathscr{B}_\alpha$ be an α-base of $\mathscr{D}_\alpha$.

Step 2. Obviously, $f_\alpha \in \mathscr{B}$ satisfies (3) and (6) in Theorem 8. Furthermore, for $r, s > 0$, $A \in L^X$, if there is $B_{[\alpha]} \not\subset (f_\alpha^r \odot f_\alpha^s(A))_{[\alpha]}$. Then there is $C \in L^X$, such that $B_{[\alpha]} \not\subset f_\alpha^r(C)_{[\alpha]}$ and $C_{[\alpha]} \not\subset f_\alpha^s(A)_{[\alpha]}$. So there are $\sum_{i=1}^{k} r_i = r$ and $\sum_{i=1}^{m} s_i = s$, such that

$$B_{[\alpha]} \not\subset (\varphi_\alpha^{r_1} \odot \varphi_\alpha^{r_2} \odot \cdots \odot \varphi_\alpha^{r_k}(C))_{[\alpha]}$$

and

$$C_{[\alpha]} \not\subset (\varphi_\alpha^{s_1} \odot \varphi_\alpha^{s_2} \odot \cdots \odot \varphi_\alpha^{s_m}(A))_{[\alpha]}.$$

Thus

$$B_{[\alpha]} \not\subset (\varphi_\alpha^{r_1} \odot \varphi_\alpha^{r_2} \odot \cdots \odot \varphi_\alpha^{r_k}) \odot (\varphi_\alpha^{s_1} \odot \varphi_\alpha^{s_2} \odot \cdots \odot \varphi_\alpha^{s_m})(A)_{[\alpha]}.$$

As a result, $B_{[\alpha]} \not\subset f_\alpha^{r+s}(A)_{[\alpha]}$. Consequently, $f_\alpha^{r+s} \leq f_\alpha^r \odot f_\alpha^s$. This is the proof of (4) in Theorem 8.

Step 3. for $r > s > 0$, $A \in L^X$,

$$\begin{aligned} f_\alpha^r(A) &= \wedge\left\{(\varphi_\alpha^{r_1} \odot \varphi_\alpha^{r_2} \odot \cdots \odot \varphi_\alpha^{r_k}(A) \mid \sum_{i=1}^{k} r_i = r\right\} \\ &\leq \wedge\left\{(\varphi_\alpha^{s_1} \odot \varphi_\alpha^{s_2} \odot \cdots \odot \varphi_\alpha^{s_m} \odot \varphi_\alpha^{r-s}(A)) \mid \sum_{i=1}^{m} s_i = m\right\} \\ &\leq \wedge\left\{(\varphi_\alpha^{s_1} \odot \varphi_\alpha^{s_2} \odot \cdots \odot \varphi_\alpha^{s_m}(A)) \mid \sum_{i=1}^{m} s_i = m\right\} = f_\alpha^s(A). \end{aligned}$$

Hence $f_\alpha^r \leq \bigwedge\limits_{s<r} f_\alpha^s$.

Conversely. Let's prove the reverse result.

If $B \in L^X$, $B_{[\alpha]} \not\subset f_\alpha^r(A)_{[\alpha]}$, then there is $\sum\limits_{i=1}^{k} r_i = r$, such that $B_{[\alpha]} \not\subset \varphi_\alpha^{r_1} \odot \varphi_\alpha^{r_2} \odot \cdots \odot \varphi_\alpha^{r_k}(A)_{[\alpha]}$. So there is $C \in L^X$, such that $C_{[\alpha]} \not\subset (\varphi_\alpha^{r_2} \odot \varphi_\alpha^{r_2} \odot \cdots \odot \varphi_\alpha^{r_k}(A)_{[\alpha]}$ and $B_{[\alpha]} \not\subset \varphi_\alpha^{r_1}(C)_{[\alpha]}$. By $\varphi_\alpha^{r_1} = \bigwedge\limits_{t<r_1} \varphi_\alpha^t$, there is $t < r_1$, such that $B_{[\alpha]} \not\subset \varphi_\alpha^t(C)_{[\alpha]}$, i.e., $B_{[\alpha]} \not\subset \varphi_\alpha^t \odot \varphi_\alpha^{r_2} \odot \cdots \odot \varphi_\alpha^{r_k}(A)_{[\alpha]}$. Let's take $s = t + \sum\limits_{i=2}^{k} r_i$, we have $s < r$ and $B_{[\alpha]} \not\subset f_\alpha^s(A)_{[\alpha]}$. Therefore $f_\alpha^r \geq \bigwedge\limits_{s<r} f_\alpha^s$. Therefore (5) in Theorem 8 holds.

Theorem 5.7 *Each α-C_{II} α-layer topological space is P.Q.-metriclizable.*

Proof Let $(L^X, D_\alpha(\delta))$ be an α-C_{II} space, $\{P_n \mid n \in N\}$ be an α-base. $\forall n \in N$, $f_\alpha^{P_n} : L^X \to L^X$ is defined as: $\forall A \in L^X$,

$$f_\alpha^{P_n}(A) = \begin{cases} 0_X, & A_{[\alpha]} \subset P_{n[\alpha]}. \\ P_n, & A_{[\alpha]} \not\subset P_{n[\alpha]}. \end{cases}$$

Let's take $\mathscr{D}^* = \{f_\alpha^{P_n} \mid n \in N\}$ and

$$\mathscr{B}_\alpha = \{f_\alpha \mid \exists f_\alpha^{P_{n_i}} \in \mathscr{D}^*, i = 1, 2, \cdots, m,\ f_\alpha = \bigvee_{i=1}^{m} f_\alpha^{P_{n_i}}\}.$$

Then $\mathscr{B}_\alpha$ is an α-base of some uniform denoted by $\mathscr{D}_\alpha$ and clearly, $\eta_\alpha = \eta_\alpha(\mathscr{D}_\alpha)$. Furthermore, since $\mathscr{B}_\alpha$ is countable, we know $(L^X, D_\alpha(\delta))$ is P.Q.-metriclizable.

Theorem 5.8 *An α-layer co-topology $D_\alpha(\delta)$ on L^X can be α-P.Q. metriclizable iff there is a Sequence of α-remote neighborhood mappings $\{f_\alpha^n\}_{n\in N}$ satisfying*

(1) $\forall n \in N, f_\alpha^n \leq f_\alpha^{n+1} \odot f_\alpha^{n+1} \odot f_\alpha^{n+1}$,
(2) $\forall a \in M^*(L^X)$, $\{f_\alpha^n(a)\}_{n\in N}$ *is the α-remote neighborhood family of a.*

Proof Necessary. If $D_\alpha(\delta)$ can be α-P.Q. metriclizable, there is an α-P.Q. metric, say d_α. Let's take $f_\alpha^n = P_\alpha^{\frac{1}{3^n}}$. By Theorem 8, it is clear that (1) and (2) hold.

Sufficiency. If $\{f_\alpha^n\}_{n\in N}$ satisfies (1) and (2). Clearly, $\{f_\alpha^n\}_{n\in N}$ is countable. So by Theorem 12, it is an α-base of some α-quasi uniform $\mathscr{D}_\alpha$. Therefore $D_\alpha(\delta) = \eta_\alpha(\mathscr{D}_\alpha)$. This means $D_\alpha(\delta)$ is α-P.Q. metriclizable.

Theorem 5.9 *An α-layer co-topology $D_\alpha(\delta)$ on L^X can be α-P.Q. metriclizable iff there is a Sequence of α-symmetric remote neighborhood mappings $\{f_\alpha^n\}_{n\in N}$ satisfying*

(1) $\forall n \in N, f_\alpha^n \leq f_\alpha^{n+1} \odot f_\alpha^{n+1} \odot f_\alpha^{n+1}$,
(2) $\forall a \in M^*(L^X)$, $\{f_\alpha^n(a)\}_{n\in N}$ *is the α-remote neighborhood family of a.*

Theorem 5.10 *An α-layer co-topology $D_\alpha(\delta)$ on L^X can be α-P.Q. metriclizable iff there is a Sequence of α-symmetric remote neighborhood mappings $\{f_\alpha^n\}_{n\in N}$ satisfying*

(1) $\forall n \in N, f_\alpha^n \leq (f_\alpha^{n+1})^\diamond \leq f_\alpha^{n+1}$,
(2) $\forall a \in M^*(L^X)$, $\{f_\alpha^n(a)\}_{n\in N}$ *is the α-remote neighborhood family of a.*

Proof Necessity. It is similar to that of Theorem 14.

Sufficiency. If $\{f_\alpha^n\}_{n\in N}$ satisfies (1) and (2). By Theorem 2, $\forall n \in N$, $f_\alpha^n \leq (f_\alpha^{n+1})^\diamond \leq (f_\alpha^{n+1})^\nabla \leq f_\alpha^{n+1}$. By Theorem 4, $\forall n \in N$, $f_\alpha^n \leq (f_\alpha^{n+1})^\diamond \leq f_\alpha^{n+1} \odot f_\alpha^{n+1}$.

Therefore $\forall n \in N$,

$$f_\alpha^n \leq f_\alpha^{n+1} \odot f_\alpha^{n+1} \leq (f_\alpha^{n+2})^\nabla \odot (f_\alpha^{n+2})^\nabla \leq (f_\alpha^{n+2})^\nabla.$$

Again, by Theorem 4,

$$\begin{aligned} f_\alpha^n &\leq (f_\alpha^{n+2})^\nabla \odot (f_\alpha^{n+2})^\nabla \leq ((f_\alpha^{n+4})^\nabla)^\nabla \odot ((f_\alpha^{n+4})^\nabla)^\nabla \\ &\leq (f_\alpha^{n+4})^\nabla \odot (f_\alpha^{n+4})^\nabla \odot (f_\alpha^{n+4})^\nabla \odot (f_\alpha^{n+4})^\nabla \\ &\leq (f_\alpha^{n+4})^\nabla \odot (f_\alpha^{n+4})^\nabla \odot (f_\alpha^{n+4})^\nabla. \end{aligned}$$

$\forall n \in N$, $g_\alpha^n = (f_\alpha^{4n-3})^\nabla$. Then $\{g_\alpha^n\}_{n\in N}$ is a family of α-symmetric remote neighborhood mappings. It is clear that $\{g_\alpha^n(a)\}_{n\in N}$ is the α-remote neighborhood family of a.

Theorem 5.11 *An α-layer co-topology $D_\alpha(\delta)$ on L^X can be α-P.Q. metriclizable iff there is a Sequence of α-remote neighborhood mappings $\{f_\alpha^n\}_{n\in N}$ satisfying*

(1) $\forall n \in N, f_\alpha^n \leq (f_\alpha^{n+1})^\nabla \leq f_\alpha^{n+1}$,

(2) $\forall a \in M^*(L^X)$, $\{f_\alpha^n(a)\}_{n \in N}$ *is the* α*-remote neighborhood family of* a.

Acknowledgments Thanks to the support by: 1. National Science Foundation (No.10971125). 2. Science Foundation of Guangdong Province (No. 01000004) 3. The construct program of the key discipline in Hunan University of Science and Engineering.

References

1. Erceg, M.A.: Metric in fuzzy set theory. J. Math. Anal. Appl. **69**, 205–230 (1979)
2. Hutton, B.: Uniformities on fuzzy topological spaces. J. Math. Anal. Appl. **58**, 557–571 (1997)
3. Peng, Y.W.: Pointwise p.q. metric and the family of induced maps on completely distributive lattices. Ann. Math. **3**, 353–359 (1992)
4. Shi, F.G.: Pointwise quasi-uniformities in fuzzy set theory. Fuzzy Sets Syst. **98**, 141–146 (1998)
5. Shi, F.G.: Pointwise Pseudo-metrics in L-fuzzy set theory. Fuzzy Sets Syst. **121**, 209–216 (2001)
6. Shi, F.G., Zhang, J., Zheng, C.Y.: L-proxinmities and totally bounded pointwise L-uniformities. Fuzzy Sets Syst. **133**, 321–331 (2003)
7. Zhang, J.: Lattice valued smooth pointwise Quasi-uniformity on completely distributive lattice. Fuzzy Syst. Math. **17**(2), 30–34 (2003)
8. Yue, Y., Fang, J.: Extension of Shi's quaso-uniformities in a Kubiak-spstal sense. Fuzzy Sets Syst. **157**, 1956–1969 (2006)
9. Yue, Y., Shi, F.G.: On (L, M)-fuzzy quasi-uniform spaces. Fuzzy Sets Syst. **158**, 1472–1485 (2007)
10. Wu, X.Y.: L-fuzzy α-quasi uniformtiy. Fuzzy Syst. Math. **3**(3), 94–99 (2012)
11. Meng, G.W., Meng, H.: D-closed sets and their applications. Fuzzy Syst. Math. **17**(1), 24–27 (2003)
12. Wang, G.J.: Theory of L-Fuzzy Topological Spaces. The Press of Shanxi normal University, Xi'an (1988)

The Complex Fuzzy Measure

Sheng-quan Ma, Mei-qin Chen and Zhi-qing Zhao

Abstract In this paper, we define the concept of complex Fuzzy measure, which is different from the concept of complex Fuzzy measure in [2], and discuss its properties and theorems. On the basis of the concept of complex Fuzzy measurable function in [2], we study its convergence theorem. It builds the certain foundation for the research of complex Fuzzy integral.

Keywords Complex fuzzy measure · Complex fuzzy measurable function

1 Introduction

In 1990–1991, Buckley [1] proposed the concept of fuzzy complex numbers and fuzzy complex-valued function. In 1997, Qiu Jiqing [2–5] firstly proposed the concept of the complex fuzzy measure on the basis of classical measure theory method. Since 2000, According to this issue, Ma shengquan [6] has done some exploratory work, and made a series of achievement in this field. The theory of fuzzy complex valued measure is an important part of fuzzy complex analysis, which has a strong background of practical application [7]. For instance it can use in the fuzzy system identification, fuzzy control, multi-classifier system design and other fields. The development of theoretical research of fuzzy complex valued measure is slow,

S. Ma(✉)
College of Mathematics and Information Science, Shanxi Normal University,
Xian,Shanxi 710062, China
e-mail: mashengquan@163.com

M. Chen
College of information science and Technology,Hainan Normal University, Haikou,
Hainan,Haikou571158, China

Z. Zhao and S. Ma
Dept. of Math, Hainan Normal University, Haikou,Hainan,571158Haikou, China

B.-Y. Cao and H. Nasseri (eds.), *Fuzzy Information & Engineering and Operations Research & Management*, Advances in Intelligent Systems and Computing 211,
DOI: 10.1007/978-3-642-38667-1_15,

because it is much more complicated than Fuzzy real-valued measure. The complex Fuzzy measure which defined in this paper is different from that in paper [2], the concept of Fuzzy measure was redefined, which distinguished between the real and imaginary parts in order to facilitate research.

2 Complex Fuzzy Measure

$\hat{R}^+$ denote positive real set, $\hat{C}^+$ denote the set of complex number on $\hat{R}^+$ [8].

Definition 2.1 *Let X be a nonempty set, F be a $\sigma-$ algebra, comprising of the subset of X, the mapping μ:$F \to \hat{\mathrm{C}}^+$ is set function, satisfying:*

(1) $\mu(\emptyset) = 0$;
(2) (monotonicity) If $A, B \in F$ and $A \subseteq B$,then $Re(\mu(A)) \le Re(\mu(B))$ and $Im(\mu(A)) \le Im(\mu(B))$. Denote $\mu(A) \le \mu(B)$
(3) if $A_n \in F(n = 1, 2, \cdots)$, $A_1 \subseteq A_2 \subseteq \cdots \subseteq A_n \subseteq \cdots$ then

$$\mu(\bigcup_{n=1}^{\infty} A_n) = \lim_{n\to\infty} \mu(A_n)$$

(4) if $A_n \in F(n = 1, 2, \ldots)$, $A_1 \supseteq A_2 \supseteq \cdots \supseteq A_n \supseteq \cdots$, and $\exists n_0$ such that $Re(\mu(A_{n_0})) < \infty, Im(\mu(A_{n_0})) < \infty$ then $\mu(\bigcap_{n=1}^{\infty} A_n) = \lim_{n\to\infty} \mu(A_n)$. Then μ is called as complex Fuzzy measure on F, (X,F,μ) is called as complex Fuzzy measure space.

Definition 2.2 *Fuzzy complex measure μ is said to be zero-additive, if for arbitrary $E, F \in F$, $\mu(F) = 0$ and $E \cap F = \varphi$, then $\mu(E \cup F) = \mu(E)$.*

Theorem 2.1 (X, F, μ) *is complex Fuzzy measure space, The following propositions are equivalence.*

(1) μ is zero-additive ;
(2) since $\mu(F) = 0$, then for arbitrary $E, F \in F$, such that $\mu(E \cup F) = \mu(E)$;
(3) since $\mu(F) = 0$, then for arbitrary $E, F \in F$, such that $\mu(E \backslash F) = \mu(E)$:

Proof. (1)$\Rightarrow$(2):
$E \cup F = E \cup (F \backslash E)$, since μ is nonnegative monotony, then for $\mu(F) = 0$, such that
$\mu(F\backslash E) \le \mu(F) = 0$, therefore $\mu(F\backslash E) = 0$. Applying the zero-additive of μ, $\mu(F\backslash E) = 0$ and $E \cap (F\backslash E) = \varphi$, then

$$\mu(E \cup F) = \mu(E \cup (F\backslash E)) = \mu(E).$$

(2)$\Rightarrow$(3):

Due to $E = (E\backslash F) \cup (E \cap F)$, and $\mu(F) = 0$, then $\mu(E \cap F) = 0$.
We know $\mu(E) = \mu((E\backslash F) \cup (E \cap F)) = \mu(E\backslash F)$ from the proposition (2).(3)⇒(1):
Due to $E \cap F = \varphi$, then $E = (E \cup F)\backslash F$.
If $\mu(F) = 0$ and $E \cap F = \varphi$, we can know

$$\mu(E) = \mu((E \cup F)\backslash F) = \mu(E \cup F)$$

from the proposition (3).

Theorem 2.2 *Suppose μ is complex Fuzzy measure of zero-additive, $A \in F$, there is a descending sequence of $\{B_n\} \subset F$, $(B_1 \supseteq B_2 \supseteq \cdots)$, if $\mu(B_n) \to 0$, then*

(1) $\mu(A\backslash B_n) \to \mu(A)$;
(2) whereupon $Re(\mu(A)) < \infty$, and $Im(\mu(A)) < \infty$, and if exists $Re(\mu(A \cup B_{n_0})) < \infty$ and $Im(\mu(A \cup B_{n_0})) < \infty$, therefore $\mu(A \cup B_n) \to \mu(A)$.

Proof. (1) since $\{A\backslash B_n\}$ is ascending sequence and if μ is lower-continuous, we can know $\lim\limits_{n\to\infty} \mu(A\backslash B_n) = \mu(\bigcup\limits_{n=1}^{\infty} (A\backslash B_n)) = \mu(A\backslash \bigcap\limits_{n=1}^{\infty} B_n)$.
Applying μ is upper-continuous, $\lim\limits_{n\to\infty} \mu(B_n) = \mu(\bigcap\limits_{n=1}^{\infty} B_n) = 0$, since μ is zero-additive, from the Theorem 1, then $\lim\limits_{n\to\infty} \mu(A\backslash B_n) = \mu(A)$.
(2) $\{A \cup B_n\}$ is descending sequence, $Re(\mu(A)) < \infty$, and $Im(\mu(A)) < \infty$, and exist $Re(\mu(A \cup B_{n_0})) < \infty$ and $Im(\mu(A \cup B_{n_0})) < \infty$, due to μ is upper-continuous,

$$\lim_{n\to\infty} \mu(A \cup B_n) = \mu(\bigcap_{n=1}^{\infty} (A \cup B_n)) = \mu(A \cup (\bigcap_{n=1}^{\infty} B_n)).$$

Since μ is zero-additive, and $\mu(\bigcap\limits_{n=1}^{\infty} B_n) = 0$, so we know

$$\lim_{n\to\infty} \mu(A \cup B_n) = \mu(A)$$

according to Theorem 1.

Definition 2.3 *Let (X,F) be measurable space, the mapping $\mu{:}F \to \hat{C}^+$ is set function, Complex fuzzy measure μ is upper-self-continuous, if for arbitrary A, $B_n \in F$, and $A \cap B_n = \Phi$, $\lim\limits_{n\to\infty} B_n = 0$, then*

$$\lim_{n\to\infty} \mu(A\backslash B_n) = \mu(A).$$

Complex fuzzy measure μ is lower-self-continuous, if for arbitrary A, $B_n \in F$ and $B_n \subseteq A$, $\lim\limits_{n\to\infty} B_n = 0$, then $\lim\limits_{n\to\infty} \mu(A\backslash B_n) = \mu(A)$.

Complex fuzzy measure μ is self-continuous, if and only if μ is not only upper-self-continuous but also lower-self-continuous.

Theorem 2.3 *Suppose μ is complex Fuzzy measure on (X,F), then*

(1) μ is upper-self-continuous if and only if $\lim_{n\to\infty} \mu(A \cup B_n) = \mu(A)$ if for arbitrary $A, B_n \in F$ and $\lim_{n\to\infty} B_n = 0$.

(2) μ is lower-self-continuous if and only if $\lim_{n\to\infty} \mu(A \backslash B_n) = \mu(A)$ if for arbitrary $A, B_n \in F$ and $\lim_{n\to\infty} B_n = 0$.

Proof. The necessity is easily proved from the definition.2.3

(1) Sufficiency: Let $E_n = B_n \backslash A$, we know $\mu(E_n) \leq \mu(B_n)$, therefore $E_n \cap A = \varphi$, and $\lim_{n\to\infty} \mu(E_n) = 0$, then $\lim_{n\to\infty} \mu(A \cup E_n) = \mu(A)$, so μ is upper-self-continuous.

(2) Sufficiency: Let $E_n = B_n \cap A$, we know $\mu(E_n) \leq \mu(B_n)$, therefore $E_n \subseteq A$, and $\lim_{n\to\infty} \mu(E_n) = 0$, then $\lim_{n\to\infty} \mu(A \backslash E_n) = \mu(A)$, so μ is lower-self-continuous

Definition 2.4 *Let (X, F) be measurable space, the mapping $\mu : F \to \hat{C}^+$ is set function,*

(1) Suppose for arbitrary $\varepsilon_i > 0$, exists $\delta_i = \delta(\varepsilon_i) > 0 (i = 1, 2)$, where $\varepsilon = \varepsilon_1 + i\varepsilon_2, \delta = \delta_1 + i\delta_2$,
Complex fuzzy measure μ is uniform-upper-self-continuous, if for arbitrary $A, B \in F$ and $\mu(B) \leq \delta$, then $\mu(A \cup B) \leq \mu(A) + \varepsilon$.

(2) Suppose for arbitrary $\varepsilon_i > 0$, exists $\delta_i = \delta(\varepsilon_i) > 0 (i = 1, 2)$, where $\varepsilon = \varepsilon_1 + i\varepsilon_2$, $\delta = \delta_1 + i\delta_2$, Complex fuzzy measure μ is uniform-lower-self- continuous, if for arbitrary $A, B \in F$ and $\mu(B) \leq \delta$, then $\mu(A) - \varepsilon \leq \mu(A \backslash B)$.

(3) Complex fuzzy measure μ is uniform-self-continuous, if and only if μ is not only uniform-upper-self-continuous but also uniform-lower-self-continuous.

Theorem 2.4 *Suppose set function μ is uniform-upper-self-continuous (uniform-lower-self-continuous), then μ is upper-self-continuous (lower-self- continuous).*

Proof. It is obvious.

Theorem 2.5 *Suppose μ is complex Fuzzy measure on (X,F), The following propositions are equivalence.*

(1) μ is uniform-self-continuous;
(2) μ is uniform-upper-self-continuous;
(3) μ is uniform-lower-self-continuous.

Proof. (1) $\Rightarrow$ (2) It is obvious.
(2) $\Rightarrow$ (3): Since μ is uniform-upper-self-continuous, so if for arbitrary $\varepsilon_i > 0, \exists \delta_i = \delta(\varepsilon_i) > 0 (i = 1, 2)$, where $\varepsilon = \varepsilon_1 + i\varepsilon_2, \delta = \delta_1 + i\delta_2$, and for arbitrary $A', B' \in$ F, $\mu(B') \leq \delta$,then $\mu(A') - \varepsilon \leq \mu(A' \cup B') \leq \mu(A') + \varepsilon$. if for arbitrary $A, B \in$ F, $\mu(B) \leq \delta$, Let$A' = A \backslash B$, $B' = A \cap B$, $\mu(B') \leq \mu(B) \leq \delta$,

if $\mu(A \backslash B) - \varepsilon \leq \mu(A' \cup B') \leq \mu(A \backslash B) + \varepsilon$, then $\mu(A) - \varepsilon \leq \mu(A \backslash B) \leq \mu(A) + \varepsilon$. It means μ is uniform-lower-self-continuous.
(3) $\Rightarrow$ (1): Since μ is uniform-lower-self-continuous, if for arbitrary $\varepsilon_i > 0, \exists \delta_i = \delta(\varepsilon_i) > 0 (i = 1, 2)$, where $\varepsilon = \varepsilon_1 + i\varepsilon_2, \delta = \delta_1 + i\delta_2$, and for arbitrary $A', B' \in$ F, $\mu(B') \leq \delta$, then $\mu(A') - \varepsilon \leq \mu(A' \backslash B') \leq \mu(A') + \varepsilon$. if for arbitrary
$A, B \in$ F, $\mu(B) \leq \delta$, Let$A' = A \cup B, B' = A \cap B, \mu(B') \leq \mu(B) \leq \delta$, therefore

$$\mu(A' \backslash B\prime) \geq \mu(A') - \varepsilon \geq \mu(A) - \varepsilon.$$

Again let $A'' = (A \cup B) \backslash (A \cap B), B'' = B \backslash A$, then $A'' \backslash B'' = A \backslash B$, and $\mu(B'') \leq \mu(B) \leq \delta$, so

$$\mu(A'') - \varepsilon \leq \mu(A'' \backslash B'') = \mu(A \backslash B) \Rightarrow \mu(A \backslash B) + \varepsilon \leq \mu(A) + \varepsilon.$$

Again let $A = E, B = F \backslash E$, then $(A \cup B) \backslash (A \cap B) = E \cup F, \mu(B) \leq \mu(F) \leq \delta$, so $\mu(E) - \varepsilon \leq \mu(E \cup F) \leq \mu(E) + \varepsilon$. It means μ is uniform-upper-self-continuous. So μ is uniform-self-continuous.

Definition 2.5 *Let (X,F) be measurable space, the mapping μ:F $\rightarrow \hat{C}^+$ is set function, If for arbitrary $\{B_n\} \subseteq A, B_1 \supseteq B_2 \supseteq \cdots$, if $\exists n_0, \forall n > n_0, Re(\mu(B_n)) < \infty$, $Im(\mu(B_n)) < \infty$ and $\bigcap_{n=1}^{\infty} B_n = \varphi$, there must be $\lim_{n \rightarrow \infty} \mu(B_n) = 0$, so μ is called zero-upper-continuous.*

Theorem 2.6 *μ is nonnegative monotonic ascending set function, and zero-upper-continuous, Then if μ is upper-self- continuous, then μ is upper- continuous; if μ is limit and lower-self- continuous, then μ is lower- continuous.*

Proof. (1) If $\{A_n\} \subseteq$ F, $A_1 \supseteq A_2 \supseteq \cdots$, exist $Re(\mu(B_n)) < \infty$, $Im(\mu(B_n)) < \infty$, let
$A = \bigcap_{n=1}^{\infty} A_n, B_n = A_n \backslash A \quad (n = 1, 2, \cdots)$,then $B_1 \supseteq B_2 \supseteq \cdots$, and $\bigcap_{n=1}^{\infty} B_n = \varphi$.if for
arbitrary

$$n > n_0, Re(\mu(B_n)) \leq Re(\mu(B_{n_0})) \leq Re(\mu(A_{n_0})) < \infty,$$

$\text{Im}(\mu(B_n)) \leq \text{Im}(\mu(B_{n_0})) \leq \text{Im}(\mu(A_{n_0})) < \infty$. Since μ is zero-upper-continuous, we know $\lim_{n \rightarrow \infty} \mu(B_n) = 0$. Due to $A_n = A \cup B_n, A \cap B_n = \varphi$, and μ is upper-self-continuous, if $\mu(A_n) = \mu(A \cup B_n)$, then $\mu(A) = \mu(\bigcap_{n=1}^{\infty} A_n)$.So μ is upper-continuous.
(2)The proof is similar to (1) above.

3 Complex Fuzzy Measurable Function

R denote real set, C denote the set of complex number on R.

Definition 3.1 *[2] Suppose* (X,F,μ) *is complex Fuzzy measure space, the mapping* $\tilde{f}: X \to C$ *is called complex Fuzzy measurable function, if for arbitrary* $a+bi \in C$, *then* $\{x \in X \,\big|\, Re[\tilde{f}(x)] \geq a, Im[\tilde{f}(x)] \geq b\} \in F$.

Definition 3.2 *Suppose* (X,F,μ) *is complex Fuzzy measure space,* $\tilde{f}_n (n = 1, 2, \cdots)$, $\tilde{f}$ *is complex fuzzy measurable function, for arbitrary* $A \in F$,

(1) $\{\tilde{f}_n\}$ *almost everywhere converge to* $\tilde{f}$ *on A, denote* $\tilde{f}_n \overset{a.e.}{\rightarrow} \tilde{f}$, *if there exists* $B \in F$, *such that* $\mu(B) = 0$, *then* $\{\tilde{f}_n\}$ *with pointwise convergence to* $\tilde{f}$ *on* $A \backslash B$.
(2) $\{\tilde{f}_n\}$ *pseudo-almost everywhere converge to* $\tilde{f}$ *on A, denote* $\tilde{f}_n \overset{p.a.e.}{\rightarrow} \tilde{f}$, *if there exists* $B \in F$, *such that* $\mu(A \backslash B) = \mu(A)$, *then* $\{\tilde{f}_n\}$ *with pointwise convergence to* $\tilde{f}$ *on* $A \backslash B$.
(3) $\{\tilde{f}_n\}$ *almost everywhere uniformly converge to* $\tilde{f}$ *on A, denote* $\tilde{f}_n \overset{a.e.u.}{\rightarrow} \tilde{f}$, *if there exists* $B \in F$, *such that* $\mu(B) = 0$, *then* $\{\tilde{f}_n\}$ *with pointwise uniform convergence to* $\tilde{f}$ *on* $A \backslash B$.
(4) $\{\tilde{f}_n\}$ *pseudo-almost everywhere uniformly converge to* $\tilde{f}$ *on A, denote* $\tilde{f}_n \overset{p.a.e.u.}{\rightarrow} \tilde{f}$, *if there exists* $B \in F$, *such that* $\mu(A \backslash B) = \mu(A)$, *then* $\{\tilde{f}_n\}$ *with pointwise uniform convergence to* $\tilde{f}$ *on* $A \backslash B$.
(5) $\{\tilde{f}_n\}$ *almost uniformly converge to* $\tilde{f}$ *on A, denote* $\tilde{f}_n \overset{a.u.}{\rightarrow} \tilde{f}$, *if there exists set sequence* $\{E_k\}$ *on F, such that* $\tilde{\mu}(E_k) \to 0$ *,and for arbitrary k, then* $\{\tilde{f}_n\}$ *with pointwise uniform convergence to* $\tilde{f}$ *on* $A \backslash E_k$.
(6) $\{\tilde{f}_n\}$ *pseudo-almost uniformly converge to* $\tilde{f}$ *on A, denote* $\tilde{f}_n \overset{p.a.u.}{\rightarrow} \tilde{f}$, *if there exists set sequence* $\{E_k\}$ *on F, such that* $\mu(A \backslash E_k) \to \mu(A)$, *and for arbitrary k, then* $\{\tilde{f}_n\}$ *with pointwise uniform convergence to* $\tilde{f}$ *on* $A \backslash E_k$.
(7) $\{\tilde{f}_n\}$ *converge in complex fuzzy measure* μ *to* $\tilde{f}$ *on A, denote* $\tilde{f}_n \overset{u.}{\rightarrow} \tilde{f}$, *if for arbitrary* $\varepsilon = \varepsilon_1 + i\varepsilon_2, \varepsilon_1, \varepsilon_2 > 0$, *such that*

$$\lim_{n\to\infty} \mu(\{x \,\Big|\, Re\left|\tilde{f}_n - \tilde{f}\right| \geq \varepsilon_1, Im\left|\tilde{f}_n - \tilde{f}\right| \geq \varepsilon_2\} \cap A) = 0.$$

(8) $\{\tilde{f}_n\}$ *converge in pseudo complex fuzzy measure* $\tilde{\mu}$ *to* $\tilde{f}$ *on A, denote* $\tilde{f}_n \overset{p.u.}{\rightarrow} \tilde{f}$, *if for arbitrary* $\varepsilon > 0$, *such that*

$$\lim_{n\to\infty} \mu(\{x \,\Big|\, Re\left|\tilde{f}_n - \tilde{f}\right| < \varepsilon_1, Im\left|\tilde{f}_n - \tilde{f}\right| < \varepsilon_2\} \cap A) = \mu(A).$$

Theorem 3.1 *Suppose* (X,F,μ) *is complex Fuzzy measure space,* $\tilde{f}_n (n = 1, 2, \cdots)$, $\tilde{f}$ *is complex fuzzy measurable function, for arbitrary* $A \in F$,

(1) $\tilde{f}_n \overset{a.e.}{\rightarrow} \tilde{f}$ *if and only if* $\{\tilde{f}_n\}$ *converge to* $\tilde{f}$ *on* $A \backslash E_k$, *if there exists set sequence* $\{E_k\}$ *on F, such that* $\mu(E_k) \to 0$, *and for arbitrary k.*

(2) $\tilde{f}_n \overset{p.a.e.}{\rightarrow} \tilde{f}$ *if and only if* $\{\tilde{f}_n\}$ *converge to* $\tilde{f}$ *on* $A\backslash E_k$, *if there exists set sequence* $\{E_k\}$ *on F, such that* $\mu(A\backslash E_k) \rightarrow \mu(A)$, *and for arbitrary* k.

(3) $\tilde{f}_n \overset{a.e.u.}{\rightarrow} \tilde{f}$ *if and only if* $\left|Re(\tilde{f} - \tilde{f}_k)\right| < \varepsilon_1$ *and* $\left|Im(\tilde{f} - \tilde{f}_k)\right| < \varepsilon_2$, *if there exists set sequence* $\{E_k\}$ *on F, such that* $\mu(E_k) \rightarrow 0$, *and for arbitrary* $\varepsilon_i > 0, (i = 1, 2)$, *where* $\varepsilon = \varepsilon_1 + i\varepsilon_2$, $\exists n_0, \forall n > n_0, \forall k, \forall x \in A\backslash E_k$.

(4) $\tilde{f}_n \overset{p.a.e.u.}{\rightarrow} \tilde{f}$ *if and only if* $\left|Re(\tilde{f} - \tilde{f}_k)\right| < \varepsilon_1$ *and* $\left|Im(\tilde{f} - \tilde{f}_k)\right| < \varepsilon_2$, *if there exists set sequence* $\{E_k\}$ *on F, such that* $\mu(A\backslash E_k) \rightarrow \mu(A)$, *and for arbitrary* $\varepsilon_i > 0, (i = 1, 2)$, *where* $\varepsilon = \varepsilon_1 + i\varepsilon_2$, $\exists n_0, \forall n > n_0, \forall k, \forall x \in A\backslash E_k$.

Proof. (1) If $\tilde{f}_n \overset{a.e.}{\rightarrow} \tilde{f}$, then exists $B \in$F, $\mu(B) = 0$, such that $\tilde{f}_n \rightarrow \tilde{f}$ on $A\backslash B$. Let $E_k = B(k = 1, 2, \cdots)$, we know $\mu(E_k) \rightarrow 0$, and $\{\tilde{f}_n\}$ converge to $\tilde{f}$ on $A\backslash E_k$, for arbitrary k.
Otherwise, if there exists $\{E_k\} \subseteq$F,$\mu(E_k) \rightarrow 0$, such that $\tilde{f}_n \rightarrow \tilde{f}$ on $A\backslash E_k$. Let $B_k = \bigcap_{i=1}^{k} E_i, B = \bigcap_{k=1}^{\infty} B_k = \bigcap_{k=1}^{\infty} E_k$, so $\mu(B_k) \leq \mu(E_k)$ and $B_1 \supseteq B_2 \supseteq \cdots$. Due to $\mu(E_k) \rightarrow 0$, there exists

$$\mathrm{Re}(\mu(B_{k_0})) \leq \mathrm{Re}(\mu(E_{k_0})) < \infty, \text{and } \mathrm{Im}(\mu(B_{k_0})) \leq \mathrm{Im}(\mu(E_{k_0})) < \infty.$$

Applying the upper-continuity of μ, we know that $\mu(B) = \lim_{k\rightarrow 0} \mu(B_k) = 0$. If for arbitrary $x \in A\backslash B = \bigcup_{k=1}^{\infty} (A\backslash E_k), \exists k_0, x \in A\backslash E_{k_0}$, then $\tilde{f}_n \rightarrow \tilde{f}$, therefore $\{\tilde{f}_n\}$ converge to $\tilde{f}$ on $A\backslash B$, denote $\tilde{f}_n \overset{a.e.}{\rightarrow} \tilde{f}$.

The proof of (2),(3),(4) is similar to (1), we omit here.
Inference 3.1 If $\tilde{f}_n \overset{a.e.u.}{\rightarrow} \tilde{f}$, then $\tilde{f}_n \overset{a.u.}{\rightarrow} \tilde{f}$;If $\tilde{f}_n \overset{p.a.e.u.}{\rightarrow} \tilde{f}$, then $\tilde{f}_n \overset{p.a.u.}{\rightarrow} \tilde{f}$.

Theorem 3.2 *Suppose complex Fuzzy measure* μ *is lower-self- continuity, for* $A \in$*F,*
If $\tilde{f}_n \overset{a.e.}{\rightarrow} \tilde{f}$, *then* $\tilde{f}_n \overset{p.a.e.}{\rightarrow} \tilde{f}$.
If $\tilde{f}_n \overset{a.e.u.}{\rightarrow} \tilde{f}$, *then* $\tilde{f}_n \overset{p.a.e.u.}{\rightarrow} \tilde{f}$.
If $\tilde{f}_n \overset{a.u.}{\rightarrow} \tilde{f}$ *then* $\tilde{f}_n \overset{p.a.u.}{\rightarrow} \tilde{f}$.

Proof. (1) If $\tilde{f}_n \overset{a.e.}{\rightarrow} \tilde{f}$ on A, then there exists $B \in$ F, $\mu(B) = 0$, such that $\{\tilde{f}_n\}$ converge to $\tilde{f}$ on $A\backslash B$, therefore $\tilde{f}_n \overset{p.a.e.}{\rightarrow} \tilde{f}$ on A.

The proof of (2),(3) is similar to (1), we omit here.

Theorem 3.3 *Suppose for arbitrary* $A \in$*F, and complex fuzzy measurable function* $\tilde{f}$ *and* $\tilde{f}_n(n = 1, 2, \cdots)$, *if* $\tilde{f}_n \overset{u.}{\rightarrow} \tilde{f}$, *then* $\tilde{f}_n \overset{p.u.}{\rightarrow} \tilde{f}$ *if and only if complex fuzzy measure* μ *is lower-self- continuity.*

Proof. (necessity) if $\tilde{f}_n \overset{u.}{\rightarrow} \tilde{f}$, then $\tilde{f}_n \overset{p.u.}{\rightarrow} \tilde{f}$, so $\lim_{n\rightarrow\infty} \mu(B_n) = 0$, for arbitrary $A \in$ F and $\{B_n\} \subseteq$F
Let

$$\tilde{f}_n(x) = \begin{cases} 1 & x \in B_n \\ 0 & x \notin B_n \end{cases}$$

So

$$\lim_{n\to\infty} \mu(\{x \left| \left|\mathrm{Re}(\tilde{f}_n - 0)\right| \geq \varepsilon_1, \left|\mathrm{Im}(\tilde{f}_n - 0)\right| \geq \varepsilon_2\} \cap A) = \lim_{n\to\infty} \mu(B_n) = 0,$$

where for arbitrary $\varepsilon_i > 0, (i = 1, 2), \varepsilon = \varepsilon_1 + i\varepsilon_2$. So on A, if $\tilde{f}_n \overset{u.}{\to} 0$, then $\tilde{f}_n \overset{p.u.}{\to} 0$.
Suppose for $\varepsilon_i < 1$, then $\mu(A \backslash B_n) = \mu(\{x \left| \left|\mathrm{Re}(\tilde{f}_n - 0)\right| < \varepsilon_1, \left|\mathrm{Im}(\tilde{f}_n - 0)\right| < \varepsilon_2\} \cap A) = \mu(A)$,
So μ is lower-self- continuity.
(Sufficiency): If $\tilde{f}_n \overset{u.}{\to} \tilde{f}$ on A, then $\lim_{n\to\infty} \mu(\{x \left| \left|Re(\tilde{f}_n - 0)\right| \geq \varepsilon_1, \left|Im(\tilde{f}_n - 0)\right| \geq \varepsilon_2\} \cap A) = 0$, where for arbitrary $\varepsilon_i > 0, (i = 1, 2), \varepsilon = \varepsilon_1 + i\varepsilon_2$. Let

$$B_n = \{x \left| \left|Re(\tilde{f}_n - \tilde{f})\right| \geq \varepsilon_1, \left|Im(\tilde{f}_n - \tilde{f})\right| \geq \varepsilon_2\} \cap A,$$

then $\{B_n\} \subseteq$ A and $\lim_{n\to\infty} \mu(B_n) = 0$. Due to μ is lower-self- continuity, so

$$\mu(A \cap \{x \left| \left|Re(\tilde{f}_n - \tilde{f})\right| < \varepsilon_1, \left|Im(\tilde{f}_n - \tilde{f})\right| < \varepsilon_2\}) = \mu(A \backslash B_n) \to \mu(A).$$

Therefore $\tilde{f}_n \overset{p.u.}{\to} \tilde{f}$ on A.

4 Conclusion

On the basis of the concept of complex Fuzzy measurable function in [2], we study its convergence theorem. It builds the certain foundation for the research of complex Fuzzy integral. Provide a strong guarantee for the complex fuzzy integral development, enrichment and development of complex fuzzy Discipline.

Acknowledgments This work is supported by International Science & Technology Cooperation Program of China (2012DFA11270), Hainan International Cooperation Key Project(GJXM201105) and Natural Science Foundation of Hainan Province (No.111007)

References

1. Buckley, J.J.: Fuzzy complex numbers. FSS **33**, 333–345 (1989)
2. Jiqing, Q., Fachao, L., Liangqing, S.: Complex fuzzy measure and complex fuzzy integral. J. Hebei Inst. Chem. Technol Light Ind. **18**(1), 1–5 (1997)
3. Wenxiu, Z.: An Introduction of Fuzzy Mathematics. Xi'an Jiaotong University Press (1991)
4. Congxin, W.: Fuzzy Analysis. pp. 24–28. National Defense Industry Press, Beijing (1991)

5. Guangquan, Z.: Fuzzy Limit theory of fuzzy Complex numbers.Fuzzy Sets and Systems **46**, 227–235 (1992)
6. Shengquan, M., Chun, C.: Fuzzy complex analysis. The Ethnic publishing house, Beijing (2001)
7. Jiqing, Q.,Ting, S.: Relation on Convergence of Measurable Functions Sequence on Fuzzy Complex Measure Space. Fuzzy Sets and Systems **21**, 92–96 (2007)
8. Shengquan, M.: The fundamental theory of fuzzy complex analysis. Science Press, Beijing (2010)

Part III
Algebras and Equation

Fuzzy α-Ideals of BCH-Algebras

Ju-ying Wu

Abstract The aim of this paper is tointroduce the notions of fuzzy α-ideals of BCH-algebras and to investigate their properties. The relations among various fuzzy ideals are discussed as well. The upper and lower rough ideals of BCH-algebras are defined. Then the properties of rough ideals are discussed. The left and the right rough fuzzy ideals are defined and researched. Finally, we characterize well BCH-algebras via fuzzy ideals.

Keywords BCH-algebra · Fuzzy ideal · Fuzzy α-ideals · Rough fuzzy ideals.

1 Introduction

BCK-algebras and BCI-algebras are two important classes of logical algebras introduced by Y. Imai and K. Iséki in 1966. Since then, a great deal of literature has been produced on the theory of BCK/BCI-algebras. The notion of BCH-algebras was introduced by Hu and Li [1] which is a generalization of BCK-algebras and BCI-algebras. The concept of a fuzzy set, which was introduced by Zadeh [2], provides a natural framework for generalizing many of the concepts of general mathematics. Since then, these fuzzy ideals and fuzzy subalgebras have been applied to other algebraic structures such as semigroups, groups, rings, ideals, modules,etc. the concept of fuzzy sets is also applied to BCK/BCI/BCH-algebras [3–11]. Wang [12] introduced the notions of fuzzy subalgebras and fuzzy ideals of BCH-algebras with respect to a t-norm T, and studied some of their properties.

In 1982, Pawlak introduced the concept of a rough set. This concept is fundamental for the examination of granularity in knowledge. It is a concept which has many

J. Wu
College of Science, Air Force Engineering University, Shaanxi,
710051 Xian, People's Republic of China
e-mail: davidbush168@sina.com

B.-Y. Cao and H. Nasseri (eds.), *Fuzzy Information & Engineering and Operations Research & Management*, Advances in Intelligent Systems and Computing 211,
DOI: 10.1007/978-3-642-38667-1_16, © Springer-Verlag Berlin Heidelberg 2014

applications in data analysis. The algebraic analysis is also an important branch in rough sets. Iwinski [13] and Pomykala [14] studied algebraic properties of rough sets. Biswas and Nanda [15] introduced the notions of rough subgroups. And then, Kuroki [16] introduced the notion of rough ideals in a semigroup. Dubois and Prade [17, 18] combined fuzzy sets and rough sets in a fruitful way by defining rough fuzzy sets and fuzzy rough sets.

In this paper, we apply the fuzzy set and rough set theory to BCH-algebras. Fuzzy α-ideals and rough fuzzy ideals of BCH-algebras are introduced and investigated.

2 Preliminaries

An algebra $(X, *, 0)$ of type (2, 0) is called a BCH-algebra [1], if it satisfies the following axioms:

(1) $x * x = 0$;
(2) $x * y = y * x = 0 \Longrightarrow x = y$;
(3) $(x * y) * z = (x * z) * y$.

for all $x, y, z \in X$. In a BCH-algebra X, we can define a partial ordering $\leq$ by putting $x \leq y$ if and only if $x * y = 0$.

A BCH-algebra $(X, *, 0)$ is called quasi-associative, if for all $x, y, z \in X$, $(x * y) * z \leq x * (y * z)$.

In this paper X always means a BCH-algebra unless otherwise specified. We recall that a nonempty subset I of a X is called an ideal, if for any $x, y \in X$:

(I_1) $0 \in I$,
(I_2) $x * y \in I$ and $y \in I$ imply $x \in I$.

A nonempty subset A of X is said to be a subalgebra if and only if for all $x, y \in A$ implies $x * y \in A$.

An ideal I is called quasi-associative [9], if for each $x \in I, 0 * x = 0 * (0 * x)$.

In this paper X always means a BCH-algebra unless otherwise specified.

A mapping $\mu : X \rightarrow [0, 1]$, where X is an arbitrary nonempty set, is called a fuzzy set in X. The complement of μ, denoted by $\bar{\mu}$, is the fuzzy set in X by $\bar{\mu}(x) = 1-\mu(x)$ for all $x \in X$.

Definition 2.1 Let X be a BCH-algebra. A fuzzy subset μ in X is said to be a fuzzy ideal, if it satisfies

(FI1) $\mu(0) \geq \mu(x)$,
(FI2) $\mu(x) \geq \min\{\mu(x * y), \mu(y)\}$, for any $x, y \in X$.

A fuzzy ideal μ of X is said to be a closed fuzzy ideal, if, for all $x \in X$, $\mu(0*x) \geq \mu(x)$.

A fuzzy set μ of X is called a fuzzy subalgebra of X, if and only if for any $x, y \in X$, $\mu(x * y) \geq min(\mu(x), \mu(y))$.

A fuzzy ideal μ in X is called quasi-associative, if for all $x, y, z \in X$, $\mu((x * y) * z) \geq \mu(x * (y * z))$.

Clearly, if X is quasi-associative, then every fuzzy ideal in X is quasi-associative.

Definition 2.2 Let μ be a fuzzy set in a set S. For $t \in [0, 1]$, the set $\mu_t = \{s \in S | \mu(s) \geq t\}$ is called a level subset of μ. The set $\tilde{\mu}_t = \{s \in S | \mu(s) > t\}$ is called a strong level subset of μ.

Theorem 2.1 *Let μ be a fuzzy ideal of X. Then μ is a closed fuzzy ideal if and only if μ is a fuzzy subalgebra of X.*

Proof If μ is a closed fuzzy ideal of X, then $\mu((x * y) * x) = \mu(0 * y) \geq \mu(y)$. Since μ be a fuzzy ideal of X, we have $\mu(x * y) \geq \min\{\mu((x * y) * x), \mu(x)\}$. It follows that $\mu(x * y) \geq \min\{\mu(x), \mu(y)\}$. Hence μ is a fuzzy subalgebra of X.

Conversely, if μ is a fuzzy subalgebra of X, then for any $x \in X$, $\mu(0 * x) \geq \min\{\mu(0), \mu(x)\}$. It follows that $\mu(0 * x) \geq \mu(x)$ as μ be a fuzzy ideal of X. Hence μ is a closed fuzzy ideal of X.

3 Fuzzy α-Ideals of BCH-Algebras

Definition 3.1 A nonempty subset I of X is called an α-ideal of X, if it satisfies

(I_1) $0 \in I$,

(I_3) $(x * z) * (0 * y) \in I$ and $z \in I$ imply $y * x \in I$.

Definition 3.2 A fuzzy subset μ in X is said to be a fuzzy α-ideal, if it satisfies

(FI1) $\mu(0) \geq \mu(x)$,

(FI3) $\mu(y * x) \geq \min\{\mu((x * z) * (0 * y)), \mu(z)\}$, for any $x, y, z \in X$.

Example 3.1 Let $X = \{0, 1, 2, 3\}$ be a BCH-algebra with Cayley table given by

*	0	1	2	3
0	0	1	2	3
1	1	0	3	2
2	2	3	0	1
3	3	2	1	0

Define $\mu : X \to [0, 1]$ by $\mu(0) = \mu(1) = t_0$, $\mu(2) = \mu(3) = t_1$, where $t_0 \geq t_1$ and $t_0, t_1 \in [0, 1]$. By routine calculation give that μ is a fuzzy α-ideal.

Theorem 3.1 *Any fuzzy α-ideal of X is a closed fuzzy ideal of X, but the converse is not true.*

Proof Suppose that μ is a fuzzy α-ideal of X. Setting $y = z = 0$ in (FI3) and combining (FI1), it follows that $\mu(0 * x) \geq \mu(x)$ for all $x \in X$.

Setting $x = z = 0$ in (FI3) and combining (FI1), it follows that $\mu(y) \geq \mu(0 * (0 * y))$ for all $y \in X$. We have

$$\mu(x) \geq \mu(0 * (0 * x)) \geq \mu(0 * x).$$

for all $x \in X$. Thus for any $x, z \in X$, from (FI3) we have

$$\mu(x) \geq \mu(0 * x) \geq \min\{\mu((x * z) * (0 * 0)), \mu(z)\} = \min\{\mu(x * z), \mu(z)\}.$$

Hence μ satisfies (FI2) and combining (FI1), μ is a fuzzy ideal of X. Since $\mu(0*x) \geq \mu(x)$ for all $x \in X$, μ is a closed fuzzy ideal of X.

To show the last part we see Example 3.2.

Example 3.2 Let $X = \{0, 1, 2, 3\}$ be a proper BCH-algebra with Cayley table given by

*	0	1	2	3
0	0	0	0	0
1	1	0	3	3
2	2	0	0	2
3	3	0	0	0

Define $\mu : X \to [0, 1]$ by $\mu(0) = \mu(1) = t_0$, $\mu(2) = \mu(3) = t_1$. Where $t_0 > t_1$ and $t_0, t_1 \in [0, 1]$. By routine calculation give that μ is a fuzzy closed ideal of X, but not a fuzzy α-ideal of X as follows:

$$\mu(1 * 2) = \mu(3) = t_1 < t_0 = \min\{\mu((2 * 1) * (0 * 1)), \mu(1)\}.$$

The proof is complete.

The following theorem give the characterization of fuzzy α-ideals.

Theorem 3.2 *Let μ be a fuzzy ideal of X. Then μ is a fuzzy α-ideal of X if and only if, for all $x, y, z \in X$, $\mu(y * (x * z)) \geq \mu((x * z) * (0 * y))$.*

Proof Assume that μ is a fuzzy α-ideal of X. By (FI3) we have

$$\mu(y * (x * z)) \geq \min\{\mu((x * z) * 0) * (0 * y)), \mu(0)\} = \mu((x * z) * (0 * y)).$$

Note that $(x * (0 * y)) * ((x * z) * (0 * y)) \leq z$. It follows that

$$\mu(x * (0 * y)) \geq \min\{\mu((x * z) * (0 * y)), \mu(z)\}.$$

If for all $x, y, z \in X$, $\mu(y*(x*z)) \geq \mu((x*z)*(0*y))$, then $\mu(y*x) \geq \mu(x*(0*y)) \geq \min\{\mu((x * z) * (0 * y)), \mu(z)\}$. Hence μ satisfies (FI3) and combining (FI1), μ is a fuzzy α-ideal of X. The proof is complete.

Theorem 3.3 *Let μ be a fuzzy ideal of X. Then μ is a fuzzy α-ideal of X if and only if, for all $x, y \in X$, $\mu(y * x) \geq \mu(x * (0 * y))$.*

Proof Assume that μ is a fuzzy α-ideal of X. By (FI3) we have

$$\mu(y * x)) \geq \min\{\mu(x * (0 * y)), \mu(0)\} = \mu(x * (0 * y)).$$

Note that $(x * (0 * y)) * ((x * z) * (0 * y)) \leq z$. It follows that

$$\mu(x * (0 * y)) \geq \min\{\mu((x * z) * (0 * y)), \mu(z)\}.$$

Thus, for all $x, y, z \in X$, $\mu(y * x) \geq \mu(x * (0 * y)) \geq \min\{\mu((x * z) * (0 * y)), \mu(z)\}$. Hence μ satisfies (FI3) and combining (FI1), μ is a fuzzy α-ideal of X. The proof is complete.

By Theorem 3.2 and Theorem 3.3, we have

Theorem 3.4 *Let μ be a fuzzy ideal of X. Then $\mu(y * (x * z)) \geq \mu((x * z) * (0 * y))$ if and only if $\mu(y * x) \geq \mu(x * (0 * y))$, for all $x, y, z \in X$.*

Next we establish the relations between fuzzy α-ideals and α-ideals of X.

Theorem 3.5 *A fuzzy set μ of X is a fuzzy α-ideal if and only if, for all $t \in [0, 1]$, the level ideal $\mu_t = \{x \in X | \mu(x) \geq t\}$ is either empty or an α-ideal of X.*

Proof Let μ be a fuzzy α-ideal of X and $\mu_t \neq \emptyset$, where $t \in [0, 1]$. If $(x * z) * (0 * y) \in \mu_t$ and $z \in \mu_t$, then $\mu((x * z) * (0 * y)) \geq t$ and $\mu(z) \geq t$. Since μ is a fuzzy α-ideal of X, we have $\mu(y * x) \geq \min\{\mu((x * z) * (0 * y)), \mu(z)\} \geq t$. Thus $y * x \in \mu_t$. It is clear $0 \in \mu_t$. Hence μ_t is an α-ideal of X.

Conversely, if μ_t is either empty or an α-ideal of X for all $t \in [0, 1]$. we show μ satisfies (FI1) and (FI3).

If (FI1) is false, then there exists $x' \in X$ such that $\mu(0) < \mu(x')$. Taking $t_0 = [\mu(x') + \mu(0)]/2$, then $\mu(0) < t_0 < \mu(x')$. Thus $x' \in \mu_{t_0}$ and so $\mu_{t_0} \neq \emptyset$. As μ_{t_0} is an α-ideal of X, we have $0 \in \mu_{t_0}$, then $\mu(0) > t_0$. This is a contradiction.

If (FI3) is false, then there exist $x', y', z' \in X$ such that $\mu(y' * z') < \min\{\mu((x' * z') * (0 * y')), \mu(z')\}$. Setting

$$t_1 = [\mu(y' * z') + \min\{\mu((x' * z') * (0 * y')), \mu(z')\}]/2.$$

Then $\mu(y' * z') < t_1 < \min\{\mu((x' * z') * (0 * y')), \mu(z')\}$. Thus $(x' * z') * (0 * y') \in \mu_{t_1}$ and $z' \in \mu_{t_1}$, but $y' * z' \notin \mu_{t_1}$. It means that μ_{t_1} is not an α-ideal of X. This is a contradiction. Hence μ be a fuzzy α-ideal of X. This completes the proof.

Corollary 3.1 *A fuzzy set μ of X is a fuzzy α-ideal if and only if, for all $t \in [0, 1]$, the level ideal $\tilde{\mu}_t = \{x \in X | \mu(x) > t\}$ is either empty or an α-ideal of X.*

Corollary 3.2 *If μ is a fuzzy α-ideal of X, $\forall x_0 \in X$, the level ideal $\mu_{x_0} = \{x \in X | \mu(x) \geq \mu(x_0)\}$ is α-ideal of X.*

Theorem 3.6 *If μ is a fuzzy α-ideal of X, then the set $\mu_0 = \{x \in X | \mu(x) = \mu(0)\}$ is α-ideal of X.*

Proof Let μ is a fuzzy α-ideal of X. If $(x * z) * (0 * y) \in \mu_0$ and $z \in \mu_0$, then $\mu((x * z) * (0 * y)) = \mu(z) = \mu(0)$. Since μ is a fuzzy α-ideal of X, by (FI3) we have $\mu(y * x) \geq \min\{\mu((x * z) * (0 * y)), \mu(z)\} = \mu(0)$. But $\mu(y * x) \leq \mu(0)$ by (FI1). Thus $\mu(y * x) = \mu(0)$ and so $y * x \in \mu_0$. It is clear $0 \in \mu_0$. Hence μ_0 is an α-ideal of X. This completes the proof.

Theorem 3.7 *The intersection of any sets of fuzzy α-ideal of X is also a fuzzy α-ideal.*

Proof Let $\{\nu_i\}$ be a family of fuzzy α-ideals of X. Since

$$\begin{aligned}(\cap\nu_i)(0) &= \inf(\nu_i(0)) \geq \inf(\nu_i(x)) = (\cap\nu_i)(x).\\ (\cap\nu_i)(y*x) &= \inf(\nu_i(y*x)) \geq \inf(\min\{\nu_i((x*z)*(0*y)), \nu_i(z)\}\\ &= \min\{\inf \nu_i((x*z)*(0*y)), \inf(\nu_i(z))\}\\ &= \min\{(\cap\nu_i)((x*z)*(0*y)), (\cap\nu_i)(z)\}.\end{aligned}$$

Hence $\cap\nu_i$ is a fuzzy α-ideals of X. This completes the proof.

Definition 3.3 Let μ and ν be fuzzy sets of S. The Cartesian product of μ and ν is defined by for all $x, y \in S$, $(\mu \times \nu)(x, y) = \min\{\mu(x), \nu(y)\}$.

Theorem 3.8 *Let μ and ν be two fuzzy α-ideals of X. Then $\mu \times \nu$ is also a fuzzy α-ideals of $X \times X$.*

Proof For any $x = (x_1, x_2), y = (y_1, y_2), z = (z_1, z_2) \in X \times X$, we have

$$\begin{aligned}(\mu \times \nu)(0) &= (\mu \times \nu)(0, 0) = \min\{\mu(0), \nu(0)\}\\ &\geq \min\{\mu(x_1), \nu(x_2)\} = (\mu \times \nu)(x_1, x_2) = (\mu \times \nu)(x).\end{aligned}$$

$$\begin{aligned}(\mu \times \nu)(y*x) &= (\mu \times \nu)(y_1 * x_1, y_2 * x_2) = \min\{\mu(y_1 * x_1), \nu(y_2 * x_2)\}\\ &\geq \min\{\min\{\mu((x_1 * z_1) * (0 * y_1)), \mu(z_1)\}, \min\{\nu((x_2 * z_2) * (0 * y_2)), \nu(z_2)\}\}\\ &= \min\{\min\{\mu((x_1 * z_1) * (0 * y_1)), \nu((x_2 * z_2) * (0 * y_2))\}, \min\{\mu(z_1), \nu(z_2)\}\}\\ &= \min\{(\mu \times \nu)((x*z)*(0*y)), (\mu \times \nu)(z)\}.\end{aligned}$$

Thus $\mu \times \nu$ is also a fuzzy α-ideals of $X \times X$. This completes the proof.

Theorem 3.9 *Assume that X and Y be two BCH-algebras. Let $f : X \to Y$ be an onto homomorphism and ν is a fuzzy α-ideal of Y. Then μ the primage of ν under f is also a fuzzy α-ideal of X.*

Proof For any $x \in X$, we have $\nu(f(x)) = \mu(x)$. As $f(x) \in Y$ and ν is a fuzzy α-ideal of Y, then $\nu(0') \geq \nu(f(x)) = \mu(x)$ for every $x \in X$, where $0'$ is the zero element of Y. But $\nu(0') = \nu(f(0)) = \mu(0)$, so $\mu(0) \geq \mu(x)$ for all $x \in X$. Hence μ satisfies (FI1).

For any $x, y, z \in X$, since ν is a fuzzy α-ideal of Y, we have

$$\begin{aligned}\mu(y*x) &= \nu(f(y*x)) = \nu(f(y) * f(x))\\ &\geq \min\{\nu((f(x) * f(z)) * (0 * f(y))), \nu(f(z))\}\\ &= \min\{\nu(f((x*z)*(0*y)), \nu(f(z))\}\\ &= \min\{\mu((x*z)*(0*y)), \mu(z)\}.\end{aligned}$$

Hence μ satisfies (FI3). It follows that μ is a fuzzy α-ideal of X and completing the proof.

We now give another characterization of fuzzy α-ideals of X.

Theorem 3.10 *A fuzzy set μ of X is a fuzzy α-ideal, then, for all $x, y \in X$, $\mu(x) \geq \mu(0 * (0 * x))$ and $\mu(x * y) \geq \mu(x * (0 * y))$.*

Proof Let μ is a fuzzy α-ideal of X. Setting $y = z = 0$ in (FI3) and combining (FI1), it follows that $\mu(0 * x) \geq \mu(x)$ for all $x \in X$. Setting $x = z = 0$ in (FI3) and combining (FI1), it follows that $\mu(y) \geq \mu(0 * (0 * y))$ for all $y \in X$.

Now we show that $\mu(x * y) \geq \mu(x * (0 * y))$. Since $(0 * (0 * (y * (0 * x)))) * (x * (0 * y)) = ((0 * (0 * y)) * (0 * x)) * (x * (0 * y)) \leq 0$, we have $\mu(0 * (0 * (y * (0 * x)))) \geq \min\{\mu(x * (0 * y)), \mu(0)\} = \mu(x * (0 * y))$. Hence $\mu(y * (0 * x)) \geq \mu(0 * (0 * (y * (0 * x)))) \geq \mu(x * (0 * y))$. It follows that $\mu(x * y) \geq \mu(y * (0 * x)) \geq \mu(x * (0 * y))$.

Theorem 3.11 *Let I be an α-ideal of X. Then there exists a fuzzy α-ideal μ of X such that $\mu_t = I$ for some $t \in [0, 1]$.*

Proof Define $\mu : X \to [0, 1]$ by

$$\mu(x) := \begin{cases} t, & \text{if} x \in I \\ 0, & \text{if} x \notin I \end{cases}$$

where t is a fixed number in $(0, 1)$. It is obvious $\mu(0) \geq \mu(x)$ for all $x \in X$. If $(x * z) * (0 * y) \notin I$ or $z \notin I$, then $\mu((x * z) * (0 * y)) = 0$ or $\mu(z) = 0$ and so $\mu(y * x) \geq 0 = \min\{\mu((x * z) * (0 * y)), \mu(z)\}$. If $(x * z) * (0 * y) \in I$ and $z \in I$, then $y * x \in I$ as I is an α-ideal and so $\mu(y * x) = t = \min\{\mu((x * z) * (0 * y)), \mu(z)\}$. Hence μ is an fuzzy α-ideal of X. Clearly, $\mu_t = I$. This completes the proof.

For a subset I of X, we call

$$\chi_I(x) := \begin{cases} 1, & \text{if} x \in I \\ 0, & \text{if} x \notin I \end{cases}$$

the characteristic function of I. Clearly χ_I is a fuzzy set of X.

Theorem 3.12 *Let I be a nonempty subset of X. χ_I is fuzzy ideal if and only if I is a ideal of X.*

Proof If χ_I is fuzzy ideal, then $\chi_I(0) \geq \chi_I(x)$ for all $x \in I$. Since I is nonempty, thus $\chi_I(0) = 1$ and so $0 \in I$. If $x * y \in I$ and $y \in I$, then $\chi_I(x * y) = 1$ and $\chi_I(y) = 1$. By χ_I is fuzzy ideal, we have $\chi_I(x) \geq \min\{\chi_I(x * y), \chi_I(y)\} = 1$. It follows that $x \in I$. Hence I is a ideal of X.

If I is a ideal of X, then $0 \in I$ and $\chi_I(0) = 1 \geq \chi_I(x)$ for all $x \in X$. If $y \in X$ but $y \notin I$, then for all $x \in X$, we have $\chi_I(x) \geq \min\{\chi_I(x * y), \chi_I(y)\} = 0$. since I is a ideal, If $x * y \in I$ and $y \in I$, then $x \in I$. Hence $\chi_I(x) \geq \min\{\chi_I(x * y), \chi_I(y)\} = 1$.

That is, for all $x, y \in X$, $\chi_I(x) \geq \min\{\chi_I(x * y), \chi_I(y)\}$. Hence χ_I is fuzzy ideal. The proof is complete.

Theorem 3.13 *Let I be a nonempty subset of X. χ_I is closed fuzzy ideal if and only if I is a closed ideal of X.*

Theorem 3.14 *Let I be a nonempty subset of X. χ_I is fuzzy α-ideal if and only if I is a α-ideal of X.*

Theorem 3.15 *Let I be a nonempty subset of X. χ_I is quasi-associative fuzzy ideal if and only if I is a quasi-associative ideal of X.*

Theorem 3.16 *Let I be a nonempty subset of X. χ_I is fuzzy subalgebra if and only if I is a subalgebra of X.*

4 Rough Fuzzy Ideals of BCH-Algebras

Let V be a set and E an equivalence relation on V and let $\mathcal{P}(V)$ denote the power set of V. For all $x \in V$, let $[x]_E$ denote the equivalence class of x with respect to E. Define the functions $E^+(S) : \mathcal{P}(V) \to \mathcal{P}(V)$ and $E^-(S) : \mathcal{P}(V) \to \mathcal{P}(V)$ as follows, for any $S \in V$,

$$E^-(S) = \{x \in V | [x]_E \subseteq S\}, \quad E^+(S) = \{x \in V | [x]_E \cap S \neq \emptyset\}.$$

The pair (V, E) is called an *approximation space*. Let S be a subset of V. Then S is said to be definable if $E^-(S) = E^+(S)$ and rough otherwise. $E^-(S)$ is called the *lower approximation* of S while $E^+(S)$ is called the *upper approximation*. Obviously, if x is in $E^+(S)$, then $[x]_E$ is contained in $E^+(S)$. Similarly, if x is in $E^-(S)$, then $[x]_E$ is contained in $E^-(S)$.

Let E be a congruence relation on a BCH-algebra X, that is, E is an equivalence relation on X such that $(x * y) \in E$ imply $(x * z, y * z) \in E$ and $(z * x, z * y) \in E$ for all $z \in X$. We denote by X/E the set of all equivalence classes of X with respect to E, that is, $X/E := \{[x]_E | x \in X\}$. Throughout this section X is a BCH-algebra, and E is a congruence relation on X.

A special role is played by relations determined by ideals, that is, relations Θ of the form

$$(x * y) \in \Theta \Leftrightarrow x * y \in U, y * x \in U,$$

where U is an ideal.

For any nonempty subsets A and B of X, we define

$$A * B = \{x * y | x \in A, y \in B\}.$$

Proposition 4.1 *Let A and B be nonempty subsets of X. Then the following hold [18]:*

(1) $E^-(A) \subseteq A \subseteq E^+(A)$,
(2) $E^+(A \cup B) = E^+(A) \cup E^+(B)$,
(3) $E^-(A \cap B) = E^-(A) \cap E^-(B)$,
(4) $A \subseteq B$ *imply* $E^-(A) \subseteq E^-(B)$, $E^+(A) \subseteq E^+(B)$,
(5) $E^+(E^+(A)) = E^+(A)$,
(6) $E^-(E^-(A)) = E^-(A)$,
(7) $E^+(A) * E^+(B) = E^+(A * B)$,
(8) $E^-(A) * E^-(B) \subseteq E^-(A * B)$ *whenever* $E^-(A) * E^-(B) \neq \emptyset$ *and* $E^-(A * B) \neq \emptyset$.

Definition 4.1 A nonempty subset S of X is called an *upper* (resp. *a lower*) *rough ideal* of X if the upper (resp. lower) approximation of S is a ideal. If S is both an upper and a lower rough ideal of X, it is said to be a rough ideal of X.

Definition 4.2 A nonempty subset S of X is called an *upper* (resp. *a lower*) *rough closed ideal* of X if the upper (resp. lower) approximation of S is a closed ideal. If S is both an upper and a lower rough closed ideal of X, it is said to be a rough closed ideal of X.

Lemma 4.1 *Let A be a subalgebra of X. Then A is a closed ideal of X if and only if* $y * x \in X \backslash A$ *whenever* $x \in A$ *and* $y \in X \backslash A$.

Theorem 4.1 *If A is a closed ideal of X, then the nonempty lower approximation of A is a closed ideal of X, that is A is a lower rough closed ideal of X.*

Proof Let A be a close ideal of X. Then A is a subalgebra of X, it follows from Theorem 3.7 that $E^-(A)$ is a subalgebra of X.

Let $x, y \in X$ be such that $x \in E^-(A)$ and $y \in X \backslash E^-(A)$. If $y * x \notin X \backslash E^-(A)$, then $y * x \in E^-(A)$, thus $[y]_E * [x]_E = [y * x]_E \subseteq A$. Let $a_y \in [y]_E$. Then for all $a_x \in [x]_E$, we have $a_y * a_x \subseteq A$. Since $a_x \in [x]_E \subseteq A$ and A is an ideal of X, it follows that $a_y \in A$. Therefore $y \in E^-(A)$. This is a contradiction, and thus $E^-(A)$ is a closed ideal of X. Hence A is a lower rough closed ideal of X.

Dubois and prade [17, 18] combined fuzzy sets and rough sets in a fruitful way by defining rough fuzzy sets and fuzzy rough sets. In this section, we define and study the rough fuzzy subalgebras and rough fuzzy ideals in BCH-algebras. First, we give some basis definitions.

Definition 4.3 Let A be a fuzzy subset of X. A is called a fuzzy left [right, two-side] ideal of X if for all $\lambda \in [0, 1]$, A_λ is a left [right, two-side] ideal of X.

Definition 4.4 Let μ be a congruence relation on X, if A is a fuzzy subset of X, then the upper approximation $\overline{\mu}(A)$ and lower approximation $\underline{\mu}(A)$ of A can define two fuzzy subsets of X, and their member functions are defined as follows:

$$\overline{\mu}(A)(x) = \sup\{A(y)|y \in [x]_\mu\}, x \in X, \quad \underline{\mu}(A)(x) = \inf\{A(y)|y \in [x]_\mu\}, x \in X.$$

Definition 4.5 Let μ be a congruence relation on X, if A is a fuzzy subset of X. A is called an upper [resp. lower] rough fuzzy subalgebra of X, if $\overline{\mu}(A)$ [resp. $\underline{\mu}(A)$] is a fuzzy subalgebra X.

Definition 4.6 Let μ be a congruence relation on X, if A is a fuzzy subset of X. A is called an upper [resp. lower] rough fuzzy left [resp. right, two-side] ideal of X, if $\overline{\mu}(A)$ [resp. $\underline{\mu}(A)$] is a fuzzy left [resp. right, two-side] ideal of X.

Theorem 4.2 *Let μ be a congruence relation on X, if A is a fuzzy subset of X, then for every $\lambda \in [0, 1]$, we have*

(1) $(\overline{\mu}(A))_\lambda = \overline{\mu}(A_\lambda)$;
(2) $(\underline{\mu}(A))_\lambda = \underline{\mu}(A_\lambda)$.

Proof (1) $x \in (\overline{\mu}(A))_\lambda \Leftrightarrow \overline{\mu}(A)(x) \geq \lambda \Leftrightarrow \sup\{A(y)|y \in [x]_\mu\} \geq \lambda$. Then $x \in (\overline{\mu}(A))_\lambda$ if and only if there exists $y \in [x]_\mu$ and $A(y) \geq \lambda$, that is, there exists $y \in [x]_\mu$ and $y \in A_\lambda$. This show that $x \in \overline{\mu}(A_\lambda)$. Therefore $(\overline{\mu}(A))_\lambda = \overline{\mu}(A_\lambda)$.

(2) $x \in (\underline{\mu}(A))_\lambda \Leftrightarrow \underline{\mu}(A) \geq \lambda \Leftrightarrow \inf\{A(y)|y \in [x]_\mu\} \geq \lambda \Leftrightarrow y \in [x]_\mu$, then $A(y) \geq \lambda \Leftrightarrow y \in [x]_\mu$, then $y \in A_\lambda \Leftrightarrow [x]_\mu \subseteq A_\lambda \Leftrightarrow x \in \underline{\mu}(A_\lambda)$. This completes the proof.

Lemma 4.2 *Let μ be a congruence relation on X, if A and B are nonempty subsets of X, then $\overline{\mu}(A) * \overline{\mu}(B) \subseteq \overline{\mu}(A * B)$[17].*

Lemma 4.3 *Let μ be a complete congruence relation on X, if A and B are nonempty subsets of X, then $\underline{\mu}(A) * \underline{\mu}(B) \subseteq \underline{\mu}(A * B)$ [17].*

Theorem 4.3 *Let μ be a congruence relation on X, if A is a fuzzy subalgebra of X, then A is an upper rough fuzzy subalgebra of X.*

Proof Let A be a fuzzy subalgebra of X, then for every $\lambda \in [0, 1]$, A_λ be a subalgebra of X. By Theorem 4.2 and Lemma 4.1, we have $(\overline{\mu}(A))_\lambda * (\overline{\mu}(A))_\lambda = \overline{\mu}(A_\lambda) * \overline{\mu}(A_\lambda) \subseteq \overline{\mu}(A_\lambda * A_\lambda) \subseteq \overline{\mu}(A_\lambda) = (\overline{\mu}(A))_\lambda$. That is, for every $\lambda \in [0, 1]$, $(\overline{\mu}(A))_\lambda$ is a subalgebra of X. Thus $\overline{\mu}(A)$ is a fuzzy subalgebra of X. Therefore A is an upper rough fuzzy subalgebra of X.

Theorem 4.4 *Let μ be a complete congruence relation on X, if A is a fuzzy subalgebra of X, then A is a lower rough fuzzy subalgebra of X.*

Proof Let A be a fuzzy subalgebra of X, then for every $\lambda \in [0, 1]$, A_λ be a subalgebra of X. By Theorem 4.2 and Lemma 4.2, we have $(\underline{\mu}(A))_\lambda * (\underline{\mu}(A))_\lambda = \underline{\mu}(A_\lambda) * \underline{\mu}(A_\lambda) \subseteq \underline{\mu}(A_\lambda * A_\lambda) \subseteq \underline{\mu}(A_\lambda) = (\underline{\mu}(A))_\lambda$. That is, for every $\lambda \in [0, 1]$, $(\underline{\mu}(A))_\lambda$ is a subalgebra of X. Thus $\underline{\mu}(A)$ is a fuzzy subalgebra of X. Therefore A is a lower rough fuzzy subalgebra of X.

From Theorem 4.2 and 4.3, we can know that the upper rough fuzzy subalgebra and lower rough fuzzy subalgebra are the generalization of the general fuzzy subalgebra.

Theorem 4.5 *Let μ be a congruence relation on X, if A is a fuzzy left [resp. right] ideal of X, then A is an upper rough fuzzy left [resp. right] ideal of X.*

Proof Let A be a fuzzy left ideal of X, then for every $\lambda \in [0, 1]$, A_λ be a left ideal of X, that is $X * A_\lambda \subseteq A_\lambda$. Since $\overline{\mu}(X) = X$, then $X * (\overline{\mu}(A))_\lambda = \overline{\mu}(X) * \overline{\mu}(A_\lambda) \subseteq \overline{\mu}(X * A_\lambda) \subseteq \overline{\mu}(A_\lambda) = (\overline{\mu}(A))_\lambda$. That is, for every $\lambda \in [0, 1]$, $(\overline{\mu}(A))_\lambda$ is a left ideal of X. Thus $\overline{\mu}(A)$ is a fuzzy left ideal of X. Therefore A is an upper rough fuzzy left ideal of X.

The other case can be seen in a similar way.

Theorem 4.6 *Let μ be a congruence relation on X, if A is a fuzzy two-side ideal of X, then A is an upper rough fuzzy two-side ideal of X.*

Theorem 4.7 *Let μ be a complete congruence relation on X, if A is a fuzzy left [resp. right] ideal of X, then A is a lower rough fuzzy left [resp. right] ideal of X.*

Proof Let A be a fuzzy left ideal of X, then for every $\lambda \in [0, 1]$, A_λ be a left ideal of X, that is $X * A_\lambda \subseteq A_\lambda$. Since $\underline{\mu}(X) = X$, then $X * (\underline{\mu}(A))_\lambda = \underline{\mu}(X) * \underline{\mu}(A_\lambda) \subseteq \underline{\mu}(X * A_\lambda) \subseteq \underline{\mu}(A_\lambda) = (\underline{\mu}(A))_\lambda$. Hence for every $\lambda \in [0, 1]$, $(\underline{\mu}(A))_\lambda$ is a left ideal of X. Thus $\underline{\mu}(A)$ is a fuzzy left ideal of X. Therefore A is a lower rough fuzzy left ideal of X.

The other case can be seen in a similar way.

Theorem 4.8 *Let μ be a complete congruence relation on X, if A is a fuzzy two-side ideal of X, then A is a lower rough fuzzy two-side ideal of X.*

From Theorem 4.7–4.8, we can know that the upper rough fuzzy left [resp. right, two-side] ideal and lower rough fuzzy [resp. right, two-side] ideal are the generalization of the general fuzzy ideal.

5 Fuzzy Ideal Characterizations of Well BCH-Algebras

A BCH-algebra is called well BCH-algebra if every ideal of X is subalgebra of X. Fuzzy ideal characterizations of associative BCH-algebras are given the following.

Theorem 5.1 *If every fuzzy ideal of X is a fuzzy subalgebra of X, then X is well BCH-algebra.*

Proof Let X is a BCH-algebra and I is a ideal of X. It follows from Theorem 3.12 that χ_I is a fuzzy ideal. Then χ_I is a fuzzy subalgebra. By Theorem 3.16, I is a subalgebra of X. Hence X is well BCH-algebra.

Theorem 5.2 *If every fuzzy ideal of X is a closed fuzzy ideal of X, then X is well BCH-algebra.*

Theorem 5.3 *If the zero fuzzy ideal $\chi_{\{0\}}$ is fuzzy α-ideal of X, then X is a quasi-associative BCH-algebra.*

Proof Assume that the zero fuzzy ideal $\chi_{\{0\}}$ is fuzzy α-ideal of X. Then $\{0\}$ is a α-ideal of X by Theorem 3.16 For any $x \in X$, as

$$((0 * (0 * x)) * 0) * (0 * (0 * x)) = (0 * (0 * x)) * (0 * (0 * x)) = 0 \in \{0\}$$

and $0 \in \{0\}$. Since $\{0\}$ is a α-ideal of X, we have $(0 * x) * (0 * (0 * x)) = 0$ and so

$$0 * ((0 * x) * (0 * (0 * x))) = (0 * (0 * x)) * (0 * x) = 0.$$

Hence $0 * x = 0 * (0 * x)$. Therefore X is a quasi-associative BCH-algebra.

References

1. Hu, Q., Li, X.: On BCH-algebras. Math. Sem. Notes **11**, 313–320 (1983)
2. Zadeh, L.A.: Fuzzy sets. Inf. Control **8**, 338–353 (1965)
3. Ougen, X.: Fuzzy BCK-algebras. Math. Japonica **36**(5), 512–517 (1991)
4. Meng, J., Guo, X.: On fuzzy ideals in BCK/BCI-algebras. Fuzzy Sets Syst. **149**, 509–525 (2005)
5. Liu, Y., Meng, J.: Fuzzy ideals in BCI-algebras, Fuzzy Sets Syst. **123**, 227–237 (2001)
6. Jun, Y.B.: Closed Fuzzy ideals in BCI-algebras. Math. Japonica **38**(1), 199–202 (1993)
7. Ahmad, B.: Fuzzy BCI-algebras. J. Fuzzy Math. **1**(2), 445–452 (1993)
8. Hu, B., He, J.J., Liu, M.: Fuzzy Ideals and Fuzzy H-ideals of BCH-algebras. Fuzzy Syst. Math. **18**, 9–15 (2004)
9. Wang, F.X.: Quasi-associative fuzzy ideals in BCH-algebras. Fuzzy Syst. Math. **21**, 33–37 (2007)
10. Wang, F.X., Chen L.: On (α,β) fuzzy subalgebras of BCH-algebras. In: 5th International Conference on Fuzzy Systems and Knowledge. Discovery, vol. 3, pp. 604–607 (2008)
11. Peng, J.Y.: Closed fuzzy ideals of BCH-algebras. Fuzzy Syst. Math. **20**, 1–7 (2006)
12. Wang F. X.: T-fuzzy subalgebras and T-fuzzy ideals of BCH-algebras. In: 3rd International Conference on Innovative Computing, Information and Control, pp. 316–319, (2008)
13. Iwinski, T.: Algebraic Approach to Rough Sets. Bull. Polish Acad. Sci. Math. **35**, 673–683 (1987)
14. Pomykala, J., Pomykala, J.: A. The Stone Algebra of Rough Sets. Bull. Polish Acad. Sci. Math. **36**, 495–508 (1998)
15. Biswas, R., Nanda, S.: Rough Groups and Rough Subgroups. Bull. Polish Acad. Sci. Math. **42**, 251–254 (1994)
16. Kuroki, N.: Rough Ideals in Semigroups. Inf. Sci. **100**, 139–163(1997)
17. Dubois, D., Prade, H.: Rough Fuzzy Sets and Fuzzy Rough Sets. Int. J. General Syst. **17**, 191–209 (1990)
18. Dubois, D., Prade, H.: Two Fold Fuzzy Sets and Rough Sets-some issues in knowledge representation. Fuzzy Sets Syst. **23**, 3–18 (1987)

Lattice-Semigroups Tree Automation's Congruence and Homomorphism

Xiao-feng Huang and Zhi-wen Mo

Abstract The partial order of lattice elements in lattice-semigroup tree automata(LSTA) is defined in this paper. We proved the existence of semilattices and also lattices formed by different types of LSTA. Finally, we investigate the congruence and homomorphism of lattice-semigroup by LSTA formed from the algebra angle, Then we obtain homomorphism fundamental theorem of the LSTA.

Keywords Lattice-semigroup tree automata · Partial order · Congruence · Homomorphism

1 Introduction

As early as in the 1950s, automata, and in particular tree automata, played an important role in the development of verification. Since Zadeh [1] created the theory of fuzzy sets, it has been actively studied by both mathematicians and computer scientists. Many applications of fuzzy set theory have arisen, for instance, fuzzy logic, fuzzy cellular neural networks, fuzzy computer, fuzzy control system, etc. Fuzzy automata on words have also long history. Fuzzy tree automata have already been studied in Ref. [2]. At the same time, automata theory based on residuated lattices are established in Refs. [3, 4]. Recently, Esik and Liu studied fuzzy tree automata with membership in a distributive lattice and an equivalence between recognizability and equationality of fuzzy tree language was established in Ref. [5]. Lu and Zheng [6] studied three different types of Lattice-valued finite state quantum automata (LQA) and four different kinds of LQA operations and proved the existence of semilattices and also lattice formed by different types of LQA. In 1995, Guo and Mo discussed

X. Huang (✉) · Z. Mo
College of Mathematics and Software Science, Sichuan Normal University,
Chengdu 610066, China
e-mail: hxf2006@163.com

B.-Y. Cao and H. Nasseri (eds.), *Fuzzy Information & Engineering and Operations Research & Management*, Advances in Intelligent Systems and Computing 211,
DOI: 10.1007/978-3-642-38667-1_17, © Springer-Verlag Berlin Heidelberg 2014

homomorphisms in lattices of quantum automata in Ref. [7]. However, the homomorphisms in lattices-semigroup of LSTA wasn't given. This paper is a generalization of Ref. [6, 7] on LSTA. The notion of lattice-semigroup tree automata (LSTA) is introduced in Ref. [8], and we study the three different kinds of LSTA. we study the LSTA form a lattice-semigroup and give some algebraic properties on LSTA. On the basis of Theorem 2.1, we prove that LSTA defined on different lattice-semigroup l forms all kinds of lattice-semigroup L-S(l, Σ, Θ). Finally, we show the congruence and homomorphism of L-S(l, Σ, Θ), lattice-semigroup homomorphism fundamental theorem is obtained.

2 Preliminaries

Definition 2.1 [8] *Let $l = (L, \bigvee, *)$ be a arbitrarily lattice-semigroup, Σ be a finite ranked alphabet. Θ be a finite state space. A lattice-semigroup tree automata (LSTA) $\mathscr{A}$ defined on (l, Σ, Θ) is a quadruple $\mathscr{A} = (A, \Sigma, \nabla, \Delta)$ consisting of*

(1) A nonempty state set $A \subseteq \Theta$.

(2) A ranked alphabet Σ such that $\Sigma_0 \neq \emptyset$.

(3) For each $n \geq 0$, ∇ is a set of l-valued predicates defined on $A^n \times A \times \Sigma$: for arbitrary $q_1, q_2, \cdots q_n \in A$, and $\sigma \in \Sigma$, $\delta_n((q_1, q_2, \cdots q_n), q, \sigma) \in \nabla$ is an element of l. The family of fuzzy sets $\delta = (\delta_n)_{n \geq 0}$ is called the transition, we usually write δ for δ_n.

(4) Δ is a set of l-valued predicates defined on A, Δ is called final state transformation set. i.e. for arbitrary $q \in A$, $\beta(q) \in \Delta$ is an element of l, β is called final state transition.

Note that in Δ only those $\delta((q_1, q_2, \cdots q_n), q, \sigma) \neq 0$ (least element of l) listed in Δ.

Definition 2.2 [8] *Let $\mathscr{A}_1 = (A_1, \Sigma, \nabla_1, \Delta_1)$ and $\mathscr{A}_2 = (A_2, \Sigma, \nabla_2, \Delta_2)$ be two LSTA on (l, Σ, Θ). The intersection $\mathscr{A}_1 \bigwedge \mathscr{A}_2$ of $\mathscr{A}_1$ and $\mathscr{A}_2$ is also a LSTA defined on (l, Σ, Θ), called the intersection automata, with*

$$\mathscr{A}_1 \bigwedge \mathscr{A}_2 = A_3 = (A_3, \Sigma, \nabla_3, \Delta_3),$$

where

$$\begin{aligned}
A_3 &= A_1 \cap A_2 \\
\nabla_3 &= \{\delta_3((q_1, q_2, \cdots q_n), q, \sigma) | q_1, q_2, \cdots q_n, q \in A_3, \sigma \in \Sigma, \delta_3((q_1, q_2, \cdots q_n), q, \sigma) \\
&= \delta_{\mathscr{A}_1}((q_1, q_2, \cdots q_n), q, \sigma) \wedge \delta_{\mathscr{A}_2}((q_1, q_2, \cdots q_n), q, \sigma) \neq 0\} \\
\Delta_3 &= \{\beta_3(q) | q \in A_3, \beta_3(q) = \beta_{\mathscr{A}_1}(q) \wedge \beta_{\mathscr{A}_2}(q) \neq 0\}.
\end{aligned}$$

Definition 2.3 [8] *Let* $\mathscr{A}_1 = (A_1, \Sigma, \nabla_1, \Delta_1)$ *and* $\mathscr{A}_2 = (A_2, \Sigma, \nabla_2, \Delta_2)$ *be two LSTA on* (l, Σ, Θ). *The union* $\mathscr{A}_1 \bigvee \mathscr{A}_2$ *of* $\mathscr{A}_1$ *and* $\mathscr{A}_2$ *is also a LSTA defined on* (l, Σ, Θ), *called the union automata, with*
$\mathscr{A}_1 \bigvee \mathscr{A}_2 = A_3 = (A_3, \Sigma, \nabla_3, \Delta_3)$, *where*

$$
\begin{aligned}
A_3 &= A_1 \cup A_2 \\
\nabla_3 &= \{\delta_3((q_1, q_2, \cdots q_n), q, \sigma) | q_1, q_2, \cdots q_n, q \in A_3, \sigma \in \Sigma, \\
&\quad \delta_3((q_1, q_2, \cdots q_n), q, \sigma) \\
&= \delta_{\mathscr{A}_1}((q_1, q_2, \cdots q_n), q, \sigma) \vee \delta_{\mathscr{A}_2}((q_1, q_2, \cdots q_n), q, \sigma) \neq 0\} \\
\Delta_3 &= \{\beta_3(q) | q \in A_3, \beta_3(q) = \beta_{\mathscr{A}_1}(q) \vee \beta_{\mathscr{A}_2}(q) \neq 0\}.
\end{aligned}
$$

Definition 2.4 [8] *Let* $\mathscr{A}_1 = (A_1, \Sigma, \nabla_1, \Delta_1)$ *and* $\mathscr{A}_2 = (A_2, \Sigma, \nabla_2, \Delta_2)$ *be two LSTA on* (l, Σ, Θ). *The union* $\mathscr{A}_1 * \mathscr{A}_2$ *of* $\mathscr{A}_1$ *and* $\mathscr{A}_2$ *is also a LSTA defined on* (l, Σ, Θ), *called the union automata, with*
$\mathscr{A}_1 * \mathscr{A}_2 = A_3 = (A_3, \Sigma, \nabla_3, \Delta_3)$, *where*

$$
\begin{aligned}
A_3 &= A_1 \times A_2 \\
\Sigma_3 &= \Sigma \times \Sigma \\
\nabla_3 &= \{\delta_3(((q_1, q_1'), (q_2, q_2'), \cdots (q_n, q_n')), (q, q'), (\sigma, \sigma')) \\
&\quad | q_1, q_2, \cdots q_n, q \in A_1, q_1', q_2', \cdots q_n', q' \in A_2, \sigma, \sigma' \in \Sigma, \\
&\quad \delta_3(((q_1, q_1'), (q_2, q_2'), \cdots (q_n, q_n')), (q, q'), (\sigma, \sigma')) \\
&\quad = \delta_{\mathscr{A}_1}((q_1, q_2, \cdots q_n), q, \sigma) * \delta_{\mathscr{A}_2}((q_1', q_2', \cdots q_n'), q', \sigma') \neq 0\} \\
\Delta_3 &= \{\beta_3(q, q') | q \in A_1, q' \in A_2, \beta_3(q, q') = \beta_{\mathscr{A}_1}(q) * \beta_{\mathscr{A}_2}(q') \neq 0\}.
\end{aligned}
$$

Theorem 2.1 [8] *The lattice-semigroup finite-state tree automata defined on* (l, Σ, Θ) *form a lattice-semigroup. (we call it L-S*(l, Σ, Θ)*)*

As usually,we can use the two operations$\wedge$ *and* $\vee$ *to define the partial order of lattice elements in L-S*(l, Σ, Θ).

3 Main Conclusion

3.1 Different Types Lattice-Semigroup

Definition 3.5 *For arbitrary LSTA* $\mathscr{A}_1$ *and* $\mathscr{A}_2$, *we define*
$\mathscr{A}_1 \leq \mathscr{A}_2$ *if and only if* $\mathscr{A}_1 \wedge \mathscr{A}_2 = \mathscr{A}_1$

Corollary 3.1 *"*$\leq$*" is a partial order.*

Proof Let $\mathscr{A}_1$, $\mathscr{A}_2$ and $\mathscr{A}_3$ be three arbitrary LSTA defined on (l, Σ, Θ)

(1) *Reflexive* Since $\mathscr{A}_1 \wedge \mathscr{A}_2 = \mathscr{A}_1$, hence $\mathscr{A}_1 \leq \mathscr{A}_2$

(2) *Antisymmetric* If $\mathscr{A}_1 \leq \mathscr{A}_2$ and $\mathscr{A}_2 \leq \mathscr{A}_1$, then from Definition 2.1 we know that $\mathscr{A}_1 = \mathscr{A}_1 \wedge \mathscr{A}_2 = \mathscr{A}_2 \wedge \mathscr{A}_1 = \mathscr{A}_2$, i.e.$\mathscr{A}_1 = \mathscr{A}_2$
(3) *Transitive* If $\mathscr{A}_1 \leq \mathscr{A}_2, \mathscr{A}_2 \leq \mathscr{A}_3$, then $\mathscr{A}_1 \wedge \mathscr{A}_2 = \mathscr{A}_1, \mathscr{A}_2 \wedge \mathscr{A}_3 = \mathscr{A}_2$, hence $\mathscr{A}_1 \wedge \mathscr{A}_3 = (\mathscr{A}_1 \wedge \mathscr{A}_2) \wedge \mathscr{A}_3 = \mathscr{A}_1 \wedge (\mathscr{A}_2 \wedge \mathscr{A}_3) = \mathscr{A}_1 \wedge \mathscr{A}_2 = \mathscr{A}_1$, i.e.$\mathscr{A}_1 \wedge \mathscr{A}_3 = \mathscr{A}_1$,

then $\mathscr{A}_1 \leq \mathscr{A}_3$

Hence "$\leq$" is a partial order.

It is easy to prove that Definition 3.5 is equivalent to the following form:

Definition 3.6 *For arbitrary LSTA $\mathscr{A}_1$ and $\mathscr{A}_2$, we define $\mathscr{A}_1 \leq \mathscr{A}_2$ if and only if $\mathscr{A}_1 \vee \mathscr{A}_2 = \mathscr{A}_2$ Similarly, we can prove "$\leq$" is also a partial order.*

Proposition 3.1 *L-S(l, Σ, Θ) is a lattice ordered semigroup.*

Proof For arbitrary $\mathscr{A}, \mathscr{B}, \mathscr{C} \in$ L-S(l, Σ, Θ),if $\mathscr{A} \leq \mathscr{B}$, then $\mathscr{A} \wedge \mathscr{B} = \mathscr{A}$. $(\mathscr{C} * \mathscr{A}) \wedge (\mathscr{C} * \mathscr{B}) = \mathscr{C} * (\mathscr{A} \wedge \mathscr{B}) = \mathscr{C} \wedge \mathscr{A}$ Then $\mathscr{C} * \mathscr{A} \leq \mathscr{C} * \mathscr{B}$

Similarly, we can prove $\mathscr{A} * \mathscr{C} \leq \mathscr{B} * \mathscr{C}$.

Hence L-S(l, Σ, Θ) is a lattice ordered semigroup.

Proposition 3.2 *L-S(l, Σ, Θ) is a complete lattice-semigroup if and only if l is a complete lattice-semigroup.*

Proof Let $\mathscr{A}, \mathscr{A}_i, i = 1, 2, \cdots, n$ be arbitrary LSTA defined on (l, Σ, Θ), then

$$\mathscr{A} * \left(\bigvee_{i=1}^{\infty} \mathscr{A}_i\right) = (A * \left(\bigcup_{i=1}^{\infty} A_i\right), \Sigma \times \Sigma, \{\delta(((q_1, q_1'), \cdots (q_n, q_n')), (q, q'),$$

$$(\sigma, \sigma'))|q_1, \cdots q_n, q \in A, q_1', \cdots q_n', q' \in \bigcup_{i=1}^{\infty} A_i, \sigma, \sigma' \in \Sigma,$$

$$\delta(((q_1, q_1'), \cdots (q_n, q_n')), (q, q'), (\sigma, \sigma'))$$

$$= \delta_{\mathscr{A}}((q_1, \cdots q_n), q, \sigma) * \left(\bigvee_{i=1}^{\infty} \delta_{\mathscr{A}_i}((q_1', \cdots q_n'), q', \sigma')\right)\},$$

$$\{\beta(q, q')|q \in A, q' \in \left(\bigvee_{i=1}^{\infty} A_i\right), \beta(q, q')$$

$$= \beta_{\mathscr{A}}(q) * \left(\bigvee_{i=1}^{\infty} \beta_{\mathscr{A}_i}(q')\right)\}).$$

$$= \left(\bigcup_{i=1}^{\infty}(A * A_i), \Sigma \times \Sigma, \{\delta(((q_1, q_1'), \cdots (q_n, q_n')), (q, q'), (\sigma, \sigma'))\right)$$

$$|q_1, \cdots q_n, q \in A, q_1', \cdots q_n', q' \in \bigcup_{i=1}^{\infty} A_i, \sigma, \sigma' \in \Sigma,$$

$$\delta(((q_1, q_1'), \cdots (q_n, q_n')), (q, q'), (\sigma, \sigma'))$$

$$= \bigvee_{i=1}^{\infty} (\delta_{\mathscr{A}}((q_1, \cdots q_n), q, \sigma) * \delta_{\mathscr{A}_i}((q_1', \cdots q_n'), q', \sigma'))\},$$

$$\{\beta(q, q') | q \in A, q'' \in \left(\bigcup_{i=1}^{\infty} A_i\right), \beta(q, q')$$

$$= \bigvee_{i=1}^{\infty} (\beta_{\mathscr{A}}(q) * \beta_{\mathscr{A}_i}(q'))\}).$$

$$= \bigvee_{i=1}^{\infty} (\mathscr{A} * \mathscr{A}_i)$$

$$= \mathscr{A}_{\infty}(A_{\infty}, \Sigma, \nabla_{\infty}, \Delta_{\infty}) = \mathscr{A}_{\infty}$$

Note that:

(1) $A_{\infty} = A * (\bigcup_{i=1}^{\infty} A_i) = \bigcup_{i=1}^{\infty} (A * A_i)$
(2) Since l is a complete lattice-semigroup, $\delta(((q_1, q_1'), \cdots (q_n, q_n')), (q, q'), (\sigma, \sigma'))$ is also an element of l. Therefore, $\mathscr{A}_{\infty}$ is an element of L-S(l, Σ, Θ). L-S(l, Σ, Θ) is a complete lattice-semigroup.
(3) Similarly, we can prove $(\bigvee_{i=1}^{\infty} \mathscr{A}_i) * \mathscr{A} = \bigvee_{i=1}^{\infty} (\mathscr{A}_i * \mathscr{A})$.

On the other hand, if L-S(l, Σ, Θ) is a complete lattice-semigroup, then $\mathscr{A}_{\infty}$ is an element of L-S(l, Σ, Θ). Thus, $\delta(((q_1, q_1'), \cdots (q_n, q_n')), (q, q'), (\sigma, \sigma'))$ must be also an element of l. This shows l is a complete lattice-semigroup.

Proposition 3.3 *L-S(l, Σ, Θ) is a distributive lattice-semigroup if and only if l is a distributive lattice-semigroup.*

Proof Let $\mathscr{A}_i = (A_i, \Sigma, \nabla_i, \Delta_i), i = 1, 2, 3.$ be three LSTA defined on (l, Σ, Θ). Assume that l is distributive lattice-semigroup. Then

$$\mathscr{A}_1 \wedge (\mathscr{A}_2 \vee \mathscr{A}_3) = (A_1 \cap (A_2 \cup A_3), \Sigma, \{\delta((q_1, q_2, \cdots q_n), q, \sigma)$$
$$|q_1, q_2, \cdots q_n, q \in A_1 \cap (A_2 \cup A_3), \sigma \in \Sigma, \delta((q_1, q_2, \cdots q_n), q, \sigma)$$
$$= \delta_{\mathscr{A}_1}((q_1, q_2, \cdots q_n), q, \sigma) \wedge [\delta_{\mathscr{A}_2}((q_1, q_2, \cdots q_n), q, \sigma)$$
$$\vee \delta_{\mathscr{A}_3}((q_1, q_2, \cdots q_n), q, \sigma)]\}, \{\beta(q) | q \in A_1 \cap (A_2 \cup A_3), \beta(q)$$
$$= \beta_{\mathscr{A}_1}(q) \wedge [\beta_{\mathscr{A}_2}(q) \vee \beta_{\mathscr{A}_3}(q)]\})$$
$$= ((A_1 \cap A_2) \cup (A_1 \cap A_3), \Sigma, \{\delta((q_1, q_2, \cdots q_n), q, \sigma)$$
$$|q_1, q_2, \cdots q_n, q \in (A_1 \cap A_2) \cup (A_1 \cap A_3), \sigma \in \Sigma,$$

$$\begin{aligned}
&\delta((q_1, q_2, \cdots q_n), q, \sigma) \\
&= [\delta_{\mathscr{A}_1}((q_1, q_2, \cdots q_n), q, \sigma) \wedge \delta_{\mathscr{A}_2} \\
&((q_1, q_2, \cdots q_n), q, \sigma)] \vee [\delta_{\mathscr{A}_1} \\
&((q_1, q_2, \cdots q_n), q, \sigma) \wedge \delta_{\mathscr{A}_3} \\
&((q_1, q_2, \cdots q_n), q, \sigma)]\}, \{\beta(q) | q \in (A_1 \cap A_2) \cup (A_1 \cap A_3), \\
&\beta(q) = [\beta_{\mathscr{A}_1}(q) \wedge \beta_{\mathscr{A}_2}(q)] \\
&\vee [\beta_{\mathscr{A}_1}(q) \wedge \beta_{\mathscr{A}_3}(q)]\}) \\
&= (\mathscr{A}_1 \wedge \mathscr{A}_2) \vee (\mathscr{A}_1 \wedge \mathscr{A}_3)
\end{aligned}$$

This chain of reasoning can be also done backwards.

Therefore $\mathscr{A}_1 \wedge (\mathscr{A}_2 \vee \mathscr{A}_3) = (\mathscr{A}_1 \wedge \mathscr{A}_2) \vee (\mathscr{A}_1 \wedge \mathscr{A}_3)$ if and only if $a \wedge (b \vee c) = (a \wedge b) \vee (a \wedge c)$ for any element a, b and c of l.

Similarly we can prove that $\mathscr{A}_1 \vee (\mathscr{A}_2 \wedge \mathscr{A}_3) = (\mathscr{A}_1 \vee \mathscr{A}_2) \wedge (\mathscr{A}_1 \vee \mathscr{A}_3)$ if and only if $a \vee (b \wedge c) = (a \vee b) \wedge (a \vee c)$ for any element a, b and c of l.

Thus we complete the proof.

Proposition 3.4 *L-S*(l, Σ, Θ) *is a modular lattice-semigroup if and only if* l *is a modular lattice-semigroup.*

Proof Let $\mathscr{A}_i = (A_i, \Sigma, \nabla_i, \Delta_i), i = 1, 2, 3.$ be three LSTA defined on (l, Σ, Θ). Assume that l is modular lattice-semigroup. Then

$$\begin{aligned}
\mathscr{A}_1 \vee (\mathscr{A}_2 \wedge (\mathscr{A}_1 \vee \mathscr{A}_3)) &= (A_1 \cup (A_2 \cap (A_1 \cup A_3)), \Sigma, \{\delta((q_1, q_2, \cdots q_n), q, \sigma) \\
&\quad |q_1, q_2, \cdots q_n, q \in A_1 \cup (A_2 \cap (A_1 \cup A_3)), \sigma \in \Sigma, \\
&\quad \delta((q_1, q_2, \cdots q_n), q, \sigma) \\
&= \delta_{\mathscr{A}_1}((q_1, q_2, \cdots q_n), q, \sigma) \\
&\quad \vee [\delta_{\mathscr{A}_2}((q_1, q_2, \cdots q_n), q, \sigma) \wedge (\delta_{\mathscr{A}_1} \\
&\quad ((q_1, q_2, \cdots q_n), q, \sigma) \vee \delta_{\mathscr{A}_3}((q_1, q_2, \cdots q_n), q, \sigma))]\}, \\
&\quad \{\beta(q) | q \in A_1 \cup (A_2 \cap (A_1 \cup A_3)), \beta(q) = \beta_{\mathscr{A}_1}(q) \\
&\quad \vee [\beta_{\mathscr{A}_2}(q) \wedge (\beta_{\mathscr{A}_1}(q) \vee \beta_{\mathscr{A}_3}(q))]\}) \\
&= ((A_1 \cup A_2) \cap (A_1 \cup A_3), \Sigma, \{\delta((q_1, q_2, \cdots q_n), q, \sigma) \\
&\quad |q_1, q_2, \cdots q_n, q \in (A_1 \cup A_2) \cap (A_1 \cup A_3)), \sigma \in \Sigma, \\
&\quad [\delta_{\mathscr{A}_1}((q_1, q_2, \cdots q_n), q, \sigma) \vee \delta_{\mathscr{A}_2} \\
&\quad ((q_1, q_2, \cdots q_n), q, \sigma)] \wedge [\delta_{\mathscr{A}_1} \\
&\quad ((q_1, q_2, \cdots q_n), q, \sigma) \vee \delta_{\mathscr{A}_3} \\
&\quad ((q_1, q_2, \cdots q_n), q, \sigma)]\}, \{\beta(q) | q \in (A_1 \cup A_2) \\
&\quad \cap (A_1 \cup A_3), \beta(q) = [\beta_{\mathscr{A}_1}(q) \vee \beta_{\mathscr{A}_2}(q)] \\
&\quad \wedge [\beta_{\mathscr{A}_1}(q) \vee \beta_{\mathscr{A}_3}(q)]\}) \\
&= (\mathscr{A}_1 \vee \mathscr{A}_2 \wedge (\mathscr{A}_1 \vee \mathscr{A}_3)
\end{aligned}$$

This chain of reasoning can be also done backwards.

Therefore$\mathscr{A}_1 \vee (\mathscr{A}_2 \wedge (\mathscr{A}_1 \vee \mathscr{A}_3)) = (\mathscr{A}_1 \vee \mathscr{A}_2) \wedge (\mathscr{A}_1 \vee \mathscr{A}_3)$ if and only if $a \vee (b \wedge (a \vee c) = (a \vee b) \wedge (a \vee c)$ for any element a, b and c of l.

Thus we complete the proof.

3.2 The Congruence and Homomorphism of L-$S(l, \Sigma, \Theta)$

Definition 3.7 [9] *Let $\equiv$ be both a semigroup congruence and lattice congruence on lattice-semigroup S. Then $\equiv$ is called a lattice-semigroup congruence.*

Definition 3.8 [9] *Let S and T be two lattice-semigroup. φ is a mapping from S and T, if*

$$\varphi(x * y) = \varphi(x) * \varphi(y)$$
$$\varphi(x \vee y) = \varphi(x) \vee \varphi(y)$$
$$\varphi(x \wedge y) = \varphi(x) \wedge \varphi(y)$$

Then φ is called a lattice-semigroup homomorphism of S into T. If φ is one-one, then φ is called a monomorphism. If φ is onto T, then φ is called an epimorphism. If φ is both a monomorphism and epimorphism, then φ is called an lattice-semigroup isomorphism, and S and T are said to be isomorphic.

Note that let l be a lattice-semigroup, $\equiv$ is a lattice-semigroup congruence. Let $[a]$ is a set of all element equivalent to a in l, then $l/\equiv = \{[a] | a \in L\}$. We define $[a] * [b] = [a * b]$, $[a] \vee [b] = [a \vee b]$, $[a] \wedge [b] = [a \wedge b]$ is validity, for arbitrary $[a], [b] \in l/\equiv$.

Obviously, $l/\equiv$ still form a semigroup for binary operation $*$, and constitute a lattice for $\vee$ and $\wedge$. Thus $l/\equiv$ is called a quotient semigroup.

Proposition 3.5 *If $\equiv$ is a lattice-semigroup congruence on L_T =L-S(l, Σ, Θ), then quotient semigroup $L_T/\equiv$ is also a lattice-semigroup.*

Proof Let L_T = L-S(l, Σ, Θ) be a set of LSTA defined on (l, Σ, Θ). Since $\equiv$ is a lattice-semigroup congruence on L_T, let $[\mathscr{A}]$ is a set of all element equivalent to $\mathscr{A}$ in L_T. Then $L_T/\equiv = \{[\mathscr{A}] | \mathscr{A} \in L_T\}$.

Let arbitrary $[\mathscr{A}_i] \in L_T/\equiv, i = 1, 2, 3$, obviously, $L_T/\equiv$ is a lattice, hence, $([\mathscr{A}_1] \vee [\mathscr{A}_2]) * [\mathscr{A}_3] = [\mathscr{A}_1 \vee \mathscr{A}_2] * [\mathscr{A}_3] = [(\mathscr{A}_1 \vee \mathscr{A}_2) * \mathscr{A}_3] = [(\mathscr{A}_1 * \mathscr{A}_3) \vee (\mathscr{A}_2 * \mathscr{A}_3)] = [\mathscr{A}_1 * \mathscr{A}_3] \vee [\mathscr{A}_2 * \mathscr{A}_3] = ([\mathscr{A}_1] * [\mathscr{A}_3]) \vee ([\mathscr{A}_2] * [\mathscr{A}_3])$

Similarly, we can prove $[\mathscr{A}_3] * ([\mathscr{A}_1] \vee [\mathscr{A}_2]) = ([\mathscr{A}_3] * [\mathscr{A}_1]) \vee ([\mathscr{A}_3] * [\mathscr{A}_2])$

Thus, quotient semigroup $L_T/\equiv$ is also a lattice-semigroup.

Proposition 3.6 *Let L_T = L-S(l, Σ, Θ) and M_T = L-S(l', Σ', Θ'), be two arbitrary lattice-semigroup set which is formed by LSTA defined on (l, Σ, Θ) and*

(l', Σ', Θ'), *φ is a lattice-semigroup homomorphism from L_T onto M_T, define a relation $\equiv$ on L_T by $\forall \mathscr{A}, \mathscr{B} \in L_T$, $\mathscr{A} \equiv \mathscr{B}$ if and only if $\varphi(\mathscr{A}) = \varphi(\mathscr{B})$. Then $\equiv$ is a lattice-semigroup congruence relation on L_T.*

Proof Clearly $\equiv$ is an equivalence relation on L_T. For $\forall \mathscr{A}, \mathscr{B}, \mathscr{C} \in L_T$, if $\mathscr{A} \equiv \mathscr{B}$, then $\varphi(\mathscr{A}) = \varphi(\mathscr{B})$. Since φ is a lattice-semigroup homomorphism from L_T onto M_T. Then

$$\varphi(\mathscr{A} * \mathscr{B}) = \varphi(\mathscr{A}) * \varphi(\mathscr{B})$$
$$\varphi(\mathscr{A} \vee \mathscr{B}) = \varphi(\mathscr{A}) \vee \varphi(\mathscr{B})$$
$$\varphi(\mathscr{A} \wedge \mathscr{B}) = \varphi(\mathscr{A}) \wedge \varphi(\mathscr{B})$$

Hence $\varphi(\mathscr{A} * \mathscr{C}) = \varphi(\mathscr{A}) * \varphi(\mathscr{C}) = \varphi(\mathscr{B}) * \varphi(\mathscr{C}) = \varphi(\mathscr{B} * \mathscr{C})$, then $\mathscr{A} * \mathscr{C} \equiv \mathscr{B} * \mathscr{C}$.

Similarly, we can prove $\mathscr{C} * \mathscr{A} \equiv \mathscr{C} * \mathscr{B}$.

If arbitrary $\mathscr{A}_i, \mathscr{B}_i \in L_T$, $\mathscr{A}_i \equiv \mathscr{B}_i$, $i = 1, 2$, Then $\varphi(\mathscr{A}_i) = \varphi(\mathscr{B}_i)$.

Hence $\varphi(\mathscr{A}_1 \vee \mathscr{A}_2) = \varphi(\mathscr{A}_1) \vee \varphi(\mathscr{A}_2) = \varphi(\mathscr{B}_1) \vee \varphi(\mathscr{B}_2) = \varphi(\mathscr{B}_1 \vee \mathscr{B}_2)$. Then $\mathscr{A}_1 \vee \mathscr{A}_2 \equiv \mathscr{B}_1 \vee \mathscr{B}_2$. Similarly, we can prove $\mathscr{A}_1 \wedge \mathscr{A}_2 \equiv \mathscr{B}_1 \wedge \mathscr{B}_2$.

Thus $\equiv$ is a lattice-semigroup congruence relation on L_T.

Corollary 3.2 *(1) Let L_T = L-S(l, Σ, Θ), M_T = L-S(l', Σ', Θ'), be two arbitrary lattice-semigroup set which is formed by LSTA defined on (l, Σ, Θ) and (l', Σ', Θ'), φ is a lattice-semigroup homomorphism from L_T onto M_T, $\equiv_M$ is a congruence relation on M_T. Define a relation $\equiv_L$ on L_T by $\forall \mathscr{A}, \mathscr{B} \in L_T$, $\mathscr{A} \equiv_L \mathscr{B}$ if and only if $\varphi(\mathscr{A}) \equiv_M \varphi(\mathscr{B})$. Then $\equiv_L$ is a lattice-semigroup congruence relation on L_T.*

(2) Let L_T = L-S(l, Σ, Θ), M_T = L-S(l', Σ', Θ'), be two arbitrary lattice-semigroup set which is formed by LSTA defined on (l, Σ, Θ) and (l', Σ', Θ'), φ is a lattice-semigroup homomorphism from L_T onto M_T, $\equiv_L$ is a congruence relation on L_T. Define a relation $\equiv_M$ on M_T by $\forall \mathscr{B}_1, \mathscr{B}_2 \in M_T$, $\mathscr{B}_1 \equiv_M \mathscr{B}_2$ if and only if there exists $\mathscr{A}_2, \mathscr{A}_2 \in L_T$,such that $\varphi(\mathscr{A}_1) = \mathscr{B}_1, \varphi(\mathscr{A}_2) = \mathscr{B}_2$, and$\mathscr{A}_1 \equiv_L \mathscr{A}_2$ Then $\equiv_M$ is a lattice-semigroup congruence relation on M_T.

Proof The proof is similar to that of Proposition 3.6

Proposition 3.7 *Let L_T = L-S (l, Σ, Θ), M_T = L-S(l', Σ', Θ'), be two arbitrary lattice-semigroup set which is formed by LSTA defined on (l, Σ, Θ) and (l', Σ', Θ'), φ is a lattice-semigroup epimorphism from L_T onto M_T, $\equiv_L$ is a congruence relation on L_T. $\equiv_M$ is a congruence relation on M_T. Then $L_T/\equiv_L$ and $M_T/\equiv_M$ are isomorphic as lattice-semigroup.*

Proof Let $L_T/\equiv_L = \{[\mathscr{A}]_L | \mathscr{A} \in L_T\}$, $M_T/\equiv_M = \{[\mathscr{B}]_M | \mathscr{B} \in M_T\}$. Define $\psi : L_T/\equiv_L \to M_T/\equiv_M$ by $\psi([\mathscr{A}]_L) = [\varphi(\mathscr{A})]_M$, $\forall [\mathscr{A}]_L \in L_T/\equiv_L$.

If $[\mathscr{A}_1]_L = [\mathscr{A}_2]_L$ for $\forall[\mathscr{A}_1]_L, [\mathscr{A}_2]_L \in L_T/\equiv_L$, then $\mathscr{A}_1 \equiv_L \mathscr{A}_2$. Hence $\varphi(\mathscr{A}_1) \equiv_M \varphi(\mathscr{A}_2)$, then $[\varphi(\mathscr{A}_1)]_M = [\varphi(\mathscr{A}_2)]_M$. Thus ψ is a mapping.

Since φ is a epimorphism, then there exists $\mathscr{A} \in L_T$ such that $\varphi(\mathscr{A}) = \mathscr{B}$ for $\forall[\mathscr{B}]_M \in M_T/\equiv_M$. Hence there exists $[\mathscr{A}]_L \in L_T/\equiv_L$such that $\psi([\mathscr{A}]_L) = [\varphi(\mathscr{A})]_M = [\mathscr{B}]_M$.Thus ψ is a epimorphism.

If $[\mathscr{A}_1]_L \neq [\mathscr{A}_2]_L$ for $\forall[\mathscr{A}_1]_L, [\mathscr{A}_2]_L \in L_T/\equiv_L$, then $\mathscr{A}_1$ and $\mathscr{A}_2$ is not equivalence about $\equiv_L$, hence $\varphi(\mathscr{A}_1)$ and $\varphi(\mathscr{A}_2)$ is not equivalence about $\equiv_M$. Thus $[\varphi(\mathscr{A}_1)]_M \neq [\varphi(\mathscr{A}_2)]_M$.i.e.$\psi([\mathscr{A}_1]_L) \neq \psi([\mathscr{A}_2]_L)$, so ψ is monomorphism.

For $\forall[\mathscr{A}_1]_L, [\mathscr{A}_2]_L \in L_T/\equiv_L$,$\psi([\mathscr{A}_1]_L * [\mathscr{A}_2]_L) = \psi([\mathscr{A}_1 * \mathscr{A}_2]_L) = [\varphi(\mathscr{A}_1 * \mathscr{A}_2)]_M = [\varphi(\mathscr{A}_1) * \varphi(\mathscr{A}_2)]_M = [\varphi(\mathscr{A}_1)]_M * [\varphi(\mathscr{A}_2)]_M = \psi([\mathscr{A}_1]_L) * \psi([\mathscr{A}_2]_L)$.

$\psi([\mathscr{A}_1]_L \vee [\mathscr{A}_2]_L) = \psi([\mathscr{A}_1 \vee \mathscr{A}_2]_L) = [\varphi(\mathscr{A}_1 \vee \mathscr{A}_2)]_M = [\varphi(\mathscr{A}_1) \vee \varphi(\mathscr{A}_2)]_M = [\varphi(\mathscr{A}_1)]_M \vee [\varphi(\mathscr{A}_2)]_M = \psi([\mathscr{A}_1]_L) \vee \psi([\mathscr{A}_2]_L)$.

Similarly, we can prove $\psi([\mathscr{A}_1]_L \wedge [\mathscr{A}_2]_L) = \psi([\mathscr{A}_1]_L) \wedge \psi([\mathscr{A}_2]_L)$.

Thus ψ is a isomorphic from $L_T/\equiv_L$ onto $M_T/\equiv_M$.

Theorem 3.2 *(lattice-semigroup homomorphism fundamental theorem) Let* $L_T =$ *L-S*(l, Σ, Θ)*,* $M_T =$ *L-S*(l', Σ', Θ')*, be two arbitrary lattice-semigroup set which is formed by LSTA defined on* (l, Σ, Θ) *and* (l', Σ', Θ')

(1) Let $\equiv$ *is a lattice-semigroup congruence relation on* $L_T =$ *L-S*(l, Σ, Θ)*, then* $\equiv^\sharp$: $L_T \to L_T/\equiv$ *be a lattice-semigroup homomorphism.*

(2) Let φ *be a lattice-semigroup homomorphism from* $L_T =$ *L-S*(l, Σ, Θ) *onto* $M_T =$ *L-S*(l', Σ', Θ')*. Define* $\ker\varphi = \{(\mathscr{A}, \mathscr{B}) \in L_T \times L_T | \varphi(\mathscr{A}) = \varphi(\mathscr{B}\}$*. Then* $\ker\varphi$ *is a lattice-semigroup congruence on* L_T *and there exists a monomorphism* $\psi : L_T/ker\varphi \to M_T$ *such that* $\varphi = \psi \circ (\ker\varphi)^\sharp$.

Proof (1) Define $\equiv^\sharp$: $L_T \to L_T/\equiv$ by $\equiv^\sharp(\mathscr{A}) = [\mathscr{A}], \forall\mathscr{A} \in L_T$, where $L_T/\equiv = \{[\mathscr{A}] | \mathscr{A} \in L_T\}$.If $\mathscr{A} = \mathscr{B}$ for $\forall\mathscr{A}, \mathscr{B} \in L_T$,then $\mathscr{A} \equiv \mathscr{B}$,Hence$[\mathscr{A}] = [\mathscr{B}]$ i.e.$\equiv^\sharp(\mathscr{A}) = \equiv^\sharp(\mathscr{B})$. Thus $\equiv^\sharp$ is a mapping.

For $\forall\mathscr{A}, \mathscr{B} \in L_T$

$$\equiv^\sharp(\mathscr{A} * \mathscr{B}) = [\mathscr{A} * \mathscr{B}] = [\mathscr{A}] * [\mathscr{B}] = \equiv^\sharp(\mathscr{A}) * \equiv^\sharp(\mathscr{B})$$
$$\equiv^\sharp(\mathscr{A} \vee \mathscr{B}) = [\mathscr{A} \vee \mathscr{B}] = [\mathscr{A}] \vee [\mathscr{B}] = \equiv^\sharp(\mathscr{A}) \vee \equiv^\sharp(\mathscr{B})$$
$$\equiv^\sharp(\mathscr{A} \wedge \mathscr{B}) = [\mathscr{A} \wedge \mathscr{B}] = [\mathscr{A}] \wedge [\mathscr{B}] = \equiv^\sharp(\mathscr{A}) \wedge \equiv^\sharp(\mathscr{B})$$

Thus $\equiv^\sharp$ is a lattice-semigroup homomorphism from L_T onto $L_T/\equiv$.

(2)Clearly $\ker\varphi$ is a equivalence relation. For $\forall\mathscr{A}, \mathscr{B} \in L_T$, Since φ is a lattice-semigroup homomorphism from L_T onto M_T.Then

$$\varphi(\mathscr{A} * \mathscr{B}) = \varphi(\mathscr{A}) * \varphi(\mathscr{B})$$
$$\varphi(\mathscr{A} \vee \mathscr{B}) = \varphi(\mathscr{A}) \vee \varphi(\mathscr{B})$$
$$\varphi(\mathscr{A} \wedge \mathscr{B}) = \varphi(\mathscr{A}) \wedge \varphi(\mathscr{B})$$

For $\forall(\mathscr{A}, \mathscr{B}) \in \ker\varphi$, we have that $\varphi(\mathscr{A}) = \varphi(\mathscr{B})$

Then $\varphi(\mathscr{A} * \mathscr{C}) = \varphi(\mathscr{A}) * \varphi(\mathscr{C}) = \varphi(\mathscr{B}) * \varphi(\mathscr{C}) = \varphi(\mathscr{B} * \mathscr{C})$, hence $(\mathscr{A} * \mathscr{C}, \mathscr{B} * \mathscr{C}) \in \ker\varphi$.

Similarly, we can prove $(\mathscr{C} * \mathscr{A}, \mathscr{C} * \mathscr{B}) \in \ker\varphi$.

If arbitrary $(\mathscr{A}_i, \mathscr{B}_i) \in\in \ker\varphi, i = 1, 2$, we have that $\varphi(\mathscr{A}_i) = \varphi(\mathscr{B}_i)$. Then $\varphi(\mathscr{A}_1 \vee \mathscr{A}_2) = \varphi(\mathscr{A}_1) \vee \varphi(\mathscr{A}_2) = \varphi(\mathscr{B}_1) \vee \varphi(\mathscr{B}_2) = \varphi(\mathscr{B}_1 \vee \mathscr{B}_2)$. Hence $(\mathscr{A}_1 \vee \mathscr{A}_2, \mathscr{B}_1 \vee \mathscr{B}_2) \in \ker\varphi$. Similarly, we can prove$(\mathscr{A}_1 \wedge \mathscr{A}_2, \mathscr{B}_1 \wedge \mathscr{B}_2) \in \ker\varphi$.

Thus, $\ker\varphi$ is a lattice-semigroup congruence on L_T.

Define $\psi : L_T/\ker\varphi \to M_T$ by $\psi([\mathscr{A}]) = \varphi(\mathscr{A}), \forall[\mathscr{A}] \in L_T/\ker\varphi$. For $\forall[\mathscr{A}], [\mathscr{B}] \in L_T/\ker\varphi$. If $[\mathscr{A}] = [\mathscr{B}]$, then $(\mathscr{A}, \mathscr{B}) \in \ker\varphi$, we can know that $\varphi(\mathscr{A}) = \varphi(\mathscr{B})$,i.e.$\psi([\mathscr{A}]) = \psi([\mathscr{B}])$ from definition of $\ker\varphi$. Thus ψ is a mapping.

For $\forall[\mathscr{A}], [\mathscr{B}] \in L_T/\ker\varphi$, If $[\mathscr{A}] \neq [\mathscr{B}]$, then $(\mathscr{A}, \mathscr{B}) \subseteq \ker\varphi$,hence $\varphi(\mathscr{A}) \neq \varphi(\mathscr{B})$, i.e.$\psi([\mathscr{A}]) \neq \psi([\mathscr{B}])$. Thus ψ is a monomorphism.

For $\forall[\mathscr{A}], [\mathscr{B}] \in L_T/\ker\varphi$

$\psi([\mathscr{A}]*[\mathscr{B}]) = \psi([\mathscr{A}*\mathscr{B}]) = \varphi(\mathscr{A}*\mathscr{B}) = \varphi(\mathscr{A})*\varphi(\mathscr{B}) = \psi([\mathscr{A}])*\psi([\mathscr{B}])$

$\psi([\mathscr{A}] \vee [\mathscr{B}]) = \psi([\mathscr{A} \vee \mathscr{B}]) = \varphi(\mathscr{A} \vee \mathscr{B}) = \varphi(\mathscr{A}) \vee \varphi(\mathscr{B}) = \psi([\mathscr{A}]) \vee \psi([\mathscr{B}])$

$\psi([\mathscr{A}] \wedge [\mathscr{B}]) = \psi([\mathscr{A} \wedge \mathscr{B}]) = \varphi(\mathscr{A} \wedge \mathscr{B}) = \varphi(\mathscr{A}) \wedge \varphi(\mathscr{B}) = \psi([\mathscr{A}]) \wedge \psi([\mathscr{B}])$

Thus ψ is a lattice-semigroup homomorphism, and $\psi \circ (\ker\varphi)^\sharp(\mathscr{A}) = \psi([\mathscr{A}]) = \varphi(\mathscr{A})$ for $\forall\mathscr{A} \in L_T$.i.e.there exists a monomorphism $\varphi = \psi \circ (\ker\varphi)^\sharp$.

Corollary 3.3 *If φ is a epimorphism from L_T onto M_T, then* $\ker\varphi$ *is a lattice-semigroup congruence on L_T and there exists a isomorphic. i.e. $L_T/\ker\varphi \cong ran(\varphi)$.*

Definition 3.9 *Let S be a lattice-semigroup, ρ is a lattice-semigroup congruence on S. Then ρ is called a* regular congruence *if there exist a partial relation "$\leq$" on quotient semigroup such that*

(1) $(S/\rho, *, \leq, \vee, \wedge)$ is a lattice ordered semigroup.

(2) $\varphi : S \to S/\rho | x \to (x)_\rho$ is a mapping which hold partial relation, i.e. there exist a lattice-semigroup homomorphism from S onto S/ρ.

Proposition 3.8 *Let L-S(l', Σ', Θ') be a lattice-semigroup set which is formed by LSTA defined on (l, Σ, Θ), $\equiv$ is a lattice-semigroup congruence relation. Then $\equiv$ is a regular congruence.*

Proof Clearly $L_T/\equiv$ is a lattice-semigroup by Proposition 3.5. We define a binary relation of similar Definition 3.5 on S/ρ:

$\leq := \{((\mathscr{A})_\equiv, (\mathscr{B})_\equiv) | (\mathscr{A})_\equiv = (\mathscr{A})_\equiv \wedge (\mathscr{B})_\equiv\}$

We know that "$\leq$" is a partial relation by Corollary 3.1, for $\forall\mathscr{A}, \mathscr{B}, \mathscr{C} \in L_T$, if $(\mathscr{A})_\equiv \leq (\mathscr{B})_\equiv$,then $(\mathscr{A})_\equiv = (\mathscr{A})_\equiv \wedge (\mathscr{B})_\equiv$.

$[(\mathscr{C})_\equiv * (\mathscr{A})_\equiv] \wedge [(\mathscr{C})_\equiv * (\mathscr{B})_\equiv] = (\mathscr{C} * \mathscr{A})_\equiv \wedge (\mathscr{C} * \mathscr{B})_\equiv = [(\mathscr{C} * \mathscr{A}) \wedge (\mathscr{C} * \mathscr{B})]_\equiv = [\mathscr{C} * (\mathscr{A} \wedge \mathscr{B})]_\equiv = (\mathscr{C})_\equiv * (\mathscr{A} \wedge \mathscr{B})_\equiv = (\mathscr{C})_\equiv * (\mathscr{A})_\equiv$, then $(\mathscr{C})_\equiv * (\mathscr{A})_\equiv \leq (\mathscr{C})_\equiv * (\mathscr{B})_\equiv$.

Similarly, we can prove $(\mathscr{A})_\equiv * (\mathscr{C})_\equiv \leq (\mathscr{B})_\equiv * (\mathscr{C})_\equiv$.

Thus $L_T/\equiv$ is a lattice ordered semigroup, and $\equiv^\sharp: L_T \to L_T/\equiv$ be a lattice-semigroup homomorphism by Theorem 3.2, (1).

Thus$\equiv$ is a regular congruence on L_T.

Proposition 3.9 *L-S(l', Σ', Θ') be a lattice-semigroup set which is formed by LSTA defined on (l, Σ, Θ), $\equiv$ is a lattice-semigroup congruence relation. Then $\equiv$ is a regular congruence if and only if there exist a lattice ordered semigroup M_T = L-S(l', Σ', Θ') and a homomorphism $\varphi : L_T \to M_T$ such that*

$$\equiv = \{(\mathscr{A}, \mathscr{B}) | \varphi(\mathscr{A}) = \varphi(\mathscr{B})\}.$$

Proof ($\Rightarrow$) Let $\equiv$ is a regular congruence, there exist a lattice ordered semigroup $L_T/\equiv$ such that $\varphi : L_T \to M_T | \mathscr{A} \to (\mathscr{A})_\equiv$ is a homomorphism by Definition 3.9, clearly

$$\equiv = \{(\mathscr{A}, \mathscr{B}) | \varphi(\mathscr{A}) = \varphi(\mathscr{B})\}.$$

($\Leftarrow$) Let exist a lattice ordered semigroup M_T = L-S(l', Σ', Θ') and a homomorphism $\varphi : L_T \to M_T$ such that $\equiv = \{(\mathscr{A}, \mathscr{B}) | \varphi(\mathscr{A}) = \varphi(\mathscr{B})\}$. Then define $\equiv = \ker \varphi$, $M_T = L_T/\ker \varphi$, clearly $\equiv$ is a regular congruence.

Acknowledgments The authors are very grateful to the anonymous referee for his/her careful review and constructive suggestions.This work is supported by the National Natural Science Foundation of China (Grant No.11071178), the Research Foundation of the Education Department of Sichuan Province (Grant No. 12ZB106) and the Research Foundation of Sichuan Normal University (Grant No. 10MSL06).

References

1. Zadeh, L.A.: Fuzzy sets. Inf. Control **8**, 338–353 (1965)
2. Inagaki, Y., Fukumura, T.: On the description of fuzzy meaning of context-free language. In: Fuzzy Sets and their Applications to Cognitive and Decision Processes. Proceedings of the U.S.-Japan Seminar, University of California, Berkeley, CA, 1974. Academic Press, New York, pp. 301–328 (1975)
3. Qiu, D.W.: Automata theory based on complete residuated lattice-valued logic. Sci. China(F) **45**(6), 442–452 (2002)
4. Qiu, D.W.: Pumping lemma in automata theory based on complete residuated lattice-valued logic: a note. Fuzzy Sets Syst. 2128–2138 (2006)
5. Esik, Z., Liu, G.W.: Fuzzy tree automata. Fuzzy Sets Syst. **158**, 1450–1460 (2007)
6. Lu, R.Q., Zheng, H.: Lattices of quantum automata. Int. J. Theor. Phys. 1425–1449 (2003)
7. GUO, X.H., Mo, Z.W.: Homomorphisms in lattices of quantum automata. J. Sichuan Normal Univ. (Nat. Sci.) 635–638 (2005)
8. Huang, X.F., Mo, Z.W.: Some algebraic properties on lattice semigroup tree automata. Fuzzy Syst. Math. (To appear)
9. Xie, X.Y.: Ordered semigroup introduce. Science Press, pp. 9–10 (2001)

$(\in, \in \vee q_{(\lambda,\mu)})$-Fuzzy Completely Semiprime Ideals of Semigroups

Zu-hua Liao, Li-hua Yi, Ying-ying Fan and Zhen-yu Liao

Abstract We introduce a new kind of generalized fuzzy completely ideal of a semigroup called $(\in, \in \vee q_{(\lambda,\mu)})$-fuzzy completely semiprime ideals. These generalized fuzzy completely semiprime ideals are characterized.

Keywords Fuzzy algebra · Fuzzy points · $(\in, \in \vee q_{(\lambda,\mu)})$-fuzzy completely semiprime ideals · Level subsets · Completely semiprime ideals

1 Introduction

Fuzzy semigroup theory plays a prominent role in mathematics with ranging applications in many disciplines such as control engineering, information sciences, fuzzy coding theory, fuzzy finite state machines, fuzzy automata, fuzzy languages.

Using the notion of a fuzzy set introduced by Zadeh [1] in 1965, which laid the foundation of fuzzy set theory, Rosenfeld [2] inspired the fuzzification of algebraic structures and introduced the notion of fuzzy subgroups. Since then fuzzy algebra came into being. Bhakat and Das gave the concepts of fuzzy subgroups by using the "belongs to" relation ($\in$) and "quasi-coincident with" relation (q) between a fuzzy point and a fuzzy set, and introduced the concept of a $(\in, \in \vee q)$-fuzzy subgroup [3–6]. It is worth to point out that the ideal of quasi-coincident of a fuzzy point with a fuzzy set, which is mentioned in [7], played a vital role to generate some different types of fuzzy subgroups. In particular, $(\in, \in \vee q)$-fuzzy subgroup is an important and useful generalization of Rosenfeld's fuzzy subgroup, which provides sufficient motivation to researchers to review various concepts and results from the realm of

Z. Liao (✉) · Y. Fan · L. Yi
School of Science, Jiangnan University, 214122 Wuxi, China
e-mail: liaozuhua57@yahoo.cn

Z. Liao
Department of Mathematics, State University of New York at Buffalo, 14260 New York, USA

B.-Y. Cao and H. Nasseri (eds.), *Fuzzy Information & Engineering and Operations Research & Management*, Advances in Intelligent Systems and Computing 211,
DOI: 10.1007/978-3-642-38667-1_18,

abstract algebra in the broader framework of fuzzy setting. Zhan [8], Jun et al. [9] introduced the notion of $(\in, \in \vee q)$-fuzzy interior ideals of a semigroup. Davvaz [10–13] defined $(\in, \in \vee q)$-fuzzy subnear-rings and characterized H_ν-fuzzy submodules, R-fuzzy semigroups using the relation $(\in, \in \vee q)$.

Later, the definition of a generalized fuzzy subgroup was introduced by Yuan [13]. Based on it, Liao [14] expanded common "quasi-coincident with" relationship to generalized "quasi-coincident with" relationship, which is the generalization of Rosenfeld's fuzzy algebra and Bhakat and Das's fuzzy algebra. And a series results were gotten by using generalized "quasi-coincident with" relationship [15–18]. When $\lambda = 0$ and $\mu = 1$ we get common fuzzy algebra by Rosenfeld and When $\lambda = 0$ and $\mu = 0.5$ we get the $(\in, \in \vee q)$-fuzzy algebra defined by Bhakat and Das and when $\lambda = 0$ and $\mu = 0.5$ we get the $(\overline{\in}, \overline{\in} \vee \overline{q})$-fuzzy algebra.

The concept of a fuzzy ideal in semigroups was developed by Kuroki. He studied fuzzy ideals, fuzzy bi-ideals and fuzzy semiprime ideals in semigroups [19–21]. Fuzzy ideals, generated by fuzzy sets in semigroups, are considered by Mo and Wang [22]. After that Bhakat and Das [23] investigated fuzzy subrings and several types of ideals, including fuzzy prime ideals and $(\in, \in \vee q)$-fuzzy prime ideals. Jun et al.[24–26] studied L-fuzzy ideals in semigroups, fuzzy h-ideals in hemirings,fuzzy ideals in inclines. Besides, Bahushri [27] did some research on c-prime fuzzy ideals in nearrings. It is now natural to investigate similar type of generalization of the existing fuzzy subsystems of some algebraic structures. Our aim in this paper is to introduce and study $(\in, \in \vee q_{(\lambda,\mu)})$-fuzzy completely semiprime ideals, and obtain some properties: an $(\in, \in \vee q_{(\lambda,\mu)})$-fuzzy ideal is an $(\in, \in \vee q_{(\lambda,\mu)})$-fuzzy completely semiprime ideal, if and only if $A_t (\neq \emptyset)$ is a completely semiprime ideal, $\forall t \in (\lambda, \mu)$. This showed that $(\in, \in \vee q_{(\lambda,\mu)})$-fuzzy completely semiprime ideals are generalizations of the existing concepts of two types of fuzzy ideals.

2 Preliminaries

Throughout the paper we always consider S as a semigroup.

A mapping from A to $[0, 1]$ is said to be a fuzzy subset of S.

A fuzzy subset A of S of the form $A(y) = \begin{cases} \lambda(\neq 0), & y = x \\ 0, & y \neq x \end{cases}$ is said to be a fuzzy point support x and value λ is denoted by x_t.

Definition 2.1 *Let A be a fuzzy subset of S, for all $t, \lambda, \mu \in [0, 1]$ and $\lambda < \mu$, a fuzzy point x_t is called belonging to A if $A(x) \geq t$, denoted by $x_t \in A$; A fuzzy point x_t is said to be generalized quasi-coincident with A if $t > \lambda$ and $A(x) + t > 2\mu$, denoted by $x_t q_{(\lambda,\mu)} A$. If $x_t \in A$ or $x_t q_{(\lambda,\mu)} A$, then denoted by $x_t \in \vee q_{(\lambda,\mu)} A$.*

Definition 2.2 *[15] A fuzzy subset A of S is said to be an $(\in, \in \vee q_{(\lambda,\mu)})$-fuzzy subsemigroup if for all $x, y \in S, t_1, t_2 \in (\lambda, 1]$, $x_{t_1}, y_{t_2} \in A$ implies $(xy)_{t_1 \wedge t_2} \in \vee q_{(\lambda,\mu)} A$.*

Theorem 2.3 *[15] A fuzzy subset A of S is an $(\in, \in \vee q_{(\lambda,\mu)})$-fuzzy subsemigroup if and only if $A(xy) \vee \lambda \geq A(x) \wedge A(y) \wedge \mu$, for all $x, y \in S$.*

Definition 2.4 *[15] A fuzzy subset A of S is an $(\in, \in \vee q_{(\lambda,\mu)})$-fuzzy left (right) ideal if (i) A is an $(\in, \in \vee q_{(\lambda,\mu)})$-fuzzy subsemigroup of S;*
(ii) For all $x_t \in A, y \in S$, implies $(yx)_t \in \vee q_{(\lambda,\mu)} A$ $((xy)_t \in \vee q_{(\lambda,\mu)} A)$.
If A is both an $(\in, \in \vee q_{(\lambda,\mu)})$-fuzzy left ideal and an $(\in, \in \vee q_{(\lambda,\mu)})$-fuzzy right ideal, then A is said to be an $(\in, \in \vee q_{(\lambda,\mu)})$-fuzzy ideal.

Theorem 2.5 *[15] A fuzzy subset A of S is an $(\in, \in \vee q_{(\lambda,\mu)})$-fuzzy ideal if and only if for all $t \in (\lambda, \mu]$, the non-empty set A_t is an ideal.*

Theorem 2.6 *[15] A fuzzy subset A of S is an $(\in, \in \vee q_{(\lambda,\mu)})$-fuzzy left (right) ideal if and only if for all $x, y \in S$, (i) $A(xy) \vee \lambda \geq A(x) \wedge A(y) \wedge \mu$; (ii)$A(xy) \vee \lambda \geq A(y) \wedge \mu (A(x) \wedge \mu)$.*

Definition 2.7 *[15] An ideal I of S is said to be a completely semiprime ideal, if for all $x \in S, x^2 \in I$ implies $x \in I$.*

Based on [23], in a semigroup we have the following definitions and theorems:

Definition 2.8 *A fuzzy ideal A of S is said to be a fuzzy completely semiprime ideal,if for all $x \in S, t \in (0, 1]$, $(x^2)_t \in A$ implies $x_t \in A$.*

Theorem 2.9 *A fuzzy ideal A of S is a fuzzy completely semiprime ideal, if and only if $A(x^2) = A(x)$, for all $x \in S$.*

Definition 2.10 *An $(\in, \in \vee q)$-fuzzy ideal A of S is said to be an $(\in, \in \vee q)$-fuzzy completely semiprime ideal, if for all $x \in S, t \in (0, 1]$, $(x^2)_t \in A$ implies $x_t \in \vee q A$.*

Theorem 2.11 *An $(\in, \in \vee q)$-fuzzy ideal of S is an $(\in, \in \vee q)$-fuzzy completely semiprime ideal if and only if $A(x) \geq A(x^2) \wedge 0.5$, for all $x \in S$.*

Lemma 2.12 *Let $\{H_t | t \in I \subset [0, 1]\}$ be a family of completely semiprime ideals of S such that for all $s, t \in I, t < s$, $H_s \subset H_t$. Then $\cup_{t \in I} H_t, \cap_{t \in I} H_t$ are completely semiprime ideals of S.*

3 $(\in, \in \vee q_{(\lambda,\mu)})$-Fuzzy Completely Semiprime Ideals

In this section, we give the new definition of an $(\in, \in \vee q_{(\lambda,\mu)})$-fuzzy completely semiprime ideal of semigroups. Then some equivalent descriptions and properties of it are discussed.

Definition 3.1 *An* $(\in, \in \vee q_{(\lambda,\mu)})$*-fuzzy ideal A of S is said to be an* $(\in, \in \vee q_{(\lambda,\mu)})$*-fuzzy completely semiprime ideal, if for all* $x \in S, t \in (\lambda, 1]$, $(x^2)_t \in A$ *implies* $x_t \in \vee q_{(\lambda,\mu)} A$.

Theorem 3.2 *An* $(\in, \in \vee q_{(\lambda,\mu)})$*-fuzzy ideal of S is an* $(\in, \in \vee q_{(\lambda,\mu)})$*-fuzzy completely semiprime ideal if and only if* $A(x) \vee \lambda \geq A(x^2) \wedge \mu$*, for all* $x \in S$.

Proof $\Rightarrow$ Let A be an $(\in, \in \vee q_{(\lambda,\mu)})$-fuzzy completely semiprime ideal of S.Assume that there exists x_0 such that $A(x_0) \vee \lambda < A(x_0^2) \wedge \mu$. Choose t to satisfy $A(x_0) \vee \lambda < t < A(x_0^2) \wedge \mu$. Then we have $A(x_0^2) > t, \lambda < t < \mu, A(x_0) < t$ and $A(x_0) + t < 2\mu$. So $(x_0^2)_t \in A$. But $(x_0)_t \overline{\in \vee q_{(\lambda,\mu)}} A$,a contradiction.
$\Leftarrow$ For all $x \in S, t \in (\lambda, 1]$ and $(x^2)_t \in A$, then $A(x^2) \geq t$ and $\lambda < t$. So $A(x) \vee \lambda \geq A(x^2) \wedge \mu \geq t \wedge \mu$. Since $\lambda < \mu$, then $A(x) \geq t \wedge \mu$.
If $t \geq \mu$, then $A(x) \geq \mu$, we have $A(x) + t > \mu + \mu = 2\mu$, so $x_t q_{(\lambda,\mu)} A$.
If $t < \mu$, then $A(x) \geq t$. So $x_t \in A$.

Hence, $x_t \in \vee q_{(\lambda,\mu)} A$. That is to say, A is an $(\in, \in \vee q_{(\lambda,\mu)})$-fuzzy completely semiprime ideal.

Theorem 3.3 *A non-empty subset* S_1 *of S is a completely semiprime ideal if and only if* χ_{S_1} *is an* $(\in, \in \vee q_{(\lambda,\mu)})$*-fuzzy completely semiprime ideal of S.*

Proof $\Rightarrow$ Let S_1 be a completely semiprime ideal, then χ_{S_1} is an $(\in, \in \vee q_{(\lambda,\mu)})$-fuzzy ideal of S. If $(x^2)_t \in \chi_{S_1}$, then $\chi_{S_1}(x^2) \geq t > 0$, so $x^2 \in S_1$. Since S_1 is a completely semiprime ideal, we have $x \in S_1$, then $x_t \in \vee q_{(\lambda,\mu)} \chi_{S_1}$. Thus χ_{S_1} is an $(\in, \in \vee q_{(\lambda,\mu)})-$ fuzzy completely semiprime ideal.
$\Leftarrow$ Let χ_{S_1} be an $(\in, \in \vee q_{(\lambda,\mu)})$-fuzzy completely semiprime ideal, then S_1 is an ideal of S. Now $x^2 \in S_1$, then $\chi_{S_1}(x^2) = 1$. Since χ_{S_1} is an $(\in, \in \vee q_{(\lambda,\mu)})$-fuzzy completely semiprime ideal, by the Theorem 3.2, we have $\chi_{S_1}(x) \vee \lambda \geq \chi_{S_1}(x^2) \wedge \mu$ $= \mu$ Then $\chi_{S_1}(x) \geq \mu > 0$. So $\chi_{S_1}(x) = 1$,i.e.,$x \in S_1$. Therefore, S_1 is a completely semiprime ideal.

Remark When $\lambda = 0, \mu = 1$, we can obtain the corresponding results in the sense of Rosenfeld; When $\lambda = 0, \mu = 0.5$, we can get the corresponding results in the sense of Bhakat and Das.

Theorem 3.4 *An* $(\in, \in \vee q_{(\lambda,\mu)})$*-fuzzy ideal A of S is an* $(\in, \in \vee q_{(\lambda,\mu)})$*-fuzzy completely semiprime ideal if and only if non-empty set* A_t *is a completely semiprime ideal for all* $t \in (\lambda, \mu]$.

Proof $\Rightarrow$ Let A be an $(\in, \in \vee q_{(\lambda,\mu)})$-fuzzy completely semiprime ideal of S, then the non-empty set A_t is an ideal, for all $t \in (\lambda, \mu]$. Let $x^2 \in A_t$, then $A(x^2) \geq t$. By Theorem3.2,we have $A(x) \vee \lambda \geq A(x^2) \wedge \mu \geq t \wedge \mu = t$. So $x \in A_t$. Hence A_t is a completely semiprime ideal.
$\Leftarrow$Let A_t be a completely semiprime ideal of S, by Theorem2.5, we have that A is an $(\in, \in \vee q_{(\lambda,\mu)})$-fuzzy ideal. Suppose that A is not an $(\in, \in \vee q_{(\lambda,\mu)})$-fuzzy completely semiprime ideal. By Theorem3.2, there exists x_0 such that $A(x_0) \vee \lambda < A(x_0^2) \wedge \mu$. Choose t such that $A(x_0) \vee \lambda < t < A(x_0^2) \wedge \mu$. Then we have $A(x_0^2) > t$, $A(x_0) <$

$t, \lambda < t < \mu$ and $x_0^2 \in A_t$. Since A_t is completely semiprime, we have $x_0 \in A_t$, a contradiction.

Therefore A is an $(\in, \in \vee q_{(\lambda,\mu)})$-fuzzy completely semiprime ideal.

Corollary 3.5 A fuzzy set A of S is an $(\in, \in \vee q_{(\lambda,\mu)})$-fuzzy completely semiprime, ideal of S if and only if non-empty set A_t is a completely semiprime ideal, for all $t \in (\lambda, \mu]$.

Theorem 3.6 *Let I be any completely semiprime ideal of S. There exists an $(\in, \in \vee q_{(\lambda,\mu)})$-fuzzy completely semiprime ideal A of S such that $A_t = I$ for some $t \in (\lambda, \mu]$.*

Proof If we define a fuzzy set in S by

$$A(x) = \begin{cases} t, & if \quad x \in I \\ 0, & otherwise \end{cases} \quad \text{for some } t \in (\lambda, \mu].$$

Then it follows that $A_t = I$.

For given $r \in (\lambda, \mu]$, we have

$$A_r = \begin{cases} A_t(= I), & if \quad r \le t \\ \emptyset, & if \quad t < r < \mu \end{cases}$$

Since I itself is a completely semiprime ideal of S, it follows that every non-empty level subset A_r of S is a completely semiprime ideal of S. By Corollary3.5, A is an $(\in, \in \vee q_{(\lambda,\mu)})$-fuzzy completely semiprime ideal of S, which satisfies the conditions of the Theorem.

Theorem 3.7 *Let A be an $(\in, \in \vee q_{(\lambda,\mu)})$-fuzzy completely semiprime ideal of S such that $A(x) \le \mu$ for all $x \in S$. Then A is a fuzzy completely semiprime ideal of S.*

Proof Let $x \in S$, $t \in (\lambda, 1]$ and $(x^2)_t \in A$. It follows that $x_t \in \vee q_{(\lambda,\mu)} A$ from Definition3.1.

By known conditions, we have $t \le A(x^2) \le \mu$ and $t < \mu$. Thus $A(x) + t \le \mu + \mu = 2\mu$, *i.e.*, $x_t \overline{q_{(\lambda,\mu)}} A$. Hence $x_t \in A$. Therefore A is a fuzzy completely semiprime ideal of S.

Theorem 3.8 *Let A be an $(\in, \in \vee q_{(\lambda,\mu)})$-fuzzy ideal of S and B be an $(\in, \in \vee q_{(\lambda,\mu)})$-fuzzy completely semiprime ideal of S. Then $A \cap B$ is an $(\in, \in \vee q_{(\lambda,\mu)})$-fuzzy completely semiprime ideal of A_μ.*

Proof Let $x \in A_\mu$ and $(x^2)_t \in A \cap B$,then $(A \cap B)(x^2) \ge t$. So $A(x^2) \ge t$ and $B(x^2) \ge t$. Thus $(x^2)_t \in A$ and $(x^2)_t \in B$. Since B is an $(\in, \in \vee q_{(\lambda,\mu)})$-fuzzy completely semiprime ideal of S, we have $x_t \in \vee q_{(\lambda,\mu)} B$, so $x_t \in B$ or $x_t q_{(\lambda,\mu)} B$.

Assume $x_t \in B$ which implies $B(x) \ge t$. If $t \le \mu$, then $A(x) \ge \mu \ge t$. So $(A \cap B)(x) = A(x) \wedge B(x) \ge t$. Therefore $x_t \in (A \cap B)$ and $x_t \in \vee q_{(\lambda,\mu)}(A \cap B)$; If $t > \mu$, then $A(x) + t > \mu + \mu = 2\mu$ and $B(x) + t \ge t + t > 2\mu$, then $(A \cap B)(x) + t = (A(x) \wedge B(x)) + t > 2\mu$. So $x_t q_{(\lambda,\mu)}(A \cap B)$.

Assume $x_t q_{(\lambda,\mu)} B$ which implies $B(x) + t > 2\mu$. If $t \leq \mu$, then $B(x) > 2\mu - t \geq \mu \geq t$ and $A(x) \geq \mu \geq t$, so $x_t \in A$ and $x_t \in B$. Thus $x_t \in A \cap B$ and $x_t \in \vee q_{(\lambda,\mu)}(A \cap B)$. If $t > \mu$, then $A(x) + t > 2\mu$ and $(A \cap B)(x) + t = (A(x) \wedge B(x)) + t > 2\mu$. So $x_t q_{(\lambda,\mu)}(A \cap B)$ and $x_t \in \vee q_{(\lambda,\mu)}(A \cap B)$.

Therefore, $A \cap B$ is an $(\in, \in \vee q_{(\lambda,\mu)})$-fuzzy completely semiprime ideal of A_μ.

Theorem 3.9 *Let $\{A_i \mid i \in I\}$ be a family of $(\in, \in \vee q_{(\lambda,\mu)})$-fuzzy completely semiprime ideals of S such that $A_i \subseteq A_j$ or $A_j \subseteq A_i$ for all $i, j \in I$. Then $A = \cup_{i \in I} A_i$ is an $(\in, \in \vee q_{(\lambda,\mu)})$-fuzzy completely semiprime ideal of S.*

Proof For all $x, y \in S$,

$$\begin{aligned} A(xy) \vee \lambda &= \left(\cup_{i \in I} A_i\right)(xy) \vee \lambda = \left(\vee_{i \in I} A_i(xy)\right) \vee \lambda \\ &= \vee_{i \in I} \left(A_i(xy) \vee \lambda\right) \\ &\geq \vee_{i \in I} \left(A_i(x) \wedge A_i(y) \wedge \mu\right) \\ &= \left(\vee_{i \in I} A_i(x)\right) \wedge \left(\vee_{i \in I} A_i(y)\right) \wedge \mu \\ &= \left(\cup_{i \in I} A_i(x)\right) \wedge \left(\cup_{i \in I} A_i(y)\right) \wedge \mu \\ &= A(x) \wedge A(y) \wedge \mu \end{aligned}$$

In the following we show that $\vee_{i \in I} (A_i(x) \wedge A_i(y) \wedge \mu) = (\vee_{i \in I} A_i(x)) \wedge (\vee_{i \in I} A_i(y)) \wedge \mu$ holds. It is clear that $\vee_{i \in I} (A_i(x) \wedge A_i(y) \wedge \mu) \leq (\vee_{i \in I} A_i(x)) \wedge (\vee_{i \in I} A_i(y)) \wedge \mu$. If possible, let $\vee_{i \in I} (A_i(x) \wedge A_i(y) \wedge \mu) \neq (\vee_{i \in I} A_i(x)) \wedge (\vee_{i \in I} A_i(y)) \wedge \mu$.

Then there exists t such that $\vee_{i \in I} (A_i(x) \wedge A_i(y) \wedge \mu) < t < (\vee_{i \in I} A_i(x)) \wedge (\vee_{i \in I} A_i(y)) \wedge \mu$. Since $A_i \subseteq A_j$ or $A_j \subseteq A_i$ for all $i, j \in I$, there exists $k \in I$ such that $t < A_k(x) \wedge A_k(y) \wedge \mu$. On the other hand, $A_i(x) \wedge A_i(y) \wedge \mu < t$ for all $i \in I$, a contradiction. Hence, $\vee_{i \in I} (A_i(x) \wedge A_i(y) \wedge \mu) = (\vee_{i \in I} A_i(x)) \wedge (\vee_{i \in I} A_i(y)) \wedge \mu$. $\forall x, y \in S$, $A(xy) \vee \lambda = (\cup_{i \in I} A_i(xy)) \vee \lambda = \vee_{i \in I} (A_i(xy) \vee \lambda) \geq \vee_{i \in I} (A_i(x) \wedge \mu)$
$= (\vee_{i \in I} A_i(x)) \wedge \mu = A(x) \wedge \mu$.

Similarly prove $A(xy) \vee \lambda \geq A(y) \wedge \mu$, for all $x, y \in S$.
$A(x) \vee \lambda = (\cup_{i \in I} A_i(x)) \vee \lambda = \cup_{i \in I} (A_i(x) \vee \lambda) \geq \vee_{i \in I} \left(A_i(x^2) \wedge \mu\right) = \left(\vee_{i \in I} A_i(x^2)\right) \wedge \mu = A(x^2) \wedge \mu$.

By Theorem3.2, A is an $(\in, \in \vee q_{(\lambda,\mu)})$-fuzzy completely semiprime ideal of S.

Theorem 3.10 *Let $\{A_i \mid i \in I\}$ be a family of $(\in, \in \vee q_{(\lambda,\mu)})$-fuzzy completely semiprime ideals of S such that $A_i \subseteq A_j$ or $A_j \subseteq A_i$ for all $i, j \in I$. Then $A = \cap_{i \in I} A_i$ is an $(\in, \in \vee q_{(\lambda,\mu)})$-fuzzy completely semiprime ideal of S.*

Proof For any $x, y \in S$,

$$\begin{aligned} A(xy) \vee \lambda &= (\cap_{i\in I} A_i)(xy) \vee \lambda = (\wedge_{i\in I} A_i(xy)) \vee \lambda \\ &= \wedge_{i\in I}(A_i(xy) \vee \lambda) \geq \wedge_{i\in I} (A_i(x) \wedge A_i(y) \wedge \mu) \\ &\geq (\wedge_{i\in I} A_i(x)) \wedge (\wedge_{i\in I} A_i(y)) \wedge \mu = A(x) \wedge A(y) \wedge \mu \\ A(xy) \vee \lambda &= (\cap_{i\in I} A_i)(xy) \vee \lambda = (\wedge_{i\in I} A_i(xy)) \vee \lambda = \wedge_{i\in I} (A_i(xy) \vee \lambda) \\ &\geq \wedge_{i\in I} (A_i(x) \wedge \mu) = (\wedge_{i\in I} A_i(x)) \wedge \mu = A(x) \wedge \mu \end{aligned}$$

Similarly prove $A(xy) \vee \lambda \geq A(y) \wedge \mu, \forall x, y \in S$.

$$\begin{aligned} A(x) \vee \lambda &= (\cap_{i\in I} A_i)(x) \vee \lambda \\ &= (\wedge_{i\in I} A_i)(x) \vee \lambda \\ &= \wedge_{i\in I} (A_i(x) \vee \lambda) \\ &\geq \wedge_{i\in I} \left(A_i(x^2) \wedge \mu\right) \end{aligned}$$

In the following we show that $(\wedge_{i\in I} A_i)(x) \vee \lambda = \wedge_{i\in I} (A_i(x) \vee \lambda)$ holds. It is clear that $\wedge_{i\in I} (A_i(x) \vee \lambda) \geq (\wedge_{i\in I} A_i(x)) \vee \lambda$.

If possible, let $\wedge_{i\in I} (A_i(x) \vee \lambda) > (\wedge_{i\in I} A_i(x)) \vee \lambda$. Then there exists t such that $\wedge_{i\in I} (A_i(x) \vee \lambda) > t > (\wedge_{i\in I} A_i(x)) \vee \lambda$. Since $A_i \subseteq A_j$ or $A_j \subseteq A_i$ for all $i, j \in I$, there exists $k \in I$ such that $t > A_k(x) \vee \mu$. On the other hand, $A_i(x) \vee \mu > t$ for all $i \in I$, a contradiction.

Hence, $\wedge_{i\in I} (A_i(x) \vee \lambda) = (\wedge_{i\in I} A_i(x)) \vee \lambda$.

Here we finish the proof of the theorem.

Theorem 3.11 *Let S and S' be semigroups and $f : S \to S'$ be an onto homomorphism. Let A and B be $(\in, \in \vee q_{(\lambda,\mu)})$-fuzzy completely semiprime ideals of S and S', respectively. Then*

(i) *$f(A)$ is an $(\in, \in \vee q_{(\lambda,\mu)})$-fuzzy completely semiprime ideal of S';*
(ii) *$f^{-1}(B)$ is an $(\in, \in \vee q_{(\lambda,\mu)})$-fuzzy completely semiprime ideal of S.*

Proof (i) For any $x' \in S$, then

$$\begin{aligned} f(A)(x'^2) \vee \lambda &= \vee\{A(z) \mid z \in S, f(z) = x'^2\} \vee \lambda \\ &\geq \vee\{A(x^2) \mid x \in S, f(x) = x'\} \vee \lambda \\ &= \vee\{A(x^2) \vee \lambda \mid x \in S, f(x) = x'\} \\ &\geq \vee\{A(x) \wedge \mu \mid x \in S, f(x) = x'\} \\ &= f(A)(x') \wedge \mu. \end{aligned}$$

Therefore $f(A)$ is an $(\in, \in \vee q_{(\lambda,\mu)})$-fuzzy completely semiprime ideal of S'.

(ii) For all $x, y \in S$, $f^{-1}(B)(x^2) \vee \lambda = B(f(x^2)) \vee \lambda = B(f(x)^2) \vee \lambda \geq B(f(x)) \wedge \mu$. So $f^{-1}(B)$ is an $(\in, \in \vee q_{(\lambda,\mu)})$-fuzzy completely semiprime ideal of S.

4 Conclusion

In the study of fuzzy algebraic system, we notice that fuzzy ideals with special properties always play an important role. In this paper, we give the new definition of $(\in, \in \vee q_{(\lambda,\mu)})$-fuzzy completely semiprime ideals of semigroups. Using inequalities, characteristic functions and level sets, we consider its equivalent descriptions. Apart from those, the properties of the union, intersection, homomorphic image and homomorphic preimage of $(\in, \in \vee q_{(\lambda,\mu)})$-fuzzy completely semiprime ideal of semigroups are investigated. Those results extend the corresponding theories of fuzzy completely semiprime ideals and enrich the study of fuzzy algebra. At present, although a series of work on this aspect have been done, there is much room for further study.

Acknowledgments This work is supported by Program for Innovative Research Team of Jiangnan University(No:200902).

References

1. Zadeh, L.A.: Fuzzy sets. Inf. Control **8**, 338–353 (1965)
2. Rosenfeld, A.: Fuzzy groups. J. Math. Anal. Appl. **35**, 512–517 (1971)
3. Bhakat, S.K., Das, P.: (α, β)-fuzzy mapping. Fuzzy Sets Syst. **56**, 89–95 (1993)
4. Bhakat, S.K.: $(\in, \in \vee q)$-level subsets. Fuzzy Sets Syst. **103**, 529–533 (1999)
5. Bhakat, S.K., Das, P.: On the definition of a fuzzy subgroup. Fuzzy Sets Syst. **51**, 235–241 (1992)
6. Bhakat, S.K., Das, P.: $(\in, \in \vee q)$-fuzzy subgroups. Fuzzy Sets Syst. **80**, 359–368 (1996)
7. Pu, P.M., Liu, Y.M.: Fuzzy topologyI: neighbourhood structure of a fuzzy point and Moore-Smith convergence. J. Math. Anal. Appl. **76**, 571–599 (1980)
8. Zhan, J.M., Ma, X.L.: On fuzzy interior ideals in semigroups. J. Math. Res. Exposition **28**, 103–110 (2008)
9. Jun, Y.B., Song, S.Z.: Generalized fuzzy interior ideals in semigroups. Inf. Sci. **176**, 3079–3093 (2006)
10. Davvaz, B.: $(\in, \in \vee q)$-fuzzy subnear-rings and ideals. Soft Comput. -A Fusion Found. Metodologies Appli. **10**, 206–211 (2006)
11. Davvaz, B., Corsini, P.: Redefined fuzzy H_v-submodules and many valued implication. Inf. Sci. **177**, 865–875 (2007)
12. Davvaz, B.: Fuzzy R-subgroups with thresholds of near rings ang implication oqerations. Soft Comput. A Fusion Found. Metodologies Appli. **12**, 875–979 (2008)
13. Yuan, X.H., Zhang, C., Ren, Y.H.: Generalized fuzzy subgroups and many-valued implications. Fuzzy Sets Syst. **138**, 205–211 (2003)

14. Liao, Z.H., Chen, Y.X., Lu, J.H., Gu, H.: Generalized fuzzy subsemigroup and generalized fuzzy completely regular semigroup. Fuzzy Syst. Math. 18, 81–84 (2004)
15. Chen, M., Liao, Z.H.: ($\in, \in \vee q_{(\lambda,\mu)}$)-fuzzy subsemigroups and ($\in, \in \vee q_{(\lambda,\mu)}$)-fuzzy ideals of semigroups. In:The sixth International Conference on Information and Management Sciences, Springer, Lhasa, Tibet, pp. 575–579 (2007)
16. Du, J., Liao, Z.H.: ($\in, \in \vee q_{(\lambda_1,\lambda_2)}$)-generalized fuzzy ideal of BCK-Algebra. In: IEEE FSKD 2007, 1,294–299(2007)
17. Liao, Z.H., Gu, H.: ($\in, \in \vee q_{(\lambda,\mu)}$)-fuzzy normal subgroup. Fuzzy Syst. Math. **20**, 47–53 (2006)
18. Zhou, J., Liao, Z.H.: Generalized (α, β)-convex fuzzy cones. In:The 2008 IEEE Internation Conference on Fuzzy Systems, Hongkong, pp. 439–442 (2008)
19. Kuroki N.: On fuzzy semigroups. Information Sciences, 53, 203–236 (1991)
20. Kuroki, N.: Fuzzy generalized bi-ideals in semigroups. Inf. sci. **66**, 235–243 (1992)
21. Kuroki, N.: Fuzzy semiprime quasi-ideals in semigroups. Inf. Sci. **75**, 202–211 (1993)
22. Mo, Z.W.: Wang X.P.:Fuzzy ideals generated by fuzzy sets in semigroups. Inf. Sci. **86**, 203–210 (1995)
23. Bhakat, S.K., Das, P.: Fuzzy subrings and ideals refined. Fuzzy Sets Syst. **81**, 383–393 (1996)
24. Jun, Y.B., Neggers, J., Kim, H.S.: On L-fuzzy ideals in semirings I. Czechoslovak Math. J. **48**, 669–675 (1998)
25. Jun, Y.B., Öztürk, M.A., Song, S.Z.: On fuzzy h-ideals in hemirings. Inf. Sci. **162**, 211–226 (2004)
26. Jun, Y.B., Ahn, S.S., Kim, H.S.: Fuzzy subclines(ideals) of incline algebras. Fuzzy Sets Syst. **123**, 217–225 (2001)
27. Babushri, S.k., Syam, P.K., Satyanarayana B.: Equivalent, 3-prime and c-prime fuzzy ideals of nearrings. Soft Computing-A Fusion of foundations, Methodologies and Applications, 13, 933–944 (2009)

Duality Theorem and Model Based on Fuzzy Inequality

Xin Liu

Abstract This article reports a study on duality theorem and model with fuzzy approaches. The study focuses on the economical interpretation of duality theorem as well as on solving duality problems in a fuzzy mathematical perspective. Besides the regular duality concepts, this article puts forward the methods of drawing non-symmetric fuzzy duality programming from that of symmetric fuzzy duality and drawing symmetric fuzzy duality programming from that of non-symmetric fuzzy duality. It sums up the general rules of forming fuzzy duality programming and proves symmetric duality theorem of fuzzy inequality type.

Keywords Fuzzy linear programming · Dual fuzzy theorem · Symmetric fuzzy duality model · Non-symmetric fuzzy duality model

1 Introduction

Fuzzy linear programming is such problems which objective function is linear and constraints are "linear approximation" [1]. As in the actual mathematical programming problems, most of the constraints or objectives are vague and, therefore, we defined those problems which contained fuzzy constraints and objectives as fuzzy linear programming. Recently, with the efforts of many scholars, fuzzy programming, especially the fuzzy linear programming, have developed rapidly and a range of methods for solving fuzzy linear programming have also been obtained [2]. But

Foundation Item: NSFC:71201019 Author: Liu Xin (1961-), female(Fuxin, Liaoning), professor, mainly work on the study of fuzzy logic and economy optimization.

X. Liu (✉)
School of Mathematics and Quantitative Economics,
Dongbei University of Finance and Economics, Dalian 116025, China
e-mail: liuxin060@dufe.edu.cn

B.-Y. Cao and H. Nasseri (eds.), *Fuzzy Information & Engineering and Operations Research & Management*, Advances in Intelligent Systems and Computing 211,
DOI: 10.1007/978-3-642-38667-1_19, © Springer-Verlag Berlin Heidelberg 2014

there are few studies of duality theorem and model with fuzzy approach, especially those contained inequality. So it caused many scholars to pay a great deal of attention to this issues [3].

Common fuzzy linear programming model can be summarized into the following types [4]:
1.1FLP(I): Classic coefficients type, that is A, b, C constant type.
1.1.1FLP(I-a): $\leq$ Fuzzy type (fuzzy type$\leq$)

$$\max Z = CX$$
$$s.t. \begin{cases} AX \lesssim b \\ X \geq 0 \end{cases}$$

1.1.2 FLP(I-b): Fuzzy objective and fuzzy type $\leq$

$$m\tilde{a}x Z = CX$$
$$s.t. \begin{cases} AX \lesssim b \\ X \geq 0 \end{cases}$$

1.2 FLP(II): Fuzzy coefficients type, that is A, b, C fuzzy type
1.2.1 FLP(II-a): Fuzzy right-side coefficients, that is $\tilde{b}$ type

$$\max Z = CX$$
$$s.t. \begin{cases} AX \leq \tilde{b} \\ X \geq 0 \end{cases}$$

1.2.2 FLP(II-b): Fuzzy objective function, that is $\tilde{C}$ type

$$\max Z = \tilde{C}X$$
$$s.t. \begin{cases} AX \leq b \\ X \geq 0 \end{cases}$$

1.2.3 FLP(II-c): Fuzzy constraint coefficients, that is $\tilde{A}, \tilde{b}$ type

$$\max Z = CX$$
$$s.t. \begin{cases} \tilde{A}X \leq \tilde{b} \\ X \geq 0 \end{cases}$$

1.2.4 FLP(II-d): All coefficients are fuzzy, that is $\tilde{A}, \tilde{b}, \tilde{C}$ type

$$\max Z = \tilde{C}X$$
$$s.t. \begin{cases} \tilde{A}X \leq \tilde{b} \\ X \geq 0 \end{cases}$$

where $\tilde{C} = (\tilde{c}_1, \tilde{c}_2, \ldots, \tilde{c}_n)$ $\tilde{A} = (\tilde{a}_{ij})_{m\times n}$ $\tilde{b} = (\tilde{b}_1, \tilde{b}_2, \ldots, \tilde{b}_m)^{\mathrm{T}}$

We can use different methods to study the above fuzzy linear programming model from the symmetric and non-symmetric. This paper only focuses on those programming models which contain fuzzy inequalities and puts forward its dual models and the related theorems.

2 The Proposal of Linear Programming Dual Problem of Fuzzy Inequality type

E.g. Machine A can run a monthly maximum of about 400 hours while Machiine B can run a monthly maximum of about 250 h , An hour's work of A cost (maintenance, depreciation, etc.) 3 yuan whit a net profit of 7 yuan, but B costs 2 yuan, whit a net profit of 3 yuan. A and B's cost per month may not exceed the sum of 1,500 yuan. How to organize production in order to gain the maximum profit?

x_1 and x_2 are the run hours of A and B respectively [5], and this is fuzzy linear programming problem

$$\begin{cases} \max f = 7x_1 + 3x_2, & \\ 3x_1 + 2x_2 \tilde{\leq} 1500, & (1_1) \\ x_1 \tilde{\leq} 400, & (1_2) \\ x_2 \tilde{\leq} 250, & (1_3) \\ x_1 \geq 0, x_2 \geq 0. & \end{cases} \tag{1}$$

the flexible targets of the corresponding constraints (1_1), (1_2), (1_3) are 50 (yuan) and 5(hours), 5(hours) respectively. Now, we discuss this problem from another side. If the factory does not arrange for the machine to produce, but rent it, how to price can make the factory profit higher than the production gains and how can they make the pricing competitive?

Assumes that rental machines designed to run about 1 hour, renting Machine A priced at 2 yuan per working hour, B is 3 yuan, the profit under these pricing arrangements should not be less than the gains from production, otherwise, the factory would produce rather than to rent. Thus, the cost of this factory renting A and working hours should be about the total value of not less than 7 yuan, in mathematical terms:

$$3y_1 + y_2 \tilde{\geq} 7$$

for B

$$2y_1 + y_3 \tilde{\geq} 3$$

and the rental income of two machines

$$w = 1500y_1 + 400y_2 + 250y_3$$

Table 1 The relationship between food amount and nutrients

Nutrients		Food				
		x_1		x_2	$\ldots x_n$	
		1		2	$\ldots n$	Minimum requirements of various nutrients
y_1	1	a_{11}		a_{12}	$\ldots a_{1n}$	b_1
y_2	2	a_{21}		a_{22}	$\ldots a_{2n}$	b_2
…	…		…	…		…
y_m	m	a_{m1}		a_{m2}	$\ldots a_{mm}$	b_m
Unit price		c_1		c_2	$\ldots c_n$	

For the decision-makers, of course, the bigger w is, the better. But for the receiver, the less his pay is, the better, therefore, in order to be competitive pricing, decision-makers can only make the total income as small as possible, under the condition of not less than the profits of all products, so the objective function can describe as follow:

$$\min w = 1500y_1 + 400y_2 + 250y_3$$

Linking the above expression

$$\begin{cases} \min w = 1500y_1 + 400y_2 + 250y_3 \\ 3y_1 + y_2 \tilde{\geq} 7 & (2_1) \\ 2y_1 + y_3 \tilde{\geq} 3 & (2_2) \\ y_1 \geq 0, y_2 \geq 0, y_3 \geq 0 \end{cases} \quad (2)$$

the flexible targets of the corresponding constraints (2_1), (2_2) are $7/30$ (yuan) and $1/10$ (yuan) respectively.

In general, such as nutrition, there are n kinds of food, each containing m kinds of nutrients, a_{ij} represents the nutrient i of each unit j, b_i represents the minimum amount that each person in need of $i$nutrients per day, c_j is the price of j, how should consumers buy food in order to satisfy their basic needs at the minimum cost? $x_j (j = 1, 2, \ldots, n)$ is the amount of j, and their relationship show in Table 1 [6].

In general, one can only roughly estimate the need for nutrients, therefore, we obtain the following forms of linear programming problems in the fuzzy constraint conditions.

$$\begin{cases} \min CX \\ AX \tilde{\geq} b \quad (L) \\ X \geq 0 \end{cases}$$

$$\text{where } A = \begin{bmatrix} a_{11} & a_{12} & \cdots & a_{1n} \\ a_{21} & a_{22} & \cdots & a_{2n} \\ \cdots & \cdots & \cdots & \cdots \\ a_{m1} & a_{m2} & \cdots & a_{mn} \end{bmatrix}, b = \begin{bmatrix} b_1 \\ b_2 \\ \cdots \\ b_m \end{bmatrix},$$

$$C = \begin{bmatrix} c_1 \ c_2 \ \dots \ c_n \end{bmatrix}, X = \begin{bmatrix} x_1 \\ x_2 \\ \dots \\ x_n \end{bmatrix}.$$

the flexible targets of the corresponding constraints $AX \tilde{\geq} b$ may

$$d = (d_1, d_2, \dots, d_m)^{\mathrm{T}} \geq 0$$

Now we raise a question from another side Manufacturer A want to produce m kinds of pills to replace the above-mentioned n kinds of food, then how to determine the price of each pill in order to benefit the most?
We still can use the above table, assuming y_i is the price of pill i, and $Y = [\,y_1 \ y_2 \ \cdots \ y_m\,]$, in order to more sales, the price of pills generally should not exceed the price of food equivalent. So, there should be $YA \tilde{\leq} C$, and this problem becomes

$$\begin{cases} \max Yb \\ YA \tilde{\leq} C \quad (D) \\ Y \geq 0 \end{cases}$$

The flexible targets of the corresponding constraints

$$YA \tilde{\leq} C \quad \text{and} \quad d' = (d'_1, d'_2, \dots, d'_m)^{\mathrm{T}} \geq 0$$

Accordingly, we have the following definition:
We defined those fuzzy linear programming problems as (L) and (D) are required as mutually dual problem . If one is called the original problem, the other is called the dual problem, also known as a group of symmetric fuzzy dual programming.
Their relationship can be described in Table 2.

Table 2 The relationship of fuzzy dual programming

y_i	x_j					
	x_1	x_2	$\dots$	x_n	Original relation	$\max Yb$
y_1	a_{11}	a_{12}	$\dots$	a_{1n}	$\tilde{\geq}$	b_1
y_2	a_{21}	a_{22}	$\dots$	a_{2n}	$\tilde{\geq}$	b_2
$\dots$						
y_m	$\dots$	$\dots$	$\dots$	$\dots$	$\dots$	
	a_{m1}	a_{m2}	$\dots$	a_{mm}	$\tilde{\geq}$	b_m
Duality relation	$\tilde{\leq}$	$\tilde{\leq}$	$\dots$	$\tilde{\leq}$		
Min CX	c_1	c_2	$\dots$	c_n		

3 The Relationship Between Symmetric and Non-Symmetric Fuzzy Dual Programming

3.1 Symmetric Fuzzy Dual Programming Models

$$\text{We defined} \quad \begin{cases} \min CX \\ AX \tilde{\geq} b \quad (L) \\ X \geq 0 \end{cases} \quad \text{and} \quad \begin{cases} \max Yb \\ YA \tilde{\leq} C \quad (D) \\ Y \geq 0 \end{cases}$$

as symmetric fuzzy dual programming.

3.2 Non-Symmetric Fuzzy Dual Programming Models

Generally in fuzzy linear programming, when we experience constraints with fuzzy equations constraints (adjust the parameters and make it into equations), its fuzzy linear programming takes the following form:

$$\begin{cases} \min CX \\ AX \approx b \\ X \geq 0 \end{cases}$$

Then what about duality?
We defined

$$(L') \quad \begin{cases} \min CX \\ AX \approx b \\ X \geq 0 \end{cases} \quad \text{and} \quad (D') \quad \begin{cases} \max Yb \\ YA \tilde{\leq} C \end{cases}$$

as a group of non- symmetric fuzzy dual programming. Symmetric and non-symmetric fuzzy dual programming is introduced to each other, and it can be demonstrated as follows.

3.3 Reduced Non-Symmetric Fuzzy Dual Programming Models from Symmetric Fuzzy Dual Programming Models

Considering

$$AX \approx b \Leftrightarrow \begin{cases} AX \tilde{\leq} b \\ AX \tilde{\geq} b \end{cases}$$

now (L') becomes

$$\begin{cases} \min CX \\ \begin{pmatrix} A \\ -A \end{pmatrix} X \tilde{\geq} \begin{pmatrix} b \\ -b \end{pmatrix} \\ X \geq 0 \end{cases}$$

and this is (L) style, according to the definition, its fuzzy dual programming models can be described as follows:

$$\begin{cases} \max\,(Y_1, Y_2) \begin{pmatrix} b \\ -b \end{pmatrix} \\ (Y_1, Y_2) \begin{pmatrix} A \\ -A \end{pmatrix} \tilde{\leq} C \\ (Y_1, Y_2) \geq 0 \end{cases}$$

alternatively, it can be expressed as:

$$\begin{cases} \max(Y_1 - Y_2)b \\ (Y_1 - Y_2)A \tilde{\leq} C \\ (Y_1, Y_2) \geq 0 \end{cases}$$

We defined $Y^* = Y_1 - Y_2$, and there has no non-negative restrictions for W^*, the above expression becomes:

$$\begin{cases} \max Y^* b \\ Y^* A \tilde{\leq} C \end{cases}$$

and this is the form of (D'), it shows that the fuzzy dual programming of (L') is (D').

3.4 Reduced Symmetric Fuzzy Dual Programming Models from Non-Symmetric Fuzzy Dual Programming ones

If the duality of (L') is (D'), we also can certify that the duality of (L) is (D).

For (L), we introduce the remaining variables W, $W = (w_1, w_2, \ldots, w_m)^{\mathrm{T}} \geq 0$, the form of (L) becomes:

$$\begin{cases} \min(C\ 0) \begin{pmatrix} X \\ W \end{pmatrix} \\ (A\ -I) \begin{pmatrix} X \\ W \end{pmatrix} \approx b \\ X \geq 0,\ W \geq 0 \end{cases}$$

the form of its duality:

$$\begin{cases} \max Yb \\ Y(A\ -I) \tilde{\leq} (C\ 0) \end{cases}$$

it can also be expressed as:

$$\begin{cases} \max Yb \\ YA \tilde{\lesssim} C \\ Y \geq 0 \end{cases}$$

and this is (D).

4 Mixed Symmetric and Non-Symmetric Dual Programming Model of the Fuzzy

Except the symmetric and non-symmetric fuzzy dual programming, there is another form which mixed the two, for example, we assume that:

$$A = \begin{bmatrix} A_{11} & A_{12} \\ A_{21} & A_{22} \end{bmatrix}, X = \begin{pmatrix} X^1 \\ X^2 \end{pmatrix}$$

A_{ij} is an m_i by m matrix, $i, j = 1, 2$. $m_1 + m_2 = m$, $n_1 + n_2 = n$, X^1 and X^2 the n_1 and n_2 dimensional vector respectively.

If the original problem is

$$\begin{cases} \min(C^1X^1 + C^2X^2) \\ A_{11}X^1 + A_{12}X^2 \tilde{\gtrsim} b^1 \\ A_{21}X^1 + A_{22}X^2 \approx b^2 \\ X^1 \geq 0 \end{cases}$$

where X^2 has no restricts, it's duality is

$$\begin{cases} \max(Y^1b^1 + Y^2b^2) \\ Y^1A_{11} + Y^2A_{12} \tilde{\lesssim} C^1 \\ Y^1A_{21} + Y^2A_{22} \approx C^2 \\ Y^1 \geq 0 \end{cases}$$

and Y^2 also has no limits.

5 The General Rules of Constituting the Dual Programming Using Fuzzy Inequality

In summary, the rules of constituting a general fuzzy dual programming can be summarized as follows:

1. When constraints in the original problems is unified into $\tilde{\geq}$ *and* $\approx$ (or $\tilde{\leq}$ *and* $\approx$), the objective function is minimize (or maximum).
2. A line constraint of the original problems corresponds to a variable y_i, where $y_i \geq 0$, if the line constraint is fuzzy inequality, otherwise, y_i has no sign restrictions.
3. Each variable of the original problem, x_j corresponds to a line constraint, where $\sum_{i=1}^{n} y_i a_{ij} \tilde{\leq} c_j$(or $\sum_{i=1}^{n} y_j a_{ij} \tilde{\geq} c_j$), if $x_j \geq 0$(if $x_j \leq 0$, we can rewrite it to $x_j = -x_j'$); and $\sum_{i=1}^{n} y_j a_{ij} \approx c_j$ (c_j is the coefficient of the objective function in original problem), if x_j has no restrictions.
4. If the Original objective function CX is minimum requirement, then it corresponds to an objective function with maximum requirement, Yb. (where b is the column of constant in the original problem constraints.)
   e.g. the original problem

$$\begin{cases} x_1 + 2x_2 - x_3 - x_4 \approx -7 \\ 6x_1 - 3x_2 + x_3 - 7x_4 \tilde{\geq} 14 \\ -28x_1 - 17x_2 + 4x_3 + 2x_4 \tilde{\leq} -3 \\ x_1 \geq 0, x_2 \geq 0 \\ \min(5x_1 - 6x_2 + 7x_3 + 4x_4) \end{cases}$$

where x_3 and x_4 has no restrictions, calculate its duality.

Unified the fuzzy in equality into $\tilde{\geq}$, and then described it in the Table 3.

So we can obtained its duality directly:

$$\begin{cases} y_1 + 6y_2 + 28y_3 \tilde{\leq} 5 \\ 2y_1 - 3y_2 + 17y_3 \tilde{\leq} -6 \\ -y_1 + y_2 - 4y_3 \approx 7 \\ -y_1 - 7y_2 - 2y_3 \approx 4 \\ y_2, y_3 \geq 0 \\ \max(-7y_1 + 14y_2 + 3y_3) \end{cases}$$

where y_1 has no restrictions.

Table 3 Dual programming of fuzzy inequality

	x_1	x_2	x_3	x_4	
y_1	1	2	−1	−1	≈ -7
y_2	6	−3	1	−7	$\tilde{\geq}$ 14
y_3	28	17	−4	−2	$\tilde{\geq}$ 3
	5	−6	7	4	$= \min f$
	$x_1 \geq 0, x_2 \geq 0, x_3, x_4$ with no restrictions				

6 Symmetric Theorem of DFLP

(L) is a FLP problem, and it duality is (D), then we know that (L) is the duality of (D).

And this can be demonstrated as follows:

Assumes that: the original problem (L):

$$\begin{cases} \min W = CX \\ AX \tilde{\geq} b \\ X \geq 0 \end{cases}$$

and its duality (D):

$$\begin{cases} \max Z = Yb \\ YA \tilde{\leq} C \\ Y \geq 0 \end{cases}$$

both sides of this expression becomes negative, knowing that $\max Z = -\min(-W)$, this expression becomes:

$$\begin{cases} \min(-W) = -Yb \\ -YA \tilde{\geq} -C \\ Y \geq 0 \end{cases}$$

according to the symmetry transformation, we obtained its duality as follows:

$$\begin{cases} \max(-W') = -CX \\ -AX \tilde{\leq} -b \\ X \geq 0 \end{cases}$$

and because $\max(-W') = -\min(W')$, so that $Z = W'$, and it becomes:

$$\begin{cases} -\min(W') = -\min Z = -CX \\ AX \tilde{\geq} b \\ X \geq 0 \end{cases}$$

this expression can also be expressed as:

$$\begin{cases} \min Z = CX \\ AX \tilde{\geq} b \\ X \geq 0 \end{cases}$$

this is just the original problem.

7 Conclusions

This article summarized some common models of the fuzzy linear programming [7], and then studied the fuzzy inequality-type linear programming problem of duality theory and its models; given a general definition and the economic interpretation of inequality-type fuzzy linear programming dual problem; constructs the symmetric and non-symmetric model of the fuzzy dual problem; summarized the general rule of how to construct a fuzzy dual programming model; and proved the symmetry duality theorem for the fuzzy inequality-type.

There are many other theorem about fuzzy duality theory. Taking into account the length of this paper, the remaining contents will be described in other papers.

References

1. hen, S., Li, J., Wang X.: Fuzzy Set Theory and its applications Beijing, China, Science Press, (2004)
2. Li, A., Zhang, Z,, etc., Fuzzy Mathematics and its applications, Beijing, China, Metallurgical Industry Press, (1994)
3. Cao, Bingyuan: Appli. Fuzzy Math. Syst. Science Press, Beijing, China (2005)
4. Masahiro Inuiguchi, Jaroslar Ramik, Tetsuzo Tanino, ctal. Satisfying solutions and duality in interval and fuzzy linear programming. Fuzzy sets syst. 135, 151–177 (2003)
5. Bactor, C.R., Chandra S.: On duality in linear programming under fuzzy environment. Fuzzy sets and systems, 125, pp. 317–325 (2002)
6. Xia, S.: Operations Research, Beiing, China, QingHua University Press, (2005)
7. Li, C., Xu, X., Zhan, D.: Joint replenishment Problem with fuzzy resource constraint, Comput. Int. Manuf. Syst. vol. 14 No.1 pp. 113–117 (2008)
8. Li, F., Jin, C., Liu, L,. Quasi-linear fuzzy number and its application in fuzzy programming, Syst. Eng.-Theory Practice. Vol. 29, No, 4. pp. 120–127(2009)
9. Andre, L., Walmir, C., Femando Gomide. Multivariable gaussian evolving fuzzy modeling system, IEEE Trans. Fuzzy Syst. 19(1). pp. 91-104. (2011)

$(\in, \in \vee q_{(\lambda,\mu)})$-Fuzzy Completely Prime Ideals of Semigroups

Zu-hua Liao, Yi-xuan Cao, Li-hua Yi, Ying-ying Fan and Zhen-yu Liao

Abstract We introduce a new kind of generalized fuzzy completely prime ideal of a semigroup called $(\in, \in \vee q_{(\lambda,\mu)})$-fuzzy completely prime ideals. These generalized fuzzy completely prime ideals are characterized. We also discuss the equivalence relationship between $(\in, \in \vee q_{(\lambda,\mu)})$-fuzzy completely prime ideals and $A_{t \vee q_{(\lambda,\mu)}}$.

Keywords $(\in, \in \vee q_{(\lambda,\mu)})$-fuzzy completely prime ideals · Completely prime ideals · Level subsets · Homomorphism

1 Introduction

Fuzzy semigroup theory plays a prominent role in mathematics with wide applications in many disciplines such as control engineering, information sciences, fuzzy coding theory, fuzzy finite state machines, fuzzy automata and fuzzy languages.

Using the notion of a fuzzy set introduced by Zadeh [1] in 1965, which laid the foundation of fuzzy set theory, Rosenfeld [2] inspired the fuzzification of algebraic structures and introduced the notion of fuzzy subgroups. Since then fuzzy algebra came into being. Bhakat and Das gave the concepts of fuzzy subgroups by using the "belongs to" relation ($\in$) and "quasi-coincident with" relation (q) between a fuzzy point and a fuzzy set, and introduced the concept of a $(\in, \in \vee q)$-fuzzy subgroup [3–6]. It is worth to point out that the ideal of quasi-coincident of a fuzzy point with a fuzzy set, which is mentioned in [7], played a vital role to generate some different

Z. Liao (✉) · Y. Fan · L. Yi
School of Science, Jiangnan University, Wuxi 214122, China
e-mail: liaozuhua57@yahoo.cn

Z. Liao
Department of Mathematics, State University of New York at Buffalo, NY 14260, USA

Z. Liao · Y. Cao
Engineering of Internet of Things, Jiangnan University, Wuxi 214122, China

B.-Y. Cao and H. Nasseri (eds.), *Fuzzy Information & Engineering and Operations Research & Management*, Advances in Intelligent Systems and Computing 211,
DOI: 10.1007/978-3-642-38667-1_20,

types of fuzzy subgroups. In particular, $(\in, \in \vee q)$-fuzzy subgroup is an important and useful generalization of Rosenfeld's fuzzy subgroup, which provides sufficient motivation to researchers to review various concepts and results from the realm of abstract algebra in the broader framework of fuzzy setting. Zhanjianming [8], Young Bae Jun [9] et al. introduced the notion of $(\in, \in \vee q)$-fuzzy interior ideals of a semigroup. Bavvaz [10–12] defined $(\in, \in \vee q)$-fuzzy subnear-rings and characterized H_ν-fuzzy submodules, R-fuzzy semigroups using the relation $(\in, \in \vee q)$.

Later, the definition of a generalized fuzzy subgroup was introduced by Yuan Xuehai [13]. Based on it, Liao Zuhua [14] expanded common "quasi-coincident with" relationship to generalized "quasi-coincident with" relationship, which is the generalization of Rosenfeld's fuzzy algebra and Bhakat and Das's fuzzy algebra. And a series of results were obtained by using generalized "quasi-coincident with" relationship [15–18]. When $\lambda = 0$ and $\mu = 1$ we get common fuzzy algebra by Rosenfeld and when $\lambda = 0$ and $\mu = 0.5$ we get the $(\in, \in \vee q)$-fuzzy algebra defined by Bhakat and Das.

The concept of a fuzzy ideal in semigroups was developed by Kuroki. He studied fuzzy ideals, fuzzy bi-ideals and fuzzy semiprime ideals in semigroups [19–21]. Fuzzy ideals, generated by fuzzy sets in semigroups, were considered by Mo and Wang [22]. After that Bhakat and Das [23] investigated fuzzy subrings and several types of ideals, including fuzzy prime ideals and $(\in, \in \vee q)$-fuzzy prime ideals. Young Bae Jun et al. [24–26] studied L-fuzzy ideals in semigroups, fuzzy h-ideals in hemirings, fuzzy ideals in inclines. Besides, Bahushri [27] did some research on c-prime fuzzy ideals in nearrings. It is now natural to investigate similar type of generalization of the existing fuzzy subsystems of some algebraic structures. Our aim in this paper is to introduce and study $(\in, \in \vee q_{(\lambda,\mu)})$-fuzzy completely prime ideals, and obtain some properties: an $(\in, \in \vee q_{(\lambda,\mu)})$-fuzzy ideal is an $(\in, \in \vee q_{(\lambda,\mu)})$-fuzzy completely prime ideal, if and only if $A_t (\neq \emptyset)$ is a completely prime ideal, $\forall t \in (\lambda, \mu]$. This showed that $(\in, \in \vee q_{(\lambda,\mu)})$-fuzzy completely prime ideals are generalizations of the existing concepts of two types of fuzzy ideals. We also discuss the equivalence relationship between $(\in, \in \vee q_{(\lambda,\mu)})$-fuzzy completely prime ideals and $A_{t \vee q_{(\lambda,\mu)}}$.

2 Preliminaries

Throughout the paper we always consider S as a semigroup.

A mapping from A to $[0, 1]$ is said to be a fuzzy subset of S.

A fuzzy subset A of S of the form $A(y) = \begin{cases} \lambda(\neq 0), & y = x \\ 0, & y \neq x \end{cases}$ is said to be a fuzzy point supporting x and value λ, denoted by x_t.

Definition 2.1 *Let A be a fuzzy subset of S, for all $t, \lambda, \mu \in [0, 1]$ and $\lambda < \mu$, a fuzzy point x_t is called belonging to A if $A(x) \geq t$, denoted by $x_t \in A$; A fuzzy point*

x_t is said to be generalized quasi-coincident with A if $t > \lambda$ and $A(x) + t > 2\mu$, denoted by $x_t q_{(\lambda,\mu)} A$. If $x_t \in A$ or $x_t q_{(\lambda,\mu)} A$, then denoted by $x_t \in \vee q_{(\lambda,\mu)} A$.

Definition 2.2 [15] *A fuzzy subset A of S is said to be an $(\in, \in \vee q_{(\lambda,\mu)})$-fuzzy subsemigroup if for all $x, y \in S, t_1, t_2 \in (\lambda, 1]$, $x_{t_1}, y_{t_2} \in A$ implies $(xy)_{t_1 \wedge t_2} \in \vee q_{(\lambda,\mu)} A$.*

Theorem 2.3 [15] *A fuzzy subset A of S is an $(\in, \in \vee q_{(\lambda,\mu)})$-fuzzy subsemigroup if and only if $A(xy) \vee \lambda \geq A(x) \wedge A(y) \wedge \mu$, for all $x, y \in S$.*

Definition 2.4 [15] *A fuzzy subset A of S is an $(\in, \in \vee q_{(\lambda,\mu)})$-fuzzy left (right) ideal if (1) A is an $(\in, \in \vee q_{(\lambda,\mu)})$-fuzzy subsemigroup of S;*
(2) For all $x_t \in A$, $y \in S$, implies $(yx)_t \in \vee q_{(\lambda,\mu)} A$ $((xy)_t \in \vee q_{(\lambda,\mu)} A)$.
If A is both an $(\in, \in \vee q_{(\lambda,\mu)})$-fuzzy left ideal and an $(\in, \in \vee q_{(\lambda,\mu)})$-fuzzy right ideal, then A is said to be an $(\in, \in \vee q_{(\lambda,\mu)})$-fuzzy ideal.

Theorem 2.5 [15] *A fuzzy subset A of S is an $(\in, \in \vee q_{(\lambda,\mu)})$-fuzzy ideal if and only if for all $t \in (\lambda, \mu]$, the non-empty set A_t is an ideal.*

Theorem 2.6 [15] *A fuzzy subset A of S is an $(\in, \in \vee q_{(\lambda,\mu)})$-fuzzy left (right) ideal if and only if for all $x, y \in S$, (1) $A(xy) \vee \lambda \geq A(x) \wedge A(y) \wedge \mu$; (2) $A(xy) \vee \lambda \geq A(y) \wedge \mu (A(x) \wedge \mu)$.*

Definition 2.7 *An ideal I of S is said to be a completely prime ideal, if for all $x, y \in S, xy \in I$ implies $x \in I$ or $y \in I$.*

Based on [23], in a semigroup we have the following definitions and theorems:

Definition 2.8 *A fuzzy ideal A of S is said to be a fuzzy completely prime ideal, if for all $x, y \in S, t \in (0, 1]$, $(xy)_t \in A$ implies $x_t \in A$ or $y_t \in A$.*

Theorem 2.9 *A fuzzy ideal A of S is a fuzzy completely prime, if and only if $A(xy) = A(x)$ or $A(xy) = A(y)$, for all $x, y \in S$.*

Definition 2.10 *An $(\in, \in \vee q)$-fuzzy ideal A of S is said to be an $(\in, \in \vee q)$-fuzzy completely prime ideal, if for all $x, y \in S, t \in (0, 1]$, $(xy)_t \in A$ implies $x_t \in \vee q A$ or $y_t \in \vee q A$.*

Theorem 2.11 *An $(\in, \in \vee q)$-fuzzy ideal of S is an $(\in, \in \vee q)$-fuzzy completely prime ideal if and only if $A(x) \vee A(y) \geq A(xy) \wedge 0.5$, for all $x, y \in S$.*

3 $(\in, \in \vee q_{(\lambda,\mu)})$-Fuzzy Completely Prime Ideals

In this section, we give the new definition of an $(\in, \in \vee q_{(\lambda,\mu)})$-fuzzy completely prime ideal of semigroups. Then some equivalent descriptions and properties of it are discussed.

Definition 3.1 *An $(\in, \in \vee q_{(\lambda,\mu)})$-fuzzy ideal A of S is said to be an $(\in, \in \vee q_{(\lambda,\mu)})$-fuzzy completely prime ideal, if for all $x, y \in S, t \in (\lambda, 1], (xy)_t \in A$ implies $x_t \in \vee q_{(\lambda,\mu)} A$ or $y_t \in \vee q_{(\lambda,\mu)} A$.*

Theorem 3.2 *An $(\in, \in \vee q_{(\lambda,\mu)})$-fuzzy ideal of S is an $(\in, \in \vee q_{(\lambda,\mu)})$-fuzzy completely prime ideal if and only if $A(x) \vee A(y) \vee \lambda \geq A(xy) \wedge \mu$, for all $x, y \in S$.*

Proof: $\Rightarrow$ Let A be an $(\in, \in \vee q_{(\lambda,\mu)})$-fuzzy completely prime ideal of S. Assume that there exist x_0, y_0 such that $A(x_0) \vee A(y_0) \vee \lambda < A(x_0y_0) \wedge \mu$. Choose t to satisfy $A(x_0) \vee A(y_0) \vee \lambda < t < A(x_0y_0) \wedge \mu$. Then we have $A(x_0y_0) > t, \lambda < t < \mu, A(x_0) < t, A(y_0) < t, A(x_0) + t < 2\mu$ and $A(y_0) + t < 2\mu$. So $(x_0y_0)_t \in A$. But $(x_0)_t \overline{\in \vee q_{(\lambda,\mu)}} A$ and $(y_0)_t \overline{\in \vee q_{(\lambda,\mu)}} A$, a contradiction.
$\Leftarrow$ For all $x, y \in S, t \in (\lambda, 1)$ and $(xy)_t \in A$, then $A(xy) \geq t$ and $\lambda < t$. So $A(x) \vee A(y) \vee \lambda \geq A(xy) \wedge \mu \geq t \wedge \mu$. Since $\lambda < \mu$, then $A(x) \vee A(y) \geq t \wedge \mu$.

If $t \geq \mu$, then $A(x) \vee A(y) \geq \mu$, so $A(x) \geq \mu$ or $A(y) \geq \mu$.

Case 1: Suppose $A(x) \geq \mu$, we have $A(x) + t > \mu + \mu = 2\mu$, so $x_t q_{(\lambda,\mu)} A$.

Case 2: Suppose $A(y) \geq \mu$, we have $A(y) + t > \mu + \mu = 2\mu$, so $y_t q_{(\lambda,\mu)} A$. Then $x_t q_{(\lambda,\mu)} A$ or $y_t q_{(\lambda,\mu)} A$.

If $t < \mu$, then $A(x) \vee A(y) \geq t$. So $x_t \in A$ or $y_t \in A$.

Hence, $x_t \in \vee q_{(\lambda,\mu)} A$ or $y_t \in \vee q_{(\lambda,\mu)} A$, that is to say, A is an $(\in, \in \vee q_{(\lambda,\mu)})$-fuzzy completely prime ideal.

Theorem 3.3 *A non-empty subset S_1 of S is a completely prime ideal if and only if χ_{S_1} is an $(\in, \in \vee q_{(\lambda,\mu)})$-fuzzy completely prime ideal of S.*

Proof: $\Rightarrow$ Let S_1 be a completely prime ideal, then χ_{S_1} is an $(\in, \in \vee q_{(\lambda,\mu)})$-fuzzy ideal of S. If $(xy)_t \in \chi_{S_1}$, then $\chi_{S_1}(xy) \geq t > 0$, so $xy \in S_1$. Since S_1 is a completely prime ideal, we have $x \in S_1$ or $y \in S_1$, then $x_t \in \vee q_{(\lambda,\mu)} \chi_{S_1}$ or $y_t \in \vee q_{(\lambda,\mu)} \chi_{S_1}$.

Thus χ_{S_1} is an $(\in, \in \vee q_{(\lambda,\mu)})-$fuzzy completely prime ideal.

$\Leftarrow$ Let χ_{S_1} be an $(\in, \in \vee q_{(\lambda,\mu)})$-fuzzy completely prime ideal, then S_1 is an ideal of S. Now $xy \in S_1$, then $\chi_{S_1}(xy) = 1$. Since χ_{S_1} is an $(\in, \in \vee q_{(\lambda,\mu)})$-fuzzy completely prime ideal. By the Theorem 3.2, we have $\chi_{S_1}(x) \vee \chi_{S_1}(y) \vee \lambda \geq \chi_{S_1}(xy) \wedge \mu = \mu$. Then $\chi_{S_1}(x) \vee \chi_{S_1}(y) \geq \mu > 0$. So $\chi_{S_1}(x) = 1$ or $\chi_{S_1}(y) = 1, i.e., x \in S_1$ or $y \in S_1$.

Therefore, S_1 is a completely prime ideal.

Remark When $\lambda = 0, \mu = 1$, we can obtain the corresponding results in the sense of Rosenfeld; When $\lambda = 0, \mu = 0.5$, we can get the corresponding results in the sense of Bhakat and Das.

Theorem 3.4 *An $(\in, \in \vee q_{(\lambda,\mu)})$-fuzzy ideal A of S is an $(\in, \in \vee q_{(\lambda,\mu)})$-fuzzy completely prime ideal if and only if non-empty set A_t is a completely prime ideal for all $t \in (\lambda, \mu]$.*

Proof: $\Rightarrow$ Let A be an $(\in, \in \vee q_{(\lambda,\mu)})$-fuzzy completely prime ideal of S, then the non-empty set A_t is an ideal, for all $t \in (\lambda, \mu]$. Let $xy \in A_t$, then $A(xy) \geq t$. By

Theorem 3.2,we have $A(x) \vee A(y) \vee \lambda \geq A(xy) \wedge \mu \geq t \wedge \mu = t$. So $x \in A_t$ or $y \in A_t$.

Hence A_t is a completely prime ideal.

$\Leftarrow$ Let A_t be a completely prime ideal of S, by Theorem 2.5, we have that A is an $(\in, \in \vee q_{(\lambda,\mu)})$-fuzzy ideal. Suppose that A is not an $(\in, \in \vee q_{(\lambda,\mu)})$-fuzzy completely prime ideal. By Theorem 3.2, there exist x_0, y_0 such that $A(x_0) \vee A(y_0) \vee \lambda < A(x_0y_0) \wedge \mu$. Choose t such that $A(x_0) \vee A(y_0) \vee \lambda < t < A(x_0y_0) \wedge \mu$. Then we have $A(x_0y_0) > t$, $A(x_0) < t$, $A(y_0) < t$, $\lambda < t < \mu$ and $x_0y_0 \in A_t$. Since A_t is a completely prime ideal, we have $x_0 \in A_t$ or $y_0 \in A_t$, a contradiction.

Therefore A is an $(\in, \in \vee q_{(\lambda,\mu)})$-fuzzy completely prime ideal.

Corollary 3.5 A fuzzy set A of S is an $(\in, \in \vee q_{(\lambda,\mu)})$-fuzzy completely prime ideal of S if and only if non-empty set A_t is a completely prime ideal, for all $t \in (\lambda, \mu]$.

Theorem 3.6 *Let I be any completely prime ideal of S. There exists an $(\in, \in \vee q_{(\lambda,\mu)})$-fuzzy completely prime ideal A of S such that $A_t = I$ for some $t \in (\lambda, \mu]$.*

Proof: If we define a fuzzy set in S by

$$A(x) = \begin{cases} t, & \text{if } x \in I \\ 0, & otherwise \end{cases} \quad \text{for some } t \in (\lambda, \mu].$$

Then it follows that $A_t = I$.

For given $r \in (\lambda, \mu]$, we have

$$A_r = \begin{cases} A_t(= I), & \text{if } r \leq t \\ \emptyset, & \text{if } t < r < \mu \end{cases}$$

Since I itself is a completely prime ideal of S, it follows that every non-empty level subset A_r of S is a completely prime ideal of S. By Corollary 3.5, A is an $(\in, \in \vee q_{(\lambda,\mu)})$-fuzzy completely prime ideal of S, which satisfies the conditions of the theorem.

Theorem 3.7 *Let A be an $(\in, \in \vee q_{(\lambda,\mu)})$-fuzzy completely prime ideal of S such that $A(x) \leq \mu$ for all $x \in S$. Then A is a fuzzy completely prime ideal of S.*

Proof: Let $x, y \in S$, $t \in (\lambda, 1]$ and $(xy)_t \in A$. It follows that $x_t \in \vee q_{(\lambda,\mu)} A$ or $y_t \in \vee q_{(\lambda,\mu)} A$ from Definition 3.1.

Without loss of generality ,we assume that $x_t \in \vee q_{(\lambda,\mu)} A$ holds. By known conditions, we have $t \leq A(xy) \leq \mu$ and $t < \mu$. Thus $A(x) + t \leq \mu + \mu = 2\mu$, *i.e.*, $x_t \overline{q_{(\lambda,\mu)}} A$. Hence $x_t \in A$. Therefore A is a fuzzy completely prime ideal of S.

Theorem 3.8 *Let A be an $(\in, \in \vee q_{(\lambda,\mu)})$-fuzzy ideal of S and B be an $(\in, \in \vee q_{(\lambda,\mu)})$-fuzzy completely prime ideal of S. Then $A \cap B$ is an $(\in, \in \vee q_{(\lambda,\mu)})$-fuzzy completely prime ideal of A_μ.*

Proof: Let $x, y \in A_\mu$ and $(xy)_t \in A \cap B$, then $(A \cap B)(xy) \geq t$. So $A(xy) \geq t$ and $B(xy) \geq t$. Thus $(xy)_t \in A$ and $(xy)_t \in B$. Since B is an $(\in, \in \vee q_{(\lambda,\mu)})$-fuzzy completely prime ideal of S, we have $x_t \in \vee q_{(\lambda,\mu)} B$ or $y_t \in \vee q_{(\lambda,\mu)} B$.

Case 1: If $x_t \in \vee q_{(\lambda,\mu)} B$, then $x_t \in B$ or $x_t q_{(\lambda,\mu)} B$.

Assume $x_t \in B$ implies $B(x) \geq t$. If $t \leq \mu$, then $A(x) \geq \mu \geq t$. So $(A \cap B)(x) = A(x) \wedge B(x) \geq t$. Therefore $x_t \in (A \cap B)$ and $x_t \in \vee q_{(\lambda,\mu)}(A \cap B)$; If $t > \mu$, then $A(x) + t > \mu + \mu = 2\mu$ and $B(x) + t \geq t + t > 2\mu$, then $(A \cap B)(x) + t = (A(x) \wedge B(x)) + t > 2\mu$. So $x_t q_{(\lambda,\mu)}(A \cap B)$ and $x_t \in \vee q_{(\lambda,\mu)}(A \cap B)$.

Assume $x_t q_{(\lambda,\mu)} B$ implies $B(x) + t > 2\mu$. If $t \leq \mu$, then $B(x) > 2\mu - t \geq \mu \geq t$ and $A(x) \geq \mu \geq t$, so $x_t \in A$ and $x_t \in B$. Thus $x_t \in A \cap B$ and $x_t \in \vee q_{(\lambda,\mu)}(A \cap B)$. If $t > \mu$, then $A(x) + t > 2\mu$, then $(A \cap B)(x) + t = (A(x) \wedge B(x)) + t > 2\mu$. So $x_t q_{(\lambda,\mu)}(A \cap B)$ and $x_t \in \vee q_{(\lambda,\mu)}(A \cap B)$.

Case 2: Suppose $y_t \in \vee q_{(\lambda,\mu)} B$, then similarly prove $y_t \in \vee q_{(\lambda,\mu)}(A \cap B)$.

Hence $A \cap B$ is an $(\in, \in \vee q_{(\lambda,\mu)})$-fuzzy completely prime ideal of A_μ.

Theorem 3.9 *Let $\{A_i \mid i \in I\}$ be a family of $(\in, \in \vee q_{(\lambda,\mu)})$-fuzzy completely prime ideals of S such that $A_i \subseteq A_j$ or $A_j \subseteq A_i$ for all $i, j \in I$. Then $A = \cap_{i \in I} A_i$ is an $(\in, \in \vee q_{(\lambda,\mu)})$-fuzzy completely prime ideal of S.*

Proof: For any $x, y \in S$,

$$\begin{aligned} A(xy) \vee \lambda &= (\cap_{i \in I} A_i)(xy) \vee \lambda = (\wedge_{i \in I} A_i(xy)) \vee \lambda \\ &= \wedge_{i \in I}(A_i(xy) \vee \lambda) \geq \wedge_{i \in I}(A_i(x) \wedge A_i(y) \wedge \mu) \\ &\geq (\wedge_{i \in I} A_i(x)) \wedge (\wedge_{i \in I} A_i(y)) \wedge \mu = A(x) \wedge A(y) \wedge \mu \\ A(xy) \vee \lambda &= (\cap_{i \in I} A_i)(xy) \vee \lambda = (\wedge_{i \in I} A_i(xy)) \vee \lambda = \wedge_{i \in I}(A_i(xy) \vee \lambda) \\ &\geq \wedge_{i \in I}(A_i(x) \wedge \mu) = (\wedge_{i \in I} A_i(x)) \wedge \mu = A(x) \wedge \mu \end{aligned}$$

Similarly prove $A(xy) \vee \lambda \geq A(y) \wedge \mu, \forall x, y \in S$.

$$\begin{aligned} A(x) \vee A(y) \vee \lambda &= (\cap_{i \in I} A_i)(x) \vee (\cap_{i \in I} A_i)(y) \vee \lambda \\ &= (\wedge_{i \in I} A_i)(x) \vee (\wedge_{i \in I} A_i)(y) \vee \lambda \\ &= \wedge_{i \in I}(A_i(x) \vee A_i(y) \vee \lambda) \qquad (1) \\ &\geq \wedge_{i \in I}(A_i(xy) \wedge \mu) \\ &= A(xy) \wedge \mu \end{aligned}$$

In the following we show that Eq. (1) holds. It is clear that $\wedge_{i \in I}(A_i(x) \vee A_i(y) \vee \lambda) \geq (\wedge_{i \in I} A_i(x)) \vee (\wedge_{i \in I} A_i(y)) \vee \lambda$.

If possible, let $\wedge_{i \in I}(A_i(x) \vee A_i(y) \vee \lambda) > (\wedge_{i \in I} A_i(x)) \vee (\wedge_{i \in I} A_i(y)) \vee \lambda$. Then there exists t such that $\wedge_{i \in I}(A_i(x) \vee A_i(y) \vee \lambda) > t > (\wedge_{i \in I} A_i(x)) \vee (\wedge_{i \in I} A_i(y)) \vee \lambda$. Since $A_i \subseteq A_j$ or $A_j \subseteq A_i$ for all $i, j \in I$, there exists $k \in I$ such that $t > A_k(x) \vee A_k(y) \vee \mu$. On the other hand, $A_i(x) \vee A_i(y) \vee \mu > t$ for all $i \in I$, a contradiction.

Hence, $\wedge_{i\in I}(A_i(x) \vee A_i(y) \vee \lambda) = (\wedge_{i\in I} A_i(x)) \vee (\wedge_{i\in I} A_i(y)) \vee \lambda$.
Here we finish the prove of the theorem.

Theorem 3.10 *Let $\{A_i \mid i \in I\}$ be a family of $(\in, \in \vee q_{(\lambda,\mu)})$-fuzzy completely prime ideals of S such that $A_i \subseteq A_j$ or $A_j \subseteq A_i$ for all $i, j \in I$. Then $A = \cup_{i\in I} A_i$ is an $(\in, \in \vee q_{(\lambda,\mu)})$-fuzzy completely prime ideal of S.*

Proof: For all $x, y \in S$,

$$\begin{aligned} A(xy) \vee \lambda &= (\cup_{i\in I} A_i)(xy) \vee \lambda = (\vee_{i\in I} A_i(xy)) \vee \lambda \\ &= \vee_{i\in I}(A_i(xy) \vee \lambda) \\ &\geq \vee_{i\in I}(A_i(x) \wedge A_i(y) \wedge \mu) \\ &= (\vee_{i\in I} A_i(x)) \wedge (\vee_{i\in I} A_i(y)) \wedge \mu \qquad (2) \\ &= (\cup_{i\in I} A_i(x)) \wedge (\cup_{i\in I} A_i(y)) \wedge \mu \\ &= A(x) \wedge A(y) \wedge \mu \end{aligned}$$

In the following we show that Eq. (2) holds. It is clear that $\vee_{i\in I}(A_i(x) \wedge A_i(y) \wedge \mu) \leq (\vee_{i\in I} A_i(x)) \wedge (\vee_{i\in I} A_i(y)) \wedge \mu$. If possible, let $\vee_{i\in I}(A_i(x) \wedge A_i(y) \wedge \mu) \neq (\vee_{i\in I} A_i(x)) \wedge (\vee_{i\in I} A_i(y)) \wedge \mu$. Then there exists t such that $\vee_{i\in I}(A_i(x) \wedge A_i(y) \wedge \mu) < t < (\vee_{i\in I} A_i(x)) \wedge (\vee_{i\in I} A_i(y)) \wedge \mu$. Since $A_i \subseteq A_j$ or $A_j \subseteq A_i$ for all $i, j \in I$, there exists $k \in I$ such that $t < A_k(x) \wedge A_k(y) \wedge \mu$. On the other hand, $A_i(x) \wedge A_i(y) \wedge \mu < t$ for all $i \in I$, a contradiction. Hence, $\vee_{i\in I}(A_i(x) \wedge A_i(y) \wedge \mu) = (\vee_{i\in I} A_i(x)) \wedge (\vee_{i\in I} A_i(y)) \wedge \mu$.

$\forall x, y \in S$, $A(xy) \vee \lambda = (\cup_{i\in I} A_i(xy)) \vee \lambda = \vee_{i\in I}(A_i(xy) \vee \lambda) \geq \vee_{i\in I}(A_i(x) \wedge \mu) = (\vee_{i\in I} A_i(x)) \wedge \mu = A(x) \wedge \mu$.

Similarly prove $A(xy) \vee \lambda \geq A(y) \wedge \mu$, for all $x, y \in S$.

$A(x) \vee A(y) \vee \lambda = (\cup_{i\in I} A_i(x)) \vee (\cup_{i\in I} A_i(y)) \vee \lambda = \cup_{i\in I}(A_i(x) \vee A_i(y) \vee \lambda) \geq \vee_{i\in I}(A_i(xy) \wedge \mu) = (\vee_{i\in I} A_i(xy)) \wedge \mu = A(xy) \wedge \mu$.

By Theorem 3.2, A is an $(\in, \in \vee q_{(\lambda,\mu)})$-fuzzy completely prime ideal of S.

Theorem 3.11 *Let S and S′ be semigroups and $f : S \to S'$ be an onto homomorphism. Let A and B be $(\in, \in \vee q_{(\lambda,\mu)})$-fuzzy completely prime ideals of S and S′, respectively. Then*

(1) $f(A)$ is an $(\in, \in \vee q_{(\lambda,\mu)})$-fuzzy completely prime ideal of S′;
(2) $f^{-1}(B)$ is an $(\in, \in \vee q_{(\lambda,\mu)})$-fuzzy completely prime ideal of S.

Proof: For any $x', y' \in S$, then

$$\begin{aligned} f(A)(x'y') \vee \lambda &= \vee\{A(z) \mid z \in S, f(z) = x'y'\} \vee \lambda \\ &\geq \vee\{A(xy) \mid x, y \in S, f(x) = x', f(y) = y'\} \vee \lambda \\ &= \vee\{A(xy) \vee \lambda \mid x, y \in S, f(x) = x', f(y) = y'\} \\ &\geq \vee\{A(x) \wedge A(y) \wedge \mu \mid x, y \in S, f(x) = x', f(y) = y'\} \\ &\geq \{A(x) \wedge A(y) \wedge \mu \mid x, y \in S, f(x) = x', f(y) = y'\} \end{aligned}$$

then $f(A)(x'y') \vee \lambda \geq \vee_{f(x)=x'}\{A(x) \wedge A(y) \wedge \mu \mid x, y \in S, f(x) = x',$
$f(y) = y'\} = \{(\vee_{f(x)=x'} A(x)) \wedge A(y) \wedge \mu \mid y \in S, f(y) = y'\}.$

$$\begin{aligned} \text{So } f(A)(x'y') \vee \lambda &\geq \vee_{f(y)=y'}\{f(A)(x') \wedge A(y) \wedge \mu \mid y \in S, f(y) = y'\} \\ &= f(A)(x') \wedge (\vee_{f(y)=y'} A(y)) \wedge \mu \\ &= f(A)(x') \wedge f(A)(y') \wedge \mu. \\ f(A)(x'y') \vee \lambda &= \vee\{A(z) \mid z \in S, f(z) = x'y'\} \vee \lambda \\ &\geq \vee\{A(xy) \mid x, y \in S, f(x) = x', f(y) = y'\} \vee \lambda \\ &= \vee\{A(xy) \vee \lambda \mid x, y \in S, f(x) = x', f(y) = y'\} \\ &\geq \vee\{A(x) \wedge \mu \mid x \in S, f(x) = x'\} \\ &\geq \vee\{A(x) \mid x \in S, f(x) = x'\} \wedge \mu \\ &= f(A)(x') \wedge \mu. \end{aligned}$$

Similarly prove $f(A)(x'y') \vee \lambda \geq f(A)(y') \wedge \mu$, for all $x', y' \in S'$.

$$\begin{aligned} f(A)(x') \vee f(A)(y') \vee \lambda &= \{\vee\{A(x) \mid x \in S, f(x) = x'\}\} \\ &\quad \vee \{\vee\{A(y) \mid y \in S, f(y) = y'\}\} \vee \lambda \\ &= \vee\{A(x) \vee A(y) \vee \lambda \mid x, y \in S, f(x) = x', f(y) = y'\} \\ &\geq \vee\{A(xy) \wedge \mu \mid f(xy) = x'y', x, y \in S\} \\ &= \vee\{A(xy) \mid f(xy) = x'y', x, y \in S\} \wedge \mu \\ &= f(A)(x'y') \wedge \mu. \end{aligned}$$

Therefore $f(A)$ is an $(\in, \in \vee q_{(\lambda,\mu)})$-fuzzy completely prime ideal of S'.
(2) For all $x, y \in S$,

$$\begin{aligned} f^{-1}(B)(xy) \vee \lambda &= B(f(xy)) \vee \lambda = B(f(x)f(y)) \vee \lambda \geq B(f(x)) \wedge B(f(y)) \wedge \mu \\ &= f^{-1}(B)(x) \wedge f^{-1}(B)(y) \wedge \mu \\ f^{-1}(B)(xy) \vee \lambda &= B(f(xy)) \vee \lambda = B(f(x)f(y)) \vee \lambda \\ &\geq B(f(x)) \wedge \mu = f^{-1}(B)(x) \wedge \mu. \end{aligned}$$

Similarly prove $f^{-1}(B)(xy) \vee \lambda \geq f^{-1}(B)(y) \wedge \mu$, for all $x, y \in S$.

$$\begin{aligned} f^{-1}(B)(x) \vee f^{-1}(B)(y) \vee \lambda &= B(f(x)) \vee B(f(y)) \vee \lambda \geq B(f(x)f(y)) \wedge \mu \\ &= B(f(xy)) \wedge \mu = f^{-1}(B)(xy) \wedge \mu. \end{aligned}$$

So $f^{-1}(B)$ is an $(\in, \in \vee q_{(\lambda,\mu)})$-fuzzy completely prime ideal of S.

Theorem 3.12 *Let A be a fuzzy subset of S, $t \in (\lambda, 1]$ and $2\mu - \lambda > 1$, then the following statements are equivalent:*

(1) A is an $(\in, \in \vee q_{(\lambda,\mu)})$-fuzzy completely prime ideal;
(2) $A_{t\vee q_{(\lambda,\mu)}} := \{x \in S | A(x) \geq t \quad or \quad A(x) + t > 2\mu\}$ is a completely prime ideal.

Proof: (1)⇒ (2) (a) Let $xy \in A_{t\vee q_{(\lambda,\mu)}}$, we claim $x \in A_{t\vee q_{(\lambda,\mu)}}$ or $y \in A_{t\vee q_{(\lambda,\mu)}}$. In fact, we have $A(xy) \geq t$ or $A(xy) + t > 2\mu$. Since A is an $(\in, \in \vee q_{(\lambda,\mu)})$-fuzzy completely prime ideal of S. Then $A(x) \vee A(y) \vee \lambda \geq A(xy) \wedge \mu$.

Case 1: Suppose $A(xy) \geq t$. Then $A(x) \vee A(y) \vee \lambda \geq t \wedge \mu$. If $t > \mu$, then $A(x) > \mu$ or $A(y) > \mu$. We get $A(x)+t > \mu+\mu = 2\mu$ or $A(y)+t > \mu+\mu = 2\mu$. If $t \leq \mu$, then $A(x) \geq t$ or $A(y) \geq t$. So $x \in A_{t\vee q_{(\lambda,\mu)}}$ or $y \in A_{t\vee q_{(\lambda,\mu)}}$.

Case 2: Suppose $A(xy) + t > 2\mu$. If $t > \mu$ and $2\mu - \lambda > 1$, then $2\mu - t < \mu$ and $\lambda < 2\mu - t$, $A(x) \vee A(y) \vee \lambda \geq A(xy) \wedge \mu > 2\mu - t$. So $A(x) + t > 2\mu$ or $A(y)+t > 2\mu$. If $t \leq \mu$ then $A(x) \vee A(y) \vee \lambda \geq \mu \geq t$, *i.e.*, $A(x) \geq t$ or $A(y) \geq t$. Hence $x \in A_{t\vee q_{(\lambda,\mu)}}$ or $y \in A_{t\vee q_{(\lambda,\mu)}}$.

Similarly prove the following:

(b) If $x, y \in A_{t\vee q_{(\lambda,\mu)}}$, then $xy \in A_{t\vee q_{(\lambda,\mu)}}$;

(c) If $x \in A_{t\vee q_{(\lambda,\mu)}}$, $y \in S$, then $xy \in A_{t\vee q_{(\lambda,\mu)}}$;

(d) If $x \in S$, $y \in A_{t\vee q_{(\lambda,\mu)}}$, then $xy \in A_{t\vee q_{(\lambda,\mu)}}$.

(2)⇐(1) (a1) We will prove $A(x) \vee A(y) \vee \lambda \geq A(xy) \wedge \mu$. If the (a1) does not hold, then there exist x_0, $y_0 \in S$ such that $A(x_0) \vee A(y_0) \vee \lambda < A(x_0 y_0) \wedge \mu$. Choose t to satisfy $A(x_0) \vee A(y_0) \vee \lambda < t < A(x_0 y_0) \wedge \mu$. Note that $\lambda < t < \mu$, $A(x_0) < t$ and $A(y_0) < t$. Then $A(x_0) + t < t + t < 2\mu$ and $A(y_0) + t < t + t < 2\mu$. This implies $x_0 \overline{\in} A_{t\vee q_{(\lambda,\mu)}}$ and $y_0 \overline{\in} A_{t\vee q_{(\lambda,\mu)}}$. Since $A(x_0 y_0) \wedge \mu > t$, we have $A(x_0 y_0) > t$. Thus we get $x_0 y_0 \in A_{t\vee q_{(\lambda,\mu)}}$. But $x_0 \overline{\in} A_{t\vee q_{(\lambda,\mu)}}$ and $y_0 \overline{\in} A_{t\vee q_{(\lambda,\mu)}}$. This is a contradiction to the fact that $A_{t\vee q_{(\lambda,\mu)}}$ is a completely prime ideal.

Similarly for all $x, y \in S$, we can prove the following:

(b1) $A(xy) \vee \lambda \geq A(x) \wedge A(y) \wedge \mu$

(c1) $A(xy) \vee \lambda \geq A(x) \wedge \mu$

(d1) $A(xy) \vee \lambda \geq A(y) \wedge \mu$

Using (a1)(b1)(c1)(d1), A is an $(\in, \in \vee q_{(\lambda,\mu)})$-fuzzy completely prime ideal of S.

4 Conclusion

In the study of fuzzy algebraic system, we notice that fuzzy ideals with special properties always play an important role. In this paper, we give the new definition of $(\in, \in \vee q_{(\lambda,\mu)})$-fuzzy completely prime ideals of semigroups. Using inequalities, characteristic functions and level sets, we consider its equivalent descriptions. Apart from those, the properties of the intersection, union, homomorphic image and homomorphic preimage of $(\in, \in \vee q_{(\lambda,\mu)})$-fuzzy completely prime ideal of semigroups are investigated. The equivalence relationship between $(\in, \in \vee q_{(\lambda,\mu)}))$-fuzzy completely prime ideals and $A_{t\vee q_{(\lambda,\mu)}}$ is also discussed. Those results extend the corresponding theories of fuzzy completely prime ideals and enrich the study of fuzzy algebra. At present, although a series of work on this aspect have been done, there is much room for further study.

Acknowledgments This work is supported by Program for Innovative Research Team of Jiangnan University(No:200902).

References

1. Zadeh, L.A.: Fuzzy sets. Inf. Control **8**, 338–353 (1965)
2. Rosenfeld, A.: Fuzzy groups. J. Math. Anal. Appl. **35**, 512–517 (1971)
3. Bhakat, S.K., Das, P.: (α, β)-fuzzy mapping. Fuzzy Sets Syst. **56**, 89–95 (1993)
4. Bhakat, S.K.: $(\in, \in \vee q)$-level subsets. Fuzzy Sets Syst. **103**, 529–533 (1999)
5. Bhakat, S.K., Das, P.: On the definition of a fuzzy subgroup. Fuzzy Sets Syst. **51**, 235–241 (1992)
6. Bhakat, S.K., Das, P.: $(\in, \in \vee q)$-fuzzy subgroups. Fuzzy Sets Syst. **80**, 359–368 (1996)
7. Pu, P.M., Liu, Y.M.: Fuzzy topologyI: neighbourhood structure of a fuzzy point and Moore-Smith convergence. J. Math. Anal. Appl. **76**, 571–599 (1980)
8. Zhan, J.M., Ma, X.L.: On fuzzy interior ideals in semigroups. J. Math. Res. Exposition **28**, 103–110 (2008)
9. Jun, Y.B., Song, S.Z.: Generalized fuzzy interior ideals in semigroups. Inf. Sci. **176**, 3079–3093 (2006)
10. Davvaz, B.: $(\in, \in \vee q)$-fuzzy subnear-rings and ideals. Soft Comput. Fusion Found. Metodologies Appl. **10**, 206–211 (2006)
11. Davvaz, B., Corsini, P.: Redefined fuzzy H_v-submodules and many valued implication. Inf. Sci. **177**, 865–875 (2007)
12. Davvaz, B.: Fuzzy R-subgroups with thresholds of near rings ang implication oqerations. Soft Comput. Fusion Found. Metodologies Appl. **12**, 875–979 (2008)
13. Yuan, X.H., Zhang, C., Ren, Y.H.: Generalized fuzzy subgroups and many-valued implications. Fuzzy Sets Syst. **138**, 205–211 (2003)
14. Liao, Z.H., Chen, Y.X., Lu, J.G., Gu, h: Generalized fuzzy subsemigroup and generalized fuzzy completely regular semigroup. Fuzzy Syst. Math. **18**, 81–84 (2004)
15. Chen, M., Liao, Z.H.: $(\in, \in \vee q_{(\lambda,\mu)})$-fuzzy subsemigroups and $(\in, \in \vee q_{(\lambda,\mu)})$-fuzzy ideals of semigroups. In:The sixth International Conference on Information and Management Sciences, pp. 575–579, Springer, Lhasa, Tibet (2007)
16. Du, J., Liao, Z.H.: $(\in, \in \vee q_{(\lambda_1,\lambda_2)})$-generalized fuzzy ideal of BCK-Algebra. In: IEEE FSKD 2007, 1, pp. 294–299 (2007)
17. Liao, Z.H., Gu, H.: $(\in, \in \vee q_{(\lambda,\mu)})$-fuzzy normal subgroup. Fuzzy Syst. Math. **20**, 47–53 (2006)
18. Zhou, J., Liao, Z.H.: Generalized (α, β)-convex fuzzy cones. In:The 2008 IEEE Internation Conference on Fuzzy Systems, Hongkong, pp. 439–442 (2008)
19. Kuroki, N.: On fuzzy semigroups. Inf. Sci. **53**, 203–236 (1991)
20. Kuroki, N.: Fuzzy generalized bi-ideals in semigroups. Inf. Sci. **66**, 235–243 (1992)
21. Kuroki, N.: Fuzzy semiprime quasi-ideals in semigroups. Inf. Sci. **75**, 202–211 (1993)
22. Mo, Z.W., Wang, X.P.: Fuzzy ideals generated by fuzzy sets in semigroups. Inf. Sci. **86**, 203–210 (1995)
23. Bhakat, S.K., Das, P.: Fuzzy subrings and ideals refined. Fuzzy Sets Syst. **81**, 383–393 (1996)
24. Jun, Y.B., Neggers, J., Kim, H.S.: On L-fuzzy ideals in semirings I. Czechoslovak Math. J. **48**, 669–675 (1998)
25. Jun, Y.B., Öztürk, M.A., Song, S.Z.: On fuzzy h-ideals in hemirings. Inf. Sci. **162**, 211–226 (2004)
26. Jun, Y.B., Ahn, S.S., Kim, H.S.: Fuzzy subclines(ideals) of incline algebras. Fuzzy Sets Syst. **123**, 217–225 (2001)
27. Babushri, S.K., Syam, P.K., Satyanarayana, B.: Equivalent, 3-prime and c-prime fuzzy ideals of nearrings. Soft Comput. Fusion Found. Methodol. Appl. **13**, 933–944 (2009)

Soft Relation and Fuzzy Soft Relation

Yu-hong Zhang and Xue-hai Yuan

Abstract Molodtsov introduced the concept of soft sets as a general mathematical tool for dealing with uncertainties. Research on soft set theory has been made progress in recent years. This paper firstly introduce the concept of soft relation and fuzzy soft relation; secondly we propose the concept of the projection and the section of fuzzy soft relation and study some properties; finally we give the concept of fuzzy soft linear transformation and get some conclusions.

Keywords Soft set · Soft relation · Fuzzy soft relation · Fuzzy soft linear transformation

1 Introduction

Soft set theory was initiated by Molodtsov [1] in 1999 as a general mathematical tool for dealing with uncertainty, fuzzy, not clearly defined objects. Since then,several special soft sets such as the fuzzy soft set [2], the generalized fuzzy soft set [3], the interval-valued fuzzy soft sets [4], the interval-valued intuitionistic fuzzy soft set [5], the bijective soft set [6] and the exclusive disjunctive soft set [7] have been proposed. In recent years, great progress, both in theory and in application, has been made by many authors. For example, the theoretical aspect of operations on soft sets [8–10],

Y. Zhang (✉)
School of Mathematical Sciences, Dalian University of Technology,
Dalian 116024, People's Republic of China
e-mail: yuhz@dlut.edu.cn

Y. Zhang
Department of Basic Education, City Institute, Dalian University of Technology,
Dalian 116000, People's Republic of China

X. Yuan
Faculty of Electronic Information and Electrical Engineering, Dalian University of Technology,
Dalian 116024, People's Republic of China
e-mail: yuanxh@dlut.edu.cn

B.-Y. Cao and H. Nasseri (eds.), *Fuzzy Information & Engineering and Operations Research & Management*, Advances in Intelligent Systems and Computing 211,
DOI: 10.1007/978-3-642-38667-1_21,

the soft topology on a soft set [11–13], the algebraic structure of soft set [14–18], soft set relations and functions [19, 20] and the applications of soft sets in decision making problem [21, 22].

Majumdar et al. [23] gave an idea of soft mapping. Based on that mapping and relation are intimately connected, we introduce the notion of soft relation which is the main purpose of this paper.

The organization of this paper is as follows: In Sect. 2, some preliminaries are given. In Sect. 3, the definitions of soft relation and fuzzy soft relation are given and some properties are studied. Section 4 present some conclusions from the paper.

2 Preliminaries

Definition 2.1. *[1] Let U be an initial universe set and E be a set of parameters. Let $\mathcal{P}(U)$ denote the power set of U and $A \subset E$. A pair (F, A) is called a soft set over U iff F is a mapping given by $F : A \to \mathcal{P}(U)$.*

Example 2.2. Suppose a soft set (F, E) describes the attractiveness of the shirts which the authors are going to wear.

$U =$ the set of all shirts under consideration $= \{x_1, x_2, x_3, x_4, x_5\}$

$E =$ {colorful,bright,cheap,warm} $= \{e_1, e_2, e_3, e_4\}$

Let $F(e_1) = \{x_1, x_2\}$, $F(e_2) = \{x_1, x_2, x_3\}$, $F(e_3) = \{x_4\}$, $F(e_4) = \{x_2, x_5\}$

So, the soft set (F, E) is a family $\{F(e_i), i = 1, 2, 3, 4\}$ of $\mathcal{P}(U)$.

Definition 2.3. *[23] Let A, B be two non-empty set and E be a parameter set. Then the mapping $F : E \to \mathcal{P}(B^A)$ is called a soft mapping from A to B under E, where B^A is the collection of all mappings from A to B.*

Actually a soft mapping F from A to B under E is a soft set over B^A.

The next definitions and results are from [24].

Let $R \in \mathcal{F}(X \times Y)$, $x \in X$, $y \in Y$, we set

$(R\,|_x)(y) = R(x, y)$, $(R\,|_y)(x) = R(x, y)$.

$(R\,|_{[x]})(y) = 1 - R(x, y)$, $(R\,|_{[y]})(x) = 1 - R(x, y)$.

$(R_X)(x) = \bigvee_{y \in Y} R(x, y)$, $(R_Y)(y) = \bigvee_{x \in X} R(x, y)$.

$(R_{[X]})(x) = \bigwedge_{y \in Y} R(x, y)$, $(R_{[Y]})(y) = \bigwedge_{x \in X} R(x, y)$.

Property 2.4.

1. $R_{[X]} \subseteq R_X$.
2. $R\,|_{[x]} = R^c\,|_x$, $R^c\,|_{[x]} = (R\,|_{[x]})^c$, $R^c\,|_x = (R\,|_x)^c$.
3. $R_{[X]} = ((R^c)_X)^c$.
4. $(R \cup S)_X = R_X \cup S_X$, $(R \cap S)_X \subseteq R_X \cap S_X$
   $(\bigcup_{t \in T} R_t)_X = \bigcup_{t \in T} (R_t)_X$, $(\bigcap_{t \in T} R_t)_X \subseteq \bigcap_{t \in T} (R_t)_X$.

5. $(R \cup S)_{[X]} \supseteq R_{[X]} \cup S_{[X]}, (R \cap S)_{[X]} = R_{[X]} \cap S_{[X]}$.
   $(\bigcup_{t \in T} R_t)_{[X]} \supseteq \bigcup_{t \in T} (R_t)_{[X]}, (\bigcap_{t \in T} R_t)_{[X]} = \bigcap_{t \in T} (R_t)_{[X]}$.
6. $(\bigcup_{t \in T} R_t) \mid_x = \bigcup_{t \in T} (R_t) \mid_x, (\bigcap_{t \in T} R_t) \mid_x = \bigcap_{t \in T} (R_t) \mid_x$.
7. $(\bigcup_{t \in T} R_t) \mid_{[x]} = \bigcap_{t \in T} (R_t) \mid_{[x]}, (\bigcap_{t \in T} R_t) \mid_{[x]} = \bigcup_{t \in T} (R_t) \mid_{[x]}$.

Let R, S, T and Q be fuzzy relations, then we have that

1. $(R \circ S) \circ T = R \circ (S \circ T)$.
2. $(R \cup S) \circ T = (R \circ T) \cup (S \circ T), T \circ (R \cup S) = (T \circ R) \cup (T \circ S)$.
   $(\bigcup_{t \in T} R_t) \circ T = \bigcup_{t \in T} (R_t \circ T), S \circ (\bigcup_{t \in T} R_t) = \bigcup_{t \in T} (S \circ R_t)$.
3. $(\lambda R) \circ S = \lambda(R \circ S) = R \circ (\lambda S)$ $(\lambda \in [0, 1])$
   $(\bigcup_{r \in \Gamma} \lambda_r R^{(r)}) \circ T = \bigcup_{r \in \Gamma} \lambda_r (R^{(r)} \circ T), S \circ (\bigcup_{r \in \Gamma} \lambda_r R^{(r)}) = \bigcup_{r \in \Gamma} \lambda_r (S \circ R^{(r)})$.
4. $Q \subseteq R \Rightarrow Q \circ T \subseteq R \circ T, S \circ Q \subseteq S \circ R, Q^n \subseteq R^n$,
   $(\bigcap_{r \in \Gamma} R^{(r)}) \circ T \subseteq \bigcap_{r \in \Gamma} (R^{(r)} \circ T)$.

3 Soft Relation and Fuzzy Soft Relation

In this section we introduce the notion of soft relation and fuzzy soft relation and study their properties. Let X be the universal set and E be a parameter set. Then the pair (X, E) will be called a soft universe. Throughout this section we assume that (X, E) is our soft universe.

Definition 3.1. *Let A, B be two non-empty set and E be a parameter set. Then the mapping*

$$F : E \to \mathcal{P}(A \times B)$$

is called a soft relation from A to B under E, where $\mathcal{P}(A \times B)$ is the set of all relations from A to B.
Note: Actually a soft relation F from A to B under E is a soft set over $A \times B$, which is different from the definition of soft set relation in [19]. In fact, let (F, A) and (G, B) be two soft sets over U, then a soft set relation from (F, A) to (G, B) is a soft subset of $(F, A) \times (G, B)$, where $(F, A) \times (G, B) = (H, A \times B)$, where $H : A \times B \to \mathcal{P}(U \times U)$ and $H(a, b) = F(a) \times G(b)$, where $(a, b) \in A \times B$.

Example 3.2. Let $R : A \times B \to [0, 1]$ is a fuzzy relation on A and B. Then the mapping

$$\begin{aligned} F : E = [0, 1] &\to \mathcal{P}(A \times B) \\ \alpha &\mapsto R_{\underline{\alpha}} = \{(x, y) \in A \times B \mid R(x, y) > \alpha\} \end{aligned}$$

is a soft relation from A to B under E.

Definition 3.3. *Let A, B be tow non-empty set and E be a parameter set. Then the mapping*

$$F : E \to \mathcal{F}(A \times B)$$

is called a fuzzy soft relation from A to B under E, where $\mathcal{F}(A \times B)$ is the set of all fuzzy relations from A to B.

Example 3.4. Let $E = \{e_1, e_2\}$, $A = \{a_1, a_2\}$, $B = \{b_1, b_2, b_3\}$.
Let $R_1, R_2 \in \mathcal{F}(A \times B)$ be defined as follows:

$$R_1 = \begin{pmatrix} 0.3 & 0.7 & 0.5 \\ 0.6 & 1 & 0 \end{pmatrix}, R_2 = \begin{pmatrix} 0.7 & 1 & 0.6 \\ 0.5 & 0 & 0.4 \end{pmatrix}$$

Let $F : E \to \mathcal{F}(A \times B)$ be defined as follows:

$$F(e_1) = R_1, F(e_2) = R_2$$

Then F is a fuzzy soft relation from A to B under E.

Definition 3.5. *Let F be a fuzzy soft relation from A to A under E, we say that*

1. *F is reflexive$\Leftrightarrow$ $F(e)$ is reflexive,$\forall e \in E$*
2. *F is symmetric$\Leftrightarrow$ $F(e)$ is symmetric,$\forall e \in E$*
3. *F is transitive$\Leftrightarrow$ $F(e)$ is transitive,$\forall e \in E$*

A fuzzy soft relation F on A is called a fuzzy soft equivalence relation if it is reflexive,symmetric and transitive.

Definition 3.6. *Let F_1 be a fuzzy soft relation from A to B under E, F_2 be a fuzzy soft relation from B to C under E.Then a new fuzzy soft relation,the composition of F_1 and F_2 expressed as $F_1 \circ F_2$ from A to C under E is defined as follows:*

$$(F_1 \circ F_2)(e) = F_1(e) \circ F_2(e)$$

Let F and G be fuzzy soft relations from A to B under E,we set
$F \leq G \Leftrightarrow \forall e \in E, F(e) \subseteq G(e)$; $F = G \Leftrightarrow \forall e \in E, F(e) = G(e)$;
$F^c(e) = (F(e))^c$; $(F \cup G)(e) = F(e) \cup G(e)$;
$(F \cap G)(e) = F(e) \cap G(e)$; $(\lambda F)(e) = \lambda F(e)$.

Property 3.7. *Let F, G and S be fuzzy soft relations,*

1. $(F \circ G) \circ S = F \circ (G \circ S)$.
2. $(F \cup G) \circ S = (F \circ S) \cup (G \circ S), S \circ (F \cup G) = (S \circ F) \cup (S \circ G)$.
   $(\bigcup_{t \in T} F_t) \circ G = \bigcup_{t \in T} (F_t \circ G), S \circ (\bigcup_{t \in T} F_t) = \bigcup_{t \in T} (S \circ F_t)$.
3. $(\lambda F) \circ G = \lambda(F \circ G) = F \circ (\lambda G)$ $(\lambda \in [0, 1])$
   $(\bigcup_{r \in \Gamma} \lambda_r F^{(r)}) \circ G = \bigcup_{r \in \Gamma} \lambda_r (F^{(r)} \circ G), S \circ (\bigcup_{r \in \Gamma} \lambda_r F^{(r)}) = \bigcup_{r \in \Gamma} \lambda_r (S \circ F^{(r)})$.

4. $F_1 \leq F_2 \Rightarrow F_1 \circ G \leq F_2 \circ G$, $S \circ F_1 \leq S \circ F_2$, $F_1^n \leq F_2^n$,
$(\bigcap_{r\in\Gamma} F^{(r)}) \circ G \leq \bigcap_{r\in\Gamma} (F^{(r)} \circ G)$.

Definition 3.8. *Let F be a fuzzy soft relation from A to B under E. For $a \in A$ and $b \in B$, we set*

$$\begin{aligned}
F\mid_a : E &\to \mathcal{P}(B) \\
e &\mapsto F\mid_a(e) = F(e)\mid_a \\
F\mid_b : E &\to \mathcal{P}(A) \\
e &\mapsto F\mid_b(e) = F(e)\mid_b \\
F\mid_{[a]} : E &\to \mathcal{P}(B) \\
e &\mapsto F\mid_{[a]}(e) = F(e)\mid_{[a]} \\
F\mid_{[b]} : E &\to \mathcal{P}(A) \\
e &\mapsto F\mid_{[b]}(e) = F(e)\mid_{[b]} \\
F_A : E &\to \mathcal{P}(A) \\
e &\mapsto F_A(e) = (F(e))_A \\
F_B : E &\to \mathcal{P}(B) \\
e &\mapsto F_B(e) = (F(e))_B \\
F_{[A]} : E &\to \mathcal{P}(A) \\
e &\mapsto F_{[A]}(e) = (F(e))_{[A]} \\
F_{[B]} : E &\to \mathcal{P}(B) \\
e &\mapsto F_{[B]}(e) = (F(e))_{[B]}
\end{aligned}$$

Property 3.9. *Let F be a fuzzy soft relation from A to B under E, then we have that*

1. $F_{[A]} \leq F_A$.
2. $F\mid_{[a]} = F^c\mid_a$, $F^c\mid_{[a]} = (F\mid_{[a]})^c$, $F^c\mid_a = (F\mid_a)^c$.
3. $F_{[A]} = ((F^c)_A)^c$.
4. $(F \cup G)_A = F_A \cup G_A$, $(F \cap G)_A \leq F_A \cap G_A$.
$(\bigcup_{t\in T} F_t)_A = \bigcup_{t\in T} (F_t)_A$, $(\bigcap_{t\in T} F_t)_A \leq \bigcap_{t\in T} (F_t)_A$.
5. $(F \cup G)_{[A]} \geq F_{[A]} \cup G_{[A]}$, $(F \cap G)_{[A]} = F_{[A]} \cap G_{[A]}$.
$(\bigcup_{t\in T} F_t)_{[A]} \geq \bigcup_{t\in T} (F_t)_{[A]}$, $(\bigcap_{t\in T} F_t)_{[A]} = \bigcap_{t\in T} (F_t)_{[A]}$.
6. $(\bigcup_{t\in T} F_t)\mid_a = \bigcup_{t\in T} (F_t)\mid_a$, $(\bigcap_{t\in T} F_t)\mid_a = \bigcap_{t\in T} (F_t)\mid_a$.
7. $(\bigcup_{t\in T} F_t)\mid_{[a]} = \bigcap_{t\in T} (F_t)\mid_{[a]}$, $(\bigcap_{t\in T} F_t)\mid_{[a]} = \bigcup_{t\in T} (F_t)\mid_{[a]}$.

Definition 3.10. *Let $F : E \to \mathcal{F}(A \times B)$ and $T : E \to \mathcal{F}(A)$, we set*

$$\begin{aligned} T \circ F : E \to \mathcal{F}(B) & \\ & e \mapsto (T \circ F)(e) = T(e) \circ F(e) \end{aligned}$$

Let $T_1 : E \to \mathcal{F}(A)$ and $T_2 : E \to \mathcal{F}(A)$,we set

$$\begin{aligned} T_1 \cup T_2 : E \to \mathcal{F}(A) & \\ & e \mapsto (T_1 \cup T_2)(e) = T_1(e) \cup T_2(e) \end{aligned}$$

Let $T : E \to \mathcal{F}(A)$,we set

$$\begin{aligned} \lambda T : E \to \mathcal{F}(A) & \\ & e \mapsto (\lambda T)(e) = \lambda T(e) \end{aligned}$$

Let $F : E \to \mathcal{F}(A \times B)$ and $T : E \to \mathcal{F}(A)$,we set

$$\begin{aligned} \lambda(T \circ F) : E \to \mathcal{F}(B) & \\ & e \mapsto [\lambda(T \circ F)](e) = \lambda(T(e) \circ F(e)) \end{aligned}$$

Theorem 3.11. *Let $T_1 : E \to \mathcal{F}(A), T_2 : E \to \mathcal{F}(A)$ and $F : E \to \mathcal{F}(A \times B)$, then*

(1) $(T_1 \cup T_2) \circ F = (T_1 \circ F) \cup (T_2 \circ F)$.
(2) $(\lambda T) \circ F = \lambda(T \circ F)$.
(3) $T_1(e) \subseteq T_2(e) \Rightarrow T_1 \circ F \leq T_2 \circ F$.

Proof. 1. $[(T_1 \cup T_2) \circ F](e) = (T_1 \cup T_2)(e) \circ F(e) = [T_1(e) \cup T_2(e)] \circ F(e) = [T_1(e) \circ F(e)] \cup [T_2(e) \circ F(e)] = [(T_1 \circ F)(e)] \cup [(T_2 \circ F)(e)] = [(T_1 \circ F) \cup (T_2 \circ F](e)$. 2. $[(\lambda T) \circ F](e) = (\lambda T)(e) \circ F(e) = \lambda[T(e) \circ F(e)] = [\lambda(T \circ F)](e)$.

Definition 3.12. *Let $\mathcal{F}_A = \{T_A \mid T_A : E \to \mathcal{F}(A)\}$, $\mathcal{F}_B = \{T_B \mid T_B : E \to \mathcal{F}(B)\}$.*
Let $\Gamma : \mathcal{F}_A \to \mathcal{F}_B$ is a mapping. If for $\lambda_t \in [0, 1]$, ${T_A}^{(t)} \in \mathcal{F}_A$ $(t \in T)$, we have that

$$\Gamma(\bigcup_{t \in T} \lambda_t {T_A}^{(t)}) = \bigcup_{t \in T} \lambda_t \Gamma({T_A}^{(t)})$$

then Γ is called a fuzzy soft linear transformation from A to B.

Property 3.10. *(1)* $\Gamma({T_A}^{(1)} \cup {T_A}^{(2)}) = \Gamma({T_A}^{(1)}) \cup \Gamma({T_A}^{(2)})$.
(2) $\Gamma(\lambda T_A) = \lambda \Gamma(T_A)$.
(3) ${T_A}^{(1)} \leq {T_A}^{(2)} \Rightarrow \Gamma({T_A}^{(1)}) \leq \Gamma({T_A}^{(2)})$.
i.e. ${T_A}^{(1)}(e) \subseteq {T_A}^{(2)}(e) \Rightarrow \Gamma({T_A}^{(1)})(e) \subseteq \Gamma({T_A}^{(2)})(e)$.

Theorem 3.13. *Let* $\Gamma : \mathcal{F}_A \to \mathcal{F}_B$ *be a fuzzy soft linear transformation from A to B, then there is an unique fuzzy soft relation* $F_\Gamma : E \to \mathcal{F}(A \times B)$ *such that* $\Gamma(T_A) = T_A \circ F_\Gamma$*; On the contrary, let* $F : E \to \mathcal{F}(A \times B)$ *be a fuzzy soft relation, then there is an unique fuzzy soft linear transformation* $\Gamma_F : \mathcal{F}_A \to \mathcal{F}_B$ *such that* $\Gamma_F(T_A) = T_A \circ F$.

Proof. We first show that $\{T_A(e) \mid T_A \in \mathcal{F}_A\} = \mathcal{F}(A)$, for any $e \in E$.

In fact,for $H \in \mathcal{F}(A)$, we set ${T_A}^H : E \to \mathcal{F}(A)$ and ${T_A}^H(e) \equiv H$. Then $\mathcal{F}(A) \supseteq \{T_A(e) \mid T_A \in \mathcal{F}_A\} \supseteq \{{T_A}^H(e) \mid H \in \mathcal{F}(A)\} = \mathcal{F}(A)$. It follows that $\{T_A(e) \mid T_A \in \mathcal{F}_A\} = \mathcal{F}(A), \forall e \in E$.

Let $e \in E$ and $f_e : \mathcal{F}(A) \to \mathcal{F}(B)$ be a mapping such that $f_e(T_A(e)) = \Gamma(T_A)(e)$, then

$$f_e(\bigcup_{t\in T} \lambda_t {T_A}^{(t)}(e)) = f_e((\bigcup_{t\in T} \lambda_t {T_A}^{(t)})(e)) = \Gamma(\bigcup_{t\in T} \lambda_t {T_A}^{(t)})(e)$$
$$= (\bigcup_{t\in T} \lambda_t \Gamma({T_A}^{(t)}))(e) = \bigcup_{t\in T} \lambda_t \Gamma({T_A}^{(t)})(e) = \bigcup_{t\in T} \lambda_t f_e({T_A}^{(t)}(e)).$$

Then f_e is a fuzzy linear transformation from A to B.

Let $F_\Gamma : E \to \mathcal{F}(A \times B)$ be a mapping and

$$F_\Gamma(e)(x, y) = \Gamma({T_A}^{\{x\}})(e)(y) = f_e({T_A}^{\{x\}}(e))(y) = f_e(\{x\})(y).$$

Then $\Gamma(T_A)(e) = f_e(T_A(e)) = f_e(\bigcup_{x\in A} {\lambda_x}^e\{x\})$, where ${\lambda_x}^e = T_A(e)(x)$.

Then $\Gamma(T_A)(e) = \bigcup_{x\in A} {\lambda_x}^e f_e(\{x\}) = \bigcup_{x\in A} {\lambda_x}^e f_e({T_A}^{\{x\}}(e))$.

Then $\Gamma(T_A)(e)(y) = \bigvee_{x\in A} ({\lambda_x}^e \wedge f_e({T_A}^{\{x\}}(e))(y))$

$$= \bigvee_{x\in A} (T_A(e)(x) \wedge F_\Gamma(e)(x, y)) = (T_A(e) \circ F_\Gamma(e))(y).$$

It follows that $\Gamma(T_A)(e) = T_A(e) \circ F_\Gamma(e)$ and consequently $\Gamma(T_A) = T_A \circ F_\Gamma$.

We need to prove that F_Γ is unique.

In fact, if the soft relation $F : E \to \mathcal{F}(A \times B)$ satisfy $\Gamma(T_A) = T_A \circ F$, then

$$F_\Gamma(e)(x, y) = \Gamma({T_A}^{\{x\}})(e)(y) = ({T_A}^{\{x\}}(e) \circ F(e))(y)$$
$$= (\{x\} \circ F(e))(y) = \bigvee_{x'\in A} (\{x\}(x') \wedge F(e)(x', y)) = F(e)(x, y)$$

Hence, $F(e) = F_\Gamma(e)$. It follows that $F = F_\Gamma$.

On the country, let $F : E \to \mathcal{F}(A \times B)$ be a soft relation.

Let $\Gamma_F : \mathcal{F}_A \to \mathcal{F}_B$

$$T_A \mapsto \Gamma_F(T_A) = T_A \circ F$$

then Γ_F is a fuzzy soft linear transformation. In fact,

$$\Gamma(\bigcup_{t\in T} \lambda_t {T_A}^{(t)})(e) = (\bigcup_{t\in T} \lambda_t {T_A}^{(t)})(e) \circ F(e) = (\bigcup_{t\in T} \lambda_t {T_A}^{(t)}(e)) \circ F(e)$$
$$= \bigcup_{t\in T} (\lambda_t {T_A}^{(t)}(e)) \circ F(e) = \bigcup_{t\in T} \lambda_t ({T_A}^{(t)} \circ F)(e) = (\bigcup_{t\in T} \lambda_t ({T_A}^{(t)} \circ F))(e)$$

$= (\bigcup_{t \in T} \lambda_t \Gamma(T_A{}^{(t)}))(e).$

It follows that $\Gamma(\bigcup_{t \in T} \lambda_t T_A{}^{(t)}) = \bigcup_{t \in T} \lambda_t \Gamma(T_A{}^{(t)})$.

Since $\Gamma_F(T_A) = T_A \circ F$, Γ_F is unique.

4 Conclusion

In this paper, we first introduced the concept of soft relation and fuzzy soft relation; In the second, we proposed the concept of the projection and the section of fuzzy soft relation and study their properties; Finally, we gave the concept of fuzzy soft linear transformation and get some conclusions.

References

1. Molodtsov, D.: Soft set theory-first results. Comput. Math. Appl. **37**, 19–31 (1999)
2. Maji, P.K., et al.: Fuzzy soft-sets. J. Fuzzy Math. **9**(3), 589–602 (2001)
3. Majumdar, Pinaki, Samanta, S.K.: Generalised fuzzy soft sets. Comp. Math. Appl. **59**, 1425–1432 (2010)
4. Feng, Feng, Li, Yongming: Violeta Leoreanu-Fotea: Application of level soft sets in decision making based on interval-valued fuzzy soft sets. Comput. Math. Appl. **60**, 1756–1767 (2010)
5. Jiang, Y., Tang, Y., Chen, Q., Liu, H.: Jianchao Tang: Interval-valued intuitionistic fuzzy soft sets and their properties. Comput. Math. Appl. **60**, 906–918 (2010)
6. Gong, K., Xiao, Z., Zhang, X.: The bijective soft set with its operations. Comput. Math. Appl. **60**, 2270–2278 (2010)
7. Xiao, Z., Gong, K., Xia, S., Zou, Y.: Exclusive disjunctive soft sets. Comput. Math. Appl. **59**, 2128–2137 (2010)
8. Maji, P.K., Biswas, R., Roy, A.R.: Soft set theory. Comput. Math. Appl. **45**, 555–562 (2003)
9. Irfan Ali, M., Feng, F., Liu, X., keun Min, W., Shabir, M.: On some new operations in soft set theory. Comput. Math. Appl. **57**, 1547–1553 (2009)
10. Sezgin, Aslıhan, Atagün, A., Osman, K.: On operations of soft sets. Comput. Math. Appl. **61**, 1457–1467 (2011)
11. Çagman, N., Karataş, S.: Serdar Enginoglu:Soft topology. Comput. Math. Appl. **62**, 351–358 (2011)
12. Shabir, Muhammad, Naz, M.: On soft topological spaces. Comput. Math. Appl. **61**, 1786–1799 (2011)
13. Tanay, B., Burç Kandemir, M.: Topological structure of fuzzy soft sets. Comput. Math. Appl. **61**, 2952–2957 (2011)
14. Aktas, H., Cagman, N.: Soft sets and soft groups. Inf. Sci. **177**, 2726–2735 (2007)
15. Aygünogu, A., Aygün, H.: Introduction to fuzzy soft groups. Comput. Math. Appl. **58**, 1279–1286 (2009)
16. Acar, Ummahan, Koyuncu, Fatih, Tanay, B.: Soft sets and soft rings. Comput. Math. Appl. **59**, 3458–3463 (2010)
17. Irfan Ali, M., Shabir, M., Naz, M.: Algebraic structures of soft sets associated with new operations. Comput. Math. Appl. **61**, 2647–2654 (2011)
18. Jun, Y.B., Lee, K.J., Park, C.H.: Fuzzy soft set theory applied to BCK/BCI-algebras. Comput. Math. Appl. **59**, 3180–3192 (2010)

19. Babitha, K.V., Sunil, J.J.: Soft set relations and functions. Comput. Math. Appl. **60**, 1840–1849 (2010)
20. Yang, Hl, Guo, Z.L.: Kernels and closures of soft set relations, and soft set relation mappings. Comput. Math. Appl. **61**, 651–662 (2011)
21. Maji, P.K., Roy, A.R.: An application of soft sets in a decision making problem. Comput. Math. Appl. **44**, 1077–1083 (2002)
22. Çagman, Naim, Enginoglu, S.: Soft matrix theory and its decision making. Comput. Math. Appl. **59**, 3308–3314 (2010)
23. Majumdar, Pinaki, Samanta, S.K.: On soft mappings. Comput. Math. Appl. **60**, 2666–2672 (2010)
24. Luo, C.Z.: Introduction to Fuzzy Sets(1)(In Chinese), pp. 1–486. Beijing Normal University Press, Beijing (1989)

Existence of Solutions for the Fuzzy Functional Differential Equations

Ya-bin Shao, Huan-huan Zhang and Guo-liang Xue

Abstract In this paper, we consider the existence theorems of solution for fuzzy functional differential equations under the compactness-type conditions and dissipative type conditions, via the properties of the embedding mapping from fuzzy number to Banach space.

Keywords Fuzzy number · Fuzzy functional differential equations · Initial value problems · Compactness-type conditions · Dissipative-type conditions

1 Introduction

The fuzzy differential equation was first introduced by Kandel and Byatt [1]. Since 1987, the Cauchy problems for fuzzy differential equations have been extensively investigated by several authors [2–11] on the metric space (E^n, D) of normal fuzzy convex set with the distance D given by the maximum of the Hausdorff distance between the corresponding level sets. In particular, Kaleva [2] studied the initial value problem

$$x'(t) = f(t, x(t)), \quad x(t_0) = x_0$$

where $f : T \times E^n \to E^n$ is a continuous fuzzy mapping, $T = [a, b]$, E^n is a fuzzy number space and $x_0 \in E^n$. The result was obtained as follows

Theorem 1. *Let $f : T \times E^n \to E^n$ be continuous and assume that there exist a $k > 0$ such that*

$$D(f(t, x), f(t, y)) \le kD(x, y)$$

for all $t \in T$, $x, y \in E^n$. Then the above initial value problem has a unique solution on T.

Y. Shao (✉) · H. Zhang · G. Xue
College of Mathematics and Computer Science, Northwest University for Nationalities, Lanzhou 730030, China
e-mail: yb-shao@163.com

B.-Y. Cao and H. Nasseri (eds.), *Fuzzy Information & Engineering and Operations Research & Management*, Advances in Intelligent Systems and Computing 211,
DOI: 10.1007/978-3-642-38667-1_22,

Since then, many authors studied the existence and uniqueness of solutions for the initial value problem for fuzzy differential equations under kinds of conditions and obtained several meaningful results. However, those research findings were not satisfactory. Until 1997, Nieto [3] proved the initial value problem for fuzzy differential equations has solutions if f is a continuous and bounded function. It is well known that the Lipschtz conditions can not be educed if f is a continuous and bounded function. That is to say, the results in [3] were perfect complement for the Theorem 1. What's more, Wu and Song [5, 6] and Song, Wu and Xue [7] changed the initial value problem of fuzzy differential equations into a abstract differential equations on a closed convex cone in a Banach space by the operator j that is the isometric embedding from (E^n, D) onto its range in the Banach space X. They established the relationship between a solution and its approximate solutions to fuzzy differential equations. Furthermore, they obtained the local existence theorem under the compactness-type and dissipative-type conditions. Park and Han [8] obtained the global existence and uniqueness of fuzzy solution of fuzzy differential equation using the the properties of Hasegawa's function and successive approximation.

There exists an extensive theory for function differential equations with includes qualitative behavior of classes of such equations and application to biological and engineering processes, for details, see [12, 13]. However, the concrete example is the radio-cardiogram, where the two compartments correspond to the left and right ventricles of the pulmonary and systematic circulation. Pipes coming out from and returning into the same compartment may represent shunts, and the equation representing this model is a nonlinear neutral Volterra integrodifferential equation in [14]. This classes of equations also arise, for example, in the study of problems such as heat conduction in materials with memory or population dynamics for spatially distributed populations, see [15, 16]. It was well known that fuzzy functional differential equations are more exactly describe the objective worlds than fuzzy differential equations, so the deeply researching of fuzzy functional equations is needed. In [18], the existence and uniqueness of a fuzzy solution for the nonlinear fuzzy neutral functional differential equation are established via Banach fixed point analysis approach.

In this paper, we investigate the Cauchy problems for fuzzy functional differential equation

$$x'(t) = f(t, x_t), \quad x(t_0) = \varphi_0$$

in fuzzy number space (E^n, D) by using the Radstorm embedding results [19].

2 Preliminaries

Let $P_k(R^n)$ denote the family of all nonempty compact convex subset of R^n and define the addition and scalar multiplication in $P_k(R^n)$ as usual. Let A and B be two nonempty bounded subset of R^n. The distance between A and B is defined by the Hausdorff metric [20]:

$$d_H(A, B) = \max\{\sup_{a \in A} \inf_{b \in B} \parallel a - b \parallel, \sup_{b \in B} \inf_{a \in A} \parallel b - a \parallel\}.$$

Denote $E^n = \{u : R^n \rightarrow [0, 1] | u$ satisfies (1)–(4) below $\}$ is a fuzzy number space. where

1. u is normal, i.e. there exists an $x_0 \in R^n$ such that $u(x_0) = 1$,
2. u is fuzzy convex, i.e. $u(\lambda x + (1 - \lambda)y) \geq \min\{u(x), u(y)\}$ for any $x, y \in R^n$ and $0 \leq \lambda \leq 1$,
3. u is upper semi-continuous,
4. $[u]^0 = cl\{x \in R^n | u(x) > 0\}$ is compact.

For $0 < \alpha \leq 1$, denote $[u]^\alpha = \{x \in R^n | u(x) \geq \alpha$. Then from (1)–(4), it follows that the α-level set $[u]^\alpha \in P_k(R^n)$ for all $0 \leq \alpha < 1$.

According to Zadeh's extension principle, we have addition and scalar multiplication in fuzzy number space E^n as follows: $[u+v]^\alpha = [u]^\alpha + [v]^\alpha$, $[ku]^\alpha = k[u]^\alpha$, where $u, v \in E^n$ and $0 \leq \alpha \leq 1$.

Define $D : E^n \times E^n \rightarrow [0, \infty)$

$$D(u, v) = \sup\{d_H([u]^\alpha, [v]^\alpha) : \alpha \in [0, 1]\},$$

where d_H is the Hausdorff metric defined in $P_k(R^n)$. Then it is easy to see that D is a metric in E^n. We know that [20]:

1. (E^n, D) is a complete metric space,
2. $D(u + w, v + w) = D(u, v)$ for all $u, v, w \in E^n$,
3. $D(\lambda u, \lambda v) = |\lambda| D(u, v)$ for all $u, v, w \in E^n$ and $\lambda \in R$.

The metric space (E^n, D) has a linear structure, it can be imbedded isomorphically as a cone in a Banach space of function $u^* : I \times S^{n-1} \longrightarrow R$, where S^{n-1} is the unit sphere in R^n, with an imbedding function $u^* = j(u)$ defined by $u^*(r, x) = \sup_{\alpha \in [u]^\alpha} \langle \alpha, x \rangle$ for all $\langle r, x \rangle \in I \times S^{n-1}$ [20]. For more detailed results about embedding, we refer the read to [21]. For a development, see Ma [22].

Theorem 2. *[19] There exist a real Banach space X such that E^n can be imbedding as a convex cone C with vertex 0 into X. Furthermore the following conditions hold true:*

1. *the imbedding j is isometric,*
2. *addition in X induces addition in E^n,*
3. *multiplication by nonnegative real number in X induces the corresponding operation in E^n,*
4. *$C - C$ is dense in X,*
5. *C is closed.*

In the following, we recall some main concepts and properties of integrability and H-differentiability for the fuzzy number function [2, 21, 23–25].

Let $x, y \in E^n$. If there exist a $z \in E^n$ such that $x = y + z$, z is called the H-difference of x and y. That is denoted $x - y$. For brevity, we always assume that it satisfies the H-difference when dealing with the operation of subtraction of fuzzy numbers throughout this paper.

Definition 1. *[23] Let $T = [t_0, t_0 + p] \subset R (p > 0)$ be compact interval. A mapping $F : T \to E^n$ is differentiable at $t_0 \in T$, if there exist a $F^{'}(t_0) \in E^n$, such that the limits*

$$\lim_{h \to 0^+} \frac{F(t_0 + h) - F(t_0)}{h}, \quad \lim_{h \to 0^+} \frac{F(t_0) - F(t_0 + h)}{h}$$

exist and are equal to $F^{'}(t_0)$.

Here the limits are taken in the metric space (E^n, D). At the endpoint of T, we consider only one-side fuzzy derivatives. If $F : T \to E^n$ is differentiable at $t_0 \in T$, then we say that $F^{'}(t_0)$ is the fuzzy derivative of $F(t)$ at the point t_0.

The fuzzy mapping $F : T \to E^n$ is called strong measurable, if for all $\alpha \in [0, 1]$ the set-valued mapping $F_\alpha : T \to P_k(R^n)$ defined by $F_\alpha(t) = [F(t)]^\alpha$ is Lebesgue measurable, where $P_k(R^n)$ is endowed with the topology generated by the Hausdorff metric d.

A mapping $F_\alpha : T \to E^n$ is called integrable bounded, if there exist an integrable function h such that $\| x \| \leq h(t)$ for all $x \in F_0(t)$.

Definition 2. *[24] Let $F : T \to E^n$. The integral of F over T, denote by $\int_T F(t)\mathrm{d}t$, is defined level-wise by the equation*

$$\left[\int_T F(t)\mathrm{d}t\right]^\alpha = \int_T F_\alpha(t)\mathrm{d}t$$

$$= \{\int_T f(t)\mathrm{d}t \,|\, f \; is \; a \; measurable \; selection \; for \; F_\alpha\}$$

for all $0 < \alpha \leq 1$.

A strongly measurable and integrable mapping $F_\alpha : T \to E^n$ is said to be integrable over T if $\int_T F(t)dt \in E^n$. From [23], we know that if $F_\alpha : T \to E^n$ is continuous, then it is integrable.

Theorem 3. *[25] If $F, G : T \to E^n$ be integrable and $\lambda \in R$, then*

1. *$\int_T (F(t) + G(t))\mathrm{d}t = \int_T F(t)\mathrm{d}t + \int_T G(t)\mathrm{d}t$,*
2. *$\int_T \lambda F(t)\mathrm{d}t = \lambda \int_T F(t)\mathrm{d}t$,*
3. *$D(F(t), G(t))$ is integrable,*
4. *$D(\int_T F(t)\mathrm{d}t, \int_T G(t)\mathrm{d}t) \leq \int_T D(F(t) + G(t))\mathrm{d}t$.*

Furthermore, we go on to list the contents on the Kuratowski's measure of non-compactness and the Ascoli-Arzela theorem in [26] on Banach space.

Definition 3. *[26] Let S be an arbitrary bounded subset of real Banach space X, the Kuratowski's measure of non-compactness is defined as:*

$$\alpha(S) = \inf\{\delta > 0 : S = \bigcup_{i=1}^{m} S_i, d(S_i) \leq \delta\}.$$

For $T = [t_0, t_0+p](p > 0)$, $C[T, X]$ denotes the space of the abstract continuous function from T to X, the norm $\|x(t)\| < \max_{t\in T} \|x(t)\|, \forall x \in C[T, X]$, we denote $H(t) = \{x(t) : x \in H\} \subset X$.

Theorem 4. *[26] A set $H \subset C(T, X)$ is a relative compact set if and only if H is equi- continuous and fo any $t \in T$, $H(t)$ is relative compact set in X.*

In the following we list several comparison theorems on classical ordinary differential equations as follows:

Theorem 5. *[26] Let $G \subset R^2$ be an open set $g \in C[G, R^1]$, $(t_0, u_0) \in G$. Suppose that the maximum solution of initial value problem*

$$u'(t) = g(t, u), \quad u(t_0) = u_0$$

and its largest interval of existence of right solution is $[t_0, t_0 + a)$, If $[t_0, t_1] \subset [t_0, t_0 + a)$, then there exists an $\varepsilon_0 > 0$ such that the maximum solution $r(t, \varepsilon)$ to the initial value problem

$$u'(t) = g(t, u) + \varepsilon, \quad u(t_0) = u_0 + \varepsilon$$

exists on $[t_0, t_1]$ whenever $0 < \varepsilon < \varepsilon_0$,and $r(t, \varepsilon)$ uniformly converges to $r(t)$ on $[t_0, t_1]$ as $\varepsilon \to 0^+$.

3 Main Results

In this section, by using theorem of ε_n-approximate solutions and the embedding results on fuzzy number space E^n, we give the existence and uniqueness theorems under dissipative-type conditions for the initial problem of the functional differential equations:

$$x'(t) = f(t, x_t), \quad x(t_0) = \varphi_0 \in E^n. \tag{1}$$

Assume that (E^n, D) is a fuzzy number space, there exists a real number $\tau > 0$, let $C = C[[-\tau, 0], E^n]$. For $\varphi \in C$, we define

$$D(\varphi, \check{0})_0 = \sup_{t\in[-\tau,0]} D(\varphi(t), \check{0}),$$

where $D(\cdot)_0$ denoted the metric in E^n. Let F is a nonempty closed subset of (E^n, D), we assume that $C_F = \{\varphi \in C|\varphi(0) \in F\}$, $\tilde{C_F} = \{\varphi \in C|\varphi(0) \in F, \varphi(t)\}$ for all $t \in [-\tau, 0]\}$. Obviously C_F and $\tilde{C_F}$ are the closed subset in C. Let $f \in C[R_+ \times C_F, E^n]$. From [27], we can get that there exist $b > 0$ such that f is bounded in $C_0(b) = \cup_{t\in[t_0,t_0+a]}(\{t\} \times C_F^t(b))$.

Firstly, we list the basic definition of dissipative-type conditions as follows:

Definition 4. *[7] Assume that $V : [t_0, t_0 + a] \times B(x_0, b) \times B(x_0, b) \to R_+$ is a continuous function provided*

1. *$V(t, x, x) = 0$, $V(t, x, y) > 0 (x \neq y)$ for all $t \in [t_0, t_0 + a]$, $x, y \in B(x_0, b)$ and that $\lim\limits_{n\to\infty} V(t, x_n, y_n)$ implies $\lim\limits_{n\to\infty} D(x_n, y_n) = 0$ whenever $\{x_n\} \subset B(x_0, b)$, $\{y_n\} \subset B(x_0, b)$,*
2. $$|V(t, x, y) - V(t, x_1, y_1)| \leq L \cdot (D(x - x_1) + D(y - y_1))$$

   for every $t \in [t_0, t_0 + a]$, $x, y, x_1, y_1 \in B(x_0, b)$, where $L > 0$ is a constant. Then V is called a function of $L - D$ type.

Definition 5. *[7] Assume that $f \in C[R_0, E^n]$ and there exist a function V of $L - D$ type such that*

$$D_+V(t, x, y) \leq g(t, V(t, x, y))$$

for all $t \in [t_0, t_0 + a]$, $x, y \in B(x_0, b)$, where

$$D_+V(t, x, y) = \lim_{h\to 0_+} \frac{1}{h}[V(t + h, x + hf(t, x), y + hf(t, y)) - V(t, x, y)]$$

and $g \in C[[t_0, t_0 + a] \times R_+, R]$, $g(t, 0) = 0$ and maximum solution of the scalar differential equation

$$u' = g(t, u), \quad u(t_0) = 0,$$

is $u = 0$ on $[t_0, t_0+a]$. Then f is said to satisfy Lyapunov dissipative-type conditions.

Next we listed the conditions for using in this paper:

(H_1) $f \in C[[t_0, t_0+a]\times C_F, E^n]$ is a fuzzy number value function. $t_0 \in R_+$, $\varphi_0 \in C_F$, a, b and $M (M \leq 1)$ satisfies $D(f(t, \varphi), \check{0}) \leq M - 1$ for any $(t, \varphi) \in C_0(b)$,

(H_2) For all $(t, \varphi) \in [t_0, t_0 + a] \times C_F$, we have

$$\lim_{h\to 0^+} \frac{1}{\lambda} \inf_{z\in F} D(\varphi_0 + hf(t, \varphi), z) = 0,$$

(H_3) For all $t \in [t_0, t_0 + a]$ and $\phi^t \subset C_F^t(b)$, if $\alpha(j\phi^t(s)) \leq \alpha(j\phi^t(0)) (\forall s \in [-\tau, 0])$ we have

$$\lim_{h\to 0^+} \frac{1}{h}[\alpha(j\phi^t(0)) - \alpha\{j\varphi(0) - jhf(t,\varphi)|\varphi \in \phi^t]$$
$$\le g(t, \alpha(j\phi^t(0)))$$

where $\alpha(\cdot)$ is the Kuratowski's measure of non-compactness and j is the embedding mapping from fuzzy number space E^n to Banach space X,

(H_4)

$$D(f(t,\varphi), f(t,\psi)) \le D(\varphi(0), \psi(0)) \quad (2)$$

for $t \in [t_0, t_0 + a]$, where $\varphi, \psi \in C_F^t$, such that

$$D(\varphi(T), \psi(T)) \le D(\varphi(0), \psi(0)), \quad \forall T \in [-\tau, 0]$$

$(H_5) g \in C[[t_0, t_0 + a] \times [0, 2b], R_+]$ and $g(t, 0) = 0$, the initial problems

$$u' = g(t, u), \quad u(t_0) = 0$$

has only solution $u(t) = 0$.

Theorem 6. *[27] Let* $r = \min\{a, \frac{b}{M}\}$ *and satisfies conditions* (H_1), (H_2), *then for all* $0 < \varepsilon < 1$, *the initial problem (1) has such* $\varepsilon-$*approximate solution* $x(t)$: $[t_0 - \tau, t_0 + r] \to E^n$ *satisfies:*

(a) *There exist* $\{\sigma_i | i = 0, 1, \cdots\} \subset [t_0, t_0 + r]$ *such that* $\sigma_0 = t_0, \sigma_i < \sigma_{i+1}, \sigma_i < t_0 + r, \sigma_{i+1} - \sigma_i \le \varepsilon$ *and implies* $\lim_{i\to\infty} \sigma_i = t_0 + r$,

(b) *For all* $t \in [t_0 - \tau, t_0]$, *we have* $x(t) = \varphi_0(t - t_0)$ *and*

$$D(x(t), x(s)) \le M|t - s|, \quad \forall t, s \in [t_0, t_0 + r),$$

(c) *For all* $i \ge 0$, $(\sigma_i, x_{\sigma_i}) \in C_0(b)$, *and it is linear on every intervals* $[\sigma_i, \sigma_{i+1}]$,

(d) *For all* $t \in (\sigma_i, \sigma_{i+1})$, *we have* $\|jx'(t) - jf(\sigma_i, x_{\sigma_i})\| \le \varepsilon$.

Next, we give the main results of this section.

Theorem 7. *Assume that* F *is a nonempty closed convex subset of* E^n, *and satisfies conditions* (H_1), (H_2), (H_3) *and* (H_4), *then the initial problem (1) has unique solution* $x(t) \in F$ *for* $\forall t \in [t_0, t_0 + r]$, *where* $r = \min\{a, \frac{b}{M}\}$.

Proof Let $\{\varepsilon_n\} \subset (0, 1)$, and ε_n is monotone converges decreasingly to 0. Let x_m, x_n are the ε_m-approximate solution and ε_n-approximate solution of initial problem (1) which satisfies properties in Theorem 6. Since F is the convex set we have

$$x_n(t) \in F, \qquad x_m \in F. \quad (3)$$

for $t \in [t_0, t_0 + r]$. We define $m(t)$ by

$$m(t) = D(x_m(t) - x_n(t)), \quad t \in [t_0, t_0 + r]$$

When $t \in (\sigma_i^n, \sigma_{i+1}^n) \cap (\sigma_j^n, \sigma_{j+1}^n)$, we get

$$\begin{aligned} D^+m(t) \le \|jx_n'(t) - jx_m'(t)\| &\le \|jx_n(t) - jf(\sigma_i^n, x_{n,\sigma_i^n})\| \\ &+ \|jx_m'(t) - jf(\sigma_j^m, x_{m,\sigma_i^m})\| \\ &+ \|jf(\sigma_i^n, x_{n,\sigma_i^n}) - jf(t, x_{n,t})\| \\ &+ \|jf(\sigma_j^m, x_{m,\sigma_i^m}) - jf(t, x_{m,t})\| \\ &+ \|jf(t, x_{n,t}) - jf(t, x_{m,t})\|. \end{aligned} \tag{4}$$

From Theorem 6, we can obtain

$$D(f(t, \varphi), f(\sigma_i^n, x_{n,\sigma_i^n})) \le \varepsilon_n$$

when $(t, \varphi) \in [\sigma_i^n, \sigma_{i+1}^n] \times C_F, D(\varphi - x_{n,\sigma_i^n}) \le 2M\delta_i^n, |t - \sigma_i^n| \le \delta_i^n$. From the properties (d) in Theorem 6 and Eq. (4), we have

$$D^+m(t) \le 2(\varepsilon_n + \varepsilon_m) + \|jf(t, x_{n,t}) - jf(t, x_{m,t})\|.$$

If t satisfies $D(x_{n,t}(T) - x_{m,t}(T)) \le D(x_m(t) - x_n(t))$, we apply condition (H_4)

$$D^+m(t) \le 2(\varepsilon_n + \varepsilon_m) + g(t, m(t)) \tag{5}$$

is true. Then, $m(t) : [t_0 - \tau, t_0 + r] \to R_+$ is continuous and satisfies Eq. (5) for $\forall t \in [t_0, t_0 + r] \backslash S$(where S is a at most countable set that consisted of t and t satisfies $m_t(T) \le m(t)$). According to Theorem 5, we have

$$m(t) \le r_{n,m(t,t_0,0)}, \quad t \in [t_0, t_0 + r],$$

where $r_{n,m(t,t_0,0)}$ is the maximum solution of initial problem

$$u' = g(t, u) + 2(\varepsilon_n + \varepsilon_m), \quad u(t_0) = 0.$$

According to Theorem 4, $r_{n,m}(t, t_0)$ converges uniformly to $r(t, t_0, 0)$ for $t \in [t_0, t_0 + r]$, where $r(t, t_0, 0)$ is the maximum solution of initial problem

$$u' = g(t, u), \quad u(t_0) = 0.$$

From (H_4), we get $r(t, t_0, 0) = 0$. This indicated that

$$\lim_{n,m \to \infty} D(x_n, x_m) = 0.$$

In other words, $\{x_n(t)\}$ converges uniformly to $x(t)$ on $[t_0-\tau, t_0+r]$. From Theorem 6, $x(t)$ is the solution of initial problem (1). From Eq. (3), we know that $x(t) \in F$ for $t \in [t_0, t_0+r]$.

Finally, we prove uniqueness. Suppose $\tilde{x}(t)$ is another solution of initial problem (1) such that

$$x(t), \tilde{x}(t) \in F \cap B[\varphi_0(0), b], \quad \forall t \in [t_0, t_0+r],$$

where $B[\varphi_0(0), b] = \{x \in E^n | D(x, \varphi_0(0)) \le b\}$. Let $m(t) = D(x(t), \tilde{x}(t))$. Then similar to the proof of Eq. (4), we have

$$D^+m(t) \le D(f(t, x_t), f(t, \tilde{x}_t)) \le g(t, m(t))$$

for $t > t_0, m_t(T) \le m(t)$. Since $m_{t_0} = 0$, according to Theorem 5 and condition (H_4), we know $m(t) = 0$, that is to say $x(t) = \tilde{x}_t$. The proof is completed.

At the last, we give the existence theorems under the compactness-type conditions for fuzzy functional differential equations.

Theorem 8. *Assume that condition (H_1), (H_2), (H_3) and (H_5) satisfied. f is continuous uniformly on $[t_0, t_0+a] \times C_F$. Let $r = \min\{a, \frac{b}{M}\}$. Then the initial problems (1) has at least one solution $x(t)$ on $[t_0-\tau, t_0+r]$ and $x(t) \in F, \forall t \in [t_0, t_0+r]$.*

Proof Let $\{\varepsilon_n\} \subset (0, 1)$, ε_n is monotonous decreasingly converges to 0. Let $\{\varepsilon_n\}$ is the ε_n-approximate solution sequence of initial problem (1) and satisfied the properties (a)–(d) in Theorem 6. In fact, we need to prove that $\{x_n\}$ has a subsequence and it converges uniformly on $[t_0-\tau, t_0+r]$. In fact, applying to Theorem 4, we need to prove that $\{x_n\}$ is a relative compact set in E^n for $\forall t \in [t_0, t_0+r]$. (According to Theorem 6, we already know $\{x_n\}$ is uniformly bounded and equi-continuous.)

Define

$$p(t) = \alpha(j\{x_n\}) \quad \forall t \in [t_0, t_0+r].$$

We easily know, for any natural number k,

$$p(t) = \alpha(j\{x_n(t)|n \ge k\}) = \alpha(j\{x_n(\tau_n(t))|n \ge k\}), \tag{6}$$

and

$$\alpha(j\{x_n - hf(\tau_n(t), x_{n,\tau_n(t)})\}) = \alpha(j\{x_n(\tau_n(t)) - hf(\tau_n(t), x_{n,\tau_n(t)})\}). \tag{7}$$

Let $t \in [t_0, t_0+r]$. Combining Eqs. (6) and (7) if $t \in (\sigma_i^n, \sigma_{i+1}^n)$, we get

$$\begin{aligned}\frac{p(t)-p(t-h)}{h} &= \frac{1}{h}[\alpha(j\{x_n(\tau_n(t))\}) - \alpha(j\{x_n(t-h)\})] \\ &\le \frac{1}{h}[\alpha(j\{x_n(\tau_n(t))\}) - \alpha(j\{x_n(t) - hf(\tau_n(t), x_{n,\tau_n(t)})\})]\end{aligned}$$

$$\begin{aligned}
&+\frac{1}{h}[\alpha(j\{x_n(t)-x_n(t-h)-hf(\tau_n(t),x_{n,\tau_n(t)})\})] \\
&=\frac{1}{h}[\alpha(j\{x_n(\tau_n(t))\})-\alpha(j\{x_n(\tau_n(t))-hf(\tau_n(t),x_{n,\tau_n(t)})\})] \\
&+\frac{1}{h}[\alpha(j\{x_n(t)-x_n(t-h)-hf(\tau_n(t),x_{n,\tau_n(t)})\})]. \qquad (8)
\end{aligned}$$

According to the properties (d) in Theorem 6, we have

$$\begin{aligned}
D(x_n(t),x_n(t-h)+hf(\tau_n(t),x_{n,\tau_n(t)})) &\le \int_{t-h}^{t} D(x_n'(s),f(\tau_n(s),x_{n,\tau_n(s)}))\mathrm{d}s \\
&+\int_{t-h}^{t} D(f(\tau_n(s),x_{n,\tau_n(s)}),f(\tau_n(t),x_{n,\tau_n(t)}))\mathrm{d}s \\
&\le \varepsilon_n h+\int_{t-h}^{t} D(f(\tau_n(s),x_{n,\tau_n(s)}),f(\tau_n(t),x_{n,\tau_n(t)}))\mathrm{d}s
\end{aligned} \qquad (9)$$

Since $f(t,x)$ is continuous uniformly, there exists $\delta=\delta(\eta)>0$ such that

$$|t_n(s)-\tau_n(t)|<\delta,\quad D(x_{n,\tau_n(s)},x_{n,\tau_n(t)})_0<\delta$$

for $\forall\eta>0$. We could obtain

$$D(f(t_n(s),x_{n,\tau_n(s)}),f(t_n(t),x_{n,\tau_n(t)}))<\eta.$$

On the other hand, we have

$$|t_n(s)-\tau_n(t)|<|t-s|+\varepsilon_n\le h+\varepsilon_n,$$

and $D(x_n(\tau_n(s)+T),x_n(\tau_n(t)+T))=$

$$\begin{cases}
M|\tau_n(s)-\tau_n(t)|\le M(h+\varepsilon_n), & t_0\le\tau_s+T\\
D(\varphi_0(\tau_n(s)+T-t_0),\varphi_0(\tau_n(t)+T-t_0)), & \tau_n(t)+T\le t_0\\
D(\varphi(\tau_n(s)+T-t_0),\varphi_0(0))+D(x_n(t_0),x_n(\tau_n(t)+T)), & t_0-\tau_n(T)<T<t_0-\tau_n(s).
\end{cases}$$

Since φ_0 is continuous uniformly on $[-\tau,0]$, there exists $\bar{\delta}>0$, such that $\tau_s\in[t_0,t_0+r]$, $\tau_n(t)\in[t_0,t_0+r]$, and $|\tau_n(s)-\tau_n(t)|\le h+\varepsilon_n<\bar{\delta}$. We have $D(\varphi_0(\tau_n(s),\varphi_0(\tau_n(t))_0<\frac{\delta}{2}$.

Hence

$$D(x_n(\tau_n(s)+T),x_n(\tau_n(t)+T))\le\begin{cases}
M(h+\varepsilon_n), & T\ge t_0-\tau_n(s)\\
\frac{\delta}{2}, & T\le t_0-\tau_n(t)\\
\frac{\delta}{2}+M(h+\varepsilon_n), & t_0-\tau_n(t)\le T\le t_0-\tau_n(s).
\end{cases}$$

We make $m = m(\delta)$ big enough and $\tilde{h}$ small enough. When $n \geq m(\delta)$, and $h < \tilde{h}$, we get

$$\varepsilon_n < \eta, h + \varepsilon_n < \bar{\delta}, M(h + \varepsilon_n) \leq \frac{\delta}{2}.$$

Then

$$\frac{1}{h} D(x_n(t), x_n(t-h) + hf(\tau_n(t), x_{n,\tau_n(t)})) \leq \varepsilon_n + \eta < 2\eta.$$

For this reason, we have

$$\frac{1}{h}\alpha(j\{x_n(t) - x_n(t-h) - hf(\tau_n(t), x_{n,\tau_n(t)})\})$$

$$= \alpha(j\{\frac{x_n(t) - x_n(t-h)}{h} - f(\tau_n(t), x_{n,\tau_n(t)}) | n \geq m(\eta)\}) \leq 2(2\eta).$$

We can obtain

$$\lim_{h \to 0^+} \frac{1}{h}\alpha(j\{x_n(t) - x_n(t-h) - hf(\tau_n(t), x_{n,\tau_n(t)})\}) \leq 4\eta.$$

From the arbitrariness of η, we know that

$$\lim_{h \to 0^+} \frac{1}{h}\alpha(j\{x_n(t) - x_n(t-h) - hf(\tau_n(t), x_{n,\tau_n(t)})\}) = 0.$$

According to Eq. (8), we have

$$D_{-}p(t) \leq \lim_{h \to 0^+} \frac{1}{h}[\alpha(j\{x_n(\tau_n(t))\} - \alpha(j\{x_n(\tau_n(t)) - hf(\tau_n(t), x_{n,\tau_n(t)})\})].$$

For any natural number N, we define $\Phi_N^t = \{x_{n,\tau_n(t)} | n \geq N\}$. Next, when N big enough, we will prove

$$\Phi_N^t \in C_F^t(b). \tag{10}$$

In fact, we have $D(x_n(\tau_n(t) + T), y(t + T)) =$

$$\begin{cases} D(x_n(\tau_n(t) + T), x_n(t_0)) \leq Mr \leq b, & \tau_n(t) + T \geq t_0 \\ D(\varphi_0(\tau_n(t) + T - t_0), \varphi_0(t + T - t_0)), & t + T \leq t_0 \\ D(\varphi_0(\tau_n(t) + T - t_0), \varphi_0(0)), & \tau_n(t) + T \leq t_0 \leq t_0 + T. \end{cases}$$

Let $\delta_{\varphi_0}(b) > 0$, such that $|t - \tau_n(t)| < \delta_{\varphi_0}(b)$, we can obtain $D(x_{n,\tau_n(t)}, y_t)_0 \leq b$. We make N big enough, when $n \geq N$, we get

$$|t - \tau_n(t)| \leq \varepsilon_n \leq \varepsilon_N < \delta_{\varphi_0}(b).$$

Thus, $D(x_{n,\tau_n(t)}, y_t)_0 \leq b$, in other words, Eq. (10) holds.

From condition (H_3), we can easily prove $D_{-}p(t) \leq g(t, p(t))$ hold true. So, according to Theorem 5, we have $p(t) \leq r(t, t_0, 0)$, where $r(t, t_0, 0)$ is the maximum solution of

$$u' = g(t, u), \quad u(t_0) = 0$$

From (H_4), we know $p(t) \equiv 0$. This completes the proof.

4 Conclusion

In [18], the existence and uniqueness of a fuzzy solution for the nonlinear fuzzy neutral functional differential equation

$$\frac{\mathrm{d}}{\mathrm{d}t}[x(t) - f(t, x_t)] = Ax(t) + g(t, x_t)$$

where A is a fuzzy coefficient and f and g are continuous function, were established via Banach fixed point analysis approach. In this paper, we investigate the Cauchy problems for fuzzy functional differential equation

$$x'(t) = f(t, x_t), \quad x(t_0) = \varphi_0$$

in fuzzy number space (E^n, D) by using the Radstorm embedding results [19]. Our results improve and extend the fuzzy functional differential equations and some relevant results in ordinary fuzzy differential equations.

Acknowledgments Thanks to the support by National Natural Science Foundation of China (No. 11161041) and 2012 Scientific Research Fund for State Ethnic Affairs Commission of China; Fundamental Research Fund for the Central Universities(No. Zyz2012079, No. Zyz2012081, No. 31920130009 and No. 31920130010).

References

1. Kandel, A., Byatt, W.J.: Fuzzy differential equations. In: Proceedings of International Conference on Cybernatics and Society, Tokyo (1978)
2. Kaleva, O.: Fuzzy diffrential equations. Fuzzy Sets Syst. **24**, 301–319 (1987)
3. Nieto, J.J.: The Cauchy problem for fuzzy differential equations. Fuzzy Sets Syst. **102**, 259–262 (1999)
4. Kaleva, O.: The Cauchy problem for fuzzy differential equations. Fuzzy Sets Syst. **35**, 389–396 (1990)
5. Wu, C.X., Song, S.J.: Existence theorem to the Cauchy problem of fuzzy diffrential equations under compactness-type conditions. Inf. Sci. **108**, 123–134 (1998)
6. Wu, C.X., Song, S.J., Lee, E.S.: Approximate solutions and existence and uniqueness theorem to the Cauchy problem of fuzzy diffrential equations. J. Math. Anal. Appl. **202**, 629–644 (1996)

7. Song, S.J., Wu, C.X., Xue, X.P.: Existence and uniqueness theorem to the Cauchy problem of fuzzy diffrential equations under dissipative conditions. Comput. Math. Appl. **51**, 1483–1492 (2006)
8. Park, J.Y., Han, H.K.: Fuzzy differential equations. Fuzzy Sets Syst. **110**, 69–77 (2000)
9. Song, S.J., Guo, L., Feng, C.B.: Global existence of solution to fuzzy diffrential equations. Fuzzy Sets Syst. **115**, 371–376 (2000)
10. Ding, Z.H., Ma, M., Kandel, A.: Existence of the solutions of fuzzy differential equations with parameters. Inf. Sci. **99**, 205–217 (1997)
11. Seikkala, S.: On the initial value problem. Fuzzy Sets Syst. **24**, 319–330 (1987)
12. Hale, J.K., Verduyn Lunel, S.M.: Introduction to Functional Differential Equations. Springer, New York (1993)
13. Kolmonovskii, V.B., Nosov, V.R.: Stability of Functional Differential Equations. Academic Press, New York (1986)
14. Gyori, I., Wu, J.: A neutral equation arising from compartmental systems with pipes. J. Dyn. Differ. Equ. **3**, 289–311 (1991)
15. Gopalsamy, K.: Stability and Oscillations in Delay Differential Equations of Population Dynamics. Kluwer Academic, Dordrecht (1992)
16. Guo, M.S., Xue, X.P., Li, R.L.: Impulsive functional differential inclusions and fuzzy population models. Fuzzy Sets Syst. **138**, 601–615 (2003)
17. Kloeden, P.E.: Fuzzy dynamical systems. Fuzzy Sets Syst. **7**, 275–296 (1982)
18. Balasubramaniam, P., Muralisankar, S.: Existence and uniqueness of a fuzzy solution for the nonlinear fuzzy neutral functional differential equation. Comput. Math. Appl. **42**, 961–967 (2001)
19. Radstrom, H.: An embedding theorem for spaces of convex set. Proc. Amer. Math. soc. **3**, 165–169 (1959)
20. Kloeden, P.P.: Characterization of compact subsets of fuzzy sets. Fuzzy Sets Syst. **29**, 341–348 (1989)
21. Puri, M.L., Ralescu, D.A.: Differentials of fuzzy functions. J. Math. Anal. Appl. **91**, 552–558 (1983)
22. Ma, M.: On embedding problems of fuzzy number space: part 5. Fuzzy Sets Syst. **55**, 313–318 (1989)
23. Dubois, D., Prade, H.: Towards fuzzy differential calculus: part 1: integration of fuzzy mappings. Fuzzy Sets Syst. **8**, 1–17 (1982)
24. Aumann, R.J.: Integrals of set-valued fuctions. J. Math. Anal. Appl. **12**, 1–12 (1965)
25. Puri, M.L., Ralescu, D.A.: Fuzzy random variables. J. Math. Anal. Appl. **114**, 409–422 (1986)
26. Lakshmikantham, V., Leela, S.: Nonlinear Diffrential Equations in Abstract Spaces. Pergamon Press, New York (1981)
27. Shao, Y.B., Gong, Z.T.: The existence of ε-approximate solutions to fuzzy functional differential equations, The 9th Intenational Conference on Fuzzy Systems and Knowledge Discovery, Chongqing, 181–184 (2012)

The Generalized Solution for Initial Problems of Fuzzy Discontinuous Differential Equations

Ya-bin Shao, Huan-huan Zhang and Zeng-tai Gong

Abstract In this paper, we generalized the existence theorems of Caratheodory solution for initial problems of fuzzy discontinuous differential equation by the definition of the $\omega - ACG^*$ for a fuzzy-number-valued function and the nonabsolute fuzzy integral and its controlled convergence theorem.

Keywords Fuzzy number · Fuzzy-number-valued function · Fuzzy Henstock integral · Controlled convergence theorem · Discontinuous fuzzy differential equation · Generalized solutions

1 Introduction

The Henstock integral is designed to integrate highly oscillatory functions which the Lebesgue integral fails to do. It is known as nonabsolute integration and is a powerful tool. It is well-known that the Henstock integral includes the Riemann, improper Riemann, Lebesgue and Newton integrals [1, 2]. Though such an integral was defined by Denjoy in 1912 and also by Perron in 1914, it was difficult to handle using their definitions. But with the Riemann-type definition introduced more recently by Henstock [1] in 1963 and also independently by Kurzweil [2], the definition is now simple and furthermore the proof involving the integral also turns out to be easy. For more detailed results about the Henstock integral, we refer to [3]. Since the concept of fuzzy sets [4] was first introduced by Zadeh in 1965, it has been studied extensively

Y. Shao (✉) · H. Zhang
College of Mathematics and Computer Science, Northwest University for Nationalities, Lanzhou 730030, China
e-mail: yb-shao@163.com

Z. Gong
College of Mathematics and Information Science, Northwest Normal University, Lanzhou 730070, China

B.-Y. Cao and H. Nasseri (eds.), *Fuzzy Information & Engineering and Operations Research & Management*, Advances in Intelligent Systems and Computing 211,
DOI: 10.1007/978-3-642-38667-1_23, © Springer-Verlag Berlin Heidelberg 2014

from many different aspects of the theory and applications, especially in information science, such as linguistic information system and approximate reasoning [5–8], fuzzy topology [9], fuzzy analysis [10], fuzzy decision making, fuzzy logic [11, 12] and so on. Recently, Wu and Gong [13, 14] have combined the above theories and discussed the fuzzy Henstock integrals of fuzzy-number-valued functions which extended Kaleva [15] integration. In order to complete the theory of fuzzy calculus and to meet the solving need of transferring a fuzzy differential equation into a fuzzy integral equation, we [16, 17] has defined the strong fuzzy Henstock integrals and discussed some of their properties and the controlled convergence theorem.

On the other hand, the characterization of the derivatives, in both real and fuzzy analysis, is an important problem. Bede and Gal [18] have subsequently introduced a more general definition of a derivative for fuzzy-number-valued function enlarging the class of differentiable fuzzy-number-valued functions. Following this idea, Chalco-Cano and Roman-Flores [19] have defined the lateral H-derivative for a fuzzy-number-valued function.

The Cauchy problems for fuzzy differential equations have been studied by several authors [15, 20–24] on the metric space (E^n, D) of normal fuzzy convex set with the distance D given by the maximum of the Hausdorff distance between the corresponding level sets. In [22], the author has been proved the Cauchy problem has a uniqueness result if f was continuous and bounded. In [15, 21], the authors presented a uniqueness result when f satisfies a Lipschitz condition. For a general reference to fuzzy differential equations, see a recent book by Lakshmikantham and Mohapatra [25] and references therein. In 2002, Xue and Fu [26] established solutions to fuzzy differential equations with right-hand side functions satisfying Caratheodory conditions on a class of Lipschitz fuzzy sets.

However, there are discontinuous systems in which the right-hand side functions $f : [a, b] \times E^n \to E^n$ are not integrable in the sense of Kaleva [15] on certain intervals and their solutions are not absolute continuous functions. To illustrate, we consider the following example:

Example 1. Consider the following discontinuous system

$$x'(t) = h(t), x(0) = \tilde{A},$$

$$g(t) = \begin{cases} 2t \sin \frac{1}{t^2} - \frac{2}{t} \cos \frac{1}{t^2}, & t \neq 0, \\ 0, & t = 0. \end{cases}$$

$$\tilde{A}(s) = \begin{cases} s, & 0 \leq s \leq 1, \\ 2 - s, & 1 < s \leq 2, \\ 0, & others. \end{cases}$$

$$h(t) = \chi_{|g(t)|} + \tilde{A}.$$

Then $h(t) = \chi_{|g(t)|} + \tilde{A}$ is not integrable in sense of Kaleva. However, above system has the following solution:

$$x(t) = \chi_{|G(t)|} + \tilde{A}t,$$

where

$$G(t) = \begin{cases} t^2 \sin \frac{1}{t^2}, & t \neq 0, \\ 0, & t = 0. \end{cases}$$

In this paper, according to the idea of [27] and using the concept of generalized differentiability [18], we shall prove the controlled convergence theorems for the fuzzy Henstock integrals, which will be foundational significance for studying the existence and uniqueness of solutions to the fuzzy discontinuous systems. As we know, we inevitably use the controlled convergence theorems for solving the numerical solutions of differential equations. As the main outcomes, we will deal with the Cauchy problem of discontinuous fuzzy systems as following:

$$\begin{cases} x'(t) = f(t, x) \\ x(\tau) = \xi \in E^n, \end{cases} \tag{1}$$

where $f : U \to E^n$ is a fuzzy Henstock integrable function, and

$$U = \{(t, x) : |t - \tau| \leq a, x \in E^n, D(x, \xi) \leq b\}.$$

To make our analysis possible, we will first recall some basic results of fuzzy numbers and give some definitions of absolutely continuous of fuzzy-number-valued function. In addition, we present the concept of generalized differentiability. In Sect. 3, we present the concept of fuzzy Henstock integrals and we prove a controlled convergence theorem for the fuzzy Henstock integrals. In Sect. 4, we deal with the Cauchy problem of discontinuous fuzzy systems. And in Sect. 5, we present some concluding remarks.

2 Preliminaries

Let $P_k(R^n)$ denote the family of all nonempty compact convex subset of R^n and define the addition and scalar multiplication in $P_k(R^n)$ as usual. Let A and B be two nonempty bounded subset of R^n. The distance between A and B is defined by the Hausdorff metric [28]:

$$d_H(A, B) = \max\{\sup_{a \in A} \inf_{b \in B} \| a - b \|, \sup_{b \in B} \inf_{a \in A} \| b - a \|\}.$$

Denote $E^n = \{u : R^n \to [0, 1] | u$ satisfies (1)–(4) below$\}$ is a fuzzy number space. where

(1) u is normal, i.e. there exists an $x_0 \in R^n$ such that $u(x_0) = 1$,

(2) u is fuzzy convex, i.e. $u(\lambda x + (1-\lambda)y) \geq \min\{u(x), u(y)\}$ for any $x, y \in R^n$ and $0 \leq \lambda \leq 1$,
(3) u is upper semi-continuous,
(4) $[u]^0 = cl\{x \in R^n | u(x) > 0\}$ is compact.

For $0 < \alpha \leq 1$, denote $[u]^\alpha = \{x \in R^n | u(x) \geq \alpha\}$. Then from above (1)–(4), it follows that the α-level set $[u]^\alpha \in P_k(R^n)$ for all $0 \leq \alpha < 1$.

According to Zadeh's extension principle, we have addition and scalar multiplication in fuzzy number space E^n as follows [28]:

$$[u+v]^\alpha = [u]^\alpha + [v]^\alpha, \quad [ku]^\alpha = k[u]^\alpha,$$

where $u, v \in E^n$ and $0 \leq \alpha \leq 1$.

Define $D : E^n \times E^n \to [0, \infty)$

$$D(u, v) = \sup\{d_H([u]^\alpha, [v]^\alpha) : \alpha \in [0, 1]\},$$

where d is the Hausdorff metric defined in $P_k(R^n)$. Then it is easy see that D is a metric in E^n. Using the results [29], we know that

(1) (E^n, D) is a complete metric space,
(2) $D(u + w, v + w) = D(u, v)$ for all $u, v, w \in E^n$,
(3) $D(\lambda u, \lambda v) = |\lambda| D(u, v)$ for all $u, v, w \in E^n$ and $\lambda \in R$.

A fuzzy-number-valued function $f : [a, b] \to R_{\mathcal{F}}$ is said to satisfy the condition (H) on $[a, b]$, if for any $x_1 < x_2 \in [a, b]$ there exists $u \in R_{\mathcal{F}}$ such that $f(x_2) = f(x_1) + u$. We call u is the H-difference of $f(x_2)$ and $f(x_1)$, denoted $f(x_2) -_H f(x_1)$ [15].

For brevity, we always assume that it satisfies the condition (H) when dealing with the operation of subtraction of fuzzy numbers throughout this paper.

It is well-known that the H-derivative for fuzzy-number-functions was initially introduced by Puri and Ralescu [23] and it is based in the condition (H) of sets. We note that this definition is fairly strong, because the family of fuzzy-number-valued functions H-differentiable is very restrictive. For example, the fuzzy-number-valued function $f : [a, b] \to R_{\mathcal{F}}$ defined by $f(x) = C \cdot g(x)$, where C is a fuzzy number, $\cdot$ is the scalar multiplication (in the fuzzy context) and $g : [a, b] \to R^+$, with $g'(t_0) < 0$, is not H-differentiable in t_0 (see [18]). To avoid the above difficulty, in this paper we consider a more general definition of a derivative for fuzzy-number-valued functions enlarging the class of differentiable fuzzy-number-valued functions, which has been introduced in [18].

Definition 1. [18] Let $f : (a, b) \to R_{\mathcal{F}}$ and $x_0 \in (a, b)$. We say that f is differentiable at x_0, if there exists an element $f'(t_0) \in R_{\mathcal{F}}$, such that

(1) for all $h > 0$ sufficiently small, there exists $f(x_0 + h) -_H f(x_0)$, $f(x_0) -_H f(x_0 - h)$ and the limits (in the metric D)

$$\lim_{h \to 0} \frac{f(x_0 + h) -_H f(x_0)}{h} = \lim_{h \to 0} \frac{f(x_0) -_H f(x_0 - h)}{h} = f'(x_0)$$

or

(2) for all $h > 0$ sufficiently small, there exists $f(x_0) -_H f(x_0+h)$, $f(x_0-h) -_H f(x_0)$ and the limits

$$\lim_{h\to 0} \frac{f(x_0) -_H f(x_0+h)}{-h} = \lim_{h\to 0} \frac{f(x_0-h) -_H f(x_0)}{-h} = f'(x_0)$$

or

(3) for all $h > 0$ sufficiently small, there exists $f(x_0+h) -_H f(x_0)$, $f(x_0-h) -_H f(x_0)$ and the limits

$$\lim_{h\to 0} \frac{f(x_0+h) -_H f(x_0)}{h} = \lim_{h\to 0} \frac{f(x_0-h) -_H f(x_0)}{-h} = f'(x_0)$$

or

(4) for all $h > 0$ sufficiently small, there exists $f(x_0) -_H f(x_0+h)$, $f(x_0) -_H f(x_0-h)$ and the limits

$$\lim_{h\to 0} \frac{f(x_0) -_H f(x_0+h)}{-h} = \lim_{h\to 0} \frac{f(x_0) -_H f(x_0-h)}{h} = f'(x_0)$$

(h and $-h$ at denominators mean $\frac{1}{h}\cdot$ and $-\frac{1}{h}\cdot$, respectively).

3 The Fuzzy Henstock Integral and Its Controlled Convergence Theorem

In this section we shall give the definition of the Henstock integral for fuzzy-number-valued functions [13, 14] on a finite interval, which is an extension of the usual fuzzy Kaleva integral in [15]. In addition, we define the properties of $\omega - AC$, $\omega - AC^*$ and $\omega - ACG^*$ for fuzzy-number-valued functions. In particular, we shall prove a controlled convergence theorems for the fuzzy Henstock integrals.

Definition 2. [1, 3] Let $\delta(x)$ be a positive function defined on the interval $[a, b]$. A division $P = \{[x_{i-1}, x_i] : \xi_i\}$ is said to be $\delta-$ fine if the following conditions are satisfied:

(1) $a - x_0 < x_1 < \cdots < x_n = b$;
(2) $\xi_i \in [x_{i-1}, x_i] \subset (\xi_i - \delta(\xi_i), \xi_i + \delta(\xi_i))$.
For brevity, we write $P = \{[u, v]; \xi\}$

Definition 3. [13, 14] Let $f : [a, b] \to E^n$ be a fuzzy-number-valued function. Then $f(x)$ is said to be Henstock integrable to A on the interval $[a, b]$ if for every $\varepsilon > 0$, there exists a function $\delta(\xi) > 0$ such that for any $\delta-$ fine division $P = \{[x_{i-1}, x_i] : \xi_i\}$, we have

$$D(\sum_{i=1}^{n} f(\xi_i)(x_{i-1}, x_i), A) < \varepsilon.$$

We write $(FH) \int_a^b f(x)\mathrm{d}x = A$ or $f \in FH[a, b]$.

Definition 4. [16] A fuzzy-number-valued function F is said to be absolutely continuous on $[a, b]$ if for every $\varepsilon > 0$ such that for every finite sequence of non-overlapping intervals $\{[c_i, d_i]\}$, with $\sum_{i=1}^{n} |b_i - a_i| < \eta$ we have

$$\sum D(F(d_i), F(c_i)) < \varepsilon.$$

We write $F \in AC[a, b]$.

Definition 5. Let $X \subset [a, b]$. A fuzzy-number-valued function f defined on X is said to be $\omega - AC(X)$ if for every $\varepsilon > 0$ such that for every finite sequence of non-overlapping intervals $\{[a_i, b_i]\}$, with $\Sigma_{i=1}^{n} |b_i - a_i| < \eta$ we have

$$\sum D(f(b_i), f(a_i)) < \varepsilon$$

where the endpoints $a_i, b_i \in X$ for all i.

Definition 6. A fuzzy-number-valued function f defined on $X \subset [a, b]$ is said to be $\omega - AC^*(X)$ if for every $\varepsilon > 0$ there exists $\eta > 0$ such that for every finite sequence of non-overlapping intervals $\{[a_i, b_i]\}$, satisfying $\Sigma_{i=1}^{n} |b_i - a_i| < \eta$ where $a_i, b_i \in X$ for all i we have

$$\sum \mathcal{O}(f, [a_i, b_i]) < \varepsilon$$

where $\mathcal{O}$ denotes the oscillation of f over $[a_i, b_i]$,i.e.,

$$\mathcal{O}(f, [a_i, b_i]) = \sup\{D(f(x), f(y)); x, y \in [a_i, b_i]\}.$$

Definition 7. A fuzzy-number-valued function f is said to be $\omega - ACG^*$ on X if X is the union of a sequence of closed sets $\{X_i\}$ such that on each X_i, f is $AC^*(X_i)$.

Theorem 1. *If f is fuzzy Henstock integrable on $[a, b]$, then its primitive F is $\omega - ACG^*$ on $[a, b]$.*

Proof For every $\varepsilon > 0$, there is a function $\delta(\xi) > 0$ such that for any δ-fine partial division $P = \{[u, v], \xi\}$ in $[a, b]$, we have

$$\sum D(F([u, v]), f(\xi)(v - u)) < \varepsilon.$$

We assume that $\delta(\xi) \leq 1$. Let

$$X_{n,i} = \{x \in [a,b] : D(f(x), \tilde{0}) \leq n, \frac{1}{n} < \delta(x) \leq \frac{1}{n-1}, x \in [a + \frac{i-1}{n}, a + \frac{i}{n})\}$$

for $n = 2, 3, \cdots, i = 1, 2, \cdots$. Fixed $X_{n,i}$ and let $\{[a_k, b_k]\}$ be any finite sequence of non-overlapping intervals with $a_k, b_k \in X_{n,i}$ for all k. Then $\{([a_k, b_k], a_k)\}$ is a δ -fine partial division of $[a, b]$. Furthermore, if $a_k \leq u_k \leq v_k \leq b_k$, then $\{([a_k, u_k], a_k)\}$, $\{([a_k, v_k], a_k)\}$ are δ-fine partial division of $[a, b]$. Thus

$$\begin{aligned}
&\sum D(F(u_k), F(v_k)) \\
&\leq \sum D(F(a_k), F(u_k)) + \sum D(F(b_k), F(v_k)) + \sum D(F(a_k), F(b_k)) \\
&\leq 3\varepsilon + \sum D(f(a_k)(u_k - a_k), \tilde{0}) + \sum D(f(b_k)(b_k - v_k), \tilde{0}) \\
&+ \sum D(f(a_k)(b_k - a_k), \tilde{0}) \leq 3\varepsilon + 3n \sum (b_k - a_k).
\end{aligned}$$

Choose $\eta \leq \frac{\varepsilon}{3n}$ and $\sum(b_k - a_k) < \eta$. Then

$$\sum \mathcal{O}(F, [a_k, b_k]) \leq 3\varepsilon + \varepsilon.$$

Therefore, F is $\omega - AC^*(X_{n,i})$. Consequently, F is $\omega - ACG^*$ on $[a, b]$. This completes the proof.

Theorem 2. *[14] If f is fuzzy Henstock integrable on $[a, b]$, then its primitive $\tilde{F}$ is differentiable a.e. and $\tilde{F}'(x) = \tilde{f}(x)$ a.e. on $[a, b]$.*

Theorem 3. *If there exists a fuzzy-number-valued function F is continuous and $\omega - ACG^*$ on $[a, b]$ such that $F'(x) = f(x)$ a.e. in $[a, b]$, then f is fuzzy Henstock integrable on $[a, b]$ with primitive F.*

Proof Let F be the primitive of f and $F'(x) = f(x)$ for $x \in [a, b] \setminus S$ where S is of measure zero. For $\xi \in [a, b] \setminus S$, given $\varepsilon > 0$ there is a $\delta(\xi) > 0$ such that whenever $\xi \in [u, v] \subset (\xi - \delta(\xi), \xi + \delta(\xi))$ we have

$$D(F([u, v]), f(\xi)(v - u)) \leq \varepsilon |v - u|.$$

Since F is continuous and $\omega - ACG^*$ on $[a, b]$, there is a sequence of closed sets $\{X_i\}$ such that $\cup_i X_i = [a, b]$ and F is $\omega - AC^*(X_i)$ for each i. Let $Y_1 = X_1$, $Y_i = X_i \setminus (X_1 \cup X_2 \cdots \cup X_{i-1})$ for $i = 1, 2, \cdots$ and S_{ij} denote the set of points $x \in S \cap Y_i$ such that $j - 1 \leq D(f, \tilde{0}) < j$. Obviously, S_{ij} are pairwise disjointed and their union is the set S. Since F is also $\omega - AC^*(S_{ij})$, there is a $\eta_{ij} < \varepsilon 2^{-i-j} j^{-1}$ such that for any sequence of non-overlapping intervals $\{I_k\}$ with at least one endpoint of I_k belonging to S_{ij} and satisfying $\sum_k |I_k| < \eta_{ij}$ we have $\sum_k D(F(I_k), \tilde{0}) < \varepsilon 2^{-i-j}$. Again, $F(I)$ denotes $F(v) -_H F(u)$ where $I = [u, v]$. Choose G_{ij} to be the union of a sequence of open intervals such that $|G_{ij}| < \eta_{ij}$ and $G_{ij} \supset S_{ij}$ where $|G_{ij}|$

denotes the total length of G_{ij}. Now for $\xi \in S_{ij}$, put $(\xi - \delta(\xi), \xi + \delta(\xi)) \subset G_{ij}$. Hence we have defined a positive function $\delta(\xi)$.

Take any δ−fine division $P = \{[u, v]; \xi\}$. Split the $\sum$ over P into partial sums $\sum_1$ and $\sum_2$ in which $\xi \bar{\in} S$ and $\xi \in S$ respectively and we obtain

$$
\begin{aligned}
D(f(\xi)(v-u), F([a,b])) &\le \sum_1 D(f(\xi)(v-u), F([a,b])) \\
&\quad + \sum_2 D(F([a,b]), \tilde{0}) + \sum_2 D(f(\xi)(v-u), \tilde{0}) \\
&< \varepsilon(b-a) + \sum_{i,j} \varepsilon 2^{-i-j} + \sum_2 j\eta_{ij} \\
&< \varepsilon(b-a) + 2\varepsilon.
\end{aligned}
$$

That is to say, f is fuzzy Henstock integrable to F on $[a, b]$.

Theorem 4. *(Controlled Convergence Theorem) If a sequence of fuzzy Henstock integrable* $\{f_n\}$ *satisfies the following conditions:*

(1) $f_n(x) \to f(x)$ *almost everywhere in* $[a, b]$ *as* $n \to \infty$*;*
(2) the primitives $F_n(x) = (FH)\int_a^x f_n(s)\mathrm{d}x$ *of* f_n *are* $\omega - ACG^*$ *uniformly in* n*;*
(3) the primitives $F_n(x)$ *are equicontinuous on* $[a, b]$*,*
then $f(x)$ *is fuzzy Henstock integrable on* $[a, b]$ *and we have*

$$\lim_{n\to\infty} (FH)\int_a^b f_n(x)\mathrm{d}x = (FH)\int_a^b f(x)\mathrm{d}x.$$

If condition (1) and (2) are replaced by condition (4):
(4) $g(x) \le f(x) \le h(x)$ *almost everywhere on* $[a, b]$*, where* $g(x)$ *and* $h(x)$ *are fuzzy Henstock integrable.*

Proof In view of condition (3), $F(x)$ exist as the limit of $F_n(x)$ and is continuous. In fact, for $\forall\lambda \in [0, 1]$, $(F_n(x))_\lambda^-$ and $(F_n(x))_\lambda^+$ is uniformly ACG^* on $[a, b]$. By the Controlled Convergence Theorem of real valued Henstock integral ([3]Theorem 7.6), $F(x)$ is continuous. Because $F_\lambda^-(x)$ and $F_\lambda^+(x)$ is Henstock integrable on $[a, b]$, it follows condition (2) that F is $\omega - ACG^*$. From Theorem 3.2, it remains to show that $F'(x) = f(x)$ almost everywhere. Hence we obtain $f(x)$ is Fuzzy Henstock integrable on $[a, b]$.

Next, we put $G(x) = (FH)\int_a^x F(t)\mathrm{d}t$, in view of condition (3), for $\forall\lambda \in [0, 1]$, we have

$$\lim_{n\to\infty} (F_n(x))_\lambda^- = G_\lambda^-(x) = F_\lambda^-(x)$$

and

$$\lim_{n\to\infty} (F_n(x))_\lambda^+ = G_\lambda^+(x) = F_\lambda^+(x).$$

So, let $x = b$, we have

$$\lim_{n\to\infty}(FH)\int_a^b f_n(x)\mathrm{d}x=(FH)\int_a^b f(x)\mathrm{d}x.$$

This completes the proof.

4 The Generalized Solutions of Discontinuous Fuzzy Differential Equations

In this section, let τ and ξ be fixed and let a fuzzy-number-valued function $f(t,x)$ be a Carathéodoy function defined on a rectangle $U:|t-\tau|\leq a,\ D(x,\xi)\leq b$, i.e., f is continuous in x for almost all t and measurable in t for each fixed x.

Theorem 5. *Let a fuzzy-number-valued function $f(t,x)$ be a function as given above, then there exist two fuzzy Henstock integrable functions h and g defined on $|t-\tau|\leq a$ such that $g(x)\leq f(x)\leq h(x)$ for all $(t,x)\in U$.*

Proof Note that $f(t,x)$ be a Carathéodoy function. Thus, there exist two measurable functions $u(t)$ and $v(t)$ defined on $|t-\tau|\leq a$ with values in $D(x,\xi)\leq b$ such that

$$f(t,u(t))\leq f(t,x)\leq f(t,v(t))$$

for all $(t,x)\in U$. Next, we shall show that $f(t,u(t))$ and $f(t,v(t))$ are fuzzy Henstock integrable by using the Controlled Convergence Theorem 4. First, there exists a sequence $\{k_n(t)\}$ of step functions defined on $|t-\tau|\leq a$ with values in $D(x,\xi)\leq b$ such that $k_n(t)\to u(t)$ almost everywhere as $n\to\infty$. Let

$$F_n(t)=\int_\tau^t f(s,k_n(s))\mathrm{d}s.$$

Then $\{F_n(t)\}$ is $\omega-ACG^*$ uniformly in n and equicontinuous. By the Controlled Convergence Theorem 4, $f(t,u(t))$ is fuzzy Henstock integrable. Similarly, $f(t,v(t))$ is fuzzy Henstock integrable.

Theorem 6. *Let a fuzzy-number-valued function $f(t,x)$ be a Carathéodoy function defined on a rectangle $U:|t-\tau|\leq a,\ D(x,\xi)\leq b$. Suppose that there exist two fuzzy Henstock integrable functions $g(t)$ and $h(t)$ on the interval $|t-\tau|\leq a$ such that*

$$g(t)\leq f(t,x)\leq h(t)$$

for all x and all most t with $(t,x)\in U$. Then there exist a solution ϕ of $x'=f(t,x)$ on some interval $|t-\tau|\leq\beta,\ \beta>0$ satisfying $\phi(\tau)=\xi$.

Proof Given $g(t)\leq f(t,x)\leq h(t)$ for all x and almost all t with $(t,x)\in U$, we get $0\leq f(t,x)-_H g(t)\leq h(t)-_H g(t)$. Let

$$F(t,x) = f(t, x + \int_{\tau}^{t} g(s)\mathrm{d}s) -_H g(t).$$

Then F is a Carathéodoy function. Furthermore, $0 \leq F(t,x) \leq h(t)_H g(t)$ for all $(t,x) \in U'$, where U' is a subrectangle U, such that $D(x + \int_{\tau}^{t} g(s)\mathrm{d}s, \xi) \leq b$ for all $(t,x) \in U'$. Then, there is a fuzzy-number-valued function $\psi'(t) = F(t, \psi(t))$ a.e. in this interval and $\psi(\tau) = \xi$. Let$\phi(t) = \psi(t) + \int_{\tau}^{t} g(s)\mathrm{d}s$.Then for almost all t, we have

$$\begin{aligned}\phi'(t) &= \psi'(t) + g(t) = F(t, \psi(t)) + g(t)\\ &= f(t, \psi(t) + \int_{\tau}^{t} g(s)\mathrm{d}s) -_H g(t) + g(t)\\ &= f(t, \phi(t))\end{aligned}$$

and

$$\phi(\tau) = \psi(\tau) + \int_{\tau}^{t} g(s)\mathrm{d}s = \xi.$$

The proof is complete.

Now we shall state another existence theorem. The hypotheses involved in this theorem are motivated by Theorem 4.

Theorem 7. *Let a fuzzy-number-valued function $f(t,x)$ be a Carathéodoy function defined on a rectangle $U : |t-\tau| \leq a$, $D(x,\xi) \leq b$. Let $f(t,u(t))$ be fuzzy Henstock integrable on $|t-\tau| \leq a$ for any step function $u(t)$ defined on $|t-\tau| \leq a$ with values in $D(x,\xi) \leq b$. Denote*

$$F_u(t) = \int_{\tau}^{t} f(s, u(s))\mathrm{d}s.$$

If $\{F_u : u$ is a step function$\}$ is $\omega - ACG^*$ uniformly in u and equicontinuous on $|t-\tau| \leq a$, then there exist a solution ϕ of $x' = f(t,x)$ on some interval $|t-\tau| \leq \beta$ with $\phi(\tau) = \xi$.

Proof By Theorem 5, let $g(t) = f(t,u(t))$, $h(t) = f(t,v(t))$, then by the condition (4) in Theorem, we have a solution ϕ of $x' = f(t,x)$ on some interval $|t-\tau| \leq \beta$ with $\phi(\tau) = \xi$. The proof is completed.

5 Conclusion

In this paper, we give the definition of the $\omega - ACG^*$ for a fuzzy-number-valued function and the nonabsolute fuzzy integral and its controlled convergence theorem. In addition, we deal with the Cauchy problem of discontinuous fuzzy differential equations involving the fuzzy Henstock integral in fuzzy number space. The function governing the equations is supposed to be discontinuous with respect to some variables and satisfy nonabsolute fuzzy integrablility. Our result improves the result given in Refs. [15, 18, 26] and [27] (where uniform continuity was required), as well as those referred therein.

Acknowledgments Thanks to the support by National Natural Science Foundation of China (No.11161041 and No.71061013) and 2012 Scientific Research Fund for State Ethnic Affairs Commission of China; Fundamental Research Fund for the Central Universities (No. zyz2012081, No. 31920130009 and No. 31920130010).

References

1. Henstock, R.: Theory of Integration. Butterworth, London (1963)
2. Kurzweil, J.: Generalized ordinary differential equations and continuous dependence on a parameter. Czechoslovak Math. J. **7**, 418–446 (1957)
3. Lee, P.Y.: Lanzhou Lectures on Henstock Integration. World Scientific, Singapore (1989)
4. Zadeh, L.A.: Fuzzy sets. Inf. Control **3**, 338–353 (1965)
5. Zadeh, L.A.: The concept of a linguistic variable and its application to approximate reasoning-I. Inf. Sci. **8**(3), 199–249 (1975)
6. Zadeh, L.A.: The concept of a linguistic variable and its application to approximate reasoning-II. Inf. Sci. **8**(4), 301–357 (1975)
7. Zadeh, L.A.: The concept of a linguistic variable and its application to approximate reasoning-III. Inf. Sci. **9**(1), 43–80 (1975)
8. Zadeh, L.A.: A note on Z-numbers. Inf. Sci. **181**, 2923–2932 (2011)
9. Liu, Y.M., Luo, M.K.: Fuzzy Topology. World Scientific, Singapore (1999)
10. Negoita, C.V., Ralescu, D.: Applications of Fuzzy Sets to Systems Analysis. Wiley, New York (1975)
11. Zadeh, L.A.: Toward a generalized theory of uncertainty (GTU)-an outline. Inf. Sci. **172**, 1–40 (2005)
12. Zadeh, L.A.: Is there a need for fuzzy logic? Inf. Sci. **178**, 2751–2779 (2008)
13. Wu, C., Gong, Z.: On Henstock intergrals of interval-valued and fuzzy-number-valued functions. Fuzzy Sets Syst. **115**, 377–391 (2000)
14. Wu, C., Gong, Z.: On Henstock intergrals of fuzzy-valued functions (I). Fuzzy Sets Syst. **120**, 523–532 (2001)
15. Kaleva, O.: Fuzzy differential equations. Fuzzy Sets Syst. **24**, 301–319 (1987)
16. Gong, Z.: On the problem of characterizing derivatives for the fuzzy-valued functions (II): almost everywhere differentiability and strong Henstock integral. Fuzzy Sets Syst. **145**, 381–393 (2004)
17. Gong, Z., Shao, Y.: The controlled convergence theorems for the strong Henstock integrals of fuzzy-number-valued functions. Fuzzy Sets Syst. **160**, 1528–1546 (2009)
18. Bede, B., Gal, S.: Generalizations of the differentiability of fuzzy-number-valued functions with applications to fuzzy differential equation. Fuzzy Sets Syst. **151**, 581–599 (2005)
19. Chalco-Cano, Y., Roman-Flores, H.: On the new solution of fuzzy differential equations. Chaos Solition Fractals **38**, 112–119 (2008)
20. Gong, Z, Shao, Y.: Global existence and uniqueness of solutions for fuzzy differential equations under dissipative-type conditions. Comput. Math. Appl. **56**, 2716–2723 (2008)
21. Kaleva, O.: The Cauchy problem for fuzzy differential equations. Fuzzy Sets Syst. **35**, 389–396 (1990)

22. Nieto, J.J.: The Cauchy problem for continuous fuzzy differential equations. Fuzzy Sets Syst. **102**, 259–262 (1999)
23. Puri, M.L., Ralescu, D.A.: Differentials of fuzzy functions. J. Math. Anal. Appl. **91**, 552–558 (1983)
24. Seikkala, S.: On the fuzzy initial value problem. Fuzzy Sets Syst. **24**, 319–330 (1987)
25. Lakshmikantham, V., Mohapatra, R.N.: Theory of Fuzzy Differential Equations and Inclusions, Taylor & Francis, London (2003)
26. Xue, X., Fu, Y.: Caratheodory solution of fuzzy differential equations. Fuzzy Sets Syst. **125**, 239–243 (2002)
27. Chew, T.S., Franciso, F.: on $x' = f(t, x)$ and Henstock-Kurzwell integrals. Differ. Integr. Equ. **4**, 861–868 (1991)
28. Dubois, D., Prade, H.: Towards fuzzy differential calculus, Part 1. Integration of fuzzy mappings. Fuzzy Sets Syst. **8**, 1–17 (1982)
29. Diamond, P., Kloeden, P.: Metric Space of Fuzzy Sets: theory and Applications. World Scientific, Singapore (1994)
30. Gong, Z.T., Wang, L.L.: The Henstock-Stieltjes integral for fuzzy-number-valued functions. Inf. Sci. **188**, 276–297 (2012)

Part IV
Forecasting, Clustering and Recongnition

A Revised Grey Model for Fluctuating Interval Fuzzy Series Forecasting

Xiang-yan Zeng, Lan Shu and Gui-min Huang

Abstract Grey models are built on the basis of the real number series, not the fuzzy number series. In this paper, GM (1, 1), a basic prediction model of grey models, is generalized to the fuzzy number series. GM (1, 1) based on interval fuzzy number series [IFGM (1, 1)] is proposed firstly, which is suitable for the prediction of interval number series with weak fluctuation. In order to extend its applicable range, a revised model is then proposed. Markov prediction theory is applied to revising IFGM (1, 1) to make it suitable for the fluctuating interval number series. The general development tendency of the raw series is embodied by grey model and the random fluctuation is reflected by Markov prediction. The first practical example has shown IFGM (1, 1) is effective for small sample and weak fluctuating series. The consumer price indexes of China from 1996 to 2009 are taken as an example of fluctuating interval number series. Revised IFGM (1, 1), IFGM (1, 1), double exponential smoothing method and autoregressive moving average (ARMA) are all applied to it. Comparison of the results has shown the revised IFGM (1, 1)has the best precision than the other models.

Keywords GM (1, 1) model · Interval fuzzy number · Fluctuation · Markov prediction.

X. Zeng (✉) · L. Shu
School of Mathematics and Computational Science, Research Center on Data Science and Social Computing, University of Electronic Technology, Guilin 541004, People's Republic of China
e-mail: zengxyhbyc@163.com

X. Zeng
School of Mathematical Sciences, University of Electronic Science and Technology of China, Chengdu 611731, People's Republic of China

G. Huang
School of Information and Communication, Research Center on Data Science and Social Computing,University of Electronic Technology, Guilin 541004, People's Republic of China

B.-Y. Cao and H. Nasseri (eds.), *Fuzzy Information & Engineering and Operations Research & Management*, Advances in Intelligent Systems and Computing 211,
DOI: 10.1007/978-3-642-38667-1_24,

1 Introduction

The grey system is an uncertainty system focusing on small sample and deficient information [1, 2]. GM (1, 1), a basic prediction model of grey models, can be established with only four items of data. Some researches have presented that this model has high simulation accuracy when it is applied in the small sample time sequence [3, 4]. There are many applications with small sample, such as practical examples in [5–10]. In addition, as circumstances change, some old information should be removed in order to make the prediction more accurate. However, most of the statistic prediction methods need a large sample [11]. Therefore, GM (1, 1) model has been widely used in many fields since it was proposed.

GM (1, 1) is based on the real number series, not the fuzzy number series [1, 2]. In many applications, the acquired attribute values often change in an interval or a vicinity of a core number, so that the data are presented by the fuzzy numbers. Therefore, it is significant to study the prediction models based on the fuzzy number series. In this chapter, the grey prediction model of the interval fuzzy number series will be researched.

The modeling mechanism of GM (1, 1) model is to use an exponential type of curve to fit the raw data [2, 3]. So, before establishing the model, the sequence of raw data should meet the grey exponential law or pass the class ratio test. Otherwise, the model should be revised. The geometric figure of the prediction results of GM (1, 1) is a smooth curve, which describes the general development trend of the raw series. The random fluctuation of the development, however, is not taken into account. In reality, there are many non-stationary random processes, for example, the annual airline plane crash toll, the road traffic safety accidents, the commodity market demand and the consumer price index. In this chapter, the GM (1, 1) model based on the interval fuzzy number series [IFGM (1, 1)] is proposed firstly. Then, for the long-term and strongly fluctuating series, the Markov prediction theory is applied to revise the model.

The Markov prediction can describe a dynamic variation of a random time series. It suits the prediction of the strongly fluctuating series. The grey prediction and the Markov prediction are mutually complementary. The general development tendency of the series is embodied by the grey prediction and the random fluctuation of the series is reflected by the Markov prediction. In this chapter, the two prediction methods are integrated. It makes full use of the information of the raw data and improves the forecast precision of the fluctuating series.

2 Definition Equation of GM (1, 1)

Let a real-number sequence be $x^{(0)}(t) = \{x^{(0)}(1), x^{(0)}(2), \cdots, x^{(0)}(n)\}$, where t_i is the corresponding time. Let $x^{(1)}(i) = \sum_{k=1}^{i} x^{(0)}(k)$, $i = 1, 2, \cdots, n$, which is called the accumulated generating operation series (AGO series) in the grey system theory.

Based on the AGO series $x^{(1)}(t) = \{x^{(1)}(i)\}$, the following equation is called the white differential equation of GM (1, 1): $\frac{dx^{(1)}(t)}{dt} + ax^{(1)}(t) = b$. Make a definite integral on the interval: $[i-1, i]$, then

$$\int_{i-1}^{i} dx^{(1)}(t) + a\int_{i-1}^{i} x^{(1)}(t)dt = b\int_{i-1}^{i} dt \tag{1}$$

where $\int_{i-1}^{i} dx^{(1)}(t) = x^{(1)}(i) - x^{(1)}(i-1) = x^{(0)}(i)$. Let $z^{(1)}(i)$ be the white background value of $x^{(1)}(t)$ in the interval: $[i-1, i]$. It is often adopted as the mean value: $z^{(1)}(i) = 0.5(x^{(1)}(i-1) + x^{(1)}(i)), i = 2, 3, \cdots, n$. Then $a\int_{i-1}^{i} x^{(1)}(t)dt = az^{(1)}(i)$. Thus, the white differential equation of GM (1, 1) becomes as follows:

$$x^{(0)}(i) + az^{(1)}(i) = b, i = 2, 3, \cdots, n \tag{2}$$

Equation (2) is called the grey differential equation (or definition equation) of GM (1, 1), where a is called the developing coefficient and b is called the grey input.

3 Interval Fuzzy GM (1, 1) [IFGM (1, 1)]

The modeling process of GM (1, 1) based on interval fuzzy number series [IFGM (1, 1)] is given as follows.

Let the interval fuzzy number series be $\tilde{x}^{(0)} = \{\tilde{x}^{(0)}(1), \tilde{x}^{(0)}(2), \cdots, \tilde{x}^{(0)}(n)\}$, where $\tilde{x}^{(0)}(i) = [x_L^{(0)}(i), x_R^{(0)}(i)]$.

Definition 3.1 *The accumulated generating operation series (AGO series) of* $\tilde{x}^{(0)}$ *is* $\tilde{x}^{(1)} = \{\tilde{x}^{(1)}(1), \tilde{x}^{(1)}(2), \cdots, \tilde{x}^{(1)}(n)\}$, *where*

$$\tilde{x}^{(1)}(i) = \sum_{k=1}^{i} \tilde{x}^{(0)}(k) = \left[\sum_{k=1}^{i} x_L^{(0)}(k), \sum_{k=1}^{i} x_R^{(0)}(k)\right] = \left[x_L^{(1)}(i), x_R^{(1)}(i)\right], i = 1, 2, \cdots, n \tag{3}$$

Definition 3.2 *The mean generating operation series of* $\tilde{x}^{(1)}$ *is* $\tilde{z}^{(1)} = \{\tilde{z}^{(1)}(2), \tilde{z}^{(1)}(3), \cdots, \tilde{z}^{(1)}(n)\}$, *where*

$$\tilde{z}^{(1)}(i) = 0.5(\tilde{x}^{(1)}(i-1) + \tilde{x}^{(1)}(i)) = [0.5\left(\sum_{k=1}^{i-1} x_L^{(0)}(k) + \sum_{k=1}^{i} x_L^{(0)}(k)\right), 0.5\left(\sum_{k=1}^{i-1} x_R^{(0)}(k) + \sum_{k=1}^{i} x_R^{(0)}(k)\right)] = [z_L^{(1)}(i), z_R^{(1)}(i)], \quad i = 2, 3, \cdots, n \tag{4}$$

Definition 3.3 *The grey differential equation of the interval fuzzy number GM (1, 1) model [IFGM (1, 1)] is*

$$\tilde{x}^{(0)}(i) + a\tilde{z}^{(1)}(i) = \tilde{b} \tag{5}$$

where a is said to be the integral development coefficient and taken as an exact number. $\tilde{b}$ is said to be the grey input and taken as an interval fuzzy number: $\tilde{b} = [b_L, b_R]$.

The integral development coefficient (a) in Eq. (1) is taken as the weighted mean of the development coefficients of the left and right boundary points of the interval fuzzy number because it represents the integral development tendency of the interval fuzzy number series, not the tendency of a single boundary point. Furthermore, if a is taken as an interval fuzzy number and the development trends of the left and right boundaries have big differences, the relative positions of the left and right boundaries of the prediction values will be disordered, which will result in the failure of the prediction. The grey input ($\tilde{b}$) is taken as a fuzzy number because it represents the overlaying of information of the fuzzy number series. That meets the connotative definition of it.

Next, the parameter estimation of IFGM (1, 1) is given. Firstly, values ($\tilde{x}^{(0)}$ and $\tilde{z}^{(1)}$) are put into (1).

$$\tilde{x}^{(0)}(2) + a\tilde{z}^{(1)}(2) = \tilde{b}, \tilde{x}^{(0)}(3) + a\tilde{z}^{(1)}(3) = \tilde{b}, \cdots, \tilde{x}^{(0)}(n) + a\tilde{z}^{(1)}(n) = \tilde{b}.$$

According to the computing criterion of the interval fuzzy number, the above equations are equal to the following two equations sets.

$$x_L^{(0)}(2) + a'z_L^{(1)}(2) = b_L, x_L^{(0)}(3) + a'z_L^{(1)}(3) = b_L, \cdots, x_L^{(0)}(n) + a'z_L^{(1)}(n) = b_L.$$

$$x_R^{(0)}(2) + a''z_R^{(1)}(2) = b_R, x_R^{(0)}(3) + a''z_R^{(1)}(3) = b_R, \cdots, x_R^{(0)}(n) + a''z_R^{(1)}(n) = b_R.$$

By the least square method, we have

$$\begin{pmatrix} a' \\ b_L \end{pmatrix} = (A^T A)^{-1} A^T Y_L, \begin{pmatrix} a'' \\ b_R \end{pmatrix} = (B^T B)^{-1} B^T Y_R, \tag{6}$$

where

$$A = \begin{pmatrix} z_L^{(1)}(2) & -1 \\ z_L^{(1)}(3) & -1 \\ \cdots & \cdots \\ z_L^{(1)}(n) & -1 \end{pmatrix}, \quad Y_L = \begin{pmatrix} -x_L^{(0)}(2) \\ -x_L^{(0)}(3) \\ \cdots \\ -x_L^{(0)}(n) \end{pmatrix}, \quad B = \begin{pmatrix} z_R^{(1)}(2) & -1 \\ z_R^{(1)}(3) & -1 \\ \cdots & \cdots \\ z_R^{(1)}(n) & -1 \end{pmatrix}, \quad Y_R = \begin{pmatrix} -x_R^{(0)}(2) \\ -x_R^{(0)}(3) \\ \cdots \\ -x_R^{(0)}(n) \end{pmatrix}.$$

The integral development coefficient (a) in IFGM (1, 1) is taken as the weighted mean of a' and a'', i.e. $a = (1 - \beta)a' + \beta a''$, where if $\beta > 0.5$, the decision maker prefers to the development trend of the upper limit of the interval fuzzy number series. If $\beta < 0.5$, the decision maker prefers the development trend of the lower limit of the interval fuzzy number series.

After the parameter estimation, the time response of IFGM (1, 1) is given as follows.

Theorem 3.1 *The time response series of IFGM (1, 1) are*

$$\hat{\tilde{x}}^{(0)}(i) = \frac{2(2-a)^{i-2}(\tilde{b} - a\tilde{x}^{(0)}(1))}{(2+a)^{i} - 1}, i = 2, 3, \cdots, n. \tag{7}$$

Let predicted values be $\hat{\tilde{x}}^{(0)}(i) = [\hat{x}_L^{(0)}(i), \hat{x}_R^{(0)}(i)]$, *then*

$$\hat{x}_L^{(0)}(i) = \frac{2(2-a)^{i-2}(b_L - ax_L^{(0)}(1))}{(2+a)^{i-1}}, \hat{x}_R^{(0)}(i) = \frac{2(2-a)^{i-2}(b_R - ax_R^{(0)}(1))}{(2+a)^{i-1}}$$

Proof Due to Eq. (1) and $\tilde{z}^{(1)}(i) = 0.5(\tilde{x}^{(1)}(i-1) + \tilde{x}^{(1)}(i))$, we have

$$\tilde{x}^{(0)}(i) + 0.5a(\tilde{x}^{(1)}(i-1) + \tilde{x}^{(1)}(i)) = \tilde{b},$$

$\tilde{x}^{(1)}(i) = \sum_{k=1}^{i} \tilde{x}^{(0)}(k)$, we obtain $\tilde{x}^{(0)}(i) + 0.5a(2\tilde{x}^{(1)}(i-1) + \tilde{x}^{(0)}(i)) = \tilde{b}$. Then

$$\begin{aligned}
\tilde{x}^{(0)}(i) &= \frac{\tilde{b} - a\tilde{x}^{(1)}(i-1)}{1+0.5a} = \frac{\tilde{b} - a(\tilde{x}^{(0)}(i-1) + \tilde{x}^{(1)}(i-2))}{1+0.5a} \\
&= \frac{\tilde{b} - a\tilde{x}^{(1)}(i-2)}{1+0.5a} - \frac{a\tilde{x}^{(0)}(i-1)}{1+0.5a} = \tilde{x}^{(0)}(i-1) - \frac{a\tilde{x}^{(0)}(i-1)}{1+0.5a} \\
&= \frac{2-a}{2+a}\tilde{x}^{(0)}(i-1) = \left(\frac{2-a}{2+a}\right)^{i-2}\tilde{x}^{(0)}(2) = \left(\frac{2-a}{2+a}\right)^{i-2}\frac{\tilde{b} - a\tilde{x}^{(0)}(1)}{1+0.5a}
\end{aligned}$$

Thus

$$\tilde{x}^{(0)}(i) = \frac{2(2-a)^{i-2}(\tilde{b} - a\tilde{x}^{(0)}(1))}{(2+a)^{i-1}}, \quad i = 2,\ 3,\ \cdots,\ n.$$

After the parameter estimation and taking $\hat{\tilde{x}}^{(0)}(1) = \tilde{x}^{(0)}(1)$, Eq. (3) becomes the time response of IFGM (1, 1) directly.

4 Modeling Condition for IFGM (1, 1)

Due to the modeling mechanism of GM (1, 1), the grey differential equation of GM (1, 1): $x^{(0)}(i) + az^{(1)}(i) = b$, where $z^{(1)}(i) = 0.5(x^{(1)}(i-1) + x^{(1)}(i))$, is the difference form of the white differential equation of GM (1, 1): $\frac{dx^{(1)}}{dt} + ax^{(1)} = b$, where $x^{(1)}$ is the accumulated generating operation series (AGO series) of the raw series $x^{(0)}$. The solution of $\frac{dx^{(1)}}{dt}+ax^{(1)}=b$ is $x^{(1)}(t)=(x^{(0)}(1)-\frac{b}{a})e^{-a(t-1)}+\frac{b}{a}$, which is an exponential type of curve. Therefore, the modeling mechanism of GM (1, 1) model is to use an exponential type of curve to fit the AGO series. So we have the modeling condition for GM (1, 1) is that the AGO series should meet the grey exponential law or pass the class ratio test.

Definition 4.1 *Let a non-negative sequence be* $x = \{x(1), x(2), \cdots, x(n)\}$. *The class ratio of x at point k is*

$$\sigma(k) = \frac{x(k)}{x(k-1)}, k = 2, 3, \cdots, n. \tag{8}$$

Definition 4.2 *Let a non-negative sequence be* $x = \{x(1), x(2), \cdots, x(n)\}$. *Then*

- *If* $\forall k, \sigma(k) \in (0, 1)$, *then x is said to have negative grey exponential law;*
- *If* $\forall k, \sigma(k) \in (1, b)$, *then x is said to have positive grey exponential law;*
- *If* $\forall k, \sigma(k) \in (a, b), b - a = \delta$, *then x is said to have grey exponential law with absolutely grey degree* δ;
- *If* $\delta < 0.5$, *then x is said to have quasi-exponential law.*

For common non-negative sequences, by the accumulated generating operation (AGO), the quasi-exponential law will appear. The smoother the sequence, the more obvious the exponential law is. Then the precision of GM (1, 1) is better. This is the theoretical basis for GM (1, 1) establishing.

From the above analysis, we have the modeling condition for IFGM (1, 1). It is that the AGO series of the left and right boundary point series both have the quasi-exponential law, namely,

$$x^{(1)}(i+1) \Big/ x^{(1)}(i) = \sum_{k=1}^{i+1} x^{(0)}(k) \Big/ \sum_{k=1}^{i} x^{(0)}(k) \in [a, b]$$

where $b - a = \delta < 0.5$.

5 Revised IFGM (1, 1)

The geometric figure of the prediction results of GM (1, 1) is a smooth curve, which describes the general development trend of the raw data. The random fluctuation of the development is not taken into account. Therefore, for the strongly fluctuating series, which normally can not meet the quasi-exponential law, the forecast precision of GM (1, 1) is not high. The Markov prediction can describe a dynamic variation of a random time series. It is suitable for the prediction of the strongly fluctuating series. In this paper, the two prediction methods are integrated. The grey prediction is applied to indicate the general development tendency of the raw series and the Markov prediction is applied to reflect the random fluctuation of the series. In this way, the forecast precision of the fluctuating series is improved.

The simulated values of IFGM (1, 1) are $\hat{\tilde{x}}^{(0)}(t) = [\hat{x}_L^{(0)}(t), \hat{x}_R^{(0)}(t)], t = 1, 2, \cdots, n$, including the left and right boundaries (i.e. $\hat{x}_L^{(0)}(t)$ and $\hat{x}_R^{(0)}(t)$). Next, the revised process for the left boundary point ($\hat{x}_L^{(0)}(t)$) is proposed. It is similar for the right boundary point ($\hat{x}_R^{(0)}(t)$).

Step 1: (Status partition) According to the relative ratios of initial values against simulation values of IFGM (1, 1), i.e. $x_L^{(0)}(t) \Big/ \hat{x}_L^{(0)}(t)$, the system is divided into m statuses, which is denoted as $E_i \in [A_i, B_i], i = 1, 2, \cdots, m$. The number of statuses depends on the actual situation. The more raw data, the more statuses needed in order to get proper transition frequencies between statuses.

Step 2: (Establish the transition probability matrix) Let $M_{ij}(k)$ be the number of raw data transferred from E_i to E_j by k steps. Let M_i be the occurrence number of E_i. Then the transition probability from E_i to E_j by k steps is $P_{ij}(k) = \frac{M_{ij}(k)}{M_i}$, $i,\ j = 1, 2, \cdots, m$. The transition probability matrix is $P(k) = (P_{ij}(k))_{m \times m}$. In practice, only the one-step transition probability matrix ($P(1)$) is considered.

Step 3: (Predicted statuses identification) Let the predicted value at the $(n+1)^{\text{th}}$ moment of IFGM (1, 1) be $\hat{x}_L(n+1)$. If $\hat{x}_L(n)$ is in the status of E_k. From the kth row of $P(1)$, the revised prediction value at the $(n+1)^{\text{th}}$ moment ($\hat{x}'_L(n+1)$) is taken as:

$$\hat{x}'_L(n+1) = \hat{x}_L(n+1) \times \sum_{j=1}^{m} \frac{1}{2} P_{kj}(A_j + B_j) \tag{9}$$

Step 4: (Multi-step prediction) If the predicted value at the $(n+1)^{\text{th}}$ moment ($\hat{x}'_L(n+1)$) is added into the raw series and step 1-4 are repeated, we can obtain the predicted value at the $(n+2)^{\text{th}}$ moment ($\hat{x}'_L(n+2)$).

6 Practical Examples

Example 1 [10] From 1998 to 2002, a change sequence of a sort of energy price in a region is: $\tilde{x}^{(0)}$={[1.05, 1.09], [1.05, 1.10], [1.09, 1.15], [1.10, 1.20], [1.15, 1.25]}. Based on this series we establish the IFGM (1,1) model. By the software MATLAB, the simulating process is as follows.

Step 1: From Eq. (3), the AGO series of $\tilde{x}^{(1)}$ is:

$$\tilde{x}^{(1)} = \{[1.05, 1.09], [2.10, 2.19], [3.19, 3.34], [4.29, 4.54], [5.44, 5.79]\}.$$

Step 2: Quasi-exponential law test is made for $\tilde{x}^{(1)}$. From Eq. (8), we have, for $k \geq 3$, $\sigma_L^{(1)}(k) = \{1.52, 1.34, 1.27\}$, $\sigma_R^{(1)}(k) = \{1.52, 1.36, 1.28\}$, then $\delta_L = 0.25$,

Table 1 Simulation results of IFGM (1, 1)

Raw data	Simulation data	Relative errors (%)
[1.05, 1.10]	[1.0628, 1.0892]	1.22, 0.98
[1.09, 1.15]	[1.1011, 1.1285]	1.02, 1.87
[1.10, 1.20]	[1.1408, 1.1692]	3.71, 2.57
[1.15, 1.25]	[1.1819, 1.2113]	2.78, 3.09
Average relative error		2.15

$\delta_R = 0.26$. So $\tilde{x}^{(1)}$ has the quasi-exponential law and the smoothness is good. IFGM (1, 1) can be established.

Step 3: The mean generating operation series is:

$$\tilde{z}^{(1)} = \{[1.575, 1.64], [2.645, 2.765], [3.74, 3.94], [4.865, 5.165]\}.$$

Step 4: From Eq. (6), the parameter estimation is:

$$\begin{pmatrix} a' \\ b_L \end{pmatrix} = \begin{pmatrix} -0.0283 \\ 1.0068 \end{pmatrix}, \begin{pmatrix} a'' \\ b_R \end{pmatrix} = \begin{pmatrix} -0.0425 \\ 1.0313 \end{pmatrix}.$$

The integral developing parameter is taken as: $a = 0.5a' + 0.5a'' = -0.0354$.

Step 5: From Eq. (7), the data of simulation are obtained. Let the relative errors are $e(i) = \left|(x^{(0)}(i) - \hat{x}^{(0)}(i))/x^{(0)}(i)\right|, i = 2,\ 3, 4, 5$. The results are shown in Table 1.

The relative errors of the left and right bounds of the simulation data have respectively been shown in Table 1. The average relative error of all is 2.15 %. In [10], the average relative error is 2.52 % and the method of forecasting is more complex than IFGM (1, 1). In addition, because the sample size is very small, the forecasting methods of statistics are normally unsuitable.

Example 2 The Consumer Price Index (CPI) is the price change index of goods and services bought by consumers. It is one of important indexes related to the inflation. The State Statistics Bureau of China provides the CPI records of each month from 1996 to 2011. From the data of twelve months, the interval of the CPI in one year is obtained. The values from 1996 to 2009 are taken as the raw series to establish the revised IFGM (1, 1) and the value of 2010 will be predicted.

Through the class ratio test, we get

$$\sigma_L(k) \in [1.0750, 1.4752], \sigma_R(k) \in [1.0753, 1.4586], for\ k \geq 3.$$

Then $\delta_L = 0.4002$, $\delta_R = 0.3833$. From the test, we know the smoothness of the raw series is not very good. The series has fluctuation.

Table 2 The simulation data of IFGM (1, 1) and the relative ratios (%) of raw data against simulation data

Year	1996	1997	1998
Raw data	[106.9, 109.8]	[100.4, 105.9]	[98.5, 100.7]
Simulation data	[106.9, 109.8]	[99.23, 101.12]	[99.47, 101.36]
Ratios (%)	100,100	101.18,104.73	99.03, 99.35
1999	2000	2001	2002
[97.8, 99.4]	[99.7, 101.5]	[99.4, 101.7]	[98.7, 100.0]
[99.70, 101.59]	[99.93, 101.83]	[100.17, 102.07]	[100.41, 102.31]
98.09, 97.84	99.76, 99.67	99.23, 99.64	98.30, 97.74
2003	2004	2005	2006
[100.2, 103.2]	[102.1, 105.3]	[100.9, 103.9]	[100.8, 102.8]
[100.64, 102.55]	[100.88, 102.79]	[101.11, 103.03]	[101.35, 103.28]
99.56, 100.63	101.21, 102.44	99.79, 100.84	99.46, 99.54
2007	2008	2009	2010
[102.2, 106.9]	[101.2, 108.7]	[98.2, 101.9]	[101.5, 105.1]
[101.59, 103.52]	[101.83, 103.76]	[102.07, 104.01]	
100.60, 103.27	99.38, 104.76	96.21, 97.98	

Firstly, the IFGM (1, 1) is established. The parameter estimation of IFGM (1, 1) is: $a' = -0.0015, a'' = -0.0032, b_L = 98.8664, b_R = 100.7415$. The integral development coefficient ($a = (1-\beta)a' + \beta a''$, where let $\beta = 0.5$) is calculated as: $a = -0.0023$. Thus the prediction formula of IFGM (1, 1) is as follows:

$$\hat{\tilde{x}}^{(0)}(i) = \frac{2(2+0.0023)^{i-2}([98.8664, 100.7415] + 0.0023 \times [106.9, 109.8])}{(2-0.0023)^{i-1}},$$

$$i = 2,\ 3,\ \cdots,\ n.$$

Next, the simulation data of IFGM (1, 1) ($[\hat{x}_L^{(0)}(t), \hat{x}_R^{(0)}(t)]$) and the relative ratios of raw data against simulation data ($x_L^{(0)}(t)/\hat{x}_L^{(0)}(t)$ and $x_R^{(0)}(t)/\hat{x}_R^{(0)}(t)$) are shown in Table 2. The results show that the prediction result of IFGM (1, 1) is a smooth curve, which indicates the general development tendency of the raw series. But the fluctuation is not shown. Next, the revised process is as follows.

From Table 2, the ratio of $x_L^{(0)}$ against $\hat{x}_L^{(0)}$, i.e. $x_L^{(0)}(t)/\hat{x}_L^{(0)}(t)$, changes in the interval: [96.21, 101.21 %]. The ratio of $x_R^{(0)}$ against $\hat{x}_R^{(0)}$ is in [97.74, 104.76 %]. Based on these, we give the status partitions of left bounds and right bounds respectively in Table 3.

Due to the status partition, the 1-step transition probability matrixes of left and right bounds are respectively as follows.

Table 3 Status partitions

Left bounds		Right bounds	
Status	Relative ratios (%)	Status	Relative ratios (%)
E1	96–98.5	E1	97–99.6
E2	98.5–100.5	E2	99.6–102.4
E3	100.5–102	E3	102.4–105

$$P_L(1) = \begin{pmatrix} 0 & 1 & 0 \\ 3/8 & 2/8 & 3/8 \\ 0 & 1 & 0 \end{pmatrix}, \ P_R(1) = \begin{pmatrix} 1/4 & 2/4 & 1/4 \\ 2/5 & 1/5 & 2/5 \\ 2/4 & 1/4 & 1/4 \end{pmatrix}$$

In 2009, the left and right boundaries are both in the status of E1. The forecast value of IFGM (1, 1) in 2010 are $\hat{x}_L(n+1) = 102.0682$ and $\hat{x}_R(n+1) = 104.0061$. Due to the first row of $P_L(1)$ and $P_R(1)$, we calculate the forecast value in 2010 according to (9).

$$\begin{aligned} \hat{x}'_L(n+1) =& 102.0682 \times \frac{1}{2} \times (98.5 + 100.5\,\%) = 101.7965, \\ \hat{x}'_R(n+1) =& 104.0061 \times \frac{1}{2} \times [\frac{1}{4} \times (97 + 99.6\,\%) + \frac{2}{4} \times (99.6 + 102.4\,\%) \\ &+ \frac{1}{4} \times (102.4 + 105\,\%)] = 105.2930 \end{aligned}$$

Next, we will apply the exponential smoothing method and the autoregressive moving average model (ARMA) to the example in order to show the compares of these models with grey models. These models can not be established based on the two bounds of the interval number directly because it destroys the integrity of the fuzzy number and may cause the lower and upper bounds of the interval number disordered. Therefore, we establish these models based on two transformed series and then deduce the forecast values of the interval number series through simple calculations. The two transformed series are the mid-point and length series of the raw interval number series, denoted respectively as $(x_L^{(0)} + x_R^{(0)})/2$ and $x_R^{(0)} - x_L^{(0)}$, which are

$$\begin{aligned} (x_L^{(0)} + x_R^{(0)})/2 =& \{108.35, 103.15, 99.6, 98.6, 100.6, 100.55, 99.35, \\ & 101.7, 103.7, 102.4, 101.8, 104.55, 104.95, 100.05\} \\ x_R^{(0)} - x_L^{(0)} =& \{2.9, 5.5, 2.2, 1.6, 1.8, 2.3, 1.3, 3, 3.2, 3, 2, 4.7, 7.5, 3.7\} \end{aligned}$$

By applying the software EVIEWS, we draw the correlograms of the two series and make the model order estimation. Then we establish ARMA (2, 2) for $(x_L^{(0)} + x_R^{(0)})/2$ and ARMA (2, 1) for $x_R^{(0)} - x_L^{(0)}$. For the exponential smoothing method, we adopt

Table 4 Forecast results of 2010

Revised IFGM (1, 1)		IFGM (1, 1)	
Forecast data	Relative errors (%)	Forecast data	Relative errors (%)
[101.7965, 105.2930]	0.29, 0.18	[102.3080, 104.2505]	0.80, 0.81
Double exponential smooth		ARMA	
[99.7944, 104.5905]	1.68, 0.48	[100.5146, 104.0258]	0.97, 1.02

double exponential smoothing method for the two series. The results of four models are shown in Table 4.

From Table 4, we can see the relative errors of the left and right forecast bounds of the revised IFGM (1, 1) (0.29 and 0.18%) are smaller than the ones of IFGM (1, 1) and is the smallest of the four models, which indicates that the revised IFGM (1, 1) model is effective as for the fluctuating series.

7 Conclusion

In this paper, we improved the definition equation of GM (1, 1) to make it suitable for the interval fuzzy number series. IFGM (1, 1) and revised IFGM (1, 1) have been proposed. The models have generalized the applicable range of grey prediction. IFGM (1, 1) is suitable for interval fuzzy number series with weak fluctuation. Revised IFNGM (1, 1) has integrated advantages of grey prediction and Markov prediction. For the fluctuating series, the prediction precision has been improved. The prediction of the fuzzy number series is a new issue in applications. There are many problems needed to be studied. We will research further.

Acknowledgments This research is supported by the National Science Foundation of China (No.11071178 and No.11162004) and the Research Foundation of Humanity and Social Science of Ministry of Education of China (No.11YJAZH131). Authors would like to thank referees for their helpful comments.

References

1. Deng, J.L.: Introduction to grey system theory. J. Grey Syst. **1**(1), 1–24 (1989)
2. Deng, J.L.: Basis of Grey Theory. Huazhong University of Science and Technology Press, Wuhan (2002)
3. Yao, T.X., Liu, S.F., Xie, N.M.: On the properties of small sample of GM(1,1) model. Appl. Math. Model. **33**, 1894–1903 (2009)
4. Liu, S.F., Deng, J.L.: The range suitable for GM (1,1). J. Grey Syst. **11**(1), 131–138 (1999)
5. Wang, Y.F.: Predicting stock price using fuzzy grey prediction system. Expert Syst. Appl. **22**(1), 33–38 (2002)

6. Xie, N.M., Liu, S.F.: Discrete grey forecasting model and its optimization. Appl. Math. Model. **33**(1), 1173–1186 (2009)
7. Tan, G.J.: The structure method and application of background value in grey system GM (1, 1) model. J. Syst. Eng. Theory Pract. **5**, 126–127 (2000)
8. Zhang, X.X., Xiao, X.P.: Study on the connotation of parameters in GM (1, 1) model. J. Grey Syst. **18**, 26–32 (2006)
9. Zeng, X.Y., Xiao, X.P.: Study on generalization for GM (1, 1) model and its application. Control Decis. **24**(7), 1092–1096 (2009)
10. Fang, Z.G., Liu, S.F.: Study on GM(11) model based on interval grey number (GMBIGN (11)). Chin. J. Manage. Sci. **12**(10), 130–134 (2004)
11. Vapnik, V.N.: Statistical Learning Theory. Wiley, New York (1998)

Exponential Forecasting Model with Fuzzy Number in Long-Distance Call Quantity

Bing-yuan Cao

Abstract An exponential forecasting model, in this paper, is established with fuzzy numbers in long-distance call quantity, and a determining method is given in comparison with examples mentioned in the paper. Besides, the model is verified by example comparison, and the effectiveness of its method.

Keywords Fuzzy number · Exponential model · Call quantity · Forecast.

1 Introduction

It proves by practice that the technical features of the long distance call quantity meets index law, at the same time, the long distance call quantity statistics is quite complex because there exists a lot of uncertainty. In this paper, we consider fuzzy uncertainty, put forward the phone quantity forecasting model with fuzzy numbers. In Sect. 2, we introduce the concept of triangular fuzzy numbers. In Sect. 3, we establish an exponential forecasting model containing fuzzy numbers in the call quantity, and give a method to determine it. In Sect. 4, through examples, we testify the model and effectiveness of the proposed method. In Sect. 5, in comparison with the examples and judgment, we obtain some regularity of results.

B. Cao (✉)
School of Mathematics and Information Science; Key Laboratory of Mathematics and Interdisciplinary Sciences of Guangdong Higher Education Institutes, Guangzhou University, ZIP 510006, People's Republic of China
e-mail: caobingy@163.com

B.-Y. Cao and H. Nasseri (eds.), *Fuzzy Information & Engineering and Operations Research & Management*, Advances in Intelligent Systems and Computing 211,
DOI: 10.1007/978-3-642-38667-1_25, © Springer-Verlag Berlin Heidelberg 2014

2 Concept of Triangular Fuzzy Numbers

Definition 1 [1, 2] Suppose $\widetilde{A} \in F(X)$, As for $\alpha \in [0, 1]$, we write

$$A_\alpha = \left\{x \in X \left| \mu_{\widetilde{A}}(x) \geq \alpha \right.\right\},$$

calling A_α the α-cut sets of $\widetilde{A}$, while calling α a threshold or a confidence level. Again we write down

$$A_\alpha = \left\{x \in X \left| \mu_{\widetilde{A}}(x) > \alpha \right.\right\},$$

calling A_α the α-strong cut sets of $\widetilde{A}$.

Definition 2 [2, 3] Suppose $\widetilde{A}$ to be a fuzzy set in a real axis R, if it meets

1. $\exists x_0 \in R$, such that $\tilde{A}(x_0) = 1$;
2. $\forall \lambda \in [0, 1]$, A_λ represents a finite closed interval in R,

we said $\tilde{A}$ to be a fuzzy number in R.

By Definition 1 and 2, we know that α-cut sets of fuzzy number $\widetilde{A}$ can be expressed in $A_\alpha = [A_\alpha^L, A_\alpha^R]$, where

$$A_\alpha^L = \inf \left\{x \in X \left| \mu_{\tilde{A}}(x) \geq \alpha \right.\right\},$$
$$A_\alpha^R = \sup \left\{x \in X \left| \mu_{\tilde{A}}(x) \geq \alpha \right.\right\}, \alpha \in [0, 1].$$

Definition 3 [3, 4] Call fuzzy sets in real axis R a fuzzy number written down as

$$\tilde{A} = \int_{x \in R} \frac{\tilde{A}(x)}{x} \text{ or } \tilde{A} \Leftrightarrow \tilde{A}(x) \in [0, 1],$$

where $\tilde{A}(x)$ is a membership function in $\tilde{A}$.

Definition 4 [5, 6] Call fuzzy number $\tilde{A}$ a triangular fuzzy number, if its membership function $\mu_{\tilde{A}}(x)$ meets

$$\mu_{\tilde{A}}(x) = \begin{cases} \dfrac{x - a^L}{a_M - a^L}, & a^L \leq x \leq a_M, \\ 0, & others, \\ \dfrac{x - a^R}{a_M - a^R}, & a_M \leq x \leq a^R, \end{cases}$$

where $[a^L, a^R]$ is a supporting interval and point $[a_M, 1]$ is a peak value.

Definition 5 [5, 6] Call fuzzy number $\tilde{A}$ a symmetric triangular fuzzy number in the center, if its membership function $\mu_{\tilde{A}}(x)$ meets

$$\mu_{\tilde{A}}(x) = \begin{cases} 2\frac{x-a^L}{a^R-a^L} & a^L \le x \le \frac{a^L+a^R}{2}, \\ 0, & others, \\ 2\frac{x-a^R}{a^L-a^R} & \frac{a^L+a^R}{2} \le x \le a^R. \end{cases}$$

Write $\tilde{A}$ a symmetric triangular fuzzy number as (m, a), where m represents the median value of three fuzzy numbers; a represents width of the left and right of the triangular fuzzy number.

In particular, at $a = 0$, $\tilde{A}$ degenerates to ordinary real numbers.

Definition 6 [4] We define

$$E(\tilde{A}) = \int\limits_{x\in R} x \frac{\mu_{\tilde{A}}(x)}{\int\limits_{x\in R} \mu_{\tilde{A}}(x)dx} dx$$

as expectation value of fuzzy number $\tilde{A}$, where $\frac{\mu_{\tilde{A}}(x)}{\int\limits_{x\in R} \mu_{\tilde{A}}(x)dx}$ is a relative membership degree of fuzzy number $\tilde{A}$.

Definition 7 [4] Suppose $\tilde{A} = (a_1, a_2, a_3)$ to be a triangular fuzzy number, then its expectation value is defined as

$$E(\tilde{A}) = \frac{a_1 + a_2 + a_3}{3}. \tag{1}$$

3 Exponential Model with Fuzzy Numbers

Due to a variety of uncertain factors, telephone traffic often produces a certain fluctuation. If a backlog of fuzzy parameter in Eq. (1) is a fuzzy number, the model will contain more information. Following that , we will fuzzify coefficients in Model Eq. (1), and establish an exponential prediction model containing fuzzy coefficient.

We consider

$$\tilde{Y} = \tilde{A}_1 \tilde{A}_2^t, \tag{2}$$

where A_1, A_2 is triangular fuzzy numbers, $\tilde{A}_i = (a_i^L, a_i, a_i^R)$, calling Eq. (2) an exponential model containing triangular fuzzy numbers. They are parameters to be estimated, $\hat{Y}(t)$ denotes the evaluation in telephone quantity during t years, and telephone quantity fluctuates with various indeterminable factors.

Theorem 1 Problem Eq. (1) is equivalent to

$$E(Y) = E(A_1)E(A_2^t). \tag{3}$$

Proof Obviously.
1. Non-fuzzify Eq. (2) by use of Theorem 1 and we obtained Eq. (3). Linearize Eq. (3) by taking logarithm and we can get

$$\ln E(Y) = \ln E(A_1) + t \ln E(A_2).$$

Coming next, we estimate parameters A_1 and A_2.

As for the given sample set $\{Y(t_1), Y(t_2), \ldots, Y(t_n)\} \in [0, 1]$, we take any two adjacent sample points $t_k, t_{k+1} (k = 1, 2, \ldots, N-1)$, the least square estimations for parameters A_1, A_2 are

$$\ln E(\hat{Y}(t_k)) = \ln E(A_1) + t_k \ln E(A_2), \tag{4}$$

$$\ln E(\hat{Y}(t_{k+1})) = \ln E(A_1) + t_{k+1} \ln E(A_2). \tag{5}$$

(4) $\times\, t_{k+1}$ − (5) $\times\, t_k$, and then we have

$$(t_{k+1} - t_k) \ln E(A_1) = t_{k+1} \ln E(\hat{Y}(t_k)) - t_k \ln E(Y(t_{k+1})). \tag{6}$$

2. Applying the least square method, we build an objective function by Eq. (6) as

$$J_1 = \sum_{k=1}^{N-1} [t_{k+1} \ln E(\hat{Y}(t_k)) - t_k \ln E(Y(t_{k+1})) - (t_{k+1} - t_k) \ln E(A_1)]^2.$$

Again, applying the method, we build an objective function by Eq. (4) as

$$J_2 = \sum_{k=1}^{N} [\ln E(Y(t_k)) - \ln E(\hat{Y}(t_k))]^2 = \sum_{k=1}^{N} [\ln E(Y(t_k)) - (\ln E(A_1) + t_k \ln E(A_2))]^2.$$

Let $\frac{\partial J_1}{\partial \ln A_1} = 0, \frac{\partial J_2}{\partial \ln A_2} = 0$. Write down $\Delta t_k = t_{k+1} - t_k$, and we obtained

$$\sum_{k=1}^{N-1} [t_{k+1} \ln E(\hat{Y}(t_k)) - t_k \ln E(Y(t_{k+1}))] \Delta t_k = \sum_{k=1}^{N-1} \Delta t_k^2 \ln E(A_1), \tag{7}$$

$$2 \sum_{k=1}^{N} [\ln E(Y(t_k)) - \ln E(A_1) - t_k \ln E(A_2))] t_k = 0. \tag{8}$$

Solve Eqs. (7 and 8) and we get

$$A = \exp\left\{\frac{\sum_{k=1}^{N-1}[\Delta t_k \ln E(\hat{Y}(t_k)) - \sum_{k=1}^{N-1} t_k \ln \frac{E(Y(t_{k+1}))}{E(Y(t_k))}}{\sum_{k=1}^{N-1} \Delta t_k^2}\right\}, \tag{9}$$

$$B = \exp\left\{\frac{\sum_{k=1}^{N-1} t_k \ln E(Y(t_k)) - \ln E(A_1) \sum_{k=1}^{N} t_k}{\sum_{k=1}^{N} t_k^2}\right\}. \tag{10}$$

Especially, at $t_k = k(k = 1, 2, \ldots, N)$, we have $E(Y(t_k)) = E(Y(k))$, where $\Delta t_k = 1$, and we change Eqs. (9), (10) into

$$A = \exp\left\{\frac{\sum_{k=1}^{N-1}[\ln E(\hat{Y}(k)) - \sum_{k=1}^{N-1} k \ln \frac{E(Y(k+1))}{E(Y(k))}}{N-1}\right\}, \tag{11}$$

$$B = \exp\left\{\frac{6\sum_{k=1}^{N} k \ln E(Y(k)) - 3N(N+1) \ln E(A_1)}{N(N+1)(2N+1)}\right\}. \tag{12}$$

Its deterministic model refers to

$$E(Y(k)) = E(A_1)E(A_2^k). \tag{13}$$

Again by formula

$$S = \sqrt{\frac{\sum_{k=1}^{N} [E(Y(t_k)) - E(\hat{Y}(t_k))]^2}{N}}, \tag{14}$$

$$e\,\% = \frac{1}{N}\sum_{k=1}^{N} |1 - \frac{E(Y(t_k))}{E(\hat{Y}(t_k))}| \times 100\,\%, \tag{15}$$

after finding a standard deviation S in a forecasting error and an average relative error percentage $e\,\%$, we determine a fitting degree for forecasting models.

Its deterministic model refers to

$$\hat{Y}(k) = \hat{A}_1 \hat{A}_2^k, \tag{16}$$

Therefore, the fuzzy exponential model corresponding to Eq. (2) is

$$\tilde{Y}(k) = \tilde{A}_1 \tilde{A}_2^k. \tag{17}$$

4 Practical Example

Example 1 Long-distance call in China during 2000–2010 as follows (Table 1):

Forecast time for telephone by applying an exponential Model Eq. (16) and we have (Table 2)

If we use Eqs. (11) and (12), we can get parameters $A_1 = 12{,}380$, $A_2 = 1.2069$, then

$$E(Y(k)) = 12{,}380 \times 1.2069^k. \tag{18}$$

By a standard deviation formula (14)

$$S = \sqrt{\frac{\sum_{k=1}^{11} [E(Y(k)) - E(\hat{Y}(k))]^2}{11}},$$

we can obtain S = 4019. Again, from formula (15) of a percentage error

Table 1 Quantity of long-distance call

Year	No	Practical date	Year	No	Practical date
2000	1	[21000, 21404, 21808]	2006	7	[42200, 42303, 42402]
2001	2	[18000, 22049, 22080]	2007	8	[51000, 51525, 52046]
2002	3	[21000, 23574, 24334]	2008	9	[64000, 64617, 65232]
2003	4	[26200, 26556, 26618]	2009	10	[78000, 78462, 78920]
2004	5	[31500, 31553, 51602]	2010	11	[97832, 106291, 106391]
2005	6	[38000, 38254, 38504]			

Table 2 Quantity of long-distance call

Year	No	Practical date	Year	No	Practical date
2000	1	21404	2006	7	42303
2001	2	22049	2007	8	51525
2002	3	23574	2008	9	64617
2003	4	26556	2009	10	78462
2004	5	31553	2010	11	106291
2005	6	38254			

$$e\,\% = \frac{1}{11}\sum_{k=1}^{11}|1-\frac{E(Y(k))}{E(\hat{Y}(k))}| \times 100\,\%,$$

we can get an average relative error to be 8.21 %. While, by the aid of geometric average, we obtain S = 9,405, e = 19.78 %, and S = 4,811, e = 9.74 %$ by average value exponential curve. Therefore, the fuzzy exponential forecast method mentioned here is superior to the above two [7].

Under the fiducial degree of 95 %, the long-distance call in China vary at the following interval $\hat{Y} \pm S$. Hence, their forecast quantity between 2000–2010 shows below (Table 3):

Table 3 Forecast quantity of long-distance call in China

Year	No	Practical date	Year	No	Practical date
2000	1	[21000, 21404, 21808]	2006	7	[42200, 42303, 42402]
2001	2	[18000, 22049, 22080]	2007	8	[51000, 51525, 52046]
2002	3	[21000, 23574, 24334]	2008	9	[64000, 64617, 65232]
2003	4	[26200, 26556, 26618]	2009	10	[78000, 78462, 78920]
2004	5	[31500, 31553, 51602]	2010	11	[97832, 106291, 106391]
2005	6	[38000, 38254, 38504]			

5 Conclusion

The method in this paper is an extension of exponential forecast model. A forecasting value is obtained for a linearized model respectively by adopting a two-step of the least square method. Each forecast value $\overset{\frown}{Y}(t)$ fluctuates in the band region composed of Y, which presents us more information. It is pointed out that the model here can be still expanded to situation with various fuzzy coefficients and evens with fuzzy variables. [1, 8–11].

Acknowledgments Thanks to the support by National Natural Science Foundation of China under Grant No. 70771030.

References

1. Cao, B.-Y.: Study on non-distinct self-regression forecast model. Chin. Sci. Bull. **35**(13), 1057–1062 (1990). (also in Kexue Tongbao 34(17),1291–1294, 1989)
2. Cao, B.-Y.: Nonlinear regression forecasting model with T-fuzzy data to be linearized. J. Fuzzy Syst. and Math. **7**(2), 43–53 (1993)

3. Cao, B.-Y.: Optimal Models and Methods with Fuzzy Quantities. Springer Heidelberg, Berlin, Heidelberg (2010)
4. Song, Z., Zhang, D., Gong, L.: Ranking fuzzy numbers based on relative membership degree. J. Air Force Eng. Univ. **7**(6), 81–83 (2006). (Natural Science Edition)
5. Li, F.: Fuzzy Information Processing System. Piking University Press, Beijing (1998)
6. Tran, L., Duckstein, L.: Comparison of fuzzy numbers using a fuzzy distance measure. Fuzzy Sets Syst. **130**, 331–341 (2002)
7. Zheng, X.C.: Prospect forecast in the quantity of long distance telephone in China. Forecast **3** (1992)
8. Cao, B.-Y.: Extended fuzzy geometric programming. J Fuzzy Math. **1**(2), 285–293 (1993)
9. Tanaka, H., Uejima, S., Asai, K.: Linear regression analysis with fuzzy model. IEEE Trans. Syst. Man Cybernetics SMC **12**(6), 903–907 (1982)
10. Tanaka, H., Uejima, S., Asai, K.: Linear regression analysis with fuzzy model. IEEE Trans. Syst. Man Cybernetics SMC **12**(6) (1982)
11. Zadeh, L.A., Chen, G.Q.: (Translation): Fuzzy Sets. Language Variables and Fuzzy Logics. Science in China Press, Piking (1982)

Contingent Valuation of Non-Market Goods Based on Intuitionistic Fuzzy Clustering: Part I

Zeng-tai Gong and Bi-qian Wu

Abstract In order to value the non-market goods, we consider the uncertain preference of the respondents for non-market goods, individual often have trouble trading off the good or amenity against a monetary measure. Valuation in these situations can best be described as fuzzy in terms of the amenity being valued. We move away from a probabilistic representation of uncertainty and propose the use of intuitionistic fuzzy contingent valuation. That is to say we could apply intuitionistic fuzzy logic to contingent valuation. Since intuitionistic fuzzy sets could provide the information of the membership degree and the nonmembership degree, it has more expression and flexibility better than traditional fuzzy sets in processing uncertain information data. In this paper, we apply intuitionistic fuzzy logic to contingent, developing an intuitionistic fuzzy clustering and interval intuitionistic fuzzy clustering approach for combining preference uncertainty. We develop an intuitionistic fuzzy random utility maximization framework where the perceived utility of each individual is intuitionistic fuzzy in the sense that an individual's utility belong to each cluster to some degree. Both the willingness to pay (WTP) and willingness not to pay (WNTP) measures we obtain using intuitionistic fuzzy approach are below those using standard probability methods.

Keywords Random utility maximization · Contingent valuation · Preference uncertainty · Intuitionistic fuzzy c-means clustering

---

Z. Gong (✉) · B. Wu
College of Mathematics and Information Science, Northwest Normal University, Lanzhou 730070, People's Republic of China
e-mail: gongzt@nwnu.edu.cn

B.-Y. Cao and H. Nasseri (eds.), *Fuzzy Information & Engineering and Operations Research & Management*, Advances in Intelligent Systems and Computing 211,
DOI: 10.1007/978-3-642-38667-1_26, © Springer-Verlag Berlin Heidelberg 2014

1 Introduction

The contingent valuation method (CVM) is a widely used technique to estimate economic values for all kinds of ecosystem and environmental services. It can be used to estimate both use and non-use values, and it is the most widely used method for estimating non-use values. Most CV surveys rely on a dichotomous choice question to elicit willingness to pay (WTP) and willingness not to pay (WNTP). Calculation of the Hicksian compensating or equivalent welfare measure is based on the assumption that the survey respondent knows her utility function with certainty [1]. The assumption of preference certainty is a strong one because CV seeks to elicit values for environmental resources from respondents who may lack the experience to make such assessments. While Hanemann [1] provide an explanation of what preference uncertainty means in the context of the CV method, several authors have adopted varying but complex approaches for dealing with preference uncertainty in non-market valuation [2]. Apparent precision of standard WTP and WNTP estimates faces the underlying uncertainty of preferences and may lead to bias outcomes. In the random utility maximization (RUM) framework popularized by Hanemann [1], an individuals utility is modeled as consisting of a deterministic component plus an unobservable random error. However, there are some uncertain which is unsure about a respondent's preference [3]. In the tradition contingent valuation, response have two choice take it or leave it, we only know that an individual's compensating surplus is greater or less than the offered payment, but not by how much. Attempts have been made to hone in on the compensating surplus using a follow-up question, as in the double-bounded approach that seeks to increase the confidence of the estimated welfare measure. Our view is that the valuation can be best be described as fuzzy in terms of perceptions about the property rights to be good, the amenity being valued, and the actual tradeoffs between the amenity and money metric. In the area of non-market valuation Cornelis Van Kooten [4] may have been the first apply fuzzy logic to the context of non-market valuation, the researchers found that fuzzy as fuzzy as opposed to conventional regression significantly improved the mean squared error, fuzzy set theory [5] provides a useful alternative for interpreting preference and analyzing willingness to pay respondences in the CV framework. Fuzzy logic addresses both imprecision about what is to be valued and uncertainty about values that are actually measured. In the article of sun and Van Kooten [6] they use the fuzzy cluster approach to fuzzify respondent utility functions from the beginning. Cluster analysis is commonly used for pattern recognition, soft learning, and other engineering applications. In economics, it is mainly used to segment markets by incorporating heterogeneous preferences. Thus, in modeling choice of shopping trips. In the context of non-market valuation, Boxall and Adamowicz [7] applied latent segmentation to the choice of wilderness recreation sites, identifying latent classes by incorporating motivation, perceptions and individual characteristics. They found significant differences in welfare measures with the segment model. Fuzzy clustering analysis provides an alternative to the latent segmentation model that addresses non-linearity in a flexible way and avoids identification problems.

Intuitionistic fuzzy sets are defined by Atanassov in 1986 just for deal with uncertainty information. Since intuitionistic fuzzy sets could provide the information of the membership degree and the non-membership degree, it has more expression and flexibility better than traditional fuzzy sets in processing uncertain information data.

In our article we apply intuitionistic fuzzy cluster and interval intuitionistic fuzzy cluster approach to fuzzify respondent utility functions, from Takagi-Sugeno fuzzy inference, we get Takagi-Sugeno [8] intuitionistic fuzzy inference. We develop a intuitionistic fuzzy random utility maximization (IFRUM) framework for analyzing follow-up certainty responses to a dichotomous valuation question. The intuitionistic fuzzy results are compared with those obtained from a traditional RUM approach and an intuitionistic fuzzy approach. The paper is organized as follows. In the next section, we present the economic principle of contingent valuation, our empirical model is described in Sects. 3, 4, 5 and 6. We introduce intuitionistic fuzzy clustering, interval intuitionistic fuzzy clustering, Takagi-Sugeno intuitionistic fuzzy inference and intuitionistic fuzzy random utility maximization principle.

2 The Economic Principle of Contingent Valuation

The individual's preference across a bundle of market commodities c and environment quality q can be described by a utility function $\mu(c, q)$. In general the utility function is assumed to be strictly increasing, continuous, and strictly quasi-concave. Given income m prices p for market commodities c, and the environment quality q, the maximum attainable utility function $v(p, q, m)$. A dual problem is to solve for the minimum expenditure that would sustain a certain utility level, given prices and environment quality taking the inverse of $v(p, q, m)$ with respect to m. We have the expenditure function $e(p, q, \mu) = v^{-1}(p, q, \mu)$, at initial price p^0, initial environment quality q^0, and initial income m^0, initial utility is $\mu^0 = v(p^0, q^0, m^0)$. At the same price and environment quality, and it is obvious that initial income maintains initial utility $m^0 = e(p^0, q^0, \mu^0)$, if the environment quality is changed to a new level q^1, the minimum expenditure required to restored the initial utility would be $e(p^0, q^1, \mu^0)$. By the definition, the expenditure difference constitutes the compensating values of environment change $y(q^1, q^0, m^0) = m^0 - e(p^0, q^1, \mu^0)$, since the expenditure function is strictly decreasing and convex in the environment quality q, given assumption on utility. The Hicksian value above is positive for an environment improvement, and negative for an environment deterioration.

3 Intuitionistic Fuzzy Clustering

The theoretical foundations of fuzzy in [5] and intuitionistic fuzzy sets are described in [9], and the section briefly outlines the related notions used in this paper.

Definition 3.1 [5] *Let X be a universe of discourse. A fuzzy set on X is an object A of the form* $A = \{(x, \mu_A(x))|x \in X\}$, *where* $\mu_A : X \to [0, 1]$ *defines the degree of membership of the element* $x \in X$ *to the set* $A \subseteq X$, *for every element* $x \in X$, *and* $0 \leq \mu_A(x) \leq 1$.

In real-life situations, when a person is asked to express his/her preference degree to an object, there usually exists an uncertainty or hesitation about the degree, and there is no means to incorporate the uncertainty or hesitation in a fuzzy set. To solve this issue, Atanassov generalized Zadeh's fuzzy set to intuitionistic fuzzy set (IFS) by adding an uncertainty (or hesitation) degree. IFS is defined as follow.

Definition 3.2 [10]. *An intuitionistic fuzzy set A is an object of the form* $A = \{x, \mu_A(x), \nu_A(x) \mid x \in X\}$, *which is characterized by a membership degree* $\mu_A(x)$ *and a non-membership degree* $\nu_A(x)$, *where*

$$\mu_A(x) : X \to [0, 1], x \in X \to \mu_A(x) \in [0, 1],$$

$$\nu_A(x) : X \to [0, 1], x \in X \to \nu_A(x) \in [0, 1]$$

with the condition $\mu_A(x) + \nu_A(x) \leq 1$ *for all* $x \in X$. *For each IFS A in X, if write* $\pi_A(x) = 1 - \mu_A(x) - \nu_A(x)$, *then* $\pi_A(x)$ *is called the degree of uncertainty (or hesitation) of x to A. Especially, if* $\pi_A(x) = 0$, *then the IFSA is reduced to a fuzzy set, and it is obvious that* $0 \leq \pi_A(x) \leq 1$.

Definition 3.3 [10] *Let* $A = \{x_i, \mu_A(x_i), \nu_A(x_i), i = 1, 2, \ldots, n\}$, $B = \{x_i, \mu_A(x_i), \nu_A(x_i), i = 1, 2 \ldots n\}$ *be two intuitionistic fuzzy sets (IFS). The Euclidean distance between two intuitionistic fuzzy sets is defined by*

$$d_{IFS}(A, B) = \left(\frac{1}{2n}\right)\left(\sum_{i=1}^{n}[\mu_A(x_i) - \mu_B(x_i)]^2 + [\nu_A(x_i) - \nu_B(x_i)]^2\right.$$
$$\left. + [\pi_A(x_i) - \pi_B(x_i)]^2\right)^{\frac{1}{2}}.$$

The distance between A and B fulfill the following conditions:

1. $0 \leq d_{IFS}(A, B) \leq 1$;
2. $d_{IFS}(A, B) = 0$ if A=B;
3. $d_{IFS}(A, B) = d_{IFS}(B, A)$;
4. if $A \subseteq B \subseteq C$, then $d_{IFS}(A, C) \geq d_{IFS}(A, B)$ and $d_{IFS}(A, C) \geq d_{IFS}(B, C)$.

In the following section, the algorithm of intuitionistic fuzzy clustering will be investigated. To implement the concept of fuzzy inference and classify inputs into c categories. We employ a fuzzy c-mean clustering (FCMC) algorithm proposed by Bezdek [7], which is considered an improvement over an earlier 'hard' c-mean algorithm. In contrast to the crisp classification of 'hard' c-mean clustering, fuzzy c-mean clustering allows each date point belong to a cluster to a degree specified by a grade of membership and a single data point to be a member of more that one cluster. Now we extend the fuzzy c-mean cluster to intuitionistic fuzzy c-mean cluster, we called IFCMC. The objective of the IFCMC algorithm is to partition a collective sets or clusters $(A_1, A_2, \ldots, A_c)$ in a way that best fits the structure of the data. Let $\mu_{A_i}(x_k)$ be the degree of membership of data point x_k in cluster A_i, $\nu_{A_i}(x_k)$ the degree of nonmembership of data point x_k, and $D = \{D_1, D_2, \ldots, D_n\}$ a effective sample of the observation (every sample has s attributes). $D_j = \{(\mu_{D_j}(x_1), \nu_{D_j}(x_1)), (\mu_{D_j}(x_2), \nu_{D_j}(x_2)), \ldots, (\mu_{D_j}(x_s), \nu_{D_j}(x_s))\}$ is an intuitionistic fuzzy set, j $= 1, 2, \ldots,$ n , $P = \{P_1, P_2, \ldots, P_c\}$, $P_1, P_2, \ldots, P_c$ are c cluster centers, c indicates the category of cluster. Every cluster center has s attributes, then P_i is an intuitionistic fuzzy set described by

$$P_i = \{(\mu_{P_i}(x_1), \nu_{P_i}(x_1)), (\mu_{P_i}(x_2), \nu_{P_i}(x_2)), \ldots, (\mu_{P_i}(x_s), \nu_{P_i}(x_s))\}.$$

For the partition matrix $U = (u_{ij})_{c\times n}$, u_{ij} indicates the membership of the ith cluster center of the jth sample. Then the partition matrix U can be indicated by

$$U = \begin{pmatrix} u_{11} & u_{12} & \ldots & u_{1c} \\ u_{21} & u_{22} & \ldots & u_{2c} \\ \ldots & \ldots & \ldots & \ldots \\ u_{n1} & u_{n2} & \ldots & u_{nc} \end{pmatrix}.$$

The objective function is then to minimize the criterion function

$$J_m(U, P) = \sum_{i=1}^{c}\sum_{j=1}^{n}(u_{ij})^m (d_{IFS}(D_j, P_i))^2$$

satisfying the conditions $u_{ij} \in [0, 1]$, $0 \le \sum_{j=1}^{n} u_{ij} \le n$, $\sum_{i=1}^{c} u_{ij} = 1$. The Lagrange function can be constructed as:

$$F = \sum_{i=1}^{c}\sum_{j=1}^{n}(u_{ij})^m (d_{IFS}(D_j, P_i))^2 + \lambda(\sum_{i=1}^{c} u_{ij} - 1) \tag{1}$$

where λ is the Lagrange multiplier. Differentiating (1) with respect to u_{ij}, and setting these partial derivatives to zero, the following equation is obtained.

$$u_{ij} = \frac{1}{\sum_{h=1}^{c}[\frac{(d_{IFS}(D_j,P_i))^2}{(d_{IFS}(D_j,P_h))^2}]^{\frac{1}{m-1}}}.$$

4 Interval Intuitionistic Fuzzy Clustering

Consider that, sometimes, it is not approximate to assume that the membership degrees for certain elements to of an IFS are exactly defined, but a value range can be given. In this situation, Atanassov [10] introduced the notion of IVIFS, which is characterized by a membership function and a nonmembership function, whose values are intervals rather than exact numbers.

Definition 4.1 [10] *Let X be a fixed non-empty universe set. We call* $A = \{x, \mu_A(x), \nu_A(x) \mid x \in X\}$ *an interval intuitionistic fuzzy set (IIFS), which is characterized by a membership degree* $\mu_A(x)$ *and a non-membership degree* $\nu_A(x)$*, with the condition* $\sup(\mu_A) + \sup(\nu_A) \leq 1$ *for all* $x \in X$.

Furthermore, $\mu_A(x)$ and $\nu_A(x)$ are denoted by $[\mu_A^L(x), \mu_A^U(x)], [\nu_A^L(x), \nu_A^U(x)]$, respectively. There for an other equivalent ways to express IVIFS A is:

$$A = \{x, \langle[\mu_A^L(x), \mu_A^U(x)], [\nu_A^L(x), \nu_A^U(x)] \mid x \in X\},$$

where $\mu_A^U(x) + \nu_A^U(x) \leq 1, 0 \leq \mu_A^L(x) \leq \mu_A^U(x) \leq 1, 0 \leq \nu_A^L(x) \leq \nu_A^U(x) \leq 1$. For each IIFS A in X, if we write $\pi_A(x) = 1 - \mu_A(x) - \nu_A(x)$ then $\pi_A(x)$ is called the degree of uncertainty (or hesitation) of x to A. Especially, if $\pi_A(x) = 0$ then the IIFS is reduced to a fuzzy set, and it is obvious that $0 \leq \pi_A(x) \leq 1$. In the contingent valuation, we know that individual's willingness to pay can be indicates by a interval and willingness no to pay can be indicated by a interval also.

Definition 4.2 *Let* A, B *be two interval intuitionistic fuzzy sets. The normalization Euclidean distance between two interval intuitionistic fuzzy sets is defined by*

$$\begin{aligned} d_{IIFS}(A, B) = (\frac{1}{4n})(\sum_{i=1}^{s}[\mu_A^L(x_i) - \mu_B^L(x_i)]^2 + [\mu_A^U(x_i) - \mu_B^U(x_i)]^2 \\ + [\nu_A^L(x_i) - \nu_B^L(x_i)]^2 + [\nu_A^U(x_i) - \nu_B^U(x_i)]^2 \\ + [\pi_A^L(x_i) - \pi_B^L(x_i)]^2 + [\pi_A^U(x_i) - \pi_B^U(x_i)]^2)^{\frac{1}{2}}. \end{aligned}$$

The distance between two interval intuitionistic fuzzy sets fulfill the the following conditions:

1. $0 \leq d_{IIFS}(A, B) \leq 1$;
2. $d_{IIFS}(A, B) = 0$ if $A = B$;
3. $d_{IIFS}(A, B) = d_{IIFS}(B, A)$;
4. if $A \subseteq B \subseteq C$, then $d_{IIFS}(A, C) \geq d_{IIFS}(A, B)$ and $d_{IIFS}(A, C) \geq d_{IIFS}(B, C)$.

In the following section, the algorithm of intuitionistic fuzzy clustering will be investigated. Let $D = D_1, D_2, \ldots, D_n$ be a effective sample of observation. Every sample has s attributes. $D_j = ([\mu_{D_j}^L(x_1), \mu_{D_j}^U(x_1)], [\nu_{D_j}^L(x_1), \nu_{D_j}^U(x_1)], [\mu_{D_j}^L(x_2), \mu_{D_j}^U(x_2)], [\nu_{D_j}^L(x_2), \nu_{D_j}^U(x_2)], \ldots, [\mu_{D_j}^L(x_s), \mu_{D_j}^U(x_s)], [\nu_{D_j}^L(x_s), \nu_{D_j}^U(x_s)])$ is an intuitionistic fuzzy set, $P = \{P_1, P_2, \ldots, P_c\}$, $P_1, P_2, \ldots, P_c$ are c cluster centers, c indicates the category of cluster. Every cluster center has s attributes, then P_i can be indicated by $P = ([\mu_{P_i}^L(x_1), \mu_{P_i}^U(x_1)], [(\nu_{P_i}^L(x_1)), \nu_{P_i}^U(x_1))], [\mu_{P_i}^L(x_2), \mu_{P_i}^U(x_2)], [(\nu_{P_i}^L(x_2)), \nu_{P_i}^U(x_2))], \ldots, [\mu_{P_i}^L(x_s), \mu_{P_i}^U(x_s)], [(\nu_{P_i}^L(x_s)), \nu_{P_i}^U(x_s))]$. For the partition matrix $U = (u_{ij})_{c \times n}$, u_{ij} indicates the membership of the ith cluster center about the attribute of jth. Then the partition matrix U can be indicated by

$$U = \begin{pmatrix} u_{11} & u_{12} & \ldots & u_{1c} \\ u_{21} & u_{22} & \ldots & u_{2c} \\ \ldots & \ldots & \ldots & \ldots \\ u_{n1} & u_{n2} & \ldots & u_{nc} \end{pmatrix}.$$

The objective function is then to minimize the criterion function

$$J_m(U, P) = \sum_{i=1}^{c} \sum_{j=1}^{n} (u_{ij})^m (d_{IIFS}(D_j, P_i))^2$$

fulfilling the conditions $u_{ij} \in [0, 1]$, $0 \le \sum_{j=1}^{n} u_{ij} \le n$, $\sum_{i=1}^{c} u_{ij} = 1$.

The Lagrange function can be constructed as:

$$F = \sum_{i=1}^{c} \sum_{j=1}^{n} (u_{ij})^m (d_{IIFS}(D_j, P_i))^2 + \lambda(\sum_{i=1}^{c} u_{ij} - 1) \tag{2}$$

where λ is the lagrange multiplier, differentiating(2) with respect to u_{ij}, and setting these partial derivatives to zero, and the follow equation is obtained:

$$u_{ij} = \frac{1}{\sum_{h=1}^{c} [\frac{(d_{IIFS}(D_i, P_j))^2}{(d_{IIFS}(D_i, P_h))^2}]^{\frac{1}{m-1}}}.$$

Since $d_{IIFS}(D_i, P_j)$ may be zero, so we should make some adjust, if $d_{IIFS}(D_i, P_j) = 0$ then $u_{ij} = 1$. Differentiating (3) with respect to cluster centers P, attribute membership function and non-membership function, and setting these partial derivatives to zero. That is, for

$$\begin{aligned} d_{IIFS}(D, P) = (\frac{1}{4n})(\sum_{l=1}^{s} [\mu_{D_j}^L(x_l) - \mu_{P_i}^L(x_l)]^2 + [\mu_{D_j}^U(x_l) - \mu_{P_i}(x_l)^U]^2 \\ + [\nu_{D_j}^L(x_l) - \nu_{P_i}^L(x_l)]^2 + [\nu_{D_j}^U(x_l) - \nu_{P_i}^U(x_l)]^2 \\ + [\pi_{D_j}^L(x_l) - \pi_{P_i}^L(x_l)]^2 + [\pi_{D_j}^U(x_l) - \pi_{P_i}^U(x_l)]^2) \end{aligned} \tag{3}$$

we can get

$$u^L_{P_i(x_l)} = \frac{\sum_{j=1}^n u_{ij}^m \mu^L_{D_j}(x_l)}{\sum_{j=1}^n u_{ij}^m}, u^U_{P_i(x_l)} = \frac{\sum_{j=1}^n u_{ij}^m \mu^U_{D_j}(x_l)}{\sum_{j=1}^n u_{ij}^m},$$

$$v^L_{P_i(x_l)} = \frac{\sum_{j=1}^n u_{ij}^m \nu^L_{D_j}(x_l)}{\sum_{j=1}^n u_{ij}^m}, v^U_{P_i(x_l)} = \frac{\sum_{j=1}^n u_{ij}^m \nu^U_{D_j}(x_l)}{\sum_{j=1}^n u_{ij}^m}.$$

The IIFCMC algorithm consisting of iterations altertaing, it contain the following steps:

1. Fix the number of cluster c $(2 \leq c \leq n)$, and threshold level ε;
2. Initialize the cluster center P^{i+1};
3. Computer the membership Matrix U;
4. Update the cluster centers by calculating P^{i+1};
5. Calculate the defect measure $D = |u^{i+1} - u^i|$;
6. Stop if $D \leq \varepsilon$;
7. Defuzzify the results by assigning every observation to that cluster for which it has maximum membership value the 'home' cluster.

5 Takagi-Sugeno Intuitionistic Fuzzy Inference

Next consider Takagi-Sugeno fuzzy inference [8]. Suppose that we can classify inputs x into c fuzzy sets, $A_1, A_2, \ldots, A_c$ with associated membership functions $\mu_{A_1}(x)$, $\mu_{A_2}(x), \ldots, \mu_{A_c}(x)$. Further, suppose that we can assign crisp functions to each of the clusters such that, if $x \in A_i$, then $y = f_i(x)$. Then, according to Takagi-Sugeno fuzzy inference, the combined effect is represented by:

$$y = \frac{\sum_{i=1}^c \mu_{A_i(x)} f_i(x)}{\sum_{i=1}^c \mu_{A_i(x)}}. \tag{4}$$

Now we consider Takagi-Sugeno [8] intuitionistic fuzzy inference, suppose that we can classify inputs x into c fuzzy sets, $A_1, A_2, \ldots, A_c$ with associated membership functions $\mu_{A_1}(x), \mu_{A_2}(x), \ldots, \mu_{A_c}(x)$ and non-membership functions $\nu_{A_1}(x)$, $\nu_{A_2}(x), \ldots, \nu_{A_c}(x)$. Further, supposed that we can assign crisp functions to each of the clusters such that, if $x \in A_i$, then $y = f_i(x)$, (4) well be get. If $x \notin A_i$, then $y = g_i(x)$. And according to Takagi-Sugeno fuzzy inference, the combined effect is represented by:

$$y = \frac{\sum_{i=1}^c \nu_{A_i(x)} g_i(x)}{\sum_{i=1}^c \nu_{A_i(x)}}, \tag{5}$$

we get Takagi and Sugeno intuitionistic fuzzy inference. The conjunction of the intuitionistic fuzzy c means clustering method with Takai-Sugeno intuitionistic fuzzy inference is enable to obtain the construction of models in a flexible way. In fact, Giles and Draeseke [11] employ this method to model econometric relationships.

6 Intuitionistic Fuzzy Random Utility Maximization Principle

We share the sample observations into clusters based on information from the follow-up certainty confidence question using the intuitionistic fuzzy c-mean clustering method. That is, individual's with similar certainty confidence are grouped into the 'home' cluster. These clusters have intuitionistic fuzzy boundaries because each observation can, at the same time, belong to other clusters to some degree smaller than their membership in the 'home' cluster. The intuitionistic fuzzy random utility maximization model is in much the same way as the standard RUM model [1], and individual k's intuitionistic fuzzy utility function μ_k and ν_k can be specified as a function of a fuzzy deterministic component ω_k and a crisp additive stochastic component ε_k.

$$\mu_k(z, y, s) = \omega_k(z, y, s) + \varepsilon_{z,k}, \quad (6)$$

$$\nu_k(z, y, s) = \omega_k(z, y, s) + \varepsilon_{z,k}, \quad (7)$$

where $z \in \{0, 1\}$ is an indicator variable that takes on the value 1 if the individual accepts the proposed change in the amenity and 0 otherwise, y is income, s is a vector of the respondent attributes, and ε is the stochastic disturbance arising from uncertainty on the part of the observer. Each individual's utility function is intuitionistic fuzzy in the sense that it belongs to every cluster to some degree. The probability of saying 'yes' and saying 'no' for each observation is then,

$$\begin{aligned} Pr_k(yes) &= Pr(\omega_k(1, y, s) + \varepsilon_{1k} > \omega_k(0, y, s) + \varepsilon_{0k}) \\ &= Pr\{(\varepsilon_{1k} - \varepsilon_{0k}) > -[\omega_k(1, y, s) - \omega_k(0, y, s)]\}, \end{aligned} \quad (8)$$

replacing $[\omega_k(1, y, s) - \omega_k(0, y, s)]/\sigma$ with $\Delta\omega_k$ and $(\varepsilon_{1k} - \varepsilon_{0k})/\sigma$ with ε_k, where ε_k is i.i.d, because ε_{1k} and ε_{0k} are i.i.d. Yields the intuitionistic fuzzy model:

$$Pr_k(yes) = Pr(\varepsilon_k > - \Delta\, \omega_k) = F_\varepsilon(\Delta\omega_k^1), \quad (9)$$

$$Pr_k(no) = Pr(\varepsilon_k \leq - \Delta\, \omega_k) = F_\varepsilon(\Delta\omega_k^0) = 1 - F_\varepsilon(\Delta\omega_k^1), \quad (10)$$

where F_ε is the cumulative distribution function. If $\varepsilon_k \sim N(0, 1)$, then (9) and (10) would be probit models. Assuming a linear utility function, the change in thedeterministic part of the utility function between the two states is then given as:

$$\Delta\, \omega_k^z = \alpha_k^z + \beta_k^z P_k + \gamma_k^z s_k, z \in \{0, 1\} \quad (11)$$

which is estimated based on the information from each cluster. Once the sample observations are proportioned into c intuitionistic fuzzy clusters, we can use the data for each fuzzy cluster separately and specify each individual's utility at the 'home' cluster as

$$\mu_{ij}^1 = \omega_{ij}^1 + \varepsilon_{ij}^1; \quad j = 1, \ldots, n_i; i = 1, \ldots, c, \tag{12}$$

$$\nu_{ij}^0 = \omega_{ij}^0 + \varepsilon_{ij}^0; \quad j = 1, \ldots, n_i; i = 1, \ldots, c. \tag{13}$$

Note that an individual's utility is intuitionistic fuzzy, since it is estimated from the coefficient estimates for each cluster, but that utility is assumed to be crisp within each cluster, so that it is possible to employ a standard probability framework within each cluster. A linear specification of the indirect utility function can be assumed (as in RUM) and the change in the deterministic parts of the utility functions between the two states is then given as

$$\omega_{ij}^1 = \alpha_i^1 + \beta_i^1 P_{ij} + \gamma_i^1 S_{ij} + \varepsilon_{ij}^1, \tag{14}$$

$$\omega_{ij}^0 = \alpha_i^0 + \beta_i^0 P_{ij} + \gamma_i^0 S_{ij} + \varepsilon_{ij}^0, \tag{15}$$

where P_{ij} is the bid, S_{ij} is a vector of observable attributes, ε_{ij}^1 and ε_{ij}^0 are random components, and $\alpha^1, \alpha^0, \beta^1, \beta^0$ and γ^1, γ^0 constitute parameters to be estimated. A standard probit (or logit) model can be estimated within each cluster. Using Takagi and Sugeno intuitionistic fuzzy inference, the intuitionistic fuzzy indirect is then:

$$\begin{aligned}\Delta\omega_k^1 &= \alpha_k^1 + \beta_k^1 P_k + \gamma_k^1 s_k \\ &= \frac{\sum_{i=1}^c \alpha_i^1 \mu_{A_i}(x_k)}{\sum_{i=1}^c \mu_{A_i}(x_k)} + \frac{\sum_{i=1}^c \beta_i^1 \mu_{A_i}(x_k)}{\sum_{i=1}^c \mu_{A_i}(x_k)} P_k + \frac{\sum_{i=1}^c \gamma_i^1 \mu_{A_i}(x_k)}{\sum_{i=1}^c \mu_{A_i}(x_k)} S_k,\end{aligned} \tag{16}$$

$$\begin{aligned}\Delta\omega_k^0 &= \alpha_k^0 + \beta_k^0 y_k + \gamma_k^0 s_k \\ &= \frac{\sum_{i=1}^c \alpha_i^0 \nu_{A_i}(x_k)}{\sum_{i=1}^c \nu_{A_i}(x_k)} + \frac{\sum_{i=1}^c \beta_i^0 \nu_{A_i}(x_k)}{\sum_{i=1}^c \nu_{A_i}(x_k)} P_k + \frac{\sum_{i=1}^c \gamma_i^0 \nu_{A_i}(x_k)}{\sum_{i=1}^c \nu_{A_i}(x_k)} S_k,\end{aligned} \tag{17}$$

where P_{ij} is the bid, S_{ij} is a vector of observable attributes, ε_{ij} is a random component, and $\alpha^1, \alpha^0, \beta^1, \beta^0, \gamma^1$ and γ^0 constitute parameters to be estimated, where $k = 1, \ldots, n$. Probability of saying yes and saying no for each observation can be rewritten as

$$Pr(yes) = F_\varepsilon(\alpha_k^1 + \beta_k^1 P_k + \gamma_k^1 s_k) = F_\varepsilon[\frac{\sum_{i=1}^c (\alpha_k^1 + \beta_k^1 P_k + \gamma_k^1 s_k)\mu_{A_i}(x_k)}{\sum_{i=1}^c \mu_{A_i}(x_k)}], \tag{18}$$

$$Pr(no) = F_\varepsilon(\alpha_k^0 + \beta_k^0 P_k + \gamma_k^0 s_k) = F_\varepsilon[\frac{\sum_{i=1}^c (\alpha_k^0 + \beta_k^0 P_k + \gamma_k^0 s_k)\nu_{A_i}(x_k)}{\sum_{i=1}^c \nu_{A_i}(x_k)}], \tag{19}$$

where $k = 1, \ldots, n$ and $F(.)$ is the cumulative distribution function of the stochastic term. The median of each individual based on FRUM is then given as

$$WTP_k = \frac{(-\alpha_i^1 - \gamma_i^1 s_k)}{\beta_i^1} = \frac{-\sum_{i=1}^{c} \alpha_i^1 \mu_{A_i}(x_k) - \sum_{i=1}^{c} \gamma_i^1 \mu_{A_i}(x_k)}{\sum_{i=1}^{c} \beta_i^1 \mu_{A_i}(x_k)}, \quad (20)$$

$$WNTP_k = \frac{(-\alpha_i^0 - \gamma_i^0 s_k)}{\beta_i^0} = \frac{-\sum_{i=1}^{c} \alpha_i^0 \nu_{A_i}(x_k) - \sum_{i=1}^{c} \gamma_i^0 \nu_{A_i}(x_k)}{\sum_{i=1}^{c} \beta_i^0 \nu_{A_i}(x_k)}, \quad (21)$$

$k = 1, \ldots, n$. That is, based on the FRUM model, the predicted probability or median WTP and median WNTP are certain form of weighted average information for the intuitionistic fuzzy clusters, with weights varying continuously throughout the sample. This is different from the traditional RUM model, where median WNTP and median WTP is derived from a homogeneous model underlying assumption that utility is crisp. With the contingent valuation question, respondents can choose to answer an open-ended question by giving an interval. Interpretation of the resulting valuation uncertainty is more straightforward with this type of question than other types of valuation question. It is assumed that the intervals reflect the respondents' uncertainty around a point value, i.e. that individuals state an interval because they only know that their valuation is within the stated range. The observed relationship between the different WTPs is encouraging in terms of the potential utility and validity of the new WTP estimates and we need to buttress this finding in future studies, i.e. further investigation is needed to explain why some people express valuation uncertainty.

7 Conclusion

We move away from a probabilistic representation of uncertainty and propose the use of intuitionistic fuzzy contingent valuation. Extended the fuzzy clustering to intuititonistic fuzzy clustering and interval intuitionistic fuzzy clustering, and proposed Takagi and Sugeno intuitionistic fuzzy inference. Applied the intutionistic fuzzy clustering and interval intuitionistic fuzzy clustering to contingent valuation non-market goods.

Acknowledgments Thanks to the support by National Natural Science Foundation of China (71061013, 61262022) the Scientific Research Project of Northwest Normal University (NWNU-KJCXGC-03-61).

References

1. Hanemann, W.M.: Welfare evaluation in contingent valuation experiment with discrete responses. Am. J. Agric. Econ. **66**, 332–341 (1984)
2. Ready, R.C., Whitehead, J.C.: Alternative approaches for incorporating respondent uncertainty when estimating willingness-to-pay: the Case of the Mexican Spotted Owl. Ecol. Econ. **27**, 29–41 (1998)
3. Alberini, A., Boyle, K.: Analysis of contingent valuation data with multiple bids and response options allowing respondents to express uncertainty. J. Environ. Econ. Manage. **45**, 40–62 (2003)
4. Van Kooten, G.C., Krcmar, E.: Preference uncertainty in non-market valuation: a fuzzy approach. Am. J. Agric. Econ. **83**, 487–500 (2001)
5. Zadeh, L.: Fuzzy sets. Inf. Control **8**, 338–353 (1965)
6. Sun, L., Van Kooten, G.C.: Comparing fuzzy and probabilistic approaches to preference uncertainty in non-market valuation. Environ. Resour. Econ. **42**, 471–489 (2009)
7. Boxall, P.W., Adamowicz, W.: A comparison of stated preference approaches to the measurement of environmental values. Ecol. Econ. **18**, 243–253 (1996)
8. Takagi, T., Sugeno, M.: Fuzzy identification of systems and its application to modelling and control. IEEE Trans. Syst. Man Cybern. **15**, 116–132 (1985)
9. Atanassov, K.: Intuitionistic fuzzy set. Fuzzy Set Syst. **20**, 86–96 (1986)
10. Atanassov, K.: Intuitionistic Fuzzy Sets: Theory and Applications. Springer, Heidelberg (1999)
11. Giles, D.E., Draeseke, R.: Econometric modelling using pattern recognition via the fuzzy *c* means algorithm. Comput. Aided Econ. 407–450 (2003)
12. Boxall, P.C., Adamowicz, W.L.: Understanding heterogeneous preferences in random utility models: a latent class approach. Environ. Resour. Econ. **23**, 421–446 (2002)
13. Cooper, J.C.: A comparison of approaches to calculating confindence intervals for benfit measure from dichotomous choice contingent valuation surveys. Land Econ. **37**, 202–209 (1994)
14. Aadland, D., Caplan, A.J.: Hypothetical surveys and real economic commitments. Land Econ. **70**, 145–154 (1994)

Forecasting Crude Oil Price with an Autoregressive Integrated Moving Average (ARIMA) Model

Chun-lan Zhao and Bing Wang

Abstract In this chapter, an autoregressive integrated moving average (ARIMA) model is proposed to predict world crude oil. Data from 1970 to 2006 is used for model development. We find that the model is able to describe and predict the average annual price of world crude oil with the aid of SAS software. The mean absolute percentage error (MAPE) is 4.059 %. Experiment shows the model have the preferable approach ability and predication performance, particularly for the short - term forecast.

Keywords Crude oil price forecast · ARIMA model · SAS software.

1 Introduction

Crude oil has been playing an increasingly important role in the world economy since nearly two-third of the world's energy demands is met from crude oil [1]. For example, central banks and private sector forecasters view the price of oil as one of the key variables in generating macroeconomic projections and in assessing macroeconomic risks.

Forecast of the price of oil, play a role in generating projections of energy use, in modeling investment decisions in the energy sector, in predicting carbon emissions and climate change, and in designing regulatory policies such as automotive fuel standards or gasoline taxes [2]. However, crude oil price forecasting is a very important topic, albeit an extremely hard one, due to its intrinsic difficulties and high volatility [3]. The average annual price of world crude oil series can be considered as

C. Zhao (✉) · B. Wang
School of Science, Southwest Petroleum University, Chengdu 610500, China
e-mail: your_wang1119@sina.com

B. Wang
e-mail: w9521423@sina.com

B.-Y. Cao and H. Nasseri (eds.), *Fuzzy Information & Engineering and Operations Research & Management*, Advances in Intelligent Systems and Computing 211,
DOI: 10.1007/978-3-642-38667-1_27, © Springer-Verlag Berlin Heidelberg 2014

a nonlinear and non-stationary time series, which is interactively affected by many factors, predicting it accurately is rather challenging.

In the past decades, traditional statistical and econometric techniques, such as linear regression (LinR), co-integration analysis, GARCH models, naive random walk, vector auto-regression (VAR) and error correction models (ECM) have been widely applied to crude oil price forecasting [1].

In 1994, Huntington applied a sophisticated econometric model to predict crude oil prices in the 1980s [4]. In 1995, Abramson and Finizza utilized a probabilistic model for predicting oil prices [5]. In 2001, Morana suggested a semiparametric statistical method for short-term oil price forecasting based on the GARCH properties of crude oil price [6]. Similarly, in 1998, Barone-Adesi et al. suggested a semi-parametric approach for oil price forecasting [7]. In 1988, Gulen used co-integration analysis to predict the West Texas Intermediate (WTI) price [8]. In 2002, 2005 and 2006, Ye, M. et al presented a simple econometric model of WTI prices, using OECD petroleuminventory levels, relative inventories, and high-inventory and low-inventory variables [9–11]. In 2004, Mirmirani and Li used the VAR model to predict U.S. oil price [12]. In 2005, Lanza et al. Investigated crude oil and oil products' prices [13]. However, until now, no one had been used the autoregressive integrated moving average (ARIMA) model to predict world crude oil.

In this paper, we find that the $ARIMA(4, 3, 0)$ model is able to describe and predict the average annual price of world crude oil with the aid of SAS software. We also give a practical example in the end. It is successfully solved, and the computational results are presented. The rest of this chapter is organized as follows. Section 2 provides the ARIMA model. In Sects. 3 and 4, we describe the proposed validity in predict oil price and the experimental results are shown. We conclude our remarks in Sect. 5.

2 ARIMA Model

An autoregressive integrated moving average (ARIMA) model [14] generalization of an autoregressive moving average (ARMA) model. These models are fitted to time-series data either to better understand the data or to predict future points in the series (forecasting). They are applied in some cases where data show evidence of non-stationary, where an initial differencing step (corresponding to the "integrated" part of the model) can be applied to remove the non-stationary.

The model is generally referred to as a $ARIMA(p, d, q)$ model where p, d, and q are non-negative integers that refer to the order of the autoregressive, integrated, and moving average parts of the model respectively. ARIMA models form an important part of the Box-Jenkins approach to time-series modeling.

Given a time series of data X_t where t is an integer index and theX_t are real numbers, then an $ARMA(p, q)$ model is given by:

$$(1-\sum_{i=1}^{p}\alpha_i L^i)X_t = (1+\sum_{i=1}^{q}\theta_i L^i)\varepsilon_t$$

where L is the lag operator, the α_i are the parameters of the autoregressive part of the model, the θ_i are the parameters of the moving average part and the ε_t are error terms. The error terms ε_t are generally assumed to be independent, dentically distributed variables sampled from a normal distribution with zero mean.

Assume now that the polynomial:

$$1-\sum_{i=1}^{p}\alpha_i L^i$$

has a unitary root of multiplicity d. Then it can be rewritten as:

$$(1-\sum_{i=1}^{p}\alpha_i L^i)X_t = (1-\sum_{i=1}^{p-d}\phi_i L^i)(1-L)^d$$

An $ARIMA(p, d, q)$ process expresses this polynomial factorization property, and is given by:

$$(1-\sum_{i=1}^{p}\phi_i L^i)(1-L)^d X_t = (1+\sum_{i=1}^{q}\theta_i L^i)\varepsilon_t$$

and thus can be thought as a particular case of an $ARMA(p+d, q)$ process having the auto-regressive polynomial with some roots in the unity. For this reason every ARIMA model with d > 0 is not wide sense stationary. The main steps of the ARIMA model is as follows [15].

- Step1: Using the unit root test: to test whether a time series variable is non-stationary, if not, though the mathematic, such as the differential transform or logarithmic differential transform, to let it be the stationary;
- Step2: Pattern Recognition: to calculate some statistic, such as the AIC and SBC to make sure the APIMA model
- Step3: Parameters estimation: the unknown parameters are needed to estimate with the estimate model, and to test them with the T statistic, and to get to rationality.
- Step 4: Model test: to test residual sequence of the fitted value and the real value whether white noise or not
- Step 5: Forecasting: all processes have been done and then to have a predict time series.

3 ARIMA Model Used for Forecasting the Oil Price

In this section, we will establish an ARIMA model to forecast the average annual price of world crude oil with the aid of SAS software.

A. Date Collection

The data from 1970 to 2004 are used to establish the model; the data from 2002 to 2006 are used to test. In the end, we use the model to forecast the oil price from 2007 to 2011. The world oil price in 2006 is 55.3 dollars a barrel. Table 1 and Fig. 1 show the oil price from 1970 to 2005.

B. The Stationary Test in Oil Prices

Next, we test the autocorrelation and partial autocorrelation of the sample data by SAS software, and the results are shown in Figs. 2 and 3.

Table 1 The oil price

Year	Actual price (s\barrel)	Year	Actual price (s\barrel)	Year	Actual price (s\barrel)
1970	17.37	1982	65.62	1994	21
1971	17.68	1983	58.06	1995	21.86
1972	18.95	1984	55.03	1996	25.92
1973	21.1	1985	49.75	1997	23.09
1974	37.64	1986	26.18	1998	14.53
1975	45.08	1987	31.05	1999	19.72
1976	47.21	1988	25.03	2000	31.61
1977	45.77	1989	29.38	2001	25.83
1978	47.21	1990	35.18	2002	25.19
1979	68.11	1991	29.49	2003	29.93
1980	90.39	1992	27.27	2004	39.61
1981	78.27	1993	23.06	2005	50.15

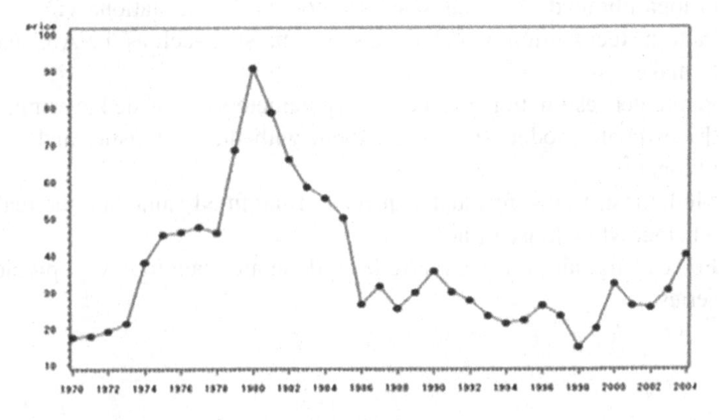

Fig. 1 The oil price figure

```
                         Name of Variable = price

                    Mean of Working Series     36.18714
                    Standard Deviation         18.17701
                    Number of Observations           35


                              Autocorrelations

Lag    Covariance    Correlation    -1 9 8 7 6 5 4 3 2 1 0 1 2 3 4 5 6 7 8 9 1      Std Error

  0       330.404        1.00000    |                    |********************|             0
  1       283.903        0.85926    |                .   |*****************   |      0.169031
  2       220.366        0.66696    |             .      |*************       |      0.266011
  3       169.461        0.51289    |            .       |********** .        |      0.310131
  4       122.747        0.37151    |           .        |*******     .       |      0.333486
  5     75.877620        0.22965    |          .         |*****        .      |      0.345108
  6     13.776567        0.04170    |          .         |*            .      |      0.349447
  7    -36.575587        -.11070    |          .       **|             .      |      0.349589
  8    -70.897861        -.21458    |          .     ****|             .      |      0.350589
  9    -97.027938        -.29366    |          .   ******|             .      |      0.354322
 10      -102.955        -.31160    |          .   ******|             .      |      0.361209
 11      -104.270        -.31558    |         .    ******|              .     |      0.368809
 12      -105.011        -.31783    |         .    ******|              .     |      0.376445
 13      -101.403        -.30691    |         .    ******|              .     |      0.384036
 14    -97.258493        -.29436    |        .     ******|               .    |      0.390980
 15    -81.035603        -.24526    |        .      *****|               .    |      0.397262

                        "." marks two standard errors
```

Fig. 2 Autocorrelation analysis

```
                          Partial Autocorrelations

     Lag    Correlation    -1 9 8 7 6 5 4 3 2 1 0 1 2 3 4 5 6 7 8 9 1

       1        0.85926    |                .   |*****************   |
       2       -0.27274    |                .***|**      .           |
       3        0.07369    |                .   |*       .           |
       4       -0.11361    |                .  **|        .           |
       5       -0.08966    |                .  **|        .           |
       6       -0.30960    |                .******|      .           |
       7        0.06350    |                .   |*       .           |
       8       -0.09111    |                .  **|        .           |
       9       -0.03533    |                .   *|        .           |
      10        0.12463    |                .   |**      .           |
      11       -0.06985    |                .   *|        .           |
      12       -0.07581    |                .  **|        .           |
      13       -0.04526    |                .   *|        .           |
      14       -0.07680    |                .  **|        .           |
      15        0.02987    |                .   |*       .           |
```

Fig. 3 Partial autocorrelation analysis

From the Figs. 2 and 3, it is easily seen that the coefficients are decreasing and the rapids deceleration with the increasing of lag phase. And the partial autocorrelations have no cut off and tailing property.

C. Use the differential transform to change the time series

From the Fig. 4, it get a stationary time. To deepen the stationary time series, used the unit root test.

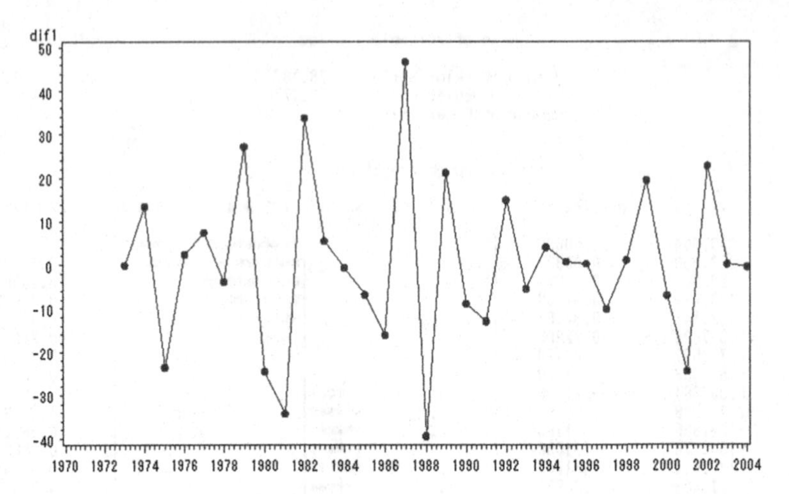

Fig. 4 The series figure of the three differences

Augmented Dickey-Fuller Unit Root Tests

Type	Lags	Rho	Pr < Rho	Tau	Pr < Tau	F	Pr > F
Zero Mean	0	-46.2363	<.0001	-9.38	<.0001		
	1	-126.124	0.0001	-7.78	<.0001		
Single Mean	0	-46.2388	0.0002	-9.22	0.0002	42.54	0.0010
	1	-126.087	0.0001	-7.64	0.0002	29.19	0.0010
Trend	0	-46.2532	<.0001	-9.07	0.0001	41.12	0.0010
	1	-127.166	0.0001	-7.56	0.0001	28.58	0.0010

Fig. 5 Unit root test

D. Unit root test

From Fig. 5, it is easy to see that $P < 0.01$, $\tau = -9.38$, and the Null Hypothesis need to be rejected under 99 % confidence level. Therefore, the series have no unit root and is a stationary level.

E. White noise test

From the Fig. 6, it is seen that the probability $P < 0.01$ of the LB statics under

Let level a = 0.01 and DF = 6. Then it shows itself isn't a white noise by Fig. 6.

Autocorrelation Check for White Noise

To Lag	Chi-Square	DF	Pr > ChiSq	Autocorrelations					
6	18.48	6	0.0051	-0.492	-0.120	0.131	-0.179	0.365	-0.230

Fig. 6 White noise test

4 Build an ARIMA Model to Forecast World Oil Average Price

A. Identify the Model of oil price and order number
From the Figs. 7 and 8, we can set several models to forecast the world oil average price, and these models just as the first consideration, such as $AR(1, 2, 4)$, $ARIMA((1, 2, 4), 3, 0)$, $ARIMA(4, 3, 0)$, $ARIMA(1, 3, 4)$, $ARIMA(4, 3, 1)$.
B. Chose the best model as the oil price forecasting model
In this section, we will use the standards to balance the models such as, detect its correlation, white noise and AIC and SBC Figs. 8, 9 and 10 are the models $AR(1, 2, 4)$, $ARIMA((1, 2, 4), 3, 0)$

```
                              Name of Variable = x

             Period(s) of Differencing                          1,1,1
             Mean of Working Series                          0.124375
             Standard Deviation                              18.65919
             Number of Observations                                32
             Observation(s) eliminated by differencing              3

                                Autocorrelations

Lag    Covariance    Correlation    -1 9 8 7 6 5 4 3 2 1 0 1 2 3 4 5 6 7 8 9 1      Std Error

  0       348.165        1.00000    |                    |********************|              0
  1      -171.144        -.49156    |          **********|      .             |       0.176777
  2    -41.648165        -.11962    |           .      **|        .           |       0.215295
  3     45.688309        0.13123    |           .        |***     .           |       0.217362
  4    -62.363819        -.17912    |           .    ****|        .           |       0.219824
  5       127.217        0.36539    |           .        |******* .           |       0.224338
  6    -80.048957        -.22992    |          .    *****|         .          |       0.242224
  7     -9.064457        -.02603    |          .        *|         .          |       0.248950
  8     34.309212        0.09854    |          .         |**       .          |       0.249035
  9    -63.876522        -.18347    |          .     ****|         .          |       0.250251
 10     91.515462        0.26285    |          .         |*****    .          |       0.254419

                          "." marks two standard errors
```

Fig. 7 Autocorrelation analysis of the three difference oil average price

```
                           The ARIMA Procedure

                        Partial Autocorrelations

         Lag    Correlation    -1 9 8 7 6 5 4 3 2 1 0 1 2 3 4 5 6 7 8 9 1

           1       -0.49156    |          **********|        .           |
           2       -0.47635    |          **********|        .           |
           3       -0.32363    |             .******|        .           |
           4       -0.58646    |        ************|        .           |
           5       -0.19416    |           .    ****|        .           |
           6       -0.24224    |           .   *****|        .           |
           7       -0.10816    |           .      **|        .           |
           8       -0.00202    |           .        |        .           |
           9       -0.16999    |           .     ***|        .           |
          10       -0.08528    |           .      **|        .           |
```

Fig. 8 Partial autocorrelation analysis of the three difference oil average price

The ARIMA Procedure

Conditional Least Squares Estimation

Parameter	Estimate	Standard Error	t Value	Approx Pr > |t|	Lag
MU	0.11008	1.15066	0.10	0.9245	0
AR1,1	-0.71054	0.16355	-4.34	0.0002	1
AR1,2	-0.48789	0.16328	-2.99	0.0058	2
AR1,3	-0.16048	0.15132	-1.06	0.2980	4

Constant Estimate	0.259669
Variance Estimate	224.2658
Std Error Estimate	14.97551
AIC	267.7497
SBC	273.6126
Number of Residuals	32

* AIC and SBC do not include log determinant.

Fig. 9 Fitting process of the $AR(1,2,4)$

The ARIMA Procedure

Conditional Least Squares Estimation

Parameter	Estimate	Standard Error	t Value	Approx Pr > |t|	Lag
AR1,1	-0.72561	0.16053	-4.52	<.0001	1
AR1,2	-0.47635	0.16053	-2.97	0.0059	2

Variance Estimate	217.7646
Std Error Estimate	14.75685
AIC	265.0161
SBC	267.9476
Number of Residuals	32

* AIC and SBC do not include log determinant.

Fig. 10 White noise test of the $ARIMA((1, 2, 4), 3, 0)$

Figure 9 shows some coefficients are weak correlation, they are not all passed. Figure 10 the estimate result of the model $ARIMA((1, 2, 4), 3, 0)$ which its mean is zero.

From the result of the Fig. 10, we can see that the coefficients are correlation. Figure 11 the test result of the model $ARIMA((1, 2, 4), 3, 0)$ which its mean is zero.

Autocorrelation Check of Residuals

To Lag	Chi-Square	DF	Pr > ChiSq	Autocorrelations					
6	16.21	4	0.0028	-0.154	-0.349	-0.256	0.088	0.431	-0.132
12	20.81	10	0.0224	-0.170	-0.061	-0.007	0.230	-0.005	-0.101
18	31.52	16	0.0115	0.025	-0.243	0.202	0.125	0.018	-0.210
24	39.91	22	0.0111	-0.090	0.231	0.080	-0.091	-0.098	-0.036

Fig. 11 Test result of the model $ARIMA((1, 2, 4), 3, 0)$

Table 2 The comparison of three models which are fitted

MODEL	Fitting results	*AIC*	*SBC*	Standard deviation
ARIMA (4,3,0)	All parameters are outstanding	250.1847	256.0477	11.38
ARIMA (1,3,4)	All parameters are outstanding	251.5808	254.5123	11.96
ARIMA (4,3,1)	One parameter is not outstanding	244.8771	252.2058	10.34

This shows us P < 0.05, and then it is a whit noise. Since it is not passed too, we could give up this model.

Next, we use the same method to test the rest models. The following Table 2 shows us the results.

From the Table2, it is easily seen AIC and SBC from the $ARIMA(4, 3, 1)$ are the minimum. And by the SAS, we get that residual error series of $ARIMA(4, 3, 1)$ is a white noise and parameters are not outstanding. But the $ARIMA(4, 3, 0)$ and $ARIMA(1, 3, 4)$ have passed the test. Especially, the standard deviation which is from the $ARIMA(4, 3, 0)$ is 11.38 and the minimum one. In conclude, we choose the $ARIMA(4, 3, 0)$ as the most reasonable model to forecast the world oil average price.
C. Estimate parameters of the $ARIMA(4, 3, 0)$ MODEL

From the Figs. 12 and 13, we see the results of the estimation of parameters and get that P < 0.01 after T-test. So we can conclude the parameters are outstanding.
D. Test the oil price model series

From Fig. 14, it is seen that the p>0.05 from all the lags 6, 12, 18 and 24. It is proved that the residual error series are white noise and is good to forecasting.

Conditional Least Squares Estimation

Parameter	Estimate	Standard Error	t Value	Approx Pr > |t|	Lag
MU	-0.02198	0.44413	-0.05	0.9609	0
AR1,1	-1.09406	0.15319	-7.14	<.0001	1
AR1,2	-1.17701	0.20001	-5.88	<.0001	2
AR1,3	-0.90257	0.20310	-4.44	0.0001	3
AR1,4	-0.63859	0.15904	-4.02	0.0004	4

Constant Estimate	-0.10577
Variance Estimate	134.3172
Std Error Estimate	11.58953
AIC	252.1818
SBC	259.5105
Number of Residuals	32

* AIC and SBC do not include log determinant.

Fig. 12 Estimate parameters of the $ARIMA(4, 3, 0)$

Conditional Least Squares Estimation

Parameter	Estimate	Standard Error	t Value	Approx Pr > |t|	Lag
AR1,1	-1.09392	0.15042	-7.27	<.0001	1
AR1,2	-1.17673	0.19635	-5.99	<.0001	2
AR1,3	-0.90215	0.19928	-4.53	0.0001	3
AR1,4	-0.63844	0.15615	-4.09	0.0003	4

Variance Estimate	129.5319
Std Error Estimate	11.38121
AIC	250.1847
SBC	256.0477
Number of Residuals	32

* AIC and SBC do not include log determinant.

Fig. 13 Fitting process of the *ARIMA*(4, 3, 0)

E. the world oil average price model —*ARIMA*(4, 3, 0), let:

$$Y_t = \nabla^3 X_t$$

$$\nabla X_t = X_t - X_{t-1}$$

$$Y_t = \nabla^3 X_t = \nabla^2 X_t - \nabla^2 X_{t-1}$$

$$Y_t = X_t - 3X_{t-1} + 3X_{t-2} - X_{t-3}$$

So the final model is:

$$Y_t = -1.09398Y_{t-1} - 1.17673Y_{t-2} - 0.9015Y_{t-3} - 0.63844Y_{t-4} + \varepsilon_t$$

Check Fig. 15, it is easily seen that the outcome is good, when the predicted value is under 95 % confidence level. The black dot is the real value and the red curve is fitting value. We just see the black have a wave surround red one.

5 Forecast the Oil Price by Model

Now we use the data of the world oil average price from 2003 to 2006 to test the model, if the outcome is perfect and we'll use it to forecasting the price after the year 2006.

Based on the ARIMA model, the model's result and the real value are given by the following Table 3.

From the Table 3, the results of the average error rate is just 4.095 %, and the relative error just close to 5 % in every year .so we think the model is good. We also use the model to predict the oil price from 2007 to 2011. The results are given in the following Table 4 and Fig. 14.

It is easy to see that the price is higher than the last year.

Table 3 The error between the model and real value

Year	Real value	Prediction value	Error	Relative error
2003	29.93	27.34	2.59	0.0865
2004	39.61	39.43	0.18	0.0045
2005	50.15	49.47	0.68	0.0135
2006	55.3	52.02	3.28	0.0593
	The average error rate		0.04095	

Table 4 The predicted value in world oil average price

Year	2007	2008	2009	2010	2011
Predicted value	60.13	73.59	86.94	99.96	110.47

Autocorrelation Check of Residuals

To Lag	Chi-Square	DF	Pr > ChiSq	Autocorrelations					
6	2.77	2	0.2499	-0.124	-0.171	-0.128	-0.036	-0.095	0.061
12	5.79	8	0.6710	0.025	-0.142	0.057	0.107	0.028	-0.155
18	15.23	14	0.3624	0.133	-0.193	0.066	0.153	0.090	-0.219
24	17.71	20	0.6068	-0.025	0.143	-0.053	-0.032	0.012	-0.033

Fig. 14 White noise test of the *ARIMA*(4, 3, 0)

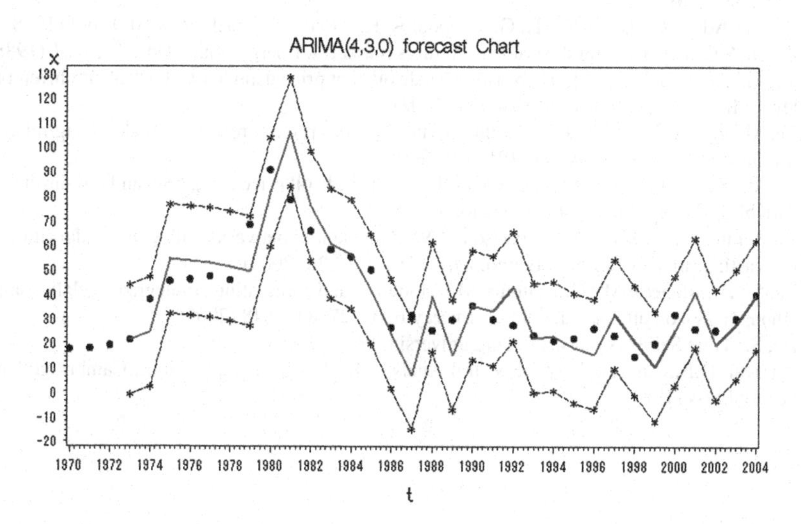

Fig. 15 The *ARIMA*(4, 3, 0) forecast chart

6 Conclusion

The movement of oil price seems to be uncertain and arbitrary, since its influential factors are complex. However, in this chapter, we use data from 1970 to 2006 to develop a reasonable ARIMA model to describe and predict the average annual price of world crude oil with the aid of *SAS* software by extracting the statistic features. We also give a practical example. By comparison, the value of prediction is good to worth with actual data. The model we established has the preferable approach ability and predication performance, particularly for the short-term forecast.

References

1. Yu, L., Wang, S., Lai, K.K.: Forecasting crude oil price with an EMD-based neural network ensemble learning paradigm. Energy Econ. **30**, 2623–2635 (2008)
2. Dargay, J.M., Gately, D.: World oil demand's shift toward faster growing and less price-responsive products and regions. Energy Policy **38**, 6261–6277 (2010)
3. Wang, S.Y., Yu, L., Lai, K.K.: Crude oil price forecasting with TEII methodology. J. Syst. Sci. Complexity **18**(2), 145–166 (2005)
4. Huntington, H.G.: Oil price forecasting in the 1980s: What went wrong? Energy J. **15**(2), 1–22 (1994)
5. Abramson, B., Finizza, A.: Using belief networks to forecast oil prices. Int. J. Forecast. **7**(3), 299–315 (1991)
6. Morana, C.: A semiparametric approach to short-term oil price forecasting. Energy Econ. **23**(3), 325–338 (2001)
7. Barone-Adesi, G., Bourgoin, F., Giannopoulos, K. :Don't look back. Risk, 100–103 (1998)
8. Gulen, S.G.:Efficiency in the crude oil futures market. J. Energy Finan. Dev. 3, 13–21 (1988)
9. Ye, M., Zyren, J., Shore, J.: Forecasting crude oil spot price using OECD petroleum inventory levels. Int. Adv. Econ. Res. **8**, 324–334 (2002)
10. Ye, M., Zyren, J., Shore, J.: A monthly crude oil spot price forecasting model using relative inventories. Int. J. Forecast. **21**, 491–501 (2005)
11. Ye, M., Zyren, J., Shore, J.: Forecasting short-run crude oil price using high and low-inventory variables. Energy Policy **34**, 2736–2743 (2006)
12. Mirmirani, S., Li, H.C.: A comparison of VAR and neural networks with genetic algorithm in forecasting price of oil. Adv. Econometrics **19**, 203–223 (2004)
13. Lanza, A., Manera, M., Giovannini, M.: Modeling and forecasting cointegrated relationships among heavy oil and product prices. Energy Econ. **27**, 831–848 (2005)
14. He, S. :Time Series Analysis, Peking university Press (2007)
15. Percival, D.B., Andrew, T.W. : Spectral Analysis for Physical Applications. Cambridge University Press (1993)

Use of Partial Supervised Model of Fuzzy Clustering Iteration to Mid- and Long-Term Hydrological Forecasting

Yu Guo and Xiao-qing Lai

Abstract Most operational hydrological forecasting systems produce deterministic forecasts and most research in operational hydrology has been devoted to finding the "best" forecasts rather than quantifying the predictive uncertainty. With the complex non-linear relation between forecasting indicators and forecasting object, it's difficult to get forecasting results with high quality from satisfied forecasting model. Therefore, based on fuzzy clustering algorithm, a partial supervised model of fuzzy clustering iteration is presented with the history data supervised and the forecasting precision is improved. The forecasting model is distinct in mathematic and physical conception, and is of good dispersion. A case study of Yamadu station in Xinjiang, China, is given to show the effectiveness of the model in mid-and long- term hydrological forecasting.

Keywords Fuzzy pattern recognition · Fuzzy clustering iteration · Indicator weights · Partial supervision

1 Introduction

Uncertainty about future events is the reason for forecasting. In general, a forecast does not eliminate uncertainty; it only reduces uncertainty [1]. The scientific and technological advances of the past century have been harnessed to increase the spatial and temporal resolution, the accuracy, and the lead time of forecasts for all principal hydrological varieties (precipitation, temperature, and runoff). Still, for many events, especially for mid- and long- term hydrological forecasting, forecasts remain far from perfect, and on occasions fall short of society's rising expectations for timely

Y. Guo (✉) · X. Lai
Pearl River Hydraulic Research Institute, Ministry of Water Resources, Guangzhou 51061, People's Republic of China
e-mail: gy_mail@sohu.com

B.-Y. Cao and H. Nasseri (eds.), *Fuzzy Information & Engineering and Operations Research & Management*, Advances in Intelligent Systems and Computing 211,
DOI: 10.1007/978-3-642-38667-1_28,

and reliable warnings. There are theoretical, technological, and budgetary limits to predictability. But there is also a largely untapped source of additional information, and hence potential benefits, that hydrological forecast could bring to society. This source is information about the predictive uncertainty. Research and development efforts should be directed to quantify the predictive uncertainty and to communicate that uncertainty to the users of forecasts.

Occurrence and development of hydrological phenomenon is a result worked by multi-factors, and there are very complicated nonlinear relations between forecast objectives and multi-factors, so if forecast models set up directly by relations between forecast objectives and multi-factors, its forecast precision can't satisfy actual demand, which means that effective employing known knowledge to obtain higher precision will be key point of forecast study. After Pedrycz et al. [2] developed fuzzy clustering method (FCM) to partial supervised FCM (SFCM), SFCM algorithm applied widely in signal processing [3, 4], pattern recognition [5, 6] and other fields. Then in this chapter, based on fuzzy clustering iterative model [7] that the partial supervised item was introduced into the SFCM model and forecast clustering process was supervised by clustering results of known data, the forecast precision is effective improved and a new worthy groping approach is presented for improving precision of mid- and long- term hydrological forecast.

2 Partial Supervised Model of Fuzzy Clustering Iteration

Assume that forecasting objective Y has n samples, viz., $Y = \{y_1, y_1, \cdots, y_n\}$, according to cause analysis and historical records that each sample has m forecast factors, thus we have forecast factors vector $X = \{X_1, X_2, \cdots, X_n\} \in R^m$. For different physical dimensions of m forecast factors that the forecast factors should be normalized to eliminate effect of dimensions difference, viz., normalize forecast factors matrix into relative membership degree (RMD) matrix $R_{m\times n} = (r_{ij})_{m\times n}$, here r_{ij} refers to RMD and $0 \le r_{ij} \le 1$.

Generally, n samples recognize regarding to m forecast factors, so we set RMD matrix of samples regarding grades is: $U = (u_{hj})_{c\times n}$, here u_{hj} stands for RMD of jth sample to hth grade and $h = 1, 2, \cdots, c$, it satisfies conditions $\sum_{h=1}^{c} u_{hj} = 1; 0 \le u_{hj} \le 1; \sum_{j=1}^{n} u_{hj} > 0; \forall j, \forall h$. Standard feature values matrix of hth grade to m forecast factors is: $S = (s_{ih})_{m\times c}$, here s_{ih} stands for standard feature values of hth grade to ith forecast factor and $0 \le s_{ih} \le 1$.

Taking different effects of different forecast factors on hydrological phenomena into account that weights vector of forecast factors can be set as: $W = (w_1, w_2, \cdots, w_m)$, and it satisfies condition $\sum_{i=1}^{m} w_i = 1$, $0 < w_i < 1$. Therefore difference of jth sample to hth grade can be expressed as generalized Euclidean weight distance: $d_{hj} = \sqrt{\sum_{i=1}^{m} [w_i(r_{ij} - s_{ih})]^2}$. For completely describing difference between jth sample and hth grade, we take RMD u_{hj} of jth sample to hth grade as weight of generalized Euclidean weight distance and we have:

$D_{hj} = u_{hj} \cdot d_{hj} = u_{hj}\sqrt{\sum_{i=1}^{m}[w_i(r_{ij} - s_{ih})]^2}$, then we can construct objective function as follows:

$$J = \min\left\{F(U, S, W) = \sum_{j=1}^{n}\sum_{h=1}^{c}\left[u_{hj}^2\sum_{i=1}^{m}\left(w_i(r_{ij} - s_{ih})\right)^2\right]\right\} \tag{1}$$

We introduce supervised item into objective function just like SFCM algorithm:

$$J_\alpha = \min\left\{F_\alpha(U, S, W) = \sum_{j=1}^{n}\sum_{h=1}^{c}\left[(1-\alpha)u_{hj}^2\sum_{i=1}^{m}\left(w_i(r_{ij} - s_{ih})\right)^2 + \alpha\cdot(u_{hj} - f_{hj})^2\sum_{i=1}^{m}\left(w_i(r_{ij} - s_{ih})\right)^2\right]\right\} \tag{2}$$

It satisfies conditions :

$$\begin{cases} \sum_{i}^{m} w_i = 1, 0 < w_i < 1 \\ 0 \le s_{ih} \le 1 \\ \sum_{h=1}^{c} u_{hj} = 1 \sum_{j=1}^{n} u_{hj} > 0 \; 0 \le u_{hj} \le 1 \end{cases} \quad \begin{matrix} i = 1, 2, \cdots, m \\ h = 1, 2, \cdots, c, \\ j = 1, 2, \cdots, n \end{matrix} \tag{3}$$

where s_{ih} refers to grade center value of hth grade to ith index, u_{hj} to RMD of jth sample to hth grade, $w_i$to weight of ith index; Supervised RMD matrix $F = (f_{hj})_{c\times n}$, here $\sum_{h=1}^{c} f_{hj} \le 1$ and $\alpha(\alpha \ge 0)$ is supervised proportion factor.

And we have Lagrange function:

$$\begin{aligned} L_\alpha(U, S, W, \lambda_w, \lambda_j) = \sum_{j=1}^{n}\sum_{h=1}^{c}\Big[(1-\alpha)u_{hj}^2\sum_{i=1}^{m}\left(w_i(r_{ij} - s_{ih})\right)^2 \\ +\alpha\cdot(u_{hj} - f_{hj})^2\sum_{i=1}^{m}\left(w_i(r_{ij} - s_{ih})\right)^2\Big] \\ -\lambda_w\left(\sum_{i=1}^{m} w_i - 1\right) - \lambda_j\left(\sum_{h=1}^{c} u_{hj} - 1\right) \end{aligned}$$

Let

$$\frac{\partial L_\alpha}{\partial u_{hj}} = 0, \quad \frac{\partial L_\alpha}{\partial w_i} = 0, \quad \frac{\partial L_\alpha}{\partial \lambda_w} = 0, \quad \frac{\partial L_\alpha}{\partial \lambda_j} = 0, \quad j = 1, 2, \cdots, n \tag{4}$$

Therefore we can obtain partial supervised model of fuzzy clustering iteration as:

$$s_{ih} = \frac{\sum_{j=1}^{n}\left[(1-\alpha)u_{hj}^2 + \alpha\cdot\left(u_{hj} - f_{hj}\right)^2\right]\cdot r_{ij}}{\sum_{j=1}^{n}\left[(1-\alpha)u_{hj}^2 + \alpha\cdot\left(u_{hj} - f_{hj}\right)^2\right]} \tag{5}$$

$$u_{hj} = \alpha f_{hj} + \left(1 - \alpha \sum_{k=1}^{c} f_{kj}\right) \left[\sum_{k=1}^{c} \frac{\sum_{i=1}^{m} [w_i (r_{ij} - s_{ih})]^2}{\sum_{i=1}^{m} [w_i (r_{ij} - s_{ik})]^2} \right]^{-1} \tag{6}$$

$$w_i = \left\{ \sum_{k=1}^{p} \frac{\sum_{h=1}^{c} \sum_{j=1}^{n} \left\{ \left[(1-\alpha) u_{hj}^2 + \alpha \cdot (u_{hj} - f_{hj})^2 \right] (r_{ij} - s_{ih})^2 \right\}}{\sum_{h=1}^{c} \sum_{j=1}^{n} \left\{ \left[(1-\alpha) u_{hj}^2 + \alpha \cdot (u_{hj} - f_{hj})^2 \right] (r_{kj} - s_{kh})^2 \right\}} \right\}^{-1} \tag{7}$$

So we can take below steps to solve partial supervised model of fuzzy clustering iteration (5), (6) and (7):

(a) Giving iteration precisions ε_1,ε_2,ε_3 and grades number c.
(b) Setting initial weights vector (w_i^l), fuzzy standard feature matrix (s_{ik}^l) and samples RMD matrix$\left(u_{hj}^l\right)$, here l=1.
(c) Using Eqs. (5), (6) and (7) to calculate$\left(s_{ik}^{l+1}\right)$, $\left(u_{hj}^{l+1}\right)$ and $\left(w_i^{l+1}\right)$ respectively.
(d) If results satisfy: max $\left|w_i^l - w_i^{l-1}\right| \le \varepsilon_1$, max $\left|u_{hj}^l - u_{hj}^{l-1}\right| \le \varepsilon_2$ and max $\left|s_{ih}^l - s_{ih}^{l-1}\right| \le \varepsilon_3$, then iteration end, and matrixes (s_{ik}^l), $\left(u_{hj}^l\right)$ and (w_i^l) can be employed as optimal indexes weight vector (w_i^*), optimal fuzzy clustering matrix(u_{hj}^*) and optimal fuzzy clustering center matrix(s_{ih}^*)that satisfy calculating precisionε_1, ε_2, ε_3; else, let $l \Rightarrow l+1$ and turn to step (c) to continue iterated calculation.

After iteration that the RMD matrix of samples set regarding to grades can be obtained, and according to inapplicability principle of maximal RMD under grades conditions [8], grade feature values of jth sample are:

$$H_j = \sum_{h=1}^{c} h \cdot u_{hj}. \tag{8}$$

3 Fuzzy Pattern Recognition Forecast

Assume that forecast objective Y has n_0 known samples$Y = \{y_1, y_2, \cdots, y_{n_0}\}$, so we use known samples to recognize five grades of low, little low, medium, little high and high and obtain RMD matrix $Z = \left(z_{hj}\right)_{c \times n_0}$, here z_{hj}refers to RMD of $jth(1 \le j \le n_0)$ sample to hth grade. Owing to single index recognition that the fuzzy recognition model [7] can be simplified as:

$$z_{hj} = \left\{ \sum_{k=1}^{c} \frac{(y_j - v_h)^2}{(y_j - v_k)^2} \right\}^{-1}, \tag{9}$$

where c is number of clustering grades; v_h is standard feature value of hth grade and we often have

$$v_h = \min_j y_j + (h-1) \cdot \frac{\max_j y_j - \min_j y_j}{c-1} \quad . \tag{10}$$

From RMD matrix $Z = \left(z_{hj}\right)_{c \times n_0}$ of known forecast samples we get supervised matrix:

$$F = \begin{pmatrix} z_{11} & z_{12} & \cdots & z_{1n_0} & 0 & \cdots 0 \\ z_{21} & z_{22} & \cdots & z_{2n_0} & 0 & \cdots 0 \\ \cdots & \cdots & \cdots & \cdots & \cdots & \cdots 0 \\ z_{c1} & z_{c1} & \cdots & z_{cn_0} & 0 & \cdots 0 \end{pmatrix} \quad . \tag{11}$$

Therefore we normalize vector$X$of n forecast factors feature values and obtain corresponding RMD matrix$R$, then apply partial supervised model of fuzzy clustering iteration to get RMD matrix U and grades feature values $H$of forecast factors.

Accordingly the forecast values of forecast objective k are:

$$Y_k = \min_j y_j + (H_k - 1) \cdot \frac{\max_j y_j - \min_j y_j}{c-1} \quad k = n_0+1, n_0+2, \cdots, n \quad . \tag{12}$$

4 Case Study

Observed annual runoff quantities of 23 years in Yamadu and its relevant 4 prophase feature values of forecast factors are listed in Table 1 [9]. Where factor $x_1$is aggregate rainfall volume from last Dec. to this Mar. in Yili weather station; factor x_2 is lunar average zonal circulation index in last Aug. at Europe and Asia region; factorx_3 is meridianal index of Europe and Asia region in last May; factor x_4 is 2,800 MHz sun radio jet stream of last Jun.

According to necessary data number of clustering and forecast recognition test, we take former 17 years data as clustering supervision, latter 6 years as forecast test. Then we divide annual runoffs of former 17 years into five grades as low, little low, medium, little high and high and obtain RMD matrix of 17 samples:

$$Z = \begin{pmatrix} 0.03 & 0.02 & 0.00 & 0.00 & 0.70 & 0.00 & 0.00 & 0.01 & 0.01 & 0.03 & 0.04 & 0.01 & 0.70 & 0.00 & 0.00 & 0.95 & 0.00 \\ 0.87 & 0.05 & 0.00 & 0.00 & 0.24 & 0.01 & 0.00 & 0.01 & 0.96 & 0.95 & 0.33 & 0.01 & 0.24 & 0.01 & 1.00 & 0.04 & 0.01 \\ 0.08 & 0.59 & 0.99 & 0.01 & 0.04 & 0.03 & 0.00 & 0.03 & 0.02 & 0.02 & 0.55 & 0.04 & 0.04 & 0.03 & 0.00 & 0.01 & 0.01 \\ 0.01 & 0.30 & 0.00 & 0.98 & 0.01 & 0.89 & 0.00 & 0.17 & 0.01 & 0.00 & 0.05 & 0.83 & 0.01 & 0.94 & 0.00 & 0.00 & 0.07 \\ 0.01 & 0.04 & 0.00 & 0.01 & 0.01 & 0.07 & 1.00 & 0.79 & 0.00 & 0.00 & 0.02 & 0.12 & 0.01 & 0.02 & 0.00 & 0.00 & 0.90 \end{pmatrix}$$

From matrix Z we have supervised matrix of 23 samples:

Table 1 Observed annual runoff quantities and factors feature values of Yamadu Hydrographic Sation at Yili River, Xijiang

No.	x_1	x_2	x_3	x_4	Annual runoff y	No.	x_1	x_2	x_3	x_4	Annual runoff y
1	114.6	1.10	0.71	85	346	13	55.3	0.96	0.40	69	300
2	132.4	0.97	0.54	73	410	14	152.1	1.04	0.49	77	433
3	103.5	0.96	0.66	67	385	15	81.0	1.08	0.54	96	336
4	179.3	0.88	0.57	87	446	16	29.8	0.83	0.49	120	289
5	92.7	1.15	0.44	154	300	17	248.6	0.79	0.50	147	483
6	115.0	0.74	0.65	252	453	18	64.9	0.59	0.50	167	402
7	163.6	0.85	0.58	220	495	19	95.7	1.02	0.48	160	384
8	139.5	0.70	0.59	217	478	20	89.9	0.96	0.39	105	314
9	76.7	0.95	0.51	162	341	21	121.8	0.83	0.60	140	401
10	42.1	1.08	0.47	110	326	22	78.5	0.89	0.44	94	280
11	77.8	1.19	0.57	91	364	23	90.0	0.95	0.43	89	301
12	100.6	0.82	0.59	83	456						

$$F = \begin{pmatrix} 0.03 & 0.02 & 0.00 & 0.00 & 0.70 & 0.00 & 0.00 & 0.01 & 0.01 & 0.03 & 0.04 & 0.01 & 0.70 & 0.00 & 0.00 & 0.95 & 0.00 & 0.0 & 0.0 & 0.0 & 0.0 & 0.0 & 0.0 \\ 0.87 & 0.05 & 0.00 & 0.00 & 0.24 & 0.01 & 0.00 & 0.01 & 0.96 & 0.95 & 0.33 & 0.01 & 0.24 & 0.01 & 1.00 & 0.04 & 0.01 & 0.0 & 0.0 & 0.0 & 0.0 & 0.0 & 0.0 \\ 0.08 & 0.59 & 0.99 & 0.01 & 0.04 & 0.03 & 0.00 & 0.03 & 0.02 & 0.02 & 0.55 & 0.04 & 0.04 & 0.03 & 0.00 & 0.01 & 0.01 & 0.0 & 0.0 & 0.0 & 0.0 & 0.0 & 0.0 \\ 0.01 & 0.30 & 0.00 & 0.98 & 0.01 & 0.89 & 0.00 & 0.17 & 0.01 & 0.00 & 0.05 & 0.83 & 0.01 & 0.94 & 0.00 & 0.00 & 0.07 & 0.0 & 0.0 & 0.0 & 0.0 & 0.0 & 0.0 \\ 0.01 & 0.04 & 0.00 & 0.01 & 0.01 & 0.07 & 1.00 & 0.79 & 0.00 & 0.00 & 0.02 & 0.12 & 0.01 & 0.02 & 0.00 & 0.00 & 0.90 & 0.0 & 0.0 & 0.0 & 0.0 & 0.0 & 0.0 \end{pmatrix}$$

After normalized forecast factors feature values we get RMD matrix:

$$R = \begin{pmatrix} 0.39 & 0.47 & 0.34 & 0.68 & 0.29 & 0.39 & 0.61 & 0.50 & 0.21 & 0.06 & 0.22 & 0.32 & 0.12 & 0.56 & 0.23 & 0.00 & 1.00 & 0.16 & 0.30 & 0.27 & 0.42 & 0.22 & 0.28 \\ 0.82 & 0.55 & 0.53 & 0.37 & 0.92 & 0.08 & 0.31 & 0.00 & 0.51 & 0.78 & 1.00 & 0.24 & 0.53 & 0.69 & 0.78 & 0.27 & 0.18 & 0.00 & 0.65 & 0.53 & 0.27 & 0.39 & 0.51 \\ 1.00 & 0.47 & 0.84 & 0.56 & 0.16 & 0.81 & 0.59 & 0.63 & 0.38 & 0.25 & 0.56 & 0.63 & 0.03 & 0.31 & 0.47 & 0.31 & 0.34 & 0.34 & 0.28 & 0.00 & 0.66 & 0.16 & 0.13 \\ 0.10 & 0.03 & 0.00 & 0.11 & 0.47 & 1.00 & 0.83 & 0.81 & 0.51 & 0.23 & 0.13 & 0.09 & 0.01 & 0.05 & 0.16 & 0.29 & 0.43 & 0.54 & 0.50 & 0.21 & 0.39 & 0.15 & 0.12 \end{pmatrix}$$

Taking supervised weight $\alpha = 0.5$, i.e., subjective weights and objective weights have same important, and using Eqs. (5), (6) and (7) that we have samples RMD matrix:

$$U = \begin{pmatrix} 0.07 & 0.06 & 0.04 & 0.03 & 0.50 & 0.05 & 0.02 & 0.04 & 0.14 & 0.19 & 0.08 & 0.03 & 0.68 & 0.08 & 0.02 & 0.66 & 0.04 & 0.16 & 0.28 & 0.78 & 0.03 & 0.90 & 0.85 \\ 0.55 & 0.10 & 0.08 & 0.03 & 0.32 & 0.07 & 0.02 & 0.05 & 0.69 & 0.68 & 0.46 & 0.05 & 0.20 & 0.09 & 0.94 & 0.15 & 0.04 & 0.12 & 0.41 & 0.09 & 0.04 & 0.04 & 0.07 \\ 0.20 & 0.43 & 0.75 & 0.06 & 0.09 & 0.13 & 0.04 & 0.09 & 0.12 & 0.08 & 0.36 & 0.34 & 0.06 & 0.11 & 0.03 & 0.11 & 0.06 & 0.54 & 0.17 & 0.06 & 0.15 & 0.03 & 0.04 \\ 0.13 & 0.36 & 0.10 & 0.65 & 0.06 & 0.61 & 0.11 & 0.28 & 0.04 & 0.04 & 0.07 & 0.50 & 0.03 & 0.63 & 0.01 & 0.05 & 0.13 & 0.14 & 0.10 & 0.04 & 0.74 & 0.02 & 0.03 \\ 0.05 & 0.06 & 0.02 & 0.23 & 0.03 & 0.13 & 0.81 & 0.55 & 0.01 & 0.02 & 0.02 & 0.07 & 0.02 & 0.09 & 0.00 & 0.03 & 0.72 & 0.05 & 0.04 & 0.02 & 0.04 & 0.01 & 0.01 \end{pmatrix}$$

Weight vector of forecast factors is: $W = (0.438\ 0.192\ 0.227\ 0.143)$.

Standard feature values matrix of forecast factors is:

$$S = \begin{pmatrix} 0.223 & 0.229 & 0.307 & 0.472 & 0.707 \\ 0.484 & 0.721 & 0.436 & 0.333 & 0.213 \\ 0.131 & 0.451 & 0.602 & 0.594 & 0.519 \\ 0.186 & 0.270 & 0.196 & 0.343 & 0.645 \end{pmatrix}.$$

According to above samples RMD matrix we obtain grade feature values of samples from 18th to 23th are: $H = (2.8\ 2.2\ 1.4\ 3.7\ 1.2\ 1.3)$, then forecast values of annual runoff is: $Y = (376.4\ 344.7\ 302.6\ 426.1\ 289.6\ 294.6)$, observed values is:$Y_o$ =(402 384 314 401 280 301), and relative forecast error is:$E(\%) =$ (6.4 10.2 3.6 6.2 3.4 2.1). The forecast results show that, forecast precision of annual forecast in this chapter has great improved than $E(\%) =$ (10.0 7.8 7.3 3.7 14.6 12.3) in Chen et al. [7], and if taking complicacy and randomicity of mid- and long- term hydrological forecast and shortage of hydrological data into account, the chapter's forecast results are reasonable and perfect.

5 Conclusion

The partial supervised model of fuzzy clustering iteration that presented in this chapter can supervise clustering forecast process by known experience and data, and in entire forecast process, the model need not qualitative analysis and judgment. The model has clear concept in mathematical and physical, so it suitable for general forecast and control system with multi-factor input-single index output, and it also offers a new deserved groping approach for forecast and control of complicated system and other fields.

References

1. Krzysztofowicz, R.: The case for probabilistic forecasting in hydrology. J. Hydrology **249**, 2–9 (2000)
2. Pedrycz, W.: Fuzzy clustering with partial supervision. IEEE Trans. Syst. Man Cybern. **27**(5), 787–795 (1997)
3. Eom, K.B.: Unsupervised segmentation of space borne passive radar images. Pattern Recogn. Lett. **20**, 485–494 (1999)
4. Marcelloni, F.: Recognition of olfactory signals based on supervised fuzzy C-means and k-NN algorithms. Pattern Recogn. Lett. **22**, 1007–1019 (2001)
5. Shukhat, B.: Supervised fuzzy pattern recognition. Fuzzy Sets Syst. **100**, 257–265 (1998)
6. Zahid, N., Abouelala, O., Limouri, M., et al.: Unsupervised fuzzy clustering. Pattern Recogn. Lett. **20**(2), 123–129 (1999)
7. Yu, G., Chen, W., Chen, S.: Intelligent forecasting mode and its application for mid and long term hydrological forecasting. Oper. Res. Manage. Sci. Fuzziness **7**, 1–12 (2007)
8. Yu, G., Chen, W., Chen, S.: Use of fuzzy preference matrix for multi-objective fuzzy optimization decision-making model. Proceedings of the 7th World Congress on, Intelligent Control and Automation, WCICA'08, p 7108–7112(2008).
9. Shouyu, C.: Relative membership function and new frame of fuzzy sets theory for pattern recognition. J. Fuzzy Math. **5**(2), 401–411 (1997)

A Fuzzy Clustering Approach for TS Fuzzy Model Identification

Mei-jiao Lin and Shui-li Chen

Abstract In this paper, a fuzzy clustering approach for TS fuzzy model identification is presented. In the proposed method, the modified mountainx clustering algorithm is employed to determine the number of clusters. Secondly, the fuzzy c-regression model (FCRM) algorithm is used to obtain an optimal fuzzy partition matrix. As a result, the initial parameters can be determined by the optimal fuzzy partition. Finally, gradient descent algorithm is adopted to precisely adjust premise parameters and consequent parameters simultaneously. The simulation results reveal that the proposed algorithm can model an unknown system with a small number of fuzzy rules.

Keywords Fuzzy modeling · FCRM algorithm · Modified mountain clustering algorithm · Gradient descent method.

1 Introduction

In recent years, Takagi-Sugeno fuzzy model has attracted the most attention of the fuzzy modeling community since TS-based method can express a highly nonlinear functional relation with a few fuzzy rules than any other type of fuzzy models [1]. Fuzzy model identification consists of structure identification and parameter estimation [2, 3]. Structure identification is concerned with the determination of the number of rules, while parameter estimation concerns the calculation of the appropriate fuzzy set parameters that provide a reliable system description [4, 5].

M. Lin (✉)
College of Mathematics and Computer Science, Fuzhou University, Fuzhou 350108, China
e-mail: lintingkiko@163.com

S. Chen
School of Science, Jimei University, Xiamen 361021, China
e-mail: sgzx@jmu.edu.cn

B.-Y. Cao and H. Nasseri (eds.), *Fuzzy Information & Engineering and Operations Research & Management*, Advances in Intelligent Systems and Computing 211,
DOI: 10.1007/978-3-642-38667-1_29, © Springer-Verlag Berlin Heidelberg 2014

At present there are many approaches for TS fuzzy model identification. These approaches consist of two problems. On the one handfuzzy clustering algorithms were used to extract fuzzy rules from available data. However, in order to obtain optimal number of fuzzy rules, some algorithms need to process repeatedly by using validity function, which increases the calculation. On the other hand fuzzy clustering algorithms were used to obtain optimal fuzzy partition and the premise parameters, and then the consequent parameters can be identify by some optimization algorithm. Whereas, consequent parameters are optimized under the condition which premise parameters are not optimal, which result in precision can no be improved [6].

In this paper, a fuzzy clustering approach for TS fuzzy model identification is proposed to solve the aforementioned problems. Firstly, the number of fuzzy rules can be determined automatically as long as the modified mountain clustering algorithm is processed only once [7]. As a result, the time complexity has been reduced. Secondly, it has been pointed out that the FCRM algorithm [8] develops hyper-plane-shaped clusters and it is suitable for obtaining linear equations. Therefore, FCRM is deployed to obtain the optimal fuzzy partition matrix. In the next step, the initial values of system parameters can be obtained via the optimal fuzzy partition matrix. Finally, the premise parameters and consequent parameters are fine adjusted simultaneously by the gradient descent algorithm [9], which solve the problem of parameters' asynchronous optimization.

2 TS Fuzzy Model

TS fuzzy model is able to approximate a nonlinear system by using a number of linear subsystems [2]. The TS fuzzy model discussed in this chapter is expressed as the following form:

$$\begin{aligned} R^i &: If\ x_1\ is\ A_j^1\ and\ x_2\ is\ A_j^2\ and\ \cdots and\ x_m\ is A_j^m \\ &Then\ y^i = a_0^i + a_1^i x_1 + \cdots + a_m^i x_m. \end{aligned} \tag{1}$$

The inferred output of the TS fuzzy model is calculated as:

$$\hat{y} = (\sum_{i=1}^{c} w^i y^i)/(\sum_{i=1}^{c} w^i),\ w^i = MIN_{j=1}^{m} A_j^i(x_j). \tag{2}$$

where c is the number of fuzzy rules, x is the m dimensionality input vector $x = [x_1, x_2, \ldots, x_m]$ and $\hat{y}$ is the output of the model. $A_j^i(1 \le i \le c, 1 \le j \le m)$ are fuzzy sets, which in this paper have bell-typed membership functions $A_j^i = exp\{-[(x - p_{j1}^i)/p_{j2}^i]^2\}$, the parameters p_{j1}^i and p_{j2}^i denote the mean and standard deviation of the jth membership function of the ith fuzzy rule, respectively. $a_j^i(1 \le i \le c, 0 \le j \le m)$ denote the consequent parameters.

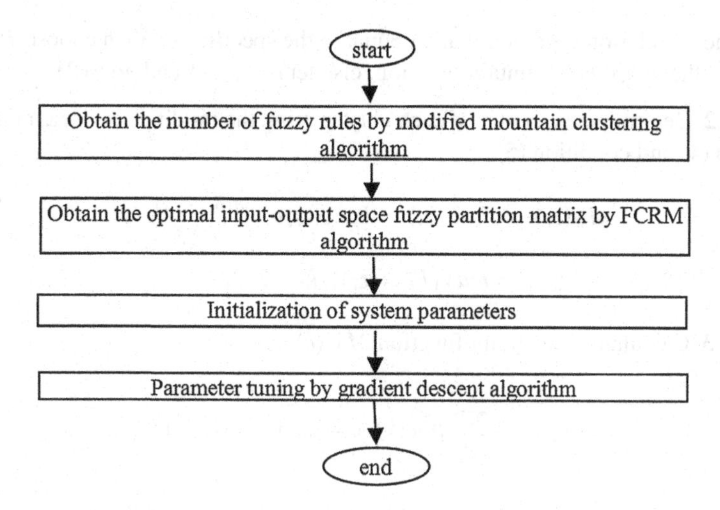

Fig. 1 The flow sheet of the proposed algorithm

3 The Proposed Algorithm

The proposed algorithm consists of four steps and the flow sheet is shown in Fig. 1.

3.1 Obtain the Number of Fuzzy Rules by the Modified Mountain Clustering Algorithm

As a clustering tool to a grouped dataset, the modified mountain algorithm provides a good cluster number estimate via the new proposed validity measure function [7, 10].

Step 1 Acquire the modified mountain function using the correlation self-comparison algorithm.

(1) Set $m_0 = 1, m_l = 5l, l = 1$ and $w = 0.99$.
(2) Calculate $P_1^{m(l-1)}(z_i)$ and $P_1^{m_l}(z_i)$, $1 \le i \le n$.

$$P_1^{m_0}(z_i) = \sum_{j=1}^{n} e^{-m_0 \rho d(z_i,z_j)},\ P_1^{m_l}(z_i) = \sum_{j=1}^{n} e^{-m_l \rho d(z_i,z_j)}. \tag{3}$$

where $z_k = (x_k, y_k), \rho = [(\sum_{j=1}^{n} \|z_j - \bar{z}\|^2)/n]^{-1}, \bar{z} = (\sum_{j=1}^{n} z_j)/n. d(z_i, z_j)$ is the Euclidean distance between z_i and z_j.

(3) Calculate the correlation between $P_1^{m(l-1)}$ and $P_1^{m_l}$.

(4) If the correlation is greater than or equal to the specified w, then choose $P_1^{m(l-1)}$ to be the modified mountain function; else let $l = l + 1$ and go to 2).

Step 2 Determine the center of each cluster using the modified revised mountain function (4) and condition (5).

$$P_k(z_i) = P_{k-1}(z_i) - P_{k-1}(z_i)e^{-\rho d(v_{k-1}, z_j)}. \tag{4}$$

$$v_{k-1} = max\{P_{k-1}(z_i)\}, k = 2, 3, \cdots. \tag{5}$$

Step 3 Calculate the validity function $MV(c)$.

$$MV(c) = \sum_{k=2}^{c} pot(k), c = 2, 3, \ldots (n-1). \tag{6}$$

$$pot(k) = P_1(v_k)\frac{P_1(v_k)}{P_1(v_1)} - ne^{-m(d_k/\rho)^2}. \tag{7}$$

$$d_k = min\{d(v_k, v_{k-1}), d(v_k, v_{k-2}), \ldots, d(v_k, v_1)\}. \tag{8}$$

The function $pot(k)$ is the potential of kth cluster center v_k, d_k is the minimum distance among v_k and all (k-1) previous identified cluster centers.

Step 4 Choose the cluster number estimate with maximum value of $MV(c)$ and select these c extracted cluster centers.

Step 5 Construct the initial input-output space fuzzy partition

$$\begin{gathered} u_{ij}^0 = 1/\sum_{k=1}^{n}(d(z_j, v_i)/d(z_k, v_i))^2, \ d(z_j), v_i \neq 0 \\ u_{ij}^0 = 1, \ otherwise \\ 1 \leq i \leq c, 1 \leq j \leq n. \end{gathered} \tag{9}$$

3.2 *Obtain the Optimal Input-Output Space Fuzzy Partition Matrix by FCRM Algorithm*

3.2.1 FCRM Algorithm

The FCRM algorithm was introduced by Hathaway and Bezdek [8] and belongs to the range of clustering algorithms with linear prototype. It develops hyper-plane-shaped clusters and is suitable to describe the structure of data space of TS fuzzy model. Therefore, in this subsection, FCRM algorithm is adopted to obtain an optimal input-output space fuzzy partition matrix.

Suppose that a set of n sample data denoted by (x_k, y_k), $1 \le k \le n$,$X = [1, x_1, \ldots, x_m]^T$,$P^i = [a_0^i, a_1^i, \ldots, a_m^i]$. The representative of ith cluster is expressed as follows:

$$y^i = a_0^i + a_1^i x_1 + \cdots + a_m^i x_m = X^T P^i, i = 1, 2, \ldots c. \tag{10}$$

Step 1 Calculate initial cluster representatives $P^i(1)$ by applying U^0 and WRLS algorithm.

Step 2 At the lth iteration, assign fuzzily each sample data (X_k, y_k), to each cluster with the representative being $y^i = X^T P^i(l)$, and then modify matrix $U(l)$ by Eq. (11).

$$I_k \equiv \{i|1 \le i \le c, d_k^i \equiv \|y_k - X^T P^i(l)\| = 0\}$$

$$\bar{I}_k \equiv \{1, 2, \ldots, c\} - I_k$$

$$\begin{cases} u_k^i = 1/[\sum_{j=1}^{c} (d_k^i, d_k^j)^{2/(m-1)}], \ if I_k = \phi \\ u_k^i = 0 i \in \bar{I}_k \ \ and \sum_{i \in I_k} u_k^i = 1, \ other \end{cases} \tag{11}$$

Step 3 If $\|U(l) - U(l-1)\| \le \varepsilon$, ($\varepsilon$ is equal to 10^{-5} in this chapter), then stop; otherwise go to step 4.

Step 4 Using u_k^i obtained by step 2 and WRLS algorithm to calculated new representative $P^i(l+1)$ at the $(l+1)$ iteration.

Step 5 Let $l = l + 1$ and go to step 2.

3.2.2 Weighted Recursive Least Squared Algorithm

Step 1 Initial values of the algorithm are determined as follows:

$$P_0^i = 0, \quad S_0 = \alpha I, \quad k = 0. \tag{12}$$

where I is the identity matrix, α is a sufficiently large number and is equal to 10^6 in this paper.

Step 2 Get new data (X_{k+1}, y_{k+1}) and then P_{k+1}^i is updated via Eqs. (13)–(15).

$$K_k = S_k X_{k+1}/[(1/u_{k+1}^i) + X_{k+1}^T S_k X_{k+1}]. \tag{13}$$

$$P_{k+1}^i = P_k^i + K_k[y_{k+1} - X_{k+1}^T P_k^i]. \tag{14}$$

$$S_{k+1} = [I - K_k X_{k+1}^T] S_k. \tag{15}$$

Step 3 Let $k = k + 1$, if $k \leq n$, then go to step 2, otherwise stop. Then, the representative for the ith cluster is given by $P^i = P_n^i (i \leq i \leq c)$.

3.3 Initialization of System Parameters

The initial consequent parameters are equal to P^i and initial premise parameters can be easily calculated by Eqs. (16)–(18).

$$A_j^i = exp\{-[(x - p_{j1}^i)/p_{j2}^i]^2\}. \tag{16}$$

$$p_{j1}^i = (\sum_{k=1}^{n} u_k^i x_{kj})/(\sum_{k=1}^{n} u_k^i). \tag{17}$$

$$p_{j2}^i/\sqrt{2} = \sqrt{[\sum_{k=1}^{n} u_k^i (x_{kj} - p_{j1}^i)^2]/(\sum_{k=1}^{n} u_k^i)}. \tag{18}$$

3.4 Parameter Tuning by Gradient Descent Algorithm

In this step, the system parameters are further adjusted by the well-known gradient descent algorithm [9]. The objective function is given as the following equation:

$$e_h = 0.5(y_h - \hat{y}_h)^2, \quad 1 \leq h \leq n. \tag{19}$$

The premise parameters and consequent parameters should be adjusted to minimize e_h and they are adjusted by Eqs.(20)–(22), respectively.

$$\Delta P_{jk}^i = -\beta \frac{\partial e_h}{\partial P_{jk}^i} = \beta(y_h - \hat{y}_h)(y_h^i - \hat{y}_h) \frac{1}{\sum_{i=1}^{c} w_h^i} \frac{\partial w_h^i}{\partial P_{jk}^i}. \tag{20}$$

$$\Delta a_j^i = -\gamma \frac{\partial e_h}{\partial a_j^i} = \gamma(y_h - \hat{y}_h) \frac{1}{\sum_{i=1}^{c} w_h^i} w_h^i x_{hj}. \tag{21}$$

$$\begin{cases} \dfrac{\partial w_h^i}{\partial P_{j1}^i} = \dfrac{2}{P_{j2}^i} \dfrac{x_{hj} - p_{j1}^i}{p_{j2}^i} exp\{-(\dfrac{x_{hj} - p_{j1}^i}{p_{j2}^i})^2\} \\ \dfrac{\partial w_h^i}{\partial P_{j2}^i} = \dfrac{2}{P_{j2}^i} (\dfrac{x_{hj} - p_{j1}^i}{p_{j2}^i})^2 exp\{-(\dfrac{x_{hj} - p_{j1}^i}{p_{j2}^i})^2\} \end{cases}, w_h^i = MIN_{j=1}^m A_j^i(x_{hj}) \\ \dfrac{\partial w_h^i}{\partial P_{j1}^i} = \dfrac{\partial w_h^i}{\partial P_{j2}^i} = 0, otherwise \quad (22)$$

where y_h is a desired output, $\hat{y_h}$ is an output of model, β is a premise learning rate, γ is a consequent learning rate.In this chapter, β and γ are equal to 0.3.

4 Simulation Examples

In this section, two examples are given to illustrate the validity of the proposed algorithm. The famous Box-Jenkins data set is a benchmark example used to check the effectiveness of nonlinear system modeling method, which consists of 296 input-output measurements of a gas furnace process [11]. x(k), x(k-1), x(k-2), x(k), y(k-1), y(k-2), y(k-3) are chose as the inputs and y(k) as the output of the system model. In order to compare with other methods, two experimental cases(case1 and case 2) are performed. The mean squared error (MSE) is used as a performance Index (PI).

$$PI = MSE = \bar{e}^2 = \frac{1}{n}\sum_{i=1}^{n}[y(i) - \hat{y}(i)]^2. \quad (23)$$

In case 1, all the data are used as training data which has 293 valid data pairs. By applying the proposed algorithm, three fuzzy rules are obtained and shows as follows

R^1: if $x(k)$ is A_1^1 and $x(k-1)$ is A_2^1 and $x(k-2)$ is A_3^1 and $y(k-1)$ is A_4^1 and $y(k-2)$ is A_5^1 and $y(k-3)$ is A_6^1
then $y^1(k) = 1.5716 - 0.7824x(k) + 2.5580x(k-1) - 1.8949x(k-2)$
$+2.6203y(k-1) - 2.5729y(k-2) + 0.9193y(k-3)$

R^2: if $x(k)$ is A_1^2 and $x(k-1)$ is A_2^2 and $x(k-2)$ is A_3^2 and $y(k-1)$ is A_4^2 and $y(k-2)$ is A_5^2 and $y(k-3)$ is A_6^2
then $y^2(k) = 11.8985 - 0.5997x(k) + 0.9541x(k-1) - 0.8401x(k-2)$
$+1.1714y(k-1) + 0.1651y(k-2) - 0.5509y(k-3)$

R^3: if $x(k)$ is A_1^3 and $x(k-1)$ is A_2^3 and $x(k-2)$ is A_3^3 and $y(k-1)$ is A_4^3 and $y(k-2)$ is A_5^3 and $y(k-3)$ is A_6^3
then $y^3(k) = 2.5391 + 0.6820x(k) - 1.0932x(k-1) + 0.0933x(k-2)$
$+1.5006y(k-1) - 1.0061y(k-2) + 0.4517y(k-3)$

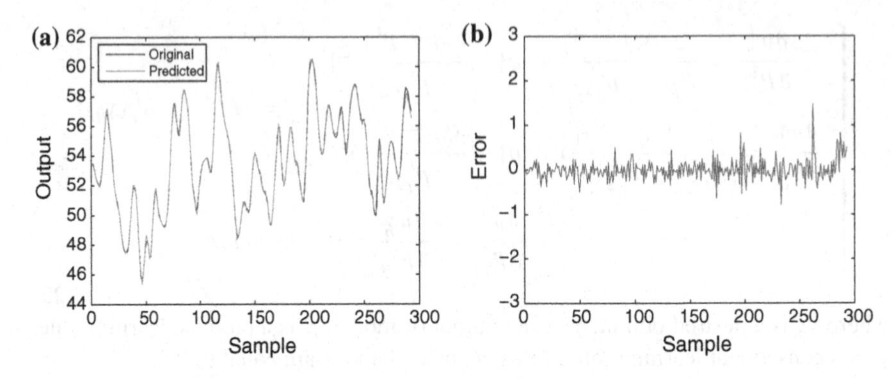

Fig. 2 **a** Original and predicted outputs **b** the respective errors (case 1)

Table 1 Comparison of performance (case 1)

Model	Number of inputs	Number of rules	MSE
Reference [2]	6	2	0.0598
Reference [12]	2	90	0.090
Reference [13]	2	4	0.123
Reference [14]	6	2	0.057
Reference [15]	2	4	0.148
Reference [16]	6	8	0.075
Reference [17]	6	9	0.055
Reference [18]	6	2	0.055
our model	6	3	0.055

Figure 2 gives the original and predicted output values for the training data and the respective errors. In Table 1, the MSE and the number of rules of the suggested fuzzy model are compared with those of prior models. The MSE of our model is equal to 0.055.

In case 2, the first 148 input-output data are used as training data and the remaining data as test data. By applying the proposed algorithm, 6 fuzzy rules are obtained and given as follows

R^1: if $x(k)$ is A_1^1 and $x(k-1)$ is A_2^1 and $x(k-2)$ is A_3^1 and $y(k-1)$ is A_4^1 and $y(k-2)$ is A_5^1 and $y(k-3)$ is A_6^1
then $y^1(k) = 8.1083 + 0.1005x(k) - 0.1750x(k-1) - 0.4588x(k-2) + 1.1832y(k-1) - 0.3169y(k-2) - 0.0259y(k-3)$

R^2: if $x(k)$ is A_1^2 and $x(k-1)$ is A_2^2 and $x(k-2)$ is A_3^2 and $y(k-1)$ is A_4^2 and $y(k-2)$ is A_5^2 and $y(k-3)$ is A_6^2
then $y^2(k) = 5.6566 - 0.4220x(k) + 0.8714x(k-1) - 0.7741x(k-2) + 1.7127y(k-1) - 1.0695y(k-2) + 0.2471y(k-3)$

R^3: if $x(k)$ is A_1^3 and $x(k-1)$ is A_2^3 and $x(k-2)$ is A_3^3 and $y(k-1)$ is A_4^3 and $y(k-2)$ is A_5^3 and $y(k-3)$ is A_6^3

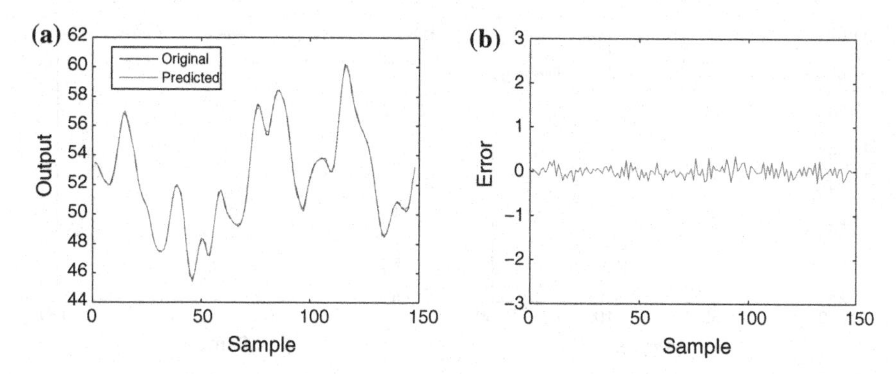

Fig. 3 **a** Original and predicted outputs **b** the respective errors (training data)

Table 2 Comparison of performance (case 2)

Model	Number of inputs	Number of rules	MSE (training)	MSE (validation)
Reference [2]	6	2	0.0164	0.145
Reference [4]	3	12	0.5072	0.2447
Reference [13]	2	4	0.02	0.271
Reference [18]	6	2	0.034	0.244
Reference [19]	2	75	0.0374	0.0403
Reference [20]	6	8	0.6184	0.2037
our model	6	6	0.0163	0.1425

then $y^3(k) = 9.5591 - 0.0289x(k) + 1.0124x(k-1) - 1.4921x(k-2)$
$+1.3675y(k-1) - 0.6457y(k-2) + 0.1018y(k-3)$

R^4: if $x(k)$ is A_1^4 and $x(k-1)$ is A_2^4 and $x(k-2)$ is A_3^4 and $y(k-1)$ is A_4^4 and $y(k-2)$ is A_5^4 and $y(k-3)$ is A_6^4
then $y^4(k) = 9.1171 + 0.1487x(k) + 0.0603x(k-1) - 0.7107x(k-2)$
$+1.2994y(k-1) - 0.6262y(k-2) + 0.1596y(k-3)$

R^5: if $x(k)$ is A_1^5 and $x(k-1)$ is A_2^5 and $x(k-2)$ is A_3^5 and $y(k-1)$ is A_4^5 and $y(k-2)$ is A_5^5 and $y(k-3)$ is A_6^5
then $y^5(k) = 8.6447 - 0.1719x(k) + 0.6213x(k-1) - 0.8157x(k-2)$
$+1.3234y(k-1) - 0.3809y(k-2) - 0.1037y(k-3)$

R^6: if $x(k)$ is A_1^6 and $x(k-1)$ is A_2^6 and $x(k-2)$ is A_3^6 and $y(k-1)$ is A_4^6 and $y(k-2)$ is A_5^6 and $y(k-3)$ is A_6^6
then $y^6(k) = 12.2225 + 0.1729x(k) - 0.2575x(k-1) - 0.8213x(k-2)$
$+0.7859y(k-1) + 0.1467y(k-2) - 0.1627y(k-3)$

Figure 3a illustrates the original and predicted output values for the training data and Fig. 3b shows the respective errors. Figure 4a depicts the original and predicted output values for the test data and Fig. 4b shows the respective errors. Table 2 gives the comparison of the results with those of prior models. In our model, the MSE for training data and test data are equal to 0.0163 and 0.1425, respectively.

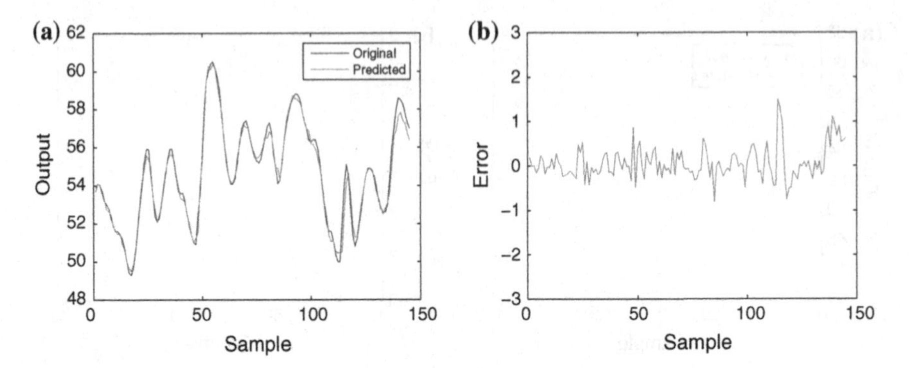

Fig. 4 **a** Original and predicted outputs **b** the respective errors (test data)

From the given Tables 1 and 2, it is obvious that our model gives good performances and have high precision.

5 Conclusion

In this chapter, a new fuzzy clustering approach for TS model identification is proposed. Firstly, the optimal number of fuzzy rules can be obtained by the modified mountain clustering algorithm. Secondly, the optimal fuzzy partition matrix can be obtained via the FCRM algorithm. Finally, the gradient descent algorithm is used to fine tune system parameters. Compared with other fuzzy modeling methods, the proposed algorithm has the advantages of simplicity, high precision, high accuracy and can be handled by an automatic procedure. But it may suffer from no convergence and is very sensitive to noise.

Acknowledgments The work was supported by Natural Science Foundation of Fujian Province of China (No.2011J01013), and Special Fund of Science, Technology of Fujian Provincial University of China (JK2010013) and Fund of Science, Technology of Xiamen (No. 3502Z20123022), The Projects of Education Department of Fujian Province (JK2010031, JA10196).

References

1. Jang, J.S.R., Sun, C.T., Mizutan, E.I.: Neuro-Fuzzy and Soft Computing: A Computational Approach to learning and Machine Intelligence. Prentice Hall, New York (1997)
2. Tsekouras, G., Sarimveis, H., Kavakli, E.: A hierarchical fuzzy-clustering approach to fuzzy modeling. Fuzzy Sets Syst. **150**, 245–266 (2005)
3. Takagi, T., Sugeno, M.: Fuzzy identification of systems and its application to modeling and control. IEEE Trans. Syst. Man Cybern. **15**(1), 116–132 (1985)

4. Sugeno, M., Yasukawa, T.: A fuzzy-logic-based approach to qualitative modeling. IEEE Trans. Fuzzy Syst. **1**(1), 7–31 (1993)
5. Kroll, A.: Identification of functional fuzzy models using multidimensional reference fuzzy sets. Fuzzy Sets Syst. **80**(2), 149–158 (1996)
6. Hong-qian, Lu, Song, Q.-N.: A novel hybird T-S model identification algorithm. J. Harbin Inst. Technol. **43**(9), 1–6 (2011)
7. Yang, M.-S., Wu, K.-L.: A modified mountain clustering algorithm. Pattern Anal. Appl. **8**, 125–138 (2005)
8. Hathaway, R., Bezdek, J.C.: Switching regression models and fuzzy clustering. IEEE Trans. Fuzzy Syst. **1**(3), 7–31 (1993)
9. Wang, L.: A Course in Fuzzy Systems and Control. Tsinghua University press, Beijing (2003)
10. Yager, R., Filev, D.: Generation of fuzzy rules by mountain clustering. J. Intell. Fuzzy Syst. **2**, 209–219 (1994)
11. Box, G.E.P., Jenkins, G.M.: Time Series Analysis, Forecasting and Control. Holden Day, San Francisco (1970)
12. Evsukoff, A., Branco, A.C.S., Galichet, S.: Structure identification and parameter optimization for non-linear fuzzy modeling. Fuzzy Sets Syst. **132**(2), 173–188 (2002)
13. Oh, S., Pedrycz, W.: Identification of fuzzy systems by means of an auto-tuning algorithm and its application to nonlinear systems. Fuzzy Sets Syst. **115**(2), 205–230 (2000)
14. Guo, Y., Lv, J.: Fuzzy modeling based on TS model and its application for thermal process. J. Syst. Simul. **22**(1), 210–215 (2010)
15. Bagis, A.: Fuzzy rule base design using tabu search algorithm for nonlinear system modelling. ISA Trans. **47**(1), 32–44 (2008)
16. Tsekouras, G.E.: On the use of the weighted fuzzy c-means in fuzzy modeling. Adv. Eng. Softw. **36**(5), 287–300 (2005)
17. Jun, H., Pan, W.: A denclue based approach to neuro-fuzzy system modelling. Adv. Comput. Control **4**, 42–46 (2010)
18. Kim, E., Park, M., Ji, S.: A new approach to fuzzy modeling. IEEE Trans. Fuzzy Syst. **5**(3), 328–337 (1997)
19. Farag, W.A., Quinatana, V.H., Lambert-Torres, G.: A genetic-based neuro-fuzzy approach for modeling and control of dynamical systems. IEEE Trans. Neural Networks **9**(5), 756–767 (1998)
20. Wang, S.-D., Lee, C.-H.: Fuzzy system modeling using linear distance rules. Fuzzy Sets Syst. **108**(2), 179–191 (1999)

Model Recognition Orientated at Small Sample Data

Jun-ling Yang, Yan-ling Luo and Ying-ying Su

Abstract System modeling is a prerequisite to understand object properties, while the chances are that an industrial site may come along with restricted conditions, leading to less experimental data acquired about the objects. In this case, the application of traditional statistical law of large numbers for modeling certainly will influence the identification precision. Aiming at the problems in recognition of small samples being not that high in precision, it is proposed to introduce the bootstrap-based re-sampling technique, upon which the original small sample data are expanded in order to meet the requirements of the statistical recognition method for sample quantity, so as to meet the requirements of precision. The simulation results showed that the extended sample model recognition accuracy is substantially higher than that of the original small sample. This illustrates the validity of the bootstrap-based re-sampling technique, working as an effective way for small sample data processing.

Keywords Small sample · Bootstrap method · Expansion · Data processing

1 Introduction

As a spurt of progress has been made in the development of information technique, data modeling is getting increasingly important for us to research the method of complex object. Abnormalities in certain condition, however, are complicated by the industrial site condition that may result in fewer experimental data about the object. Concerning the research of certain system objects, less number of process samples may not be able to bring out accurate judgment and analysis for object models. A better application of the law of large numbers in the typical statistical recognition

J. Yang (✉) · Y. Luo · Y. Su
School of Electric and Information Engineering, Chongqing University of Science and Technology, Chongqing 401331, China
e-mail: junlingyang@126.com

B.-Y. Cao and H. Nasseri (eds.), *Fuzzy Information & Engineering and Operations Research & Management*, Advances in Intelligent Systems and Computing 211,
DOI: 10.1007/978-3-642-38667-1_30,

method contributes to the establishment of small sample and improved recognition accuracy. To do this, the research on the data processing method for small sample data extension is viewed as a very important issue that requires special attention.

There are many commonly used methods being applied to look into the small sample issue in industry, such as Bayes, Bootstrap and Bayes Bootstrap, BP neural network, Monte Carlo, as well as the gray system, support vector machine, the fuzzy analysis and physical simulation [1–5]. Bayes is a combination of a priori knowledge of the system and the existing knowledge that conducts statistical inference on the future actions of the system. It is a method proven to be effective in estimating parameters in conformity with both the normal distribution and non-normal distribution, thus being taken as a more reasonable statistical analysis method for small samples or single simulation running. As one of the commonly used statistical inference methods, Bootstrap features non-priori and merely requires the actual observation data in the calculation process, it is quite convenient when used in the actual processing of the data, and also available for validation approximate convergence of the data model. Relative to other methods rooted from the weighted concept, this non-parametric statistical method does not involve any assumptions about the distribution of the unknown population. Rather, it is performed by means of the computer to complete the sampling of the original data, able to transform the small sample issue into a large sample one to simulate unknown distribution. The Bootstrap-based re-sampling technique works with computers in simulation to replace the deviation, variance, and other statistics as a complex and not-so-accurate approximate analysis method. Thus, the Bootstrap method for conducting statistical inference on small samples is mainly used in the probabilistic model of unknown parameters with too complex derivation being theoretically infeasible. It may also be found present in optimizing the inference effect with inaccurate statistical models or adequate statistical information.

In summary, this paper presents a bootstrap re-sampling technique for the expansion of sample data, aiming to build a relatively large number of virtual samples. On this basis, the use of the research methods for the traditional model recognition may improve the accuracy of object recognition.

2 Bootstrap-Based Re-sampling Technique for Small Sample Expansion

2.1 Principle

Let us set an unknown small sample population $x(n) = (x_1, x_2, \ldots, x_n)$, n is smaller, and then get the unknown parameter estimates, such as the expectation of this unknown population (also known as the Mean). Traditionally, the mean of such n samples $\frac{1}{n}\sum_{i=1}^{n} x_i$ is used to estimate it. Yet with a very small n, such an

Table 1 Distribution law of discrete sample

Value x1	x_1	x_2	...	x_n
Probability	1/n	1/n	...	1/n

expected value (mean) of unknown distribution shows a poor effect. Here, Bootstrap provides an effective solution, which is completed in the following steps:

Step 1: Collect the small samples $(x_1, x_2 \ldots, x_n)$, here n is smaller.

Step 2: Get the empirical distribution functions of this sample, which are discrete, and then can be written by the following distribution law, as shown in Table 1.

Step 3: Select n samples from this empirical distribution. According to the above distribution law, the successful selection of $x_i, i = 1, 2, \ldots, n$ has an average probability of $\frac{1}{n}$. And then there comes the successful selection of the following set of samples $(x_1, x_1 \ldots, x_1)$, i.e. all is x_1, or $(x_n, x_n \ldots, x_n)$; all is x_n, or $(x_1, x_1 \ldots, x_n)$, that means there is individual repeat, but the total number of samples is n. In short, the taken samples can be any combination of the original samples. This is because the distribution law has every value to be shown in probability as $\frac{1}{n}$. Here we might as well have the picked n samples marked as $(x_1^*, x_2^*, \ldots, x_n^*)$.

Step 4: Calculate $\mu^* = \frac{1}{n}\sum_{i=1}^{n} x_i^*$, the average value of the small samples picked at step 3.

Step 5: Repeat step 3, step 4 for K times (K can be a very large figure, usually K=100, 000 times), and then each time there can be obtained with an average, taken as $\mu_i^*, i = 1, 2, \ldots, K$.

Step 6: Calculate $\mu' = \frac{1}{K}\sum_{i=1}^{K} \mu_i^*$, here$\mu_i^*, i = 1, 2, \ldots, K$ is the average of the samples from the selected repetition at i times. Alternatively, the average of these K samples obtained at K times is averaged one more time. μ' is the desired estimate of the unknown population.

2.2 Re-sampling Technique Based on the Bootstrap Method

According to the principle of bootstrap methods as indicated in 2.1, the following re-sampling technique was designed to obtain the expanded samples for research need.

Step 1–2. Same as step 1–2 for the principle of the bootstrap methods

Step 3. The distribution law was obtained from step 2 as Table 1.

In turn, extraction was repeated B times from the discrete distribution (usually $B \gg n$, appropriate when meeting the needs of the sample size), ending up with the obtained B samples that constitute the final expansion of the sample. As B was larger, it is generally believed that the expanded B samples contain the original n small samples. Based on the above analysis of USRC elements, according to the establishment principles of index system [6–8], the evaluation index system can be established, as shown in Table 1.

3 Small Sample and the Simulation of Sample Model Recognition After Expansion

Regarding the recognition problem on the small sample model, the traditional BP neural network was placed to construct a classifier for the original small sample and the sample after the expansion using the bootstrap-based re-sampling technique, respectively, in order to verify its applicability.

3.1 Simulation Model Constructed for Small-Sample Model Recognition

The simulation model for small-sample model recognition was constructed as a Gaussian mixture model that generates "Swiss Roll":

$$Y=1, \text{当} \begin{bmatrix} Z_1 \\ Z_2 \end{bmatrix} \sim N_2(\mu_1, \Sigma) \quad Y=2, \text{当} \begin{bmatrix} Z_1 \\ Z_2 \end{bmatrix} \sim N_2(\mu_2, \Sigma),$$

$$Y=3, \text{当} \begin{bmatrix} Z_1 \\ Z_2 \end{bmatrix} \sim N_2(\mu_3, \Sigma) \quad Y=4, \text{当} \begin{bmatrix} Z_1 \\ Z_2 \end{bmatrix} \sim N_2(\mu_4, \Sigma),$$

$$\mu_1 = \begin{bmatrix} 7.5 \\ 7.5 \end{bmatrix}, \mu_2 = \begin{bmatrix} 7.5 \\ 12.5 \end{bmatrix}, \mu_3 = \begin{bmatrix} 12.5 \\ 7.5 \end{bmatrix}, \mu_4 = \begin{bmatrix} 12.5 \\ 12.5 \end{bmatrix},$$

the covariance matrix

$$\Sigma = \begin{bmatrix} 1 & 0 \\ 0 & 1 \end{bmatrix}.$$

(1) Original small sample model

Based on the constructed Gaussian mixture model that features the Swiss Roll, 32 samples were randomly generated (8 samples each category, a total of 32 samples for four categories), with the original small sample model in Fig. 1.

(2) Sample model after the expansion using the bootstrap-based re-sampling technique

With the help of the bootstrap-based re-sampling technique, the original 32 small samples were expanded to hit 300 samples, as shown in Fig. 2.

Comparing Figs. 1 and 2, the expanded number of samples was increased visually.

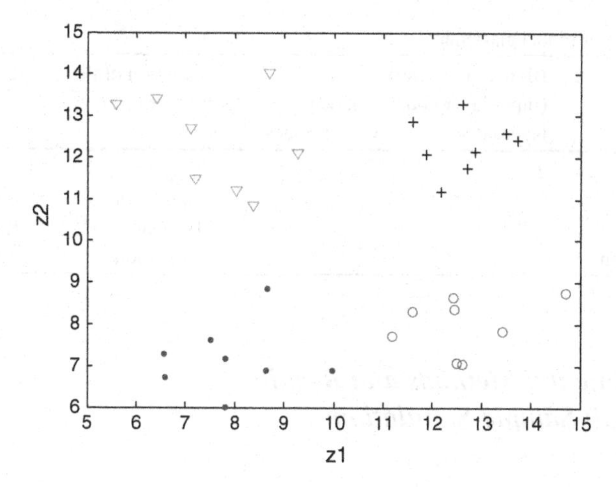

Fig. 1 Original small sample model

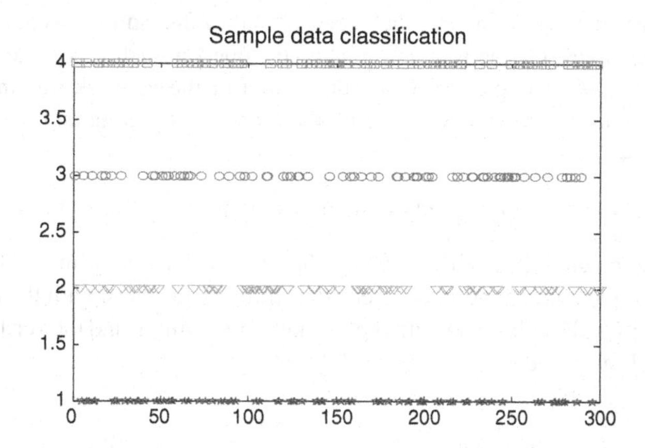

Fig. 2 Sample model after the expansion using the bootstrap-based re-sampling technique

3.2 Classifier Designed for Small-Sample Model Recognition

Targeting at the original small sample of the model, while taking care of the characteristics of Gaussian mixture model, a BP neural network structure was designed to perform model recognition experiments in line with the pertinent guidelines, as shown in Table 2.

Table 2 Design of model classifier

Small sample	BP neural network structure (Input layer nodes - hidden layer nodes - output layer nodes)	Function of the output layer	Guidelines
Original sample	2-5-1	Tangent function	Relative error less than 5 %
Small sample after expansion	2-5-1	Tangent function	Relative error less than 2 %

3.3 Experimental Methods and Results on Small Sample Simulation

(1) Original small sample simulation:

The original model was given seven samples each category, a total of 32 small samples, where the first 7 for each category, a total of 28 samples were selected as the training samples. Of the remaining samples, one for each type, a total of four were selected as test samples to obtain the output of the network training and the relative error map, as shown in Fig. 3, with the relative test output and the error map shown in Fig. 4.

(2) Expanded small sample based on the bootstrap re-sampling technique:

There were expanded 300 samples using the bootstrap re-sampling technique, of which 270 samples were randomly selected as training samples, with the remaining 30 as test samples. The obtained output of the network training and the verified output are shown in Figs. 5 and 6.

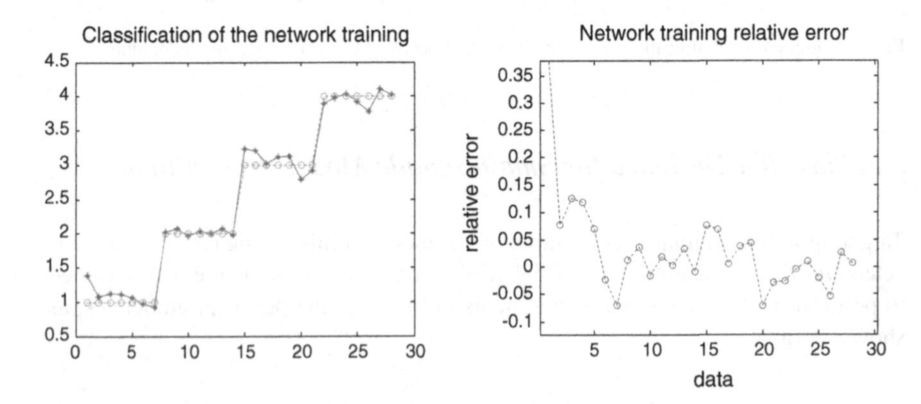

Fig. 3 The network training and the relative error

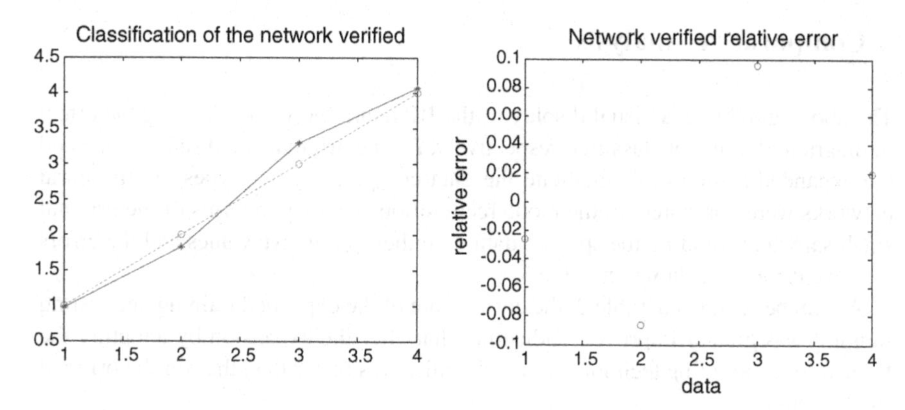

Fig. 4 The relative test output and the error map

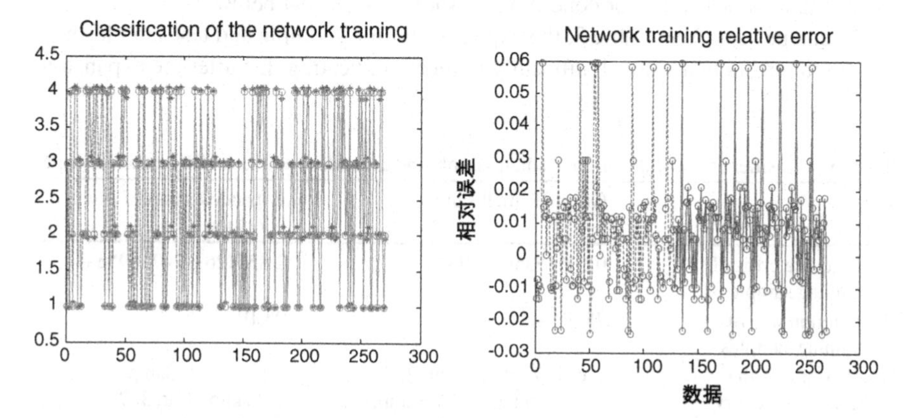

Fig. 5 The obtained output of the network training

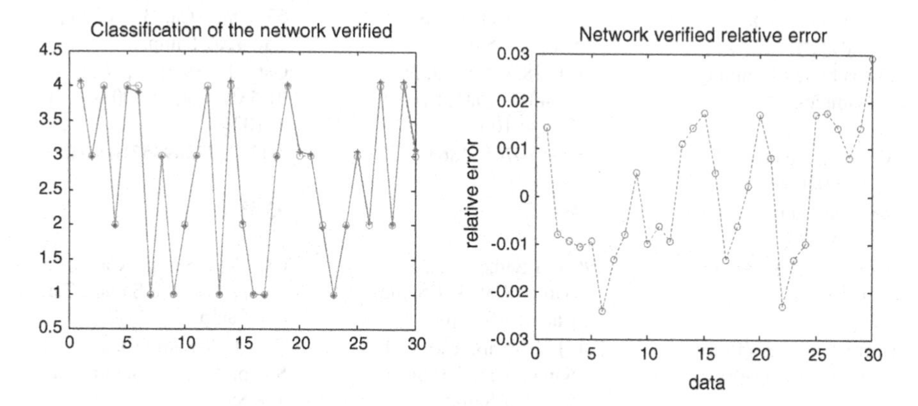

Fig. 6 The verified output of the network training

4 Comparative Analysis

The above graphic is a visual display of the BP neural network, showing the effect of diagram of a model classifier, respectively, for the original small sample size and the expanded samples. To facilitate the data comparison, two types of BP neural networks were compared on the mode recognition accuracy in terms of the original small sample, providing the specific data with their predictive values and the errors (relative errors), as shown in Table 3.

As can be seen from Table 2, the correct rate of the expanded training and testing samples was greatly improved, indicating that after the expansion by adopting the Bootstrap re-sampling technology, the classifier was better than that for the original small sample.

The original small sample and the expanded sample in BP neural network showed their output values and error conditions as shown in Table 4 below.

As can be seen from Table 3, the expanded training sample and test samples had the output errors that were significantly reduced, indicating that after the expansion

Table 3 Classifier effect of the original small sample and expanded samples in BP neural network

	Original small sample	Expanded samples using bootstrap
Mean square error of training samples	0.006553567539420	2.834126992471895e-04
Recognition rate of training samples	75 %	95.19 %
Number of training samples category	Cat. 1: 7 Samp.; Cat. 2: 7 Samp.; Cat. 3: 7 Samp.; Cat. 4: 7 Samp.	Cat. 1: 67 Samp.; Cat. 2: 71 Samp.; Cat. 3: 73 Samp.; Cat. 4: 59 Samp.
Correct number of training samples category	Cat. 1: 2 Samp.; Cat. 2: 5 Samp.; Cat. 3: 7 Samp.; Cat. 4: 7 Samp.	Cat. 1: 60 Samp.; Cat. 2: 65 Samp.; Cat. 3: 73 Samp.; Cat. 4: 59 Samp.
Correct rate of training samples category	Cat. 1: 28.57 %; Cat. 2: 71.43 %; Cat. 3: 100 %; Cat. 4: 100 %	Cat. 1: 89.55 %; Cat. 2: 91.55 % Cat. 3: 100 %; Cat. 4: 100 %
Mean square error of test samples	0.067451076534863	3.478027726935210e-04
Recognition rate of test samples	50 %	93.33 %
Number of test samples category	Cat. 1: 1 Samp.; Cat. 2: 1 Samp.; Cat. 3: 1 Samp.; Cat. 4: 1 Samp.	Cat. 1: 9 Samp.; Cat. 2: 5 Samp.; Cat. 3: 8 Samp.; Cat. 4: 8 Samp.
Correct number of test samples category	Cat. 1: 1 Samp.; Cat. 2: 0 Samp.; Cat. 3: 0 Samp.; Cat. 4: 1 Samp.	Cat. 1: 9 Samp.; Cat. 2: 3 Samp.; Cat. 3: 8 Samp.; Cat. 4: 8 Samp.
Correct rate of test samples category	Cat. 1: 100 %; Cat. 2: 0 %; Cat. 3: 0 %; Cat. 4: 100 %	Cat. 1: 100 %; Cat. 2: 60 %; Cat. 3: 100 %; Cat. 4: 100 %

Table 4 The original small sample and the expanded sample in BP neural network showed their output values and error conditions

	The original small sample				The expanded sample			
	Sequence of samples	True value	Output values	Absolute error (10^-2)	Sequence of samples	True value	Output values	Absolute error (10^-2)
Training sample	1	1	0.91876	8.12393	1	1	0.99221	0.77907
	2	1	0.98089	1.91084	2	2	1.98055	1.94477
	3	1	0.98599	1.40055	3	4	3.98072	1.92751
	4	1	1.12220	12.21996	4	1	1.02538	2.53836
	...	...	...	...	...	...	...	...
	25	4	4.03473	3.47282	267	3	2.99233	0.76750
	26	4	4.11668	11.66820	268	1	0.99221	0.77907
	27	4	4.00009	0.00936	269	3	2.99445	0.55490
	28	4	3.97737	2.26296	270	4	4.02254	2.25426
Test sample	29	1	0.95533	4.46721	271	1	4.02254	2.25426
	30	2	2.12924	12.92396	272	2	0.98545	1.45501
	31	3	2.50399	49.60053	...	...	...	...
	32	4	3.92869	7.13062	299	3	0.99449	0.55131
					300	4	3.98918	1.08225

by adopting the Bootstrap re-sampling technology, the classifier was better than that for the original small sample.

Through the above tabular data comparison, we can get the following conclusions:

(1) Original small sample expanded by using the bootstrap re-sampling method, worked as the expanded small sample;
(2) The bootstrap-based re-sampling method to expand the original sample may greatly improve the model classifier in the BP neural network on its recognition accuracy.

5 Conclusion

As illustrated in the above analysis and simulation, this chapter presented the bootstrap-based method of re-sampling techniques for the expansion of small samples, which effectively extracted more data from small samples. The resulting sampling samples met the requirements of the traditional recognition method from the statistical theory for the number of samples. And the expanded samples, compared with the original small sample mode, showed higher recognition accuracy. This thus proved the correctness of bootstrap re-sampling method for the expansion of the original small sample.

Acknowledgments The work is supported by Chongqing Educational Committee Science and Technology Research Project No.KJ091402, No. KJ111417, and the Natural Science Foundation of Chongqing University of Science & Technology No.CK2011Z01.

References

1. Xu, L.: Neural Network Control. Publishing House of Electronics Industry, Beijing (2003)
2. Saridis, G.N.: Entropy formation of optimal and adaptive control. IEEE Trans. Autom. Control **33**(8), 713–721 (1988)
3. Lin, J.-H., Isik, C.: Maximum entropy adaptive control of chaotic systems. In: Procecdings of IEEE ISIC/CIRA/ISAS joint conference, pp. 243–246. Gaithersburg (1998).
4. Xiaoqun, Y.: Intelligent control processes based on information entropy. South China University of Technology (2004)
5. Masory, O., Koren, Y.: Adaptive control system for tuming. Ann. CIRP **29**(1), 281–284 (1980)
6. Hyvarinen, A., Karhunen, J., Oja, E.: Independent Component Analysis. Wiley, New York (2001)
7. Shuang, C.: Neural Network Theory Oriented at MATLAB toolbox and its Applications (Version 2). China University of Science and Technology Press (2003)
8. Shi, F., Wang, X., Yu, L., Li, Y.: Studies on 30 Cases of MATLAB Neural Network. Beijing University of Aeronautics and Astronautics Press, Beijing (2010)

A New Fuzzy Clustering Validity Index with Strong Robustness

Xiang-Jun Xie, Yong Wang and Li Zhong

Abstract Cluster validity has been widely used to evaluate the fitness of partitions produced by fuzzy c-means (FCM) clustering algorithm. Many validity functions have been proposed for evaluating clustering results. Most of these popular validity measures do not work well for clusters with different fuzzy weighting exponent m and data with outliers at the same time. In this paper, we propose a new validity index for fuzzy clustering. This validity index is based on the compactness and separation measure. The compactness is defined by fuzzy Z-membership function based on the gold dividing point and separation is described by monotone linear function. The contrasting experimental results show that the proposed index works well.

Keywords Clustering validity function · Z-membership function · Monotone linear function

1 Introduction

Fuzzy c-means (FCM) clustering algorithm [1] proposed by Bezdek has become the most popular fuzzy clustering approach in recent decades. But this algorithm requires several parameters, and the most significant one affecting the performances is known as the number of cluster c. These problems are cluster validity problems. So, designing an effective validity function to find the best number of cluster c for a given data set is quite important.

So far, many fuzzy cluster functions have been proposed for evaluating fuzzy partitions. The first proposed fuzzy cluster validity functions are the partition coefficient V_{PC} [2] and partition entropy V_{PE} [3] by Bezdek. The separation coefficient proposed by Gunderson was the first validity index that explicitly takes into account

X. Xie (✉) · Y. Wang · L. Zhong
School of sciences, Southwest Petroleum University, Chengdu 610500, P.R. China
e-mail: xiangjunxie@126.com

B.-Y. Cao and H. Nasseri (eds.), *Fuzzy Information & Engineering and Operations Research & Management*, Advances in Intelligent Systems and Computing 211,
DOI: 10.1007/978-3-642-38667-1_31,

the data geometrical properties. XB index was proposed by Xie and Beni [4] and Fukuyama and Sugeno proposed FS index V_{FS} [5]. The fuzzy hypervolume (FHV) and partition density indices PD were proposed by Gath and Geva [5]. A new validity index for fuzzy clustering is introduced by Mohamed Bouguessa and Sheng-Rui Wang in 2004 [6]. Kuo-Lung Wu and Miin-Shen Yang propose a new index for fuzzy clustering called a partition coefficient and exponential separation (PCAES) index in 2004 [5]. In 2010, a new cluster validity index (CS(c)) is proposed to evaluate the fitness of clusters obtained by FCM by Horng-Lin Shieh and Po-Lun Chang [7].

Although so many indices have been proposed, almost no one has discussed the robustness of parameter m and data sets with noisy points at the same time. Thus, we present a new validity index for fuzzy clustering. The rest of this paper is organized as follows. Section 2 provides the FCM clustering algorithm. In Sect. 3, we describe the proposed validity index in detail. In Sect. 4, the experimental results are shown. Sect. 5 is the conclusion.

2 Fuzzy C-Means Clustering

Fuzzy c-means is an unsupervised clustering algorithm that has been applied successfully to a number of problems involving feature analysis, clustering and classifier design. Let $X = (x_1, x_2, \cdots, x_n)$ be an n points data set in a p-dimensional Euclidean space R^p with its usual Euclidean norm $\|\cdot\|$. The fuzzy c-means (FCM) algorithm partitions X into c clusters by minimizing the following objective function (1):

$$J_m(U, V) = \sum_{i=1}^{c} \sum_{j=1}^{n} (u_{ij})^m \left\| x_j - v_i \right\|^2 (m \neq 1), \tag{1}$$

where c is the number of cluster. $V = \{v_1, v_2, \cdots, v_c\}$ is a vector of unknown cluster prototype $v_k \in R^p$. U $(U = (u_{ij})_{c \times n})$ is a fuzzy partition matrix composed of the membership of each feature vector x_j in each cluster i, where u_{ij} should satisfy $\sum_{i=1}^{c} u_{ij} = 1 (j = 1, 2, \cdots, n), 0 \leq u_{ij} \leq 1 (i = 1, 2, \cdots, c, j = 1, 2, \cdots, n)$, and $0 < \sum_{j=1}^{n} u_{ij} < n$ $(i = 1, 2, \cdots, c)$. The exponent $m (m > 1)$ is a fuzzifier, which can control the fuzzy degree of the clustering result.

The steps for FCM based algorithm are as follows:

Step 1: Suppose a preselected number of cluster c, a chosen value of m, an initial partition matrix U and a threshold value ε.

Step 2: Compute the cluster centers V for $i = 1, 2, \cdots, c$ using (2),

$$v_i = \frac{\sum\limits_{k=1}^{n} (u_{ik})^m \cdot x_k}{\sum\limits_{k=1}^{n} (u_{ik})^m}, \tag{2}$$

Step 3: Update fuzzy membership matrix U according to following criterion. For any of the j or k, if $d_{jk} > 0$, U is updated according to Eq. (3); elseif $d_{ik} = 0$ for i and k, $u_{ik} = 1$; otherwise $u_{ik} = 0$.

$$u_{ik} = \{\sum_{j=1}^{c} [(\frac{d_{ik}}{d_{jk}})^{\frac{2}{m-1}}]\}^{-1}, \tag{3}$$

where d_{ik} is Euclidean distance between point x_k and cluster center v_i.

Step 4: Check the termination conditions, if they are satisfied, then halt; otherwise go back to step 2.

The FCM algorithm has an important drawback that is FCM needs to know the number of clusters before practical applications of the FCM algorithm. However, the user usually does not know the exact number of clusters in the data set. The performance of clustering algorithms in terms of the clustering results can be affected significantly if the number of clusters given is not accurate. So, designing an effective validity function to detect the best number of cluster c for a given data set is very important.

The validation procedure used to find the best number of cluster c is as follows: run FCM algorithm over the range $c = 2, \cdots, c_{\max} \approx \sqrt{n}$ and compute the validity function for each partition (U, V) generated by FCM. Choose the best cluster number which has the optimal validity value.

3 Cluster Validity

3.1 Some Validities Which Have Been Proposed

Many cluster validity indices have been proposed for fuzzy clustering and are used to establish what partition best explains the unknown cluster structure in a given data set.

Bezdek proposed two cluster validity indices, the partition coefficient (PC) [2] and partition entropy (PE) [3] as following (4) and (5). Both PC and PE possess monotonic evolution tendency with c. In general, The optimal partition is obtained by maximizing V_{PC} (or minimizing V_{PE}) with respect to $c = 2, \cdots, c_{\max}$.

$$V_{PC} = \frac{\sum_{j=1}^{n} \sum_{i=1}^{c} (u_{ij}^2)}{n}, \tag{4}$$

$$V_{PE} = -\frac{1}{n} \sum_{j=1}^{n} \sum_{i=1}^{c} [u_{ij} \log(u_{ij})] \tag{5}$$

Xie and Beni [4] proposed a validity index that focused on two properties compactness and separation, defined as (6). In general, an optimal c^* is found by solving max (V_{XB}) to produce a best clustering performance for the data set.

$$V_{XB} = \frac{\sum_{j=1}^{n}\sum_{i=1}^{c} u_{ij}^2 \left\|x_j - v_i\right\|^2}{n(\min_{i \neq k} \|v_i - v_k\|)^2} \tag{6}$$

Bensaid et al. [8] gave a partition index V_{SC} as (7) which is the ration of the sum of compactness and separation of the clusters. V_{SC} is very useful when different partitions having equal number of clusters are compared. In general, an optimal c^* is found by solving min (V_{SC}) to produce a best clustering performance for the data set.

$$V_{SC} = \sum_{i=1}^{c} \frac{\sum_{j=1}^{n} u_{ij}^m \left\|x_j - v_i\right\|^2}{n_i \sum_{k=1}^{c} (\|v_k - v_i\|)^2} \tag{7}$$

Then, a new cluster validity has been proposed by Fukayama and Sugno [5], denoted by FS index as (8), for evaluating fuzzy c-partitions by exploiting the concepts of compactness and separation. The optimal partition is obtained by minimizing V_{FS} with respect to $c = 2, \cdots, c_{\max}$.

$$V_{FS} = \sum_{j=1}^{n}\sum_{i=1}^{c} u_{ij}^m \left\|x_j - v_i\right\|^2 - \sum_{j=1}^{n}\sum_{i=1}^{c} u_{ij}^m \left\|v_i - \bar{v}\right\|^2. \tag{8}$$

3.2 A New Cluster Validity

Membership function is used to indicate whether an element belongs to fuzzy set. So far, some important membership functions have been proposed, such as: triangle membership function, bell membership function, trapezoidal membership function [9].

Fuzzy clustering validity function is composed of two important factors: the compactness and separation for the data structure. In order to reduce the outlier to affect the clustering, Z-type membership function [9] w_{ij} is used to depict the total compactness and is the membership degree of x_j belonging to center v_i as (9). The total separation of the data set is measured by monotone linear function m_{ik} as (10). It is the separation degree between different center v_i and v_k.

At the same time, the gold dividing point has the strict proportionality, artistic quality and harmonious quality, and carries rich aesthetic value. What is more, it has been widely applied in many fields, especially in mathematic. So, we adopt the gold dividing point to establish the function w_{ij}.

$$w_{ij} = \begin{cases} 1, & d \leq 0.372d_{\min}; \\ 1 - \frac{d}{d_{\max}}, & 0.372d_{\min} < d \leq 0.628d_{\min}; \\ 0, & d > 0.628d_{\min}. \end{cases} \tag{9}$$

$$m_{ik} = \frac{\|v_i - v_k\|}{d_{\max}} \tag{10}$$

where, $\|\cdot\|$ denotes Euclidean distance,

$$d_{\min} = \min_{i \neq k}(\|v_i - v_k\|)(i, k = 1, 2, \cdots, c),$$

$$d_{\max} = \max_{i \neq k}(\|v_i - v_k\|)(i, k = 1, 2, \cdots, c)$$

Thus, through the linearly combined $\sum_{j=1}^{n} \sum_{i=1}^{c} (w_{ij})^2$and $\sum_{i=1}^{c-1} \sum_{k=i+1}^{c} (m_{ik})^2$ by parameter α, a robust validity index is defined as (11)

$$V_{ZL} = \alpha \sum_{j=1}^{n} \sum_{i=1}^{c} (w_{ij})^2 + (1 - \alpha) \sum_{i=1}^{c-1} \sum_{k=i+1}^{c} (m_{ik})^2. \tag{11}$$

where, $0 \leq \alpha \leq 1$.

V_{ZL} is a bounded function, $0 \leq V_{ZL} \leq c(n+c-1)$, and it is very simple. A large value of V_{ZL} indicates that the fuzzy c-partition is characterized by well-separated fuzzy clusters.

4 Experiments

The experiments are done to demonstrate the effectiveness and superiority of the new index for noisy data and robustness of m. In this section, we implement the FCM clustering algorithm on data set with the cluster number $c = 2, \cdots, c_{\max} \approx \sqrt{n}$. Three groups of experiments are carried out and their results show that the validity index proposed in this paper outperforms other indices in the literature.

4.1 Robustness to Data Set with Noisy Points

Our purpose is to show the new index V_{ZL} has strong robustness when faced with some noisy points. Two artificial datasets and one real dataset are used in this experiment. Data 1 is an artificial data set which contains 56 points with three noisy points, as shown in Fig. 1. It is a two dimensional dataset consisting of 2 classes. Data 2

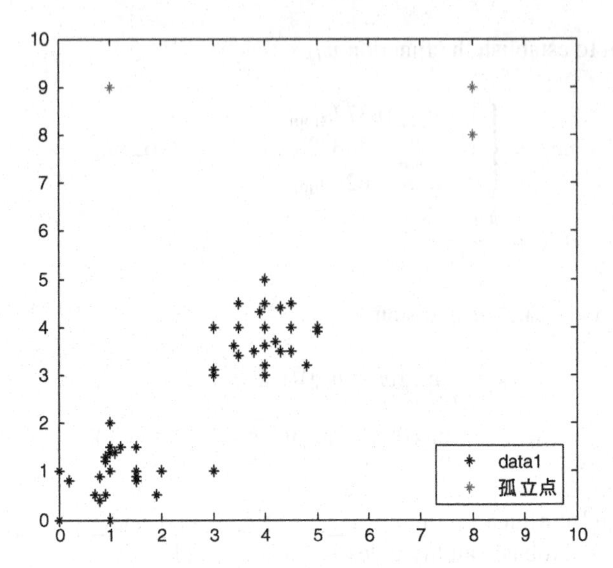

Fig. 1 Data 1 with noisy points

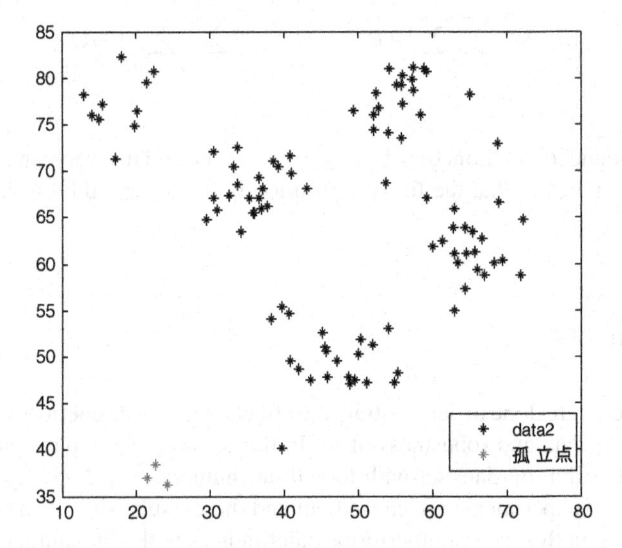

Fig. 2 Synthetic data 2

is also an artificial data set which consisting of 5 classes with three noisy points, as shown in Fig. 2. Its noisy points make validity index harder to determine the number of clusters. Data 3 is the IRIS data set which has three classes (Iris Setosa, Iris Versicolor and Iris virginica) or two classes. And it is the representative data set with some outliers.

Table 1 The value of V_{ZL}on different data sets

Data	$c = 2$	$c = 3$	$c = 4$	$c = 5$	$c = 6$	$c = 7$
Data1	**21.2865**	13.6090	16.3630	16.5561	18.1044	19.2461
Data2	45.4083	70.1990	78.9861	**96.5681**	57.7550	59.0643
IRIS	**55.0530**	*50.3896*	44.8709	41.3916	29.1647	40.1474

Table 1 reports the results obtained for the new validity index on $m = 2.0$, $\alpha = 0.6$. The new validity index V_{ZL} always gives the right number of clusters at any given time.

4.2 Compared with Other Indices

Second, several numerical examples are presented to compare the proposed V_{ZL} index with four other validity indices: V_{PC},V_{PE},V_{XB} and V_{CS}. One artificial dataset and four real datasets are used in this experiment. The artificial is data 2 and three real datasets are respectively iris data, BENSAID data and seeds data set. BENSAID data was described by Bensaid et al. [8], as shown in Fig. 3. It includes 49 data points in two dimensional measurement spaces, and consists of three clusters [10]. Seeds data set contains 210 data points with three classes (Kama, Rosa and Canadian). It is from Machine Learning Repository.

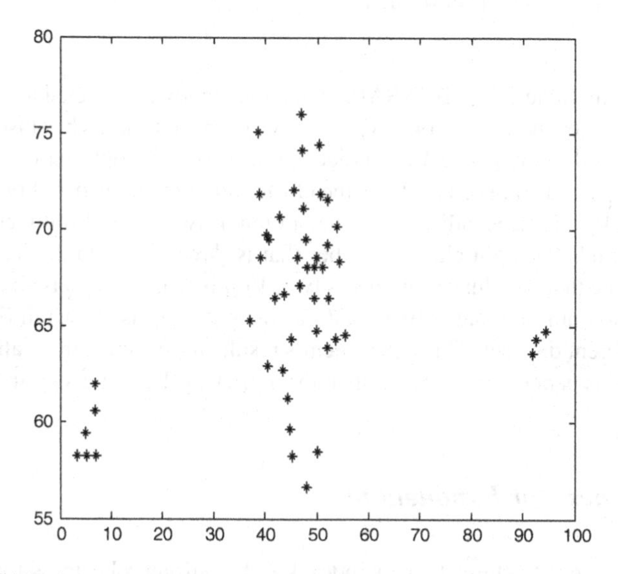

Fig. 3 BENSAID data

Table 2 The cluster number obtained by different validity indices

Validity function	BENSAID	IRIS	Seeds	Data 2
V_{PC}	2	2	2	2
V_{PE}	2	2	2	2
V_{XB}	7	12	14	10
V_{ZL}	**3**	**2,3**	**3**	**5**
V_{CS}	3	2	2	6

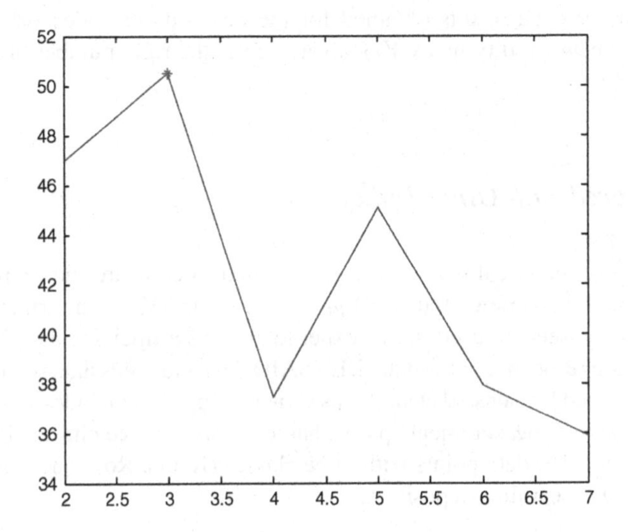

Fig. 4 The value of V_{ZL}on BENSAID data

As shown in Table 2, for BENSAID data, V_{PC} index indicates that the optimal is two, V_{PE} is seven, V_{XB} is two, V_{CS} and V_{ZL}can find the right cluster number three. For IRIS data, V_{PC} and V_{PE} indices indicate that the optimal is two, V_{XB} is eleven, and V_{CS} is also two. V_{ZL} finds the cluster number that is two. For seeds data set, V_{PC} and V_{PE} indices indicate that the optimal is two, V_{XB} is fourteen, and V_{CS} is two. V_{ZL} finds the right cluster number that is three. For data 2, V_{PC} and V_{PE} indices indicate that the cluster number is two, V_{XB} is ten, and V_{CS} is six. V_{ZL} finds the right cluster number that is five. And the value of V_{ZL} is shown in Figs. 4, 5, 6 and 7 on different data set. This experiment's results have shown that validity index proposed in this paper outperformed than the V_{PC},V_{PE},V_{XB} and V_{CS} indices.

4.3 Robustness for Exponent m

Last, our purpose is to show the new index V_{ZL} has strong robustness for exponent m. Seeds data and IRIS data are used in the experiments.

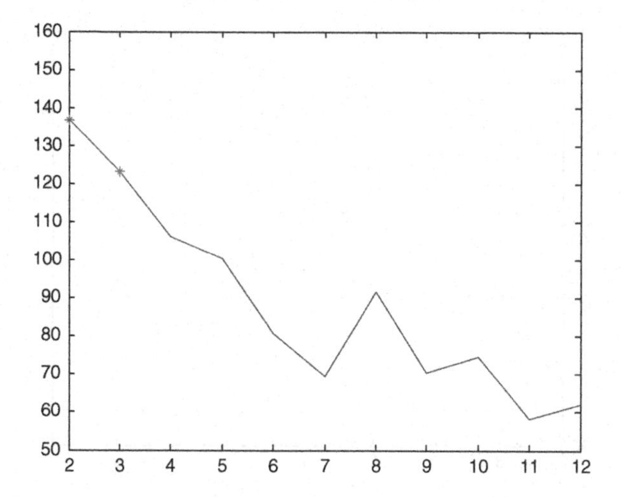

Fig. 5 The value of V_{ZL} on IRIS data

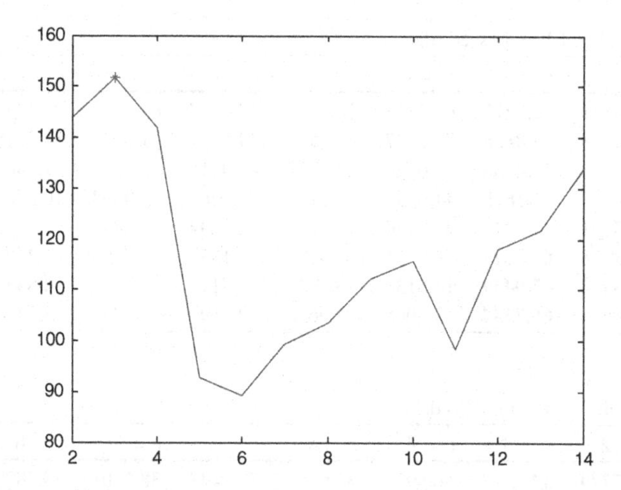

Fig. 6 The value of V_{ZL} on seeds data

As Tables 3 and 4 show, on various values of $m (m = 1.2, m = 2, m = 3, m = 4, m = 6, m = 8, m = 10, m = 12)$, V_{ZL} can find the optimal number of cluster three for seeds data and the optimal number of cluster two or three for IRIS data.

From these experiments, the proposed index V_{ZL} can correctly recognize the optimal c^* on different data set with noisy points. Furthermore, on various values of m, V_{ZL} can correctly recognize the optimal c^*, even $m = 12$. Through the different data sets, the proposed index V_{ZL} has high ability to produce a good cluster number.

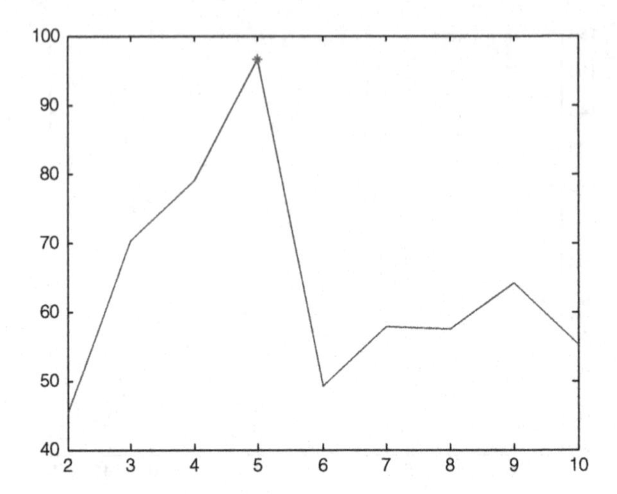

Fig. 7 The value of V_{ZL} on data 2

Table 3 The value of V_{ZL} on seeds data

	$c=2$	$c=3$	$c=4$	$c=5$	$c=6$	$c=7$	$c=8$	$c=9$
$m=1.2$	70.6561	**76.9277**	62.2338	56.1791	58.9467	67.7101	71.4378	70.9161
$m=2$	71.9818	**75.8838**	70.9837	46.3947	44.5817	49.6729	51.7318	56.0699
$m=3$	71.2123	**74.4536**	64.6746	29.1977	28.9315	32.0267	31.9466	32.4741
$m=4$	70.0373	**72.1899**	41.6797	16.1123	17.8263	10.9093	10.6767	8.7589
$m=6$	65.3121	**69.5245**	20.0180	6.7783	7.5694	3.8843	8.8120	6.4058
$m=8$	62.8408	**67.8621**	3.9431	7.4931	5.1337	2.5886	4.3199	4.2239
$m=10$	61.3454	**67.0459**	10.6134	3.8756	1.7114	2.7537	4.8333	3.3796
$m=12$	58.9493	**66.7354**	3.3065	13.9635	2.2590	2.9728	3.9464	6.2200

Table 4 The value of V_{ZL} on IRIS data

	$c=2$	$c=3$	$c=4$	$c=5$	$c=6$	$c=7$	$c=8$	$c=9$
$m=1.2$	**68.7779**	*62.3513*	56.5978	38.5257	51.1148	38.2301	40.2879	45.2736
$m=2$	**68.2835**	*61.5637*	53.1354	35.9002	43.4019	34.7778	45.8766	32.5361
$m=3$	**68.1357**	*59.1015*	24.2753	39.3635	26.4340	21.2548	17.0658	5.4151
$m=4$	**67.8236**	*56.3635*	46.6811	17.7812	26.1044	20.1522	11.8655	6.1299
$m=6$	**67.7235**	*50.7925*	43.4952	13.7001	3.2347	3.6853	4.3485	6.7215
$m=8$	**67.7004**	*47.2231*	35.6833	16.7906	2.9151	3.4913	4.7591	6.6698
$m=10$	**67.6819**	*46.0136*	6.2793	13.1404	2.5431	3.3387	4.1656	5.7321
$m=12$	**67.6653**	*42.5963*	2.2489	2.0596	1.8784	4.0825	4.9803	4.7946

5 Conclusion

The cluster validity index has been used to search for the optimal number of clusters when the number of clusters is not known a priori. In this paper, a new cluster validity index V_{ZL} for the fuzzy c-means algorithm has been proposed. It is defined by Z-type membership function and monotone linear function. Through the different data sets and both fixed m and varying m, the results of experimental tests show that it has stronger robustness of m than other indices, and it can correctly deal with some data with outliers.

References

1. Xingbo, G.: Fuzzy clustering analysis and application. Xian university of electronic science and technolog press, Xian (2004)
2. Zadeh, L.A.: Inf. control. Fuzzy sets **15**(8), 338–353 (1965)
3. Fukuyama, Y., Sugeno, M.: A new method of choosing the number of clusters for the Fuzzy C-means method. Fuzzy sets Sys. **15**(9), 12–19 (1989)
4. Xie, X.L., Beni, G.: A validity measure for fuzzy clustering. Pattern Anal. Mach. Intell. **13**(8), 841–847 (1991)
5. Wu, K.L., Yang, M.S.: A cluster validity index for fuzzy clustering. Pattern Recogn. Lett. **25**(9), 1275–1291 (2004)
6. Zahid, N., Limouri, M., Essaid, A.: A new cluster-validity for fuzzy clustering. Pattern Recogn. **32**(6), 1089–1097 (1999)
7. Shieh, H.L., Chang P.L.: A new robust validity index for fuzzy clustering algorithm. International conference on Industrial Engineering and, Engineering Management, pp. 767–771 (2010)
8. Bensaid, A.M., Hall, L.O., Bezdek, J.C., Clarke, L.P., Silbiger, M.L., Arrington, J.A., Murtagh, R.F.: Validity-guided (Re) Clustering with applications to image segmentation. IEEE Trans. Fuzzy Sys. **4**, 112–123 (1996)
9. Li, B.N.: Fuzzy mathematics, pp. 56–207. Hefei industry press, Hefei (2007)
10. Kim, D.W., Lee, K.H.: Fuzzy cluster validity index based on inter-cluster proximity. Pattern Recogn. Lett. **24**, 2561–2574 (2003)

Regional Economic Forecasting Combination Model Based on RAR+SVR

Da-rong Luo, Kai-zhong Guo and Hao-ran Huang

Abstract Regional economy has become a critical part in national economy system. Mastering its change is important for national economic decision-making. Yet many economic variables in regional economy system have characteristics of nonlinearity and instability, the result of forecasting achieved by traditional linear modeling and predicting technology doesn't reach the demand of accuracy. Thus, a combination model based on Residual Auto-regressive and Support Vector Regression is proposed in this chapter. In the model the linear part of time series will be fitted by means of Residual Auto-regressive first, then the nonlinear part included in the residual will be draw by means of Support Vector Regression. The combination model helps to increase the accuracy of forecasting in regional economy system. At the end, a prediction of GDP in Guangdong province shows the efficiency of the model.

Keywords Regional economy · Forecasting · Support vector regression · Time series analysis.

1 Introduction

The regional economy is a kind of complex economic system. The prediction of its developing trend has important guiding significance for the sustainable development of the regional economy, and has become a hot area of economic research [1–4]. Compared with macro-economy, the variables of regional economy system have characteristics of high nonlinearity and great fluctuation, lead to a lot of traditional linear modeling and forecasting technology are difficult to be applicable.

D. Luo (✉)
College of Economics and Management, Wuyi University, Jiangmen 529020, People's Republic of China
e-mail: dr_luo@126.com

K. Guo · H. Huang
School of Management, Guangdong University of Technology, Guangzhou 510520, People's Republic of China

B.-Y. Cao and H. Nasseri (eds.), *Fuzzy Information & Engineering and Operations Research & Management*, Advances in Intelligent Systems and Computing 211,
DOI: 10.1007/978-3-642-38667-1_32,

Time series analysis (TSA) [5–8] is an important forecasting method in complex economic system, but usually TSA fits and estimates data on the basis of the linear model, and seems unable to reach the need of fitting time series which has non-linear characteristic. To make up for the deficiency that TSA has, a forecast method based on TSA combining with neural network has been proposed [9]. The method fits the linear part in a time series by TSA model and fits the non-linear part by neural network that compensates the error from linear part fitting, thus has improved the accuracy of the fitting. However, the method really has a stronger ability for portraying the non-linear characteristic of a time series by using neural network, but it has defects of over-estimated and local optimization, and also sets up on the principle of empirical risk minimization, so as to lay particular emphasis on fitting but not its generalization ability when forecasting.

On the other hand, Support Vector Regression (SVR) is a method of machine learning with high generalization ability [1, 10, 11]. The method is an algorithm based on structural risk minimization principle, overcoming the deficiency of neural network method, having better ability for fitting non-linear part of time series, and having higher generalization ability. For this reason, considering the characteristics of high nonlinearity and great volatility that variables in regional economy system have, this chapter proposes a regional economic forecasting combination model based on RAR and SVR. The model adopts RAR fitting linear part of time series, SVR fitting non-linear part of time series, thus compensates the error of fitting by RAR.

2 Residual Auto-Regressive Model

RAR model is a kind of common model in time series analysis. Compared with other models in time series analysis, it has advantages that model can be interpreted easily and visually and can make the most of the residual, the form of RAR is:

$$\begin{cases} x_t = T_t + S_t + \gamma_t, \\ \gamma_t = \sum\limits_{j=1}^{p} \varphi_j \gamma_{t-j} + a_t, \\ E(a_t) = 0, Var(a_t) = \sigma^2, Cov(a_t, a_{t-i}) = 0, \forall i \geq 1, \end{cases} \tag{1}$$

where T_t means the fitting of the trend that time series has, S_t means the season value given in advance, γ_t is the residual. If a trend exists in time series, $T_t = \beta_0 + \sum_{j=1}^{k} \beta_i x_{t-i}$; Otherwise, if a season period exists in time series, $T_t = \alpha_0 + \sum_{j=1}^{k} \alpha_i x_{t-im}$, and is the constant period.

RAR model supposes that time series is composed of linear part T_t and non-linear part γ_t. T_t is the main part of time series. γ_t represents the non-linear part. To improve the accuracy of fitting, RAR extracts the information among γ_t again using similar linear way.

3 SVR Model

SVR is a kind of machine learning method based on statistical theory. It has very strong ability of non-linear modeling and high performance generalization. The basic thought [11] of SVR is that data in input space is mapped upon to the feature space of high-dimensional through particular nonlinear function $\phi(\cdot)$, thus the optimal classification hyperplane can be achieved to classify the transformed data in feature space and the regression function $f(x) = w^T \phi(x) + b$ can be obtained. If is insensitive loss function, $f(x)$ can be turned into the optimization problem as follows:

$$\min_{w,b,\xi,\xi'} \left(\frac{1}{2} \|w\|^2 + C \sum_{i=1}^{n} (\xi_i + \xi_i') \right), \tag{2}$$

$$s.t. \begin{cases} < w, \phi(x_i) > +b - y_i \leq \varepsilon + \xi_i, i = 1, 2, \ldots, n, \\ - < w, \phi(x_i) > -b + y_i \leq \varepsilon + \xi_i, i = 1, 2, \ldots, n, \\ \xi_i \geq 0, \xi_i' \geq 0, i = 1, 2, \ldots, n \end{cases} \tag{3}$$

where is the weight vector of the hyperplane, C is the penalty factor, b is the threshold value, ε is a parameter value of insensitive loss function, is the amount of sample data, $\xi_i$and $\xi_i{}'$ are slack variables, $< x, y >$ means the inner product between vector x and vector y. The optimization question can be transformed to a quadratic optimization problem with Lagrangian [1]:

$$\max_{\alpha,\alpha'} \left(-\frac{1}{2} \sum_{i,j=1}^{n} (\alpha_i' - \alpha_i) \left(\alpha_j' - \alpha_j\right) K(x_i, x_j) - \varepsilon \sum_{i=1}^{n} (\alpha_i' + \alpha_i) + \sum_{i=1}^{n} y_i (\alpha_i' - \alpha_i) \right), \tag{4}$$

$$s.t. \begin{cases} \sum_{i=1}^{n} (\alpha_i' - \alpha_i) = 0, \\ 0 < \alpha_i, \alpha_i \leq C, i = 1, 2, \ldots, n, \end{cases} \tag{5}$$

where the nuclear function $K(x_1, x_j)$ is $< \phi(x_i), \phi(x_i) >$. After solving the optimization question mentioned above, α_i and α_i' can be achieved. Hence the optimum solution of original problem can be obtained and the final regression function is as follow:

$$f(x) = \sum_{i=1}^{n} \left(\alpha_i - \alpha_i'\right) K(x_i, x_i) + b, \tag{6}$$

when $\alpha_i \in (0, C)$, $b = y_i - \sum_{j=1^n} \left(\alpha_i - \alpha_i'\right) K(x_i, x_i) - \varepsilon$; when $\alpha_i{}' \in (0, C)$, $b = y_i - \sum_{j=1^n} \left(\alpha_i - \alpha_i'\right) K(x_i, x_i) + \varepsilon$.

4 Forecasting Model and Algorithm Based on RAR and SVR

4.1 Forecasting Model

In traditional RAR model, the linear method is still adopted while fitting residual series. Due to the residual series includes non-linear part of original time series $\{x_{i-1}, x_{i-2}, \ldots, x_{i-p}\}$, it is obvious that the non-linear information of residual series can't be draw completely by linear fitting method. Therefore SVR that possesses higher exploring non-linear information ability is adopted in place of the linear fitting method. Supposing there is a time series, forecasting model combined with RAR and SVR is set up as follow:

$$\begin{cases} x_t = T_t + S_t + \gamma_t, \\ \gamma_t = \gamma_t^(SVR) + a_t, \end{cases} \tag{7}$$

where is the fitting value at the moment t; If time series has obvious trend feature, $T_t = \beta_0 + \sum_{i=1}^{k} \beta_i x_{t-i}$; If time series has season period feature, $T_t = \alpha_0 + \sum_{j=1}^{l} \alpha_i x_{t-im}$, the value of the error, will be replaced by the fitting value obtained by SVR method :

$$\gamma_t^{(SVR)} = \sum_{i=t-p}^{t-1} \left(\alpha_i - \alpha_i'\right) K\left(x_i, x\right) + b, \tag{8}$$

where, if $\alpha_i \in (0, C)$, $b = y_i - \sum_{j=t-p}^{t-1} \left(\alpha_i - \alpha_i'\right) K\left(x_j, x_i\right) - \varepsilon$; and if $\alpha_i' \in (0, C)$, $b = y_i - \sum_{j=t-p}^{t-1} \left(\alpha_i - \alpha_i'\right) K\left(x_j, x_i\right) + \varepsilon$.

4.2 Forecasting Algorithm Based on RAR and SVR

Supposing there is a time series $\{x_{i-1}, x_{i-2}, \ldots, x_{i-p}\}$, after sample data set $\{x_{i-1}, x_{i-2}, \ldots, x_{i-p}\}$ is input, the forecasting algorithm procedure can be described as follows:

Step 1: Fit time series by RAR model using Maximum likelihood estimation and initial model, $\widehat{x_t} = T_t + S_t + \widehat{\gamma_t}$ can be obtained, then calculate the residual $\{\widehat{\gamma_t}\}$;
Step 2: Examine de-noising of residual after first fitting, if the residual is white noise, stop algorithms, output forecasting models; Otherwise, go to Step 3;
Step 3: Calculate residual fitting value using SVR method, update forecasting model as $\widehat{x_t} = T_t + S_t + \widehat{\gamma_t}^{(SVR)}$.

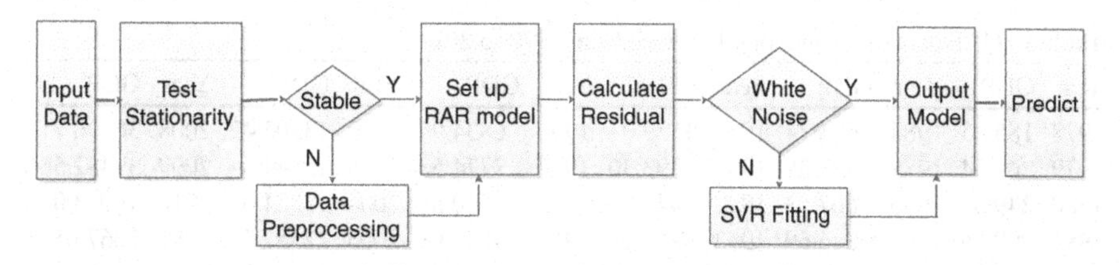

Fig. 1 Flow chart of the forecasting algorithm.

Export: Output the fitting value $\widehat{x}_t$.

The flow chart of the algorithm is shown as in Fig. 1:

4.3 Performance Evaluation of Forecasting Model

To evaluate the performance of the forecasting model, mean absolute error (MAE) is adopted while assessing, the definition of MAE is as follows:

$$MAE = \frac{1}{n}\sum_{i=1}^{n} |\widehat{x}_i - x_i|, \tag{9}$$

where is the actual value and $\widehat{x}_i$ is the fitting value. MAE reflects how big the error value is. the more little its value , the more accurate the predicting result is.

5 Regional Economic Forecasting Model Empirical Analysis

In order to verify the efficiency of the forecasting model mentioned above, a prediction of Guangdong Province GDP will be conducted in this chapter. Test data includes GDP data from 1978 to 2011 in Guangdong Province is shown in Table 1.

5.1 Data Preprocessing

Usually a lot of time series in the field of regional economy are obviously non-stationary. In order to avoid the non-stationary effect bringing more complexity of model estimating and predicting, data will usually be tested and then be smoothed by logarithmic or difference way.

According to Table 1, data map and autocorrelogram can be obtained in Figs. 2 and 3. On the basis of analyzing these two figures, we can conclude that the series does not meet the conditions that values of mean and variance are constant, and its fluc-

Table 1 GDP data of Guangdong Province from 1978 to 2011

Year	GDP	Year	GDP	Year	GDP	Year	GDP	Year	GDP	Year	GDP
1978	185.85	1984	458.74	1990	1559.03	1996	6834.97	2002	13502.42	2008	36796.71
1979	209.34	1985	577.38	1991	1893.30	1997	7774.53	2003	15844.64	2009	39482.56
1980	249.65	1986	667.53	1992	2447.54	1998	8530.88	2004	18864.62	2010	46013.06
1981	290.36	1987	846.69	1993	3469.28	1999	9250.68	2005	22557.37	2011	52673.59
1982	339.92	1988	1155.37	1994	4619.02	2000	10741.25	2006	26587.76		
1983	368.75	1989	1381.39	1995	5933.05	2001	12039.25	2007	31777.01		

Unit: 100 million yuan

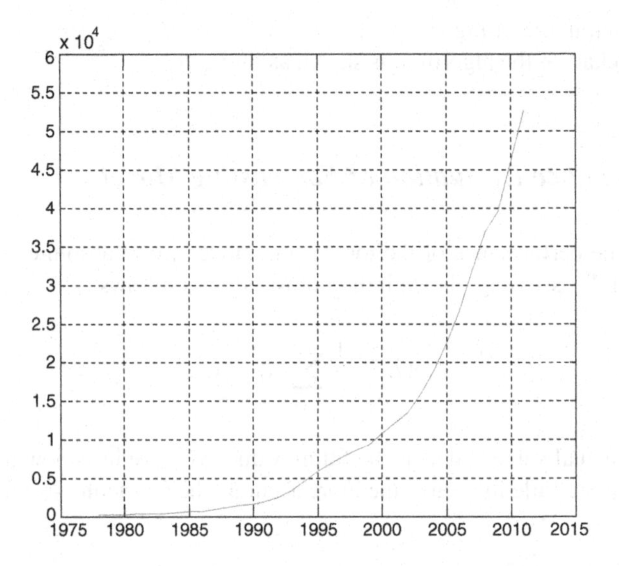

Fig. 2 Data map of GDP data.

tuating range is relatively wide. So, logarithmic smoothing and then first difference is adopted, the graph of the processed data $\{x_t\}$ is in Fig. 4. From autocorrelogram of $\{x_t\}$ in Fig. 5, we can see that it possesses strong short time relevance and can be thought to be stationary.

5.2 Predicting by Combination Model

We take datas from 1978 to 2007 as training data and datas from 2008 to 2011 as test data. Obviously GDP data has no season period feature, so $S_t = 0$. The parameter T_t can be obtained by maximum likelihood method, thus the fitting model is $x_t = 1.020175469x_{t-1} + \gamma_t$.

The model can be intuitively interpreted that each logarithmic value of GDP is 1.020175469 times of last year. At the same time, influenced by a great deal of

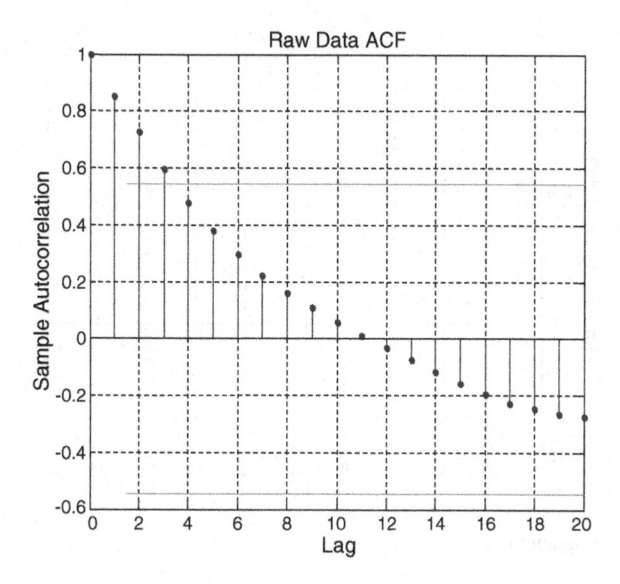

Fig. 3 Autocorrelogram of GDP data

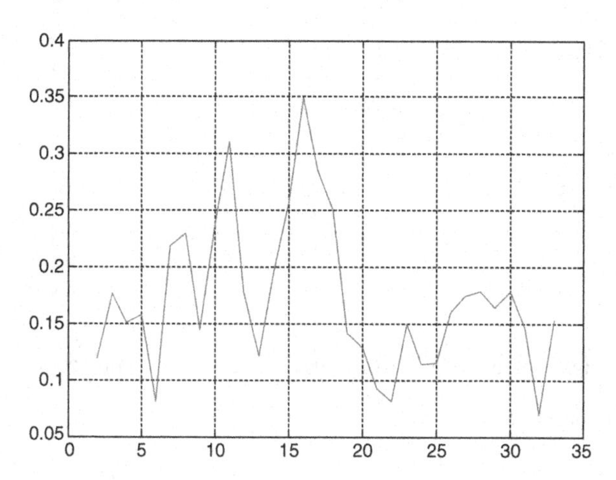

Fig. 4 GDP data after smoothing and difference

factors, logarithm series of GDP is autocorrelative. It can be proved by white noise test of residual. Then, fitting residual by SVR method is the next step, the result is as shown in Fig. 6. the final predicting value compensated with SVR is shown in Fig. 7. In Fig.7, the full line shows the actual data, the asterisk shows the predicting data. In this chapter radial basis function $K\left(x_i, x_j\right) = exp\left\{-\frac{|x_i - x_j|^2}{\sigma^2}\right\}$ is used as the nuclear function of SVR algorithm and parameters C and σ have the value of 100 and 0.001, respectively.

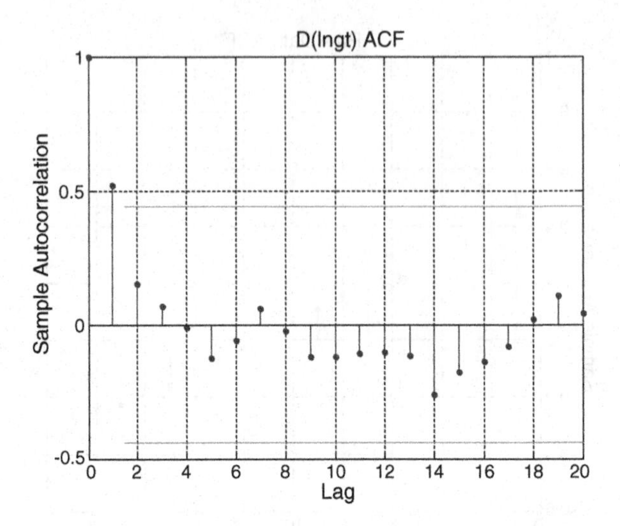

Fig. 5 Autocorrelogram of x_t

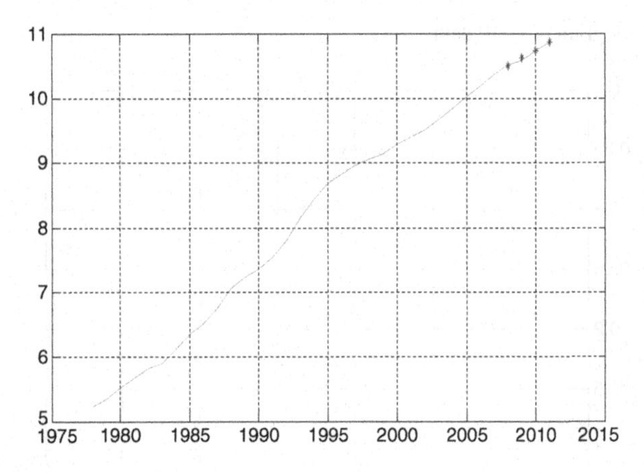

Fig. 6 Fitting residual by SVR

5.3 Result Analysis

In order to further illustrate the efficiency of the algorithm, results of AR, RAR and SVR model are compared, the logarithmic predicting value of each model and logarithmic actual value are shown in Table 2.

From Table 2, we can know that the overall predicting accuracy of combination model based on RAR and SVR is more than 98.8 %, higher than the accuracy that is obtained simply by RAR or SVR model. And the predicting accuracy of 2010, 2011 is up to 99.8 %. By comparison of the results, when economic variable is of non-linearity and fluctuation, predicting using traditional model doesn't work anymore.

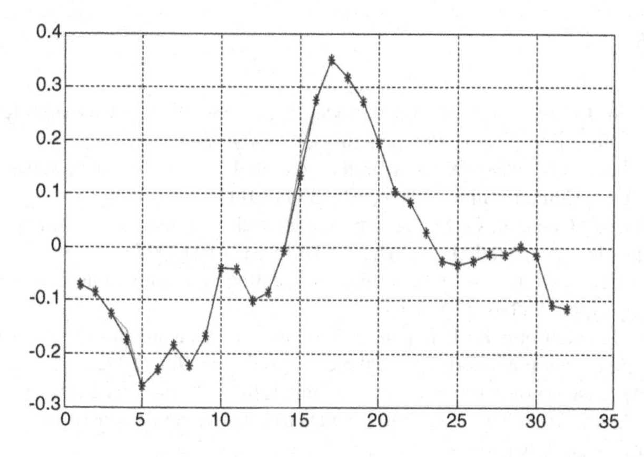

Fig. 7 Actual and predicting value

Table 2 MAE comparison between AR, RAR, SVR and SVR+RAR

Year	Actual value	AR	RAR	SVR	SVR+RAR
2008	10.513	9.513	9.531	10.537	10.503
2009	10.584	9.705	9.723	10.681	10.621
2010	10.737	9.901	9.919	10.812	10.737
2011	10.872	10.101	10.119	10.944	10.872
MAE		0.8713	0.8532	0.0669	0.0119

This proves the efficiency and superiority of the combination model in this chapter. Forecasting using the RAR+SVR combination model, GDP of Guangdong Province from 2012 to 2014 will be 57414.2131, 62696.3207,68276.2933 respectively.

6 Conclusion

Considering the characteristics of strong nonlinearity and great fluctuation that a lot of economic variables in regional economy field have, this chapter has proposed the combination model based on SVR and RAR, keeping the advantage of RAR model that can be explained easily and intuitively, improving the ability of fully excavating non-linear information at the same time. Example and comparison shows that such model has relatively higher predicting accuracy. In addition, this model is suitable for not only short-term but also long-term forecasting. For combined with SVR, the more nonlinearity series have, the more effectively the model predicts.

Acknowledgments Thanks to the support by Guangdong Provincial Natural Science Foundation of China (No.S2011010006103) and 2012 Jiangmen Industrial Technology Research and Development Projects (No.[2012]156).

References

1. Xiao, J.H., Lin, J., Liu, J.: A SVR-based model for regional economy medium-term and long-term forecast. Sys. Eng. Theory Practice **26**(4), 97–103 (2006)
2. Zhou, Z.Y., Duan, J.N., Chen, X.: Research on regional economy prediction based on support vector machines. Comput. Simul. **28**(4), 375–378 (2011)
3. Fei, Z.H., Li, B., Chen, X.X.: Mathematic model and case analysis of strategy for regional economic development. Math. Pract. Theory **37**(15), 12–19 (2007)
4. Deng, H.Z., Chi, Y., Tan, Y.J.: The nonlinear modeling and simulation in economic system. Comput. Eng. Appl. **37**(18), 7–9 (2001)
5. Chakraborty, K., Mehrotra, K., Kmohan, C., Ranka, S.: Forecasting the behavior of multivariate time series using neural networks. Neural Networks **5**, 961–970 (1992)
6. Wang, Y.: Application time series analysis. China Renmin University Press, Beijing (2008)
7. Chen, X.J.: Application of TSA in GDP predicts in Guangdong Province. Sun Yat-sen University, Master Dissertation (2009)
8. Zhu, B.Z., Lin, J.: A novel nonlinear ensemble forecasting model incorporating ARIMA and LSSVM. Math. Prac. Theory **39**(12), 34–40 (2009)
9. Hu, S.S., Zhang, Z.D.: Fault prediction for nonlinear time series based on neural network. Acta Automatica Sinica **33**(7), 744–748 (2007)
10. Vapnik, V.: The nature of statistical learning theory. Springer, New York (1995)
11. Muller, K.R., Mika, S., Ratsch, G., et al.: An tntroduction to kernel-based learning algorithms. IEEE Trans. on Neural Networks **12**(2), 181–201 (2001)

Part V
Systems and Algorithm

On Cycle Controllable Matrices over Antirings

Jing Jiang, Xin-an Tian and Lan Shu

Abstract In this paper, cycle controllable matrices are defined over an arbitrary commutative antiring L. Some properties for cycle controllable matrices are established, a necessary and sufficient condition for a cycle controllable matrix which has a given nilpotent index is obtained. Finally, expressions for a cycle controllable matrix as a sum of square-zero matrices are shown.

Keywords Antiring · Cycle controllable matrix · Nilpotent matrix

1 Introduction

In the field of applications, semirings are useful tools in diverse domains such as design of switching circuits, automata theory, information systems, dynamic programming, decision theory, and so on [1]. But the techniques of matrices over semirings play a very important role in optimization theory, models, models of discrete event networks, grapy theory, and so on. Therefore, the study of matrices over semi-

J. Jiang (✉)
Mathematics and Statistics College, Chongqing University of Arts and Sciences, Chongqing 402160, China
e-mail: jiangjingstu@163.com

J. Jiang
School of Mathematical Sciences, University of Electronic Science and Technology of China, Chengdu 611731, Sichuan , China

X. Tian
Modern Education Technology Center, Chongqing University of Arts and Sciences, Chongqing 402160, China

L. Shu
Mathematics and Statistics College, Chongqing University of Arts and Sciences, Chongqing 402160, China

B.-Y. Cao and H. Nasseri (eds.), *Fuzzy Information & Engineering and Operations Research & Management*, Advances in Intelligent Systems and Computing 211,
DOI: 10.1007/978-3-642-38667-1_33,

rings has a long history. Especially, many authors have studied nilpotent matrix over some special semirings such as antiring, path algebra, lattice and incline. Nilpotent matrices represent acyclic fuzzy graphs used to represent consistent systems, and in general, acyclic graphs are important in the representation of precedence relations (see e.g. [2, 6, 7]). Thus, the study of the nilpotent matrix is valuable.

In [3, 4], Tan considered the nilpotency of matrices over commutative antirings. From Tan‘s paper, many properties obtained were discussed in terms of the property that $a_{i_1i_2}a_{i_2i_3}\cdots a_{i_mi_1}$ is nilpotent for any positive integer m and any $i_1, i_2, \cdots, i_m \in N$. According to this, these matrices with the property are defined as cycle controllable matrices in this chapter. Some properties for cycle controllable matrices are established, a necessary and sufficient condition for a cycle controllable matrix which has a given nilpotent index is obtained. Finally, expressions for a cycle controllable matrix as a sum of square-zero matrices are shown.

2 Definitions and Preliminaries

In this section, we will give some definitions and lemmas. For convenience, we use N to denote the set $\{1, 2, \cdots, n\}$ and use Z^+ to denote the set of positive integers.

A semiring is an algebraic system $(L, +, \bullet)$ in which $(L, +)$ is a commutative monoid with identity element 0 and $(L, \bullet)$ is another monoid with identity element 1, connected by ring-like distributivity. Also, $0r = r0 = 0$ for all $r \in L$ and $0 \neq 1$. A semiring L is called an antiring if it is zerosumfree, that is, if the condition $a+b = 0$ implies that $a = b = 0$ for all $a, b \in L$. Antirings are abundant: for examples, every Boolean algebra, every distributive lattice and any incline are commutative antirings. An antirings is called entire if $ab = 0$ implies that either $a = 0$ or $b = 0$. For example, the set Z^+ of nonnegative integers with the usual operations of addition and multiplication of integers is a commutative entire antiring.

In this paper, the semiring $(L, +, \bullet)$ is always supposed to be a commutative antiring. Let $M_{m\times n}(L)$ be the set of all $m \times n$ matrices over L. Especially, put $M_n(L) = M_{n\times n}(L)$. For any $A \in M_{m\times n}(L)$, we denote by a_{ij} or A_{ij} the element of L which stands in the (i, j) -entry of$A$. Also, we denote by $E(A)$ the set that contains all nonzero entries in matrix A and by $|E(A)|$ the number of the set$E(A)$.

For any $A, B \in M_{m\times n}(L)$ and$C \in M_{n\times l}(L)$, we define:

$$A + B = (a_{ij} + b_{ij})_{m\times n}; \tag{1}$$

$$AC = \left(\sum_{k=1}^{n} a_{ik}c_{kj}\right)_{m\times l}. \tag{2}$$

The $(i, j)th$ entry of A^l is denoted by $a_{ij}^{(l)} (l \in Z^+)$, and obviously

$$a_{ij}^{(l)} = \sum_{1 \le i_1, i_2, \cdots, i_{l-1} \le n} a_{ii_1} a_{i_1 i_2} \cdots a_{i_{l-1} j}. \tag{3}$$

It is easy to see that $(M_n(L), +, \bullet)$ is a semiring. In particular, if L is a commutative antiring then $(M_n(L), +, \bullet)$ is an antiring.

Let L be a semiring, an element a in L is said to be nilpotent if $a^k = 0$ for some positive integer k. Let $A \in M_n(L)$, A is said to be nilpotent if there exists a positive integer k such that $A^k = 0$. The least positive integer k satisfying $A^k = 0$ is called the nilpotent index of A, and denoted by $h(A)$. For any $A \in M_n(L)$, we denote by det A the permanent of A (the definition of the permanent for A can be seen in Definition 2.2 [3]) and by $D(A)$ the directed graph of A (the definition of the directed graph $D(A)$ for A can be found in [5]). Also, some definitions about the principal submatrix, the permanental adjoint matrix $adjA$, the submatrix $A(U/V)$, the submatrix $A[U/V]$ and $A(i \Rightarrow j)$ can be found in [4].

Definition 2.1. A matrix $R = (r_{ij}) \in M_n(L)$ is said to be a cycle controllable matrix, if $r_{ii}^{(k)}$ is nilpotent for any $i \in N$ and any $k \in Z^+$.

Definition 2.2. A semiring L is called join-irreducible if $a = a_1 + a_2$ implies that $a_1 = a$ and $a_2 = a$ for all $a, a_1, a_2 \in L$.

Remark. It is clear that if a semiring L is called join-irreducible then L is an antiring.

Lemma 2.1 Let [4]. Let L be a commutative semiring, $r, r_1, r_2, \cdots, r_m \in L$. Then

(1) if $r$is nilpotent then kr and rk are nilpotent for any k in L;
(2) if L is a commutative antiring, then $\sum_{i=1}^{m} r_i$ is nilpotent if and only if $r, r_1, r_2, \cdots, r_m$ are nilpotent.

Lemma 2.2 Let [4]. Let L be a commutative semiring and $A, B \in M_n(L)$. Then

(1) A is nilpotent if and only if $PAP^T$is nilpotent for any $n \times n$ permutation matrix P, and $h(A) = h(PAP^T)$;
(2) if L is a commutative antiring and $A + B$ is nilpotent, then A and B are nilpotent.

3 Basic Properties of Cycle Controllable Matrices

In this section, some properties of cycle controllable matrices over a commutative antiring L with nonzero nilpotent elements are posed.

Proposition 3.1. If $R = (r_{ij}) \in M_n(L)$ is a cycle controllable matrix, then $R$is nilpotent.

Proof. Let $T = r_{ii_1} r_{i_1 i_2} \cdots r_{i_{n-1} j}$ be any term of $r_{ij}^{(n)}$, where $1 \le i, i_1, i_2, \cdots, i_{n-1}, j \le n$. Since the number of indices in T is greater than n, there must be two indices i_u and i_v such that $i_u = i_v$ for some $u, v (u < v)$, and so $r_{i_u i_{u+1}} \cdots r_{i_{v-1} i_v} =$

$r_{i_u i_{u+1}} \cdots r_{i_{v-1} i_u}$. But $r_{i_u i_{u+1}} \cdots r_{i_{v-1} i_u}$ is a term of $r_{i_u i_u}^{(v-u)}$ and $r_{i_u i_u}^{(v-u)}$ is nilpotent (since R is a cycle controllable matrix), we have $r_{i_u i_{u+1}} \cdots r_{i_{v-1} i_u}$ is nilpotent (by Lemma 2.1(2)). Then T is nilpotent (by Lemma 2.1(1)), and so $r_{ij}^{(n)}$ is nilpotent (by Lemma 2.1(2)). Therefore, R^n is nilpotent, and so R is nilpotent. This proves the proposition.

Proposition 3.2. If $R = (r_{ij}) \in M_n(L)$ is a cycle controllable matrix, then all principal submatrices of R are cycle controllable matrices.

Proof. By Proposition 3.1, R is nilpotent. Let B be any principal submatrix with order k of R. Then, there exists an $n \times n$ permutation matrix P such that

$$PRP^T = \begin{pmatrix} B & C \\ E & D \end{pmatrix} = \begin{pmatrix} B & 0 \\ 0 & 0 \end{pmatrix} + \begin{pmatrix} 0 & C \\ E & D \end{pmatrix}$$

where $C \in M_{k \times (n-k)}(L)$, $E \in M_{(n-k) \times k}(L)$ and $D \in M_{n-k}(L)$. Since PRP^T is nilpotent (by Lemma 2.2(1)), the matrix $\begin{pmatrix} B & 0 \\ 0 & 0 \end{pmatrix}$ is nilpotent (by Lemma 2.2(2)), and so B is nilpotent. Then, $B^l = 0$ for some positive integer l, and so $(B^m)^l = B^{ml} = 0$ for any positive integer m. Therefore, $B_{ii}^{(ml)} = 0$ for any $i \in N$. But $(B^m)_{ii}^{(l)}$ is some term of $B_{ii}^{(ml)}$, we have that $(B^m)_{ii}^{(l)} = 0$ (because L is a commutative antiring), that is, $B_{ii}^{(m)}$ is nilpotent for any $i \in N$ and any $m \in Z^+$. Therefore, B is a cycle controllable matrix. This proves the proposition.

Proposition 3.3. Let $R = (r_{ij}) \in M_n(L)$. If every element of R is nilpotent, then the matrix R is a cycle controllable matrix.

Proof. Since every element of R is nilpotent, we have $r_{ii}^{(k)}$ is nilpotent for any $i \in N$ and any $k \in Z^+$ (by Lemma 2.1). Thus, R is a cycle controllable matrix. This proves the proposition.

Proposition 3.4. Let $R = (r_{ij}) \in M_n(L)$ is a cycle controllable matrix. Then the matrix $Radj R$ is a cycle controllable matrix.

Proof. Let $B = Radj R$. Then $b_{ij} = \sum_{k \in N} r_{ik} \det R(j/k) = \det R(i \Rightarrow j)$ for all $i, j \in N$.

In the following, we will show that b_{ij} is nilpotent.

(1) For any $i \in N$, we have $b_{ii} = \det R(i \Rightarrow i) = \det R$, and we consider any term $T_\sigma = r_{1\sigma(1)} r_{2\sigma(2)} \cdots r_{n\sigma(n)}$ of $\det R$, where $\sigma \in s_n$ and s_n denotes the symmetric group of the set N. Since $\sigma^k(1) \in N$ for all positive integer k, there exist $r, s \in \{0, 1, \cdots, n\}$ such that $\sigma^r(1) = \sigma^s(1) (r < s)$ and $\sigma^r(1), \sigma^{r+1}(1), \cdots, \sigma^{s-1}(1)$ are mutually different (note that $\sigma^0(1) = 1$). Then $r_{\sigma^r(1)\sigma^{r+1}(1)} \cdots r_{\sigma^{s-1}(1)\sigma^s(1)} = r_{\sigma^r(1)\sigma^{r+1}(1)} \cdots r_{\sigma^{s-1}(1)\sigma^r(1)}$ is a factor of T_σ. But $r_{\sigma^r(1)\sigma^{r+1}(1)} \cdots r_{\sigma^{s-1}(1)\sigma^r(1)}$ is nilpotent (because R is a cycle controllable matrix), we have T_σ is nilpotent for any $\sigma \in s_n$ (by Lemma 2.1(1)). Thus, $\det R = b_{ii}$ is nilpotent (by Lemma 2.1(2)).

(2) For any $i, j \in N$ with $i \neq j$, we have $b_{ij} = \det R(i \Rightarrow j) = \sum_{\sigma \in s_n} r_{1\sigma(1)} \cdots r_{i\sigma(i)} \cdots r_{i\sigma(j)} \cdots r_{n\sigma(n)}$. For any $\sigma \in s_n$, if $\sigma^l(i) \neq j$ for all $l \geq 1$, then there must be a k such that $\sigma^k(i) = i$ with $1 \leq k \leq n$ and $i, \sigma(i), \cdots, \sigma^{k-1}(i)$ are pairwise different. Thus $r_{i\sigma(i)}r_{\sigma(i)\sigma^2(i)} \cdots r_{\sigma^{k-1}(i)i}$ is nilpotent (because R is a cycle controllable matrix). Since the product $r_{i\sigma(i)}r_{\sigma(i)\sigma^2(i)} \cdots r_{\sigma^{k-1}(i)i}$ is a factor of the product $r_{1\sigma(1)} \cdots r_{i\sigma(i)} \cdots r_{i\sigma(j)} \cdots r_{n\sigma(n)}$, we can get $r_{1\sigma(1)} \cdots r_{i\sigma(i)} \cdots r_{i\sigma(j)} \cdots r_{n\sigma(n)}$ is nilpotent (by Lemma 2.1(1)).
If there exists a positive integer l such that $\sigma^l(i) = j$, then there exists a positive integer k such that $i = \sigma^k(j)$ with $1 \leq k \leq n$ and $i, \sigma(j), \cdots, \sigma^{k-1}(j)$ are pairwise different. Thus, $r_{i\sigma(j)}r_{\sigma(j)\sigma^2(j)} \cdots r_{\sigma^{k-1}(j)i}$ is nilpotent (because R is a cycle controllable matrix). But the product $r_{i\sigma(j)}r_{\sigma(j)\sigma^2(j)} \cdots r_{\sigma^{k-1}(j)i}$ is a factor of the product $r_{1\sigma(1)} \cdots r_{i\sigma(i)} \cdots r_{i\sigma(j)} \cdots r_{n\sigma(n)}$, again, we have $r_{1\sigma(1)} \cdots r_{i\sigma(i)} \cdots r_{i\sigma(j)} \cdots r_{n\sigma(n)}$ is nilpotent (by Lemma 2.1(1)).
Consequently, we obtain b_{ij} is nilpotent. Therefore, the matrix $B = RadjR$ is a cycle controllable matrix (by Proposition 3.3). This proves the proposition.
In the following, a necessary and sufficient condition for a cycle controllable matrix over a commutative antiring which has a given nilpotent index is obtained.

Theorem 3.1. If $R = (r_{ij}) \in M_n(L)$ is a cycle controllable matrix, then $R^m = 0$ if and only if $B^m = 0$ for any principal submatrix B of R.

Proof. "⇒". Let B be any principal submatrix of order r of R. Then there must be an $n \times n$ permutation matrix P such that

$$PRP^T = \begin{pmatrix} B & C \\ E & D \end{pmatrix} = \begin{pmatrix} B & 0 \\ 0 & 0 \end{pmatrix} + \begin{pmatrix} 0 & C \\ E & D \end{pmatrix}$$

where $C \in M_{r\times(n-r)}(L)$, $E \in M_{(n-r)\times r}(L)$ and $D \in M_{n-r}(L)$. Since $R^m = 0$, we have $PR^mP^T = \begin{pmatrix} B & C \\ E & D \end{pmatrix}^m = 0$. But $(M_n(L), +, \bullet)$ is an antiring, we can get $B^m = 0$.

"⇐". The proof is obvious. This completes the proof.

Theorem 3.2. If $R = (r_{ij}) \in M_n(L)$ is a cycle controllable matrix, then $h\,(R) = r$ if and only if $B^r = 0$ for any principal submatrix B of$R$ and $C^{r-1} \neq 0$ for some principal submatrix $C$of R.

Proof. "⇒". Since $h\,(R) = r$, we have $R^r = 0$ and $R^{r-1} \neq 0$. By Theorem 3.1, $B^r = 0$ for any principal submatrix B of R and $C^{r-1} \neq 0$ for some principal submatrix C of R.

"⇐". By the hypothesis and Theorem 3.1, we obtain $R^r = 0$ and $R^{r-1} \neq 0$, thus $h(R) = r$. This completes the proof.

At the end of this section, expressions of a cycle controllable matrix are discussed. In [5], for any $n \times n$ nilpotent matrix $A \in M_n(L)$ over an entire antiring L without nonzero nilpotent elements,A can be written as a sum of $\lceil \log_2 n \rceil$ square-zero matrices. So there must be two matrices B_i and B_j such that $B_i = B_j$ for some matrix

A. In the following, we develop further results on the case which these square-zero matrices are pairwise different under some conditions. We use Dolzan's notation (see [5]) and let L be always a commutative semiring without nonzero nilpotent elements below. Then, for any cycle controllable matrix $A \in M_n(L)$, the diagonal entry of A must be equal to 0, and A is nilpotent.

Theorem 3.3. Let L be a join-irreducible and entire semiring. For any cycle controllable matrix $A = (a_{ij}) \in M_n(L)$ with $|E(A)| = m$, A can be written as $\sum_{t=1}^{k} A_t$, where $A_t(t = 1, \cdots, k) \in M_n(L)$ is a square-zero matrix (that is, $A_t^2 = 0$), $A_1, A_2, \cdots, A_k$ are pairwise different, and $\chi(D(A)) \leq k \leq |E(A)|$ $(\chi(D(A)) \leq \lceil \log_2 n \rceil)$.

Proof. By Lemma 4.1 [3] and Proposition 3.1, we can assume that a cycle controllable matrix $A \in M_n(L)$ is a strictly uppertriangular matrix. Let $\chi(D(A))$ be the least number of colors needed to color the edges of a graph $D(A)$ such that every path in $D(A)$ has no two incident edges of the same color. Obviously, $\chi(D(A)) \leq \lceil \log_2 n \rceil$ (the reason can be found in the proof of Theorem 11 [5]).

By Lemma 4 [5], all paths in the digraph corresponding to a square-zero matrix are of length at most 1. Thus, by Lemma 10 [5], it follows that every cycle controllable matrix can be written as a sum of at least $\chi(D(A))$ square-zero matrices. Since the semiring L is join-irreducible, and the most number of colors needed to color the edges of a graph $D(A)$ such that no vertex is a source and a sink of two edges of the same color is $|E(A)|$, we have any cycle controllable matrix can be written as a sum of at most $|E(A)|$ square-zero matrices. This completes the proof.

Remark 3.1. For any controllable matrix $A \in M_n(L)$ with L being a join-irreducible and entire semiring, A cannot be written as a sum of t $(t < \chi(D(A)) \leq \lceil \log_2 n \rceil$ or $t > |E(A)|)$ square-zero matrices.

Example 3.1. Let $A = \begin{pmatrix} 0 & a & b & 0 \\ 0 & 0 & c & 0 \\ 0 & 0 & 0 & d \\ 0 & 0 & 0 & 0 \end{pmatrix}$ be a 4×4 cycle controllable matrix over a join-irreducible and entire semiring. Then $\chi(D(A)) = 2$ and $|E(A)| = 4$, and so A can be written as follows,

$$A = \begin{pmatrix} 0 & a & 0 & 0 \\ 0 & 0 & 0 & 0 \\ 0 & 0 & 0 & d \\ 0 & 0 & 0 & 0 \end{pmatrix} + \begin{pmatrix} 0 & 0 & b & 0 \\ 0 & 0 & c & 0 \\ 0 & 0 & 0 & 0 \\ 0 & 0 & 0 & 0 \end{pmatrix} \quad \text{or}$$

$$A = \begin{pmatrix} 0 & a & 0 & 0 \\ 0 & 0 & 0 & 0 \\ 0 & 0 & 0 & 0 \\ 0 & 0 & 0 & 0 \end{pmatrix} + \begin{pmatrix} 0 & 0 & b & 0 \\ 0 & 0 & 0 & 0 \\ 0 & 0 & 0 & 0 \\ 0 & 0 & 0 & 0 \end{pmatrix} + \begin{pmatrix} 0 & 0 & 0 & 0 \\ 0 & 0 & c & 0 \\ 0 & 0 & 0 & 0 \\ 0 & 0 & 0 & 0 \end{pmatrix} + \begin{pmatrix} 0 & 0 & 0 & 0 \\ 0 & 0 & 0 & 0 \\ 0 & 0 & 0 & d \\ 0 & 0 & 0 & 0 \end{pmatrix}$$

At last, expressions of an $n \times n$ trace-zero matrix are further investigated.

Theorem 3.4. Let L be a join-irreducible semiring without nonzero nilpotent elements, and $A = (a_{ij}) \in M_n(L)$ be an $n \times n$ trace-zero matrix. Then A can be written as $\sum_{t=1}^{k} A_t$, where $A_t (t = 1, \cdots, k) \in M_n(L)$ is a square-zero matrix (that is, $A_t^2 = 0$), $A_1, A_2, \cdots, A_k$ are pairwise different, and $\chi(D(A)) \leq k \leq |E(A)|$ $(\chi(D(A)) \leq N(n))$.

Proof. Similarly, let $\chi(D(A))$ be the least number of colors needed to color the edges of a graph $D(A)$ such that every path in $D(A)$ has no two incident edges of the same color. Obviously, $\chi(D(A)) \leq N(n)$ (the reason can be found in Theorem 13 [5]).

From the proof of Theorem 3, we also have every trace-zero matrix can be written as a sum of at least $\chi(D(A))$ square-zero matrices and at most $|E(A)|$ square-zero matrices.

Remark 3.2. For any trace-zero matrix $A \in M_n(L)$ with L being a join-irreducible and no nonzero nilpotent elements, A cannot be written as a sum of t $(t < \chi(D(A)) \leq N(n)$ or $t > |E(A)|)$ square-zero matrices.

Example 3.2. Let $A = \begin{pmatrix} 0 & 0 & b \\ c & 0 & d \\ e & f & 0 \end{pmatrix}$ be a 3×3 trace-zero matrix over a join-irreducible semiring without nonzero nilpotent elements. Then $\chi(D(A)) = 3$ and $|E(A)| = 5$. Therefore, A can be written as follows,

$$A = \begin{pmatrix} 0 & 0 & 0 \\ c & 0 & d \\ 0 & 0 & 0 \end{pmatrix} + \begin{pmatrix} 0 & 0 & b \\ 0 & 0 & 0 \\ 0 & 0 & 0 \end{pmatrix} + \begin{pmatrix} 0 & 0 & 0 \\ 0 & 0 & 0 \\ e & f & 0 \end{pmatrix} \quad \text{or}$$

$$A = \begin{pmatrix} 0 & 0 & b \\ 0 & 0 & 0 \\ 0 & 0 & 0 \end{pmatrix} + \begin{pmatrix} 0 & 0 & 0 \\ c & 0 & 0 \\ 0 & 0 & 0 \end{pmatrix} + \begin{pmatrix} 0 & 0 & 0 \\ 0 & 0 & d \\ 0 & 0 & 0 \end{pmatrix} + \begin{pmatrix} 0 & 0 & 0 \\ 0 & 0 & 0 \\ e & 0 & 0 \end{pmatrix} + \begin{pmatrix} 0 & 0 & 0 \\ 0 & 0 & 0 \\ 0 & f & 0 \end{pmatrix}$$

4 Conclusion

In this paper, cycle controllable matrices are defined over an arbitrary commutative antiring. Some properties for cycle controllable matrices are discussed. The main results in the present chapter are the generalizations of the nilpotent matrices. Therefore, the study of cycle controlla-ble matrices is valuable.

Acknowledgments This work was supported by the Foundation of National Nature Science of China (Grant No. 11071 178) and the Fundamental Research Funds for the Central Universities.

References

1. Carre, B.: Groups and networks. Clarendon Press, Oxford (1979)
2. Harary, F.: On the consistency of precedence matrices. J. Assoc. Comput. Math **7**, 255–259 (1960)
3. Tan, Y.J.: On nilpotent matrices over antirings. Linear Algebra Appl. **429**, 1243–1253 (2008)
4. Tan, Y.J.: On nilpotency of matrices over antirings. Linear Algebra Appl. **433**, 1541–1554 (2010)
5. Dolzan, D., Oblak, P.: Invertible and nilpotent matrices over antirings. Linear Algebra Appl. **430**, 271–278 (2009)
6. Fisher, A.C., Liebmain, J.S., Nemhauser, G.L.: Computer construction of project networks. Commun. ACM **11**, 493–497 (1968)
7. Marimont, R.B.: A new method of checking the consistency of precedence matrices. J. Assoc. Comput. Math **6**, 162–171 (1959)

Control Strategy of Wastewater Treatment in SBR Method Based on Intelligence Fusion

Xian-kun Tan, Ren-ming Deng and Chao Xiao

Abstract To solve the puzzle of dissolved oxygen control for wastewater treatment in SBR method, the paper proposed a sort of intelligence fusion control algorithm. In the paper, it summarized up the main puzzles in control, proposed the intelligence fusion based control strategy, constructed the structure of controller, discussed the control algorithm, and made the simulation by means of intelligence fusion based control algorithm. The simulation curve demonstrated that the proposed control strategy would be stronger in robustness and suitable for the dissolved oxygen control of wastewater treatment. The research result shows that it is feasible and reasonable for wastewater treatment in SBR method.

Keywords Wastewater treatment · SBR · Dissolved oxygen · Intelligence fusion based control strategy

1 Introduction

Along with high speed development of China economy, the treatment of polluted water in city has become an important research subject in modern civilization metropolis. If the polluted water is directly sluiced into the rivers then the wild ecological environment would be suffered from serious destruction, and it would seriously endanger the subsistence of the humanity and various species. The method

X. Tan (✉)
School of Polytechnic, Chongqing Jiaotong University, Chongqing 400074, China
e-mail: txkcx11@163.com

R. Deng · C. Xiao
College of Automation, Chongqing University, Chongqing 400030, China
e-mail: dengrenming65106683@126.com

C. Xiao
e-mail: sngeet@163.com

B.-Y. Cao and H. Nasseri (eds.), *Fuzzy Information & Engineering and Operations Research & Management*, Advances in Intelligent Systems and Computing 211,
DOI: 10.1007/978-3-642-38667-1_34, © Springer-Verlag Berlin Heidelberg 2014

of sequencing batch reactor (SBR) is generally adopted in the process of polluted water treatment, and in which the aeration is the most important key node [1, 2]. Its function is to supply the oxygen for polluted water, blow off the gaseous fluid, make full agitation for polluted water, and enhance the mass transfer effect. In the process of charging oxygen in aeration, if the polluted water quantity is more then the needed oxygen quantity is always more, and therefore the aeration quantity is more. And vice versa, if the aeration wind quantity is not enough then the time of biochemistry processing would be postponed, and the quality of output water would not be ensured. The too much wind quantity of aeration will result in lots of energy waste, and therefore the correct control method is that the aeration wind quantity is proportionally to track the change of polluted water quantity automatically. The aeration control is the technical bottleneck and key technology node in the process of polluted water treatment. In view of the polluted water treatment is a complex process of biochemistry reaction, and it is very hard to be described by accurate mathematic model [3, 4], so it is a puzzle in control engineering. The following makes a certain exploration for control strategy.

2 Main Puzzles in Control Process

Because the sewerage bio-chemical disposing is a typical complex process, it is very difficult to describe the bio-chemical process characteristic by strict mathematics method, such as nonlinearity, time varying, randomness, fuzziness and instability. In general speaking, the cybernetics characteristic of waste water disposal process can be summarized up as the following [5, 6]. (1) unknown of time varying, randomness, and decentralization in process parameter, (2) uncertainty of lag time process, (3) serious nonlinearity in chemical reaction process, (4) correlativity among process parameters, (5) disturbance of process environment being unknown, diversity and randomness.

In the face of the characteristic mentioned above, the conventional control strategy is difficult to carry through the effective control for waste water disposal process. The conventional control method is the control based on mathematics model, but it is hard to establish precise mathematic model for sewerage disposing. Because the water quality and water quantity from sewerage disposing factory are variable and random, therefore the needed oxygen quantity of sewerage disposing pool is also variable. According to the technology demand, only the supplied oxygen quantity put into sewerage disposing pool is equal to the needed consuming oxygen quantity in the sewerage disposing pool in the same period, it can just ensure the water quality of factory output water. The DO parameter control is very complicated in the sewerage disposing process, therefore it is necessary to explore the control strategy corresponding to the characteristic of treatment process in waste water.

3 Control Strategy Based on Intelligence Fusion

By dint of conventional control strategy (such as classical control theory, PID control) and the method of modern control theory, it is hard to obtain better control effect for waste water treatment. The existent main puzzles are centralized to represent as following aspects. First of all it is difficult to build accurate mathematical model according to the cybernetics characteristic mentioned above, and the precondition adopting classical control theory and modern control theory is to construct the mathematical model, their analysis and design are based on the math model, and the solution is quantitative. And for waste water treatment, due to the uncertainty it is difficult to give the quantitative description, therefore it is good for nothing to adopt the conventional control strategy. In addition, if the conventional control strategy is adopted then there are other problems such as (1) system complexity, (2) semi-structured and non-structured, (3) high nonlinearity, (4) uncertainty, (5) reliability. The existent puzzle of characteristic mentioned above results in that the conventional control strategy can not carry through the effective control for sewerage disposing process, therefore it is necessary to research further the control strategy.

There are lots of control strategies those can be supplied to choose, but there still are lots of puzzle needed to be solved. The expert control system is based on the knowledge. Because it is difficult to collect the characteristic information, to express the characteristic information, and to build the maturity repository, therefore the expert control system is also difficult to realize the control of sewerage disposing process. NN control needs definite experiment samples. Due to the influence of uncertainty, it is hard to obtain the experiment samples from the known experience and aforehand experiment. In view of the method limitation, it is also difficult to realize effective control generally. The hominine control experience can carry through the summarization and description by means of hominine language, it can be depicted to fuzzy language by means of fuzzy set in fuzzy mathematics, and also it can be realized by the sentence of "If condition Then action". But because the uncertainty factors are too much the fuzzy control is unnecessarily a good choice for sewerage disposing process. The basic property of HSIC (Human Simulated Intelligent Controller) is to simulate the control behavior of control expert. Therefore its control algorithm is multi-modal control, and the material method is to execute alternate use among multi-modal control. Such a property can make contradictions of control quality demand be perfectly harmonized for control system. Therefore it is a sort of more wise choice [7]. Starting from strategy optimization in this paper, it proposes a sort of fusion control strategy based on combination with expert control system and human simulated intelligent controller. The intelligence fusion based control strategy summarizes the control skill and experience of field operators, and it can be concluded as the engineering control algorithm. All the simulation and field debugging results show that the fusion control strategy is feasible and reasonable.

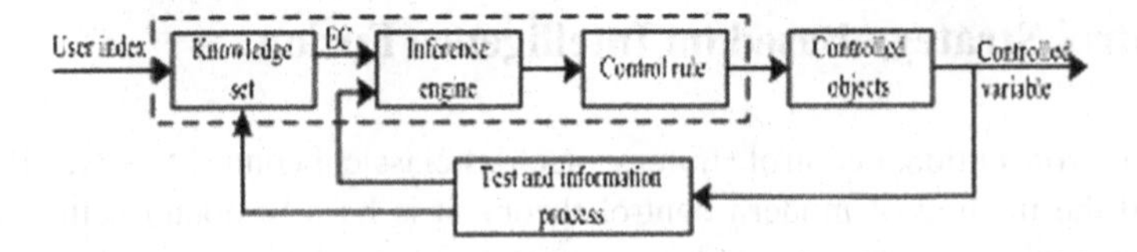

Fig. 1 Structure of control model based on fusion strategy

4 Structure of Controller and Its Algorithm

4.1 Structure of Controller

Based on the fusion of human simulated intelligent control and expert system control techniques, the structure of controller is shown as in Fig. 1.

After the expert system technique is introduced into control system, it can make the structure of control model be simplified. The expert system is an intelligent program system that can solve the problem only solved by expert in special field in terms of knowledge base and inference machine. All the knowledge base, inference machine and rule set could be integrated into the human simulated intelligent controller that it is excellent in performance, and simple in system structure, therefore it can excellently control the input in air amount (oxygen). The DO parameter system of intelligence fusion based control strategy can realize multi-modal control by use of expert system techniques in flexible mode because of the control strategy being based on the basic characteristic combined open-loop with closed-loop control. Therefore it is enhanced to the performance of the judge and inference. In fact, establishing knowledge set is how to express the obtained known-knowledge. The control system always adopts production rule to establish the knowledge set, its basic structure can be expressed as "$If < condition > Then < action >$". The outstanding advantages of system structure of control model based on fusion strategy constituted by production rule are that it is good in modularity, it can be independent in additions, deletions and modifications for each piece of control rule, and it is not direct affiliation among each piece of control rule. And also it is good in naturalness, and suitable for the peculiarity of industrial process control.

4.2 Control Algorithms

In the theory of human simulated intelligent controller, a certain relation of specified operation, which consists of control error and change rate of control error in the control system, is called as feature primitives, and it can be represented by q_i Combining with all the feature primitives forms the characteristic state of control system, it is a both qualitative and quantitative description of dynamic characteristic in intelligent control system, it can be expressed by φ_i. The characteristic model of control system is the set of characteristic state, and a sort of macro-control strategy set after partitioned dynamic information space of system can be expressed by φ.

In the prototype algorithm of human simulated intelligent control, the feature primitives are as the following.

$q_1 : e = 0;$
$q_2 : \dot{e} = 0;$
$q_3 : \dot{e} \neq 0;$
$q_4 : e \cdot \dot{e} > 0;$
$q_5 : e \cdot \dot{e} < 0;$

The characteristic states are as the following.

$\varphi_1 = \{e \cdot \dot{e} > 0 \cup e = 0 \cap \dot{e} \neq 0\}$
$\varphi_2 = \{e \cdot \dot{e} < 0 \cup \dot{e} = 0\}$

The characteristic states are as the following.

$\phi = \{\varphi_1, \varphi_2\} = \{[e \cdot \dot{e} > 0 \cup e = 0 \cup \dot{e} \neq 0], [e \cdot \dot{e} < 0 \cup \dot{e} = 0]\}$

The memory quantity of characteristic is the following.

$\lambda_1 : e_{m,i}$ is i_{th} extremum of error.
$\lambda_2 : u_{0(n-1)}$ is the holding value of control output quantity in previous period.

The prototype algorithm of human simulated intelligent controller is the control mode of alternate open loop and closed loop, and aimed at two sort of different characteristic states it selects different control decision, the algorithm is as the following.

$$U = \begin{cases} K_p \cdot e + k \cdot K_p \cdot \sum_{i=1}^{n-1} e_{m,i} \\ k \cdot K_p \cdot \sum_{i=1}^{n} e_{m,i} \end{cases}$$

The control algorithm adopts improved prototype algorithm of human simulated intelligent controller based on semi-proportion adjustor.

$$u = \begin{cases} K_p e + k K_p \sum_{i=1}^{n-1} e_{m,i} \quad (e \cdot \dot{e} > 0 \cup e = 0 \cap \dot{e} \neq 0) \\ k K_p \sum_{i=1}^{n} e_{m,i} \quad (e \cdot \dot{e} > 0 \cup \dot{e} = 0) \end{cases}$$

In which, u is the control output, K_p is the proportion coefficient, k is the restraining coefficient, e is the system error, $\dot{e}$ is the system error change rate, $e_{m,j}$ is ith error peak value. The algorithm is an algorithm of running control unit level In the control algorithm, firstly the controller identifies two sorts of simple relation characteristic of error and error change rate, then it judges two sorts of different motion state of dynamical system, finally the controller adopts respectively two sorts of different control modal. The output of quantitative control operation is decided by the relation among error peak value of characteristic memory and experienced knowledge (proportion coefficient K_p, restraining k) and current error value. Therefore the algorithm not only has qualitative decision process (identification of motion state and selection of corresponding control mode) but also has quantitative control (output of material control mode). According to the algorithm of human simulated intelligent controller based on semi-proportion adjustor, aimed at the particular situation it can form simplified control algorithm and engineering control algorithm.

For simplification and fused the control expert experience, the simplified algorithm can be described as the following.

Pattern 1 If $e_n{\cdot}_n > 0$ or $e_n = 0$, $|e_n| > 0$,

Then $u_n = u_{n-1} + k_+ * e_n$

Pattern 2 If $e_n \cdot \Delta e_n < 0$ and $|e_n| \geq M$,

Then $u_n = u_{n-1} + k_- \ * \ e_n$

Pattern 3 If $e_n \cdot \Delta e_n < 0$ and $|e_n| < M$ or $e_n = 0$,

Then $u_n = u_{n-1} + k * e_{m,n}$

In which, e_n is nth error, $e_n = e_n - e_{n-1}$, $e_{m{\cdot}n}$ is the error of n_{th} extremum, k_+ is a quickening scale coefficient and $k_+ > 1$, k_- is a suppression scale coefficient and $0 < k_- \leq 1$, k is a hold scale coefficient and $0 < k < 1$, M is the setting error boundary, n represents the order number of control cycle, un is the amount of current control output, un-1 is the amount of control output before the nth cycle.

5 Simulation Experiment

For convenience, here it only takes simplified control algorithm to make the simulation. Ignoring the influence of various factors such as nonlinearity, time varying, uncertainty and so on, the robust controller itself can ensure the control quality of waste water, and therefore the simulated math model is not so important. If the controlled system is considered as one order process with time lag, then its control process can adopt the model of inertia node with pure time lag to be described approximately.

$$W(s) = Ke^{-\tau s}/(Ts + 1)$$

In which, τ, T, K is respectively the pure lag time, time constant of rolling process, gain coefficient. Generally three parameters in the formula can be determined by means of unit step response method. For convenience, here it takes $\tau = 2, T = 1.2, K = 1$, and therefore the model is

$$W(s) = Ke^{-2s}/(1.2s + 1)$$

Under the condition of MATLAB environment, adopting unit step input signal, by means of Simulink tool box it can build system simulation model to make the simulation of control process. The simulation adopts respectively the PID and intelligence fusion based control algorithm (simplified as HSIC in the following figure) to be in charge of the same process, and the simulation curve is shown as in Fig. 2.

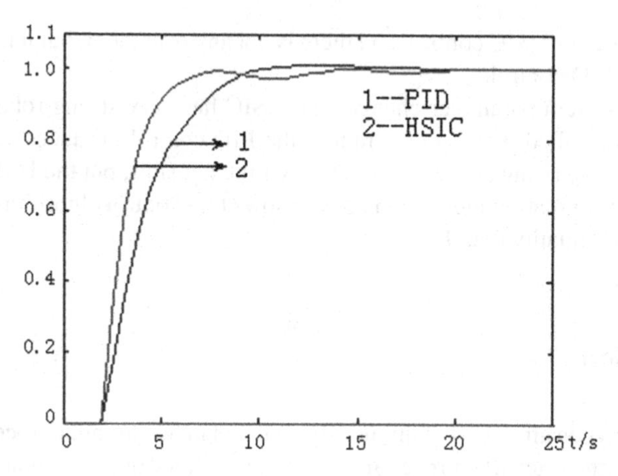

Fig. 2 Process response

In the Fig. 2, the curve 1 and curve 2 is respectively the response curve by PID and HSIC, and from Fig. 2, it can be seen that both PID and HSIC can not be in overshoot, but for the former the rising and regulating time is lower than later, therefore the later owns better control effect.

In order to validate the advantages of fusion control strategy, here it gives a comparison of robustness for parameter changing. the parameter of controller consists of K, T and τ, and for simplification, here it would add an inertia node $1/(2s + 1)$ in controlled object, namely when the transfer function of controlled object changes from $W(s) = e^{-2s}/(1.2s + 1)$ to $W_1(s) = e^{-2s}/(1.2s + 1)(2s + 1)$ the response curve is shown as in Fig. 3. From the Fig. 3, it can be seen that it hardly changes in

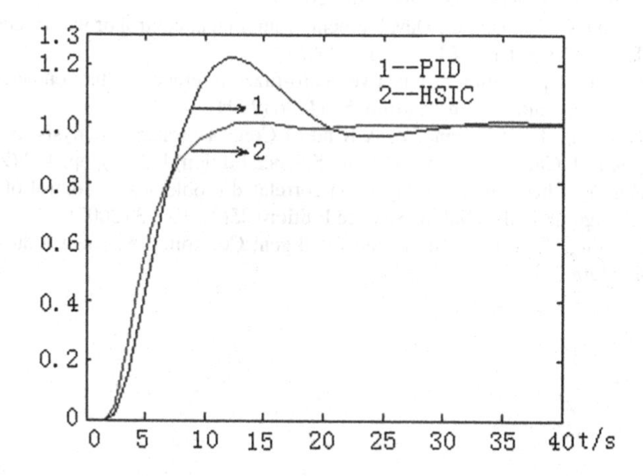

Fig. 3 Response of two-order process with pure lag

system response for HSIC control, and there is not any overshoot, but it has seriously overshoot for PID control.

When the system parameter changes the HSIC has very strong robustness than PID. In the controlled process of aeration, the PID control produces very obvious overshoot, and the rising as well as regulating time get slow, but the HSIC is hardly changing. The simulation mentioned above shows that control algorithm of HSIC is better in control quality than PID.

6 Conclusion

The needed oxygen amount is continuously changed in the air input process. In order to make the waste water after processing attain the specified output standard, the paper proposed a kind of fusion control strategy based on HSIC and expert control system. By means of the fusion control strategy, the DO parameter density can be controlled within allowable index range in the waste water disposal tank. The simulation results show that when the fusion control strategy is adopted the control system quality is rather ideal than PID, and it is simpler in system structure, and better in real time performance.

References

1. Lin, L., Tiecheng, D.: Application of intelligent control in sewerage disposing. Microcomput. Inf. **34**, 35–37 (2007)
2. Punal, A., Rocca, E.: An expert system for monitoring and diagnosis of anaerobic wastewater treatment plants. Water Res. **16**, 2656–2666 (2002)
3. Taijie, L., Lifeng, C., Zaiwen, L.: Development of intelligent control of wastewater treatment. J. Beijing Technol. Bus. Univ. **23**(3), 9–11 (2005)
4. Jiaquan, H., et al.: Application of adaptive neuro-fuzzy inference system on aeration control of wastewater treatment. Autom. Instrum. **5**, 34–36 (2004)
5. Taifu, L., Guoliang, F., Bianxiang, Z. A Kind of Control Strategy Analysis for uncertainty complex system. J. Chongqing Univ. (Nature Science Edition) l, 26(1), pp. 4–7 (2003).
6. Taifu, L., Zhi, Y., Chaojian, S.: Analysis on correlated problem with control of uncertainty system. J. Chongqing Univ. (Nature Science Edition) **25**(2), 19–23 (2002)
7. Zhusu, L., Yaqing, T.: Human Simulated Intelligent Controller. National Industry Defence Press, Beijing (2003)

Grey Bottleneck Transportation Problems

Guo-zhong Bai and Li-na Yao

Abstract In some real situations, such as transporting emergency goods when a natural disaster occurs or transporting military supplies during the war, the transport network may be destroyed, the transportation cost (time or mileage) from sources to destinations may not be deterministic, but uncertain grey number. This paper investigated a new bottleneck transportation problem called the grey bottleneck transportation problem, in which the transportation time (or mileage) from sources to destinations may be uncertain, and introduces its mathematical model and algorithms.

Keywords Bottleneck transportation problem, Natural disaster, Uncertain transportation time, Grey number

1 Introduction

In all the transportation models, such as traditional transportation problem [1, 2], bottleneck transportation problem [3, 4] and so on, the transportation cost/time from every source (supply point) to every destination (demand point) has been considered to be deterministic. However, in real life, it is not necessarily true. For example, transporting emergency goods in the event of a natural disaster or transporting military supplies in wartime, the transport network may be destroyed, the transportation time (or mileage) from certain of the sources to certain of the destinations may be uncertain, but a grey number [5, 6]. In these special cases, the transportation capacity often is poor; the optimization of the transportation project becomes even more important; the most important task is how to transport the emergency supplies (including emergency squad and so on) from supply points to demand points as quickly as possible.

G. Bai (✉) · L. Yao
Department of Mathematics, Guangdong University of Business Studies,
Gungzhou 510320 , China
e-mail: baiguozhong@163.com

B.-Y. Cao and H. Nasseri (eds.), *Fuzzy Information & Engineering and Operations Research & Management*, Advances in Intelligent Systems and Computing 211,
DOI: 10.1007/978-3-642-38667-1_35, © Springer-Verlag Berlin Heidelberg 2014

In this paper, if the transportation time (or mileage) from supply points to demand points be uncertain grey number, the bottleneck transportation problem is said to be a Grey Bottleneck Transportation Problem. Because China is a developing country and the transport network is also developing, traffic jams are frequent. The time wasted by the traffic jam is uncertain. Thus studying such bottleneck transportation problems will contribute to the society and economy.

2 Definitions and Theorems

The bottleneck transportation problem can be stated as follows: A set of supplies and a set of demands are specified such that the total supply is equal to the total demand. There is a transportation time (or mileage) associated between each supply point and each demand point. It is required to find a feasible distribution (of the supplies) which minimizes the maximum transportation time associated between a supply point and a demand point such that the distribution between the two points is positive. In this paper, we will consider only balanced transportation problems because it is not difficult to convert an unbalanced transportation problem into a balanced one [7].

It is assumption that there are m supply points $A_1, A_2, \ldots, A_m$ and n demand points $B_1, B_2, \ldots, B_n$. Let a_i be the supply of A_i, b_j be the demand of B_j, and the transportation time from A_i to B_j be t_{ij}, all i and j, then the bottleneck transportation model is

$$\begin{aligned} &\min z = \max\{t_{ij} \left| x_{ij} \neq 0, i = 1, \ldots m; j = 1, \ldots n \right\} \\ &s.t. \sum_{j=1}^{n} x_{ij} = a_i, i = 1, 2, \ldots, m \\ &\sum_{i=1}^{m} x_{ij} = b_j, j = 1, 2, \ldots, n \\ &x_{ij} \geq 0, i = 1, 2, \ldots, m; j = 1, 2, \ldots, n \end{aligned} \tag{1}$$

where $\sum_{i=1}^{m} a_i = \sum_{j=1}^{n} b_j, a_i \geq 0, b_j \geq 0$, all i and j.

In a grey bottleneck transportation problem, the transportation time from A_i to B_j may be uncertain, it is a grey number $\otimes_{ij}$ [5, 6]. The grey bottleneck transportation model is

$$\begin{aligned} &\min f = \max\{\otimes_{ij} \left| x_{ij} \neq 0, i = 1, \ldots m; j = 1, \ldots n \right\} \\ &s.t. \sum_{j=1}^{n} x_{ij} = a_i, i = 1, 2, \ldots, m \end{aligned} \tag{2}$$

$$\sum_{i=1}^{m} x_{ij} = b_j, j = 1, 2, \ldots, n$$
$$x_{ij} \geq 0, i = 1, 2, \ldots, m; j = 1, 2, \ldots, n$$

where $\otimes_{ij}$ be the grey transportation time [5] from A_i to B_j, $a_i \geq 0$, $b_j \geq 0$, $i = 1, 2, \ldots, m; j = 1, 2, \ldots, n$.

The grey bottleneck transportation problem (2) is written as *GBTP*($\otimes_{ij}$), or *GBTP* for short.

If the grey number $\otimes_{ij} \in [a_{ij}, b_{ij}]$, the following transportation model (3) is said to be a Lower Limit Transportation Problem of the grey transportation problem (2), written as $GBTP(\otimes_{ij})_a$ or for short.

$$\min f_a = \max\{a_{ij} \,|x_{ij} \neq 0, i = 1, \ldots m; j = 1, \ldots n\}$$
$$s.t. \sum_{j=1}^{n} x_{ij} = a_i, i = 1, 2, \ldots, m \tag{3}$$
$$\sum_{i=1}^{m} x_{ij} = b_j, j = 1, 2, \ldots, n$$
$$x_{ij} \geq 0, i = 1, 2, \ldots, m; j = 1, 2, \ldots, n$$

The following transportation model (4) is said to be an Upper Limit Transportation Problem of the grey transportation problem (2), written as or $GBTP^a$ for short.

$$\min f^a = \max\{b_{ij} \,|x_{ij} \neq 0, i = 1, \ldots m; j = 1, \ldots n\}$$
$$s.t. \sum_{j=1}^{n} x_{ij} = a_i, i = 1, 2, \ldots, m \tag{4}$$
$$\sum_{i=1}^{m} x_{ij} = b_j, j = 1, 2, \ldots, n$$
$$x_{ij} \geq 0, i = 1, 2, \ldots, m; j = 1, 2, \ldots, n$$

Theorem 1. *In the GBTY($\otimes_{ij}$) (2), if the optimal values of lower limit transportation problem $GBTP_a$ and upper limit transportation problem $GBTP^a$ are α and β respectively, then $\alpha \leq \beta$.*

Proof: Write the feasible solution set of the *GBTP*($\otimes_{ij}$) as

$$D = \{(x_{11}, \ldots, x_{1n}, \ldots, x_{mn}) \left| \sum_{j=1}^{n} x_{ij} = a_i, \sum_{i=1}^{m} x_{ij} = b_j, \; x_{ij} \geq 0\right\}.$$

Let $X^0 = (x_{11}^0, \ldots, x_{1n}^0, \ldots, x_{mn}^0)$ be an optimal solution of the $GBTP_a$ corresponding to the optimal value α, then

$$f_a(X^0) = \min_{X \in D} \max\{a_{ij} \left| x_{ij} \neq 0\right.\} = \max\{a_{ij} \left| x_{ij}^0 \neq 0\right.\} = \alpha,$$

and for any $Y \in D$ we have $f_a(Y) \geq \alpha$.

Let $Y^0 = (y_{11}^0, \ldots, y_{1n}^0, \ldots, y_{mn}^0)$ be an optimal solution of the $GBTP^a$ corresponding to the optimal value β, then

$$f^a(Y^0) = \min_{X \in D} \max\{b_{ij} \left| x_{ij} \neq 0\right.\} = \max\{b_{ij} \left| y_{ij}^0 \neq 0\right.\} = \beta.$$

Because of $\otimes_{ij} \in [a_{ij}, b_{ij}], 0 \leq a_{ij} \leq b_{ij}$, for any $\bar{X} = (\bar{x}_{11}, \ldots, \bar{x}_{1n}, \ldots, \bar{x}_{mn}) \in D$ we have

$$\max\{a_{ij} \left| \bar{x}_{ij} \neq 0\right.\} \leq \max\{b_{ij} \left| \bar{x}_{ij} \neq 0\right.\}. \tag{5}$$

If $\alpha > \beta$, then

$$f_a(Y^0) \geq \alpha > \beta = f^a(Y^0).$$

That is

$$\max\{a_{ij} \left| y_{ij}^0 \neq 0\right.\} > \max\{b_{ij} \left| y_{ij}^0 \neq 0\right.\}.$$

This is in contradiction with formula (5). Thus we have $\alpha \leq \beta$.

Definition 1. *In the grey bottleneck transportation model (2), if the optimal values of $GBTP_a$ and $GBTP^a$ are α and β respectively, then the grey number $\otimes \in [\alpha, \beta]$ is said to be the Grey Optimal Value of the grey bottleneck transportation model (2).*

Definition 2. *Let $X^0 = (x_{11}^0, \ldots, x_{1n}^0, \ldots, x_{mn}^0)$ and $Y^0 = (y_{11}^0, \ldots, y_{1n}^0, \ldots, y_{mn}^0)$ be the optimal solutions of $GBTP_a$ and $GBTP^a$ respectively, if $x_{ij}^0 = y_{ij}^0$, all i and j, then X^0 is said to be a Synchronal Optimal Solution of the grey bottleneck transportation model (2).*

Theorem 2. *Let α and β be the optimal values of $GBTP_a$ and $GBTP^a$ respectively, if $X^0 = (x_{11}^0, \ldots, x_{1n}^0, \ldots, x_{mn}^0)$ is a synchronal optimal solution of the grey bottleneck transportation model (2), then for every t, $0 \leq t \leq 1$, X^0 is also an optimal solution to the following bottleneck transportation problem (6):*

$$\begin{aligned} &\min z = \max\{a_{ij} + t(b_{ij} - a_{ij}) \left| x_{ij} \neq 0\right.\} \\ &s.t. \sum_{j=1}^{n} x_{ij} = a_i, i = 1, 2, \ldots, m \\ &\sum_{i=1}^{m} x_{ij} = b_j, j = 1, 2, \ldots, n \end{aligned} \tag{6}$$

$$x_{ij} \geq 0, i = 1, 2, \ldots, m; j = 1, 2, \ldots, n$$

and the optimal value of the bottleneck transportation model (6) equals $\alpha + t(\beta - \alpha)$.

Proof: Because X^0 is a synchronal optimal solution of the grey bottleneck transportation problem (2), for any $X \in D$ we have

$$\alpha = \max\{a_{ij} \left| x_{ij}^0 \neq 0\right\} \leq \max\{a_{ij} \left| x_{ij} \neq 0\right\}$$
$$\beta = \max\{b_{ij} \left| x_{ij}^0 \neq 0\right\} \leq \max\{b_{ij} \left| x_{ij} \neq 0\right\}.$$

Since $0 \leq t \leq 1, 0 \leq a_{ij} \leq b_{ij}$, we have

$$\begin{aligned}
&\alpha + t(\beta - \alpha) = (1 - t)\alpha + t\beta \\
&= (1 - t)\max\{a_{ij} \left| x_{ij}^0 \neq 0\right\} + t\max\{b_{ij} \left| x_{ij}^0 \neq 0\right\} \\
&\leq (1 - t)\max\{a_{ij} \left| x_{ij} \neq 0\right\} + t\max\{b_{ij} \left| x_{ij} \neq 0\right\} \\
&= \max\{(1 - t)a_{ij} \left| x_{ij}^0 \neq 0\right\} + \max\{tb_{ij} \left| x_{ij}^0 \neq 0\right\} \\
&= \max\{[(1 - t)a_{ij} + tb_{ij}] \left| x_{ij}^0 \neq 0\right\} \\
&= \max\{[(a_{ij} + t(b_{ij} - a_{ij})] \left| x_{ij}^0 \neq 0\right\}
\end{aligned}$$

Thus X^0 is an optimal solution of the bottleneck transportation problem (6), and the corresponding optimal value equals $\alpha + t(\beta - \alpha)$.

Definition 3. *Let the grey number* $\otimes \in [\alpha, \beta]$ *be a grey optimal value of the grey bottleneck transportation model (2). For* $m \times n$ *real numbers* γ_{ij}, $a_{ij} \leq \gamma_{ij} \leq b_{ij}$, *all* i *and* j, *there exists* $X^0 \in D$ *such that* $\alpha \leq \max\{\gamma_{ij} \left| x_{ij}^0 \neq 0\right\} \leq \beta$, *then* X^0 *is said to be a Semi-optimal Solution of the grey bottleneck transportation problem (2), and* $\max\{\gamma_{ij} \left| x_{ij}^0 \neq 0\right\}$ *is said to be the Semi-optimal Value corresponding to the semi-optimal solution.*

Because a balanced bottleneck transportation model always has optimal solutions [1, 2], and both optimal solutions of the lower limit transportation problem $GBTP_a$ and the upper limit transportation problem $GBTP^a$ may be regarded as the semi-optimal solutions of the grey bottleneck transportation problem (2), we have the following theorem.

Theorem 3. *A balanced GBTP always has semi-optimal solutions and the grey optimal values.*

3 The Methods for Solving *GBTP*

In the grey transportation problem (2), grey transportation time $\otimes_{ij}$ from$i$th supply point to jth demand point is not a definite real number, but a grey number. Sometime, we cannot obtain an optimal value of *GBTP* in the mathematical significance. According to Theorem 1 and Theorem 3, we can obtain the grey optimal value or the semi-optimal value of *GBTP*. It is regretful that the grey optimal value is not a definite real number. The following are some special methods for solving a grey bottleneck transportation problem with the grey transportation time $\otimes_{ij}$. With these we can obtain the whitened value [5, 6] of the grey optimal value of *GBTP* and the semi-optimal value.

3.1 *Time Sequence Grey Number*

In the grey bottleneck transportation model (2), if the grey transportation time $\otimes_{ij}$ is given by the time sequence, say

$$\otimes_{ij}: \quad \{t_{ij}(1), t_{ij}(2), \ldots, t_{ij}(n)\}.$$

(1) Let the average number

$$t_{ij} = \frac{1}{n}\sum_{k=1}^{n} t_{ij}(k),$$

then use t_{ij} in place of $\otimes_{ij}$.

(2) Use the method of grey forecasting [6] to calculate the forecasting value $t_{ij}(n+1)$ of the $\otimes_{ij}$, then use $t_{ij}(n+1)$ in place of $\otimes_{ij}$.

Then solve the bottleneck transportation problem, and obtain an optimal solution. The optimal solution is a semi-optimal solution to the grey bottleneck transportation problem (2), and the corresponding optimal value is a semi-optimal value of (2).

It is easy to see that the *GBTP* not only suits to study static but also suits to study dynamic bottleneck transportation problems.

3.2 *Rational Grey Number*

In the grey bottleneck transportation model (2), the grey transportation time is a rational grey number [5] $\otimes_{ij} \in [a_{ij}, b_{ij}]$.

(1) Let

$$t_{ij} = a_{ij} + t(b_{ij} - a_{ij}), \quad 0 \le t \le 1,$$

then use t_{ij} in place of $\otimes_{ij}$. Where t is said to be the risk coefficient. For each t, $0 \le t \le 1, t_{ij}$ is a definite real number. Specially, $t = 0$ is said to be the optimistic coefficient; t=1 is said to be the pessimistic coefficient. Using this method to determine the whitened value of the grey number $\otimes_{ij}$, the dependable degree may be defined as $\sqrt{2t}$ when $0 \le t \le 0.5$, or $\sqrt{2(1-t)}$ when $0.5 \le t \le 1$. The reason why we make such a definition can be found in the reference [8].

(2) Let

$$t_{ij} = a_{ij} + \lambda_{ij}(b_{ij} - a_{ij}), 0 \le \lambda_{ij} \le 1,$$

then use t_{ij} in place of $\otimes_{ij}$. Where λ_{ij} is said to be the weighted risk coefficient.

Then solve the bottleneck transportation problem, and obtain an optimal solution. The optimal solution is a semi-optimal solution of the grey bottleneck transportation problem (2), and the corresponding optimal value is a semi-optimal value of (2).

3.3 Especial Grey Number

In the grey bottleneck transportation problem (2), determining the whitened value of the grey transportation time $\otimes_{ij}$ is difficult, and has no precedent to go by. Then the following methods may be used to determine the whitened value of the grey number $\otimes_{ij}$.

(1) Three-value-estimate. At first, the following three estimates are given by the experienced policymakers, experts and concerned person: the minimal transportation time α_{ij}, the maximal transportation time β_{ij} and the most possible transportation time γ_{ij}. Let the weighted average

$$t_{ij} = \frac{\alpha_{ij} + 4\gamma_{ij} + \beta_{ij}}{6},$$

then use the weighted average t_{ij} in place of the grey transportation time $\otimes_{ij}$.

(2) Two-value-estimate. If estimating the most possible transportation time is very difficult, let the weighted average

$$t_{ij} = \frac{3\alpha_{ij} + 2\beta_{ij}}{5},$$

then use t_{ij} in place of $\otimes_{ij}$.

Then solve the bottleneck transportation problem, and obtain an optimal solution. The optimal solution is a semi-optimal solution of the grey bottleneck transportation problem (2), and the corresponding optimal value is a semi-optimal value of (2).

4 Conclusion

This paper investigated a new transportation problem called the grey bottleneck transportation problem, in which the transportation time (or mileage) from sources to destinations may be uncertain. The mathematical model is given, and some methods for solving the grey bottleneck transportation problem are introduced.

If we regard a real number a ($a \in R$) as a special grey number, namely rational grey number $a = \otimes \in [a, a]$, then a traditional bottleneck transportation problem can be considered as a special grey bottleneck transportation problem.

Many of the problems normally encountered in practice deal with grey number, such as Grey Payoff Matrix Game [9], Grey Assignment Problems [10], and so on.

References

1. James K. Strayer.: Linear Programming and Its Applications. New York: Springer-Verlag World Publishing Corp (1989)
2. Hamdy, A.: Taha: Operations Research: An Introduction. Macmillan Publishing Company, New York (1989)
3. Hammer, P.L.: Time-minimizing transportation problems. Naval Research Logistics Quarterly **16**(3), 345–35 (1989)
4. Garfinkel, R.S., Rao, M.S.: The bottleneck transportation problem. Naval Research Logistics Quarterly **18**(4), 465–472 (1991)
5. Deng, J.L.: The Tutorial to Grey System Theory. Huazhong University of Science and Technology Press, Wuhan (1990)
6. Deng, J.L.: Grey Forecasting and Decision-Making. Huazhong University of Science and Technology Press, Wuhan (1988)
7. Bai G. Z.: B-Transportation Problem and Its Applications. Systems Engineering–Theory & Practice, 17(11), 97–102 (1997).
8. Xu, G.H., Liu, Y.P., Cheng, K.: Handbook of Operations Research Fundamentals. Science Press, Beijing (1999)
9. Bai, G.Z.: The Grey Payoff Matrix Game. The Journal of Grey System **4**(4), 323–331 (1992)
10. Bai, G.Z.: Grey Assignment Problem. Chinese Journal of Operations Research **11**(2), 67–69 (1992)

A Simulated Annealing Genetic Algorithm for Solving Timetable Problems

Yi-jie Dun, Qian Wang and Ya-bin Shao

Abstract The post-enrolment course timetabling (PE-CTT) is one of the most studied timetabling problems, for which many instances and results are available. In this paper, we design a metaheuristic approach based on Simulated Annealing to solve the PE-CTT. We consider all the different variants of the problems that have been proposed in the literature and perform a comprehensive experimental analysis on all the public instances available. The outcome is that the solver, properly engineered and tuned, performs very well in all cases. Thus we provide the new best known results for many instances and state-of-the-art values for the others. An algorithm SAGA for solving timetable problem was presented by analyzing all kinds of restricting conditions and special requirements in timetable arrangement of colleges and universities. Moreover, the crossover and mutation operators in the simulated annealing genetic algorithm were improved with adaptive strategy in order to enhance its searching ability and efficiency of the algorithm. The numerical experiments showed that the algorithm was efficient and feasible.

Keywords Course timetabling · Genetic algorithms · Simulated annealing · Metaheuristics

1 Introduction

The timetabling of events [1] (such as lectures, tutorials, and seminars) at universities in order to meet the demands of its users is often a difficult problem to solve effectively. As well as wanting a timetable that can actually be used by the institution, users will also want a timetable that is "nice" to use and which doesn't overburden the people who will have to base their days' activities around it. Timetabling is also

Y. Dun (✉) · Q. Wang · Y. Shao
School of Mathematics and Computer Technology, Northwest University for Nationalities, Lanzhou 730030, China
e-mail: dunyijie@126.com

B.-Y. Cao and H. Nasseri (eds.), *Fuzzy Information & Engineering and Operations Research & Management*, Advances in Intelligent Systems and Computing 211,
DOI: 10.1007/978-3-642-38667-1_36,

a very idiosyncratic problem that can vary between different countries, different universities, and even different departments. From a computer- science perspective, it is therefore a problem that is quite difficult to study in a general way. Educational timetabling is a sub-field of timetabling that considers the scheduling of meetings between teachers and students. In the middle of 1970s, the authors demonstrated that the timetable problem is a NP complete problem [1] with other researchers. In the 1990s', Colorni et al. [2] using genetic algorithms (GAs) to solve the timetable problems. Sigeru [3] using GAs to solve the problem of university timetable arrangements by the way of adds control constraints. Zhang Chunmei [4] introduced the Adaptive genetic algorithms for solving the university timetable problem, they divided the university courses into Several categories such as compulsory, elective, minor and Solving respectively, and have achieved good results. Ye Ning et al. [5], who industry and presents an algorithm of automatically arranging courses in universities. They used genetic algorithms to set up a data model and defined a four-dimensional chromosome encoding scheme. Legierski [6] gave the application of simulated annealing algorithm to optimize the arrangement of the curriculum program, and discussed the various issues involved in the program. Schaerf et al. [7] applied one kind of mixed simulation annealing algorithm [8] to solve the class schedule problem. A large number of variants of educational timetabling problems have been proposed in the literature, which differ from each other based on the type of institution involved (university, school, or other), the type of meeting (course lectures, exams, seminars, . . .), and the constraints imposed. The university course timetabling (CTT) problem is one of the most studied educational timetabling problems and consists in scheduling a sequence of events or lectures of university courses in a prefixed period of time (typically a week), satisfying a set of various constraints on rooms and students. Many formulations have been proposed for the CTT problem over the years. Indeed, it is impossible to write a single problem formulation that suits all cases since every institution has its own rules, features, costs, and fixations. Nevertheless, two formulations have recently received more attention than others, mainly thanks to the two timetabling competitions, ITC 2002 and ITC 2007 [9] (McCollum et al. 2010), which have used them as competition ground. These are the so-called curriculum-based course timetabling (CBCTT) and post-enrolment course timetabling (PE-CTT). The main difference between the two formulations is that in the CB-CTT all constraints and objectives are related to the concept of curriculum, which is a set of courses that form the complete workload for a set of students. On the contrary, in PE-CTT this concept is absent and the constraints and objectives are based on the student enrolments to the courses. However, for the curriculum schedule problem, the search ability of the algorithm is not high and it is inefficient, what's more, the initial parameters have a big influence on it. Such as genetic algorithm, the parameters improperly is easy to fall into the "premature", and the simulated annealing algorithm is very harsh restrictions on the "retire temperature" process conditions, and the time performance is poor. Therefore this paper presents a method of using hybrid genetic algorithm–simulated annealing genetic algorithm(SAGA) to optimize the curriculum arrangement, and effectively combine genetic algorithms and simulated annealing genetic algorithm, and improve the genetic algorithms fitness function, crossover and mutation rate, so that can get better optimization effect.

This paper is organized as follows. In Sect. 2 we will review the related definitions of SAGA as the learning and inference. In Sect. 3 we will introduce our algorithm learning framework, followed by the experimental evaluations in Sect. 4. The conclusions are given in Sect. 5.

2 Relational the Basic Principle of SAGA

As discussed, the genetic algorithm is search algorithms which is based on natural selection and genetic theory, and combine the fittest rules of the process of biological evolution survival with the random information exchange mechanism within group chromosomes. It was first proposed by U.S.J. Holland. Its main characteristic is exchange information between the search group strategy and individuals; however the search does not require any prior knowledge when solving the problem itself. So it is suitable to deal with traditional search methods which are difficult to solve and complex nonlinear problems. The genetic algorithm is widely used in combinatorial optimization, machine learning, adaptive control, planning and design, and artificial life.

In the original timetabling competition, a problem model was used in which a number of "events" had to be scheduled into rooms and "timeslots" in accordance with a number of constraints. These constraints can be divided into two classes: the hard constraints and the soft constraints. The former are mandatory in their satisfaction and reflect constraints that need to be satisfied in order for the timetable to be useable; the latter are those that are to be satisfied only if possible and are intended to make a timetable "nice" for the people who were supposed to use it.

2.1 Simulated Annealing

The simulated annealing algorithm is a random optimization algorithm which is based on the Monte Carlo iterative solution strategy, The starting point is based on the similarity between the physical annealing process and the general combinatorial optimization problems. The simulated annealing algorithm is in some or other initial temperature, along with the falling of the problem parameters, combined with the probabilistic jumping feature to find the optimal solution of the objective function in the problem solution space randomly.

Paul L. Stoffa learned from the idea of simulated annealing simulated annealing genetic algorithm SAGA. Use the following fitness stretching method, at a certain temperature to adjust the individual fitness and overcome the "premature" and "stagnation" phenomenon in genetic algorithm:

$$f_i = \frac{e^{fi/T}}{\sum_{i=1}^{M} e^{fi/T}}$$

and

$$T = T_0(0.99^{g^{-1}})$$

where f_i is the individual's fitness, M is the population size, g is a hereditary algebra, T is the temperature and T_0 is the initial temperature.

2.2 *General Definition of PE-CTT*

In the PE-CTT problem it is given a set E = {1, …, E} of events, a set T = {1, …, T} of timeslots, and a set R = {1, …, R} of rooms. It is also defined a set of days D = {1, …, D}, such that each timeslot belongs to one day and each day is composed by T/D timeslots.

It is also given a set of students S and an enrolment relation $M \subseteq E \times S$, such that $(e, s) \in M$ if student s attends event e.

Furthermore, it is given a set of features F that may be available in rooms and are required by events. More precisely, we are given two relations $\Phi R \subseteq R \times F$ and $\Phi E \subseteq E \times F$, such that $(r, f) \in \Phi R$ if room r has feature f and $(e, f) \in \Phi E$ if event e requires feature f, respectively, Each room $r \in R$ has a fixed capacity Cr, expressed in terms of seats for students.

In addition, it is defined a precedence relation $\prod \subseteq \varepsilon \times \varepsilon$, such that if $(e1, e2) \in \prod$, events e1 and e2 must be scheduled in timeslots t1 and t2 such that $t1 < t2$. Finally, there is an availability relation $A \subseteq \varepsilon \times T$, stating that event e can be scheduled in timeslot t only if $(e, t) \in A$.

3 The Model of the Timetable Problem

3.1 *Description of the Problem*

There are five mutual restraint factors are involved in the curriculum which is the classroom, class, time, curriculum, teachers. The question solution process is to find a suitable teachers and classrooms-time on any course. When arranges the curriculum, meet the necessary constraints, as far as possible to meet some special optimization requirements.

The existing constraints are: (1) A teacher can only be one course at the same time; (2) a class can only be a course at the same time; (3) A classroom only for a course at the same time; (4) Student numbers cannot be greater than the current maximum capacity of the designated classroom.

Special requirements: (1) Trunk courses or examination classes are arranged in the morning, elective, or examine the lessons are arranged in the afternoon; (2) The courses which have more hours in a week should arrange dispersed to facilitate teachers in preparing lessons and students to review; (3) Keep some special time periods, such as the class meeting activities, outdoor activities; (4) Which have many hours of the same course within a week should be arranged in the same class room.

The curricula arrangement should defined as a four-group C, C = (teacher, classroom, course, class), which representing teachers, classroom, course scheduling tasks and classes. Among them, the Teacher (Tno, Tname, Cno) include the number of teachers, teachers' name and course code, the Classroom (class no, capability, type) include a number of the classroom, classroom capacity and classroom type, the Class (class no, nature_class, student no) include the class number, nature classes number, students number, the Course (Cno, Cname, Stu_num, period, distribute) include the course number, course name, number of who select it, how many hours per week and its distribution codes.

The rules of hours per week distributed coding are: 1 stands for the week hours is 2 and should take one class; 2 stands for the week hours is 3 and should take two classes; 3 stands for should take one class in single week; 4 stands for should take one class in double week; 0 shows that more than 24 hours interval between the two classes of the course, and if there are no number 0 between two numbers. Course on continuous, for example, 11 stands for 4 courses ranked in the continuous-time.

3.2 Coding and Chromosome Representation

Students should make choose courses online based on teaching plan, and collect it so that can form the original data of course arrangement, it is Course (Cno, Cname, Stu_num, period, distribute), from the course schedule we can get teacher's information: Teacher (Tno, Tname, Cno). From the database we can get the classroom's information: Classroom (classno, capability, type). From the teachers course schedule we can get a series of constraint, such as dedicated classrooms, a multimedia classroom and time requirements.

Uniform numbers of all courses in all classes of the school and can get the course number Uniform coding of all available time periods within a week can get time number; Uniform numbers of all classrooms can get the classroom number. Using matrix encoding, a matrix Y represents the possibly timetabling. The matrix's rows and columns represent the number of classroom and time. If a course was arranged in a classroom i and in time period j, y_{ij} stands for it, 0 indicates no Timetable of the classroom during that time period, -1 stands for go on class continuous. For example, there are 4 classrooms and 15 available time periods (Within a week of five days class, 3 large sections such every day). The corresponding matrix representation is as follows:

$$Y = \begin{bmatrix} 15 & 2 & 7 & 21 & 18 & 0 & 34 & 15 & 2 & 25 & -1 & 23 & 17 & 2 & 0 \\ 9 & 12 & -1 & 14 & 6 & 21 & 27 & 9 & 0 & 36 & 3 & 12 & 4 & -1 & 0 \\ 5 & 6 & 0 & 24 & 16 & 0 & 13 & 6 & -1 & 35 & 24 & 0 & 3 & 26 & 11 \\ 22 & 8 & 1 & 33 & -1 & 0 & 30 & 28 & 0 & 8 & 29 & 31 & 33 & 0 & 1 \end{bmatrix}$$

3.3 Fitness Function

Definition:

$$F = \{\alpha, \beta, \gamma, \chi, \eta, \theta\} \tag{1}$$

$$\alpha = \begin{cases} x.....(0 \le x < 1) \\ 1..........(x \ge 1) \end{cases} \tag{2}$$

Among them, x is the number of course in the timetable units divided by the capacity of classroom. When $\beta = 0$, there is no conflict about teacher's teaching time. When $\beta = 1$, there have conflict about teachers' teaching time. When $\gamma = 0$, there is no conflict about class to class time. When $\gamma = 1$, there have conflict about class to class time. When $\chi = 0$, the special courses is arranged in the designated classrooms. When $\chi = 1$, the special courses is not arranged in the designated classrooms. When $\eta = 0$, one class time interval is more than 3 segments in a week. When $\eta = 1$, one class time interval is less than 3 segments in a week. θ is the sum of fitness which meet other special requirements. (Meet a requirement , and its value plus 0, otherwise plus 1). For example, if the teachers ask to teaching in the morning, there have conflicts with students' course selection time.

According to the curriculum schedule and teacher curriculum schedule requirements, Calculate the value of fitness function after make corresponding weights based on the the important degree of a variety of constraints and special requirements:

$$f(Y) = \sum_{i=1}^{6} \omega_i F[i]. \tag{3}$$

3.4 Genetic Operator

It can be randomly selected as parents in the two rows of the matrix of a single chromosome, and can be exchange randomly part of the time. Its function is adjust the conflict between the number of elective students and classroom capacity and also can meet the special requirements of a course for classroom. However, it can also choose from two different chromosomal matrixes in the same line as parents, and using partially matched crossover method for cross operation.

1. Mutation operator. When the operation result is close to the optimal solution, using part random search capability of the mutation operator can accelerate the convergence to the optimal solution. For the selected chromosome select a row randomly, and then select a column value in the row randomly which corresponding to a course number (its value is not 0 or -1), and then transform the course number of some other time period arranged in the same day, so that can adjust the time conflict of teachers and class.
2. Selection operators. According to the size of the individual fitness function, we use the roulette selection strategy to choose the excellent course arrangement plan, and to meet the various special requirements based on the constraint condition.
3. Filtration operator. When determine the initial solution, it must satisfy the constraint conditions, at the same tine, each of the solution in the annealing process should satisfy this condition too. The filtering operation can be used to judge whether a solution is the feasible solution. It is need to scan the entire chromosome to judge when initialize population, however, we can get result only by scan the progeny of chromosome specific line of a part after cross and mutation.

3.5 Parameter Settings

1. The determination of the objective function

What the genetic algorithm's search direction guidance based on is the fitness. Because in the roulette selection, the chromosome with large fitness is more probably to be selected, so the optimization direction of the objective function associated with the direction of the fitness increased. However, the conflict of objective function arranged by Curriculum schedule, it belongs to the minimum optimization problem, so it should be adjusted and the transformation is as follows:

$$f(i) = \exp(-(f_i - f_{\min})/t) \tag{4}$$

where f_i denotes the fitness value which chromosome correspond; $f_{\min}$ is the smallest fitness value in the current evolutionary population, t stands for temperature parameters. This is a very good accelerating objective function, the acceleration is not apparent when the temperature if high otherwise it is very apparent. Thus it can achieve the fitness stretching.

2. The determination of initial temperature and the operations of retire temperature

The initial temperature choose $T_0 = K\delta$. Where K is a sufficiently large number, you can choose K = 10, 20, 100, ... experimental values; and $\delta = f_{s,\max} - f_{s,\min} f_{s,\max}$ is the largest objective function value in the initial population while $f_{s,\min}$ is the smallest. The retirement temperature function use the common form: $T_{K+1} = \alpha T_K$, with $0 < \alpha < 1$.

3. The probability of crossover and mutation

The crossover rate p_c and mutation rate p_m take a fixed value in accordance with the experience during the optimization, $p_c \in [0.25, 0.95]$, $p_m \in [0.005, 0.100]$, However, this method has a certain blindness. Srinivas et al. proposed that p_c and p_m change automatically with adaptive, it's main idea is adjust p_c and p_m dynamically based on the population evolution in order to achieve the purpose of overcoming premature convergence and to accelerate the search speed. According to its principle, we can get the following expression:

$$p_c = \begin{cases} \frac{k_1(f - f_{\min})}{f_{avg} - f_{\min}} & (f < f_{avg}) \\ k_1 & (f \geq f_{avg}) \end{cases}$$

$$p_m = \begin{cases} \frac{k_2(f - f_{\min})}{f_{avg} - f_{\min}} & (f < f_{avg}) \\ k_2 & (f \geq f_{avg}) \end{cases}$$

where k_1, k_2 is a constant and the specific value is determined according to the actual situation; f_{avg} is the average objective function value of the current generation of evolutionary groups; f is the smaller objective function value in the two cross individuals.

4. Discriminate criteria replication strategy based on the Metropolis

Adjustments made by crossover and mutation operation may make the new curriculum after exchange doesn't satisfy the constraint conditions. It means which will produce infeasible solutions, so it is needed to withdraw the exchange back to the original state. When the exchanges satisfy the constraints, we need to calculate the difference of objective function in order to carry out the judgment of the metropolis criterion. Metropolis criterion is: If $\Delta t < 0$ accepted S' as new current solution S, otherwise with the probability of $\exp(-\nabla t/T)$.

4 Algorithm Descriptions

Type: teacher data, course data (including the course schedule Course), classroom data. Output: a semester course schedule of the school

1. According to the classroom data and the available time in a week, generate a blank course schedule, and put the course numbers into it randomly with the requirements that satisfy the constraints. At the same time, it should be loaded n groups in total to form the initial population consisted by n-dimensional table, n is the number of chromosomes in the groups;
2. Choose the initial temperature coefficient K, annealing temperature coefficient α, the coefficient k_1, k_2 of crossover and mutation operations, the termination of coefficient q, let the number of iterations L = 0;

3. Calculate the objective function value of each chromosome in the population, and determine the initial temperature, make the initial optimal solution $s = f_{\min}$, and p=0;
4. Calculate the fitness of each chromosome, and using the roulette method for group selection, and reserves the high Fitness value of the chromosome;
5. Recalculates the value for the objective function of the chromosome and follow the crossover probability pc to run the crossover operation of genetic algorithm to keep the target's largest chromosomes;
6. Recalculates the value for the objective function of the chromosome and follow the Mutation probability pm to run the crossover operation of genetic algorithm to keep the target's largest chromosomes;
7. Run the replication strategy based on the Metropolis criterion to produce the next generation of population;
8. Executive temperature back operation with $T_{K+1} = \alpha T_K$ and L=L+1;
9. Calculate the objective function value of the chromosome in new populations, and $S' = f_{\min}$, the next judge the difference between S' and S to get ΔS, then according to the Metropolis criterion to decide whether to accept the new value. If it accept makes S = S', p=0, otherwise makes p=p+1;
10. Judge whether p is greater than or equal to q, if it is, output of the final solution with S and stop the calculation; otherwise, return to the third step.

5 Analyses of Algorithm Results

According to the above algorithms, choose a spring semester curriculum of a university in Lanzhou from 2001 to 2002 as an example, then use the software C++ Builder 6.0 compile the course arrangement program, after that, verify the algorithm results. There are 21 weeks in this semester in total and each week have 27 available time, and this university have 450 teachers and 183 classrooms which including 75 multimedia classrooms, 48 special classrooms, and it have 14,856 seats and 15,000 students in 96 classes.

The population size is n=60 when calculating, and each coefficient of the algorithm is determined as follows: the initial temperature coefficient K=20, the annealing temperature coefficient $\alpha = 0.8$, the coefficient of Crossover and mutation operations is $k_1 = 0.8, k_2 = 0$, The judgment condition of algorithm terminates is q=20. The algorithm runs to end at the 2875th generation and it spend 2,455 s, the results showed the conflict between the classrooms, teachers, classes and the time has been solved perfectly. However, the same example shows if the programming using a simple genetic algorithm to deal with and compare the results of multiple runs, we can get that the average time is about 1,967 s and the average operating algebra is 1,543. Although the algorithm running time is short, it should be run multiple times to get a more satisfactory solution.

Analysis of the proposed algorithm, the downside is that: Because it access to the database frequently during the calculation, and to verify whether the solution is feasible, therefore the algorithm running time longer. So the future work is to improve these deficiencies deeply in order to improve the efficiency of the algorithm.

6 Conclusions

This paper presents algorithms using simulated annealing genetic algorithm to solve the curriculum arrangement optimization problem, it combines the characteristics of genetic algorithms and simulated annealing genetic algorithm which makes the two algorithms search capabilities complement each other, what's more, it overcomes the genetic algorithm easy to fall into the "premature" with the parameter choice is undeserved and simulated annealing algorithm has very harsh restrictions to "Annealing temperature", and it also avoid the search process into a local minimum. In addition, the algorithms change the genetic algorithm crossover and mutation probability adaptively in the problem solving process. Thus we improve the algorithm's exploration ability and efficiency in the solution space.

Acknowledgments Thanks to the support by the National Scientific Fund of China (No. 11161041) and Fundamental Research Foundation for the Central Universities (No. 31920130051, No. Zyz2012077).

References

1. Chiarandini, M., Socha, K., Birattari, M., Rossi-Doria, O.: An effective hybrid approach for the university course timetabling problem. Technical Report AIDA(5), FG Intellektik, FB Informatik, TU, Darmstadt, Germany (2003)
2. Lewis, R.: Metaheuristics for university course timetabling. Napier University, Edinburgh, Scotland, Doctoral Thesis (2006)
3. Lewis, R.: A survey of metaheuristic-based techniques for university timetabling problems. (Available at http://www.cardiff.ac.uk/carbs/ quant/rhyd/rhyd.html) (2007)
4. chunmei, zhang.:The university course timetabling problem with self-adaptive genetic algorithm approach. Lecture Notes in Journal of Inner Mongolia University(Natural Science Edition) (2002)
5. Ning, Ye.: TTP algorithm based on genetic algorithm. Journal of Southeast University (Natural Science Edition) (2003)
6. Legierski W.:Search strategy for constraint-based class-teacher timetabling[A]. PATAT[C], Konstanz Germany (2000)
7. Schaerf, A.: Local search techniques for large high-School timetabling problems. IEEE Transactions on Systems Man and Cybernetics Part A (2005)
8. D S Johnson, C R Aragon, L A McGeoch, C Schevon.:Optimiza-tion by simulated annealing:an experimental evaluation, Part. Lecture Notes in Computer Science. (1999)
9. Stutzle Thomas, Hoos Holger H.:MAX-MIN ant system. Future Generation Computer Systems. (2010)

Weighted Statistical Approximation Properties of the q-Phillips Operators

Mei-ying Ren

Abstract In this paper, the q-Phillips operators which were introduced by I. Yüksel are studied. By the means of the q-integral and the concept of the statistical convergence, the weighted statistical approximation theorem of the operators is obtained. Then a convergence theorem of Korovkin type is given. Finally, a Voronovskaja-type asymptotic formulas is also investigated.

Keywords Weighted statistical approximation · q-Phillips operators · Korovich type theorem · Voronovskaja type asymptotic formulas · q-integral

1 Introduction

After Phillips [1] introduced and studied q analogue of Bernstein polynomials, the applications of q-calculus in the approximation theory become one of the main areas of research, many authors studied new classes of q-generalized operators (for instance, see [2–4]). In 2011, Yüksel [5] studied the approximation properties of the q-Phillips operators. The main aim of this paper is to study weighted statistical approximation the properties of the q-Phillips operators on the basis of [5].

Before, proceeding further, let us give some basic definitions and notations from q-calculus. Details on q-integers can be found in [6–10].

Let $q > 0$, for each nonnegative integer k, the q-integer $[\mathrm{k}]_q$ and the q-factorial $[\mathrm{k}]_q!$ are defined by

$$[k]_q := \begin{cases} \frac{1-q^k}{1-q}, & q \neq 1, \\ k, & q = 1. \end{cases} \text{ and } [k]_q! := \begin{cases} [k]_q[k-1]_q \ldots [1]_q, & k \geq 1, \\ 1, & k = 0. \end{cases} \text{ respectively.}$$

M. Ren (✉)
Department of Mathematics and Computer Science, Wuyi University, Wuyishan 354300, China
e-mail: npmeiyingr@163.com

B.-Y. Cao and H. Nasseri (eds.), *Fuzzy Information & Engineering and Operations Research & Management*, Advances in Intelligent Systems and Computing 211,
DOI: 10.1007/978-3-642-38667-1_37, © Springer-Verlag Berlin Heidelberg 2014

Then for $q > 0$ and integers $n, k, n \geq k \geq 0$, we have $[k+1]_q = 1 + q[k]_q$ and $[k]_q + q^k[n-k]_q = [n]_q$.

For the integers $n, k, n \geq k \geq 0$, the q-binomial coefficients is defined by

$$\begin{bmatrix} n \\ k \end{bmatrix}_q := \frac{[n]_q!}{[k]_q![n-k]_q!}.$$

The two q-analogue of the exponential function are defined as:

$$e_q(x) = \sum_{n=0}^{\infty} \frac{x^n}{[n]_q!} = \frac{1}{(1-(1-q)x)_q^{\infty}}, |x| < \frac{1}{1-q}, |q| < 1 \quad \text{and}$$

$$E_q(x) = \sum_{n=0}^{\infty} q^{n(n-1)/2} \frac{x^n}{[n]_q!} = (1+(1-q)x)_q^{\infty}, |q| < 1,$$

where $(1+x)_q^{\infty} = \prod_{j=0}^{\infty}(1+q^j x)$.

Also, it is known that $e_q(x)E_q(-x) = e_q(-x)E_q(x) = 1$.

For $0 < q < 1$, the q-Jackson integral in the interval [0, a] and the q-improper integral are defined as: $\int_0^a f(t)d_q(t) = a(1-q)\sum_{n=0}^{\infty} f(aq^n)q^n, a > 0$ and

$\int_0^{\infty/A} f(t)d_q(t) = (1-q)\sum_{n=-\infty}^{\infty} f(\frac{q^n}{A})\frac{q^n}{A}, A > 0$, respectively, provided where the sums converge absolutely.

For $t > 0$, q-Gamma function is defined as:

$$\Gamma_q(s) = K(A, s)\int_0^{\infty/A(1-q)} t^{s-1}e_q(-t)d_q(t),$$

where $K(A, s) = \frac{A^s}{1+A}(1 + \frac{1}{A})_q^s(1 + A)_q^{1-s}$. In particular, for $s \in N, K(A, s) = q^{s(s-1)/2}, K(A, 0) = 1$ and $\Gamma_q(s+1) = [s]_q\Gamma_q(s), \Gamma_q(1) = 1$.

For $f \in C[0, \infty), q \in (0, 1), x \in [0, \infty), n \in N$, q-Phillips operators is defined as (see [5]):

$$\begin{aligned} P_n^q(f; x) = [n]_q \sum_{k=1}^{\infty} p_{n,k}(x, q) \int_0^{\infty/A(1-q)} q^{k(k-1)} p_{n,k-1}(t, q) f(t) d_q(t) \\ + e_q(-[n]_q x) f(0), \end{aligned} \tag{1}$$

where

$$p_{n,k}(x, q) = \frac{([n]_q x)^k}{[k]_q!} e_q(-[n]_q x). \tag{2}$$

2 Some Lemmas

Lemma 1 *(see [5]) For the operators $P_n^q(f;x)$ given by (1) let $e_m(t) = t^m$, $m = 0, 1, 2, 3, 4$, then*

(i) $P_n^q(e_0;x) = 1;$ (3)

(ii) $P_n^q(e_1;x) = \frac{x}{q};$ (4)

(iii) $P_n^q(e_2;x) = \frac{x^2}{q^4} + \frac{[2]_q}{q^3[n]_q}x;$ (5)

(iv) $P_n^q(e_3;x) = \frac{x^3}{q^9} + \frac{[2]_q q + [4]_q}{q^8[n]_q}x^2 + \frac{[2]_q[3]_q}{q^6[n]_q^2}x;$

(v)
$$\begin{aligned}P_n^q(e_4;x) = &\frac{x^4}{q^{16}} + \frac{[2]_q q^2 + [4]_q q + [6]_q}{q^{15}[n]_q}x^3 \\ &+ \frac{[2]_q[3]_q q^2 + [2]_q[5]_q q + [4]_q[5]_q}{q^{13}[n]_q^2}x^2 + \frac{[2]_q[3]_q[4]_q}{q^{10}[n]_q^3}x.\end{aligned}$$

Lemma 2 *Let sequence $\{q_n\}$ satisfying $q_n \in (0,1)$, $\lim_{n\to\infty} q_n = 1$and $\lim_{n\to\infty} q_n^n = c(c < 1)$. Then for any $x \in [0,\infty)$, we have*

(i) $\lim_{n\to\infty} [n]_{q_n} P_n^{q_n}((t-x)^2;x) = 2(1-c)x^2 + 2x;$

(ii) $\lim_{n\to\infty} [n]_{q_n}^2 P_n^{q_n}((t-x)^4;x) = 12x^2 + 28(1-c)x^3 + 12(1-c)^2x^4.$

Proof In view of $[n]_{q_n} = \frac{1-q_n^n}{1-q_n}$, by the linearity of the $P_n^q(f;x)$ and Lemma 1, we have

$$\begin{aligned}\lim_{n\to\infty} [n]_{q_n} P_n^{q_n}((t-x)^2;x) &= \lim_{n\to\infty} [n]_{q_n}[(\frac{1}{q_n^4} - \frac{2}{q_n} + 1)x^2 + \frac{[2]_{q_n}}{q_n^3[n]_{q_n}}x] \\ &= 2(1-c)x^2 + 2x.\end{aligned}$$

$$\begin{aligned}P_n^{q_n}((t-x)^4;x) =& \frac{[2]_{q_n}[3]_{q_n}[4]_{q_n}}{q_n^{10}[n]_{q_n}^3}x \\ &+ \frac{x^2}{[n]_{q_n}^2}(\frac{[2]_{q_n}[3]_{q_n}q_n^2 + [2]_{q_n}[5]_{q_n}q_n + [4]_{q_n}[5]_{q_n}}{q_n^{13}} - \frac{4[2]_{q_n}[3]_{q_n}}{q_n^6}) \\ &+ \frac{x^3}{[n]_{q_n}}(\frac{[2]_{q_n}q_n^2 + [4]_{q_n}q_n + [6]_{q_n}}{q_n^{15}} - \frac{4([2]_{q_n}q_n + [4]_{q_n})}{q_n^8} + \frac{6[2]_{q_n}}{q_n^3})\end{aligned}$$

$$+(\frac{1}{q_n^{16}}-\frac{4}{q_n^9}+\frac{6}{q_n^4}-\frac{4}{q_n}+1)x^4$$
$$=: A_{n,q_n}+B_{n,q_n}+C_{n,q_n}+D_{n,q_n}.$$

In view of $q_n \in (0,1)$, $\lim_{n\to\infty} q_n = 1$, we have $[n]_{q_n} \to \infty$ as $n \to \infty$ (see [11]), so, using $[n]_{q_n} = \frac{1-q_n^n}{1-q_n}$, it is clear that

$$\lim_{n\to\infty}[n]_{q_n}^2 A_{n,q_n} = 0,\ \lim_{n\to\infty}[n]_{q_n}^2 B_{n,q_n} = 12x^2,$$

$$\lim_{n\to\infty}[n]_{q_n}^2 D_{n,q_n} = \lim_{n\to\infty}(\frac{1-q_n^n}{1-q_n})^2\frac{1-4q_n^7+6q_n^{12}-4q_n^{15}+q_n^{16}}{q_n^{16}}x^4$$
$$=\lim_{n\to\infty}\frac{(1-q_n^n)^2}{q_n^{16}}(q_n^{14}-2q_n^{13}-5q_n^{12}-8q_n^{11}-5q_n^{10}-2q_n^9$$
$$+q_n^8+4q_n^7+7q_n^6+6q_n^5+5q_n^4+4q_n^3+3q_n^2+2q_n+1)x^4$$
$$=12(1-c)^2x^4.$$

By a similar calculation, we have $\lim_{n\to\infty}[n]_{q_n}^2 C_{n,q_n} = 28(1-c)x^3$. So

$$\lim_{n\to\infty}[n]_{q_n}^2 P_n^{q_n}((t-x)^4;x) = 12x^2+28(1-c)x^3+12(1-c)^2x^4.$$

3 Weighted Statistical Approximation

Let N denote a set of all natural numbers, K be a subset of N, χ_K is the characteristic function of K, The density of K is defined by $\delta(K) = \lim_{n\to\infty}\frac{1}{n}\sum_{k=1}^{n}\chi_K(k)$, provided the limit exists (see [12]). A sequence $x = \{x_n\}$ is called statistically convergent to a number L, if for every $\varepsilon > 0$, $\delta\{n \in N : |x_n - L| \geq \varepsilon\} = 0$ (see [13]). This convergence is denoted as $st - \lim_{n\to\infty} x_n = L$. Let $A = \{a_{nk}\}, n, k = 1, 2, 3, \ldots$ be an infinite summability matrix. For a given sequence $x = \{x_k\}$, the A-transform of x, denoted by $Ax = ((Ax)_n)$, is defined as $(Ax)_n = \sum_{k=1}^{\infty} a_{nk}x_k$, provided the series converges for each n. A is said to be regular if $\lim_{n\to\infty}(Ax)_n = L$ whenever $\lim_{k\to\infty} x_k = L$ (see [14]) . Suppose that $A = \{a_{nk}\}$ is nonnegative regular summability matrix. Then $x = \{x_n\}$ is A-statistically convergent to L if for every $\varepsilon > 0$, $\lim_{n\to\infty}\sum_{k:|x_k-L|\geq\varepsilon} a_{nk} = 0$, and we write $st_A - \lim_{n\to\infty} x_n = L$ (see [15]). If $A = C_1$ is the Cesaro matrix of order one, then A-statistically convergence reduces to the statistical convergence (see [16]).

Let R is a set of all real number. $C(R)$ denotes the space of all functions f which are continuous in R.$\rho(x)$is called as weight function, if it satisfy $\rho(x) \in C(R)$, $\rho(x) \geq 1$and $\lim\limits_{|x|\to\infty} \rho(x) = \infty$.

Let $B_\rho(R) = \{f | f : R \to R, |f(x)| \leq M_f \rho(x)\}$,where $\rho(x)$ is weighted function, M_f is a positive constant depending only on f.Denoting $C_\rho(R) = \{f | f \in B_\rho(R) \cap C(R); ||f||_\rho = \sup\limits_{x\in R} \frac{|f(x)|}{\rho(x)}\}$.

Currently, Duman and Orhan proved a weighted Korovkin type theorem via A-statistical convergence. Now, we recall this theorem.

Theorem 1 *(see* [17]*) Let* $A = \{a_{nk}\}$ *be a nonnegative regular summability matrix,* $\{T_n\}$ *is a sequence of linear positive operators acting from* $C_{\rho_1}(R)$ *to* $B_{\rho_2}(R)$*, where* ρ_1 *and* ρ_2 *are weighted functions,* $\lim\limits_{|x|\to\infty} \frac{\rho_1(x)}{\rho_2(x)} = 0$. *Then for any* $f \in C_{\rho_1}(R)$, $st_A - \lim\limits_{n\to\infty} ||T_n f - f||_{\rho_2} = 0$ *if and only if* $st_A - \lim\limits_{n\to\infty} ||T_n F_v - F_v||_{\rho_1} = 0$*, where* $F_v(x) = \frac{x^v \rho_1(x)}{1+x^2}$, $v = 0, 1, 2$.

By Theorem 1, we can immediately get the following corollary.

Corollary 1 *Let* $\alpha > 0, e_m(t) = t^m, m = 0, 1, 2.\{T_n\}$ *is a sequence of linear positive operators acting from* $C_{1+x^2}(R)$ *to* $B_{1+x^{2+\alpha}}(R)$. *Then for any* $f \in C_{1+x^2}(R)$, $st - \lim\limits_{n\to\infty} ||T_n f - f||_{1+x^{2+\alpha}} = 0$ *if and only if* $st - \lim\limits_{n\to\infty} ||T_n e_m - e_m||_{1+x^2} = 0$.

Proof In Theorem 1, we take $\rho_1 = 1 + x^2$, $\rho_2 = 1 + x^{2+\alpha}$, also let $A = C_1$ be the Cesaro matrix of order one, then we can get the desired conclusion.

Let sequence $q = \{q_n\}, 0 < q_n < 1$ satisfies the condition:

$$st - \lim_{n\to\infty} q_n = 1, st - \lim_{n\to\infty} q_n^n = a(a < 1) \tag{6}$$

Let $m > 0$, denoting

$B_{1+x^m}[0, \infty) = \{f | f : [0, \infty) \to R, |f(x)| \leq M'_f(1 + x^m)\}$, where M'_f is a positive constant depending only on f. Also let $C[0, \infty)$ denote the space of all functions f which are continuous in $[0, \infty)$, denoting $C_{1+x^m}[0, \infty) = \{f | f \in B_{1+x^m}[0, \infty) \cap C[0, \infty); ||f||_{1+x^m} = \sup\limits_{x\in[0,\infty)} \frac{|f(x)|}{1+x^m}\}$, $C^*_{1+x^m}[0, \infty) = \{f | f \in C_{1+x^m}[0, \infty), \exists \lim\limits_{x\to\infty} \frac{f(x)}{1+x^m} < \infty\}$.

The next we give the weighted statistical approximation properties of the q-Phillips operators.

Theorem 2 *Let sequence* $q = \{q_n\}, 0 < q_n < 1$ *satisfy the condition (6) then for any* $\alpha > 0$ *and any* $f \in C_{1+x^2}[0, \infty)$*, we have* $st - \lim\limits_{n\to\infty} ||P_n^{q_n}(f; \cdot) - f||_{1+x^{2+\alpha}} = 0$.

Proof By Lemma 1 we have $P_n^q(1+t^2;x) \le C(1+x^2)$, so, $\{P_n^{q_n}(f;x)\}$ is a sequence of linear positive operators acting from $C_{1+x^2}[0,\infty)$ to $B_{1+x^{2+\alpha}}[0,\infty)$.

By (3), it is clear that $st-\lim_{n\to\infty} ||P_n^{q_n}(e_0;\cdot)-e_0||_{1+x^2}=0$.

By (4), we have $||P_n^{q_n}(e_1;\cdot)-e_1||_{1+x^2} = \sup_{x\in[0,\infty)} \frac{|P_n^{q_n}(e_1;x)-e_1|}{1+x^2} \le \frac{1}{q_n}-1$.

For any $\varepsilon>0$, let $U=\{k: ||P_k^{q_k}(e_1;\cdot)-e_1||_{1+x^2}\ge\varepsilon\}$, $U_1=\{k:\frac{1}{q_k}-1\ge\varepsilon\}$. It is clear that $U\subseteq U_1$, thus

$$\delta\{k\le n: ||P_k^{q_k}(e_1;\cdot)-e_1||_{1+x^2}\ge\varepsilon\} \le \delta\{k\le n: \frac{1}{q_k}-1\ge\varepsilon\}. \tag{7}$$

Since $st-\lim_{n\to\infty} q_n = 1$, so $st-\lim_{n\to\infty}(\frac{1}{q_n}-1)=0$. Thus, by (7) we have $st-\lim_{n\to\infty} ||P_n^{q_n}(e_1;\cdot)-e_1||_{1+x^2}=0$.

By (5), we have

$$||P_n^{q_n}(e_2;\cdot)-e_2||_{1+x^2} = \sup_{x\in[0,\infty)} \frac{|P_n^{q_n}(e_2;x)-e_2|}{1+x^2} \le (\frac{1}{q_n^4}-1)+\frac{[2]_{q_n}}{q_n^3[n]_{q_n}}.$$

For any $\varepsilon>0$, let
$V=\{k: ||P_k^{q_k}(e_2;\cdot)-e_2||_{1+x^2}\ge\varepsilon\}$, $V_1=\{k:\frac{1}{q_k^4}-1\ge\frac{\varepsilon}{2}\}$,
$V_2=\{k:\frac{[2]_{q_k}}{q_k^3[k]_{q_k}}\ge\frac{\varepsilon}{2}\}$. It is clear that $V\subseteq V_1\cup V_2$, so

$$\begin{aligned}&\delta\{k\le n: ||P_k^{q_k}(e_2;\cdot)-e_2||_{1+x^2}\ge\varepsilon\}\\ &\le \delta\{k\le n: \frac{1}{q_k^4}-1\ge\frac{\varepsilon}{2}\}+\delta\{k\le n:\frac{[2]_{q_k}}{q_k^3[k]_{q_k}}\ge\frac{\varepsilon}{2}\}.\end{aligned} \tag{8}$$

Since $st-\lim_{n\to\infty} q_n=1$, $st-\lim_{n\to\infty} q_n^n = a(a<1)$, so $st-\lim_{n\to\infty}(\frac{1}{q_n^4}-1)=0$, $st-\lim_{n\to\infty}\frac{[2]_{q_n}}{q_n^3[n]_{q_n}}=0$. Thus, by (8) we have $st-\lim_{n\to\infty}||P_n^{q_n}(e_2;\cdot)-e_2||_{1+x^2}=0$. So, by Corollary 1, Theorem 2 was got.

4 Korovkin Type Convergence Theorem

Theorem 3 *Let $q_n\in(0,1)$, then the sequence $\{P_n^{q_n}(f;x)\}$ converges to f uniformly on $[0,A]$ for any $f\in C^*_{1+x^2}[0,\infty)$ if and only if $\lim_{n\to\infty} q_n=1$.*

Proof Assume that $\lim_{n\to\infty} q_n = 1$. Fix $A>0$ and consider the lattice homomorphism $T_A: C[0,\infty)\to C[0,A]$ defined by $T_A(f)=f_{[0,A]}$, then for $e_m(t)=t^m$, $m=0,1,2$, by Lemma 1 we have $T_A(P_n^{q_n}(e_m;x))$ converges to $T_A(e_m(t))$ uniformly on $[0,A]$. {[18], Proposition 4.2.5 (6)} and its proof say that $C^*_{1+x^2}[0,\infty)$ is isomorphic to $C[0,1]$ and that the set $\{1,t,t^2\}$ is a Korovkin set in $C^*_{1+x^2}[0,\infty)$. So the universal Korovkin type property [see [18] Theorem 4.1.4 (vi)] implies that $\{P_n^{q_n}(f;x)\}$ converges to f uniformly on $[0, A]$, provided $f \in C^*_{1+x^2}[0,\infty)$.

On the other hand, if we assume that for any $f \in C^*_{1+x^2}[0,\infty)$, the sequence $\{P_n^{q_n}(f;x)\}$ converges to f uniformly on $[0, A]$, then $\lim_{n\to\infty} q_n = 1$. Indeed, if $\{q_n\}$ does not tend to 1, then it must contain a subsequence $\{q_{n_k}\}$, such that $q_{n_k} \in (0,1)$ and $q_{n_k} \to q_0 \in [0,1)$ as $k \to \infty$. Thus,

$$\frac{1}{[n_k]_{q_{n_k}}} = \frac{1-q_{n_k}}{1-(q_{n_k})^{n_k}} \to 1-q_0 \quad \text{as} \quad k \to \infty.$$

Taking $n = n_k, q = q_{n_k}$ in $\{P_n^{q_n}(t^2;x)\}$, by Lemma 1, we get $\lim_{k\to\infty}(P_{n_k}^{q_{n_k}}(t^2;x) - x^2) = (\frac{1}{q_0^4} - 1)x^2 + \frac{1-q_0^2}{q_0^3}x \neq 0$.

This leads to a contradiction, hence $\lim_{n\to\infty} q_n = 1$.

5 Voronovskaja Type Asymptotic Formulas

The last we give the Voronovskaja type asymptotic formulas of the q-Phillips operators.

Theorem 4 *Assume that $q_n \in (0,1)$, $\lim_{n\to\infty} q_n = 1$, $\lim_{n\to\infty} q_n^n = c (c < 1)$. For any $f \in C^*_{1+x^2}[0,\infty)$ such that $f', f'' \in C^*_{1+x^2}[0,\infty)$. Then, we have $\lim_{n\to\infty}[n]_{q_n}(P_n^{q_n}(f;x) - f(x)) = [(1-c)x^2 + x]f''(x)$ uniformly on any $[0, A]$, $A > 0$.*

Proof Let $f, f', f'' \in C^*_{1+x^2}[0,\infty)$ and $x \in [0,\infty)$ be fixed. By the Taylor formula, we have $f(t) - f(x) = f'(x)(t-x) + \frac{f''(x)}{2}(t-x)^2 + \psi(t,x)(t-x)^2$, where $\psi(\cdot,x) \in C^*_{1+x^2}[0,\infty)$, $\psi(t,x) \to 0 (t \to x)$. So, by Lemma 1 we can get

$$\begin{aligned}[n]_{q_n}(P_n^{q_n}(f;x) - f(x)) &= \frac{f''(x)}{2}[n]_{q_n}P_n^{q_n}((t-x)^2;x) \\ &\quad + [n]_{q_n}P_n^{q_n}(\psi(t,x)(t-x)^2;x). \qquad (9)\end{aligned}$$

By the Cauchy-Schwartz inequality, we have

$$\begin{aligned}[n]_{q_n}P_n^{q_n}(\psi(t,x)(t-x)^2;x) &\leq \sqrt{P_n^{q_n}(\psi^2(t,x);x)} \\ &\quad \cdot \sqrt{[n]_{q_n}^2 P_n^{q_n}((t-x)^4;x)} \qquad (10)\end{aligned}$$

Observe that $\psi^2(x,x) = 0$ and $\psi^2(\cdot,x) \in C^*_{1+x^2}[0,\infty)$, then by Theorem 3 we have $\lim_{n\to\infty} P_n^{q_n}(\psi^2(t,x);x) = \psi^2(x,x) = 0$ uniformly with respect to $x \in [0,A]$, $A > 0$. Thus, by (10) and Lemma 2, we can obtain $\lim_{n\to\infty} [n]_{q_n} P_n^{q_n}(\psi(t,x)(t-x)^2;x) = 0$ uniformly with respect to $x \in [0,A]$. Hence, by (9) and Lemma 2, we can get immediately $\lim_{n\to\infty} [n]_{q_n}(P_n^{q_n}(f;x) - f(x)) = [(1-c)x^2 + x]f''(x)$ uniformly on any $[0,A]$.

6 Conclusion

In the paper, the weighted statistical approximation theorem of the q-Phillips operators which was given by (1) is obtained. Also a convergence theorem of Korovkin type and a Voronovskaja-type asymptotic formulas are given. If we use King's approach to consider King type modification of the extension of the q-Phillips operators, we will obtain better weighted statistical approximation (cf. [19, 20]).

Acknowledgments This work is supported by the National Natural Science Foundation of China (Grant No. 61170324) and the Class A Science and Technology Project of Education Department of Fujian Province, China (Grant No. JA12324).

References

1. Phillips, G.M.: Bernstein polynomials based on the q-integers. Ann. Numer. math. **4**, 511–518 (1997)
2. Agratini, O., Nowak, G.: On a generalization of Bleimann, Butzer and Hahn operators based on q-integer. Math. Comput. Model. **53**(5–6), 699–706 (2011)
3. Doğru, O., Orkcu, M.: Statistical approximation by a modification of q-Meyer- König-Zeller operators. Appl. Math. Lett. **23**, 261–266 (2010)
4. Gupta, V., Radu, C.: Statistical approximation properties of q-Baskakov Kantorovich operators. Cent. Eur. J. Math. **7**(4), 809–818 (2009)
5. Yüksel, I.: Approximation by q-Phillips operators. Hacet. J. Math. Stat. **40**(2), 191–201 (2011)
6. De Sole, A., Kac, V.G.: On integral representations of q-Gamma and q-Beta Functions. Rend. Mat. Acc. Lincei **16**(6), 11–29 (2005)
7. Gasper, G., Rahman, M.: Basic hypergeometrik series. In: Encyclopedia of Mathematics and its Applications, vol. 35. Cambridge University Press, Cambridge (1990)
8. Jackson, F.H.: On a q-definite integrals. Q. J. Pure Appl. Math. **41**, 193–203 (1910)
9. Kac, V.G., Cheung, P.: Quantum calculus. Springer, New York (2002). Universitext
10. Koelink, H.T., Koornwinder, T.H.: q-special functions. In: A tutorial, in Deformation Theory and Quantum Groups with Applications to Mathematical Physics, Amherst, MA (1990). Contemp. Math. **134**, 141–142. Amer. Math. Soc., Providence, RI (1992)
11. Videnskii, V.S.: On q-Bernstein polynomials and related positive linear operators (in Russian). In: Problems of Modern Mathematics and Mathematical Education. St.-Pertersburg, pp. 118–126 (2004)
12. Niven I, Zuckerman H.S., Montgomery H.: An Introduction to the Theory of Numbers, 5th edn. Wiley, New york (1991)

13. Fast, H.: Sur la Convergence Statistique. Colloq. Math. **2**, 241–244 (1951)
14. Hardy, G.H.: Divergent series. Oxford University Press (1949)
15. Fridy, J.A., Miller, H.I.: A matrix characterization of statistical convergence. Analysis **11**, 59–66 (1991)
16. Duman, O., Orhan, C.: Rates of a-statistical convergence of positive linear operators. Appl. Math. Lett. **18**, 1339–1344 (2005)
17. Duman, O., Orhan, C.: Statistical approximation by positive linear operators. Studia Math. **161**(2), 187–197 (2004)
18. Altomare F, Campiti M.: Korovkin-type approximation theory and its applications. In: de Gruyter Studies in Mathematics, vol. 17. Walter de Gruyter and Co., Berlin, pp. xii+627 (1994)
19. King, J.P.: Positive linear operators which preserve x^2. Acta Math. Hungar. **99**(3), 203–208 (2003)
20. Mahmudov N.I.: q-Szász-Mirakjan operators which preserve x^2. J. Comput. Appl. Math. **235**, 4621–4628 (2011)

Consistency Adjustment Algorithm of the Reciprocal Judgment Matrix

Wei-xia Li, Cheng-yi Zhang and Hua Yang

Abstract In this paper, firstly, according to the problem of the consistency of reciprocal judgment matrix, two kinds of consistency recursive iterative adjustment algorithms were given. The algorithm is based on adjustment by order, and fixed value randomly to adjust other value, then choose the matrix as consistency matrix which is corresponding the minimum deviation value. Then give an example to adjust the reciprocal judgment matrix to be consistency by using the two kinds of recursive iterative adjustment algorithm.

Keywords Positive reciprocal judgment matrix · Consistency recursive iterative adjustment algorithm · Analytic hierarchy process (AHP)

1 Introduction

Since 1980, analytic hierarchy process (AHP) [1] was applied widely, and it solved many significant practical problems. The key problem of AHP is the consistency of judgment matrix which is based on the comparison each other of the elements. The inconsistency of judgment matrix imply the weight obtained of the elements is not in conformity with the actual situation, ultimately, it can not give the accurate sorting of each scheme. In recent years, there are many problems about the consistency checking and adjustment of the positive reciprocal judgment matrix. For example, [2–4] gave several consistency adjustment methods based on the relationship between the weight vector derived from consistency positive reciprocal judgment matrix and eigenvec-

W. Li
Department of Public Health, Hainan Medical University, Hainan 571158, China
e-mail: liweixia851019@163.com

C. Zhang (✉) · H. Yang
Department of Mathematics and Statistics, Hainan Normal University, Hainan 571158, China
e-mail: chengyizh@hainnu.edu.cn

B.-Y. Cao and H. Nasseri (eds.), *Fuzzy Information & Engineering and Operations Research & Management*, Advances in Intelligent Systems and Computing 211,
DOI: 10.1007/978-3-642-38667-1_38,

tors of judgment matrix; [5–7] adjusted the elements of positive reciprocal judgment matrix to be consistent based on probability theory and statistical knowledge; [8, 9] introduced the concept of the perturbation matrix, and analyzed relationship among the judgment matrix, the export matrix and measure matrix to adjust the judgment matrix; Based on the optimization point of view, [10–12] established the optimization model to adjust consistency of the positive reciprocal judgment matrix; [13] proposed interactive analysis method for the adjustment of the judgment matrix. However, the methods of the consistency judgment seemed more complex, and also was lack of a theoretical basis, then the consistency adjustment method maybe incomplete consistency, even the results appear serious inconsistency with the information contained in the original judgment matrix. About the consistency adjustment method of the reciprocal judgment matrix, two recursive iterations adjustment algorithm were introduced, and its essence is starting reciprocal judgment matrix by order. The two kinds of methods fixed elements values randomly of row vector to adjust other elements, then made positive reciprocal judgment matrix to be consistency by order. Then elect random element corresponding the minimal deviation value and the corresponding consistency adjustment matrix of this order, when compared the deviation value between the adjustment matrix and the original judgment matrix. So give the consistency adjustments of the reciprocal judgment matrix such by-order. The method avoids large deviation value between the adjusted consistency matrix and the original judgment matrix information, and adjusted positive reciprocal judgment matrix was complete consistency. It improves the awareness and understanding of judge information, as well as the accuracy of the adjustment.

2 Preliminaries

Definition 2.1 Let $A = (a_{ij})_{n\times n}$ be a judgment matrix, where $a_{ij} \in R$ $(i, j \in N)$, if

1. $a_{ij} > 0$, $(i, j = 1, 2, \ldots, n)$;
2. $a_{ji} = \frac{1}{a_{ij}}$, $(i, j = 1, 2, \ldots, n)$,

then A is called the positive reciprocal judgment matrix.

Definition 2.2 Let $A = (a_{ij})_{n\times n}$ be an positive reciprocal judgment matrix for $(i, j = 1, 2, \ldots, n)$, if for each k, $a_{ij} = a_{ik}a_{kj}$, then A is called consistency positive reciprocal judgment matrix.

Theorem 2.1 Let $A = (a_{ij})_{n\times n}$ be an positive reciprocal judgment matrix where $(i, j = 1, 2, \ldots, n)$, and $w = (w_i)_{1\times n}$ be the weight vector of $A = (a_{ij})_{n\times n}$, then for $\forall k = 1, 2, \ldots, n$, $w_i = a_{ik}/\sum_{i=1}^{n} a_{ik}$.

3 Consistency Recursive Iterative Adjustment Algorithm of Positive Reciprocal Judgment Matrix

3.1 Basic Definition and Theorem

Symbols are as follows:

1. Let $A = (a_{ij})_{n\times n}$ be an positive reciprocal judgment matrix, then $A^{(k)}$signify the leading principal submatrix of order K of A.
2. Let $A^{(k)}$ be the leading principal submatrix of order K of A, then $A_k^{(s)}$ signify the leading principal submatrix of order s of $A^{(k)}$ where $1 \le s \le k$.
3. Let $A^{(k)}$ be the leading principal submatrix of order K of A, then $B^{(k)}$ signify the consistency positive reciprocal judgment matrix of $A^{(k)}$.
4. Let $A = (a_{ij})_{n\times n}$ be an positive reciprocal judgment matrix, then $C_k^{(k-1)} = (c_{ij}^{(k-1)})_{k\times k}$ signify the leading principal submatrix of order K, which satisfied the leading principal submatrix of order $k-1$ is consistency positive reciprocal judgment matrix $B^{(k-1)}$ and the elements in the kth row(column) are the same as A.

Definition 3.3 Let $A = (a_{ij})_{n\times n}$ be positive reciprocal judgment matrix, then $A_1 = (a'_{ij})_{n\times n}$ is called as column normalized matrix of A, where $a'_{ij} = \frac{a_{ij}}{\sum_{i=1}^{n} a_{ij}}$.

Definition 3.4 Let $A_1 = (a'_{ij})_{n\times n}$ and $B_1 = (b'_{ij})_{n\times n}$ be column normalized matrix of positive reciprocal judgment matrixes of $A = (a_{ij})_{n\times n}$ and $B = (b_{ij})_{n\times n}$, then $E = (e_{ij})_{n\times n}$ is called the deviation matrix between A and B, where $e_{ij} = a'_{ij} - b'_{ij}$.

Definition 3.5 Let $A = (a_{ij})_{n\times n}$ be positive reciprocal judgment matrix, C be the consistency judgment matrix of $A = (a_{ij})_{n\times n}$, and $W = (w_1, w_2, \ldots, w_n)$ be the weight vector of C, then $D = (d_{ij})_{n\times n}$ is called the derived matrix of $A = (a_{ij})_{n\times n}$, such that $d_{ij} = a_{ij}\frac{w_j}{w_i}$.

Theorem 3.2 Let $A = (a_{ij})_{n\times n}$ be positive reciprocal judgment matrix, if A is consistency positive reciprocal judgment matrix, then all the value of elements of its derived matrix D are one, and that is $D = \begin{pmatrix} 1 & 1 & \ldots & 1 \\ 1 & 1 & \ldots & 1 \\ \vdots & \vdots & \vdots & \vdots \\ 1 & 1 & \ldots & 1 \end{pmatrix}$.

3.2 Consistency Recursive Iterative Adjustment Algorithm of Positive Reciprocal Judgment Matrix 1

Let $A = (a_{ij})_{n\times n}$ be positive reciprocal judgment matrix, then the consistent recursive iterative adjustment algorithm is as follows:

Step 1: $A_1 = (1)$ and $A_2 = \begin{pmatrix} 1 & a_{12} \\ a_{21} & 1 \end{pmatrix}$ that are consistency positive reciprocal judgment matrix are the leading principal sub-matrix of order one and of order two of A respectively;

Step 2: Suppose for each $k > 2$, $\forall 1 \leq r \leq k$, $C_k^{(k-1)} = (c_{ij}^{(k-1)})_{k\times k}$, let $t_{kr}^{(km)} = a_{km}c_{mr}^{(k-1)}$ and $t_{rk}^{(km)} = 1/t_{kr}^{(km)}$, that is $T_k^m = (t_{ij}^{(km)})_{k\times k}$ $(\forall 1 \leq m < k)$;

Step 3: Calculate the deviate value $E_k^m = (e_{ij}^{(km)})_{k\times k}$ of $A^{(k)}$and T_k^m;

Step 4: Determine $J_k = \{l_k | s_k^{l_k} = \min\{s_k^m\}\}$ such that $s_k^m = \sum_{j=1}^{k}\sum_{i=1}^{k}\left|e_{ij}^{(km)}\right|$, and let $B^{(k)} = \{T_k^{l_k} | l_k = \min\{J_k\}\}$;

Step 5: Let $k = k + 1$. If $k \leq n$, then go to Step 2. Otherwise, continue to Step 6.

Step 6: Let $B = B^{(k)}$, then output B.

Step 7: End.

Theorem 3.3 Let $A = (a_{ij})_{n\times n}$ be positive reciprocal judgment matrix, and $A^{(k-1)}$ is adjusted to $B^{(k-1)}$. If the elements of $B^{(k)}$ were recursive iterations adjusted such that $b_{kj}^{(k)} = a_{kl_k}b_{l_kj}^{(l_k)}$, $b_{jk}^{(k)} = 1/b_{kj}^{(k)}$, then $B^{(k)}$ is consistency positive reciprocal judgment matrix.

Proof. Firstly, since the elements of the kth row of $B^{(k)}$ satisfied $b_{ks}^{(k)} = b_{kl_k}^{(k)}b_{l_ks}^{(k)}$, $b_{sj}^{(k)} = b_{sl_k}^{(k)}b_{l_kj}^{(k)}$ for $\forall 1 \leq s \leq k$ and $b_{kj}^{(k)} = b_{kl_k}^{(k)}b_{l_kj}^{(k)}$, then $b_{kj}^{(k)} = b_{ks}^{(k)}b_{sj}^{(k)}$.

Moreover, since $b_{kj}^{(k)} = b_{kl_k}^{(k)}b_{l_kj}^{(k)}$, $b_{l_kj}^{(k)} = b_{l_ki}^{(k)}b_{ij}^{(k)}$ for each $1 \leq i, j \leq k - 1$, then $b_{ij}^{(k)} = b_{kj}^{(k)}b_{ik}^{(k)}$. Hence, $b_{ij}^{(k)} = b_{is}^{(k)}b_{sj}^{(k)}$ where $\forall 1 \leq s \leq k$. Therefore, $B^{(h)}$ is consistency positive reciprocal judgment matrix.

3.3 Consistency Recursive Iterative Adjustment Algorithm of Positive Reciprocal Judgment Matrix 2

Let $A = (a_{ij})_{n\times n}$ be positive reciprocal judgment matrix, then the consistent recursive iterative adjustment algorithm is as follows:

Step 1: $A_1 = (1)$ and $A_2 = \begin{pmatrix} 1 & a_{12} \\ a_{21} & 1 \end{pmatrix}$ that are consistency positive reciprocal judgment matrix are the leading principal sub-matrix of order one and of order two of A respectively;

Step 2: Suppose for each $k > 2, \forall 1 \le r \le k$, $C_k^{(k-1)} = (c_{ij}^{(k-1)})_{k\times k}$, let $t_{kr}^{(km)} = a_{km}c_{mr}^{(k-1)}$ and $t_{rk}^{(km)} = 1/t_{kr}^{(km)}$, that is $T_k^m = (t_{ij}^{(km)})_{k\times k}(\forall 1 \le m < k)$;

Step 3: Let $A^{(k)} = (a_1^{(k)}, a_2^{(k)}, \ldots, a_k^{(k)})$ and $T_k^m = (t_1^{km}, t_2^{km}, \ldots, t_k^{km})$ where $\forall 1 \le h \le k$, $a_h^{(k)}$ is the line vector of $A^{(k)}$ and t_h^{km} is the line vector of T_k^m, then calculate the value $\cos\theta_{mh}^{(k)} = \frac{(a_h^{(k)}, t_h^{km})}{|a_h^{(k)}||t_h^{km}|}$;

Step 4: Determine $J_k = \{l_k | \cos\theta_{l_k}^{(k)} = \max\{\cos\theta_m^{(k)}\}\}$such that $\cos\theta_m^{(k)} = \sum_{h=1}^k \cos\theta_{mh}^{(k)}$, and let $B^{(k)} = \{T_k^{l_k} | l_k = \min\{J_k\}\}$;

Step 5: Let $k = k+1$; If $k \le n$, then go to Step 2. Otherwise, continue to Step 6;

Step 6: Let $B = B^{(k)}$, then output B;

Step 7: End.

4 Case

Let $A = \begin{pmatrix} 1 & 1/9 & 3 & 1/5 \\ 9 & 1 & 5 & 2 \\ 1/3 & 1/5 & 1 & 1/2 \\ 5 & 1/2 & 2 & 1 \end{pmatrix}$, then adjust A to be consistency positive reciprocal judgment matrix by the two kinds of algorithm above and give the sorting weight vector of A.

Table 1 The results from recursive iterations adjustment Algorithm 1

k	m	T_k^m	S_k^m	l_k	$B^{(k)}$
3	1	$\begin{pmatrix} 1 & 1/9 & 3 \\ 9 & 1 & 27 \\ 1/3 & 1/27 & 1 \end{pmatrix}$	3.729	2	$\begin{pmatrix} 1 & 1/9 & 5/9 \\ 9 & 1 & 5 \\ 1.8 & 1/5 & 1 \end{pmatrix}$
	2	$\begin{pmatrix} 1 & 1/9 & 5/9 \\ 9 & 1 & 5 \\ 1.8 & 1/5 & 1 \end{pmatrix}$	2.933		
	1	$\begin{pmatrix} 1 & 1/9 & 5/9 & 1/5 \\ 9 & 1 & 5 & 1.8 \\ 1.8 & 0.2 & 1 & 0.36 \\ 5 & 5/9 & 5/18 & 1 \end{pmatrix}$	3.582	3	$\begin{pmatrix} 1 & 1/9 & 5/9 & 5/18 \\ 9 & 1 & 5 & 5/2 \\ 1.8 & 0.2 & 1 & 0.5 \\ 3.6 & 0.4 & 2 & 1 \end{pmatrix}$
4	2	$\begin{pmatrix} 1 & 1/9 & 5/9 & 2/9 \\ 9 & 1 & 5 & 2 \\ 1.8 & 0.2 & 1 & 0.4 \\ 4.5 & 1/2 & 5/2 & 1 \end{pmatrix}$	3.446		
	3	$\begin{pmatrix} 1 & 1/9 & 5/9 & 5/18 \\ 9 & 1 & 5 & 5/2 \\ 1.8 & 0.2 & 1 & 0.5 \\ 3.6 & 0.4 & 2 & 1 \end{pmatrix}$	3.2		

Table 2 The results from recursive iterations adjustment Algorithm 2

k	m	T_k^m	$\cos\theta_m^{(k)}$	l_k	$B^{(k)}$
3	1	$\begin{pmatrix} 1 & 1/9 & 3 \\ 9 & 1 & 27 \\ 1/3 & 1/27 & 1 \end{pmatrix}$	2.889	1	$\begin{pmatrix} 1 & 1/9 & 5/9 \\ 9 & 1 & 5 \\ 1.8 & 1/5 & 1 \end{pmatrix}$
	2	$\begin{pmatrix} 1 & 1/9 & 5/9 \\ 9 & 1 & 5 \\ 1.8 & 1/5 & 1 \end{pmatrix}$	2.962		
	1	$\begin{pmatrix} 1 & 1/9 & 5/9 & 1/5 \\ 9 & 1 & 5 & 1.8 \\ 1.8 & 0.2 & 1 & 0.36 \\ 5 & 5/9 & 5/18 & 1 \end{pmatrix}$	3.899		
4	2	$\begin{pmatrix} 1 & 1/9 & 5/9 & 2/9 \\ 9 & 1 & 5 & 2 \\ 1.8 & 0.2 & 1 & 0.4 \\ 4.5 & 1/2 & 5/2 & 1 \end{pmatrix}$	3.905	2	$\begin{pmatrix} 1 & 1/9 & 5/9 & 2/9 \\ 9 & 1 & 5 & 2 \\ 1.8 & 0.2 & 1 & 0.4 \\ 4.5 & 1/2 & 5/2 & 1 \end{pmatrix}$
	3	$\begin{pmatrix} 1 & 1/9 & 5/9 & 5/18 \\ 9 & 1 & 5 & 5/2 \\ 1.8 & 0.2 & 1 & 0.5 \\ 3.6 & 0.4 & 2 & 1 \end{pmatrix}$	3.895		

On the one hand, by using the recursive iterations adjustment Algorithm 1, we can obtain the consistency positive reciprocal judgment matrix of A as follows: $\begin{pmatrix} 1 & 1/9 & 5/9 & 5/18 \\ 9 & 1 & 5 & 5/2 \\ 1.8 & 0.2 & 1 & 0.5 \\ 3.6 & 0.4 & 2 & 1 \end{pmatrix}$. Then by the Theorem 1, we get the sorting weight vector $w = (0.0649, 0.5844, 0.1169, 0.2338)'$ and the results are in the Table 1 as following.

On the other hand, by using the recursive iterations adjustment Algorithm 2, we the consistency positive reciprocal judgment matrix of A as follows: $\begin{pmatrix} 1 & 1/9 & 5/9 & 5/18 \\ 9 & 1 & 5 & 5/2 \\ 1.8 & 0.2 & 1 & 0.5 \\ 3.6 & 0.4 & 2 & 1 \end{pmatrix}$. Then by the Theorem 1, we get the sorting weight vector $w = (0.0613, 0.5521, 0.1104, 0.2761)'$ and the results are in the Table 2 as following.

5 Conclusion

About the consistency of the reciprocal judgment matrix, two kinds of consistency recursive iterative adjustment algorithm are given. These two kinds of algorithm were complete consistency recursive iterative adjustment algorithm satisfying people's need. In the end the example was given to verify the practicality of the methods.

Acknowledgments Thanks to the support by NSF of China (71140008), KSTF of Hainan (090802) and (ZDXM20110047), KSTF of Haikou (2010072), Hainan special social development technology (2011SF003), Hainan Natural Science Fund 110008 and Hainan Normal University 2011 Graduate innovative research project (Hys2011-26).

References

1. Saaty, T.L.: The analytic hierarchy process. McGraw-Hill, New York (1980)
2. Saaty, T.L.: A scaling method for priorities in hierarchical structure. Math. Psychol. **15**(3), 234–281 (1977)
3. Saaty, T.L., Vargas, L.G.: Comparison of eigenvalue logarithmic least squares and least squares methods in estimating ratios. Math. Model. **5**, 309–324 (1984)
4. Xu, Z.: A practical method for improving consistency of judgement matrix. Syst. Eng. **16**(6), 61–63 (1998)
5. Liu, W.: A new method of rectifying judgement matrix. Syst. Eng. Theory Pract. **9**(9), 100–104 (1999)
6. Wu, Z., Zhang, W., Guan, X.: A statistical method to check and rectify the consistency of a judgement matrix. Syst. Eng. **20**(3), 67–71 (2002)
7. He, B., Meng, Q.: A new method of checking the consistency of a judgment matrix. J. Ind. Eng. Eng. Manage. **16**(4), 92–94 (2002)
8. Zhang, Q., Long, X.: An iterative algorithm for improving the consistency of judgement matrix in AHP. Math. Pract. Theory **31**(5), 565–568 (2001)
9. Tangsen, Z., Li, H., Lipeng, W.: A greedy algorithms to accelerating rectify judgement–matrix on AHP through measure matrix. Math. Pract Theory **34**(11), 94–97 (2004)
10. Wang, G., Liang, L.: A method of regulating judgement matrix according as consistent rule. Syst. Eng. **19**(4), 90–96 (2001)
11. Juliang, J., Yiming, W., Jinfeng, P.: Accelerating genetic algorithm for correcting judgement matrix consistency in analytic hierarchy process. Syst. Eng. Theory Pract. **24**(1), 63–69 (2004)
12. Wang, X., Dong, Y., Chen, Y.: Consistency modification of judgment matrix based on a genetic simulated annealing algorithm. J. Syst. Eng. **21**(l), 107–111 (2006)
13. Zhengqing, L.: A new method for adjusting inconsistency judgment matrix in AHP. Syst. Eng. Theory Pract. **6**, 84–92 (2004)

Wavelet Frequency Domain Adaptive Multi-Modulus Blind Equalization Algorithm Based on Fractional Lower Order Statistics

Jun Guo, Xiu-zai Zhang and Ye-cai Guo

Abstract A wavelet frequency domain adaptive β multi-modulus blind equalization algorithm based on Fractional lower order statistics (FLOS β FWTMMA) is proposed in the α-stable distribution noise environment. This proposed algorithm uses Fractional lower order statistics to restrain α-stable distribution noise, the equalizer output signal energy is optimized adaptively to obtain a joint blind equalization algorithm, its computational loads can be greatly reduced by using Fast Fourier Transform (FFT) and overlapping retention law. In the proposed algorithm, orthogonal wavelet transform is used to improve the convergence rate. The underwater acoustic channel simulation results show that the proposed algorithm has better performance.

Keywords Fractional lower order statistics · α-stable distribution noise · Frequency domain adaptive multi-modulus algorithm · Orthogonal wavelet transform

1 Introduction

In the blind equalization algorithm which has been studied, the channel noise is assumed to be Gaussian noise. However, a lot of noise in actual performance has obvious peak pulse [1], such as underwater acoustics noise, low frequency atmospheric noise. α-stable distribution can be used to describe this kind of impulse noise with a significant peak pulse waveform, therefore, the traditional blind equalization algorithm based on Gaussian noise model is no longer applicable in α-stable distribution noise environment.

Orthogonal wavelet transform is used to reduce the autocorrelation of the input signals, and the convergence rate of blind equalization algorithm is improved [2].

J. Guo · X. Zhang · Y. Guo (✉)
College of Electronic and Information Engineering, Nanjing University of Information Science and Technology, Nanjing 210044, People's Republic of China
e-mail: guo-yecai@163.com

B.-Y. Cao and H. Nasseri (eds.), *Fuzzy Information & Engineering and Operations Research & Management*, Advances in Intelligent Systems and Computing 211,
DOI: 10.1007/978-3-642-38667-1_39,

Frequency domain blind equalization algorithm is to make the traditional time domain blind equalization algorithm turn into the frequency domain equalization algorithm, and the Fast Fourier Transform and overlapping retention law are used to greatly reduce the amount of computation [3].

The energy of the equalizer output signals and Lagrangian multipliers are combined in adaptive β multi-modulus blind equalization algorithm, a new cost function is defined, and the optimal weight vector is obtained by seeking the most value of cost function.

In this chapter, the adaptive β multi-modulus algorithm is applied to the wavelet frequency domain blind equalization algorithm, and Fractional lower order

statistics [4] is used to restrain α-stable distribution noise, a wavelet frequency domain adaptive β multi-modulus blind equalization algorithm based on Fractional lower order statistics is proposed and its performance is tested by underwater acoustic channels [5].

2 α-Stable Distribution Noise

α-stable distribution noise without the given probability density function is described by following characteristic function, i.e.,

$$\varphi(t) = \begin{cases} \exp\left\{jat - \gamma\,|t|^{\alpha}\left[1 + j\beta \mathrm{sgn}(t)\tan\left(\frac{\pi\alpha}{2}\right)\right]\right\}, & \alpha \neq 1, \\ \exp\left\{jat - \gamma\,|t|^{\alpha}\left[1 + j\beta \mathrm{sgn}(t)\frac{2}{\pi}\lg|t|\right]\right\}, & \alpha = 1, \end{cases} \tag{1}$$

where, $\alpha \in (0, 2]$ is characteristic index and denotes α-stable distribution probability density function tail thickness. γ is dispersion coefficient, which is similar to the variance of Gaussian noise. $\beta \in [-1, 1]$ is symmetric parameter. When $\beta = 0$, α-stable distribution becomes symmetric α-stable distribution, which is regarded as SαS. $a \in (-\infty, \infty)$ is a location parameter and represents the mean or median of the distribution.

3 Adaptive Multi-Modulus Blind Equalization Algorithm

The cost function of adaptive β multi-modulus blind equalization algorithm in literature [6] is defined as

$$J_{\beta MMA} = E[|z(n)|^2]. \tag{2}$$

subject to

$$\begin{cases} \mathrm{fmax}(\gamma_1, |z_r(n)|) = \gamma_1, \\ \mathrm{fmax}(\gamma_1, |z_i(n)|) = \gamma_1, \end{cases} \tag{3}$$

where, $z(n)$ denotes the equalizer output, $z_r(n)$ and $z_i(n)$ are the real component and imaginary component of equalizer output signals, γ_1 is the maximum module of transmitted sequence, f is equalizer weight vector.

fmax is defined as

$$\text{fmax}(a,b) = \frac{|a+b|+|a-b|}{2} = \begin{cases} a, a \geq b \geq 0 \\ b, b \geq a \geq 0 \end{cases} \tag{4}$$

where, a and b are constant. The constellations of an M-alphabet square-QAM are contained in a square region $\Re$, if the equalizer output falls inside region $\Re$, then constraints in Eq. (3) are satisfied, the energy of equalizer output signals stays close to transmitted sequence energy.

For some $a, b \in \text{R}$, we have

$$\frac{\partial}{\partial a}\text{fmax}(|a|,|b|) = \frac{\text{sign}[a]}{2}(1+\text{sign}[|a|-|b|]). \tag{5}$$

where, sign[·] is a sign function. After the Lagrangian multipliers, λ_r and λ_i are employed, we obtain

$$\begin{aligned} J = {} & E[|z_r(n)|^2 + \lambda_r(\text{fmax}(\gamma_1, |z_r(n)|) - \gamma_1)] \\ & + E[|z_i(n)|^2 + \lambda_i(\text{fmax}(\gamma_1, |z_i(n)|) - \gamma_1)]. \end{aligned} \tag{6}$$

According to stochastic gradient method, the weight vector iteration formula are given by

$$\boldsymbol{f}_r(n+1) = \boldsymbol{f}_r(n) + \frac{\mu}{4}[\lambda_r g_r + 4z_r(n)]^* \boldsymbol{y}_r(n), \tag{7}$$

$$\boldsymbol{f}_i(n+1) = \boldsymbol{f}_i(n) + \frac{\mu}{4}[\lambda_i g_i + 4z_i(n)]^* \boldsymbol{y}_i(n), \tag{8}$$

$$g_L = \text{sign}[z_L(n)](1+\text{sign}(|z_L(n)|) - \gamma_1). \tag{9}$$

where, $\boldsymbol{f}_r(n)$ and $\boldsymbol{f}_i(n)$ represent the real part and imaginary part vector of weight vectors, L denotes r or i, $\boldsymbol{y}_r(n)$ and $\boldsymbol{y}_i(n)$ are the real component and imaginary component of equalizer input signals, μ is a step-size.

If $|z_L(n)| < \gamma_1$, then $g_L = 0$; the constraints in Eq. (3) are satisfied. If $|z_L(n)| > \gamma_1$, then $g_L = 2\text{sign}[z_L(n)]$; here, we suggest to compute λ_L in order that the Bussgang condition is satisfied. This consideration leads to

$$\underbrace{E[(0.5\lambda_L \text{sign}[z_L(n)] + z_L(n))z_L(n-i)]}_{|z_L(n)|>\gamma_1} + \underbrace{E[z_L(n)z_L(n-i)]}_{|z_L(n)|<\gamma_1} = 0|\forall i \in Z, \tag{10}$$

$$\lambda_L = -2(1+\beta)|z_L(n)|, \quad (\beta > 0), \tag{11}$$

where

$$\beta = \frac{M - 2\sqrt{M} + 3}{3\sqrt{M} - 3}. \tag{12}$$

According to the above analyses, the weight vector iteration formula and the cost function of adaptiveβmulti-modulus blind equalization algorithm (βMMA) can be defined as

$$J = E[|c_r| \cdot |z_r^2(n) - \gamma_1^2|] + E[|c_i| \cdot |z_i^2(n) - \gamma_1^2|], \tag{13}$$

$$\boldsymbol{f}_L(n+1) = \boldsymbol{f}_L(n) + \mu c_L z_L(n)^* \boldsymbol{y}_L(n), \tag{14}$$

$$c_L = \begin{cases} 1 & |z_L(n)| \le \gamma_1, \\ -\beta & |z_L(n)| > \gamma_1, \end{cases} \tag{15}$$

4 Wavelet Frequency Domain Adaptive Multi-Modulus Blind Equalization Algorithm Based on Fractional Lower Order Statistics

The frequency domain cost function of Eq. (6) can be written as

$$\begin{aligned} J = {} & E[|Z_r(n)|^p + \lambda_r(\text{fmax}(\gamma_1, |Z_r(n)|) - \gamma_1)] \\ & + E[|Z_i(n)|^p + \lambda_i(\text{fmax}(\gamma_1, |Z_i(n)|) - \gamma_1)]. \end{aligned} \tag{16}$$

where, $Z_r(n)$ and $Z_i(n)$ represent the real component and imaginary component of frequency domain equalizer output signals, in the α-stable distribution noise, the α order statistics of signals do not exit, so $p < \alpha$. According to stochastic gradient method [7], the weight vector iteration formula of frequency domain adaptive β multi-modulus blind equalization algorithm based on Fractional lower order statistics (FLOS β FMMA) can be written as

$$\boldsymbol{F}_L(n+1) = \begin{cases} \boldsymbol{F}_L(n) + \mu|Z_L(n)|^{p-1} \\ \qquad \cdot\text{sign}(Z_L(n))\boldsymbol{Y}_L^*(n) |z_L(n)| \le \gamma_1, \\ \boldsymbol{F}_L(n) + \mu[|Z_L(n)|^{p-1}\text{sign}(Z_L(n)) \\ \qquad -(1+\beta)Z_L(n)]\boldsymbol{Y}_L^*(n) |z_L(n)| > \gamma_1, \end{cases} \tag{17}$$

where, $\boldsymbol{F}_L(n)$ is equalizer frequency domain weight vector, $\boldsymbol{Y}_L(n)$ are equalizer input signal frequency domain vector.

The principle of FLOS β FWTMMA is shown in Fig. 1.

In Fig. 1, $\boldsymbol{a}(n)$ is input signal vector, $\boldsymbol{c}(n)$ is channel vector, $\boldsymbol{w}(n)$ is α-stable distribution noise, $\boldsymbol{y}(n)$ is the signals with noise, $\boldsymbol{r}_r(n)$ and $\boldsymbol{r}_i(n)$ are the real component and imaginary component of signal vector after wavelet transform, $\boldsymbol{Z}_r(n)$ and $\boldsymbol{Z}_i(n)$

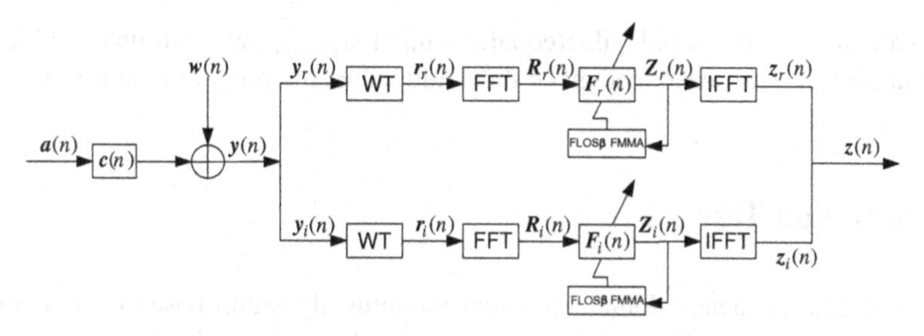

Fig. 1 The principle diagram of wavelet frequency domain adaptive β multi-modulus blind equalization algorithm based on Fractional lower order statistics

are equalizer output frequency domain signals, $z(n)$ is equalizer output vector after IFFT.

According to wavelet transform theory and Fig. 1, we have

$$\boldsymbol{r}_r(n) = \boldsymbol{V}\boldsymbol{y}_r(n), \boldsymbol{r}_i(n) = \boldsymbol{V}\boldsymbol{y}_i(n). \tag{18}$$

where, $\boldsymbol{V}$ is orthogonal wavelet transform matrix, the equalizer outputs are given by

$$\boldsymbol{Z}_r(n) = \boldsymbol{F}_r(n)\boldsymbol{R}_r(n), \boldsymbol{Z}_i(n) = \boldsymbol{F}_i(n)\boldsymbol{R}_i(n). \tag{19}$$

For wavelet frequency domain adaptive β multi-modulus blind equalization algorithm based on Fractional lower order statistics (FLOS β FWTMMA), its weight vector iteration formula can be written as

$$\boldsymbol{F}_L(n+1) = \begin{cases} \boldsymbol{F}_L(n) + \mu\hat{\boldsymbol{R}}^{-1}(n)|Z_L(n)|^{p-1} \\ \quad \cdot\text{sign}(Z_L(n))\boldsymbol{R}_L^*(n)|z_L(n)| \le \gamma_1, \\ \boldsymbol{F}_L(n) + \mu\hat{\boldsymbol{R}}^{-1}(n)[|Z_L(n)|^{p-1}\text{sign}(Z_L(n)) \\ \quad -(1+\beta)Z_L(n)]\boldsymbol{R}_L^*(n)|z_L(n)| > \gamma_1, \end{cases} \tag{20}$$

$$\hat{\boldsymbol{R}}^{-1}(n) = \text{diag}[\sigma_{j,0}^2(n), \sigma_{j,1}^2(n), \ldots, \sigma_{J,k_J-1}^2(n), \sigma_{J+1,0}^2(n), \ldots \sigma_{J+1,k_J-1}^2(n)]. \tag{21}$$

where, σ_{j,k_J}^2 and $\sigma_{j+1,k_J}^2(n)$ denote the average power estimation of r_{j,k_J} and $s_{j,k_J}(n)$. They can be given by the following recursive equations

$$\sigma_{j,k_J}^2(n+1) = \beta_\sigma\sigma_{j,k_J}^2(n) + (1-\beta_\sigma)|r_{j,k_J}(n)|^2, \tag{22}$$

$$\sigma_{j+1,k_J}^2(n+1) = \beta_\sigma\sigma_{j+1,k_J}^2(n) + (1-\beta_\sigma)|s_{j,k_J}(n)|^2. \tag{23}$$

where, $r_{j,k_J}(n)$ is wavelet transform coefficients, $s_{j,k_J}(n)$ is scale transform coefficients, β_σ is smoothing factor and $0 < \beta_\sigma < 1$.

As the α-stable distribution noise has peak pulse, we use the modified method proposed in literature [8] to suppress the abnormal value of the equalizer input.

Its idea is to set a threshold value(equalizer input signal power estimated value), if the equalizer input exceeds the given threshold value, the pretreatment is done.

5 Simulation Tests

We compared frequency domain constant modulus algorithm based on Fractional lower order statistics (FLOSFCMA), frequency domain weighted multi-modulus algorithm based on Fractional lower order statistics (FLOSFWMMA) and FLOS β FMMA with FLOS β FWTMMA in order to verify the performance of FLOS β FWTMMA, simulation experiment was done in α -stable distribution noise. In tests, the impulse response of channel was given by $c = [-0.35, 0, 0, 1]$, the characteristics index of αstable distribution noise was 1.71, $\beta = a = 0$, the step-size of the FLOSFCMA was set to 0.00003, the step-size of the FLOSFWMMA was set to 0.00013, the step-size of the FLOS β FMMA was set to 0.000012, the step-size of the FLOS β FWTMMA was set to 0.000999, the length of equalizer was 16, the ninth tap of FLOS β FMMA was initialized into 1, for FLOSFWMMA, FLOSFCMA and FLOS β FWTMMA, their tenth tap were initialized into 1, the *GSNR* [9] (generalized signal to noise ratio) was 28dB, $GSNR = 10\log_{10}(\sigma^2/\gamma)$, σ^2 was the variance of input signals. The 3000 Monte Carlo simulation results were shown in Fig. 2.

Figure 2a shows FLOS β FWTMMA has an improvement of about 2000 steps for convergence speed comparison with FLOS β FMMA, respectively. Its steady-state error has a drop of about 9 dB comparison with that of the FLOSFCMA, about 6 dB comparison with that of the FLOSFWMMA, and about 3 dB comparison with that of the FLOS β FMMA, respectively. Figure 2f shows the FLOS β FWTMMA's constellations are the clearest. Therefore, FLOS β FWTMMA has the best adaptability in α-stable distribution noise.

6 Conclusion

A wavelet frequency domain adaptive β multi-modulus blind equalization algorithm based on Fractional lower order statistics is proposed in this chapter. An adaptive β multi-modulus blind equalization algorithm based on the optimization of the energy of equalizer output signals is introduced into frequency domain blind equalization algorithm and the proposed algorithm is simulated in α-stable distribution noise. The simulation results show that the proposed FLOS β FWTMMA has favorable performance.

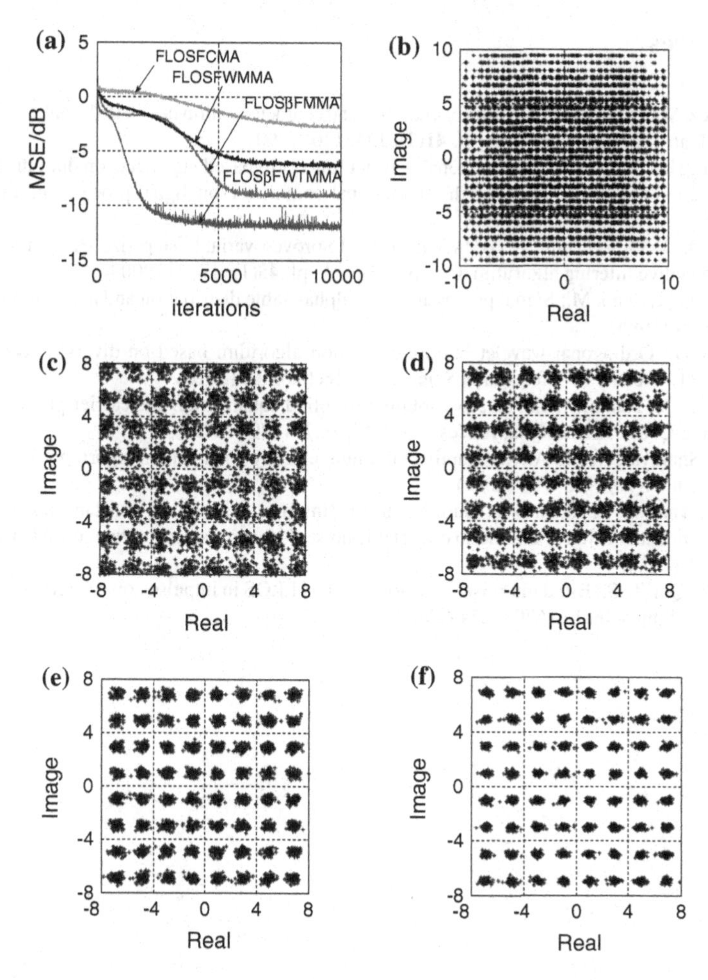

Fig. 2 Simulation results under the α -stable distribution noise. **a** Mean square error curves. **b** input of equalizer. **c** FLOSFCMA. **d** FLOSFWMMA. **e** FLOS β FMMA. **f** FLOS β FWTMMA

Acknowledgments Thanks to the support by the Author of National Excellent Doctoral Dissertation of China (200753), Natural Science Foundation of Higher Education Institution of Jiangsu Province (08KJB510010) and "the peak of six major talent" cultivate projects of Jiangsu Province (2008026), Natural Science Foundation of Higher Education Institution of Anhui Province (KJ2010A096), Natural Science Foundation of Jiangsu Province (BK2009410), Postgraduate Research and Innovation projects of Higher Education Institution of Jiangsu Province (CXLX11_0637).

References

1. Sanchez, M.G.: Impulsive noise measurements and characterization in a UHF digital TV channel. IEEE Trans. Electromagn. Compat. **41**(2), 124–136 (1999)
2. Guo, Y., Han, Y., Yang, C.: Orthogonal wavelet transform based sign decision dual-mode blind equalization algorithm. In: The 9th international conference on signal processing, pp. 80–83 (2008)
3. Yang, Q., Xiao, L., Zeng, X.-W., Wang, J.-L.: Improved variable step-size frequency domain LMS adaptive filtering algorithm. Comput. Eng. Appl. **45**(1), 17–21 (2009)
4. Nikias, C.L., Shao, M.: Signal processing with alpha-stable distribution and application. Wiley, New York (1995)
5. Xuejie, D.: Orthogonal wavelet blind equalization algorithm based on diversity technology. Nanjing University of Information Science and Technology (2010)
6. Shafayat, A., Nandi, A.K.: Adaptive solution for blind equalization and carrier-phase recovery of square-QAM. IEEE Signal Process. Lett. **17**(9), 791–794 (2010)
7. Abrar, Shafayat, Nandi, A.K.: Adaptive minimum entropy equalization algorithm. IEEE Commun. Lett. **14**(10), 966–968 (2010)
8. Zhang, Yinbing, Zhao, Junwei, Guo, Yecai, Li, Jinming: A improved constant modulus blind equalization algorithm of inhibition of α-stable noise. J Northwestern Polytechnical Univ. **28**(2), 203–206 (2010)
9. Guo, Y., Qiu, T.-S.: Blind multiuser detector based on FLOS in impulse noise environment. Acta Electron. Sinica **35**(9), 1670–1674 (2007)

The Integrated Operator Base on Complex Fuzzy Valued Intuitionist Fuzzy Sets and Its Application in the Evaluation for Hospital

Xue-ping Zhang and Sheng-quan Ma

Abstract In this paper, based on intuitionist fuzzy sets, we firstly got the integrated operator base on Complex Fuzzy valued Intuitionist Fuzzy Sets, and then applied to the evaluation of the hospital in and got very good results.

Keywords Complex fuzzy numbers · The integrated operator of Complex Fuzzy valued Intuitionist Fuzzy Sets

In 1989, Atanassov and De promote intuitionist fuzzy sets, intuitionist fuzzy sets membership and non-membership are expressed by interval-valued, but due to the uncertainty of objective things and the fuzziness of the human mind, the degree of membership and non-memberships often difficult to express by real-valued and interval-valued, so, in 2007, Chinese scholar Liu Feng, Yuan Xuehai based on the form of triangular fuzzy number, interval-valued intuitionist fuzzy sets are extended for triangular fuzzy number, finally, given the concept of fuzzy number intuitionist fuzzy set, this paper, on the basis of previous work, to further promote it and its ideological sources is:

If in a singing contest, the overall scores of the players is to design better rated projects, assign certain subjective weights for each rated item, and then the judges combine their experience and knowledge to evaluate the performance of the contestants, this is the more common method, which is simple to implement, but there is insufficient, because there is no harm, if $\{\langle x, \mu_A(x), \eta_A(x)\rangle \, | x \in X\}$ represents a competitor, $\mu_A(x)$ support this player, $\eta_A(x)$ oppose this player, the degrees of the support and the oppose with a lot of subjectivity, as some projects due to certain factors, it was generally score higher, and the weights are also great, the final result will not be able to reflect the differences of each players, so, the single scoring method sometimes cannot truly reflect the real level of the players, so, we give the concept

X. Zhang · S. Ma (✉)
School of Information and Technology, Hainan Normal University, Haikou 571158, China
e-mail: 917956085@qq.com

B.-Y. Cao and H. Nasseri (eds.), *Fuzzy Information & Engineering and Operations Research & Management*, Advances in Intelligent Systems and Computing 211,
DOI: 10.1007/978-3-642-38667-1_40,

of complex fuzzy number intuitionistic fuzzy sets base on fuzzy number intuitionist fuzzy sets and the real problem,the idea of both subjective and objective support to deal with this issue, to make sure its more reasonable and fair.

1 Complex Fuzzy Value Intuitionist Fuzzy Sets

Definition 1.1. Let the set E is non-empty and finite, the form

$$A = \left\{ \left\langle x, \tilde{M}_{TA}(x) + i\tilde{M}_{IA}(x), \tilde{N}_{TA}(x) + i\tilde{N}_{TA}(x) \right\rangle | x \in E \right\}$$

are called Complex fuzzy value intuitionist fuzzy sets on E, which

$$\tilde{M}_{TA}(x) = \left(M_{TA}^1(x), M_{TA}^2(x), M_{TA}^3(x) \right) \in F(R)$$

$$\tilde{M}_{IA}(x) = \left(M_{IA}^1(x), M_{IA}^2(x), M_{IA}^3(x) \right) \in F(R)$$

$$\tilde{N}_{TA}(x) = \left(N_{TA}^1(x), N_{TA}^2(x), N_{TA}^3(x) \right) \in F(R)$$

$$\tilde{N}_{IA}(x) = \left(N_{IA}^1(x), N_{IA}^2(x), N_{IA}^3(x) \right) \in F(R)$$

are Two triangular function in $I = [0, 1]$, and satisfied the following:

$$M_A^1(x), M_A^2(x), M_A^3(x), N_A^1(x), N_A^2(x), N_A^3(x) \in [0, 1]$$

At the same time,

$$M_{TA}^3(x) + N_{TA}^3(x) \le 1, M_{IA}^3(x) + N_{IA}^3(x) \le 1, \forall x \in E.$$

Complex number valued Intuitionist fuzzy set can be a good expression of the issues raised in the introduction,

$$A = \left\{ \left\langle x, \tilde{M}_{TA}(x) + i\tilde{M}_{IA}(x), \tilde{N}_{TA}(x) + i\tilde{N}_{TA}(x) \right\rangle | x \in E \right\}$$

represents a competitor, $\tilde{M}_{TA}(x) + i\tilde{M}_{IA}(x)$ support this player, which $\tilde{M}_{TA}(x)$ is subjective support, $\tilde{M}_{IA}(x)$ is objective support, $\tilde{N}_{TA}(x) + iN_{IA}(x)$ oppose this player, which $\tilde{N}_{TA}(x)$ are subjective opposition, $\tilde{N}_{IA}(x)$ is objective opposition.

Definition 1.2. Let X be a non-empty set,

$$A = \left\{ \left\langle x, \tilde{M}_{TA}(x) + i\tilde{M}_{IA}(x), \tilde{N}_{TA}(x) + i\tilde{N}_{IA}(x) \right\rangle | x \in E \right\},$$
$$B = \left\{ \left\langle x, \tilde{M}_{TB}(x) + i\tilde{M}_{IB}(x), \tilde{N}_{TB}(x) + i\tilde{N}_{IB}(x) \right\rangle | x \in E \right\}$$

Are Complex fuzzy value intuitionist fuzzy sets, then

$$A + B = \{< x, [\inf M_{TA\lambda}(x) + \inf M_{TB\lambda}(x) - \inf M_{TA\lambda}(x) \inf M_{TB\lambda}(x),$$
$$\sup M_{TA\lambda}(x) + \sup M_{TB\lambda}(x) - \sup M_{TA\lambda}(x) \sup M_{TB\lambda}(x)] + i$$
$$[\inf M_{IA\lambda}(x) + \inf M_{IB\lambda}(x) - \inf M_{IA\lambda}(x) \inf M_{IB\lambda}(x),$$
$$\sup M_{IA\lambda}(x) + \sup M_{IB\lambda}(x) - \sup M_{IA\lambda}(x) \sup M_{IB\lambda}(x)],$$

The integrated operator base on Complex Fuzzy valued Intuitionist Fuzzy Sets and its Application in the Evaluation for hospital

$$[\inf N_{TA\lambda}(x) \inf N_{TB\lambda}(x), \sup N_{TA\lambda}(x) \sup N_{TB\lambda}(x)] + i$$
$$[\inf N_{IA\lambda}(x) \inf N_{IB\lambda}(x), \sup N_{IA\lambda}(x) \sup N_{IA\lambda}(x)]\}_{\lambda \in [0,1]}$$

$$A{*}B = \{< x, [\inf M_{TA\lambda}(x) \inf M_{TB\lambda}(x), \sup M_{TA\lambda}(x) \sup M_{TB\lambda}(x)]$$
$$+i[\inf M_{IA\lambda}(x) \inf M_{IB\lambda}(x), \sup M_{IA\lambda}(x) \sup M_{IB\lambda}(x)],$$
$$[\inf N_{TA\lambda}(x) + \inf N_{TB\lambda}(x) - \inf N_{TA\lambda}(x) \inf N_{TB\lambda}(x),$$
$$\sup N_{TA\lambda}(x) + \sup N_{TB\lambda}(x) - \sup N_{TA\lambda}(x) \sup N_{TB\lambda}(x)]$$
$$+i[\inf N_{IA\lambda}(x) + \inf N_{IB\lambda}(x) - \inf N_{IA\lambda}(x) \inf N_{IB\lambda}(x),$$

2 Complex Fuzzy Valued Intuition Fuzzy Integrated Operator Based on Complex Fuzzy Integral

For a multi-attribute decision-making problems, the assumptions $Y = \{Y_1, Y_2, \ldots, Y_n\}$ is the program set, $G = \{G_1, G_2, \ldots, G_n\}$ is the set of attributes and

$$\varpi = \{\varpi_{T1} + i\varpi_{I1}, \varpi_{T2} + i\varpi_{I2}, \ldots, \varpi_{Tn} + i\varpi_{In}\}^T$$

$$\omega = \{\omega_{T1} + i\omega_{I1}, \omega_{T2} + i\omega_{I2}, \ldots, \omega_{Tn} + i\omega_{In}\}^T$$

Weight vector for the property, which $\varpi_{Tj} \in [0,1]$, $\varpi_{Ij} \in [0,1]$ $j = 1,2,\ldots,n$, $\sum_{j=1}^{n}\varpi_{Tj} = 1, \sum_{j=1}^{n}\varpi_{Ij} = 1, \omega_{Tj} \in [0,1], \omega_{Ij} \in [0,1]$ $j = 1,2,\ldots,n$, $\sum_{j=1}^{n}\omega_{Tj} = 1, \sum_{j=1}^{n}\omega_{Ij} = 1, \varpi = \{\varpi_{T1} + i\varpi_{I1}, \varpi_{T2} + i\varpi_{I2}, \ldots, \varpi_{Tn} + i\varpi_{In}\}^T$ is Subjective weight vector, and

$$\omega = \{\omega_{T1} + i\omega_{I1}, \omega_{T2} + i\omega_{I2}, \ldots, \omega_{Tn} + i\omega_{In}\}^T$$

is objective weight vector, And the hypothetical Scenario Y_i characteristic information are expressed by complex fuzzy numerical intuition the fuzzy set, as the following

$$Y_i = \left\{\left\langle G_i, M_{TY_i}(G_j) + M_{IY_i}(G_j), N_{TY_i}(G_j) + N_{IY_i}(G_j)\right\rangle | G_i \in G\right\} \quad (i = 1,2,\ldots,n)$$

Which

$$M_{TY_i}(G_j), M_{IY_i}(G_j), N_{TY_i}(G_j), N_{IY_i}(G_j) \in [0,1]$$

And

$$\sup M_{TY_i}(G_j) + \sup M_{IY_i}(G_j) \le 1, \sup N_{TY_i}(G_j) + \sup N_{IY_i}(G_j) \le 1$$

Option Y_i about attribute G_j' characteristic expressed by complex fuzzy values intuitionist fuzzy numbers $d_{ij} = \left(M_{Tij} + M_{Iij}, N_{Tij} + N_{Iij}\right)$, which M_{Tij} denoted the degree of the Programs Y_i subjectively satisfy the attributes G_j, M_{Iij} denoted the degree of the Programs Y_i objectively satisfy the attributes G_j, N_{Tij} denoted the degree of the Programs Y_i subjectively does not satisfy the attributes G_j, N_{Iij} denoted the degree of the Programs Y_i objectively does not satisfy the attributes G_j, All feature information of programs Y_i $(i = 1,2,\ldots,n)$ About attribute G_j $(j = 1,2,\ldots,m)$can be expressed as a complex fuzzy number valued intuitionist fuzzy decision matrix $D = \left(d_{ij}\right)_{m\times n}$, which $d_{ij} = (M_{Tij} + iM_{Iij}$, $N_{Tij} + iN_{Iij})$ to got the integration operator base on complex fuzzy number valued intuition fuzzy sets, that is

$$e_i = \left\langle \sum_{j=1}^{m}\left(M_{Tij} + iM_{Iij}\right)\left(\varpi_{Tj} + i\varpi_{Ij}\right) \sum_{j=1}^{m}\left(N_{Tij} + iN_{Iij}\right)\left(\omega_{Tj} + i\omega_{Ij}\right)\right\rangle \quad i = 1,2,\ldots,n,$$

which $\sum_{j=1}^{m}\left(M_{Tij} + iM_{Iij}\right)\left(\varpi_{Tj} + i\varpi_{Ij}\right)$ are discrete integration operator. Calculating the score as the following

$$S(e_i) = \left(\sum_{j=1}^{m}\left(M_{Tij} + iM_{Iij}\right)\left(\varpi_{Tj} + i\varpi_{Ij}\right)\right) - \left(\sum_{j=1}^{m}\left(N_{Tij} + iN_{Iij}\right)\left(\omega_{Tj} + i\omega_{Ij}\right)\right) \quad i = 1,2,\ldots,n$$

Then to sort the $S(e_i)$ and to got the decisions.

3 Application

Evaluation ranking of a city to carry out the city's hospital, in accordance with the priority requirements, evaluation process are as follows (Calculated data from the literature [4])

$$\begin{aligned}
e_1 &= \langle (0.555, 0.655, 0.755) + i(0.438, 0.559, 0.664),\\
&\quad (0.035, 0.125, 0.215) + i(0.374, 0.447, 0.650)\rangle\\
e_2 &= \langle (0.575, 0.675, 0.775) + i(0.205, 0.432, 0.582),\\
&\quad (0.025, 0.115, 0.205) + i(0.374, 0.476, 0.671)\rangle\\
e_3 &= \langle (0.515, 0.615, 0.719) + i(0.284, 0.394, 0.548),\\
&\quad (0.036, 0.126, 0.216) + i(0.43, 0.554, 0.741)\rangle\\
e_4 &= \langle (0.495, 0.595, 0.695) + i(0.298, 0.423, 0.548),\\
&\quad (0.073, 0.150, 0.227) + i(0.266, 0.380, 0.500)\rangle\\
e_5 &= \langle (0.495, 0.595, 0.695) + i(0.369, 0.513, 0.689),\\
&\quad (0.069, 0.159, 0.249) + i(0.399, 0.553, 0.705)\rangle\\
e_6 &= \langle (0.488, 0.588, 0.688) + i(0.161, 0.261, 0.424),\\
&\quad (0.056, 0.146, 0.236) + i(0.301, 0.414, 0.568)\rangle\\
e_7 &= \langle (0.449, 0.549, 0.649) + i(0.397, 0.497, 0.656),\\
&\quad (0.099, 0.189, 0.279) + i(0.286, 0.419, 0.583)\rangle
\end{aligned}$$

And obtained

$$\begin{aligned}
S(e_1) &= (0.52, 0.53, 0.54) + i\,(0.064, 0.112, 0.014) = H_1\\
S(e_2) &= (0.55, 0.56, 0.57) + i\,(0.169, 0.044, 0.089) = H_2\\
S(e_3) &= (0.479, 0.489, 0.503) + i\,(0.146, 0.16, 0.193) = H_3\\
S(e_4) &= (0.422, 0.445, 0.468) + i\,(0.032, 0.043, 0.048) = H_4\\
S(e_5) &= (0.426, 0.436, 0.446) + i\,(0.03, 0.04, 0.016) = H_5\\
S(e_6) &= (0.432, 0.442, 0.452) + i\,(0.14, 0.153, 0.144) = H_6\\
S(e_7) &= (0.35, 0.36, 0.37) + i\,(0.111, 0.078, 0.073) = H_7.
\end{aligned}$$

To sort the results and to got the decision-making program are as following

$$H_2 > H_1 > H_3 > H_4 > H_6 > H_5 > H_7$$

The results illustrate the method with complex fuzzy integral as well as subjective and objective support to deal with these problems, with the literature [4], the same result but more discrimination, such contest or decision-making on a more fair and reasonable.

4 Summary

Comparison by the results of the examples and the result of literature [4], it is clear that introduction of subjective and objective support and opposition to better deal with such problems, the results of the data show that not only a greater degree of distinction between individuals, but also with reality match. The used of the Complex fuzzy valued intuitionist fuzzy set in the appraisal decision-making than the fuzzy number intuitionist fuzzy sets even more of their superiority.

Acknowledgments This work is supported by International Science and Technology Cooperation Program of China (2012DFA11270), Hainan International Cooperation Key Project (GJXM201105) and Natural Science Foundation of Hainan Province (No.111007).

References

1. Atnassov, K., Gargov, G.: Interval-valued intuitions tic fuzzy sets [J]. Fuzzy sets and systems. **31**, 343–349 (1989)
2. Atnassov K.: Intuitionist Fuzzy sets: Theory and Applications [M]. Heidelberg; physical-verlay, 21–46(1999)
3. Liu, F., Yuan, X.: Fuzzy number intuitionistic fuzzy sets [J]. Fuzzy Systems and Mathematics **21**(1), 88–91 (2007)
4. Hu, J.: Fuzzy number intuitionistic fuzzy set theory based on structured element and its application [D], Liaoning Engineering Technology University, 2009

Part VI
Graph and Network

Fuzzy Average Tree Solution for Graph Games with Fuzzy Coalitions

Cui-ping Nie and Qiang Zhang

Abstract In this chapter, the model of graph games with fuzzy coalitions is proposed based on graph games and cooperative games with fuzzy coalitions. The fuzzy average tree solution of graph games with fuzzy coalitions is given, which can be regarded as the generalization of crisp graph games. It is shown that the fuzzy average tree solution is equal to the fuzzy Shapley value for complete graph games with fuzzy coalitions. We extend the notion of link-convexity, under which the fuzzy core is non-empty and the fuzzy average tree solution lies in this core.

Keywords Graph game · Average tree solution · Imputation · Fuzzy coalition · Link-convexity

1 Introduction

In a cooperative game, cooperation is not always possible for the players. Cooperative games with limited communication structure are called graph games introduced by Myerson [1]. The best-known single-valued solution for graph games is the Myerson value characterized by component efficiency and fairness. In [2, 3] the positional value is proposed. The value for such games is characterized by component efficiency and balanced total threats, see Slikker [4]. Herings et al. [5] defines the average tree solution for cycle-free graph games. Moreover, he generalizes this solution to the class of all graph games in [6].

C. Nie (✉) · Q. Zhang
School of Management and Economics, Beijing Institute of Technology,
Beijing 100081, People's Republic of China
e-mail: niecuiping7@163.com

C. Nie
Department of Mathematics and Information Science, Shijiazhuang College,
Shijiazhuang 050035, People's Republic of China

B.-Y. Cao and H. Nasseri (eds.), *Fuzzy Information & Engineering and Operations Research & Management*, Advances in Intelligent Systems and Computing 211,
DOI: 10.1007/978-3-642-38667-1_41, © Springer-Verlag Berlin Heidelberg 2014

There are some situations where players do not fully participate in a coalition but to a certain extent. A fuzzy coalition is introduced by Aubin [7]. Butnariu [8] defines a Shapley value and shows the explicit form of the Shapley value on a limited class of fuzzy games. Tsurumi et al. [9] gives a new class of fuzzy games with integral form. The fuzzy core for fuzzy cooperative games has been researched in [10, 11].

This chapter is organized as follows. In Sect. 2 we give preliminary notions of graph games. In Sect. 3 we discuss graph games with fuzzy coalitions represented by an undirected graph. The fuzzy average tree solution is introduced. We prove that the fuzzy average tree solution is equal to the fuzzy Shapley value for a complete graph game with fuzzy coalitions. In Sect. 4 we give the relationship between the fuzzy average tree solution and the fuzzy core.

2 Preliminaries

We consider a cooperative game with limited communication structure, called a graph game. It is represented by (N, v, L) with $N = \{1, \cdots, n\}$ a node set of players, $v : 2^N \to R$ a characteristic function and $L \subseteq \{\{i, j\} | i \neq j, i, j \in N\}$ a set of edges. Usually, a cooperative game(N, v) is thought as a complete graph game (N, v, L), i.e. $L = \{\{i, j\} | i \neq j, i, j \in N\}$.

The limited possibilities of cooperation can be represented by an graph (N, L) in which cooperation is only possible if players are directly or indirectly connected. Let $P(N)$ be the collection of all crisp coalitions. A coalition of players $K \in P(N)$ is a network of a graph (N, L) if K is connected in the graph. Furthermore, if a network K cannot form a larger network with any other player of $N \backslash K$, then K is called a component. For a graph (N, L), we denote by $C^L(N)$ the set of all networks and $\hat{C}^L(N)$ the class of all components. $(W, L(W))$ is thought as a subgraph of the graph (N, L), where $W \in P(N)$ and $L(W) = \{\{i, j\} \in L | i, j \in W\}$.

Definition 2.1. For a graph (N, v, L) a sequence of nodes $(i_1, \ldots, i_k)$ is a cycle when: 1) $k \geq 2$; 2) $i_{k+1} = i_1$; 3) $\{i_h, i_{h+1}\} \in L$, $h = 1, \ldots, k$. A graph (N, L) is said to be cycle-free if it does not contain any cycle.

Definition 2.2. Let (N, L) be a graph, then an n-tuple $B = \{B_1, \cdots B_n\}$ of n subsets of N is admissible if it satisfies:

(1) For all $i \in N$, $i \in B_i$, and for some $j \in N$, $B_j = N$;
(2) For all $i \in N$ and $K \in \hat{C}^L(B_i \backslash \{i\})$, we get $K = B_j$ and $\{i, j\} \in L$ for some $j \in N$. Let B^L be the collection of all admissible n-tuples $B = \{B_1, \cdots B_n\}$ of the graph (N, L).

Definition 2.3. For a graph game (N, v, L), the average tree solution $AT(N, v, L)$ is the payoff vector defined by

$$AT_i(N, v, L) = \frac{1}{|B^L|}\left(\sum_{B \in B^L}\left(v(B_i) - \sum_{K \in \hat{C}^L(B_i \setminus \{i\})} v(K)\right)\right), \quad i \in N. \quad (1)$$

For a cooperative game (N, v) and $W \in P(N)$,the Shapley value $Sh(v)(W)$ is an imputation given by

$$Sh_i(v)(W) = \begin{cases} \sum_{T \in P(W \setminus \{i\})} \frac{|T|!(|W|-|T|-1)!}{|W|!} \cdot [v(T \cup \{i\}) - v(T)], & i \in W \\ 0, & i \in N \setminus W \end{cases} \quad (2)$$

A fuzzy coalition $U = (U(1), \ldots, U(n))$ is a fuzzy subset of N, in which $U(i) \in [0, 1]$ and $i \in N$. $U(i)$ is the degree that i takes part in the coalition U. Let $F(N)$ be the class of all fuzzy coalitions and $d(U)$ be the cardinality of $D(U)$. We write the elements of $D(U)$ in the increasing order as $r_1 \langle \cdots \langle r_{d(U)}$ and denote $r_0 = 0$.

Definition 2.4. Let (N, tv) be a cooperative game with fuzzy coalitions, $U \subseteq F(N)$. If a function $f: F(U) \to R_+^n$ satisfies

(1) $f_i(U) = 0, \forall i \notin Supp(U)$;
(2) $\sum_{i \in Supp(U)} f_i(U) = tv(U)$;
(3) $f_i(U) \geq U(i) \cdot tv(\{i\}), \forall i \in Supp(U)$. Then we call f an imputation of (N, tv).

Note that the definition above is also applicable to crisp games. For a fuzzy game (N, tv), the crisp game (N, v) corresponding to (N, tv) is called the associated crisp game. Tsurumi et al. [9] gives the fuzzy Shapley value $f(tv)(U)$:

$$f_i(tv)(U) = \sum_{m=1}^{d(U)} Sh_i(v)\left([U]_{r_m}\right) \cdot (r_m - r_{m-1}), \ i \in N$$

where $tv(U) = \sum_{m=1}^{d(U)} v\left([U]_{r_m}\right) \cdot (r_m - r_{m-1})$ and $[U]_{r_m} = \{i \in N : U(i) \geq r_m\}$.

3 The Fuzzy Average Tree Solution on Graph Games with Fuzzy Coalitions

Now we will extend the average tree solution on the class of crisp graph games. The average tree solution is introduced as a function which derives the value from a given pair of a game and a coalition.

Definition 3.1. For a crisp graph game (N, v, L) and $W \in P(N)$, the average tree solution $AT(W, v, L(W))$ is defined by

$$AT_i(W, v, L(W)) = \begin{cases} \frac{1}{|B^{L(W)}|}\left(\sum_{B \in B^{L(W)}}\left(v(B_i) - \sum_{K \in \hat{C}^{L(W)}(B_i \setminus \{i\})} v(K), i \in W\right)\right), & i \in W, \\ 0, & i \in N \setminus W. \end{cases} \tag{3}$$

When (N, v, L) is a cycle-free graph game, let $W \in P(N)$ and $T^i(W)$ be the spanning tree with $i \in W$ as the root in the subgraph $(W, L(W))$. Obviously, we have

$$AT_i(W, v, L(W)) = \begin{cases} \frac{1}{|W|}\left(\sum_{j \in W}\left(v\left(K_i^j(W)\right) - \sum_{\{i' | (i,i') \in T^j(W)\}} v\left(K_{i'}^j(W)\right)\right)\right), & i \in W \\ 0, & i \in N \setminus W \end{cases} \tag{4}$$

where $K_j^i(W)$ is the set consisting of $j \in W$ and all its subordinates in $T^i(W)$.

Example 3.1 Let (N, v, L) be a crisp graph game with $N = \{1, 2, 3, 4\}$, $L = \{\{1, 2\}, \{2, 3\}, \{3, 4\}\}$ and v a characteristic function on N as follows: $v(\{1, 2\}) = v(\{4\}) = 1$, $v(\{2, 3\}) = v(\{3, 4\}) = 2$, $v(\{1, 2, 3\}) = 4$, $v(S) = 0$, otherwise.

Then the average tree solution of the coalition $W = \{1, 2, 3\}$ is

$$\begin{aligned} AT_1(W, v, L(W)) &= \frac{1}{3}((4-2)+0+0) = \frac{2}{3}, \\ AT_2(W, v, L(W)) &= \frac{1}{3}(2+4+1) = \frac{7}{3}, \\ AT_3(W, v, L(W)) &= \frac{1}{3}(0+0+(4-1)) = 1, \\ AT_4(W, v, L(W)) &= 0, \\ AT(W, v, L(W)) &= \left(\frac{2}{3}, \frac{7}{3}, 1, 0\right). \end{aligned}$$

Lemma 3.1 For a crisp graph game (N, v, L) and $W \in P(N)$. Then the average tree solution $AT(W, v, L(W))$ is an imputation.

Proof From Eq. (2), it is apparent that $AT_i(W, v, L(W)) = 0$, $\forall i \notin W$. Since $AT(W, v, L(W))$ is component efficient, we have

$$\begin{aligned} \sum_{i \in W} AT_i(W, v, L(W)) &= \sum_{K \in \hat{C}^{L(W)}(W)} \sum_{i \in K} AT_i(W, v, L(W)) \\ &= \sum_{K \in \hat{C}^{L(W)}(W)} v(K) = v(W). \end{aligned}$$

Because $v(B) = v\left(\left(\sum_{K \in \hat{C}^{L(W)}(B_i \setminus \{i\})} K\right) \cup i\right) \geq v(i) + \sum_{K \in \hat{C}^{L(W)}(B_i \setminus \{i\})} v(K)$ where $i \in W$, it holds that $v(B) - \sum_{K \in \hat{C}^{L(W)}(B_i \setminus \{i\})} v(K) \geq v(i)$.

Thus

$$AT_i(W, v, L(W)) = \frac{1}{\left|B^{L(W)}\right|} \sum_{B \in B^{L(W)}} \left[v(B_i) - \sum_{K \in \hat{C}^{L(W)}(B_i \setminus \{i\})} v(K) \right] \geq v(\{i\}), i \in W.$$

□

It is shown that the average tree solution $AT(N, v, L)$coincides with the Shapley value $Sh(v)(N)$ for a complete graph game (N, v, L) in [7]. Thus, it is easy to see that the average tree function $AT(W, v, L(W))$ is also equal to the Shapley value $Sh(v)(W)$, where $W \in P(N)$.

A graph game with fuzzy coalitions is a triple (N, tv, L) where $tv : F(N) \to R_+$ is the fuzzy characteristic function . Next, We discuss the fuzzy average tree solution of graph games with fuzzy coalitions.

Definition 3.2. Let (N, tv, L) be a graph game with fuzzy coalitions, $U \in F(N)$ and (N, v, L) be the associated crisp graph game of (N, tv, L). Then the fuzzy average tree solution $\tilde{AT}(U, tv, L(U))$ is defined by

$$\tilde{AT}_i(U, tv, L(U)) = \sum_{m=1}^{d(U)} AT_i\left([U]_{r_m}, v, L\left([U]_{r_m}\right)\right) \cdot (r_m - r_{m-1}), i = 1, \ldots, n. \tag{5}$$

Theorem 3.1 Let (N, tv, L) be a complete graph game with fuzzy coalitions, $U \in F(N)$. Then the fuzzy average tree solution is equal to the fuzzy Shapley value, i.e.,

$$\tilde{AT}(U, tv, L(U)) = f(tv)(U).$$

Proof For any $i \in N$, $AT_i(W, v, L(W)) = Sh_i(v)(W)$, where $W \in P(N)$ and v is a characteristic function of the associated crisp graph game.

Due to $[U]_{r_m} \in P(N)$, we get $f_i(tv)(U) = \sum_{m=1}^{d(U)} Sh_i(v)\left([U]_{r_m}\right) \cdot (r_m - r_{m-1}) = \sum_{m=1}^{d(U)} AT_i\left([U]_{r_m}, v, L\left([U]_{r_m}\right)\right) \cdot (r_m - r_{m-1}) = \tilde{AT}_i(U, tv, L(U))$.

Hence $f(tv)(U) = \tilde{AT}(U, tv, L(U))$ which completes the proof. □

4 The Relation between The Fuzzy Average Tree Solution and The Fuzzy Core

In this section, we study the relationship between the fuzzy average tree solution and the fuzzy core. Firstly, we extend the notion of link-convexity in [7].

For a graph game with fuzzy coalitions $(N, tv, L), tv(S)+tv(T) \leq tv(S \cup T)+ \sum_{K \in \hat{C}^L(S \cap T)} tv(K), \quad \forall S, T \in F(N)$ that satisfy: ①$S, T, S \backslash T, T \backslash S$and $(S \backslash T) \cup (T \backslash S)$ are non-empty network; ②$N \backslash S$or$N \backslash T$ is a network. Then (N, tv, L) is link-convex.

The fuzzy core on the game with fuzzy coalitions (N, tv) by Tsurumi is $\tilde{c}(tv)(U) = \{x \in R_+^n | \sum_{i \in N} x_i = \sum_{m=1}^{d(U)} v([U]_{rm}) \cdot (r_m - r_{m-1}), \sum_{i \in Supp(S_U)} x_i \geq \sum_{m=1}^{d(U)} v(S_{[U]_{rm}}) \cdot (r_m - r_{m-1}), \forall S \in C^L(N)\}$, where v is the associated crisp game, $Supp(S_U) = \{i \in N | S_U(i) \succ 0\}$ and $S_U(i) = \begin{cases} U(i), & i \in S \\ 0, & i \notin S \end{cases}$, for any $U \in F(N)$.

Lemma 4.1 Let (N, tv, L) be a graph game with fuzzy coalitions and $U \in F(N)$. Then the fuzzy average tree solution $\tilde{AT}(U, tv, L(U))$ is an imputation.

Proof If $i \notin Supp(U)$, then $i \notin [U]_{r_m}, m = 1, 2, \ldots, d(U)$. Consequently, $AT_i([U]_{r_m}, v, L([U]_{r_m})) = 0$, which implies that $\tilde{AT}_i(U, tv, L(U)) = \sum_{m=1}^{d(U)} AT_i([U]_{r_m}, v, L([U]_{r_m})) \cdot (r_m - r_{m-1}) = 0$, where v is the associated crisp game of tv.

Because $AT(W, v, L(W))$ is component efficient for any $W \in P(N)$, we get

$$
\begin{aligned}
\sum_{i \in SuppK} \tilde{AT}_i(U, tv, L(U)) &= \sum_{i \in SuppK} (\sum_{m=1}^{d(U)} AT_i(U_{r_m}, v, L(U_{r_m})) \cdot (r_m - r_{m-1})) \\
&= \sum_{m=1}^{d(U)} (\sum_{i \in SuppK} AT_i(U_{r_m}, v, L(U_{r_m})) \cdot (r_m - r_{m-1})) \\
&= \sum_{m=1}^{d(U)} v(K_{r_m})(r_m - r_{m-1}) \\
&= tv(K),
\end{aligned}
$$

i.e., $\tilde{AT}(U, tv, L(U))$ satisfies component efficiency. Further, we obtain that $\sum_{i \in Supp(U)} \tilde{AT}_i(U, tv, L(U)) = \sum_{K \in \hat{C}^{L(U)}(U)} \sum_{i \in Supp(K)} \tilde{AT}_i(U, tv, L(U)) = \sum_{K \in \hat{C}^{L(U)}(U)} tv(K) = tv(U)$. It remains to show that $\tilde{AT}(U, tv, L(U)) \geq U(i) \cdot tv(\{i\}), i \in Supp(U)$.

$$\widetilde{AT}_i\left(U, tv, L\left(U\right)\right) = \sum_{m=1}^{d(U)} AT_i\left(\left[U\right]_{r_m}, v, L\left(\left[U\right]_{r_m}\right)\right) \cdot \left(r_m - r_{m-1}\right)$$

$$\geq \sum_{m=1}^{d(U)} v\left(i\right)\left(r_m - r_{m-1}\right) \geq \quad U\left(i\right) \cdot tv\left(\{i\}\right)$$

The proof is completed. □

Theorem 4.1 Let a graph game with fuzzy coalitions (N, tv, L) be link-convex and $U \in F(N)$, then $\widetilde{AT}\left(U, tv, L\left(U\right)\right) \in \tilde{c}\left(tv\right)\left(U\right)$.

Proof Let (N, v, L) be the crisp graph game corresponding to (N, tv, L). Because $\widetilde{AT}\left(U, tv, L\left(U\right)\right)$ is an imputation on (N, tv, L), we have

$$\sum_{i\in N} \widetilde{AT}_i\left(U, tv, L\left(U\right)\right) = tv\left(U\right) = \sum_{m=1}^{d(U)} v\left(\left[U\right]_{r_m}\right) \cdot \left(r_m - r_{m-1}\right).$$

Next, we will show that $\sum_{i\in Supp\left(S_{[U]_{rm}}\right)} \widetilde{AT}_i\left(U, tv, L\left(U\right)\right) \geq \sum_{m=1}^{d(U)} v\left(S_{[U]_{rm}}\right) \cdot \left(r_m - r_{m-1}\right)$, for any $S \in P(N)$.

$$\sum_{i\in Supp(S_U)} \widetilde{AT}_i\left(U, tv, L\left(U\right)\right)$$

$$= \sum_{i\in Supp(S_U)} \left[\sum_{m=1}^{d(U)} AT_i\left(\left[U\right]_{r_m}, v, L\left(\left[U\right]_{r_m}\right)\right) \cdot \left(r_m - r_{m-1}\right)\right]$$

$$= \sum_{m=1}^{d(U)} \sum_{i\in Supp(S_u)} AT_i\left(\left[U\right]_{r_m}, v, L\left(\left[U\right]_{r_m}\right)\right) \cdot \left(r_m - r_{m-1}\right).$$

By the graph game with fuzzy coalitions (N, v, L) is link-convex, we can get $\sum_{i\in[U]_{rm}} AT_i\left(\left[U\right]_{r_m}, v, L\left(\left[U\right]_{r_m}\right)\right) \geq v\left(\left[U\right]_{r_m}\right)$. Then

$$\sum_{i\in Supp(S_{[U]rm})} AT_i\left(S_{[U]r_m}, v, L\left(S_{[U]r_m}\right)\right) \geq v\left(S_{[U]r_m}\right), \quad \forall S \in P\left(N\right).$$

Consequently,

$$\sum_{m=1}^{d(U)} \sum_{i \in Supp(S_{[U]r_m})} AT_i\left(S_{[U]_{r_m}}, v, L\left(S_{[U]r_m}\right)\right) \cdot (r_m - r_{m-1})$$
$$\geq \sum_{m=1}^{d(U)} v\left(S_{[U]r_m}\right) \cdot (r_m - r_{m-1}).$$

Thus, $\sum_{i \in Supp(S_{[U]r_m})} \widetilde{AT}_i\left(U, tv, L\left(U\right)\right) \geq \sum_{m=1}^{d(U)} v\left(S_{[U]r_m}\right) \cdot (r_m - r_{m-1})$.
The proof is completed □.

By Theorem 4.1 we have conclusions that when a graph game with fuzzy coalitions (N, tv, L) is link-convex the fuzzy average tree solution must exist and the fuzzy core $\tilde{c}\left(tv\right)\left(U\right)$ is non-empty.

Example 4.1 Let (N, tv, L) be a graph game with fuzzy coalitions in which $N = \{1, 2\}$, $L = \{\{1, 2\}\}$, $U\left(1\right) = 0.4$ 0.4, $U\left(2\right) = 0.6$ and $v$is a characteristic function of the crisp graph game corresponding to (N, tv, L):

$$v\left(\phi\right) = v\left(\{1\}\right) = 0, v\left(\{2\}\right) = 1, v\left(\phi\right) = v\left(\phi\right) = v\left(\{1\}\right) = 0,$$
$$v\left(\{2\}\right) = 1, v\left(\{1, 2\}\right) = 2.$$

Then, $tv\left(U\right) = U\left(1\right) \cdot v\left(\{1, 2\}\right) + \left(U\left(2\right) - U\left(1\right)\right) \cdot v\left(\{2\}\right) = 1,$

$$\tilde{c}\left(tv\right)\left(U\right)$$
$$= \left\{\left(0.4x_1^{\{1,2\}}, 0.4x_2^{\{1,2\}} + 0.2\right) | x_1^{\{1,2\}} + x_2^{\{1,2\}} = 2, x_1^{\{1,2\}} \geq 0, x_2^{\{1,2\}} \geq 1\right\}.$$

The fuzzy average tree solution of the game is

$$\widetilde{AT}_1\left(U, tv, L\left(U\right)\right) = 0.4 \times \frac{1}{2} = 0.2, \widetilde{AT}_2\left(U, tv, L\left(U\right)\right) = 0.4 \times \frac{3}{2} + 0.2 = 0.8.$$

Obviously, $(0.2, 0, 8) \in \tilde{c}\left(tv\right)\left(U\right)$.

5 Conclusion

The fuzzy average tree solution has been proposed in graph games with fuzzy coalitions. Moreover, it coincides with the fuzzy Shapley function for complete graph games with fuzzy coalitions. We have generalized the notion of link-convexity under which the fuzzy average tree solution lies in the fuzzy core. However, the fuzzy average tree solution is not unique. It will be interesting to find other kinds of fuzzy average tree solutions.

Acknowledgments Thanks to the support by the National Natural Science Foundation of China (Nos.70771010,71071018, 70801064) and Specialized Research Fund for the Doctoral Program of Higher Education (No. 20111101110036).

References

1. Myerson, R.B.: Graphs and cooperation in games. Math. Oper. Res. **2**, 225–229 (1977)
2. Borm, P., Owen, G., Tijs, S.H.: On the position value for communication situations. SIAM J. Discrete Math. **5**, 305–320 (1992)
3. Meessen, R.: Communication games. University of Nijmegen, Nijmegen, Master thesis (1988)
4. Slikker, M.: A characterization of the position value. Int. J. Game Theory **33**, 505–514 (2005)
5. Herings, P.J.J., Van der Laan, G., Talman, A.J.J.: The average tree solution for cycle-free graph games. Games Econ. Behav. **62**, 77–92 (2008)
6. Herings, P.J.J., Van der Laan, G., Talman, A.J.J., Yang, Z.: The average tree solution for cooperative games with communication structure. Games Econ. Behav. **68**, 626–633 (2010)
7. Aubin, J.P.: Mathematical Methods of Game and Economic Theory, Rev edn. North-Holland, Amsterdam (1982)
8. Butnariu, D.: Stability and Shapley value for n-persons fuzzy game. Fuzzy Sets Syst. **4**, 63–72 (1980)
9. Tsurumi, M., Tanino, T., Inuiguchi, M.: A Shapley function on a class of cooperative fuzzy games. Eur. J. Oper. Res. **129**, 596–618 (2001)
10. Tijs, S., Branzei, R., Ishihara, S., Muto, S.: On cores and stable sets for fuzzy games. Fuzzy Sets Syst. **146**, 285–296 (2004)
11. Yu, X.H., Zhang, Q.: The fuzzy core in games with fuzzy coalitions. J. Comput. Appl. Math. **230**, 173–186 (2009)
12. Baron, R., Béal, S., Rémilla, E., Solal, P.: Average tree solutions and the distribution of Harsanyi dividends. Int. J. Game Theory **40**, 331–349 (2011)
13. Gillies, D.B.: Some Theorems on n-person Games. Princeton University Press, Princeton (1953)
14. Shapley, L.S.: A value for n-persons games. Ann. Math. Stud. **28**, 307–318 (1953)
15. Talman, D., Yamamoto, Y.: Average tree solution and subcore for acyclic graph games. J. Oper. Res. Soc. Jpn **51**(3), 203–212 (2008)

Algorithm of Geometry-Feature Based Image Segmentation and Its Application in Assemblage Measure Inspect

Chao-hua Ao, Chao Xiao and Xiao-yi Yang

Abstract Aimed at the puzzle that the edge of industrial computerized tomography image is difficult to realize accurate measure for nondestructive inspection in work-piece, which is resulted from over-segmentation phenomenon when adopted traditional watershed algorithm to segment the image, the chapter proposed a sort of new improved image segmentation algorithm based fuzzy mathematical morphology. In the paper, it firstly smoothed the image by means of opening-closing algorithm based fuzzy mathematical morphology, and then it computed the gradient operators based on the mathematical morphology, after that it segmented the gradient image to get the result based on fuzzy mathematical morphology. And finally it made the assemblage measure inspect for large-complex workpiece. The result of simulation experiment shows that it is better in eliminating over segmentation phenomenon, and more applicable in image recognition.

Keywords Fuzzy mathematical morphology · Gradient operator · Image segmentation · Nondestructive inspection · Assemblage inspection.

C. Ao (✉)
Department of Automation, Chongqing Industry Polytechnic College, Chongqing 401120, China
e-mail: aochaohua@yahoo.com.cn

C. Xiao
College of Automation, Chongqing University, Chongqing 400030, China
e-mail: sngeet@163.com

X. Yang
College of Education Science, Chongqing Normal University, Chongqing 400030, China
e-mail: yangxiaoyi999@sina.com

B.-Y. Cao and H. Nasseri (eds.), *Fuzzy Information & Engineering and Operations Research & Management*, Advances in Intelligent Systems and Computing 211,
DOI: 10.1007/978-3-642-38667-1_42,

1 Introduction

In the pattern recognition of precision measurement in manufacturing, the image segmentation is a key step of image analysis and processing, and also is a sort of basic computer vision technology. Specially in assemblage measure inspection for large-scale and complex metal components, the structural characteristics of an image sometimes are very obvious, so that if the geometry features of the parts is seized when the image is processed, then it not only can reduce a great of processing time, but also can obtain a better processing result. In view of mathematical morphology considering fully the image structural character, it provided with unique advantage of structural character [1]. Watershed algorithm is an image segmentation technology based mathematical morphology, and it may get smart image edge. However, it is too sensitive to noise, and the weak noise will cause over-segmentation phenomenon. This chapter proposed a new sort of improved image segmentation algorithm based fuzzy mathematical morphology by means of technology method fusion so as to enhance the quality of image segmentation.

2 Fuzzy Mathematical Morphology

Fuzzy mathematical morphology is a kind of mathematical tool of image analysis based on morphology structural elements [2]. Its basis idea is to use morphological structural elements to measure and distill corresponding shape of an image to attain objective of analysis and recognition image, reduction image data, keep basic shape character and eliminating non-correlative structure. Mathematical morphology has four basic operations [3]. Those are the dilation, erosion, opening and closing operator. Each operation has its trait respectively in binary and gray-degree image processing. They can educe many other practical operations of mathematical morphology.

If the basic idea of fuzzy set theory is introduced into the mathematical morphology, then it is called as the fuzzy mathematical morphology. By means of itself properties, it can be used to extend the application field from binary mathematical morphology to the gray-degree image processing of pattern recognition, and widens the definition of algorithm operators of classic mathematical morphology, and therefore it obtains strong robustness in some degree, and holds good trait of classic mathematical morphology operator. Especially, it is more effective in image processing effect than by traditional morphological algorithm operator when the image includes some noise. The main idea of processing image in fuzzy morphology is to view an image as a fuzzy set because of fuzziness rooted in image itself characteristic and in process of collection and processing, so the fuzzy arithmetic operator can be introduced into the image process to make pattern recognition. Of course, the operation is different according to the different definition in fuzzy arithmetic operator. However a lot of operator can be transformed as Bloch operator to carry through each kind of operation. Before evaluating USRC, it's necessary to analyze science research capability elements of universities [4].

2.1 Fuzzy Subset

In the fuzzy theory, the fuzzy set can be educed when value range of a membership eigenfunction in classic set theory for element x is extended from open interval 0, 1 to closed interval [0, 1]. The Fuzzy subset A can be expressed as formula (1)

$$\mu_A : U \longrightarrow [0, 1], x \longrightarrow \mu_A(x) \tag{1}$$

In which, U is called as domain. The μ_A is called membership function, and $\mu_A(x)$ is called as the value of membership function. Formula (1) is any map of U over closed interval [0, 1]. Fuzzy subset A is fully described by μ_A of membership function. The membership function represents the degree that belongs to A by a value of element x over closed interval [0, 1].

2.2 Decomposition Theorem

Suppose A is a common set of domain X, $\forall\lambda \in [0, 1]$, Fuzzy set $\lambda * A$ of X can be defined, and its membership function is expressed by formula (2).

$$\mu_{\lambda * A} = \begin{cases} \lambda, x \in A \\ 0, x \notin A \end{cases} \tag{2}$$

For $\forall\breve{A} \in F(X)$, the decomposition theory form of fuzzy set is expressed by formula (3).

$$\breve{A} = \bigcup_{\lambda \in [0,1]} \lambda * A \tag{3}$$

In which, $\breve{A}$ is reflection of A.

2.3 Extension Theorem

Extension theorem presents the image structure of f(A) of fuzzy subset A in X under the common mapping relation f from domain X to domain Y. It also presented mapping rules that extend map relation between element of X and element of Y corresponding to subset or fuzzy subset element of X and subset or fuzzy subset element of Y. It can be expressed respectively by formula (4) and (5).

$$f : F(X) \to F(Y); A \to f(A) \tag{4}$$

$$f - 1 : F(Y) \to F(X); B \to f^{-1}(B) \tag{5}$$

In which, X and Y are two domains. Their mapping relation is: $f : X \to Y$ and it can educe the mapping from F(X) to F(Y) and from F(Y) to F(X). The $f(A)$ is called image of A and $f^{-1}(B)$ is called inverse image of B. their membership function is respectively expressed by formula (6) and formula (7).

$$\mu_A(y) = \bigvee_{y=f(x)} \mu_A(x) \; \forall y \in Y \tag{6}$$

$$\mu_{f^{-1}(B)}(x) = \mu_B(f(x)) \; \forall x \in X \tag{7}$$

3 Improvement of Watershed Algorithm

3.1 Watershed Algorithm Based on Immersion Simulation

Watershed algorithm is a sort of image processing tool rooted in mathematical morphology. It can be used to segment image, distill gradient image and so on. In the numerous existing sequence watershed algorithms, it is the most representative and the fastest algorithm that based on immersion simulation and its improved algorithm proposed by Vincent [5]. In this algorithm, digital image can be expressed by formula (8)

$$G = (D, E, I) \tag{8}$$

In which, (D,E) describes the image and I is the corresponding transform function of D, Each pixel p.I(p). I(p) expresses the image gray-value of each pixel p and its value range is from 0 to 255. If threshold h of image is $T = p|I(p) \leq h$, in the immersion process, then the point starts from set $T_{h_{min}}(I)$ and the point in set is the place that water reaches firstly. And these points form beginning point of iterative formula, shown as in formula (9) and (10).

$$X_{h_{min}} = \{p \in D|I(p) \leq h_{min}\} = T_{h_{min}} \tag{9}$$

$$X_{h+1} = MIN_{h+1} \cup IZ_{T_{h+1}}(X_h), h \in [h_{min}, h_{max}] \tag{10}$$

In the above, h_{min} is the minimum and h_{max} is the maximum. And the $X_{h_{min}}$ is composed of point in set I. These points are located minimum region that its altitude is the lowest. The MIN_h is union of all minimum region that their gray-values are h. Gray-value h is iterative continuously from h_{min} to h_{max}. IZ is union of measure infection region [6, 7]. In the iterative process, the minimum point district of image

I will be extended gradually. Suppose X_h is the connected discreteness of threshold set T_{h+1} under the value h of position for union of district sets started from plane position, it may be a new minimum or be located the extension region of $X(h)$. For the latter, X_{h+1}can be renewed by computing T_{h+1}. In the set D, supplementary set of $X_{h_{max}}$ is just the watershed of the image [4], expressed by formula (11).

$$Watershed(f) = D/X_{h_{max}} \tag{11}$$

According to the definition above, the gradient-value of each point of the image can be viewed as its height. Provided we drill many small holes on the bottom of each minimum region M of the image and pour water into formed ground interface, the water will be immerged gradually to the ground. So many lake-let can be formed like a catchment basin. Starting from minimum region that the altitude is the lowest, the water will immerge into all catchment basins. In the process above, if the water comes from different catchment basins it will be converged, a dam will be built at the converged edge. At the end of the immersion process, it is necessary that the dam has to surround all catchment basins and the union of dams is just corresponding watershed of the image.

3.2 Watershed Algorithm Based IFT

Image Foresting Transform (IFT) is a sort of image segmentation algorithm based on graph theory [8] and it is the shortest path first algorithm of Dijkstra in essence. It uses connectedness of graph to design image processing arithmetic operator. Its main idea is that the image will be mapped into the picture and the marked image will be obtained through computing shortest path of the picture. In the picture, the IFT algorithm defined a shortest path forest, and the nodes of the forest are pixel. The arcs between nodes are defined by adjacency relation of pixel, and the path costs are determined by path cost function [9]. The IFT algorithm regards image as a picture and its processing result is the adjacency relation of pixel. The common path cost function has additive path cost function and maximum arc path cost function. The catchment basin uses maximum arc path cost function [10], expressed by formula (12) and (13).

$$f_{max} = (< t >) = h(t) \tag{12}$$

$$f_{max}(\pi \cdot < s, t >) = max\{f_{max}(\pi), I(t)\} \tag{13}$$

In the above, A is an adjacency relation of pixels, and (s, t), s is the end node, t is the start node, $h(t)$ is its initial value of path cost started from node t and $I(t)$ is the pixel-value of t.

4 Improved Fusion Algorithm in Image Segmentation

4.1 Study on Algorithm

The idea of fusion algorithm is the following. Firstly, the optimal threshold-value [11] must be determined by auto-recognition method based on image gray-degree character for enhancing recognition efficiency and its judging criterion is to separate the goal from background farthest. Secondly, it determines optimal threshold-value of image segmentation by means of a sort of simple and nimble method based on optimal auto-recognition threshold-value. Then it restricts further path cost function of original IFT watershed algorithm according to the optimal threshold-value. In this chapter, the algorithm is to constrict search scope of optimal path of original IFT watershed in essence, so it can enhance the execution speed for operation.

Because of adding the restriction of threshold-value in the algorithm, the path cost function needs to make corresponding adjustment. The new path cost function is expressed by formula (14) and (15)

$$f_{max} = (< t >) = I(t) \tag{14}$$

$$f_{new}(\pi \cdot < s, t >) = \begin{cases} max\{f_{new}, I(t)\} \; if \; I(t) \geq T \\ +\infty \;\; otherwise \end{cases} \tag{15}$$

In the formula, T is the threshold-value. Suppose the image has N-degree gray-grade value, the steps of improved IFT watershed algorithm are as the following. Process input is respectively the image I and the template image L. Process output is the result L of each catchment-basin transformed by watershed algorithm;

Auxiliary data structure is all the node cost C (cost map), and the initial values will be set as infinite (∞);

Computing steps of algorithm are shown as the following.

(1) Do $C(p) = I(p)$ for all nodes satisfying the condition($L(p) \neq 0$), then insert node p into queue Q according to the value of C(p).
(2) Make use of auto-recognition technology to identify the threshold-value.
(3) Delete node p that its C(p) value is minimum if queue Q is not empty. For each node satisfying the condition $q \in N(p)$ and node q without inserting into the queue Q, do the following operation.

Computing $C = f_{new}\pi \cdot < p, q >$

If $C \neq +\infty$, let C(q)= C and insert node q into queue Q according to value of C(q). Then let L(p)= L(q). For the above algorithm steps, we can make the following analysis.

(1) Restriction condition of path cost function, is readjusted according to threshold-value.
(2) Seed set, any node belonged to objective.

(3) Layered queue structure Q, if the image includes N-grade gray-grade value (because there exists threshold-value restriction), the number of bucket of queue Q can be reduced to $N - T + 1$ and storage space of algorithm can be contracted to O(n+N-t+1). In the steps of original algorithm, the node inserted queue Q must be never operated by current node, so operation for the queue is different from the original algorithm.

(4) Because of adding restriction of threshold-value, the search process does not traverses all the nodes, but some nodes that their threshold-values are over threshold-value of target region of image will be visited in the target region of image. This method reduced the search area and enhanced the execution efficiency of algorithm

4.2 Implementation and Its Result Analysis for The Improved Algorithm

1. Implementation of the improved algorithm

   The algorithm can be realized under the software environment of Matlab.7.0. The improved algorithm flowchart is shown as in Fig. 1.

   The operation steps of improved algorithm are as the following.

   Step 1. To smooth image

   It adopts firstly opening-closing operation based on the fuzzy mathematical morphology to smooth image, so it can eliminate noise, save important region edge, and solve more perfectly the problems in the pretreatment process by means of morphology erosion, dilation and opening-closing operation to filter image.

   Step 2. To compute gradient

   The algorithm secondly adopts the operator of basic morphological gradient algorithm to compute gradient.

   Step 3. To achieve image

   It finally segments the image to gain the objective image by use of improved algorithm.

   It is specially worth to point out that it was chosen to the opening-closing filtering based on fuzzy morphology to carry through the image filtering in the filtering process. This kind of filtering method is based on set theory and has some smart trait, such as keeping the image edge well, distilling the signal effectively, retaining the image detail completely, and restraining noise and so on.

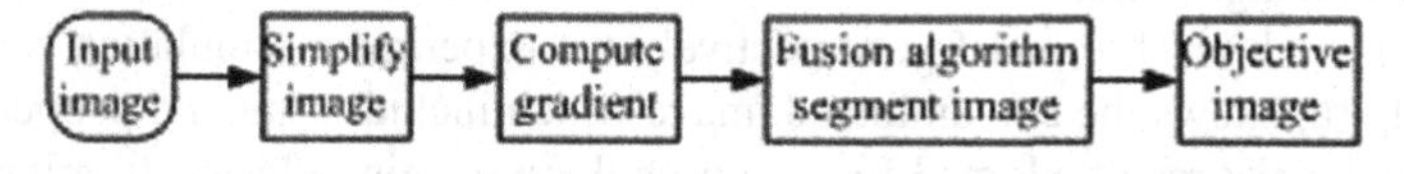

Fig. 1 Flowchart of algorithm

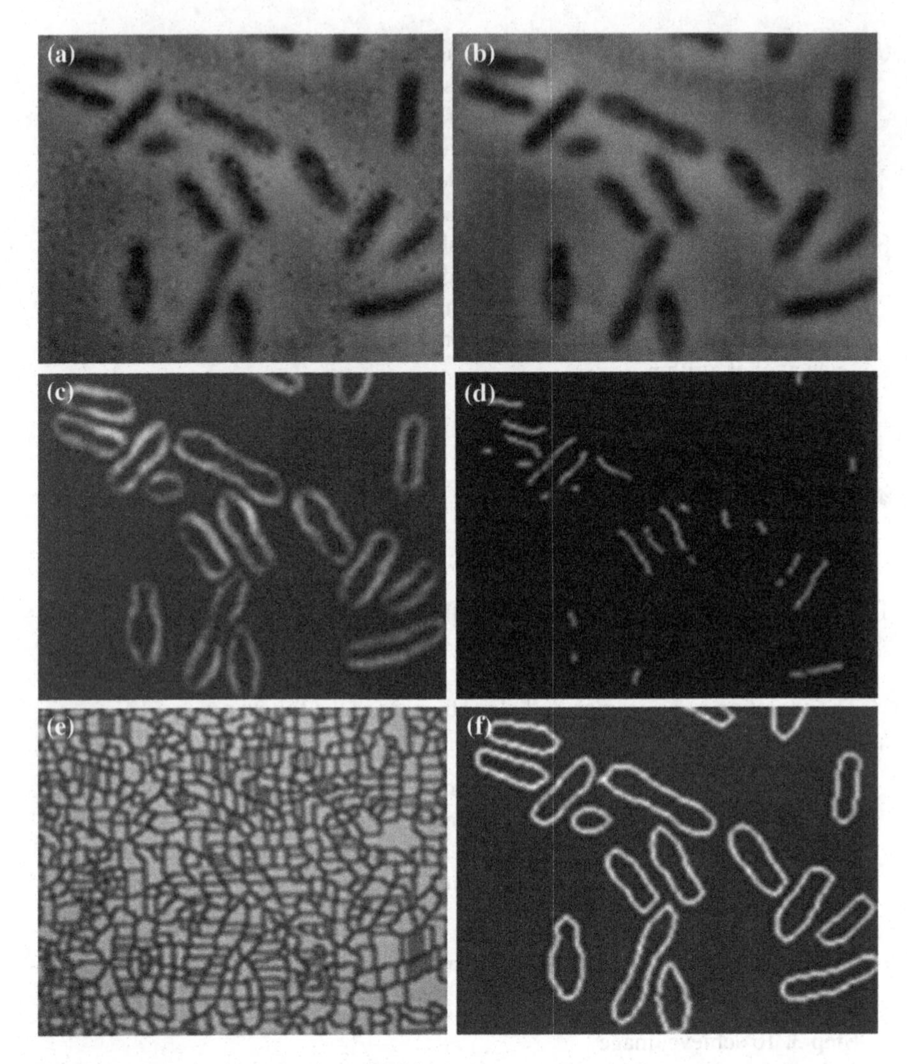

Fig. 2 Result comparison of various algorithms

2. Simulation verification In order to compare the effect of processing image, here we take the grain defect detect of grain depot as an example to judge that whether the rice is in completeness and not mixed other grain. Firstly it takes the grain image. Then it makes image processing for the taken image. Based on the software environment of Matlab.7.0, the Fig. 2 shows the results of numeric simulation of the fusion algorithm.

   In Fig. 2a, b, c, d, e and f is respectively the experiment simulation results in which, (a) shows the contaminated image which includes salt and pepper noise, (b) shows the result adopted fuzzy morphology opening-closing filtering to filter contaminated image, (c) shows the gradient image computed morphological gradient algorithm operator for filtered image, (d) shows the result adopted Prewitt algorithm to segment image, (e) shows the result used directly traditional watershed algorithm to segment image, (f) shows result adopted new improved algorithm to segment image.

3. Result analysis for experiment from the Fig. 2, it can be viewed that if it adopts directly the watershed algorithm to segment image then the result will appear over-segmentation phenomenon and if it adopts Prewitt algorithm to segment image then the following problems will be happened that its contour line will not be continuous and edge orientation is not precise. But if it adopts improved algorithm based fusion algorithm technology to segment image then it will not only be conquered to over-segmentation phenomenon produced by direct watershed in the result and get continuous and close boundary line, but also can fully save the image detail, and therefore obtain satisfactory segmentation effect and faster operation speed.

5 Use Case in Scathe-less Detect of Metal Parts Assemblage

The nondestructive inspection of metal parts assemblage has been widely applied in assemblage measure inspect for large-scale complex metal parts, and the algorithm of geometry-feature based image segmentation can make the detail of image outline of metal workpiece more clear so as to distinguish strictly the boundary of metal part defect. For example, industrial computerized tomography images based on clear image edge can make nondestructive inspection for each workpiece, and realize the location of the same section position CT image using Hausdorff distance, and also using the position information which the stencil image is located in the original image, it can judge that the assemblage is right or not. For inspect of large-scale complex metal parts, it can make image registration, image matching and localization and so on. In Fig. 3a, b and c, they are the examples of image registration, in which, (a) is original image of CT, (b) is rotating image and (c) is image with rotating and translation, and it canco make accurate localization for each workpiece.

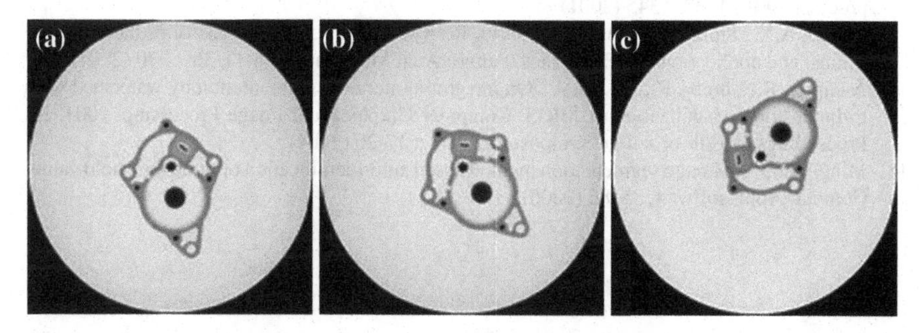

Fig. 3 Image registration

6 Conclusion

From the above effect analysis of image processing, we can make such a useful conclusion that it is a sort of method of image processing and analysis and also a solution of image segmentation based morphological technology to the fusion image segmentation algorithm based on fuzzy mathematical morphology. The method can be applied to many practical engineering field, such as military objective detective, agriculture pest image recognition, resource protection of the ocean, agriculture irrigation, environment monitoring and so on. So it has important practical engineering application value and theoretic significance. However, there are still some problems in the research process, for example, both choice of threshold-value and execution speed of algorithm need to make further improving so that it can obtain better segmentation effect and faster operation speed.

References

1. Dai, Q., Yun, Y.: Application development of mathematical morphology in image processing. Control Theor. Appl. (4):13–16 (2001)
2. Bloch, I, Maitre, H.: Fuzzy mathematical morphologies: a comparative study. Pattern Recognit. **9**, 1341–1387 (1995)
3. Bloch, I, Maitre, H.: Why Robots should use Fuzzy Mathematical Morphology. Proceeding of the 1st International ICSC-AISO Congress on Neuro-Fuzzy Technologies, La Havana, Cuba, January, pp 249–283 (2002)
4. Roerdink, B.T.M.: The watershed transform: definitions, algorithms and parallelication. Fundamenta. Informatica **41**, 197–228 (2000)
5. Vincent, L., Soille, P.: Watersheds in digital space: An efficient algorithm based on immersion simulations. Trans. Pattern anal. Mach. Intell **13**(6), 583–589 (1991)
6. Lotufo, R., Silva, W.: Minimal set of markers for the watershed transform. Proceedings of ISMM 2002. Redistribution rights reserved CSIRO Publishing, pp 359–368 (2002)
7. Hernandez, S.E., Barner K.E.: Joint region merging criteria for watershed-based image segmentation. Proceedings of international Conference on Inage Processing, vol. 2, pp 108–111 (2000)
8. Perez, D.G., Gu, C., etal.: Extensive partition operators, gray-level connected operators, and region merging/classification segmentation algorithms: theoretical links. IEEE Trams. Image process. **10**(9): 1332–1345 (2001)
9. Falcao, A.X., Stolfi, J., de Alencar Lotufo, R.: The image foresting transform: theory, algorithms, and applications. IEEEE Trans. Pattern Anal Mach Intell. **26**(1), 364–370 (2004)
10. Audigier, R., Lotufo, R., Falcao, A.: On integrating iterative segmentation by watershed with tridimensional visualization of MRIS. Computer Graphics and Image Processing. 2004. In: Proceedings of 17th Brazilian Symposium on, Oct.17-20 (2004)
11. Ming, Chen: An image segmentation method based auto-identification optimal threshold value. Comput. Appl. softw. **4**, 85–86 (2006)

The Application of GA Based on the Shortest Path in Optimization of Time Table Problem

Zhong-yuan Peng, Yu-bin Zhong and Lin Ge

Abstract Time Table Problem (*TTP*) is a constraint Combinational Optimization Problem (*COP*) with multi- objective. Based on the analysis of advantages and disadvantages of Genetic Algorithm (*GA*) and Kruskal Algorithm (*KA*), this chapter put forward to a new hybrid algorithm—the Shortest path-based Genetic Algorithm (*SPGA*), which has the advantages of both *GA* and *KA*. In this algorithm, fitness function, selection operator, crossover operator and mutation operator are studied deeply and improved greatly, so that the hybrid algorithm can be used in the actual course arrangement. The simulation results show the effectiveness of this method.

Keywords Shortest path · Genetic algorithm · Time table Problem · Kruskal algorithm

1 Introduction

GA, a self-adaptive iterative search algorithm with probability based on natural selection and genetic variation, was mainly put forward by John Holland in 1975. It includes three basic operations: selection, crossover and mutation. *GA* is parallel; it doesn't need derivative or other auxiliary knowledge, and it only needs object function and the corresponding fitness function that can affect the search direction. *GA* can be directly applied in combinational optimization, neural networks, machine learning, automatic control, planning and design, artificial life and other fields [1].

Z. Peng · L. Ge
Maoming Polytechnic, Maoming 525000, Guangdong, People's Republic of China

Y. Zhong (✉)
School of Mathematics and Information Sciences, Guangzhou University,
Guangzhou 510006, People's Republic of China
e-mail: Zhong_yb@163.com

B.-Y. Cao and H. Nasseri (eds.), *Fuzzy Information & Engineering and Operations Research & Management*, Advances in Intelligent Systems and Computing 211,
DOI: 10.1007/978-3-642-38667-1_43, © Springer-Verlag Berlin Heidelberg 2014

KA is a classic algorithm in solving Minimum Spanning Tree (*MST*) in graph theory. Compared to Prim algorithm, it is more suitable in solving *MST* in sparse graph. Suppose a connected graph $G = (V, \{E\})$, then let the initial state of *MST* be a non-connected graph $T = (V, \{\})$,which has n vertexes and no edge. In this non-connected graph T, each vertex forms a connected component. Select the lowest cost edge in E, if the vertexes to which lowest cost edge attached fall in different connected components in T, then add this edge into T; otherwise, discard this edge and select next lowest cost edge. Do this by analogy until all the vertexes in T are in the same connected component [2].

2 Principle of Hybrid Genetic Algorithm (*HGA*)

Given a weighted undirected connected graph $G = (V, E, W)$, where $w = \sum_{e \in E} w_e$ is the sum of weights on each edge. If tree $T = (V, E_T, W_T)$ contains all vertices in graph G , and makes $W_T = \sum_{e \in E_T} w_e$ the minimum, then tree T is called *MST* of graph G. According to classic algorithms for solving *MST* in graph theory, *MST* is unique. In practical problems, different forms of *MST* often represent different implementations; sometimes Sub-minimum Spanning Tree (*SMST*) may also be a better solution. If only one can be chosen, it may ignore or discard a better solution. Thus, in actual operation, we hope to choose one from a number of *MSTs* or *SMSTs*, and weigh the pros and cons of various aspects in order to get a better solution [2, 3].

In recent years, *GA* is applied to *TTP* by many researchers, and they got good results [4–7]. However, for *TTP*, the search capability and efficiency of any single search algorithm, including *GA*, is not high, and it is greatly influenced by the initial parameters. Simple genetic algorithm needs a long time and it is very easy to fall into "premature". Kruskal algorithm belongs to greedy algorithm. Although it is able to reduce time complexity to achieve local optimization in some extents, the effect of its solution is often unsatisfactory and the solution does not meet the principles of global optimization in general conditions.

Therefore, combining advantages of *GA* and *KA*, this chapter tries to put forward to a *HGA–SPGA*. Genetic algorithm based on the shortest path combine with Genetic algorithm and Kruskal algorithm. To make best use of the advantages and bypass the disadvantages. In this *HGA*, the fitness function, selection operator, crossover operator and mutation operator are improved, so better optimization results are got in the practical application of *TTP*.

3 The Application of *HGA* in *TTP*

3.1 *Representation of Graph in* TTP

In *TTP*, $V = \{v_i, i = 1, 2, 3, \cdots\}$ denotes the set of factors, where v_i indicates the ith factor, which is represented by vertexes. If there is a relationship between v_i and v_j, then they will be connected with edges (see Fig. 1). $W = \{w_i, i = 1, 2, 3, \cdots\}$ denotes the set of weights of edges, where w_i indicates ith weight of the edge. For specific questions, w_i has different meanings.

3.2 *The Design of Fitness Function*

For *TTP*, as long as we can get a *MST* or a *SMST* in Fig. 1, we can get a successful selection scheme. Of course, different selection schemes have different effects. And fitness value can be used for the measurement of different schemes. In *GA*, an individual's fitness determines the probability of being passed on: a greater fitness, a greater probability. Therefore, we define the fitness function [8] as follows:

$$F(T) = \begin{cases} K - \sum\limits_{i=1}^{n} w_i Z_i \\ 0 \end{cases}$$

Fig. 1 Representation of graph in *TTP*

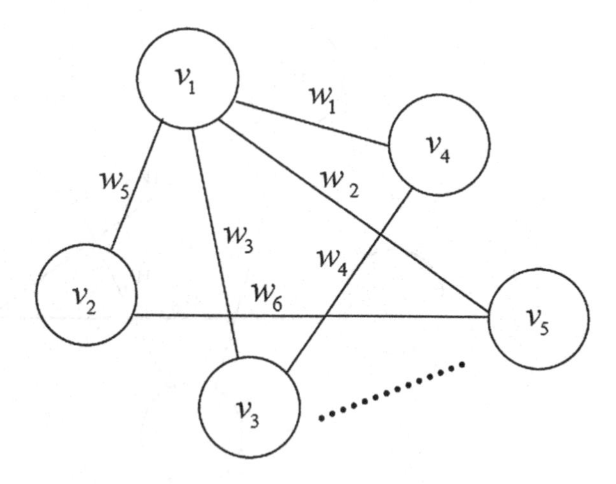

where $Z_i = \begin{cases} 1 \ e_i \in T \\ 0 \ e_i \notin T \end{cases}$; T is a spanning tree of Fig. 1; $K = (n-1)max(w_1, w_2, \cdots, w_n)$ is a larger positive constant so that it can ensure that $F(T)$ is nonnegative; w_i is the weight of ith edge.

In the process of genetic evolution, generated individuals must be tested. If all vertices can be searched, then the individual is a spanning tree of graph, an effective individual whose fitness can be calculated by the defined fitness function; otherwise, the successful selection is unable to be completed, and this individual is an invalid one. Let its fitness be zero, and it will be eliminated.

3.3 The Improvement of Genetic Operator

3.3.1 Combining *GA* with *KA*

The Standard Genetic Algorithm (*SGA*) generates new offspring individuals mainly through crossover and mutation operator; but for *MST*, the crossover operator and mutation operator of *SGA* is very easy to disrupt the basic structure of spanning tree, so it is difficult to obtain effective new individuals, namely, effective *MST* or *SMST* can not be constructed, and the search capabilities are reducing. In order to improve the efficiency of searching, combining the characteristics of *MST*, crossover operator and mutation operator are improved.

Take $i = 5$ for an example to illustrate the operating process of *SPGA*. Give fixed values to w_i, and get a *MST* by using Kruskal algorithm. For example, let $w_1 = 1$, $w_2 = 0.5$, $w_3 = 0.8$, $w_4 = 1$, $w_5 = 1$, $w_6 = 0.2$, then get a *MST* $w_2w_3w_4w_6$. A selection scheme is completed (Shown in Fig. 2). We use edge encoding method.

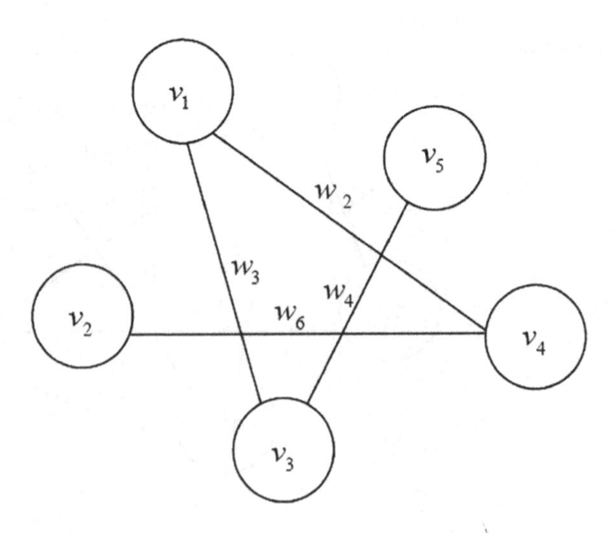

Fig. 2 *MST* $w_2w_3w_4w_6$

If the edge is selected, then let it be 1, otherwise let it be 0. For example, *MST* $w_2w_3w_4w_6$ can be represented as 011101.

3.3.2 Selection Operator

Order new generated individuals with descending sequence according to their fitness, and use random traversal sampling method as the selection strategy. Suppose S is the number of individuals to be selected, and select individuals equidistantly. The distance of selection pointer is $\frac{1}{S}$; the position of the first pointer is determined by an uniform random number in interval $\left[0, \frac{1}{S}\right]$.

3.3.3 Crossover Operator

In gene of a selected individual in the initial colony, choose any position and exchange forward and backward, and then a new individual is got, such as 011101 $\rightarrow$ 011$\dot{:}$101 $\rightarrow$ 101$\dot{:}$011 $\rightarrow$ 101011. The received new individual is 101011 that corresponds to $w_1w_3w_5w_6$, a *SMST*. It is another selection scheme, an effective individual. Using such a crossover operator can greatly improve search efficiency in the feasible solution space, and will avoid a large number of individuals that will be eliminated.

3.3.4 Mutation Operator

Choose any position in the gene of new individuals, then exchange one position before and after this position, such as 101011 $\rightarrow$ 10$\dot{:}$1011 $\rightarrow$ 11$\dot{:}$0011 $\rightarrow$ 110011. The new individual after mutation is 110011, which is an invalid individual and will be eliminated.

The improved crossover operator and mutation operator has a very prominent characteristic: in the process of obtaining progeny colony, each individual has only one parent, single-parent propagation from biological view. The advantage is that the next generation can keep characteristics of parent by the greatest extent, and it increases the possibilities of being a feasible solution as well as the search capabilities of *GA*.

3.4 Conditions for Ending Algorithm

GA needs to set an ending condition, otherwise it will do an infinite loop and will never terminate. For *TTP*, we define a generation number n as the condition of end. When iteration times of *GA* equals to this generation number, the algorithm is terminated and the present individual is the ultimate solution. Of course, n can not be too big or

Fig. 3 Flow chart for *HGA*

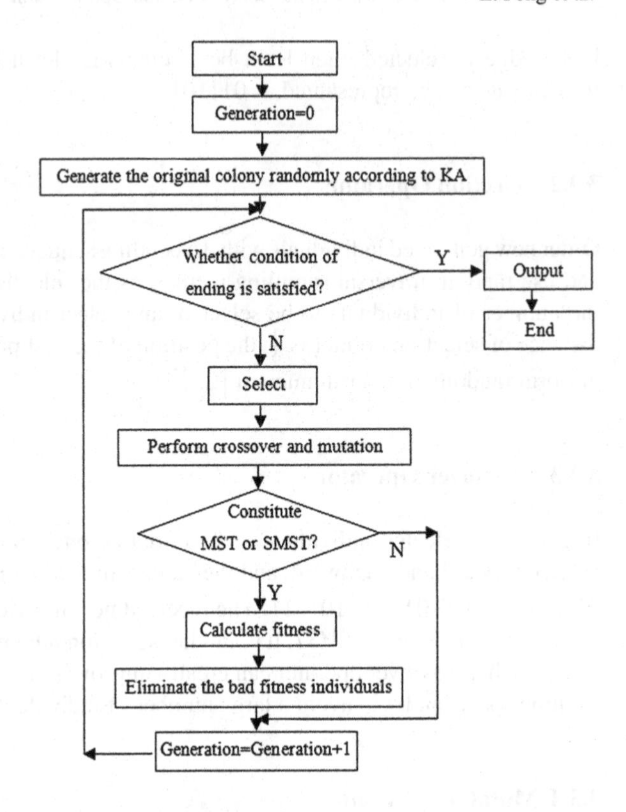

too small: if too long, the computation time is also long, then efficiency reduces; if too small, convergence effect can not reach, failing to find optimal solution.

3.5 *Flow Chart for* HGA

Improving crossover operator and mutation operator of *SGA* appropriately, we can get *SPGA*. The flow chart is shown in Fig. 3.

4 Instance of Application

The arrangement of schedule is a multi-factor optimized decision problem, and a typical problem in Combination Programming; it is mainly used to make rational utilization of time and space resources and avoid conflicts while arranging courses. A rational and scientific course schedule is very important to the work of teaching in school [9, 10].

4.1 Constraints in CSP

In the arrangement of schedule, some constraints should be met, and these constraints can be divided into soft constraints and hard constraints [9] .

4.1.1 Hard Constraint

Hard constraint must be met while arranging schedule, otherwise, teaching work will not continue normally.

(1) One teacher can not be arranged more than two different courses at the same time;
(2) One classroom can not be arranged more than two different courses at the same time;
(3) One class can not have more than two different courses at the same time;
(4) The classroom must be large enough to hold students attending classes;
(5) The type of classroom must match that of the courses.

4.1.2 Soft Constraint

Soft constraint can be met while arranging schedule and it is the standard for measuring course arrangement.

(1) Courses that have more week periods should be staggered reasonably;
(2) If one class has two consecutive courses, then the probability of changing classroom should be as small as possible, or arrange nearby classroom;
(3) A class's one week courses should be evenly distributed;
(4) A teacher's one week courses should be evenly distributed;
(5) Important courses should be arranged in a good time period as far as possible;
(6) Physical education class should be arranged in the third or fourth class, or in the seventh or eighth class;
(7) Satisfy the special requirements of a certain teacher.

4.2 The Arrangement of Courses for a Class

It is assumed that there is a numerical control class. In one semester, the required courses are: Advanced Mathematics (2 / week), Mechanical drafting and Tolerance (4 / week), College English (4 / week), Engineering Mechanics (4 / week), Basis for Information Application Technology (4 / week), Ideology Morality Accomplishment Course and Legal Basis (2 / week) and PE (2 / week). Each day is divided into four time periods, and two classes are a time period. Thursday afternoon is the learning time for teachers, so no courses are arranged. According to the steps of *HGA*, this class's courses are arranged to meet the hard constraints, and are optimized according to soft constraints. The arrangement results are shown in Table 1.

From Table 1 and we can see that the optimized timetable using Improved Genetic Algorithm (*IGA*) satisfies not only the hard constraints, but also soft constraints. From practical view, feasibility of *HGA* is verified further, and *IGA* is a feasible method of solving *CSP*.

4.3 Analysis of the Results

(1) Because of the thought of single-parent propagation, crossover operator and mutation operator of *SGA* are made appropriate changes to effectively avoid the generation of invalid individuals. The running time of this algorithm reduces greatly, and it is more efficient. (2) Both Initial Timetable Generated Randomly and Optimized Timetable Using *HGA* meet the hard constraints, and there is no conflict. Namely, one teacher, one classroom and one class are not arranged more than two different courses at the same time; the classroom is large enough to hold all students; the type of classroom matches that of the courses. (3) Optimized timetable can deal courses that require special arrangements well, such as PE. In the initial timetable, PE is arranged in 1st and 2nd class in the morning, which is not scientific. If students have class in the 3rd and 4th, they will be tired, then it would certainly affect the efficiency of class. In optimized timetable, PE is arranged in the 7th and 8th class in the afternoon. Students can have a good rest after class, achieving the purpose of physical exercise without influencing other courses. (4) The time interval must be reasonable for courses that have four or more than four classes one week. It is very necessary to maintain a certain degree of dispersion. On the one hand, teachers must make good preparations before class, such as being familiar with teaching materials, designing teaching situation, choosing teaching methods, preparing teaching plans, and correcting the exercises; if the class interval is too dense, it will affect teacher's preparation and classroom instruction. On the other hand, students need time to digest, understand and consolidate knowledge after class, and they also need some time to finish the course assignments. If the courses are too continuous, students can only be struggling to accept, and there is no time to understand and master, so the effect is bad. Optimized timetable handles reasonably at this. (5) Important courses should be arranged in better time periods as far as possible, which is based on a person's physiological characteristics and teaching rules. It plays a significant role in raising the overall classroom teaching effect. There is a clear manifestation of the arrangements of important courses in optimized timetable.

Table 1 Optimized Timetable Using *HGA*

	Monday	Tuesday	Wednesday	Thursday	Friday
1	College English Chen Shu	Engineering mechanics Xu Yan	Mechanical drafting and	Basis for information application	
2	2/303	1/301	Tolerance Gu Zhi 1/605	Technology Wu Min No.2 Computer room	
3	Basis for information application technology	Advanced mathematics Ge Lin		Engineering mechanics Xu Yan	Mechanical drafting and tolerance
4	Wu Min No.1 Computer room	1/301		2/403	Gu Zhi 1/208
5			Ideology morality accomplishment course and legal		College english Chen Shu
6			Basis Wang Feng 2/501		Comprehensive stair classroom
7		PE			
8		Liang Zhuo quan			

5 Conclusion

This chapter put forward a new hybrid algorithm - the Shortest path-based Genetic Algorithm *(SPGA)*, which has the advantages of both GA and *KA*. Combined with the actual situation of Time Table Problem, fitness function, selection operator, crossover operator and mutation operator are studied deeply and improved greatly in this algorithm. And the algorithm is used to test the course scheduling problems. Practice has shown that this is a simple, effective method. It can obtain a number of *MST* and *SMST* in a short time with high efficiency, providing a variety of schemes for course arrangement. And it is also a solid foundation for further optimization. To sum up, good results are achieved.

Acknowledgments Thanks to the Higher Education Teaching Reform Project of Guangdong Province by Prof Yu-bin Zhong.

References

1. Wang XP., Cao, LM.: Genetic algorithms-theory, application and software implementation. Xi'an Jiaotong University Press, Xi'an (2002)
2. Yan, W.M., Wu, W.M.: Data Struct. Tsinghua University Press, Beijing (1996)
3. Zhou, R., Mai, W., Lei, Y.: Minimum spanning tree algorithms based on genetic algorithms. J. ZhengZhou Univ. (Engineering Science) (2002)
4. Chu, Beasley, P.C.: A genetic algorithm for the generalized assignment problem. Eur. J. Oper. 20–22 (1995)
5. Safaai, D., Sigeru, O.: Incorporating constraint propagation in genetic for university timetable problem, Engineering Application of, Artificial Intelligence, pp. 241–253(1999)
6. Luan, F., Yao, X.: Solving real-word lecture room assignment problems by genetic algorithm, Complexity International. Electronic J. Complex Res. 15–18 (1996)
7. Colorni, A., Dorigomarco, V.: Metaheuristics for high school timetabling.Comput. Optim. Appl, Maniezzo (1998)
8. Li-juan, Z., Xiao-buo, LE.: Application of genetic algorithm in minimum spanning tree. Comput. Knowl. Technol (2007)
9. Zhao G.: Study on timetable problem of college based on genetic algorithms. Yanbian University (2006)
10. Zhao B.: Study of genetic algorithms and application in course arrangement problem. SWJTU J. (2003)

Monitoring System of Networked Gas Stations Based on Embedded Dynamic Web

Wei Huang, Kai-wen Chen and Chao Xiao

Abstract The oil is a sort of strategic material, and therefore strengthening the management of oil material has very important significance. Aimed at the difference in communication protocol among dispensers of gas stations, resulted in being difficult to realize the integration of monitoring system, this paper proposed a sort of integrated monitoring solution based on embedded Web. The core device of monitoring system selected a sort of embedded Web server based on Intel Xscale IXP-422 RISC CPU. The servers distributed in the industrial field of gas stations interconnected through industrial Ethernet, and composed a wide area network system based on Web service. The field bus of field device connected to Web server in the field local area of gas stations to complete the integrated monitoring of field device. The system adopted the architecture of distributed browser/server. By means of the approach of Apache+Html+PHP, the monitoring and management of the gas stations could be realized based on embedded Web, and the realization of dynamic Web browse could be completed by control unit. The actual test data demonstrated that it could be high in security level, stronger in anti-jamming, better in environment adaptability, and higher in real time performance. The research result shows that the proposed solution is feasible and reasonable.

Keywords Web service · IFSF · Embedded system · Protocol conversion · FCC

W. Huang (✉)
Department of Automation, Chongqing Industry Polytechnic College, Chongqing 401120, China
e-mail: huangwei051001@126.com

K. Chen · C. Xiao
College of Automation, Chongqing University, Chongqing 400030 , China
e-mail: sngeet@163.com

K. Chen
e-mail: dengrenming65106683@126.com

B.-Y. Cao and H. Nasseri (eds.), *Fuzzy Information & Engineering and Operations Research & Management*, Advances in Intelligent Systems and Computing 211,
DOI: 10.1007/978-3-642-38667-1_44,

1 Introduction

Nowadays, The Web has been widely used into the extensive industrial applications [1–3]. It has been become a hot topic to apply Web technology for industrial monitoring field. By means of running the embedded Web server of field control device in the bottom layer of industrial control system, it can be accessed and monitored to all control devices in industrial field through Internet in anywhere for using general Web browser [4–6], and gets the aim of monitoring field device expediently. Through the page layout of Web stored in the field control device, it can dynamically reflect the running state of field device and feedback information after executing the monitoring operation, and also it can collect field running real time data more accurately, make various checking analysis, control the field devices, and carry through system maintenance easily and so on. This paper takes the monitoring system of gas stations in oil product retail network as an example to explore a sort of realization of monitoring system based on embedded dynamic Web.

2 Structure of Networked Monitoring System

2.1 System Structure

The Fig. 1 shows the structure of networked monitoring system in gas stations.

The field devices in forecourt of gas stations, such as dispensers, tank level gauges and so on, are connected directly to the Intranet/Internet through embedded Web server, then by means of Ethernet port or wireless port, and it can be connected to the browsing monitoring station or moving browsing monitoring station based on Web. It is propitious to realize the data communication and monitor in real time,

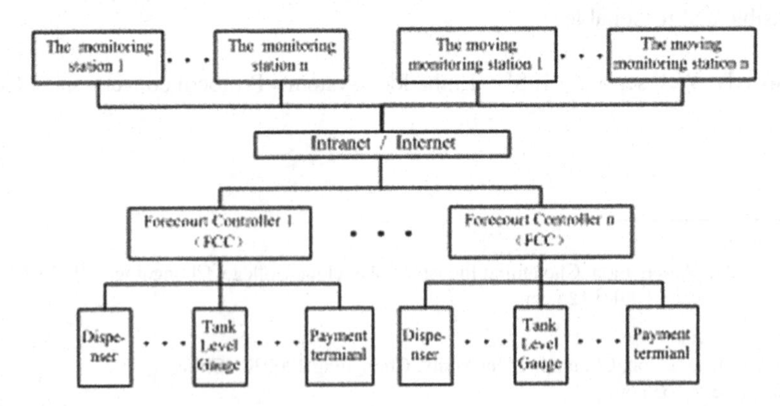

Fig. 1 Structure of networked monitoring system

network security and encrypt system for this sort of architecture, and to adopt Web with embedded technology for realizing dynamic monitoring and integrated management.The advantages mainly are incarnated as the following, such as that it is more convenient in realizing intelligent communication, configurable and controllable in device on line, realizable in store-forward of real time data, and ensures the integrality and security of data transfer.

2.2 Structure Block Diagram of Soft System

The structure block diagram is shown as in Fig. 2, and its implementation of soft system is applied by embedded dynamic Web. The soft system based on embedded dynamic Web consists of three parts which are operating system, Web server and application software. The following is the rough explaining.

(1) Realization of intelligent communication
The front-end device can carry through processing various data of heterogeneous oil device so as to be convenient for system monitoring and extending. It can adopt the embedded intelligent computer of industrial grade being suitable for field scurvy environment, and carry through structured program development of collectivity design and object oriented for communication software from systematic hierarchy. And it can design different module in terms of data communication protocol of different devices to realize the intelligent communication so as to carry through unification control and management for different oil devices.

(2) Configuration and control of device on line
The mode of system development is used by dynamic Web based on Apache + PHP + data file for Linux. Through the page layout of Web it can neatly realize the device configuration and status setting for various forecourt devices, and reflect dynamically the real time state of each device such as oil gun state of dispensers, and look over the log file of forecourt controller running and analyze the statistic log file to acquire the running statistic data that offers the base of the first hand data for erratum and correcting fault. According to the access content of purview enactment, each user owns different management purview.

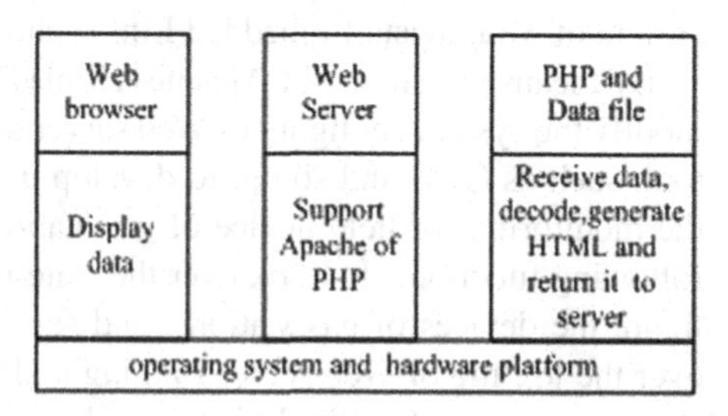

Fig. 2 Structure of embedded dynamic Web

Only the supervisor has double purview among both read and writing. It can ensure the data security transfer over network and offer the accessing function of encrypt security renew and load the price list of oil plants and oil grade in real time, realize the control for all forecourt devices such as dispensers and oil gun being on or off, it can monitor manifold oil device that will be more complex in instance and higher in control difficulty.

(3) Store-Forward of real time data
The inner ROM of front end control device is easily divided as four areas that are Boot Loader, Linux Kernel, Mini Root File System and User Root File System. When system results in breakdown rooted in user program reason, the Mini Root File System can be used as emergency root file system. When User Root File System in the Boot Loader is loaded to be failed, it can activate the Mini Root File System and introduce a sort of inner embedded mechanism to prevent system breakdown and ensure the steady reliability of system running. Once the system fault occurs in background network, the data communication of Store-Forward can be used. The data is stored into the file of JFFS2. After the fault is eliminated, the collected data will be automatically transmitted into the corresponding main computer. And it does not influence the normal work of dispensers and related devices. Therefore it ensured the integrality and security of system.

2.3 The Hardware Structure for Web Server

The Web server is an embedded computer of high performance in which a 32Mb NOR Flash ROM and a 128MB SDRAM is configured. It has rich resource such as with eight serial ports, double 10/100 Mbps Ether port, wireless communication of PCMCIA, 8-channel data input and 8-channel data output, and extended interface of CompactFlash.

3 Description of Software Function

The core part [7–9] of system is an embedded computer configured by embedded system of Montavista Embedded Linux, shown as in Fig. 3.

By means of scheme of Apache+Html+PHP, it can realize dynamic browse and modify the system configure of Web server dynamically. Also it adopts development tool, such as GCC and so on, to develop user application program so as to realize the monitoring for field device of gas stations. The Web server can implement the following function. (1) Look over the state for various dispensers in real time, configure the devices of gas stations, and set the work status of dispensers. (2) Look over the log file of Web server running and make statistic and analysis for running statistic data so as to offer the base for eliminating erratum. (3) Set accessing purview

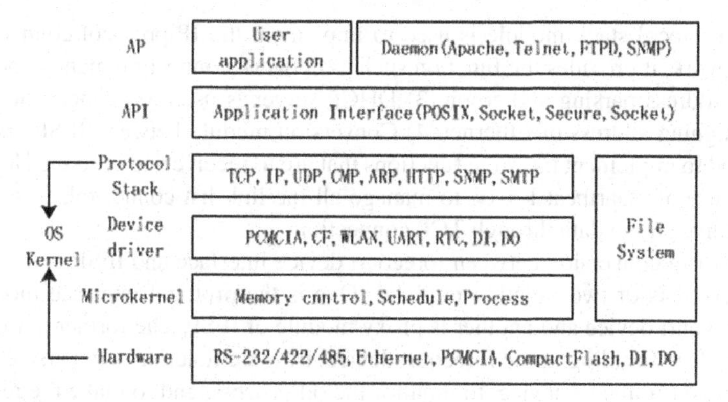

Fig. 3 System structure for Linux

so as to ensure the data being security transfers in network, and to offer the accessing function of secret security. (4) Control the parameters such as dispensers work status, setting oilcan and price brand under the condition of certain control of security level.

4 Conversion for Protocol

The function of protocol conversion is to realize information conversion from forecourt device interface protocol to IFSF protocol [10] based on TCP/IP, and to implement the real time information processing of forecourt device in the embedded Web server. The protocol conversion consists of two modules that are respectively the IFSF interface module based on TCP/IP and the conversion module between forecourt device interface protocol and IFSF protocol.

(1) IFSF protocol interface based TCP/IP
Shown as in Fig. 4, it consists of four modules. 1) Application module of IFSF is used to control the implement of application program of forecourt device. 2)

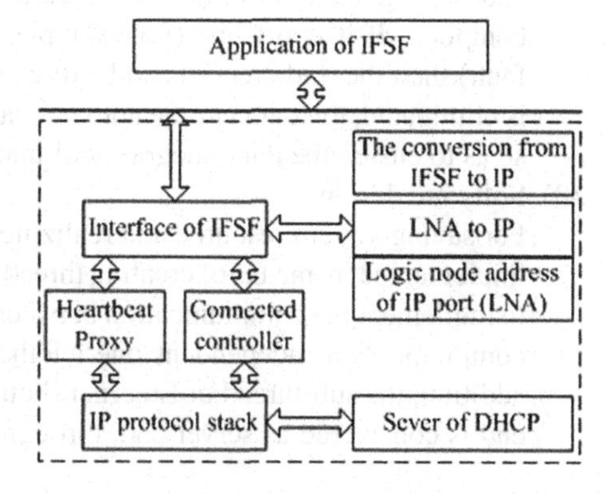

Fig. 4 Structure of IFSF based TCP/IP

IP protocol stack module is used to implement the IP protocol connected by network, it provides the function such as network connection management and IP address parsing and so on. 3) DHCP server is used to connect the device assigning address of Ethernet. 4) Conversion module between IFSF and IP is used to implement the three functions that are to receive and send the Heartbeat by use of Heartbeat Proxy, to manage all the link list connected, to send and receive all the data through TCP connection.

(2) Protocol conversion between forecourt device interface and IFSF
It consists of two function modules. One is the protocol interface module of forecourt device and another is proxy module of IFSF. The former realizes the interface for idiographic forecourt device. Its main function is to parse communication frame of device, to monitor the oil process, and to make the response for special case of dispenser. And the latter is used to respond the "write/read" for each node request of IFSF. According to the protocol and frame format of IFSF, it realizes the accessing among IFSF nodes. The communication is carried through the database among the above modules. The proxy module accesses the database that it represents to access the forecourt device. The database stored all the information of dispenser and forecourt device, including oil gun state, protocol version of dispenser, oil price version, trade record list and accumulative total of trade statistic data and so on. In all the information, the data that has higher demand of time effect is all with the time stamp. If the sent write/read information from other IFSF node wants to access the data that is time sensitive, then it can directly access the data through database. Otherwise it can send the data request order forward to forecourt device protocol interface module through another information transfer channel between device protocol interface module and IFSF proxy module, and the device protocol interface module must make response in a certain time. It can implement the monitoring and management for the function in some data file such as oil quality file and general information of gas stations of data base through integrated monitoring module. Once the module of device protocol interface apperceives that the file variety is happened, it will be automatically start up to general information query instruction in the program and go into the information renewing stage. The function of control and management is to transfer the trade record of all dispensers into the main computer. If it is off line (for example, the signal interrupting rooted in line fault) then the trade record can be directly stored in local area. When the fault is eliminated, the data will be automatically transferred into the main computer so as to ensure the data integrality of management system.

(3) Software design
For saving system resource and realizing share data, the conversion function is implemented by means of creating thread mode. Each serial port creates a thread to implement the communication between device and conversion so as to ensure communication independent one another between device and conversion. In addition, the sub-thread of Ethernet should be created. The forecourt controller end is considered as server end, through creating connection with POS, it can

Fig. 5 The system flowchart

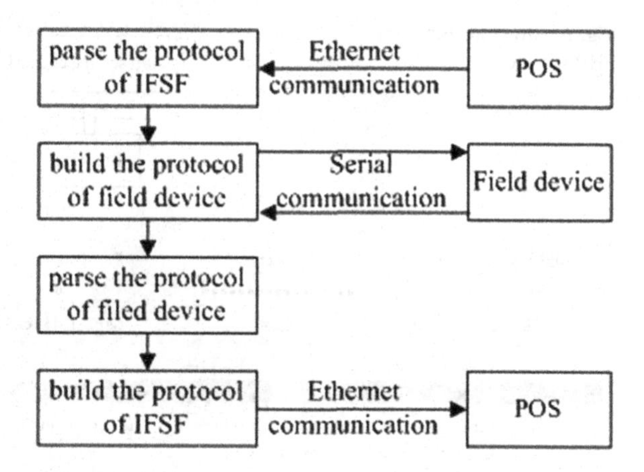

realize the communication between protocol conversion and POS. The data interaction of thread is implemented by means of sharing data storage.

The system flowchart is shown as in Fig. 5. The POS carries through control and data transfer for field device of gas stations by forecourt controller. And vice versa, the process is that the data in each forecourt device is packed in format of IFSF frame to be transferred after data being parsed through forecourt controller, finally the POS transfers data to background server.

(4) Module for serial communication

In terms of configure file, it carries through parameter needed setting for each serial port. In the program, through read in configure file it makes the serial port to complete initialization. For avoiding long time waiting, each serial port assigns a thread so as to complete time-sharing operation. Each thread assigns a private data buffer. When the data sets in, it is accepted into data buffer by use of function of recv ( ), after right checkout of CRC16, parsing packet and pick-up effective data, finally packed, and sent into the POS. When the net is off line, the oil data will be stored temporarily. In main program, the oil data stores by means of static structure array. It adopts the mutex storage of share memory to prevent the producing conflict that different thread stores data at the same time and after arriving delay time, it creates new thread to transfer data.

5 Realization for Monitoring and Its Performance Test

All the user interfaces of dynamic Web server can be implemented by the www browser, shown as in Fig. 6. This mode is the direct access between Web server and browser. The application program is put in the server, and it is unnecessary to develop client end program. There are two sorts of development methods, CGI (Common Gateway Interface) technology and embedded technology.

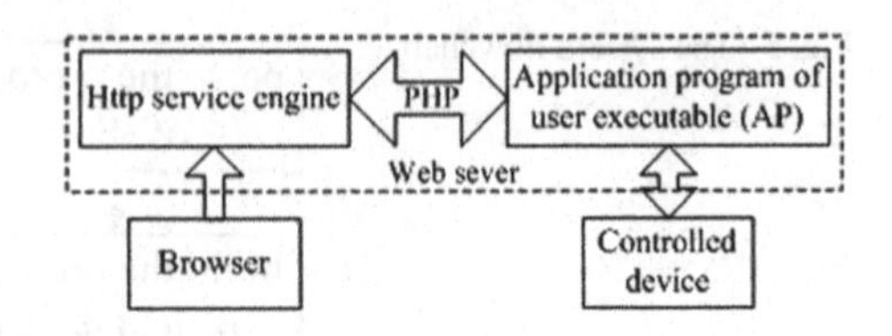

Fig. 6 Structure mode of PHP service.

Administrator name	text	Max Connection	Max COM Port
administrator	●●●●●●●●●●●●●●●●●	50	9
Server Port	Log	Serial Timeout	Support Protocol
9	9	1000	9
Ethernet			
IP of Interface 1	Netmask of Interface 1	IP of Interface 2	Netmask of Interface 2
192.168.0.127	255.255.9.0	10.1.1.131	255.255.255.0
COM			
Enable	Mode	Device File	Setting
yes	rs232	/dev/ttyM	192.168.0.127
Protocol	Broadcast		
9	9		

Fig. 7 Interface of parameter setting

The system function test carries through the environment under the condition of network. Using test signal resource being in place of analogue signal, the method is to simulate the receiving & sending data and order of POS and dispenser respectively by using two computers, by means of software server platform of Apache + PHP, after completing compiling of protocol conversion of main program and related program, it will be loaded into the development device. And at the same time, the compiling of the Web page layout will also be entered into it. Then runs the development device, we can directly view the testing results from display menu at client end. The interface of parameter setting is shown as in Fig. 7. From the Fig. 7, we can modify and look over port configuring file, and the POS can simulate to control the dispenser to oil, to set oil parameters, and to control the oil card inserting and exiting. Also it can simulate the operations of dispenser such as lifting the oil gun, hanging the oil gun and making balance etc. The testing result shows that it can complete protocol conversion of dispenser and POS, realize all the functions controlled by POS.

6 Conclusion

It is a necessary trend to integrating and communicating among different protocol for industrial automation control development in future. Through the application example above, it shows that it can better realize the monitoring in system and sharing in data information and make the maintaining easier for networked monitoring system based on embedded dynamic Web.

References

1. Cercone, N., Lijun, Hou., Keselj, V., et al.: From computational intelligence to Web intelligence. Computer, **35**(11):72–76 (2002)
2. Smimov, A., Pashkin, M., Chilov, N., et al.: Agent based Web intelligence for logistics support in networked organizations. Intelligent Transportation System. In: Proceedings. of the 7th International IEEE Conference on 3–6 Oct. pp. 922–927 (2004)
3. Butz, C.J., Hua, S., Maguire, R.B.: A Web Based intelligent tutoring system for computer programming. Web Intelligence, 2004, WI 2004. Proceedings IEEE /WIC/AMC International Conference on 20–24 Sept. pp 159–165 (2004)
4. Wei, XU., De-rong, T.: Development of control and management system based on network for gas station. J. Shandong Univ. Technol. (Sci & Tech), **18**(4), 59–63 (2004)
5. SONG Hong-wei, Zhen Ran. Design and realization of gas station central control system. Microcomputer, information. 22(6), pp. 106–108 (2006)
6. WU Q.: Development of remote monitoring system based on embedded web sever. J. Xihua Univ. (Natural Science) **25**(6), 39–41 (2006)
7. Zhuo-heng, L.: Linux network programming. China machine press, Beijing (2000)
8. Wehrle, K., Pahlke, F.: The Linux networking architecture. Prentice Hall, America (2004)
9. Matthew, N., Stones, R.: Beginning linux programming, 3rd edn. Canada, Wiley Publishing, Inc. pp 187–189 (2004)
10. International Forecourt Standard Forum. Communication on specification over TCP / IP (2002)

Research on Workflow Model Based on Petri Net with Reset Arcs

Chi Zhang, Sakirin Tam, Kai-qing Zhou and Xiao-bo Yue

Abstract To satisfy the workflow modeling requirements in the ability of powerful expression, a method by adding reset arc to extend the workflow model has been put forward, and the formal representation is proposed in this paper. Then, the soundness analysis of this method is researched by using an insurance claim model and reachability graph. Therefore, this method improved the power of describing workflow model of WF-net, especially cancellation feature which was not supported by most Petri net models.

Keywords Petri net · Workflow net · Reachability graph · Reset arcs · Soundness

1 Introduction

The concept of workflow originates from the domain of production organization and office automation, it is proposed with a fixed program activity for the routine. The workflow is designed to segment the work into well-defined tasks and roles, to perform, monitor and manage the tasks according to certain rules and procedures. The workflow can improve the work efficiency, control the procedure better, and manage business processes more effectively, etc.

C. Zhang (✉) · X. Yue
Department of Computer and Communication Engineering, Changsha University of Science and Technology, Changsha 410114, China
e-mail: czchang49@163.com

S. Tam
Faculty of Science and Technology, Yala Islamic University, A.Yarang, 94160 Pattani, Thailand

K. Zhou
Faculty of Computer Science and Information Systems, Univeristi Teknologi Malaysia, UTM Skudai, 80310 Johor, Malaysia

B.-Y. Cao and H. Nasseri (eds.), *Fuzzy Information & Engineering and Operations Research & Management*, Advances in Intelligent Systems and Computing 211,
DOI: 10.1007/978-3-642-38667-1_45, © Springer-Verlag Berlin Heidelberg 2014

At present, there are a lot of workflow modeling methods. Such as: a directed acyclic graph, process ontology, Petri net, etc [1–3]. Zhang [4] introduced the existing workflow modeling technology and its current development situation, also the existing problems, and the latest development. Among these modeling methods, some focus on describing the control relations between the tasks, the others focus on data flow description on the tasks. Petri net technology is a kind of modeling method which can not only be used for structural modeling, but carry out quantitative and qualitative analysis. It has become the main tool of workflow modeling.

Although the Petri net theory and workflow technology is mature enough, basic workflow model is still unable to express all possible workflow because of the model scale and the complexity in practical application. Therefore, it is essential to extend the model to improve the modeling ability. The existing extension methods are hierarchical, color, time, etc [5–7]. On the basis of summing up the classic Petri-net modeling theory, this paper presented the extension methods of adding reset arcs, soundness analysis of this method by using the reachability graph and an insurance claim model. By analyzing the example, we can prove that the extended method can improve workflow modeling ability.

2 Workflow Petri Net

2.1 The Definition of WFPN

Petri nets (PN) were first presented in August 1939 by Carl Adam Petri which described concurrent and asynchronous model of computer system. PN is a directed bipartite graph, in which the nodes represent transitions, places, and directed arcs. Being considered as a type of automatic theory, PN has a well-developed mathematical theory for process analysis, which makes it become the main tool of workflow modeling and analysis.

Workflow net is defined as follows:

Definition 1. *A Petri net PN= (P, T, F, M0) is called a Workflow net [8] if and only if it satisfies the following conditions:*

1. *There is a source place,* $i \in P|.i = \phi$;
2. *There is a sink place,* $o \in P|o. = \phi$;
3. *For all* $x \in P \cup T$ *are located a path from i to o.*

where $P = \{p1, p2, \ldots, pn\}$ *is a finite set of places,* $T=\{t1,t2,\ldots,tn\}$ *is a finite set of transitions, and* $F \subseteq (P\times T)\cup(T\times P)$ *is a set of arcs (flow relation). They satisfy the following conditions:*

1. $P\cap T=\phi$;
2. $P\square T\neq \phi$;

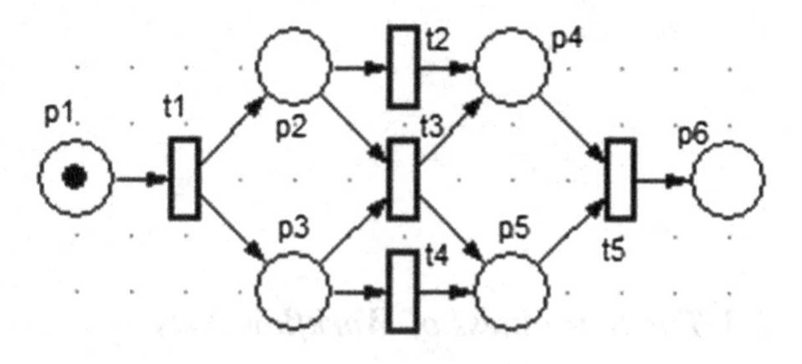

Fig. 1 A simple workflow model

(0,0,0,0,0,1) → (0,1,1,0,0,0) → (0,0,1,1,0,0)
(0,1,0,0,1,0) → (0,0,0,1,1,0) → (0,0,0,0,0,1)

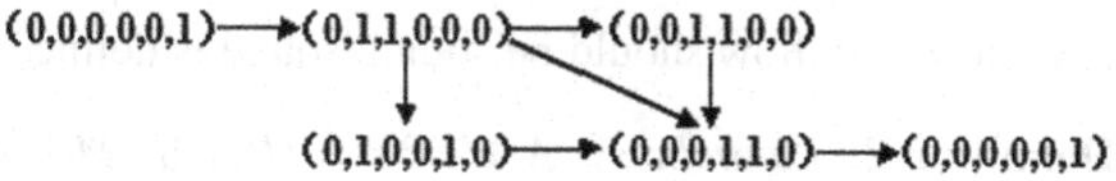

Fig. 2 WFPN reachability graph

3. *dom(F)□cod(F)=P□T Where,* $Dom\ (F)\ =\ \{x|\square y\ :\ (x, y)\square F\}, cod(F)\ =\ \{y|\square x : (x, y)\square F\}$.

M0 is the initial identification, representation of the initial state of the system. In Petri net, identification is used to indicate the token's distribution in each place at a certain hour. Place p (p□P) is called the input place of transition t (t□T), if and only if there is at least one directed arc from p to t. Place p is called the output place of transition t, if and only if there is at least one directed arc from t to p. Post (pre) set of p is the set of output (input) transitions of p, denoted by p. and .p, respectively. Symbols .t and t. have similar meaning.

Petri nets which meet the above conditions are called WFPN (Work Flow Petri Net).

2.2 Reachability Analysis

The reachability of workflow net can be described as that whether there exists a legitimate transition sequence from initial state M0 to target state, under the condition of being given workflow net and target state. Reachability analysis can detect whether there is dead task, by establishing WFPN reachability graph.

For example, a simple workflow model as shown in Fig. 1, it consists of six places (p1, p2, p3, p4, the p5, p6) and five transitions (t1, t2, t3, t4, t5).According to reachability graph generating algorithm, reachability graph can be established as shown in Fig. 2. Reachability graph is a directed graph which consists of nodes and directed arcs. Moreover the tag of each node represents a reachable state and each directed arc connecting two nodes represents a possible change of state.

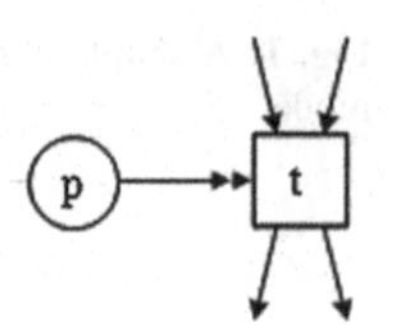

Fig. 3 Reset arcs

2.3 The Soundness of Workflow Nets

A correct workflow should be sound, sound is defined as follows:

Definition 2. *(soundness) A WFPN = (P, T, F, M0) modeling process is sound, if and only if:*

1. *For every state M reachable from the initial state M0 (there is only one token in place i), there exists a firing sequence leading from state M0 to o (there is only one token in place o).Formally:* $\forall M(i \xrightarrow{*} M) \Rightarrow (M \xrightarrow{*} o)$;
2. *State o is the only state reachable from state M0 with at least one token in place o.Formally:* $\exists M(i \xrightarrow{*} M \wedge M \geq 0) \Rightarrow (M = o)$;
3. *There is no dead transitions in (PN,i).Formally:* $\forall t \in T, \exists M, M' i \xrightarrow{*} M \xrightarrow{t} M'$ *(PN,i) represents a Petri net with initial state i.*

where: $x \xrightarrow{*} y$ *represents there is a reachable path from marker x to y in the reachability graph of the model.*

The reachability graph can be used to check if WFPN model can meet three conditions of the soundness. By contrast the first two conditions, we can analyze whether there is a node can correspond final state, and there is the only token in place o.Refered to the third condition, we can judge whether each task in workflow net corresponds with the state transition in reachability graph. If so, WFPN model in each task can be executed.

3 Extended Workflow Net

3.1 Definition of Extended Workflow Net

The basic Petri net model is very simple and is unable to express all routing constructs one may encounter in real-life workflows. Therefore, a method to extend workflow model by adding reset arc has been put forward. This method can effectively improve the modeling capabilities of workflow nets. The notion of reset arcs is illustrated in Fig. 3. Here the double-headed arcs are reset arcs.

Definition 3. *Five - tuple N=(P, T, F, M0, R) is a necessary and sufficient condition of RWFPN:*

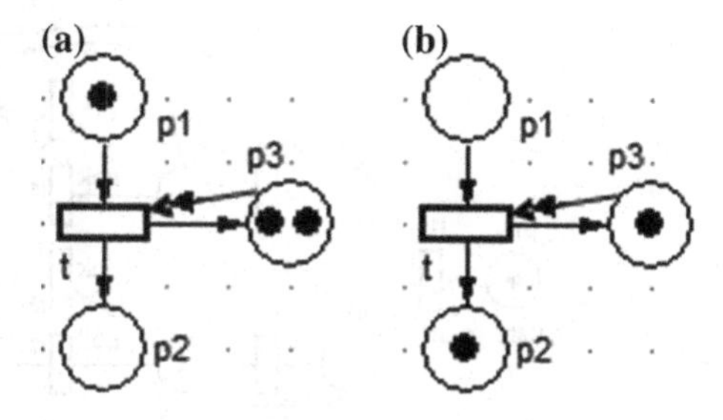

Fig. 4 An example reset net before and after firing transition t

1. *(P, T, F, M0) is a basic Petri net,*
2. *R□T→2^P is a function defining reset arcs*
3. *There is no reset arc connected to the sink place.*

Transition t is enabled at M, denoted as M [t>, if for all ∀p□·t, M(p)≥1.We denote $M \xrightarrow{N,t} M'$ *if M[t> and*

$$M' = \begin{cases} M(p) - F(p,t) + F(t,p), if p \in P \backslash R(t) \\ F(t,p), if p \in R(t) \end{cases}$$

In Fig. 4, transition t is enabled at marking p1and t may fire. When transition t fire, it removes a token from its input places p1, removes all tokens from its reset place p3, and puts one token in its output places p2. Moreover, reset arcs do not influence enabling.

3.2 Workflow Model with Adding Reset Arcs

However, in many cases a rather simple model is used (WF-nets or even less expressive) and practical features such as cancellation are missing. Many workflow languages have a feature of cancellation, e.g., Staffware has a withdraw construct, BPMN

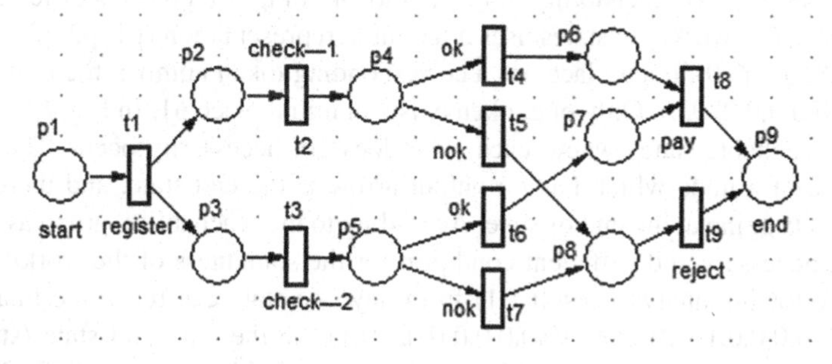

Fig. 5 Insurance claim model

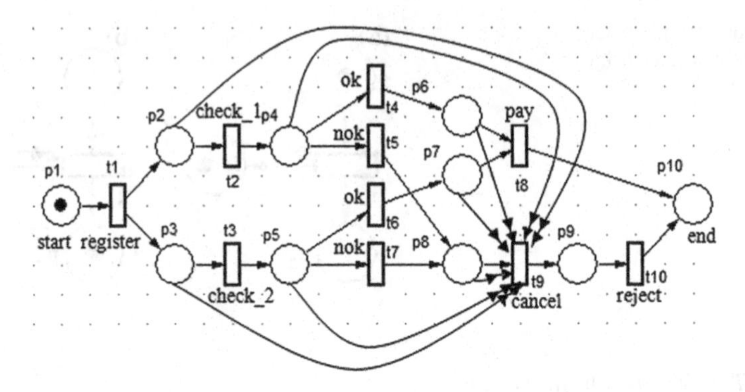

Fig. 6 Insurance claim mode with the reset arc

has cancel, compensate, and error events, etc. Since many workflow languages have cancellation features, a mapping to workflow nets is not always possible. Therefore, the workflow net with reset arcs should be taken into account.

To illustrate the problem, we can look at the example of an insurance claim in literature [8]. In order to show trigger process more intuitive, this case based on the added ok and nok of the two transitions, as shown in Fig. 5. In this example there are two review processes. The claim will be rejected if the application cannot pass any one of review process. Some researchers [9] have pointed out that, the WF-net showed in Fig. 5 is not sound. There are several deficiencies. If one check pass, the other one won't pass. The WF-net will not terminate properly because a token gets stuck in p6 or p7. If the two checks are ok, claim rejection will be executed twice because of the presence of two tokens in o with the moment of termination unclearly. However, the added reset arc insurance claim model does not have the similar situation. Added a reset arc and cancel functions of insurance claims model as shown in Fig. 6. There are seven reset arcs from the places (p2, p3, p4, p5, p6, p7, p8) to transition t9, also does not have token stuck questions in Fig. 6. When the two checks does not ok, transition t9 consume a token and remove a token, then created a token in place p9. So reject will only be executed once, end place is also only one token.

The workflow net reachability graph as shown in Fig. 7 represents different possible states. The workflow states uses a ten tuple representation (p1, p2, p3, p4, p5, p6, p7, p8, p9, p10). Each place has a corresponding token number, the initial state is (1,0,0,0,0,0,0,0, 0,0). Only one token exists in initial place p1. In Fig. 7, there are nineteen reachable states where each state does not necessarily occur. The correspondence of a node which has no output arrow is the end state, and there is no transitions trigging at the end of state. According to the reachability graph as shown in Fig. 7, necessary and sufficient condition for the soundness of the workflow net has the following analysis results: 1) from any one state can reach the final state (0,0,0,0,0,0,0,0,0,1); 2) state (0,0,0,0,0,0,0,0,0,1) are the only end state (state o), this state is reflected as a marker M = (0,0,0,0,0, 0,0,0,0,1), and M > 0; 3) Starting

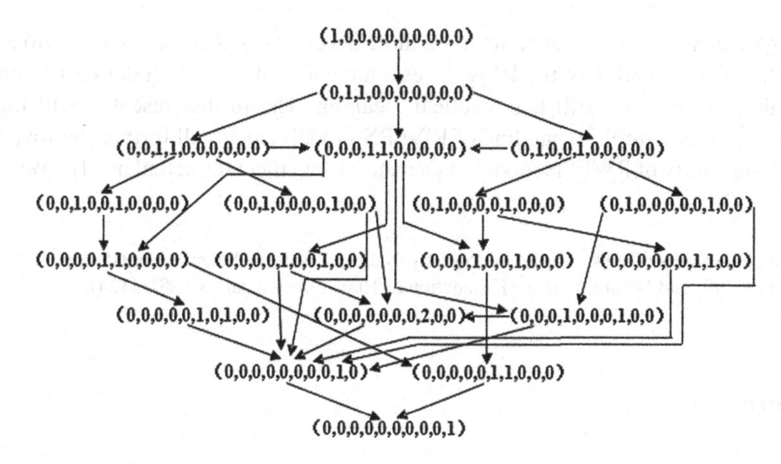

Fig. 7 Reachability graph of RWFPN

from the initial state, each transition can reach the ready state(Each transition has the opportunity to occur, there is no death transitions). Based on the above three points, insurance claim model with reset arcs meet the requirements of WF-net soundness. Therefore, the insurance claim model with adding reset arc is still sound.

4 Decidability

For a complex WF-nets, it is not easy to decide soundness, so it's more complex to decide the soundness of the extended WF-nets. Some scholars put forward that classical soundness for time extended workflow model [10] is undecidable. Unfortunately, classical soundness is undecidable for workflow models with reset arcs [11]. Therefore, for a common application, if you want to expand workflow model, you can aim at the practical model to verify them one by one only in accordance with three necessary and sufficient conditions of soundness (such as the example of 3.2 insurance claim in that way).

5 Conclusion

This article puts forward an idea of expanding WF-net with add reset arc by the analysis of Petri net which based on workflow model and then make full use of reachability graph to make a soundness analysis, so as to illustrate the validity under the common circumstances. The model example in this article is a classical workflow model. Meanwhile, we can know that it can improve the modeling ability after adding reset arc by this model. Unfortunately, classical soundness is undecidable for workflow models with reset arcs. Therefore, for a common application, if you want to expand workflow model, you can aim at the practical model to verify them one by one only in accordance with the three necessary and sufficient conditions of soundness. The workflow model with reset arc can make good modeling for many specific process, so it still has practical meaning. The further research will focus on the other structure's expanding of RWFPN model which will further improve the modeling ability of RWFPN model. As a result, it can meet the actual need of WF-net better.

Acknowledgments Thanks to the support by National Natural Science Foundation of China (No.61170199) and Natural Science Foundation of Hunan Provincial (No.08JJ3124).

References

1. Yuan, Y., Li, X., Wang, Q., et al.: Bottom level based heuristic for workflow scheduling in Grids. J. Comput. **31**(2), 282–290 (2008)
2. Lian, Y., Xu, F.: Study on modeling of process ontology for enterprise patent resources management. Comput. Eng. Appl. **46**(1), 21–27 (2010)
3. Wu, S.: Workflow of construction projects based on Petri net. Comput. Eng. Appl. **45**(30), 10–12 (2009)
4. Zhang, Z., Liu, D., Liu, W.: Review of workflow modeling technology. Microelectron. Comput. **25**(10), 69–71 (2008)
5. Hong, J., Xiangqian, D.: Modeling of hierarchical petri net-based workflow. Wuhan: Control Indus. Eng. (CCIE) **1**, 113–116 (2011)
6. Lu, H., Min, L., Wang, Y.: Aproach to master-slave workflow system and its Petri-net modeling. J. Commun. **31**(1), 92–99 (2010)
7. Shen, L., Sui, F., Bai, L., et al.: Research of workflow modeling based on extended UML activity diagrams. Appl. Res. Comput. **26**(2), 287–590 (2009)
8. van der Aalst, Wil, van Hee, Kees: Workflow Management: Model, Methods and Systems. Tsinghua Press, Beijing (2004)
9. Yuan, C.: Petri Net Theory and Application, pp. 58–66. Publishing House of Electronics Industry, Beijing (2005)
10. Tiplea, F.L., Macovei, G.I.: Soundness for S-and A-timed workflow nets is undecidable. IEEE Trans. Syst., Man Cybern. A, Syst. Humans **39**(4), 924–932 (2009)
11. Dufourd, C., Finkel, A., Schnoebelen, P.: Reset nets between decidability and undecidability, In: Larsen, K., Skyum, S., Winskel, G. (eds.) Proceedings of the 25th International Colloquium on Automata, Languages and Programming, Vol. 1443 of Lecture Notes in Computer Science, Springer-Verlag, Aalborg, Denmark, pp. 103–115 (1998).

The Closeness Centrality Analysis of Fuzzy Social Network Based on Inversely Attenuation Factor

Ren-jie Hu, Guang-yu Zhang and Li-ping Liao

Abstract Fuzzy centrality analysis is one of the most important and commonly used tools in fuzzy social network. This is a measurement concept concerning an actor's central position in the fuzzy social network, and it reflects the different positions and advantages between social network actors. In this paper we extend the notion of centrality and centralization to the fuzzy framework, propose fuzzy inversely attenuation closeness centrality, and discussed fuzzy group closeness centralization based on inversely attenuation factor in fuzzy social networks.

Keywords Fuzzy social network · Attenuation factor · Fuzzy closeness centrality · Fuzzy group closeness centralization

1 Introduction

A social network is a set of nodes representing people, groups, organizations, enterprises, etc., that are connected by links showing relations or flows between them. Social network analysis studies the implications of the restrictions of different actors in their communications and then in their opportunities of relation. The fewer constraints an actor faces, the more opportunities he/she will have, and thus he will be in a more favorable position to bargain in exchanges and to intermediate in the bargains of others that need him, increasing his influence.

R. Hu · G. Zhang (✉) · L. Liao
School of Management, Guangdong University of Technology, Guangdong 510520, China
e-mail: guangyu@gdut.edu.cn

R. Hu
e-mail: renjiehu2005@163.com

L. Liao
e-mail: liping1110@hotmail.com

B.-Y. Cao and H. Nasseri (eds.), *Fuzzy Information & Engineering and Operations Research & Management*, Advances in Intelligent Systems and Computing 211,
DOI: 10.1007/978-3-642-38667-1_46, © Springer-Verlag Berlin Heidelberg 2014

In social network analysis, the problem of determining the importance of actors in a network has been studied for a long time [1]. It is in this context that the concept of the centrality of a vertex in a network emerged. Social networks analysts consider the closely related concepts of centrality and power as fundamental properties of individuals, that inform us about aspects as who is who in the network, who is a leader, who is an intermediary, who is almost isolated, who is central, who is peripheral. Social networks researchers have developed several centrality measures. Degree, Closeness and Betweenness centralities are without doubt the three most popular ones.

Degree centrality focuses on the level of communication activity, identifying the centrality of a node with its degree [2, 3]. Closeness centrality considers the sum of the geodesic distances between a given actor and the remaining as a centrality measure in the sense that the lower this sum is, the greater the centrality [4, 5]. Closeness centrality is, then, a measure of independence in the communications, in the relations or in the bargaining, and thus, it measures the possibility to communicate with many others depending on a minimum number of intermediaries. Betweenness centrality emphasizes the value of the communication control: the possibility to intermediate in the relation of others [6, 7]. Here, all possible geodesic paths between pairs of nodes are considered. The centrality of each actor is the number of such paths in which it lies.

Centrality analysis is used extensively in social and behavioral sciences, as well as in political science, management science, economics, biology, and so on. Stephenson and Zelen [8] abandon the geodesic path as structural element in the definition of centrality, to introduce a measure based on the concept of information as it is used in the theory of statistical estimation. The defined measure uses a weighted combination of all paths between pairs of nodes, the weight of each path depending on the information contained in it. Bonacich [9, 10] suggests another concept of centrality. He proposes to measure the centrality of different nodes using the eigenvector associated with the largest characteristic eigenvalue of the adjacent matrix. The ranking of web sites as they appear in the web search engine Google was created from this measure by Brin and Page [11]. Costenbader and Valente [12] studied the stability of centrality measures when networks are sampled. A measure of betweenness centrality based on random walks can be found in Newman [13]. Zemljic and Hlebec [14] evaluate the reliability of measures of centrality and prominence of social networks among high school students. Kang [15] presents a measure of similarity between connected nodes in terms of centrality based on Euclidean distances. Kolaczyk et al. [16] provide an expansion for group betweenness in terms of increasingly higher orders of co-betweenness, in a manner analogous to the Taylor series expansion of a mathematical function in calculus. Everett and Borgatti [17] proposed a new centrality called exogenous centrality. Sohn and Kim [18] develop a robust methodology for computing zone centrality measures in an urban area. Centrality measures for complex biological networks can be found in Estrada [19]. Martı'n Gonza' lez et al. [20] discussed centrality measures and the importance of generalist species in pollination network. Kermarrec et al. [21] introduce a novel form of centrality: the second order centrality which can be computed in a distributed manner. Pozo et al. [22] define a family of centrality measures for directed social networks from a game theoretical point of view. Wang et al. [23] used complex network theory to examine the overall structure of China's air transport network and the centrality of individual cities. The research shows that the rapid development of the air transport network in China has produced a distinctive pattern. In recent years, new and small airports in China are inclined to supply direct links to the top hubs and so bypass the regional ones, resulting in underdeveloped regional centers. Qi et al. [24] propose a new centrality measure called the Laplacian centrality measure for weighted networks. Laplacian centrality is an intermediate measuring between global and local characterization of the importance (centrality) of a vertex.

There is little paper about fuzzy social networks until now. Fan et al. [25, 26] discuss structural equivalence and regular equivalence in fuzzy social networks. Data mining through fuzzy social network analysis are discussed by Premchand and Suseela [27]. Tseng [28] proposes FNBSC (fuzzy network balanced scorecard) as a performance evaluation method when the aspects and criteria are dependent and interaction is uncertain. Li-ping and Hu Ren-jie defined the concept of fuzzy social network and explores some of its basic properties [28]. The definition and relevant analysis provide the theoretical foundation for further study of the fuzzy social network.

In fuzzy social network, links represent social relationships, for instance friendships, between actors. These relationships offer benefits in terms of favors, information, etc. Moreover, actors also benefit from indirect relationships. A "friend of a friend" also results in some indirect benefits, although of a lesser value than the direct benefits that come from a "friend". The same is true of "friends of a friend of a friend", and so forth. The benefit deteriorates with the "distance" of the relationship. For instance, in the fuzzy social network where actor 1 is linked to 2, 2 is linked to 3, 3 is linked to 4, and 4 is linked to 5 in Fig. 1. Obvious, the relationship between actor 1 and actor 2 is 0.8. However, how to calculate the relationship between "friend of a friend"? How to calculate the fuzzy closeness centrality in fuzzy social network based on attenuation factor? It has not been considered formally in the literature until now, to the best of our knowledge.

Liao Li-ping and Hu Ren-jie describe the relationship between actors by fuzzy relation matrix, and use quantitative technique to define the fuzzy social network [28]. In this article, we propose some methods to calculate the relationship between "friend of a friend". In this paper we extend the notion of centrality and centralization to the fuzzy framework, propose fuzzy inversely attenuation closeness centrality, and discussed fuzzy group closeness centralization based on inversely attenuation factor in fuzzy social networks.

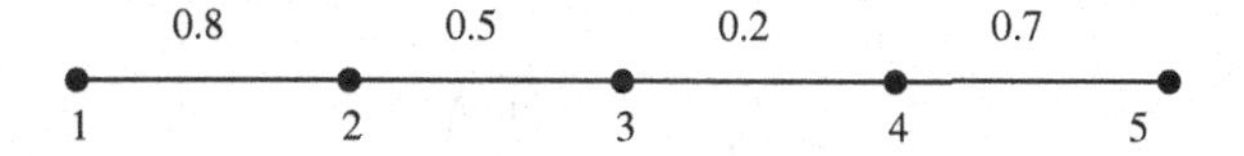

Fig. 1 A fuzzy friend's network of five actors

The organization of this paper is as follows. Section 2 contains the notation and some preliminary concepts. In Sect. 3, we propose fuzzy inversely attenuation closeness centrality, and discussed fuzzy group closeness centralization based on inversely attenuation factor in fuzzy social networks. Finally a conclusion appears in Sect. 4.

2 Preliminaries

In social networks, the relation between actors are reduced "1" and "0", 1 indicates the presence of linkage between actors, and the "0" indicates the absence of such a linkage. It can not make the relation between actors clear. Hence, how to describe the relation between actors has come into greater prominence. Premchand and Suseela [27] defined fuzzy social network as a fuzzy graph with the entities as the nodes or actors and the relations among them as the edges or links. Liao Li-ping and Hu Ren-jie defined fuzzy social network as follows:

Definition 2.1 *Fuzzy social network is defined as a fuzzy relational structure* $\widetilde{G} = (V, \widetilde{E})$, *where* $V = \{v_1, v_2, \ldots, v_n\}$ *is a non-empty set of actors or nodes, and* $\widetilde{E} = \begin{pmatrix} \widetilde{e_{11}} & \cdots & \widetilde{e_{1n}} \\ \vdots & \ddots & \vdots \\ \widetilde{e_{n1}} & \cdots & \widetilde{e_{nn}} \end{pmatrix}$ *is a fuzzy relation on* V.

In definition 1, $\widetilde{e_{ij}}$ is a fuzzy relation between v_i and v_j, $\widetilde{E}$ is called fuzzy adjacency matrix on $\widetilde{G}$.

We define intensity and connected intensity in $\widetilde{G}$ as follows:

Definition 2.2 *Assume that* $\omega = v_0 e_1 e_2 \ldots \mathbf{e}_k v_k$ *is a path of two actors* v_0 *and* v_k *in* $\widetilde{G}$,

$$\widetilde{s}(\omega) = \bigwedge_{i=1}^{k} \mu(e_i) \tag{1}$$

$s(\omega)$ *is defined as fuzzy intensity of path* ω

Definition 2.3 *If there are n path* $\omega_k, (k = 1, 2, \ldots\ldots, n)$ *connecting actor u and actor v,*

$$\widetilde{s}(u, v) = \bigvee_{k=1}^{n} \widetilde{s}(\omega_k), \tag{2}$$

$\widetilde{s}(u, v)$ *is called fuzzy connected intensity between u and v in fuzzy social network* $\widetilde{G}$.

Further, we can define fuzzy connected intensity matrix $\widetilde{R}$ in $\widetilde{G}$ as follows:

$$\widetilde{R} = \begin{pmatrix} \widetilde{s_{11}} & \cdots & \widetilde{s_{1n}} \\ \vdots & \ddots & \vdots \\ \widetilde{s_{n1}} & \cdots & \widetilde{s_{nn}} \end{pmatrix}, \tag{3}$$

where, $\widetilde{s_{ij}} = \widetilde{s}(v_i, v_j), (i, j = 1, \ldots\ldots, n)$.

From fuzzy connected intensity matrix $\widetilde{R}$ for $\widetilde{G}$, we can know the relation between any of two actors in fuzzy social network.

However, in specific fuzzy social network, the fuzzy intensity and fuzzy connected intensity between actor v_i and actor v_j are related to the total actor numbers between v_i andv_j. The more actors between v_i andv_j, the fuzzy intensity and fuzzy connected intensity will decrease more quickly. So, we further discuss fuzzy intensity, fuzzy connected intensity, fuzzy connected intensity matrix, fuzzy closeness centrality and fuzzy group closeness centralization based on attenuation factor in Sect. 3.

3 The Measures of Fuzzy Closeness Centrality About Fuzzy Social Network

Fuzzy centrality analysis is one of the most important and commonly used tools of concept in the analysis of fuzzy social network. This is a measurement concept which reflects the different positions and advantages between different actors in a fuzzy social network. Generally, according to the local difference and global difference, centrality is classified into local fuzzy centrality and global fuzzy centrality. The former, also known as fuzzy degree centrality, what it reflects is a person's dominant position in the fuzzy social network. The greater the centrality, that is, more associated with more people, the more they are in the central position. The latter refers to the relation between actors to other actors in the whole network. This reflects the closeness between actors, which is measured by the relation between different actors. Fuzzy centralization refers to the overall closeness, rather than the relative importance of certain actors.

Definition 3.1.1 *Assume that $\omega = v_0 e_1 e_2 \ldots \mathbf{e}_k v_k$ is a path of two actors v_0 and v_k in $\widetilde{G}$,*

$$\widetilde{S}^I(\omega) = \frac{1}{k} \bigwedge_{i=1}^{k} \mu(e_i), \tag{4}$$

$\widetilde{S}^I(\omega)$ is defined as fuzzy inversely attenuation intensity of path ω.

Here, $1/k$ is called inversely attenuation factor in path ω.

Definition 3.1.2 *If there are n path $\omega_k (k = 1, 2, \ldots\ldots, n)$ connecting actor u and actor v, $k_j + 1$ actors on ω_k,*

$$\widetilde{S}^I(u, v) = \bigvee_{i=1}^{n} \widetilde{S}^I(\omega_k) = \bigvee_{j=1}^{n} [\frac{1}{k_j} (\bigwedge_{i=1}^{k_j} \mu(e_i))], \tag{5}$$

$\widetilde{S}^I(u, v)$ is called fuzzy inversely attenuation connected intensity between actor u and actor v in fuzzy social network $\widetilde{G}$.

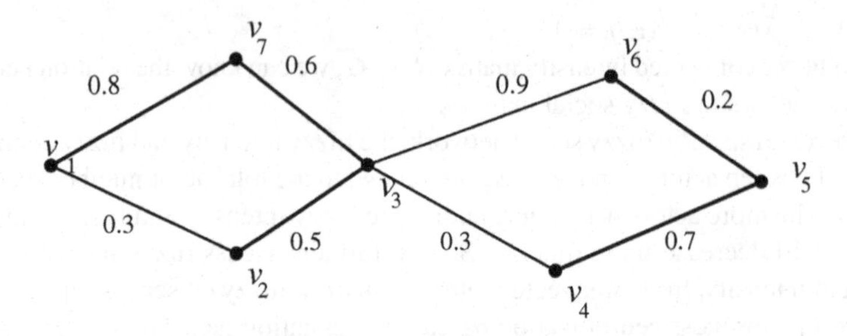

Fig. 2 A fuzzy social network consisting of seven actors

Here, $1/k_j$ is called inversely attenuation factor in ω_j.

For instance, a fuzzy social network consisting of seven actors is shown in Fig. 2.

In Fig. 2, fuzzy inversely attenuation connected intensity between actor v_1 and actor v_5 is

$$\begin{aligned}\widetilde{s}^I(v_1, v_5) &= \left[\frac{1}{4}(0.8 \wedge 0.6 \wedge 0.9 \wedge 0.2)] \vee [\frac{1}{4}(0.8 \wedge 0.6 \wedge 0.3 \wedge 0.7)\right] \\ &\vee [\frac{1}{4}(0.3 \wedge 0.5 \wedge 0.9 \wedge 0.2)] \vee [\frac{1}{4}(0.3 \wedge 0.5 \wedge 0.3 \wedge 0.7)] \\ &= 0.075.\end{aligned}$$

Further, we can define fuzzy inversely attenuation connected intensity matrix $\widetilde{R}^I$ in $\widetilde{G}$ as follows:

$$\widetilde{R}^I = \begin{pmatrix} \widetilde{s}^I_{11} & \dots & \widetilde{s}^I_{1n} \\ \vdots & \ddots & \vdots \\ \widetilde{s}^I_{n1} & \dots & \widetilde{s}^I_{nn} \end{pmatrix}. \tag{6}$$

Here, $\widetilde{S}^I_{ij} = \widetilde{s}^I(v_i, v_j)$, $(i, j = 1, \dots\dots, n)$. From fuzzy inversely attenuation connected intensity matrix $\widetilde{R}^I$ for $\widetilde{G}$, we can know the relation between any of two actors in fuzzy social network.

In Fig. 2, the fuzzy inversely attenuation connected intensity matrix is

$$\widetilde{R}^I = \begin{pmatrix} 1 & 0.3 & 0.3 & 0.1 & 0.075 & 0.2 & 0.8 \\ 0.3 & 1 & 0.5 & 0.15 & 0.1 & 0.25 & 0.25 \\ 0.3 & 0.5 & 1 & 0.3 & 0.15 & 0.9 & 0.6 \\ 0.1 & 0.15 & 0.3 & 1 & 0.7 & 0.15 & 0.15 \\ 0.075 & 0.1 & 0.15 & 0.7 & 1 & 0.2 & 0.1 \\ 0.2 & 0.25 & 0.9 & 0.15 & 0.2 & 1 & 0.3 \\ 0.8 & 0.25 & 0.6 & 0.15 & 0.1 & 0.3 & 1 \end{pmatrix}.$$

Definition 3.1.3 *Fuzzy inversely attenuation closeness centrality is the sum of fuzzy inversely attenuation connected intensity from actor v_i to the other $n-1$ actors:*

$$\widetilde{C}_C^I(v_i) = [\sum_{j=1}^{n} \widetilde{s}^I(v_i, v_j)]^{-1}. \tag{7}$$

Here $\widetilde{s}^I(v_i, v_j)$ is the fuzzy inversely attenuation connected intensity between v_i and v_j.

In Fig. 2, $\widetilde{C}_C^I(v_1) \approx 0.360$, $\widetilde{C}_C^I(v_2) \approx 0.392$, $\widetilde{C}_C^I(v_3) \approx 0.267$, $\widetilde{C}_C^I(v_4) \approx 0.392$, $\widetilde{C}_C^I(v_5) \approx 0.430$, $\widetilde{C}_C^I(v_6) \approx 0.333$, $\widetilde{C}_C^I(v_7) \approx 0.313$.

$\widetilde{C}_C^I(v_i)$ grows with decreasing relation between v_i and other actors; it is an inverse of centrality for actor v_i. Nevertheless, it is a simple measure and, since it is a sum relation, $\widetilde{C}_C^I(v_i)$ has a natural interpretation. It is, of course, only meaningful for connected fuzzy social networks.

This measure is dependent upon the number of actors in the fuzzy social network from which is calculated. We cannot, therefore, compare values of $\widetilde{C}_C^I(v_i)$ for actors drawn from fuzzy social networks of different sizes. So it would be useful to have a measure from which the impact of fuzzy social network size was removed.

In this paper, the relative fuzzy inversely attenuation closeness centrality of an actor v_i is defined as:

$$\widetilde{C}_C^{\prime I}(v_i) = \left[\frac{\sum_{j=1}^{n} \widetilde{s}^I(v_i, v_j)}{n-1}\right]^{-1} = \frac{n-1}{\sum_{j=1}^{n} \widetilde{s}^I(v_i, v_j)}. \tag{8}$$

The measures $\widetilde{C}_C^I(v_i)$ and $\widetilde{C}_C^{\prime I}(v_i)$ are both closeness-based indexes of actor centrality. Either may be used when measures based upon independence or efficiency is desired.

In Fig. 2, $\widetilde{C}_C^{\prime I}(v_1) \approx 16.667$, $\widetilde{C}_C^{\prime I}(v_2) \approx 15.306$, $\widetilde{C}_C^{\prime I}(v_3) \approx 22.472$, $\widetilde{C}_C^{\prime I}(v_4) \approx 15.306$, $\widetilde{C}_C^{\prime I}(v_5) \approx 13.953$, $\widetilde{C}_C^{\prime I}(v_6) \approx 18.018$, $\widetilde{C}_C^{\prime I}(v_7) \approx 19.169$.

From an alternative view, the centrality of an entire fuzzy social network should index the tendency of a single actor to be more central than all other actors in fuzzy social network. Measures of this type are based on differences between the centrality of the most central actor and that of all others. Thus, they are indexes of the fuzzy group closeness centralization based on inversely attenuation factor of the fuzzy social network. The measure of fuzzy inversely attenuation group closeness centralization is

$$\widetilde{C}_C^I = \frac{\sum_{i=1}^{n} (\widetilde{C}_C^{\prime I}(v^*) - \widetilde{C}_C^{\prime I}(v_i))}{\max \sum_{i=1}^{n} (\widetilde{C}_C^{\prime I}(v^*) - \widetilde{C}_C^{\prime I}(v_i))} = \frac{\sum_{i=1}^{n} (\widetilde{C}_C^{\prime I}(v^*) - \widetilde{C}_C^{\prime I}(v_i))}{(n^2 - 3n + 2)/(2n - 3)}. \tag{9}$$

Here, $\widetilde{C}_C'^I(v^*)$ = largest value of $\widetilde{C}_C'^I(v_i)$ for any actor in the fuzzy social network, $\max \sum_{i=1}^{n} (\widetilde{C}_C'^I(v^*) - \widetilde{C}_C'^I(v_i))$ = the maximum possible sum of differences in relative fuzzy closeness centrality based on inversely attenuation factor for a fuzzy social network of n actors.

In Fig. 2, $\widetilde{C}_C^I = 11.719$.

4 Conclusion

Fuzzy centrality analysis is one of the most important and commonly used tools in fuzzy social network. This is a measurement concept concerning an actor's central position in the fuzzy social network, and it reflects the different positions and advantages between social network actors. In this article, we extend the notion of centrality to the fuzzy framework, propose fuzzy inversely attenuation closeness centrality. In this paper, we extend the notion of centralization to the fuzzy framework, discussed fuzzy group closeness centralization based upon inversely attenuation factor in fuzzy social networks.

Acknowledgments Thanks to the support by National Natural Science Foundation of China (No. 71173051), ***Eleventh Five-Year Plan*** project of Philosophy and Social Sciences, Guangdong Province (No. GD10CGL07), and High-tech Zone development guide special projects of Guangdong Province.

References

1. Wasserman, S., Faust, K.: Social Network Analysis: Methods and applications. Cambridge University Press, Cambridge (1994)
2. Shaw. Communication networks. In: Berkowitz, L. (Ed.), Advances in Experimental Social Psychology. Academic Press, New York, pp. 111–147 (1954)
3. Nieminen. On the centrality in a directed graph. Social Science Research, 2,371–378(1974)
4. Beauchamp. An improved index of centrality. Behavioral Science 10, 161–163(1965)
5. Sabidussi. The centrality index of a graph. Psychometrika 31, 581–603 (1966)
6. Bavelas. A mathematical model for small group structures. Human Organization 7, 16–30 (1948)
7. Freeman. A set of measures of centrality based on betweenness. Sociometry 40, 35–41 (1977)
8. Stephenson, K., Zelen, M.: Rethinking centrality: methods and applications. Social Networks **11**, 1–37 (1989)
9. Bonacich. Factoring and weighting approaches to status scores and clique detection. Journal of Mathematical Sociology 2, 113–120 (1972)
10. Bonacich. Power and centrality: a family of measures. American Journal of Sociology 92, 1170–1182 (1987)
11. Brin, S., Page, L.: The anatomy of a large-scale hypertextual web search engine. Computer Networks and ISDN Systems **30**, 107–117 (1998)
12. Costenbader, Elizabeth, Valente, Thomas W.: The stability of centrality measures when networks are sampled. Social Networks **25**, 283–307 (2003)

13. Newman, M.E.J.: A measure of betweenness centrality based on random walks. Social Networks **27**, 39–54 (2005)
14. Barbara Zemljič and Valentina. Hlebec Reliability of measures of centrality and prominence. Social Networks 27, 73–88 (2005)
15. Soong Moon Kang. A note on measures of similarity based on centrality. Social Networks 29, 137–142 (2007)
16. Kolaczyk, Eric D., Chuab, David B., Barthélemy, Marc: Group betweenness and co-betweenness: Inter-related notions of coalition Centrality. Social Networks **31**, 190–203 (2009)
17. Everett, Martin G., Borgatti, Stephen P.: Induced, endogenous and exogenous centrality. Social Networks **32**, 339–344 (2010)
18. Sohn, Keemin, Kim, Daehyun: Zonal centrality measures and the neighborhood effect. Transportation Research Part A **44**, 733–743 (2010)
19. Estrada, Ernesto: Generalized walks-based centrality measures for complex biological networks. Journal of Theoretical Biology **263**, 556–565 (2010)
20. Ana, M.: Martı'n Gonza' lez, Bo Dalsgaard, Jens M. Olesen. Centrality measures and the importance of generalist species in pollination networks. Ecological Complexity **7**, 36–43 (2010)
21. Kermarrec, Anne-Marie, Le Merrer, Erwan, Sericola, Bruno, Trédan, Gilles: Second order centrality: Distributed assessment of nodes criticity in complex networks. Computer Communications **34**, 619–628 (2011)
22. del Pozo, Mónica, Manuel, Conrado, González-Arangüena, Enrique, Owen, Guillermo: Centrality in directed social networks. A game theoretic approach. Social Networks **33**, 191–200 (2011)
23. Wang, Jiaoe, Mo, Huihui, Wang, Fahui, Jin, Fengjun: Exploring the network structure and nodal centrality of China's air transport network: A complex network approach. Journal of Transport Geography **19**, 712–721 (2011)
24. Qi, Xingqin, Fuller, Eddie, Qin, Wu, Yezhou, Wu, Zhang, Cun-Quan: Laplacian centrality: A new centrality measure for weighted networks. Information Sciences **194**, 240–253 (2012)
25. T. F. Fan, C. J. Liau, T. Y. Lin. Positional analysis in fuzzy social networks, in: Proceedings of the Third IEEE International Conference on Granular, Computing, pp. 423–428 (2007)
26. P. S. Nair, S. Sarasamma. Data mining through fuzzy social network analysis, in: Proceedings of the 26th Annual Meeting of the North American Fuzzy Information Processing Society, pp. 251–255 (2007)
27. Tseng, Ming-Lang: Implementation and performance evaluation using the fuzzy network balanced scorecard. Computers & Educaton **55**, 188–201 (2010)
28. Liao Li-ping, Hu, Ren-jie. , : On the definition and property analysis of fuzzy social network based on fuzzy graph. Journal of Guangdong University of Technology: Social science edition **12**(3), 46–51 (2012)

Part VII
Others

Enterprise Innovation Evaluation Based on Fuzzy Language Field

Bing-ru Yang, Hui Li, Wen-bin Qian, Yu-chen Zhang and Jian-wei Guo

Abstract In this paper, we use fuzzy language field and the method of fuzzy integrative evaluation algorithm that instead of the traditional method of Analytic Hierarchy Process in the background of the enterprise innovation level rating. And develop a system of computer-assisted innovation evaluation for evaluating the enterprise innovation.

Keywords Enterprise innovation · Fuzzy language field · Fuzzy integrative evaluation · Computer-assisted innovation evaluation system

1 Introduction

With the growth of economic globalization, all enterprises have to withstand more competition. At same time, the market resource is excessive over-concentration and even be saturated. By way of the long-time existing, the enterprises must adapt with this market environment. Learning how to evolution and innovation is the best way. Now more and more researchers do lots of work around innovation action and innovation system [1–4]. The chapter designed an evaluation method based on

B. Yang (✉) · H. Li · W. Qian · Y. Zhang
School of Computer and Communication Engineering, University of Science and Technology Beijing, Beijing 100083, China
e-mail: bryang_kd@yahoo.com.cn

B. Yang · H. Li · W. Qian
Beijing Key Laboratory of Knowledge Engineering for Materials Science, Beijing 100083, China

J. Guo
School of Electronics and Information Engineering, Liaoning Technical University, Huludao 125105, China

Y. Zhang
School of Economics & Management, Tongji University, Shanghai 200092, China

B.-Y. Cao and H. Nasseri (eds.), *Fuzzy Information & Engineering and Operations Research & Management*, Advances in Intelligent Systems and Computing 211,
DOI: 10.1007/978-3-642-38667-1_47,

fuzzy integrative evaluation to measure the enterprise innovation, and developed a computer-assisted innovation evaluation system in form of web server.

2 Related Works about Fuzzy Language Field and Fuzzy Language Value Structure

2.1 Related Conception

Let us review some conceptions about fuzzy language field [5–11].

Suppose the fuzzy language variable, which describes the state or changing state, has the structure given in Fig. 1:

Definition 2.1 *Given two real intervals* L_1 *and* L_2 *, if* L_1 *and* L_2 *do not contain each other, and* $L_1 \cap L_2 \neq \phi$*, then we call* L_1 *and* L_2 *the overlapping interval pair.*

Definition 2.2 *Given a sequence of n real intervals, if every two adjacent intervals are overlapping interval pair, then we call the sequence an overlapping interval sequence.*

Obviously, all the corresponding base variable intervals of fuzzy language value X (in real domain) compose an overlapping interval sequence.

Definition 2.3 *To set D consisting of n real intervals that may compose an overlapping interval sequence, the binary relation* $\prec$ *is defined as: to any two intervals* $[x_1, x_2] \in D$ *and* $[y_1, y_2] \in D$*, we can get* $[x_1, x_2] \prec [y_1, y_2] \Leftrightarrow (x_1 \geq y_1) \wedge (x_2 \geq y_2)$.

Theorem 2.1 *The binary relation* $\prec$ *defined on D is a complete ordering relation.*

The proof is omitted.

Definition 2.4 *In the corresponding base variable region of fuzzy language variable, the dots in the middle of every overlapping subinterval (like* ξ *) and its adjacent land* ε *(*ε *is generally the allowed error value) are called standard samples (dots), the value interval* $(\xi - \varepsilon, \xi + \varepsilon)$ *taken is called standard values; any other dots are all called nonstandard samples (dots); they make up standard sample space and nonstandard sample space, respectively. The combination is called general sample space.*

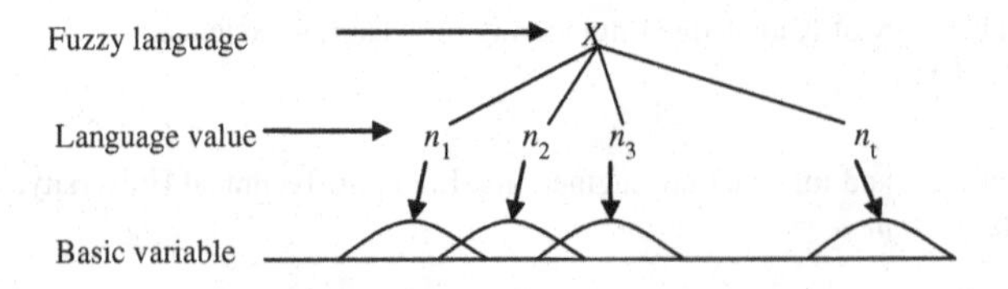

Fig. 1 Fuzzy language variable structure

Definition 2.5 $\Lambda = \langle D, I, N, \leq_N \rangle$, *if the following are satisfied:*

(1) D is a set of all overlapping intervals of base variable region on R;
(2) $N \neq \emptyset$ is a finite set of fuzzy language value;
(3) $\leq_N$ is a complete ordering relation on N;
(4) $I : N \to D$ is a standard value mapping, and satisfies isotonicity.

Then L is called a fuzzy language field.

Definition 2.6 *For the fuzzy language field* $\Lambda = \langle D, I, N, \leq_N \rangle$, $F = < \Lambda, W, K >$ *is a fuzzy language value structure of* Λ, *if*

(1) Λ satisfies Definition 2.5;
(2) K is a natural number;
(3) $W : N \to [0, 1]^K$, it satisfies the following, as in (1):

$$\begin{aligned} &\forall n_1, n_2 \in N \ (n_1 \leq_N n_2 \to W(n_1) \leq_{dic} W(n_2)), \\ &\forall n_1, n_2 \in N (n_1 \neq n_2 \to W(n_1) \neq W(n_2)), \end{aligned} \tag{1}$$

In which, $\leq_{dic}$ is a lexicographic order in $[0, 1]^K$. In F, the K dimension vector corresponding to the standard value in the subinterval of base variable region of every language value is called standard vector; or else called nonstandard vector.

Definition 2.7 *Given two fuzzy language fields Λ_1 and Λ_2, we say that Λ_1 is an expansion of Λ_2, if there is a 1-1 mapping $f : D_1 \to D_2, g : N_1 \to N_2$, satisfying, as in (2):*

1) f is monotonous;
2) $\forall n_1 \in N_1 \left(f \left(I_1 \left(n_1 \right) \right) = I_2 \left(g \left(n_1 \right) \right) \right);$ (2)

In which, $L_1 = \left\langle D_1, I_1, N_1 \leq_{N_1} \right\rangle$, $L_2 = \left\langle D_2, I_2, N_2 \leq_{N_2} \right\rangle$.

Definition 2.8 *Given fuzzy language value structures $F_1 = < \Lambda, W_1, K_1 >$ and $F_2 = < \Lambda, W_2, K_2 >$ of $\Lambda = \langle D, I, N \leq_N \rangle$, if there is a 1-1 mapping $h : [0, 1]^{K_1} \to [0, 1]^{K_2}$, that satisfies, as in (3):*

1) f is strictly monotonous in lexicography;
2) $\forall n \in N \quad (h\,(W_1(n)) = W_2\,(n));$
3) $(\exists \varepsilon \in R) \left(\forall n.n' \in N\right) \left(dis_1 \left(W_1\,(n)\,, W_1\,(n')\right) = \varepsilon \cdot dis_2 \left(W_2\,(n)\,, W_2\,(n')\right)\right).$ (3)

In which,

$$dis_1 : [0, 1]^{K_1} \times [0, 1]^{K_1} \to [0, 1]\,,$$

$$dis_2 : [0, 1]^{K_2} \times [0, 1]^{K_2} \to [0, 1]\,.$$

Then, we call F_1 and F_2 are (dis_1, dis_2) isomorphism (the abbreviation is "dis-isomorphism").

Theorem 2.2 (Expansion theorem) *Given two fuzzy language fields L_1 and L_2, L_1 is an expansion of L_2, if L_1 and L_2 are the same-type language field (it means $|N_1| = |N_2|$).*

The proof is omitted.

Theorem 2.3 (Dis-isomorphism theorem) *Suppose that F is a fuzzy language value structure of L, then F and F_{double} (the double-extension of F) are dis-isomorphic under the weighting Hamming distance.*

The proof is omitted.

From the above we can know that since the same type fuzzy language fields are not distinguished from each other in the expansion sense, the language values of other same type fuzzy language fields can be described based on the language value of natural number "large", "small" so on, and it can be known that in the dis-isomorphism sense, fuzzy language value structure can be built on different dimension space. The selection of discrete type vector corresponding to each fuzzy language value has a comparatively large free degree.

2.2 Establishment of Evaluation Index System

Based on the above analysis of USRC elements, according to the establishment principles of index system [8] and based on many interrelated reference literatures [9–11], the evaluation index system can be established, as shown in Table 1.

3 Hierarchical Structure for the Enterprise Innovation

The Fraunhofer Gesellschaft is a German research organization with 60 institutes spread throughout Germany, each focusing on different fields of applied science. They provide an excellence innovation model that classifies enterprise innovation nine fields as Strategy, Core competencies and knowledge, Process, Culture, Organizational structure and partners' network, Market development, Core Technology, Project Management and Production and service field.

On the basis of Fraunhofer Gesellschaft's research, which defines enterprise innovation into 9 innovation fields (EIf), we break down them into 26 innovation factors (EIFa) of more detailed as subcriterias, and then describe them into 77 innovation indicators (EIIn) as sub-subcriterias as below.

Table 1 Hierarchical Structure for the Enterprise Innovation model

EIf	EIFa	EIIn
Strategy field	Basic goal of strategy	A clear long-term innovation strategic aims
		Cooperative innovation strategy and overall strategy
	Pattern of strategy	Correspondence between innovation method and innovation goal
		Correspondence between innovation model and internal environment
	Implementationof strategy	Proportion of R & D sales revenue in project fund
		Proportion of research in R&D
		Proportion of research in pre-research project
		Proportion of project without implemented
Core competencies and knowledge field	Knowledge creation	Self-sufficiency rate of predominant product and main service
		Development capability in its field.
		Achievement of knowledge creation
	Knowledge application	Proportion of enterprise to use external technological achievement
		Use ratio of technological development
		Capability in enterprise dominate staff
		Effect of using new technology enterprise
	Worker knowledge structure	Proportion of R&D worker in staff
		Proportion of experienced R&D worker in staff
		Proportion of foreign R $ D worker in staff
Innovation process	Reasonability of innovation process	Reasonable degree of innovative design in workflow
		Development capability in its field
		Punctual degree around transfer of work
	Proportion of innovation resource	proportion of total investment in R & D
		proportion of transfer funds on technological achievements
		Rate of innovation and protection of information resources in a timely manner

(continued.)

Table 1 (continued.)

EIf	EIFa	EIIn
	Support of innovation	Professional research tools to meet the rate
		The ratio of projects not implemented
		R & D rate of supply of materials in a timely manner
Culture of innovation	Idea for innovation	Business leaders' degree of recognition on the importance of innovation
		Employees' degree of recognition on the importance of innovation
		Enterprises' aggressive and mental state in innovation
	Atmosphere for innovation	Tolerance degree of Failure of innovation
		The degree of innovation incentives
		Innovation team's self-management
	trust degree	Business leaders' degree of trust on the Innovation Team
		The level of trust of the person in charge of innovation and team members
		The level of trust of innovation team members
Organizational structure and network of partners	Internal coordination	The matching degree of R & D team's knowledge, ability, personality
		The coordination of R & D team and task types
		Communication network within the organization's degree of perfection
		The adequacy and timeliness of communication and exchange of knowledge information
	Ability to cope with change	Judgement for market and technological change
		Timeliness of Adjustment on innovation content
		Optimization of targeted R & D team
	Open innovation networks	Having long-term innovation partners or technology alliances
		Employment of external technical experts
		R & D institutions set up in overseas
Market Development	Insight into the potential market	Employing professional market research agencies
		Screening of market research information in a timely manner

(continued.)

Table 1 (continued.)

EIf	EIFa	EIIn
		Finding and using the market to compare with major competitors
	Analysis of customer relationships	Analysis and classification of customer Wishes or preferences of customers satisfaction Degree of customer loyalty for companies
	Use of market resources	Always listening to the views of customers or suppliers
		Use of creative of customers and suppliers in innovation
		Inviting potential customers to pre-intervention in innovation
Key Technology	Core technology of company	Number of 3rd party patents
		Number of owed patents and key techniques
	Research of core technology	Research degree for trends of production
		Research and popularization of innovative approaches
		R&D proportion from sales
Project Management	Planning and organization of project	Science of R & D period
		Reasonability for period of R & D plan
		Reasonability for organization of R&D
		Proportion of non-performing project
	Coordination and control of project	Reasonability for quality control of R & D project Reasonability for period control of R & D project Reasonability for cost control of R & D project
	Project Risk Management	Degree of risk of R & D project
		Measurement of R & D risk control
		Effectiveness of risk control
Production & service	Prospects of development	Prospect for production & service
		Competitiveness of production & service
	The basis of innovation	Ability of R & D teams from production & service
		Ability of R & D platforms from production & service
		Intellectual property of production & service
	Innovation in sustainability	Persistent plan of R & D form production & service
		Approach for dealing with suspension project

4 Language Field Description of Enterprise Innovation Degree

(1) From the angle of fuzzy language field, enterprise innovation degree can be considered as fuzzy language variable. Its language value can be very non-innovative, non-innovative, middle, innovative, very innovative. Every fuzzy language value corresponds to a fuzzy subset and can also be expressed with a corresponding vector in finite and discrete state. This enterprise innovation degree language field $\Lambda^0 = \langle D^0, I^0, N^0, \leq_{N^0} \rangle$ is called standard language field, in which N^0= {very non-innovative (R), non-innovative (P), middle (T), innovative (Q), very innovative (S)}.
The fuzzy language value structure in L^0 is called standard language value structure, given $F^0 = \langle \Lambda^0, W^0, K^0 \rangle$. According to the dis-isomorphism theorem, we might as well take K = 5 and can get enterprise innovation degree standard vector:

$$\alpha_R = (a_1, a_2, a_3, a_4, a_5),$$

$$\alpha_P = (b_1, b_2, b_3, b_4, b_5),$$

$$\alpha_T = (c_1, c_2, c_3, c_4, c_5),$$

$$\alpha_Q = (d_1, d_2, d_3, d_4, d_5),$$

$$\alpha_S = (e_1, e_2, e_3, e_4, e_5),$$

in which the concrete value of a_i, b_i, c_i, d_i, e_i $(i = 1, 2, 3, 4, 5)$ can be got one by one by using fuzzy operator operation rules from the standard vector corresponding to the alternative of "large" or "small" of natural number set {1, 2, 3, 4, 5}.

(2) There are many indicators, factors and fields influencing enterprise innovation degree such as the A clear long-term innovation strategic aims indicator being from the Basic goal of strategy factor, and this factor being from the Strategy field, and so on. The indicators are the fundamental units. According to the fuzzy language field theorem, every indicator is considered as an independent fuzzy language variable for each has base (real) variable region in itself. We will inspect A clear long-term innovation strategic aims (Acl) indicator as example in the fuzzy language field in the following.

   (i) Acl indicator fuzzy language field:

$$\Lambda_{Acl} = \left\langle D_{Acl}, I_{Acl}, N_{Acl} \leq_{N_{Acl}} \right\rangle$$

   in which D_{Acl} is the region whose base variable is "having a clear or unclear long-term aim for innovation strategic". The overlapping subinterval is divided according to its value.

N_M = {very unclear (R'), unclear (P'), middle (T'), clear (Q'), very clear (S') }.
The meanings of I_{Acl} and $\leq_{N_{Acl}}$ are the same as given by Definition 2.5.

(ii) The fuzzy language value structure in Λ_{Acl} is $F_{Acl} = \langle \Lambda_{Acl}, W_{Acl}, K_{Acl} \rangle$ in which the meaning of Λ_{Acl}, W_{Acl} and K_{Acl} are the same as given by Definition 2.6. Now take $K_{Acl} = 5$; the standard vectors corresponding to each fuzzy language value above are recorded as $\beta'_R, \beta'_P, \beta'_T, \beta'_Q, \beta'_S$, respectively.

(iii) According to the expansion theorem, Λ_{Acl} can be expanded to Λ^0 because Λ_{Acl} and Λ^0 are the same type fuzzy language field; and the five-dimensional vector in F_{Acl} can be transformed to the five-dimensional vector in F_{Acl} directly as follows, as in (4):

$$\beta'_R = \alpha_R, \beta'_P = \alpha_P, \beta'_T = \alpha_T, \beta'_Q = \alpha_Q, \beta'_S = \alpha_S. \tag{4}$$

In nature, by way of the expansion above, the transformation from the state description of the combustible degree to the one of the enterprise innovation degree is realized. Meanwhile, each indicator fuzzy language field deduced by the standard language field unitarily is to be discussed.

(3) We get corresponding conclusions completely describing and inspecting 77 innovation indicators.
(4) Operation fuzzy language field $\Lambda^0 = \langle D^0, I^0, N^0, \leq_{N^0} \rangle$ and fuzzy language value structure $F^0 = \langle \Lambda^0, W^0, K^0 \rangle$.
Since operation is distinguished from all the indicators above in quantifiable aspect, each standard vector of enterprise innovation degree can be ascertained directly according to the indicators entered by enterprise users.

5 Fuzzy Integrative Evaluation Algorithm

(1) Analyze the figure of hierarchy structure for the enterprise innovation and form a level structure as Fig. 2:
(2) Ascertain Grade 1 fuzzy synthetic judging, in which the indicator is the evaluated item.

(i) Build weight set: Grade 1 weight $A = (a_1, a_2, a_3, a_4, a_5)$; here $0 \leq a_i \leq 1$, $\sum_{i=1}^{5} a_i = 1 (i = 1, 2, \ldots, 5)$. The ascertainment of a_i can be got by the following method. The experts give the marks and then take the average value (or weighting average value) and at last converge it to get the result. Obviously, one indicator corresponds to a weight A,

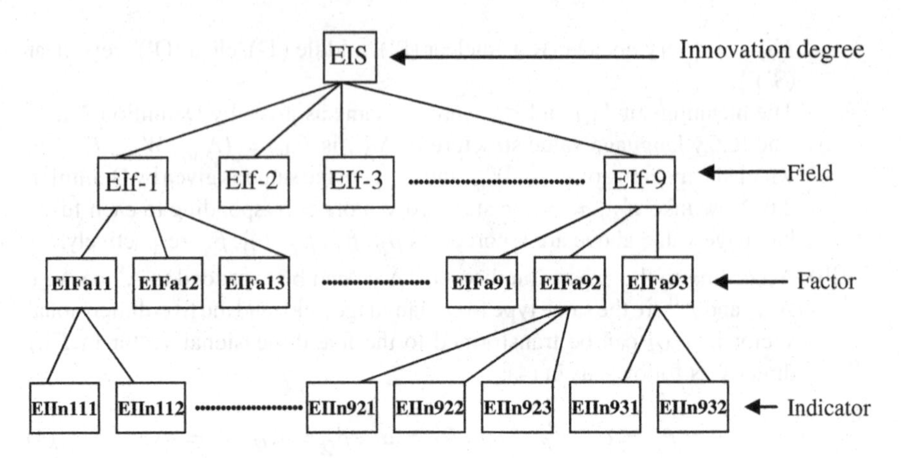

Fig. 2 Hierarchy Structure for the Enterprise Innovation

(ii) Building judge matrix, as in (5):

$$M = \begin{bmatrix} a_{11}\ a_{12}\ \ldots\ a_{15} \\ a_{21}\ a_{22}\ \ldots\ a_{25} \\ \ldots\ldots\ldots\ldots \\ a_{51}\ a_{52}\ \ldots\ a_{55} \end{bmatrix}. \tag{5}$$

The ascertainment of row vector i corresponding to the evaluated item in M is as follows:

(a) When the real input data of each factor is standard sample data, put the corresponding hazard degree standard vector into corresponding row of matrix M.

(b) When the real input data of each factor is nonstandard sample data, use the interpolation method to get the corresponding expression of hazard degree nonstandard vector and then put into corresponding row of matrix M.

The interpolation formula, as in (6):

$$\alpha_t = A_t \cdot (l - \frac{|t_i - t_i^0|}{l_i}) + A_a \cdot (\frac{|t_i - t_i^0|}{l_i}) \tag{6}$$

in which t_i is the real data of nonstandard sample in the interval NO·i; t_i^0 is the real data of standard sample in the interval NO·i; l_i is the length of the interval NO·i (when t is in the first or last interval, use 2 l_i instead of l_i); A_t is the corresponding hazard degree standard vector of the interval NO·i; A_a is the corresponding innovation degree standard vector of the left or right adjacent interval according to t.

(c) Concrete algorithm is as (iii) above.

(iii) Form result vector. Consider device l_i as an example. When the corresponding weight is A and judge matrix is M, its hazard degree result vector is $T_{I_i} = A \circ M$ (any other indicator is same).
For Factor I, given that innovation degree result vectors of P sets of devices, respectively, are $T_{I_1}, T_{I_2}, \ldots T_{I_p}$.

(3) Ascertain Grade 2 fuzzy synthetic judging, in which innovation factors are evaluated by indicators. Consider Factor I as an example. When the corresponding grade 2 weight is A' and judge matrix is $M = T_{I_1}, T_{I_2}, \ldots T_{I_p}$ its innovation degree (vector) is $T_I = A' \circ M'$. Other factors are similar, given that the innovation degree of N sets of units, respectively, are $T_I, T_{II}, \ldots T_N$.
(4) Cluster analysis. For every level of innovation degree I-V (very innovative … very non-innovative), choose the innovation degree standard vectors discussed above as clustering centre, recorded as $\text{I}_0, \text{II}_0, \text{III}_0, \text{IV}_0, \text{V}_0$ which is defined the corresponding innovation level).
Calculate the distance from $T_I, T_{II}, \ldots T_N$ to clustering centre $\text{I}_0, \text{II}_0, \ldots, \text{V}_0$, respectively, according to the formula as follows, as in (7):

$$d(T_i, J_0) = \sum_{k=1}^{5} |\mu_{Ti}(u_k) - \mu_{J0}(u_k)| (i = I, \ldots, N; J = I, \ldots, V)(HimingDistance) \quad (7)$$

For one unit T_i, according to the rule "choosing minimum", it belongs to the innovation level to that $\text{I}_0, \text{II}_0, \text{III}_0, \text{IV}_0, \text{V}_0$ whose distance from itself (absolute value) is minimal. So the result of unit fuzzy synthetic judge has been got.
Using the same algorithm we can get the judge result of grade 3 fuzzy synthetic judge, in which the Fields can be evaluated by Factors. Then, get sum enterprise innovation by grade 4 fuzzy synthetic judge that consisted of the Fields.

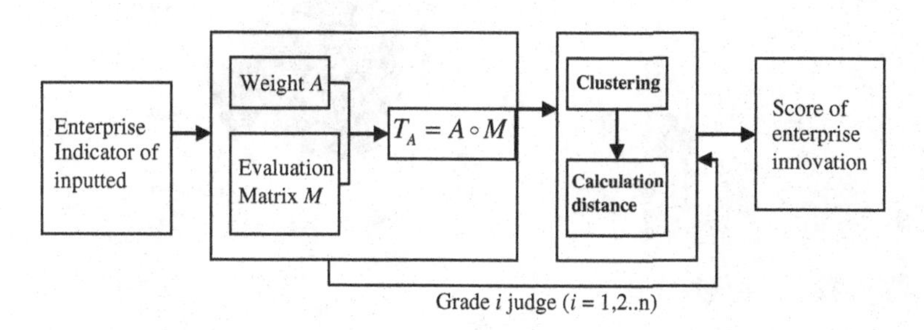

Fig. 3 System flow chat

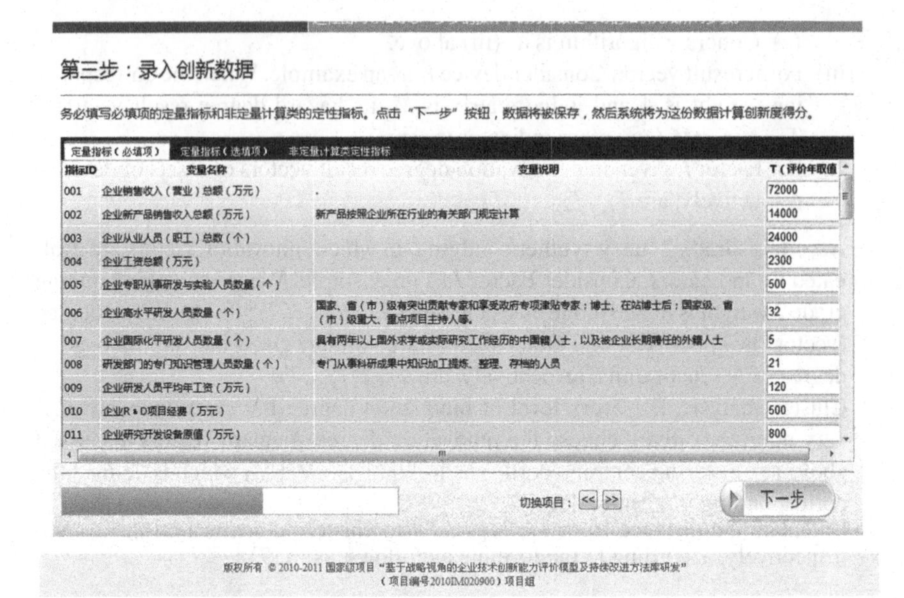

Fig. 4 Computer-assisted innovation evaluation system data input

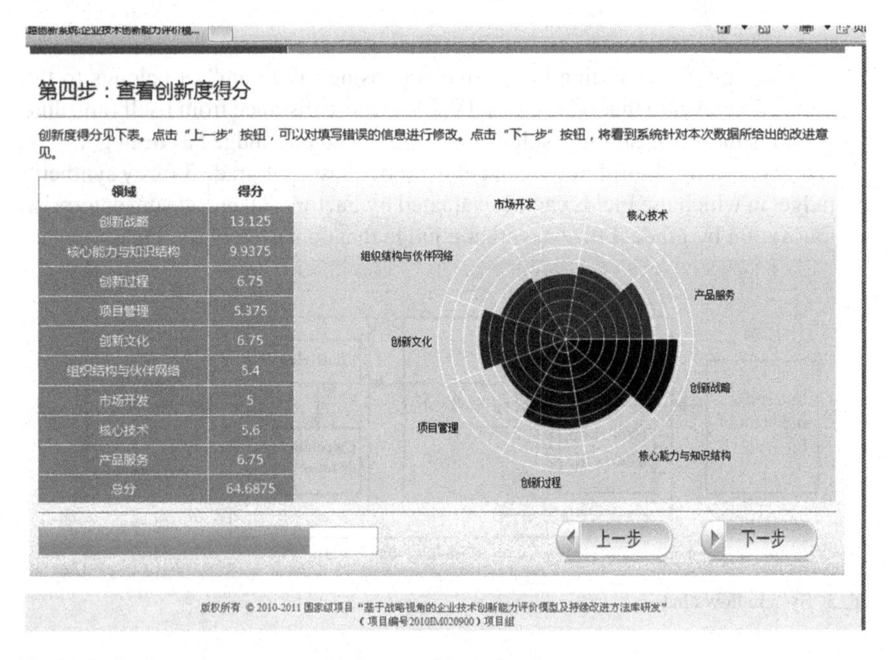

Fig. 5 Evaluation of computer-assisted innovation evaluation system

6 System and Example

(1) The enterprise user just need input enterprise information into Computer-assisted innovation evaluation system in the light of own actual conditions, then system will evaluate the degree of innovation according to above algorithm. The system flow chat is shown in Fig. 3.
(2) The Computer-assisted innovation evaluation system has been developed in web server. The procedure of evaluation is shown as above Fig. 4 and 5.
(3) The system advantage: Evaluate innovative capability of enterprise fully. From Strategy, Core competencies and knowledge, Process, Culture, Organizational structure and partners network, Market development, Core Technology, Project Management and Production & service field, system can analyze the problem precisely. On the other hand, the algorithm that this chapter proposed is more suitable for human thinking and more accurate.

7 Conclusion

This paper, fuzzy language field and the method of fuzzy integrative evaluation algorithm that instead of the traditional method of Analytic Hierarchy Process are used by evaluating enterprise innovation level rating. And develop a system of computer-assisted innovation evaluation for evaluating the enterprise innovation. The research not only solves the tasks in the enterprise innovation, but also has general adaptability to solve large classes of problems after scientific abstraction and become a powerful method.

Acknowledgments This work originated from pioneering project of the National Department of Technology (Grant No. 2010IM020900).

References

1. Asheim, B., Coenen, L.: Svensso n- Henning M. Nor-disk Industrifond, Nordic SMEs and regional innovation systems. Oslo (2003)
2. Li, X.: China's regional innovation capacity in transition: An empirical approach. Res. Policy **38**(2), 338–357 (2009)
3. Asheim, B.T., Coenen, L.: Knowledge bases and regional innovation systems: Comparing Nordic cluster s. Res. Policy **34**(8), 1173–1190 (2005)
4. Buesa, M., et al.: Regional systems o f innovation and the knowledge production function: t he Spanish case. Technovation **26**(4), 463–472 (2006)
5. Bingru, Yang: FIA and CASE based on fuzzy language field. Fuzzy sets syst. **95**(2), 83–89 (1998)
6. Bingru, Yang: Language Field and Language Value Structure - the Description Frame of Reasoning Calculation Model, pp. 38–44. Logic and Intelligence, Electronics Industry Publishing House, China (1993)

7. Bingru, Y.: The causal relation qualitative inference model based on synthetic language field. Pattern Recog. Artif. Intell. **9**(1), 31–36 (1996)
8. Yang Bingru: A multi-level inductive type intelligence inference model, inductive logic and artificial intelligence (863 Item Monograph), pp. 320–332. China Textile University Publishing House, China (1995)
9. Yang Bingru: A kind of causal relation inductive inference mechanism on fuzzy controlling of complicated system. Fuzzy Syst. Math. 10(4), (1996)
10. Halpern J.Y., Moses Y.O.: A guide to completeness and complexity for model logic of knowledge and belief. Artif. Intell. 54 (1992)
11. Turksen, I.B.: Interval valued fuzzy sets based on normal forms. Fuzzy Sets Syst. **20**, 191–210 (1986)

Integration on Heterogeneous Data with Uncertainty in Emergency System

Wei Huang, Kai-wen Chen and Chao Xiao

Abstract Aimed at the puzzle to realize the integration for heterogeneous data, this paper proposed a sort of data exchange model of heterogeneous database based on unified platform of middleware. Based on the mapping of model drive, the data exchange could be realized by a concrete model. Its thought was to express XML document being a tree composed by data object, in which each element type corresponded to an object in object pattern, namely there was mapping among patterns. The paper gave an example of power emergency processing system, the application effect shows that the speed of emergency event processing can enhance five to ten times and it is able to realize the exchange and integration for heterogeneous data easily.

Keywords Heterogeneous data · Data integration · XML · Emergency system.

1 Introduction

Along with the swift increasing of net information, the network has become a great information base composed by heterogeneous data source including being different in content and format and quality of data. The different data comes from different avenue provided by different user. Therefore it has been become a focus topic on how to make the command department of emergency center, such as fire alarm system

W. Huang (✉)
Department of Automation, Chongqing Industry Polytechnic College, Chongqing 401120, China
e-mail: huangwei051001@126.com

K. Chen · C. Xiao
College of Automation, Chongqing University, Chongqing 400030, China
e-mail: sngeet@163.com

K. Chen
e-mail: dengrenming65106683@126.com

B.-Y. Cao and H. Nasseri (eds.), *Fuzzy Information & Engineering and Operations Research & Management*, Advances in Intelligent Systems and Computing 211,
DOI: 10.1007/978-3-642-38667-1_48, © Springer-Verlag Berlin Heidelberg 2014

and electric power system, and to share data information on network better and to deal with conjuncture emergency event fleetly. The emergency processing owns great uncertainty, once it happens that a lot of heterogeneous data must be processed in real time, therefore it needs higher processing performance. Under the above background, the exchange and integration of heterogeneous data becomes more and more important, currently it has been become a hot topic for fast speed processing conjuncture emergence event.

2 Existent Puzzle of Conventional Data Integration

The essence of data integration is to realize the share of information and resource among networks which include not only management information network but also real time control network. It has essentially the following technologies to realize integration between control and information management network.

(1) DDE, each application program can share the memory to exchange the information and realize dynamic data exchange.
(2) Interconnection between control and information network implemented by gateway and router.
(3) Communication technology such as data communication by modem, and remote communication based on TCP/IP.
(4) Data access, Intranet connects into the control network through browser.
(5) Aiming at the software function integration of heterogeneous industrial control system group, the system can make seamless integration by OPC.

Now there exist two sorts of new data integration method, namely virtual database and data warehouse [1]. In fact, virtual database does not store any own data, the user's query is translated into one or multi-data resource, after that it carries through synthesis processing response of user query for those data resource, finally the processing result is returned back to user. The data warehouse method means that the data copy from several data resources is stored in the unitary database named as data warehouse. Before the data is stored into the data warehouse, it has to make pretreatment. It is difficult to adapt the situation in data happened neatly because of the rule of data conversion and integration being fused in the customizing code. Due to the offered information being always the past information, it is difficult to get the accuracy real-time information. Also it is possible to make new data isolated island because of the data integration only through making various system to form middle database or centralized database. It has to pay more cost because of the frangibility in data integration scheme, and therefore it has to solve a series of technical puzzle [2].

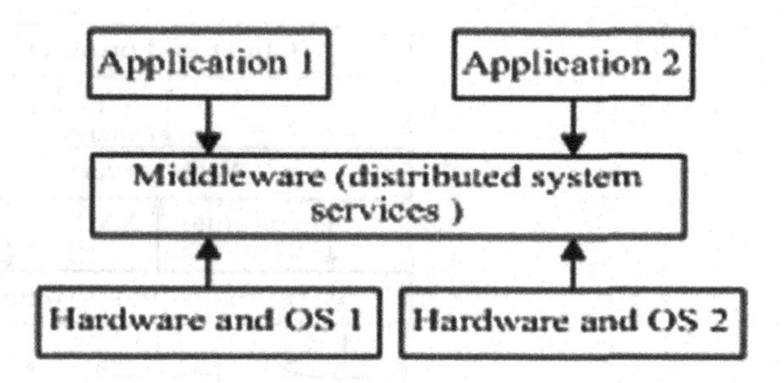

Fig. 1 Position of middleware in system structure

3 Middleware Based Data Exchange

3.1 Middleware Technology

The middleware shown in Fig. 1 is located between the application and system software. It can manage the computing resource over the operating system of client server. Its function is that it provides the environment of running and development for application software located in its upper-hierarchy, and helps user to develop and integrate the more complicated application software in agility and high efficiency, and provides the currency services located among system hardware, operating system and applications.

The middleware screened the differences of network hardware platform, heterogeneous character of operating system and network protocol. Aimed at the difference in operating system and hardware platform, it can provide lots of realization according with interface and protocol criterion and satisfy a great lot of application demand. Generally speaking, it makes the system running on various platform of hardware and operating system, supports the distributed computing and standard protocol and interface, and provides the transparence application or service interaction based on the platform by cross-network, cross-hardware and cross-operating system.

3.2 XML Based Data Exchange for Heterogeneous Database

Two basic conditions must be satisfied for going heterogeneous data exchange of non-structured database based on network application. One is compatible to variously heterogeneous data format, whether it is structured or semi-structured data. The other is easy to issue and exchange, after data exchanged it can be issued in multi-format expediently. Just the technology of XML can satisfy the above demand, not only become a sort of standard of data exchange among applications, but also be one of representation technology and an important information exchange standard in Internet [3].

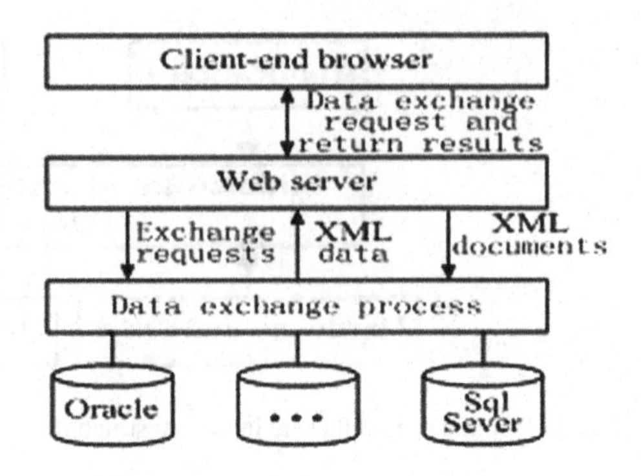

Fig. 2 System structure for platform

4 Model for Data Exchange

For the convenience of discussion, now it takes an example of data exchange among Oracle, SOL server and Access. The operating system of data exchange platform applies the Windows 2000 SP4, the development environment selects ASP.NET and Visual C, the platform structure is shown in Fig. 2.

4.1 Data Model of XML

XML is a sort of semi-structured data model. It can be used to describe anomalistic data and integrate the data from different data resources. It can unlimitedly define a set of marker, and provide independent of resolution in various feature, also it is extensible and has higher efficiency. The data model of XML separates the data from display, and modifying manner can only change the display mode. And it can display in different mode in terms of different demand [4]. Generally the file of XML includes a document type statement (DTD/Schema), the mode of representing data is truly independent of application and platform, the document is viewed as the documental database and data document. Due to being independent of platform, the document of XML is a plain text and independent of platform and application, the other system application can directly make operation for data in the file of XML described by XML itself. It is a standard pattern bestraddle by platform for data exchange and operation. And it is able to realize mutual operation of data in the heterogeneous data integration.

4.2 Database and XML

Any document of XML has its logic and physical structure. From the viewpoint of physical structure, the document is composed of entity that can quote the other entity and make it be included in the document. Logically the document consists of statement, element, notes, symbol citation and process instruction. It is needed to own the start mark and end mark for all the elements in the document of XML.

The document has to include a root element. From the viewpoint of logical structure, the document of XML is a tree with hierarchy structure, and the root element is its tree root which content is viewed as tree page [5]. Essentially the XML is the best data model to describe semi-structure data. The relative database model consists of three layer models those are the database, table (relation) and field (column). The document of XML can be located as a tree in any hierarchy. It can be used to represent the relative mode [6]. The data is always processed as the text to be treated, and the data conversion middleware transforms the text in the document of XML into the data type in the other database, and vice versa. In this way, the document of XML can be considered as a text file of unified standard.

4.3 Mapping Between XML and Database

In the conversion process between document of XML and database, it deals with the concepts such as element, sub-element, processing instruction, entity, entity citation and name field and so on. Moreover in the database it involves the concepts such as the table, record, field, keyword, main key, view and index and so on. Generally speaking, there exists mutual mapping of template and mode between document structure of XML and database. In the template, the mapping based on template drive is to embed the command of SQL. It belongs to simple hierarchy mapping based on the data result set. That is the result of the command being executed in the SQL, and does not deal with the relative mode or object mode. It can only be used to transfer data between relative database and document of XML. In the template, this sort of mapping relation only embeds the database executing processed by database conversion middleware. Therefore the template mapping provided a great agility.

When the data is transferred into the document of XML from database or into the database from the document of XML, it is realized by the model to the mapping based on model drive. The main idea is that a tree expresses the document of XML, and each element type corresponds to an object in the object model. The new mainstream database supports the technology of XML, and it includes the mode mapping of data exchange of heterogeneous database based on XML and data format mapping between the XML and relative database. But it has to pay attention that when practical heterogeneous data exchange the data exchange rule dictionary is needed.

5 Case of Application in Emergency System

The importance of electric power is evident in the modern metropolis society. Once the electric power meshwork goes wrong somewhere, it has to make emergency processing. The originated unit of accident joins to the alarm system, and informs the happened fault position. The module of alarm system will automatically create data resource of XML in terms of alarm information. The data needs to pass a series of command processing before it sends to GIS server end, including looking for

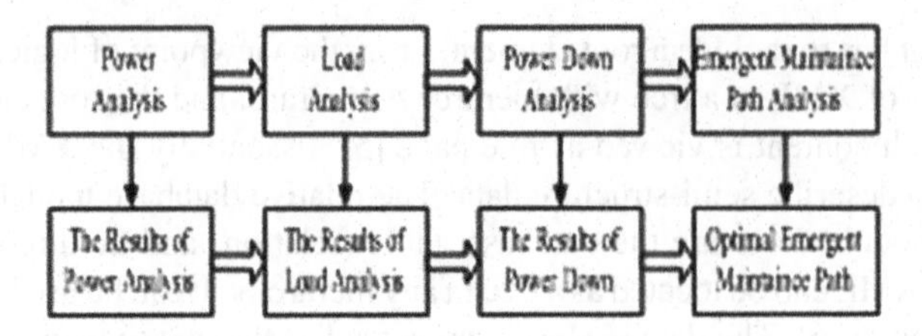

Fig. 3 Data process flowchart

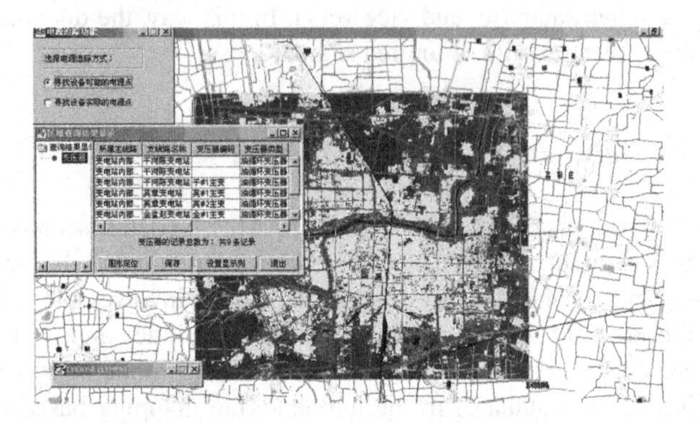

Fig. 4 Fault analysis of electric resource

the electric source, transformer load, influencing range of power supply, and best emergency maintenance solution. The information processing flowchart is shown in Fig. 3.

The electric resource analysis of fault point device can automatically switch to server interface of GIS, the red line labels the path from nonce fault device to electric resource, the path is shown in Fig. 4.

The end of GIS returns analysis data to the alarm system, and makes the space analysis for meshwork load density, and then connects all load control system to the net of GIS, the page of GIS directly displays the distribution of each load control point. By means of querying and analyzing load information of each load control point, it can analyze the singularity instance and make decision of optimal power-down isolation in least power-down range. The result is shown in Fig. 5.

When the fault happens, the system offers optimal path analysis for patrol and emergency maintenance vehicle from one point to multi-point and multi-point to one point. It does not only display the graph but also gives the dispatch path table. The emergency maintenance path is shown in Fig. 6.

The process of above interaction deals with lots of heterogeneous data system, and the alarm system can utilize the analysis data of GIS system and generate the report forms for emergency maintenance. Application of unified technical frame can easily realize exchange and integration among various isolated heterogeneous system. The better result is taken by this method for city emergency application, such as fire protection and emergency maintenance of power fault and so on. For instance,

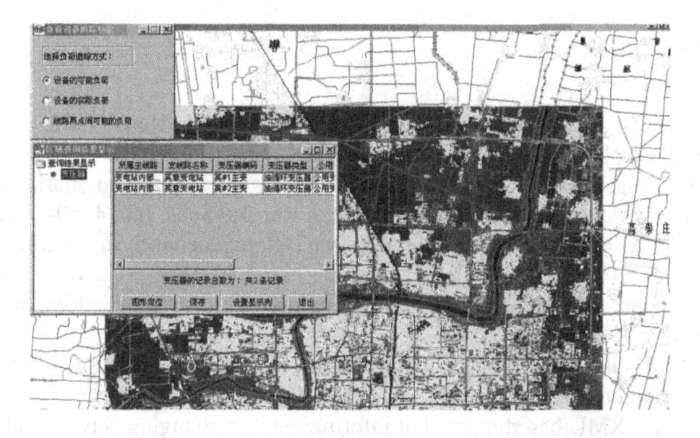

Fig. 5 Analysis of load distribution

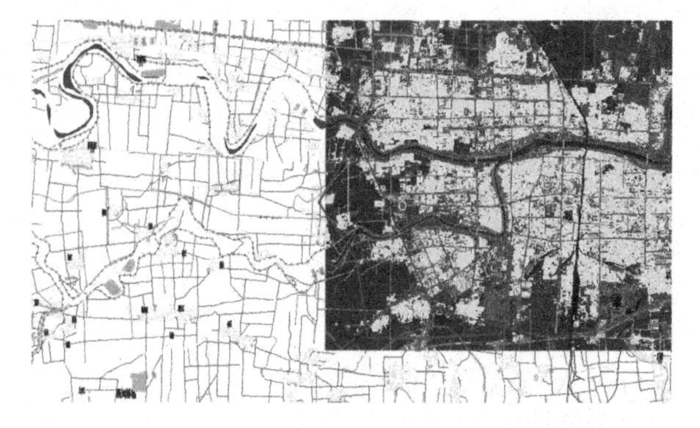

Fig. 6 Shortest path analysis

it can enhance the processing speed from five to ten times, to support dynamic appending of application and dynamic modifying of configuration parameter. Under the condition of no power-down for the whole system, it can extend the application service provided by the system. By use of data transmission platform constituted by bottom layer, it can make storing and sending data through the reliable message queue. Due to encrypt and decrypt function of data, it ensured the data transmission security. The client end can update the data without manual intervening, and be better in real time.

6 Conclusion

The actual application result shows that it is an excellent choice to take middleware system based on XML so as to realize the information exchange among heterogeneous systems. Being as a data exchange tool for XML, it has been widely used in various aspects of heterogeneous systems integration, and will be a very good application foreground in future.

References

1. Zhang, N., Jia, Z., Shi, Z.: Research on Technology of ETL in Retail Trade Data Warehouse. Comput. Eng. Appl. 24, 213–216 (2002)
2. Haibing, A., Lingkui, M., Zhiyong, L.: XML-based integrating and sharing of distributed heterogeneous geographic information databases. Remote Sens, Info 4, 50–56 (2002).
3. Dubuisson, O.: ASN.1 - Communication Between Heterogeneous Systems. Morgan Kaufmann Editor, October (2000).
4. Vakali, A., Catania, B., Maddalena, A.: XML data stores: emerging practices. Internet Comput. IEEE 9(2), 62–69 (2005).
5. Alwardt A.L.: Using XML transactions to perform closed-loop diagnostics in network centric support environments. Autotestcon 9, 707–713 (2005).
6. Gao, Y., Tan, L.: XML-based method of information interchanging between relative database and object-oriented database. Comput. Syst. Appl. 3, 196–197 (2003).

Random Fuzzy Unrepairable Warm Standby Systems

Ying Liu, Lin Wang and Xiao-zhong Li

Abstract Usually, the lifetimes of components in operation and in warm standby are assumed to be random variables. The probability distributions of the random variables have crisp parameters. In many practical situations, the parameters are difficult to determine due to uncertainties and imprecision of data. So it is appropriate to assume the parameters to be fuzzy variables. In this paper, the lifetimes of components in operation and in warm standby are assumed to have random fuzzy exponential distributions, then reliability and mean time to failure (MTTF) of the warm standby systems are given. Finally, a numerical example is presented.

Keywords Reliability · Mean time to failure · Random fuzzy variable · Random fuzzy exponential distribution · Warm standby system.

1 Introduction

Warm standby is a technique widely used to improve system reliability and availability. Warm standby means that the inactive component can fail at the standby state. Warm standby systems have been investigated extensively in the past. In classical reliability theory, the lifetimes of the components in operation or warm standby are considered as random variables. For example, Dhillon and Yang [1], Li et al. [2], Mokaddis et al. [10], Naidu and Gopalan [11], She and Pecht [12], Tan [13], Uematsu and Nishida [14], Vanderperre [15], Wang and Ke [16], Yuan and Meng [17] and so on.

In practice, the probability distribution is known except for the values of parameters. For example, the lifetime of a component is exponentially distributed variable

Y. Liu (✉) · L. Wang · X. Li
Department of Computer Sciences, Tianjin University of Science and Technology, Tianjin 300222, China
e-mail: liu@tust.edu.cn

B.-Y. Cao and H. Nasseri (eds.), *Fuzzy Information & Engineering and Operations Research & Management*, Advances in Intelligent Systems and Computing 211,
DOI: 10.1007/978-3-642-38667-1_49, © Springer-Verlag Berlin Heidelberg 2014

with parameter λ, in which λ is obtained by history data. But sometimes there is a lack of sufficient data. So it is more suitable to consider λ as fuzzy variable. Random fuzzy theory introduced by Liu [3] is one of the powerful tools to deal with this kind of phenomena. But just some researchers have paid attention to the reliability problems by using this theory. Zhao et al. [19] used random fuzzy theory into renewal process, which was a very useful tool to deal with repairable systems. Zhao and Liu [18] provided three types of system performances, in which the lifetimes of redundant systems were treated as random fuzzy variables. The lifetimes and repair times of components were assumed to have random fuzzy exponential distributions in Liu et al. [8], then the limiting availability, steady state failure frequency, mean time between failures, mean time to repair of the repairable series system were proposed. Liu et al. [9] considered a random fuzzy shock model and a random fuzzy fatal shock model, then bivariate random fuzzy exponential distribution was derived from the random fuzzy fatal shock model.

The rest of this paper is organized as follows. Section 2 introduces basic concepts related to random fuzzy reliability theory of unrepairable systems. Section 3 gives the reliability analysis of random fuzzy unrepairable warm standby systems with absolutely reliable conversion switches. Section 4 presents a numerical example.

2 Basic Concepts Related to Random Fuzzy Reliability Theory of Unrepairable Systems

Let $(\Theta, \mathcal{P}(\Theta), \text{Cr})$ be a credibility space, where Θ is a universe, $\mathcal{P}(\Theta)$ the power set of Θ and Cr a credibility measure defined on $\mathcal{P}(\Theta)$. A fuzzy variable ξ defined by Liu [4] is a function from the credibility space $(\Theta, \mathcal{P}(\Theta), \text{Cr})$ to the set of real numbers, its membership function is derived from the credibility measure by

$$\mu(x) = (2\text{Cr}\{\xi = x\}) \wedge 1, \quad x \in \Re,$$

and ξ is said to be positive if and only if $\mu(x) = 0$ for all $x \leq 0$.

Definition 1 *(Liu [3]) Let ξ be a fuzzy variable and $\alpha \in (0, 1]$. Then*

$$\xi_\alpha^L = \inf\left\{x \mid \mu(x) \geq \alpha\right\} \quad \text{and} \quad \xi_\alpha^U = \sup\left\{x \mid \mu(x) \geq \alpha\right\}$$

are called the α-pessimistic value and the α-optimistic value of ξ, respectively.

Definition 2 *(Liu and Liu [5]) Let ξ be a positive fuzzy variable. The expected value $E[\xi]$ is defined as*

$$E[\xi] = \int_0^{+\infty} \text{Cr}\{\xi \geq r\}\text{d}r$$

provided the integral is finite.

Proposition 1 (Liu and Liu [6]) *Let ξ be a fuzzy variable with finite expected value $E[\xi]$, then we have*

$$E[\xi] = \frac{1}{2}\int_0^1 \left[\xi_\alpha^L + \xi_\alpha^U\right] \mathrm{d}\alpha,$$

where ξ_α^L and ξ_α^U are the α-pessimistic value and the α-optimistic value of ξ, respectively.

Proposition 2 (Liu and Liu [6] and Zhao and Tang [19]) *Let ξ and η be two independent fuzzy variables. Then*

(i) for any $\alpha \in (0, 1]$, $(\xi + \eta)_\alpha^L = \xi_\alpha^L + \eta_\alpha^L$;

(ii) for any $\alpha \in (0, 1]$, $(\xi + \eta)_\alpha^U = \xi_\alpha^U + \eta_\alpha^U$.
Furthermore, if ξ and η are positive, then

(iii) for any $\alpha \in (0, 1]$, $(\xi \cdot \eta)_\alpha^L = \xi_\alpha^L \cdot \eta_\alpha^L$;

(iv) for any $\alpha \in (0, 1]$, $(\xi \cdot \eta)_\alpha^U = \xi_\alpha^U \cdot \eta_\alpha^U$.

The concept of the random fuzzy variable was given by Liu [3]. Let $(\Omega, \mathcal{A}, \mathrm{Pr})$ be a probability space, $\mathcal{F}$ a collection of random variables. A random fuzzy variable is defined as a function from a credibility space $(\Theta, \mathcal{P}(\Theta), \mathrm{Cr})$ to a collection of random variables $\mathcal{F}$.

Definition 3 (Liu and Liu [6]) *A random fuzzy variable ξ is said to be exponential if for each θ, $\xi(\theta)$ is an exponentially distributed random variable whose density function is defined as*

$$f_{\xi(\theta)}(t) = \begin{cases} X(\theta)\exp(-X(\theta)t), & \text{if } t \geq 0 \\ 0, & \text{if } t < 0, \end{cases}$$

where X is a positive fuzzy variable defined on Θ. An exponentially distributed random fuzzy variables is denoted by $\xi \sim \mathcal{EXP}(X)$, and the fuzziness of random fuzzy variable ξ is said to be characterized by fuzzy variable X.

Definition 4 (Liu and Liu [6]) *Let ξ be a positive random fuzzy variable defined on the credibility space $(\Theta, \mathcal{P}(\Theta), \mathrm{Cr})$. Then the expected value $E[\xi]$ is defined by*

$$E[\xi] = \int_0^{+\infty} \mathrm{Cr}\left\{\theta \in \Theta \mid E[\xi(\theta)] \geq r\right\} \mathrm{d}r$$

provided that the integral is finite.

Definition 5 (Liu and Liu [7]) Let ξ be a random fuzzy variable. Then the average chance, denoted by Ch, of random fuzzy event characterized by $\{\xi \leq 0\}$ is defined as

$$\mathrm{Ch}\{\xi \leq 0\} = \int_0^1 \mathrm{Cr}\left\{\theta \in \Theta \mid \Pr\{\xi(\theta) \leq 0\} \geq p\right\} \mathrm{d}p.$$

Remark 1 *If ξ degenerates to a random variable, then the average chance degenerates to* $\Pr\{\xi \le 0\}$, *which is just the probability of random event. If ξ degenerates to a fuzzy variable, then the average chance degenerates to* $\mathrm{Cr}\{\xi \le 0\}$, *which is just the credibility of random event.*

Consider the lifetime of an unrepairable system X be a random fuzzy variable on the credibility space $(\Theta, \mathcal{P}(\Theta), \mathrm{Cr})$. We propose the following definitions.

Definition 6 *Let X be the random fuzzy lifetime of a system, reliability of the system is defined by*

$$R(t) = \mathrm{Ch}\{X \ge t\}.$$

Definition 7 *Let X be the random fuzzy lifetime of a system, MTTF of the system is defined by*

$$\mathrm{MTTF} = \int_0^{+\infty} \mathrm{Cr}\{\theta \in \Theta \mid E[X(\theta)] \ge r\}\mathrm{d}r.$$

3 Random Fuzzy Unrepairable Warm Standby Systems with Absolutely Reliable Conversion Switches

Consider a warm standby system composed by n same type components, the lifetime of components in operation have random fuzzy exponential distribution with parameter λ defined on the credibility space $(\Theta_1, \mathcal{P}(\Theta_1), \mathrm{Cr})$. The lifetimes of components in warm standby have random fuzzy exponential distribution with parameter μ defined on the credibility space $(\Theta_2, \mathcal{P}(\Theta_2), \mathrm{Cr}')$.

We assume one component is in operation at $t = 0$, the other components are in warm standby state. When the operating component fails, another not failed component is replaced. The failure state of system occurs only when there is no operative component left. We also assume the conversion switch is absolutely reliable and the conversion is instantaneous. The lifetimes of components are independent.

Let S_i be the moment of ith failed component, $i = 1, 2, \ldots, n$, and $S_0 = 0$. So

$$S_n = \sum_{i=1}^{n} (S_i - S_{i-1})$$

is the moment of system failed. Since the random fuzzy exponential distribution has the property of memoryless, $S_i - S_{i-1}$ has random fuzzy exponential distribution with parameter $\lambda + (n-i)\mu$ defined on the product credibility space $(\Theta, \mathcal{P}(\Theta), \mathrm{Cr})$, where $\Theta = \Theta_1 \times \Theta_2$ and $\mathrm{Cr} = \mathrm{Cr}_1 \wedge \mathrm{Cr}'$.

Theorem 1 *The reliability of random fuzzy warm standby system is*

$$R(t) = \frac{1}{2}\sum_{i=0}^{n-1}\int_0^1 \left(\left[\prod_{k=0,k\neq i}^{n-1}\frac{\lambda_\alpha^L + k\mu_\alpha^L}{(k-i)\mu_\alpha^L}\right] e^{-(\lambda_\alpha^L + i\mu_\alpha^L)t} + \left[\prod_{k=0,k\neq i}^{n-1}\frac{\lambda_\alpha^U + k\mu_\alpha^U}{(k-i)\mu_\alpha^U}\right] e^{-(\lambda_\alpha^U + i\mu_\alpha^U)t}\right)\mathrm{d}\alpha.$$

Proof. By Definition 5 and Definition 6, we have

$$\begin{aligned} R(t) &= \mathrm{Ch}\{S_n \geq t\} \\ &= \int_0^1 \mathrm{Cr}\left\{\theta \in \Theta \mid \Pr\{S_n(\theta) \geq t\} \geq p\right\}\mathrm{d}p \\ &= \frac{1}{2}\int_0^1 \left(\Pr_\alpha^L\left\{\omega \in \Omega \mid S_n(\theta)(\omega) \geq t\right\} + \Pr_\alpha^U\left\{\omega \in \Omega \mid S_n(\theta)(\omega) \geq t\right\}\right)\mathrm{d}\alpha. \end{aligned} \tag{1}$$

Let $A = \{\theta_1 \in \Theta_1 \mid \mu\{\theta_1\} \geq \alpha\}$ and $B = \{\theta_2 \in \Theta_2 \mid \mu\{\theta_2\} \geq \alpha\}$. For $\forall\theta_1 \in A$ and $\forall\theta_2 \in B$, we can arrive at

$$\lambda_\alpha^L \leq \lambda(\theta_1) \leq \lambda_\alpha^U$$

and

$$\mu_\alpha^L \leq \mu(\theta_2) \leq \mu_\alpha^U.$$

We can construct three warm standby systems:

(1) The lifetime of components in operation have exponential distribution with parameter λ_α^L and the lifetimes of components in warm standby have exponential distribution with parameter μ_α^L;
(2) The lifetime of components in operation have exponential distribution with parameter $\lambda(\theta_1)$ and the lifetimes of components in warm standby have exponential distribution with parameter $\mu(\theta_2)$;
(3) The lifetime of components in operation have exponential distribution with parameter λ_α^U and the lifetimes of components in warm standby have exponential distribution with parameter μ_α^U;

It is easy to see that the system 1 and system 3 are two standard stochastic warm standby systems. For any fixed θ_1 and θ_2, system 2 is also a stochastic warm standby system. Let S_n^1, S_n^2, S_n^3 be the failed moment of system 1, system 2 and system 3, it is easy to see that

$$\Pr\{S_n^3 > t\} \leq \Pr\{S_n^2(\theta) > t\} \leq \Pr\{S_n^1 > t\}.$$

Since θ_1 and θ_2 are arbitrary points in A and B, we have

$$\Pr_\alpha^L\left\{\omega \in \Omega \mid S_n(\theta)(\omega) \geq t\right\} = \Pr\{S_n^3 > t\}$$

and

$$\Pr_\alpha^U\left\{\omega \in \Omega \mid S_n(\theta)(\omega) \geq t\right\} = \Pr\{S_n^1 > t\}.$$

From the result in classical reliability theory, we can arrive at

$$\Pr\{S_n^1 > t\} = \sum_{i=0}^{n-1}\left[\prod_{k=0,k\neq i}^{n-1}\frac{\lambda_\alpha^L + k\mu_\alpha^L}{(k-i)\mu_\alpha^L}\right]e^{-(\lambda_\alpha^L + i\mu_\alpha^L)t}$$

and

$$\Pr\{S_n^3 > t\} = \sum_{i=0}^{n-1}\left[\prod_{k=0,k\neq i}^{n-1}\frac{\lambda_\alpha^U + k\mu_\alpha^U}{(k-i)\mu_\alpha^U}\right]e^{-(\lambda_\alpha^U + i\mu_\alpha^U)t},$$

Then we have

$$\begin{aligned}
R(t) &= \mathrm{Ch}\{S_n \geq t\} \\
&= \frac{1}{2}\int_0^1\left(\sum_{i=0}^{n-1}\left[\prod_{k=0,k\neq i}^{n-1}\frac{\lambda_\alpha^L + k\mu_\alpha^L}{(k-i)\mu_\alpha^L}\right]e^{-(\lambda_\alpha^L + i\mu_\alpha^L)t}\right. \\
&\quad \left. + \sum_{i=0}^{n-1}\left[\prod_{k=0,k\neq i}^{n-1}\frac{\lambda_\alpha^U + k\mu_\alpha^U}{(k-i)\mu_\alpha^U}\right]e^{-(\lambda_\alpha^U + i\mu_\alpha^U)t}\right)\mathrm{d}\alpha \\
&= \frac{1}{2}\sum_{i=0}^{n-1}\int_0^1\left(\left[\prod_{k=0,k\neq i}^{n-1}\frac{\lambda_\alpha^L + k\mu_\alpha^L}{(k-i)\mu_\alpha^L}\right]e^{-(\lambda_\alpha^L + i\mu_\alpha^L)t}\right. \\
&\quad \left. + \left[\prod_{k=0,k\neq i}^{n-1}\frac{\lambda_\alpha^U + k\mu_\alpha^U}{(k-i)\mu_\alpha^U}\right]e^{-(\lambda_\alpha^U + i\mu_\alpha^U)t}\right)\mathrm{d}\alpha.
\end{aligned}$$

The prove is completed.

Theorem 2 MTTF *of the warm standby system is*

$$\mathrm{MTTF} = \sum_{i=0}^{n-1} E\left[\frac{1}{\lambda + i\mu}\right].$$

Proof. By Definition 7 and Proposition 1, we have

$$\begin{aligned}\text{MTTF} &= \int_0^{+\infty} \text{Cr}\{\theta \in \Theta \mid E[S_n(\theta)] \geq r\}\text{d}r \\ &= \frac{1}{2}\int_0^1 \left(E\left[S_n(\theta)\right]_\alpha^L + E\left[S_n(\theta)\right]_\alpha^U\right)\text{d}\alpha.\end{aligned} \tag{2}$$

From the three warm standby systems constructed in the proof of Theorem 1, we also can see that

$$E\left[S_n^3\right] \leq E\left[S_n(\theta)\right] \leq E\left[S_n^1\right].$$

Since θ_1 and θ_2 are arbitrary points in A and B, we have

$$E\left[S_n(\theta)\right]_\alpha^L = E\left[S_n^3\right]$$

and

$$E\left[S_n(\theta)\right]_\alpha^U = E\left[S_n^1\right].$$

From the result in classical reliability theory, we can arrive at

$$E\left[S_n^1\right] = \sum_{i=0}^{n-1} \frac{1}{\lambda_\alpha^L + i\mu_\alpha^L}$$

and

$$E\left[S_n^3\right] = \sum_{i=0}^{n-1} \frac{1}{\lambda_\alpha^U + i\mu_\alpha^U}.$$

So we have

$$\begin{aligned}\text{MTTF} &= \frac{1}{2}\int_0^1 \left(E\left[S_n(\theta)\right]_\alpha^L + E\left[S_n(\theta)\right]_\alpha^U\right)\text{d}\alpha \\ &= \frac{1}{2}\int_0^1 \left(\sum_{i=0}^{n-1} \frac{1}{\lambda_\alpha^L + i\mu_\alpha^L} + \sum_{i=0}^{n-1} \frac{1}{\lambda_\alpha^U + i\mu_\alpha^U}\right)\text{d}\alpha \\ &= \sum_{i=0}^{n-1} \frac{1}{2}\int_0^1 \left(\frac{1}{\lambda_\alpha^L + i\mu_\alpha^L} + \frac{1}{\lambda_\alpha^U + i\mu_\alpha^U}\right)\text{d}\alpha \\ &= \sum_{i=0}^{n-1} E\left[\frac{1}{\lambda + i\mu}\right].\end{aligned}$$

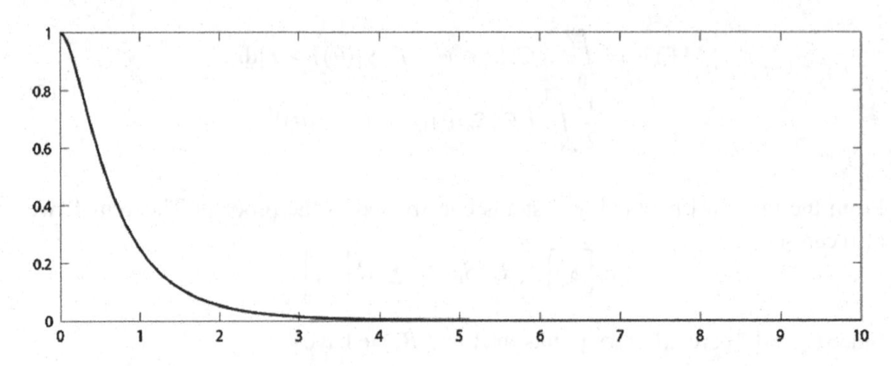

Fig. 1 The reliability of the warm standby system

4 A Numerical Example

Consider a warm standby system composed by two same type components, the lifetime of components in operation have random fuzzy exponential distribution with parameter λ and the lifetimes of components in warm standby have random fuzzy exponential distribution with parameter μ, we assume $\lambda = (1, 2, 3)$ and $\mu = (2, 3, 4)$. We can arrive at

$$\lambda_\alpha^L = 1 + \alpha, \qquad \lambda_\alpha^U = 3 - \alpha,$$

and

$$\mu_\alpha^L = 2 + \alpha, \qquad \mu_\alpha^U = 4 - \alpha.$$

It follows from Theorem 1 that

$$\begin{aligned} R(t) =& \frac{1}{2}\int_0^1 \left(\frac{\lambda_\alpha^L + \mu_\alpha^L}{\mu_\alpha^L} e^{-\lambda_\alpha^L t} + \frac{\lambda_\alpha^U + \mu_\alpha^U}{\mu_\alpha^U} e^{-\lambda_\alpha^U t} - \frac{\lambda_\alpha^L}{\mu_\alpha^L} e^{-(\lambda_\alpha^L + \mu_\alpha^L)t} - \frac{\lambda_\alpha^U}{\mu_\alpha^U} e^{-(\lambda_\alpha^U + \mu_\alpha^U)t} \right) d\alpha \\ =& \frac{1}{2}\int_0^1 \left(\frac{1+\alpha+2+\alpha}{2+\alpha} e^{-(1+\alpha)t} + \frac{3-\alpha+4-\alpha}{4-\alpha} e^{-(3-\alpha)t} - \frac{1+\alpha}{2+\alpha} e^{-(1+\alpha+2+\alpha)t} \right. \\ & \left. - \frac{3-\alpha}{4-\alpha} e^{-(3-\alpha+4-\alpha)t} \right) d\alpha \end{aligned}$$

The Fig. 1 shows a plot of the reliability of the warm standby system, we can see the reliability decrease of t.

It follows from Theorem 2 that

$$\begin{aligned} \text{MTTF} &= E\left[\frac{1}{\lambda}\right] + E\left[\frac{1}{\lambda + \mu}\right] \\ &= \frac{1}{2}\int_0^1 \left(\frac{1}{1+\alpha} + \frac{1}{3-\alpha} \right) d\alpha + \frac{1}{2}\int_0^1 \left(\frac{1}{1+\alpha+2+\alpha} + \frac{1}{3-\alpha+4-\alpha} \right) d\alpha \\ &= \frac{1}{2} ln3 - \frac{1}{4} ln\frac{7}{3} \approx 0.3375. \end{aligned}$$

5 Conclusion

As the systems become more and more complex, the probability theory is not suitable in many situations and random fuzzy methodology shows its advantages in certain circumstance. In this paper, the lifetimes of components in operation and in warm standby are assumed to have random fuzzy exponential distributions, then the expressions of reliability and mean time to failure of the warm standby systems are proposed. Further researches can pay attention to repairabe systems.

Acknowledgments This work was supported by the National Natural Science Foundation of China Grant No. 61070021 and scientific research foundation in Tianjin university of science and technology No. 20110119.

References

1. Dhillon, B.S., Yang, N.: Reliability and availability analysis of warm standby systems with common-cause failures and human errors. Microelectron. Reliab. **32**(4), 561–575 (1992)
2. Li, X.H., Yan, R.F., Zuo, M.J.: Evaluating a warm standby system with components having proportional hazard rates. Oper. Res. Lett. **37**(1), 56–60 (2009)
3. Liu, B.D.: Theory and Practice of Uncertain Programming. Physica-Verlag, Heidelberg (2002)
4. Liu, B.D.: A survey of credibility theory. Fuzzy Optimization Decis. Making **5**(4), 387–408 (2006)
5. Liu, B.D., Liu, Y.K.: Expected value of fuzzy variable and fuzzy expected value models. IEEE Trans. Fuzzy Syst. **10**(4), 445–450 (2002)
6. Liu, Y.K., Liu, B.D.: Expected value operator of random fuzzy variable and random fuzzy expected value models. Int. J. Uncertainty, Fuzziness Knowl. Based Syst. **11**(2), 195–215 (2003)
7. Liu, Y.K., Liu, B.D.: Random fuzzy programming with chance measures defined by fuzzy integrals. Math. Comput. Model. **36**(4–5), 509–524 (2002)
8. Liu, Y., Li, X.Z., Yang, G.L.: Reliability analysis of random fuzzy repairable series system. Fuzzy Inform. Eng. Adv. Soft Comput. **78**, 281–296 (2010)
9. Liu, Y., Tang, W.S., Li, X.Z.: Random fuzzy shock models and bivariate random fuzzy exponential distribution. Appl. Math. Model. **35**(5), 2408–2418 (2011)
10. Mokaddis, G.S., Labib, S.W., Ahmed, A.M.: Analysis of a two-unit warm standby system subject to degradation. Microelectron. Reliab. **37**(4), 641–647 (1997)
11. Naidu, R.S., Gopalan, M.N.: Cost-benefit analysis of a one-server two-unit warm standby system subject to different inspection strategies. Microelectron. Reliab. **23**(1), 121–128 (1983)
12. She, J., Pecht, M.G.: Reliability of a k-out-of-n warm-standby system. IEEE Trans. Reliab. **41**(1), 72–75 (1992)
13. Tan, Z.B.: Reliability and availability analysis of two-unit warm standby microcomputer systems with self-reset function and repair facility. Microelectron. Reliab. **37**(8), 1251–1253 (1997)
14. Uematsu, K., Nishida, T., Kowada, M.: Some applications of semi-regenerative processes to two-unit warm standby system. Microelectron. Reliab. **24**(5), 965–977 (1984)
15. Vanderperre, E.J.: Reliability analysis of a warm standby system with general distributions. Microelectron. Reliab. **30**(3), 487–490 (1990)
16. Wang, K.H., Ke, J.C.: Probabilistic analysis of a repairable system with warm standbys plus balking and reneging. Appl. Math. Model. **27**(4), 327–336 (2003)
17. Yuan, L., Meng, X.Y.: Reliability analysis of a warm standby repairable system with priority in use. Appl. Math. Model. **35**(9), 4295–4303 (2011)

18. Zhao, R.Q., Liu, B.D.: Redundancy optimization problems with uncertainty of combining randomness and fuzziness. Eur. J. Oper. Res. **157**(3), 716–735 (2004)
19. Zhao, R.Q., Tang, W.S., Yun, H.L.: Random fuzzy renewal process. Eur. J. Oper. Res. **169**(1), 189–201 (2006)

Separation Axioms in ω_α-*opos*

Xiu-Yun Wu, Li-Li Xie and Shi-Zhong Bai

Abstract In this paper, concepts of ω_α-T_i, ($i = 0, 1, 2, 3, 4$) sets are introduced in ω_α-order preserving spaces and their characteristic properties are studied. They consist a new complete system of separation axiom, which is proved to be a good extension of both classical and fuzzy theories. Finally, their relationships are established and their differences are discussed by some exact examples.

Keywords ω_α-Order-preserving operator · ω_α-Remote neighborhood · ω_α-T_i set · ω_α-Closed set · ω_α-Homeomorphism.

1 Introduction

Separation axiom is one of the most important parts in fuzzy topology. A lot of papers have devoted on it, There are various forms in fuzzy topological spaces [1–9]. As the concept of L-fuzzy order-preserving operator spaces was introduced by professor Chen in 2002 [10], a new type of separation axiom which can be regarded as a summary of other separation axioms, especially, in L-fuzzy topological spaces and L-fuzzy semi-topological spaces was introduced [11].

In [12], Meng introduced the concept of stratified topological space. Separation axiom was naturally introduced in it [13]. However, there is not relevant theory yet

X. Wu (✉)
Institute of Computational Mathematics, Department of Mathematics, Hunan Institute of Science and Engineering, Yongzhong 425100, China
e-mail: wuxiuyun2000@126.com

L. Xie
Department of English Teaching, Hunan Institute of Science and Engineering, Yongzhong 425100, China

S. Bai
Institute of Mathematics, Wuyi university, Jiangmen 529020, China

B.-Y. Cao and H. Nasseri (eds.), *Fuzzy Information & Engineering and Operations Research & Management*, Advances in Intelligent Systems and Computing 211,
DOI: 10.1007/978-3-642-38667-1_50, © Springer-Verlag Berlin Heidelberg 2014

has been established in stratified order-preserving spaces. In this paper, concepts of ω_α-T_i, $(i = 1, 2, 3, 4)$ sets which are the more generalized form of stratified form are introduced in ω_α-order preserving spaces [14, 15], their characteristic properties are studied. Then they're proved to be good extensions. Finally, specific examples are given to show their differences. The relationships among them are established as well.

2 Preliminaries

In this paper, X, Y will always denote nonempty crisp sets, A mapping $A : X \to L$ is called an L-fuzzy set. L^X is the set of all L-fuzzy sets on X. An element $e \in L$ is called an irreducible element in L, if $p \vee q = e$ implies $p = e$ or $q = e$, where $p, q \in L$. The set of all nonzero irreducible elements in L will be denoted by $M(L)$ (see. [12]). If $x \in X$, $\alpha \in M(L)$, then x_α is called a molecule in L^X. The set of all molecules in L^X is denoted by $M^*(L^X)$. If $A \in L^X$, $\alpha \in M(L)$, take $A_{[\alpha]} = \{x \in X \mid A(x) \geq \alpha\}$, $A^{(\alpha')} = \{x \in X \mid \alpha \nleq A(x)\}$. Clearly, $A_{[\alpha]} = (A^{(\alpha')})'$.

The following are the concepts of L-fuzzy order-preserving operator in L-fuzzy topological space and order-preserving operator in general topological space.

Let X be an nonempty set. An operator $\omega : L^X \to L^X$ is called an L-fuzzy order preserving operator in L^X, if it satisfies: (1) $\omega(1_X) = 1_X$, (2) $\forall A, B \in L^X$ and $A \leq B$ implies $\omega(A) \leq \omega(B)$. A set $A \in L^X$ is called an ω-set, if $\omega(A) = A$. The set of all ω-sets in L^X is denoted by ω. (L^X, ω) is called an order-preserving operator space(briefly, L-*opos*). A molecule $x_\alpha \in M^*(L^X)$, $P \in \omega$, P is called an ω-remote neighborhood of x_α, if $x_\alpha \nleq P$. The set of all ω-remote neighborhood of x_α is denoted by $\omega\eta(x_\alpha)$. Let $x_\alpha \in M^*(L^X)$, $A \in L^X$, x_α is called an ω-adherent point of A, if $\forall P \in \omega\eta(x_\alpha)$, $A \nleq P$. The union of all ω-adherent points of A is called the ω-closure of A, denoted by A_ω^-. A set $A \in L^X$ is called ω-closed, if $A_\omega^- = A$. The set of all ω-closed sets in L^X is denoted by Ω_ω^-. Ω_ω^- is finite union and infinite intersection preserving [10].

Let X be an nonempty set, $\mathscr{P}(X)$ be the family of all subsets of X. An operator $\sigma : X \to X$ is called an order preserving operator in X, if it satisfies: (1) $\sigma(X) = X$, (2) $\forall A \subset B \subset X$ implies $\sigma(A) \subset \sigma(B)$. A set $A \subset X$ is called an σ-set, if $\sigma(A) = A$. The set of all σ-sets in L^X is denoted by Δ. And (X, Δ) is called an order-preserving operator space on X (briefly, *opos*). Let $x \in X$, $P \in \Delta$, P is called an σ-remote neighborhood of x, if their is $Q \subset X$, such that $x \notin Q$, $P \subset Q$. The set of all σ-remote neighborhood of x is denoted by $\sigma\eta(x)$. Let $x \in X$, $A \subset X$, x is called an σ-adherent point of A, if $\forall P \in \sigma\eta(x)$, $A \not\subset P$. The union of all σ-adherent points of A is called the σ-closure of A, denoted by A_σ^-. An set $A \subset X$ is called σ-closed, if $A_\sigma^- = A$. The set of all σ-closed sets in X is denoted by Δ_σ^-. Δ_σ^- is finite union and infinite intersection preserving [11].

Let (X, Δ) be an *opos*, L be an fuzzy lattice, $\forall \alpha \in L$. An fuzzy set $A : L \to X$ is called a lower continuous function, if $\{x \in X \mid A(x) \leq \alpha\} \in \Delta_\sigma^-$. The set of all the lower continuous functions, denoted by $\omega_L(\Delta)$ consists an L-fuzzy topology in L^X. The space $(L^X, \omega_L(\Delta))$ is called the induced L-*opos* by (X, Δ) [11].

Considering the above definitions, if we say there is an L-*opos* (L^X, Ω) or an *opos* (X, Δ), we mean there is an order-preserving operator Ω on L^X, or σ on X. The two spaces are generated by them, respectively.

Let (L^X, δ) and (L^Y, τ) be two L-fuzzy topological spaces, $f^{\rightarrow} : L^X \rightarrow L^Y$ be induced by simple mapping $f : X \rightarrow Y$. If $f^{\rightarrow}$ and its reverse mapping $f^{\leftarrow}$ satisfies:

(1) $\forall A \in L^X, y \in Y, f^{\rightarrow}(A)(y) = \vee\{A(x) \mid x \in X, f(x) = y\}$;
(2) $\forall B \in L^Y, x \in X, f^{\leftarrow}(B)(x) = B(f(x))$.

Then $f^{\rightarrow}$ is called an L-fuzzy mapping.

An L-fuzzy mapping $f^{\rightarrow} L^X \rightarrow L^Y$ is called homomorphism, if it satisfies:

(1) $f^{\rightarrow}$ is union preserving. I.e., for $\{A_i \in L^X, (i \in I)\} \subset L^X$, $f^{\rightarrow}(\bigvee_{i\in I} A_i) = \bigvee_{i\in I} f^{\rightarrow}(A_i)$;
(2) $f^{\rightarrow}$ is reserving involution preserving. I.e., for $B \in L^Y$, $f^{\leftarrow}(B') = (f^{\leftarrow}(B))'$ [16].

Let (L^X, Ω) be an L-*opos*, $\alpha \in M(L)$. An operator $\omega_\alpha : L^X \rightarrow L^X$ is called ω_α order-preserving operator, (briefly, ω_α-*opo*), which is defined by: $\forall A \in L^X$,

$$\omega_\alpha(A) = \wedge\{G \in \Omega_\omega^- \mid G_{[\alpha]} \supset A_{[\alpha]}\}.$$

A set $A \in L^X$ is called ω_α-closed, if $\omega_\alpha(A)_{[\alpha]} = A_{[\alpha]}$. The set of all ω_α-closed sets in L^X is denoted by $\omega_\alpha(\Omega)$. $(L^X, \omega_\alpha(\Omega))$ is called ω_α-order-preserving operator space, (briefly, ω_α-*opos*) [14].

Let $(L^X, \omega_\alpha(\Omega))$ be an ω_α-*opos*, and $e \in M^*(L^X)$. $A \in \omega_\alpha(\Omega)$ is called an ω_α-remote neighborhood of e (briefly, ω_α-*RN* of e), if $e \not\leq \omega_\alpha(A)$. The collection of all ω_α-*RN*s of e is denoted by $\eta_{\omega_\alpha}(e)$. If $B \in L^X, \forall P \in \eta_{\omega_\alpha}(e)$, $B_{[\alpha]} \not\subset P_{[\alpha]}$, then e is called an ω_α-adherent point of B.

Lemma 2.1 *Let $(L^X, \omega_\alpha(\Omega))$ be an ω_α-opos, $A \in L^X$, $e \in M^*(L^X)$. Then e is an ω_α-adherent point of A iff $e_{[\alpha]} \subset \omega_\alpha(A)_{[\alpha]}$.*

Proof Necessity. Suppose $e_{[\alpha]} \not\subset \omega_\alpha(A)_{[\alpha]}$, then $\omega_\alpha(A) \in \eta_{\omega_\alpha}(e)$. By $A_{[\alpha]} \subset \omega_\alpha(A)_{[\alpha]}$, we have e is not an ω_α-adherent point of A. This is a contradiction.

Sufficiency. If e is not an ω_α-adherent point of A, then there is $P \in \eta_{\omega_\alpha}(e)$, such that $A_{[\alpha]} \subset P_{[\alpha]}$. Since $P \in \omega_\alpha(\Omega)$, we have $\omega_\alpha(A)_{[\alpha]} \subset \omega_\alpha(P)_{[\alpha]} = P_{[\alpha]}$. By $e_{[\alpha]} \not\subset P_{[\alpha]}$, we get $e_{[\alpha]} \not\subset \omega_\alpha(A)_{[\alpha]}$. This is a contradiction too.

Definition 2.1 *Let $(L^X, \omega_\alpha(\Omega))$ be an ω_α-opos. $\emptyset \neq Y \subset X$. $A \in L^X$, take $\Omega_{|Y} = \{A_{|Y} \mid A \in \Omega\}$ be the restriction of Ω on Y, and $(L^Y, \Omega_{|Y})$ be a subspace of (L^X, Ω). An operator $(\omega_\alpha)_{|Y} : L^Y \rightarrow L^Y$ defined by: $\forall B \in L^Y$, $(\omega_\alpha)_{|Y}(B) = \omega_\alpha(B^*)_{|Y}$. The set of all $(\omega_\alpha)_{|Y}$-closed sets is denoted by $(\omega_\alpha)_{|Y}(\Omega_{|Y})$.*

Theorem 2.1 *Let $(L^X, \omega_\alpha(\Omega))$ be an ω_α-opos. $(L^Y, \Omega_{|Y})$ be a subspace of (L^X, Ω). Then*

(1) $\forall A \in \omega_\alpha(\Omega)$, $A_{|Y} \in (\omega_\alpha)_{|Y}(\Omega_{|Y})$;
(2) $\forall B \in (\Omega_\alpha)_{|Y}(\Omega_{|Y})$, there is $D \in \omega_\alpha(\Omega)$, such that $D_{|Y} = B$.

Definition 2.2 *Let (X, Δ) be an opos, $A \subset X$. A is called a*

(1) T_0 set, if $x, y \in A$ with $x \neq y$, there is $P \in \Delta$, such that $x \in X/P$, $y \in P$. Or, there is $Q \in \Delta$, such that $y \in X/Q$, $x \in Q$.
(2) T_1 set, if $x, y \in A$ with $x \neq y$, there is $P \in \Delta$, such that $x \in X/P$, $y \in P$.
(3) T_2 set, if $x, y \in A$ with $x \neq y$, there are $P, Q \in \Delta$, such that $A \subset P \cup Q$.
(4) T_3 set, if $x \in A, B \in \Delta$ with $x \notin B \subset A$, there are $P, Q \in \Delta$, such that $A \subset P \cup Q$.
(5) T_4 set, if $B, C \in \Delta$ with $B, C \subset A$ and $B \cap C = \emptyset$, there are $P, Q \in \Delta$, such that $A \subset P \cup Q$.

Specially, (X, Δ) is called a T_i space, if X is a T_i set, (i=0,1,2,3,4).

Definition 2.3 *Let $(L^X, \omega^1_\alpha(\Omega_1))$, $(L^Y, \omega^2_\alpha(\Omega_2))$ be ω^1_α-opos and ω^2_α-opos, respectively. An L-fuzzy homeomorphism $f^{\rightarrow} : L^X \rightarrow L^Y$ is called*

(1) $(\omega^1_\alpha, \omega^2_\alpha)$-continuous, if $\forall B \in \omega^2_\alpha(\Omega_2)$, then $f^{\leftarrow}(B) \in \omega^1_\alpha(\Omega_1)$
(2) $(\Omega^1_\alpha, \Omega^2_\alpha)$-homeomorphism, if $f^{\rightarrow}$ is both $(\omega^1_\alpha, \omega^2_\alpha)$-continuous and $f^{\leftarrow}$ is$(\omega^2_\alpha, \omega^1_\alpha)$-continuous.

Notes and symbols are not mentioned here can be found in [16].

3 ω_α-T_0, T_1, T_2 sets

Definition 3.1 *Let $(L^X, \omega_\alpha(\Omega))$ be an ω_α-opos, $A \in L^X$. Then A is called an*

(1) *ω_α-T_0 set, if $\forall x, y \in A_{[\alpha]}$ with $x \neq y$, then there is $P \in \eta_{\omega_\alpha}(x_\alpha)$, such that $y_\alpha \leq P$, or there is $Q \in \eta_{\omega_\alpha}(y_\alpha)$, such that $x_\alpha \leq Q$.*
(2) *ω_α-T_1 set, if $\forall x, y \in A_{[\alpha]}$ with $x \neq y$, then there is $P \in \eta_{\omega_\alpha}(x_\alpha)$, such that $y_\alpha \leq P$.*
(3) *ω_α-T_2 set, if$\forall x, y \in A_{[\alpha]}$ with $x \neq y$, then there are $P \in \eta_{\omega_\alpha}(x_\alpha)$, $Q \in \eta_{\omega_\alpha}(y_\alpha)$, such that $A_{[\alpha]} \subset P_{[\alpha]} \cup Q_{[\alpha]}$.*
$(L^X, \omega_\alpha(\Omega))$ is called an ω_α-T_i space, if 1_X is an ω_α-T_i set, $(i = 0, 1, 2)$.

Clearly, ω_α-$T_2 \Rightarrow \omega_\alpha$-$T_1 \Rightarrow \omega_\alpha$-$T_0$. However, the converse result is not true.

Theorem 3.1 *Let $(L^X, \omega_\alpha(\Omega))$ be an ω_α-opos, $A \in L^X$, $\forall x, y \in A_{[\alpha]}$ with $x \neq y$. Then the following statements are equivalent.*

(1) A is an ω_α-T_0 set.
(2) $\eta_{\omega_\alpha}(x_\alpha) \neq \eta_{\omega_\alpha}(y_\alpha)$.
(3) $\omega_\alpha(x_\alpha)_{[\alpha]} \neq \omega_\alpha(y_\alpha)_{[\alpha]}$.

(4) $x_\alpha \not\leq \omega_\alpha(y_\alpha)$ *or* $y_\alpha \not\leq \omega_\alpha(x_\alpha)$.

Proof (1)⇒(2). Suppose A is an ω_α-T_0 set. If there is $P \in \eta_{\omega_\alpha}(x_\alpha)$, such that $y_\alpha \leq P$. Then $P \notin \eta_{\omega_\alpha}(y_\alpha)$. Or, if there is $Q \in \eta_{\omega_\alpha}(y_\alpha)$, such that $x_\alpha \leq Q$. Then $Q \notin \eta_{\Omega_\alpha}(x_\alpha)$. Therefore $\eta_{\omega_\alpha}(x_\alpha) \neq \eta_{\omega_\alpha}(y_\alpha)$.

(2)⇒(3). By (2), we get there is $P \in \eta_{\omega_\alpha}(x_\alpha)$ and $P \notin \eta_{\omega_\alpha}(y_\alpha)$, or, there is $Q \in \eta_{\omega_\alpha}(y_\alpha)$, and $Q \notin \eta_{\omega_\alpha}(x_\alpha)$. Take the former for example, we get $y \in P_{[\alpha]} = \omega_\alpha(P)_{[\alpha]}$. Since $\omega_\alpha(P) \in \Omega_\omega^-$, we have

$$\omega_\alpha(y_\alpha) = \wedge\{G \in \Omega_\omega^- \mid G_{[\alpha]} \supset (y_\alpha)_{[\alpha]} = \{y\}\} \leq \omega_\alpha(P).$$

Thus $y \in \omega_\alpha(y_\alpha)_{[\alpha]} \subset \omega_\alpha(P)_{[\alpha]} = P_{[\alpha]}$. On the other hand, as $x \in \omega_\alpha(x_\alpha)_{[\alpha]}$, $x \notin P_{[\alpha]}$. We have $x \notin \omega_\alpha(y_\alpha)_{[\alpha]}$. Therefore, $\omega_\alpha(x_\alpha)_{[\alpha]} \neq \omega_\alpha(y_\alpha)_{[\alpha]}$.

(3)⇔(4). By the proof of (2)⇒(3). Obvious.

(3)⇒(1). $\forall x, y \in A_{[\alpha]}$ with $x \neq y$. By (3), $\omega_\alpha(x_\alpha)_{[\alpha]}/\omega_\alpha(y_\alpha)_{[\alpha]} \neq \emptyset$, or $\omega_\alpha(y_\alpha)_{[\alpha]}/\omega_\alpha(x_\alpha)_{[\alpha]} \neq \emptyset$. As for the former, we prove $x \notin \omega_\alpha(y_\alpha)_{[\alpha]}$.

In fact, if

$$x \in \omega_\alpha(y_\alpha)_{[\alpha]} = \cap\{G_{[\alpha]} \mid G \in \Omega_\omega^-, G_{[\alpha]} \supset (y_\alpha)_{[\alpha]} = \{y\}\}.$$

This is to say, for any $G \in \Omega_\omega^-, \{y\} \subset G_{[\alpha]}$, then $x \in G_{[\alpha]}$. So $\omega_\alpha(x_\alpha)_{[\alpha]} \subset \omega_\alpha(y_\alpha)_{[\alpha]}$. Hence, $\omega_\alpha(x_\alpha)_{[\alpha]}/\omega_\alpha(y_\alpha)_{[\alpha]} = \emptyset$. A contradiction. Besides, as $\omega_\alpha(y_\alpha) \in \Omega_\omega^-$, we have $\omega_\alpha(y_\alpha) \in \eta_{\omega_\alpha}(x_\alpha)$, and $y_\alpha \leq \omega_\alpha(y_\alpha)$. Similarly, we can prove the later case, therefore, A is an ω_α-T_0 set.

Theorem 3.2 *Let* $(L^X, \omega_\alpha(\Omega))$ *be an* ω_α*-opos,* $\alpha \in M(L)$, $A \in L^X$. *Then* A *is an* ω_α-T_1 *set iff* $\forall x, y \in A_{[\alpha]}$ *with* $x \neq y$, *there is* $P \in \eta_{\omega_\alpha}(x_\alpha)$, *such that* $y_\alpha \leq \omega_\alpha(P)$.

Proof Suppose A is an ω_α-T_1 set. Then there is $P \in \eta_{\omega_\alpha}(x_\alpha)$, such that $y_\alpha \leq P$. So $y \in P_{[\alpha]} = \omega_\alpha(P)_{[\alpha]}$. Hence, $y_\alpha \leq \omega_\alpha(P)$.

Conversely. Suppose $\forall x, y \in A_{[\alpha]}$ with $x \neq y$, there is $P \in \eta_{\omega_\alpha}(x_\alpha)$, such that $y_\alpha \leq \omega_\alpha(P)$. So $x \notin P_{[\alpha]} = \omega_\alpha(P)_{[\alpha]}$. Thus $x_\alpha \not\leq \omega_\alpha(P)$ and $\omega_\alpha(P) \in \eta_{\omega_\alpha}(x_\alpha)$. Besides, $y_\alpha \leq \omega_\alpha(P)$ is obvious.

Theorem 3.3 *Let* $(L^X, \omega_\alpha(\Omega))$ *be an* ω_α*-opos,* $A \in L^X$. *Then* A *is an* ω_α-T_1 *set iff* $\forall x \in A_{[\alpha]}$, $x_\alpha \in \omega_\alpha(\Omega)$.

Proof Suppose A is an ω_α-T_1 set. $\forall x \in A_{[\alpha]}$, $y \in \omega_\alpha(x_\alpha)_{[\alpha]}$, if $x \neq y$, then there is $Q \in \eta_{\omega_\alpha}(y_\alpha)$, such that $x_\alpha \leq Q$, and thus, $\omega_\alpha(x_\alpha) \leq \omega_\alpha(Q)$. Thereby $y \in \omega_\alpha(x_\alpha)_{[\alpha]} \subset \omega_\alpha(Q)_{[\alpha]} = Q_{[\alpha]}$. It is a contradiction with $Q \in \eta_{\omega_\alpha}(y_\alpha)$. Therefore $\omega_\alpha(x_\alpha)_{[\alpha]} = \{x\} = (x_\alpha)_{[\alpha]}$.

Conversely. If A is not an ω_α-T_1 set, then there are $x, y \in A_{[\alpha]}$ with $x \neq y$, such that $\forall P \in \eta_{\omega_\alpha}(x_\alpha)$, $y_\alpha \not\leq \omega_\alpha(P)$. So $y \notin \omega_\alpha(P)_{[\alpha]} = P_{[\alpha]}$. Hence $\{y\} = (y_\alpha)_{[\alpha]} \not\subset P_{[\alpha]}$. It implies $x_\alpha \leq \omega_\alpha(y_\alpha)$. Therefore $x \in \omega_\alpha(y_\alpha)_{[\alpha]}$ and $(y_\alpha)_{[\alpha]} \neq \omega_\alpha(y_\alpha)_{[\alpha]}$. We get $y_\alpha \notin \omega_\alpha(\Omega)$.

Theorem 3.4 *Let $(L^X, \omega_\alpha(\Omega))$ be an ω_α-opos, $A \in L^X$. Then A is an ω_α-T_2 set iff every constant molecular net in A can not convergence to two different x_α, y_α at same time, where $x, y \in A_{[\alpha]}$ with $x \neq y$.*

Proof Let A be an ω_α-T_2 set, $S = \{x(n)_\alpha \mid n \in D\}$ be an constant value molecular net in A. If $x, y \in A_{[\alpha]}$ with $x \neq y$, $S \to \omega_\alpha$-x_α and $S \to \omega_\alpha$-y_α. Then there are $P \in \eta_{\omega_\alpha}(x_\alpha)$ and $Q \in \eta_{\omega_\alpha}(y_\alpha)$, such that $A_{[\alpha]} \subset P_{[\alpha]} \cup Q_{[\alpha]}$. Since $S \to \omega_\alpha$-x_α, there is $m_1 \in D$, such that $x_\alpha^n \not\leq P$ whenever, $n \geq m_1$. Similarly, by $S \to \omega_\alpha$-y_α, there is $m_2 \in D$, such that $x_\alpha^n \not\leq Q$ whenever, $n \geq m_2$. Take $m \in D$, and $m \geq m_1, m_2$, so $x_\alpha^m \not\leq P \vee Q$. Therefore, $(x_\alpha^m)_{[\alpha]} \not\subset P_{[\alpha]} \cup Q_{[\alpha]}$. A contradiction with $(x_\alpha^m)_{[\alpha]} \subset A_{[\alpha]}$.

Conversely. Suppose A is not an ω_α-T_1 set, then there are $x, y \in A_{[\alpha]}$ with $x \neq y$, such that $\forall P \in \eta_{\omega_\alpha}(e), \forall Q \in \eta_{\omega_\alpha}(d)$, $A_{[\alpha]} \not\subset P_{[\alpha]} \cup Q_{[\alpha]}$. Then there is $z_\alpha^{(P,Q)} \in A_{[\alpha]}$, such that $z_\alpha^{(P,Q)} \notin P_{[\alpha]} \cup Q_{[\alpha]}$. Take $D = \eta_{\omega_\alpha}(x_\alpha) \times \eta_{\omega_\alpha}(y_\alpha)$. For $(P_1, Q_1), (P_2, Q_2) \in D$, define $(P_1, Q_1) \leq (P_2, Q_2)$ iff $P_1 \leq P_2$, $Q_1 \leq Q_2$. D is a directed set. $S = \{z_\alpha^{(P,Q)} \mid (P, Q) \in D\}$, then S is a constant molecular net in A, Let's prove $S \to \Omega_\alpha$-x_α.

In fact, for each $(P_0, Q_0) \leq (P, Q)$, by $(z_\alpha^{(P,Q)})_{[\alpha]} \not\subset P_{[\alpha]} \cup Q_{[\alpha]}$, we have $(z_\alpha^{(P,Q)})_{[\alpha]} \not\subset P_{0[\alpha]}$, so $z_\alpha^{(P,Q)} \not\leq P_0$. This means $S \to \omega_\alpha$-x_α. Similarly, $S \to \omega_\alpha$-y_α. Therefore, S has two ω_α-limit points x_α, y_α with $x \neq y$. It is a contradiction.

Theorem 3.5 *Let (X, Δ) be an opos, and $(L^X, \omega_L(\Delta))$ be an L-opos induced by (X, Δ). $(L^X, \omega_\alpha(\Omega_L(\Delta)))$ be ω_α-opos. Then $(L^X, \omega_\alpha(\Omega_L(\Delta)))$ is an ω_α-T_i space iff (X, Δ) is a T_i space, $(i = 1, 2)$.*

Proof We only prove the case of $i = 2$. Suppose $(L^X, \omega_\alpha(\Omega_L(\Delta)))$ is an ω_α-T_2 space, $x, y \in 1_{X[\alpha]} = X$ with $x \neq y$. Then there are $P \in \eta_{\omega_\alpha}(x_\alpha)$, $Q \in \eta_{\omega_\alpha}(y_\alpha)$, such that $P_{[\alpha]} \cup Q_{[\alpha]} = X$. Since P are ω_α-closed sets, we have $P_{[\alpha]} \in \Delta$, and $P'^{(\alpha')} = (P_{[\alpha]})'$ is an open set in (X, Δ). As $x \notin P_{[\alpha]}$, we get $x \in P'^{(\alpha')}$, so $P'^{(\alpha')}$ is an neighborhood of x in (X, Δ). Similarly, we get $Q'^{(\alpha')}$ is an neighborhood of y. Clearly, $P'^{(\alpha')} \cap Q'^{(\alpha')} = \emptyset$. Hence, (X, Δ) is an T_2 space.

Conversely. If (X, Δ) is an T_2 space. $x, y \in X$, with $x \neq y$, then there are two closed sets $U, V \in \Delta$, such that $x \in U'$, $y \in V'$ and $U' \cap V' = \emptyset$. Take $P = \omega_\alpha(\chi_U)$, $Q = \Omega_\alpha(\chi_V)$, so $P, Q \in \omega_\alpha(\Omega)$, and $x \notin U$, $y \notin V$. Since $(L^X, \Omega_L(\Delta))$ is induced by (X, Δ). We have $P \in \eta_{\omega_\alpha}(x_\alpha)$, $Q \in \eta_{\omega_\alpha}(y_\alpha)$. Besides,

$$
\begin{aligned}
P_{[\alpha]} \cup Q_{[\alpha]} &= \omega_\alpha(\chi_U)_{[\alpha]} \cup \omega_\alpha(\chi_V)_{[\alpha]} = \omega_\alpha(\chi_U \vee \chi_V)_{[\alpha]} \\
&= \omega_\alpha(\chi_{U \cup V})_{[\alpha]} = \omega_\alpha(\chi_X)_{[\alpha]} = X.
\end{aligned}
$$

Therefore $(L^X, \omega_\alpha(\Omega_L(\Delta)))$ is an ω_α-T_2 space.

Theorem 3.6 *Let $(L^X, \omega_\alpha(\Omega))$ be an ω_α-opos, $(L^Y, \omega_{\alpha|Y}(\Omega_{|Y}))$ be its subspace. $A \in L^X$ is an ω_α-T_i set, $(i = 0, 1, 2)$, $\chi_Y \in \omega_\alpha(\Omega)$, then $A_{|Y} \in L^Y$ is an $\omega_{\alpha|Y}$-T_i set too.*

Proof We only prove the case of $i = 1$. If $A \in L^X$ is an Ω_α-T_1 set, and $x \in (A_{|Y})_{[\alpha]}$, then $x \in A_{[\alpha]}$. Since A is an ω_α-T_1 set and $x_\alpha \in M^*(L^X)$, we have $x_\alpha \in \omega_\alpha(\Omega)$. So $x_{\alpha|Y} \in M^*(L^Y)$, and $x_{\alpha|Y} \in \omega_{\alpha|Y}(\Omega_{|Y})$. Consequently, $A_{|Y}$ is an ω_α-T_i set.

Theorem 3.7 *Let* $(L^X, \omega^1_\alpha(\Omega_1))$, $(L^Y, \omega^2_\alpha(\Omega_2))$ *be* ω^1_α-*opos,* ω^2_α-*opos, respectively.* $f^{\rightarrow} : L^X \to L^Y$ *is* $(\omega^1_\alpha, \omega^2_\alpha)$-*homomorphism,* $A \in L^X$ *is an* ω^1_α-T_i *set,* $(i = 0, 1, 2)$, *then* $f^{\rightarrow}(A) \in L^Y$ *is an* ω^2_α-T_i *set, too.*

Proof Only to prove the case of $i = 0$.

$\forall y^1, y^2 \in f^{\rightarrow}(A)_{[\alpha]} = f(A_{[\alpha]})$ with $y^1 \neq y^2$, then there are $x^1, x^2 \in A_{[\alpha]}$ with $x^1 \neq x^2$, such that $f(x^1) = y^1$, $f(x^2) = y^2$. Since $A \in L^X$ is an ω^1_α-T_0 set, there is $P^1 \in \eta_{\omega^1_\alpha}(x^1_\alpha)$, such that $x^2_\alpha \leq P^1$, or there is $Q^2 \in \eta_{\omega^1_\alpha}(x^2_\alpha)$, such that $x^1_\alpha \leq Q^2$. As for the former case for example, $P^2 = f^{\rightarrow}(P^1)$. By $P^1 \in \omega^1_\alpha(\Omega_1)$, and $f^{\rightarrow}$ is $(\omega^1_\alpha, \omega^2_\alpha)$-homomorphism, so $P^2 \in \omega^2_\alpha(\Omega_2)$. Surely, $P^2 \in \eta_{\omega^1_\alpha}(y^2_\alpha)$, and $y^2_\alpha \leq P^2$. Therefore, $f^{\rightarrow}(A)$ is an ω^2_α-T_0 set.

Example 1 Let $X = \{x, y\}$, $L = \{0, 1/3, 2/3, 1\}$, $A \in L^X$ with $A(x) = a$, $A(y) = b$, we writer (a, b) instead of A. Take

$$\delta' = \{(0, 0), (1/3, 2/3), (1, 1)\}.$$

It easy to check that δ is an L-fuzzy topology on L^X. Take ω be the interior operator in δ, and $\alpha = 2/3$. Then $\omega_\alpha = D_\alpha$ [9]. Thus we have

$$\omega_\alpha(\Omega) = \{(0, 0), (0, 1/3), (0, 2/3), (0, 1), (1/3, 0), (1/3, 1/3),$$
$$(1/3, 2/3), (1/3, 1), (2/3, 2/3), (2/3, 1), (1, 2/3), (1, 1)\}.$$

Let $A = (1, 2/3)$. So $A_{[\alpha]} = \{x, y\}$, and

$$\eta_{\omega_\alpha}(x_\alpha) = \{(0, 0), (0, 1/3), (0, 2/3), (0, 1), (1/3, 0), (1/3, 1/3),$$
$$(1/3, 2/3), (1/3, 1)\}.$$
$$\eta_{\omega_\alpha}(y_\alpha) = \{(0, 0), (0, 1/3), (1/3, 0), (1/3, 1/3)\}.$$

Easily, there is $P = (0, 2/3) \in D_\alpha(\delta)$, such that $x_\alpha \not\leq P$, and $y_\alpha \leq P$. Hence A is an ω_α-T_0 set. However, $\forall Q \in \eta_{\omega_\alpha}(y_\alpha)$, $y_\alpha \not\leq Q$. Thus A is not an ω_α-T_1 set. Besides, there is not $P \in \eta_{\omega_\alpha}(x_\alpha)$, $Q \in \eta_{\omega_\alpha}(y_\alpha)$, such that $A_{[\alpha]} \subset P_{[\alpha]} \cup Q_{[\alpha]}$, which means A is not an ω_α-T_2 set, neither.

Example 2 Let X be an infinite set, $L = [0, 1]$, $\alpha \in (0, 1)$.

$$\delta' = \{A \in L^X \mid A_{[\alpha]} = X, or A_{[\alpha]}\ is\ finite\}.$$

It easy to check (L^X, δ) is an L-fuzzy topology. Take ω be the interior operator in δ, then $\Omega = \delta$. $\omega_\alpha = D_\alpha$. So $\omega_\alpha(\Omega) = \delta'$. If $A \in L^X$ and $x \in A_{[\alpha]}$. Since $(x_\alpha)_{[\alpha]} = \{x\}$ is finite, we have $x_\alpha \in \omega_\alpha(\Omega)$. Thus by Theorem 3.3, we know A is an

ω_α-T_1 set. However, for each $A \in L^X$, which satisfies $A_{[\alpha]}$ is infinite and $A_{[\alpha]} \neq X$. Furthermore, $\forall x, y \in A_{[\alpha]}$. Since $\forall P \in \omega_\alpha(\Omega)$, $P_{[\alpha]}$ is finite, or $P_{[\alpha]} = X$, then there are not $P \in \eta_{\omega_\alpha}(x_\alpha)$, $Q \in \eta_{\omega_\alpha}(y_\alpha)$, such that $A_{[\alpha]} \subset P_{[\alpha]} \cup Q_{[\alpha]}$. Therefore, A is not an ω_α-T_2 set.

4 ω_α-T_3, T_4 sets

Definition 4.1 *Let $(L^X, \omega_\alpha(\Omega))$ be an ω_α-opos. $P \in \omega_\alpha(\Omega)$ is called an ω_α-remote neighborhood of $A \in L^X$, if $\forall x \in A_{[\alpha]}$, we have $x \notin P_{[\alpha]}$. The set of all ω_α-remote neighborhood of A is denoted by $\eta_{\omega_\alpha}(A)$.*

Definition 4.2 *Let $(L^X, \omega_\alpha(\Omega))$ be an ω_α-opos. $A \in \omega_\alpha(\Omega)$ is called an*

(1) ω_α-regular set,if $\forall B \in \omega_\alpha(\Omega)$ with $B_{[\alpha]} \subset A_{[\alpha]}, x \in A_{[\alpha]}/B_{[\alpha]}$,there are $P \in \eta_{\omega_\alpha}(x_\alpha)$ and $Q \in \eta_{\omega_\alpha}(B)$, such that $A_{[\alpha]} \subset P_{[\alpha]} \cup Q_{[\alpha]}$. If A is both an ω_α-T_1 and ω_α-regular set, then A is called an ω_α-T_3 set.
(2) ω_α-normal set, if$\forall B, C \in \omega_\alpha(\Omega)$ with $\emptyset \neq B_{[\alpha]}, C_{[\alpha]} \subset A_{[\alpha]}$ and $B_{[\alpha]} \cap C_{[\alpha]} = \emptyset$, there are $P \in \eta_{\omega_\alpha}(B)$ and $Q \in \eta_{\omega_\alpha}(C)$, such that $A_{[\alpha]} \subset P_{[\alpha]} \cup Q_{[\alpha]}$.
If A is both an ω_α-T_1 and ω_α-normal set, then A is called an weakly ω_α-T_4 set.

$(L^X, \omega_\alpha(\Omega))$ is called ω_α-T_i space, if 1_X is an ω_α-T_i $(i = 3, 4)$ set.

Clearly, we have ω_α-$T_4 \Rightarrow \omega_\alpha$-$T_3 \Rightarrow \omega_\alpha$-$T_2$.

However, ω_α-regular has no relation with ω_α-normal. Besides, neither an ω_α-regular set or an ω_α-normal set is an ω_α-T_0 set.

Example 3 Let $X = \{x, y, z\}$, $L = \{0, 1/2, 1\}$, $A \in L^X$ with $A(x) = a$, $A(y) = b$, $A(z) = c$, we writer (a, b, c) instead of A. Take

$$\delta' = \{(0, 0, 0), (1, 0, 0), (0, 1, 1), (1, 1, 1)\}.$$

It easy to check δ is an L-fuzzy topology on L^X. Take $\alpha = 1/2$, $A = 1_X$, so $A_{[\alpha]} = \{x, y, z\} = X$. It is easy to have

$$\begin{aligned}\Omega_\alpha(\Omega) =& \{(0, 0, 0), (1/2, 0, 0), (1, 0, 0), (1/2, 1/2, 1/2), (1/2, 1, 1/2),\\ & (1/2, 1/2, 1), (1/2, 1, 1), (1, 1/2, 1/2), (1, 1, 1/2), (1, 1/2, 1), (0, 1/2, 1/2),\\ & (0, 1/2, 1), (0, 1, 1/2), (0, 1, 1), (1, 1, 1)\}.\end{aligned}$$

The followings are the proofs of A is both an ω_α-regular set and ω_α-normal set.

(1) If $x \in A_{[\alpha]}$, $B \in \omega_\alpha(\Omega)$ with $x \notin B_{[\alpha]} \subset A_{[\alpha]}$, then B must be one of $\{(0, 1/2, 1/2), (0, 1/2, 1), (0, 1, 1/2), (0, 1, 1)\}$. Thus, there are $P = (0, 1, 1) \in \eta_{\omega_\alpha}(x_\alpha)$, $Q = (1, 0, 0) \in \eta_{\omega_\alpha}(B)$, such that $A_{[\alpha]} \subset P_{[\alpha]} \cup Q_{[\alpha]}$.

(2) If $y \in A_{[\alpha]}$, $B \in \omega_\alpha(\Omega)$ with $y \notin B_{[\alpha]} \subset A_{[\alpha]}$, or If $z \in A_{[\alpha]}$, $B \in \omega_\alpha(\Omega)$ with $z \notin B_{[\alpha]} \subset A_{[\alpha]}$, then B must be one of $\{(1/2, 0, 0), (1, 0, 0)\}$. Thus, there are

$P = (1, 0, 0) \in \eta_{\omega_\alpha}(y_\alpha)$, $Q = (0, 1, 1) \in \eta_{\omega_\alpha}(B)$, such that $A_{[\alpha]} \subset P_{[\alpha]} \cup Q_{[\alpha]}$. Hence, A is an ω_α-regular set.

For each $B, C \in \Omega_\alpha(\Omega)$, which satisfy $B_{[\alpha]} \cap C_{[\alpha]} = \emptyset$. Then B and C must be one of $\{(1/2, 0, 0), (1, 0, 0)\}$ and $\{(0, 1/2, 1/2), (0, 1/2, 1), (0, 1/2, 1), (0, 1, 1/2), (0, 1, 1)\}$, respectively. Then there are $P = (0, 1, 1) \in \eta_{\omega_\alpha}(B)$ and $Q = (1, 0, 0) \in \eta_{\Omega_\alpha}(C)$, such that $A_{[\alpha]} \subset P_{[\alpha]} \cup Q_{[\alpha]}$. Hence, A is an ω_α-normal set.

However, as $y, z \in A_{[\alpha]}$, on one hand, there is not $P \in \eta_{\omega_\alpha}(y_\alpha)$, such that $z_\alpha \leq P$, on the other hand, there is not $Q \in \eta_{\omega_\alpha}(z_\alpha)$, such that $y_\alpha \leq Q$. Hence, A is not an ω_α-T_0 set. And therefore, A is not an ω_α-T_1 or ω_α-T_2 set, neither.

Example 4 Let $X = \{x, y\}, L = \{0, 1\}$, $A \in L^X$, with $A(x) = a, A(y) = b, A(z) = c$, we writer (a, b, c) instead of A. Take

$$\delta' = \{(0, 0, 0), (1, 0, 0), (0, 1, 1), (1, 0, 1), (0, 0, 1), (1, 1, 1)\}.$$

It easy to check δ is an L-fuzzy topology on L^X. Take $\alpha = 1$, $A = 1_X$, so $A_{[\alpha]} = \{x, y, z\} = X$. and clearly $\omega_\alpha(\Omega) = \delta'$.

For each $B, C \in \Omega_\alpha(\Omega)$, satisfy $B_{[\alpha]} \cap C_{[\alpha]} = \emptyset$. Then B must be $(1, 0, 0)$ and C must be $\{(0, 0, 1)$ or $(0, 1, 1)\}$. Then there are $P = (0, 1, 1) \in \eta_{\Omega_\alpha}(B)$ and $Q = (1, 0, 0) \in \eta_{\omega_\alpha}(C)$, such that $A_{[\alpha]} \subset P_{[\alpha]} \cup Q_{[\alpha]}$. Hence, A is an ω_α-normal set. Let $G = (0, 0, 1) \in \omega_\alpha(\Omega)$, $y \in A_{[\alpha]}$, and $y \notin G_{[\alpha]}$. Clearly, $\eta_{\omega_\alpha}(y_\alpha) = \{(1, 0, 0), (1, 0, 1), (0, 0, 1)\}$ and $\eta_{\omega_\alpha}(G) = \{(1, 0, 0)\}$. So there are not $P \in \eta_{\Omega_\alpha}(y_\alpha)$, and $Q \in \eta_{\Omega_\alpha}(G)$, such that $A_{[\alpha]} \subset P_{[\alpha]} \cup Q_{[\alpha]}$. Therefore A is not an Ω_α-regular set. $y, z \in A_{[\alpha]}$ and $\eta_{\omega_\alpha}(z_\alpha) = \{(0, 0, 0), (1, 0, 0)\}$. $\eta_{\omega_\alpha}(y_\alpha) = \{(0, 0, 0), (1, 0, 0), (1, 0, 1), (0, 0, 1)\}$. Then $\forall P \in \eta_{\omega_\alpha}(z_\alpha)$, $y_\alpha \not\leq P$. This implies A is not an ω_α-T_1 set.

Theorem 4.1 *Let $(L^X, \omega_\alpha(\Omega))$ be an ω_α-opos. Then $A \in \omega_\alpha(\Omega)$ is an ω_α-regular set iff $\forall x \in A_{[\alpha]}$, $P \in \eta_{\omega_\alpha}(x_\alpha)$ with $P_{[\alpha]} \subset A_{[\alpha]}$, there are $Q \in \eta_{\omega_\alpha}(x_\alpha)$, $R \in \eta_{\omega_\alpha}(P)$, such that $R_{[\alpha]} \cup Q_{[\alpha]} \supset A_{[\alpha]}$.*

Proof If $x \in A_{[\alpha]}$, $P \in \eta_{\Omega_\alpha}(x_\alpha)$ with $P_{[\alpha]} \subset A_{[\alpha]}$. So $P \in \omega_\alpha(\Omega)$, and $x \notin P_{[\alpha]} \subset A_{[\alpha]}$. Since $A \in \omega_\alpha(\Omega)$ is an ω_α-regular set, there are $Q \in \eta_{\omega_\alpha}(x_\alpha)$, $R \in \eta_{\omega_\alpha}(P)$, such that $A_{[\alpha]} \subset Q_{[\alpha]} \cup R_{[\alpha]}$.

Conversely. If $x \in A_{[\alpha]}$, $B \in \omega_\alpha(\Omega)$ with $x \notin B_{[\alpha]} \subset A_{[\alpha]}$. This means $B \in \eta_{\omega_\alpha}(x_\alpha)$. So there are $Q \in \eta_{\omega_\alpha}(x_\alpha)$, $R \in \eta_{\omega_\alpha}(B)$, such that $A_{[\alpha]} \subset R_{[\alpha]} \cup Q_{[\alpha]}$. Hence, A is an ω_α-regular set.

Theorem 4.2 *Let $(L^X, \omega_\alpha(\Omega))$ be an Ω_α-opos. Then $A \in \omega_\alpha(\Omega)$ is an Ω_α-normal set iff $\forall B \in \omega_\alpha(\Omega)$, with $B_{[\alpha]} \subset A_{[\alpha]}$, $\forall P \in \eta_{\omega_\alpha}(B)$ with $\emptyset \neq P_{[\alpha]} \subset A_{[\alpha]}$, there are $Q \in \eta_{\omega_\alpha}(B)$, $R \in \eta_{\omega_\alpha}(P)$, such that $Q_{[\alpha]} \cup R_{[\alpha]} = A_{[\alpha]}$.*

Theorem 4.3 *Let $(L^X, \omega_\alpha(\Omega))$ be an ω_α-opos, $(L^Y, \omega_{\alpha|Y}(\Omega_{|Y}))$ be its subspace, $\chi_Y \in \omega_\alpha(\Omega)$. If $A \in L^X$ is an ω_α-T_i set, $(i = 0, 1, 2)$, then $A_{|Y} \in L^Y$ is an $\omega_{\alpha|Y}$-T_i set too.*

Proof We only prove the case of $i = 3$. If $A \in L^X$ is an ω_α-T_3 set and $x \in (A_{|Y})_{[\alpha]}$, so $A \in \omega_\alpha(\Omega)$, $x_\alpha \in M^*(L^X)$, $x^*_\alpha \in M^*(L^X)$.

$\forall B \in \omega_{\alpha|Y}(\Omega_{|Y})$ with $x \notin B_{[\alpha]} \subset (A_{|Y})_{[\alpha]}$. By Theorem 2.1, there is $G \in \omega_\alpha(\Omega)$, such that $G_{|Y} = B$, Take $H = G \wedge \chi_Y$. Since $\chi_Y \in \omega_\alpha(\Omega)$, we get $H \in \omega_\alpha(\Omega)$and obviously, $x \notin G_{[\alpha]} = H_{[\alpha]} \subset A_{[\alpha]}$.

Since A is an ω_α-T_3 set, there are $P \in \eta_{\omega_\alpha}(x^*_\alpha)$, and $Q \in \eta_{\omega_\alpha}(H)$, such that $A_{[\alpha]} \subset P_{[\alpha]} \cup Q_{[\alpha]}$. Again, by Lemma 2.1, we have $P_{|Y}, Q_{|Y} \in \omega_{\alpha|Y}(\Omega_{|Y})$, and $P_{|Y} \in \eta_{\omega_\alpha}(x_\alpha)$, $Q_{|Y} \in \eta_{\omega_\alpha}(B)$. Clearly, $(A_{|Y})_{[\alpha]} = A_{[\alpha]} \cap Y \subset (P_{[\alpha]} \cup Q_{[\alpha]}) \cap Y = (P_{|Y})_{[\alpha]} \cup (Q_{|Y})_{[\alpha]}$. Therefore, $A_{|Y}$is an $\omega_{\alpha|Y}$-T_3 set.

Theorem 4.4 *Let (X, Δ) be an opos, $(L^X, \Omega_L(\Delta))$ be an L-opos induced by (X, Δ). $(L^X, \omega_\alpha(\Omega_L(\Delta)))$ be its ω_α-opos. Then $(L^X, \omega_\alpha(\Omega_L(\Delta)))$ is an ω_α-T_i space iff (X, Δ) is a T_i space, $(i = 3, 4)$.*

Proof We only prove the case of $i = 3$.

Suppose $(L^X, \omega_\alpha(\Omega_L(\Delta)))$ is an ω_α-T_3 space, $x \in X$, $E \in \Delta$, and $x \notin E$. So $\chi_E \in \omega_\alpha(\Omega_L(\Delta))$. Since 1_X is an ω_α-T_3 set, Then there are $P \in \eta_{\omega_\alpha}(x_\alpha)$, $Q \in \eta_{\Omega_\alpha}(\chi_E)$, such that $P_{[\alpha]} \cup Q_{[\alpha]} = X$. Hence $(P_{[\alpha]})' \cap (Q_{[\alpha]})' = \emptyset$, and $x \in (P_{[\alpha]})'$, $E \subset (Q_{[\alpha]})'$. This means $(P_{[\alpha]})'$, $(Q_{[\alpha]})'$ are the neighborhoods of x, and E, respectively. Therefore, (X, Δ) is a T_3 space.

Conversely. If (X, Δ) is an T_3 space. $x \in X$, $E \in \Delta$, with $x \notin E$, then there are two closed sets $U, V \in \Delta$, such that $x \in U'$, $E \subset V'$, and $U' \cap V' = \emptyset$. Take $P = \omega_\alpha(\chi_U)$, $Q = \omega_\alpha(\chi_V)$, so $P, Q \in \omega_\alpha(\Omega_L(\Delta))$. Since $(L^X, \Omega_L(\Delta))$ is induced by (X, Δ). We have $P \in \eta_{\omega_\alpha}(x_\alpha)$, $Q \in \eta_{\omega_\alpha}(\chi_E)$. Besides,

$$\begin{aligned} P_{[\alpha]} \cup Q_{[\alpha]} &= \omega_\alpha(\chi_U)_{[\alpha]} \cup \omega_\alpha(\chi_V)_{[\alpha]} = \omega_\alpha(\chi_U \vee \chi_V)_{[\alpha]} \\ &= \omega_\alpha(\chi_{U \cup V})_{[\alpha]} = \omega_\alpha(\chi_X)_{[\alpha]} = X. \end{aligned}$$

Therefore $(L^X, \omega_\alpha(\Omega_L(\Delta)))$ is an ω_α-T_3 space.

Theorem 4.5 *Let $(L^X, \omega^1_\alpha(\Omega_1))$, $(L^Y, \omega^2_\alpha(\Omega_2))$ be ω^1_α-opos, ω^2_α-opos, respectively. $f^\rightarrow : L^X \rightarrow L^Y$ is $(\omega^1_\alpha, \omega^2_\alpha)$-homeomorphism, $A \in L^X$ is an ω^1_α-T_i set, $(i = 3, 4)$, then $f^\rightarrow(A) \in L^Y$ is an ω^2_α-T_i set, too.*

Proof We only prove the case of $i = 3$.

If $y \in f^\rightarrow(A)_{[\alpha]} = f(A_{[\alpha]})$, $B \in \omega^2_\alpha(\Omega_2)$ with $y \notin B_{[\alpha]} \subset A_{[\alpha]}$. Since $f^\rightarrow$ is an $(\omega^1_\alpha, \omega^2_\alpha)$-homeomorphism, $f^\leftarrow(B) \in \omega^1_\alpha(\Omega_1)$ and there is $x \in A_{[\alpha]}$, such that $f(x) = y$, $x \notin f^\leftarrow(B)_{[\alpha]} \subset A_{[\alpha]}$. Because A is an ω^1_α-T_3 set, there are $P^1 \in \eta_{\omega^1_\alpha}(x_\alpha)$, $Q^1 \in \eta_{\omega^1_\alpha}(f^\leftarrow(B))$, such that $A_{[\alpha]} \subset P^1_{[\alpha]} \cup Q^1_{[\alpha]}$. Take $P^2 = f^\rightarrow(P^1)$, $Q^2 = f^\rightarrow(Q^1)$, we have $P^2 \in \eta_{\Omega^2_\alpha}(y_\alpha)$, $P^2 \in \eta_{\omega^2_\alpha}(B)$, and $f^\rightarrow(A)_{[\alpha]} \subset P^2_{[\alpha]} \cup Q^2_{[\alpha]}$. Therefore, $f^\rightarrow(A) \in L^Y$ is an ω^2_α-T_3 set.

Acknowledgments Thanks to the support by:1. National Science Foundation (No.10971125).2. Science Foundation of Guangdong Province (No. 01000004). 3. The construct program of the key discipline in Hunan University of Science and Engineering.

References

1. Bai, S.Z.: Q-convergence of nets and weak separation axioms in fuzzy lattices. Fuzzy Sets Syst. **88**(3), 379–386 (1997)
2. Bai, S.Z., Wang, W.L.: Fuzzy non-continuous mappings and fuzzy pre-semi-separation axioms. Fuzzy Sets Syst. **94**(2), 261–268 (1998)
3. El-Gayyar, M.K., Kerre, E.E., Ramadan, A.A.: On smooth topological spaces II: separation axioms. Fuzzy Sets Syst. **119**(3), 495–504 (2001)
4. El-Saady, K., Bakeir, M.Y.: Separation axioms in fuzzy topological ordered spaces. Fuzzy Sets Syst. **98**(2), 211–215 (1998)
5. Ghanim, M.H., Kerre, E.E., Mashhour, A.S.: Separation axioms, subspaces and sums in fuzzy topology. J. Math. Anal. Appl. **102**(1), 189–202 (1984)
6. Hutton, B., Reilly, I.L.: Separation axioms in fuzzy topological spaces. Fuzzy Sets Syst. **3**(1), 93–104 (1980)
7. Rodabaugh, S.E.: The Hausdorff separation axioms for fuzzy topological spaces. General Topology Appl. II **11**(2), 319–334 (1980)
8. Wuyts, Lowen R.: On separation axioms in fuzzy topological spaces, fuzzy neighborhood spaces and fuzzy uniform spaces. J. Math. Anal. Appl. **93**(4), 27–41 (1983)
9. Wang, G.J.: Separation axioms in topological molecular lattices. J. Math. Res. Exposition **12**(3), 1–16 (1983)
10. Chen, S.L., Dong, C.Q.: On L-fuzzy order-preserving operator spaces. Fuzzy Syst. Math. **16**(1), 36–41 (2002)
11. Huang, C.X.: ω-separation axioms on L-fuzzy order-preserving operator spaces. J. Math. **25**(4), 383–388 (2005)
12. Meng, G.W., Meng, H.: D_α-closed sets and their applications. Fuzzy Syst. Math. **17**(1), 24–27 (2003)
13. Sun, S.B., Meng, G.W.: D_α-separation axioms in L-fuzzy topological spaces. J. Inner Mongolia Normal Univ. **27**(2), 174–179 (2008)
14. Wu, X.Y., Xie, L.L.: Stratified order-preserving operator spaces in L-*opos*. Fuzzy Syst. Math. **25**(4), 79–82 (2011)
15. Wu, X.Y., Xie, L.L.: ω_α-connectedness in L-fuzzy Stratified order- preserving operator spaces. Fuzzy Syst. Math. **25**(5), 34–37 (2011)
16. Wang, G.J.: Theory of L-fuzzy topological spaces. The Press of Shanxi Normal University, Xi'an (1988)

The Application of FHCE Based on GA in EHVT

Ge Lin and Zhong-yuan Peng

Abstract Fuzzy Hierarchy Comprehensive Evaluation (FHCE) as a common evaluation method of the combination of qualitative analysis and quantitative analysis has been widely used in social life. At present, one of fuzzy comprehensive evaluation research difficulties is how to reasonably determine the weight of evaluation index. The main issue of analytic hierarchy process (AHP) in itself is to determine the each elements weight of judgment matrix which is artificially assigned, so it has highly subjective one-sidedness. In view of the above problems this paper attempts to propose a new model of FHCE, that is to structure judgment matrix according to interval scale of [1, 9] in AHP, and use the standard genetic algorithm (GA) to calculate each elements weight of judgment matrix.

Keywords Genetic algorithm · Analytic hierarchy process · Fuzzy comprehensive evaluation · Teaching evaluation.

1 Introduction

In recent years, our countrys higher vocational education has got rapid development, and has accounted for half of the entire higher education, but in large scale development at the same time has not been accompanied by the improvement of quality. At present, many higher vocational colleges are in the exploration to improve the quality of education, and to make scientific and objective evaluation on quality of teaching is one of effective ways to improve the quality of education. Recently, higher vocational teachers teaching quality evaluation encounters some problems, for example, most of the higher vocational colleges still use the evaluation index

G. Lin (✉) · Z. Peng
Maoming Vocational Technical College, Maoming 525000 , Guangdong,
People's Republic of China
e-mail: gelin0319@163.com

B.-Y. Cao and H. Nasseri (eds.), *Fuzzy Information & Engineering and Operations Research & Management*, Advances in Intelligent Systems and Computing 211,
DOI: 10.1007/978-3-642-38667-1_51, © Springer-Verlag Berlin Heidelberg 2014

system of ordinary colleges and universities. As the weights of this evaluation index are often worked by small number experts according to experience directly, they are lack of quantitative analysis. And this may has substantial deviation compared to actual situation, which directly influences the qualitative accuracy of evaluation results as well as quantitative accuracy, etc. To solve the above problem, this paper tries to present a comprehensive evaluation model combining GA, AHP and fuzzy comprehensive evaluation, and contact the practice of teaching quality evaluation to have some discussions.

2 Fuzzy Hierarchy Comprehensive Evaluation Based on Genetic Algorithm

2.1 Analytic Hierarchy Process

The Analytic Hierarchy Process (AHP) proposed by the United States Operations Research professor T L Saaty. It is a design-making method to make qualitative analysis and quantitative analysis on the basis of refer to the relevant elements of design-making problem is decomposed into hierarchies of objectives, guidelines, programs, and so on. The advantage of this method is qualitative and quantitative combined, with a highly logical, systematic and practical, which is an effective decision-making method aimed at the multi-level and multi-objective planning decision problem [1].

After set up a hierarchy model by using the AHP method, we can clearly see that the upper factors are determined by the underlying factors. Aim at a certain factor of the above level, to make pairwise comparisons in degrees of importance on factors which are subject to of this level, and then get the corresponding judgment matrix. Due to the complexity of the relationships among the criteria in the evaluation system, and the evaluation criteria are various, it is easy to cause the inconsistency in decision makers subjective judgment. In addition, the 1 to 9 scale of AHP method in some cases can not exactly reflect the proportional relationship between alternative schemes, and often leads judgment matrix to inconsistent. In order to better satisfy the consistency of judgment matrix, this paper takes interval scale, shown in Table 1.

Table 1 The Interval scale of judgment matrix and meaning

Scale value interval	Meaning
[1, 1]	a_i and a_j are equally important
[1, 3]	a_i is a little important than a_j
[3, 5]	a_i is obvious important than a_j
[5, 7]	a_i is strongly important than a_j
[7, 9]	a_i is extremely important than a_j
Reciprocal	If the ratio of importance of a_i and a_j is a_{ij}, then the ratio of importance of a_j and a_i is $\frac{1}{a_{ij}}$

According to the interval scale of [1, 9], we established judgment matrix through the comparison between every two indexes. This is a very crucial step. But the uncertain judgment matrix represented by element through using interval, solving its weight vector is complex, and the consistency of judgment matrix which is scaled by interval has great impact on solving the weight vectors. Better consistency can improve decision-making reliability. Only the weight vector gained in the premise that the judgment matrix has consistency that can it be used as the basis for decision-making, and to adapt to a variety of complex systems. For this reason, we need to improve the method which is used to construct judgment matrix, eliminate the inconsistency data fundamentally, cancel the consistency test, to simplify the calculation process, so that the calculation method is easier, the results are more accurate. Thus this paper introduced a genetic algorithm to calculate the weight and then structure judgment matrix. Using the genetic algorithm to calculate the weight of judgment matrix, we can minimize the subjectivity of decision-makers to meet the consistency of judgment matrix, to provide protection for the reasonable and fair judgment.

2.2 Genetic Algorithm

Genetic Algorithm (GA), is a kind of optimized search technology which is based on the biological evolution process, and develop on the basis of optimal save bad dead principle [2]. It is a kind of combination optimization algorithm which is adopted statistical heuristic search technology. When we use genetic algorithm to solve problem, each possible solution of this problem would be encoded into a "chromosome", namely the individual. Several individuals constitute the group. At the beginning of the genetic algorithm, it begins with a group of randomly generated initial population, according to the selected fitness function for each individual evaluation to make evaluation, according to certain probability to select individuals with larger fitness as parents to reproduce offspring. To make crossover, mutation on reproduced offspring to form a new generation of groups, and make re-evaluation, selection, crossover, mutation on this new generation of groups, and so on ad infinitum, so that the fitness of the best individual in the population and the average fitness continues to improve, until the fitness of the best individual to reach a certain limit or the fitness of the best individual and the average fitness of groups is no longer increase, then the iterative process convergence, the algorithm ends [3].

Thus, using genetic algorithm to calculate the index weights of each operator's design is as follows:

(1) The Generation of Initial Population

Suppose the judgment matrix that we want to obtain must satisfy:

$$\begin{pmatrix} [1,1] & [\frac{1}{5},\frac{1}{3}] & [\frac{1}{3},1] & [\frac{1}{7},\frac{1}{5}] \\ [3,5] & [1,1] & [1,3] & [\frac{1}{3},1] \\ [1,3] & [\frac{1}{3},1] & [1,1] & [\frac{1}{5},\frac{1}{3}] \\ [5,7] & [1,3] & [3,5] & [1,1] \end{pmatrix}$$

we can randomly generate several matrixes according to this requirement, make its elements satisfy the corresponding requirements, that is take the value within a predetermined range, such as matrix:

$$\begin{pmatrix} 1 & \frac{1}{3} & \frac{1}{2} & \frac{1}{5} \\ 3 & 1 & 3 & 1 \\ 2 & \frac{1}{3} & 1 & \frac{1}{4} \\ 5 & 1 & 4 & 1 \end{pmatrix}, \begin{pmatrix} 1 & \frac{1}{4} & \frac{1}{3} & \frac{1}{5} \\ 4 & 1 & 2 & \frac{1}{3} \\ 3 & \frac{1}{2} & 1 & \frac{1}{3} \\ 5 & 3 & 3 & 1 \end{pmatrix}, \begin{pmatrix} 1 & \frac{4}{15} & \frac{2}{3} & \frac{1}{5} \\ \frac{15}{4} & 1 & 3 & \frac{1}{4} \\ \frac{3}{2} & \frac{1}{3} & 1 & \frac{1}{5} \\ 5 & 4 & 5 & 1 \end{pmatrix} \cdots\cdots$$

They are all in line with the condition. And these matrixes can serve as the initial population to make operation by using GA.

(2) Coding

Judgment matrix is square matrix, and $a_{ii} = 1, a_{ij} = \frac{1}{a_{ji}}$, precisely because of judgment matrix special form of expression, in GA the coding of chromosome uses real number coding mode. Namely the judgment matrix

$$A = \begin{pmatrix} 1 & a_{12} & a_{13} & \cdots & a_{1m} \\ a_{21} & 1 & a_{23} & \cdots & a_{2m} \\ a_{31} & a_{32} & 1 & \cdots & a_{3m} \\ \vdots & \vdots & \vdots & \vdots & \vdots \\ a_{m1} & a_{m2} & a_{m3} & \cdots & 1 \end{pmatrix}$$

can be expressed as: $a_{12}a_{13} \ldots a_{1m}a_{23}a_{24} \ldots a_{2m} \ldots a_{(m-1)m}$.

(3) Selection Operator

In order to improve the efficiency of the algorithm, to ensure the effectiveness of individual which is produced by subsequent crossover and mutation operation, the algorithm takes random consistency ratio CR as the reference value. If generated initial matrixs consistency ratio $CR < 0.1$, this individual will be selected; otherwise, this individual would be eliminated. So it can ensure the judgment matrix satisfy consistency to maximum extent. In the selected effective individual, the value of consistency ratio is CR smaller, the greater the probability that the individual would be selected.

(4) Crossover Operator

Take any two matrices in the initial group, arbitrarily choosing a location in two individual genes, and then exchange, we can get two new individuals, such as: $B_1 = \begin{pmatrix} 1 & \frac{1}{4} & \frac{1}{3} & \frac{1}{6} \\ 4 & 1 & 2 & \frac{1}{3} \\ 3 & \frac{1}{2} & 1 & \frac{1}{5} \\ 6 & 3 & 5 & 1 \end{pmatrix}$ is expressed as: $\frac{1}{4}, \frac{1}{3}, \frac{1}{6}, 2, \frac{1}{3}, \frac{1}{5}$, $B_2 = \begin{pmatrix} 1 & \frac{1}{3} & \frac{2}{3} & \frac{3}{16} \\ 3 & 1 & \frac{5}{2} & \frac{2}{3} \\ \frac{3}{2} & \frac{2}{5} & 1 & \frac{3}{11} \\ \frac{16}{3} & \frac{3}{2} & \frac{11}{3} & 1 \end{pmatrix}$ is expressed as: $\frac{1}{3}, \frac{2}{3}, \frac{3}{16}, \frac{5}{2}, \frac{2}{3}, \frac{3}{11}$.

The crossover operation of B_1 and B_2 is as follows:

$$B_1 : \frac{1}{4}\ \frac{1}{3}\ \frac{1}{6} \vdots\ 2\ \frac{1}{3}\ \frac{1}{5} \qquad B_1{'} : \frac{1}{3}\ \frac{2}{3}\ \frac{3}{16} \vdots\ 2\ \frac{1}{3}\ \frac{1}{5}$$

$\updownarrow$ The location of cross $\Rightarrow$

$$B_2 : \frac{1}{3}\ \frac{2}{3}\ \frac{3}{16} \vdots\ \frac{5}{2}\ \frac{2}{3}\ \frac{3}{11} \qquad B_2{'} : \frac{1}{4}\ \frac{1}{3}\ \frac{1}{6} \vdots\ \frac{5}{2}\ \frac{2}{3}\ \frac{3}{11}$$

That the new individuals after cross are:

$$B_1{'} = \begin{pmatrix} 1 & \frac{1}{3} & \frac{2}{3} & \frac{3}{16} \\ 3 & 1 & 2 & \frac{1}{3} \\ \frac{3}{2} & \frac{1}{2} & 1 & \frac{1}{5} \\ \frac{16}{3} & 3 & 5 & 1 \end{pmatrix}, B_2{'} = \begin{pmatrix} 1 & \frac{1}{4} & \frac{1}{3} & \frac{1}{6} \\ 4 & 1 & \frac{5}{2} & \frac{2}{3} \\ 3 & \frac{2}{5} & 1 & \frac{3}{11} \\ 6 & \frac{3}{2} & \frac{11}{3} & 1 \end{pmatrix}$$

Such design of crossover operator can ensure that the individual produced after cross satisfies the requirements of judgment matrix on this level. Namely the elements in the matrix satisfy established value range, the search efficiency can be greatly improved in the feasible solution space, and it avoids the produce of a lot of individuals which will be eliminated.

(5) Mutation Operator

Arbitrary select a location in new individual gene segments, use a random number to substitute the elements on this position which in accordance with the value ranges. Namely: chromosome is: $\frac{1}{4}, \frac{1}{3}, \frac{1}{6}, 2, \frac{1}{3}, \frac{1}{5}$, the value range of the element in the fourth position is [1, 3], then inner [1, 3] randomly take a number (such as 2.5) to replace 2, the new chromosome after variation is: $\frac{1}{4}, \frac{1}{3}, \frac{1}{6}, 2.5, \frac{1}{3}, \frac{1}{5}$, and the corresponding matrix also satisfies the requirement.

2.3 Fuzzy Comprehensive Evaluation

Fuzzy Comprehensive Evaluation (FCE) provide some evaluation method to the actual comprehensive evaluation problems by the aid of the Fuzzy set theory [4]. To be specific, fuzzy comprehensive evaluation is a method which is based on fuzzy mathematics, using the principle of fuzzy relation synthesis to make quantification on some factors which have unclear border and are difficult to quantitative, and from a number of factors to make comprehensive evaluation on the membership grade condition of by evaluation of things. The comprehensive evaluation principle as follows: to determine evaluation index theory field $U = \{u_1, u_2, \ldots, u_m\}$, the fuzzy set $A = \{a_1, a_2, \ldots, a_m\}$ of U is the weight set, the rating assignment constitute a rating score set $V = \{v_1, v_2, \ldots, v_n\}$, R is the fuzzy relationship for $U \times V$, $\mu_{R(u_i,v_j)} = r_{ij}$ says the membership function of the index u_i on the comment grade v_i, namely aims at the proportion of the number that the index u_i is evaluated level v_i, R is evaluation fuzzy matrix, to make matrix multiplication on matrix A and matrix R, gain matrix $G = (g_1, g_2 \ldots, g_n)$, then introduce comment rating score matrix $V = \{v_1, v_2, \ldots, v_n\}$, make $S = \overline{G} \times V^T$, among them V^T is the transposed matrix of matrix V, S is the result of final evaluation [5].

2.4 The New Model of FHCE Based on GA (GA-FCE)

Without loss of general, the steps of mathematical modeling [6] for that fuzzy hierarchy comprehensive evaluation which is based on GA evaluating higher vocational teaching quality are as follows:

Step 1: determine evaluation index system and evaluation standards, and establish a hierarchical structure model.

Step 2: establish the judgment matrix of each index layer through choosing [1, 9] interval scale method by the AHP.

Step 3: calculate the weight value of each index in the judgment matrix by using GA, to establish weight matrix A.

Step 4: get a fuzzy evaluation matrix by statistics through the questionnaire.

Step 5: build a model $B = A \circ R$ to make single factor FCE (level 1 comprehensive evaluation and secondary comprehensive evaluation), get the normalized evaluation results; get comprehensive evaluation score by choosing the weighted average method to deal with the evaluation results [7, 8].

3 The Specific Operation of GA-FCE in HVTE

3.1 Determine Evaluation Index System and Evaluation Standards, to Establish a Hierarchical Structure Model

In this paper, we combine with the practical teaching of todays higher vocational colleges, consult relevant literature, on the basis following the four principles of objectivity and scientificalness, orientation and feasibility, structure the evaluation index system [9] Table 2.

Table 2 Quality evaluation index system of higher vocational teachers' Teaching [10]

Level 1 index	Secondary evaluation	Evaluation standard
Professional ethics X_1	Impart knowledge and educate people x_{11}	Words and deeds, whether to conduct students in education
	Stand and deliver x_{12}	Whether the teacher have uncivilized behavior in the classroom and training spaces
	Passionate in one's job x_{13}	No late and leave early, do not do things that have nothing to do with teaching in the classroom
Theory teaching ability X_2	Teaching organization x_{21}	Scientific and rational organization of classroom teaching, better interact with students
	Teaching content x_{22}	The theoretical lectures moderate enough, combined with the production of the actual implementation to teach
	Teaching method x_{23}	Classroom teaching easily understood, handle important and difficult points properly teach students in accordance with their aptitude
	Teaching effect x_{24}	Students can better master the classroom knowledge, reap is big
Practical teaching ability X_3	Teachers' professional skills x_{31}	The teacher has strong professional practice skills
	Practice activity organization x_{32}	Melt "teaching, learning and doing" as a whole, and strengthen training students' ability
	Practice process guidance x_{33}	Able to timely and effective manner to guide students in practical activities
	Skills training effect x_{34}	Take employment as the guidance, effective training post vocational ability

3.2 Establish the Judgment Matrix of Each Index Layer

According to the interval scale of [1, 9], we established judgment matrix through the comparison between every two indexes such as Table 3.

Table 3 The Judgment matrix of each index layer

U	X_1	X_2	X_3		C_1	x_{11}	x_{12}	$x13$	
X_1	[1, 1]	$[\frac{1}{5}, \frac{1}{3}]$	$[\frac{1}{3}, 1]$		x_{11}	[1, 1]	[3, 5]	[1, 3]	
X_2	[3, 5]	[1, 1]	[1, 3]		x_{12}	$[\frac{1}{5}, \frac{1}{3}]$	[1, 1]	$[\frac{1}{3}, 1]$	
X_3	[1, 3]	$[\frac{1}{3}, 1]$	[1, 1]		x_{13}	$[\frac{1}{3}, 1]$	[1, 3]	[1, 1]	
C_2	x_{21}	x_{22}	x_{23}	x_{24}	C_3	x_{31}	x_{32}	x_{33}	x_{34}
x_{21}	[1, 1]	$[\frac{1}{5}, \frac{1}{3}]$	$[\frac{1}{3}, 1]$	$[\frac{1}{7}, \frac{1}{5}]$	x_{31}	[1, 1]	[1, 3]	[3, 5]	$[\frac{1}{3}, 1]$
x_{22}	[3, 5]	[1, 1]	[1, 3]	$[\frac{1}{3}, 1]$	x_{33}	$[\frac{1}{3}, 1]$	[1, 1]	[1, 3]	$[\frac{1}{5}, \frac{1}{3}]$
x_{23}	[1, 3]	$[\frac{1}{3}, 1]$	[1, 1]	$[\frac{1}{5}, \frac{1}{3}]$	x_{23}	$[\frac{1}{5}, \frac{1}{3}]$	$[\frac{1}{3}, 1]$	[1, 1]	$[\frac{1}{7}, \frac{1}{5}]$
x_{24}	[5, 7]	[1, 3]	[3, 5]	[1, 1]	x_{34}	[1, 3]	[3, 5]	[5, 7]	[1, 1]

Table 4 The judgment matrix, weight and consistency of each Index

U	X_1	X_2	X_3		ω	C_1	x_{11}	x_{12}	x_{13}		ω
X_1	1	$\frac{1}{5}$	$\frac{1}{3}$		0.1120	x_{11}	1	$\frac{9}{2}$	$\frac{3}{2}$		0.529
X_2	5	1	$\frac{3}{2}$		0.5402	x_{12}	$\frac{2}{9}$	1	$\frac{1}{3}$		0.1176
X_3	3	$\frac{2}{3}$	1		0.3478	x_{13}	$\frac{2}{3}$	3	1		0.3529
$\lambda_{max} = 3.0012, C.R. = 0.0010 < 0.1$						$\lambda_{max} = 3.0000, C.R. = 0.0000 < 0.1$					
C_2	x_{21}	x_{22}	x_{23}	x_{24}	ω	C_3	x_{31}	x_{32}	x_{33}	x_{34}	ω
x_{21}	1	$\frac{1}{3}$	$\frac{2}{3}$	$\frac{3}{16}$	0.0914	x_{31}	1	$\frac{17}{3}$	$\frac{13}{3}$	$\frac{2}{3}$	0.322
x_{22}	3	1	2	$\frac{2}{3}$	0.3165	x_{32}	$\frac{3}{7}$	1	$\frac{3}{2}$	$\frac{3}{11}$	0.129
x_{23}	$\frac{3}{2}$	$\frac{1}{2}$	1	$\frac{3}{1}$	0.1230	x_{33}	$\frac{3}{13}$	$\frac{2}{3}$	1	$\frac{3}{16}$	0.082
x_{24}	$\frac{16}{3}$	$\frac{3}{2}$	$\frac{11}{3}$	1	0.4690	x_{34}	$\frac{3}{2}$	$\frac{13}{3}$	$\frac{16}{3}$	1	0.465
$\lambda_{max} = 4.0044, C.R. = 0.0016 < 0.1$						$\lambda_{max} = 4.0055, C.R. = 0.0020 < 0.1$					

3.3 Calculate the Weight Value of Each Index in the Judgment Matrix by Using GA, and do a Consistency Test

To make operation on judgment matrix by using GA, we can get the judgment matrix Table 4 that which has good consistency and its weight value is determined [11].

3.4 Structure Fuzzy Evaluation Matrix

Student questionnaire: we randomly select 100 students in participating class of this teacher for the teachers teaching evaluation respectively. We collect the results and make finishing statistical quantification treatment on them, combine with factor weight of evaluation index by using GA, and construct FCE Table 5 of the teaching quality evaluation.

Table 5 The FCE table of the teaching quality evaluation

Level 1 index	Weight	Secondary evaluation	Weight	Evaluation results				
				A	*B*	*C*	*D*	*E*
Professional ethics	0.1120	Impart knowledge and educate people	0.5294	0.30	0.53	0.14	0.03	0
		Stand and deliver	0.1176	0.76	0.20	0.04	0	0
		Passionate in one's job	0.3529	0.72	0.25	0.02	0.01	0
Theory teaching ability	0.5402	Teaching organization	0.0914	0.35	0.42	0.20	0.02	0.01
		Teaching content	0.3165	0.23	0.36	0.28	0.08	0.05
		Teaching method	0.1230	0.34	0.32	0.10	0.18	0.06
		Teaching effect	0.4690	0.26	0.42	0.20	0.05	0.07
Practical teaching ability	0.3478	Teachers' professional skills	0.3226	0.65	0.24	0.10	0.01	0
		Practice activity organization	0.1294	0.46	0.22	0.15	0.11	0.06
		Practice process guidance	0.0826	0.43	0.28	0.04	0.09	0.06
		Skills training effect	0.4654	0.38	0.26	0.24	0.04	0.08

3.5 FCE and Evaluation Grade Based on Model $B = A \circ R$

3.5.1 Level 1 Fuzzy Comprehensive Evaluation

(1) The single factor evaluation result $\overline{P_1}$ of professional ethics layer is:

$$\overline{P}_1 = \overline{A}_1 \circ \overline{R}_1 = \begin{pmatrix} 0.5294 & 0.1176 & 0.3529 \end{pmatrix} \circ \begin{pmatrix} 0.30 & 0.53 & 0.14 & 0.03 & 0 \\ 0.76 & 0.20 & 0.04 & 0 & 0 \\ 0.72 & 0.25 & 0.02 & 0.01 & 0 \end{pmatrix}$$

$$= \begin{pmatrix} 0.5023 & 0.3923 & 0.0859 & 0.0194 & 0 \end{pmatrix}$$

(2) The single factor evaluation result $\overline{P_2}$ of theory teaching ability layer is:

$$\overline{P}_2 = \overline{A}_2 \circ \overline{R}_2 = \begin{pmatrix} 0.0914 & 0.3165 & 0.1230 & 0.4690 \end{pmatrix} \circ \begin{pmatrix} 0.35 & 0.42 & 0.20 & 0.02 & 0.01 \\ 0.23 & 0.36 & 0.28 & 0.08 & 0.05 \\ 0.34 & 0.32 & 0.10 & 0.18 & 0.06 \\ 0.26 & 0.42 & 0.20 & 0.05 & 0.07 \end{pmatrix}$$

$$= \begin{pmatrix} 0.2685 & 0.3887 & 0.2130 & 0.0727 & 0.0569 \end{pmatrix}$$

(3) The single factor evaluation result $\overline{P_3}$ of practice teaching ability layer is:

$$\overline{P}_3 = \overline{A}_3 \circ \overline{R}_3 = \begin{pmatrix} 0.3226 & 0.1294 & 0.0826 & 0.4654 \end{pmatrix} \circ \begin{pmatrix} 0.65 & 0.24 & 0.10 & 0.01 & 0 \\ 0.46 & 0.22 & 0.15 & 0.11 & 0.06 \\ 0.43 & 0.28 & 0.04 & 0.09 & 0.06 \\ 0.38 & 0.26 & 0.24 & 0.04 & 0.08 \end{pmatrix}$$

$$= \begin{pmatrix} 0.4816 & 0.25 & 0.1667 & 0.0435 & 0.05 \end{pmatrix}$$

3.5.2 The Secondary Fuzzy Comprehensive Evaluation

Known by 3.5.1, the result $\overline{G}$ of level 1 fuzzy comprehensive evaluation is:

$$\overline{G} = \overline{A} \circ \overline{P} = \begin{pmatrix} 0.1120 & 0.5402 & 0.3478 \end{pmatrix} \circ \begin{pmatrix} 0.5023 & 0.3923 & 0.0859 & 0.0194 & 0 \\ 0.2685 & 0.3887 & 0.2130 & 0.0727 & 0.0569 \\ 0.4816 & 0.25 & 0.1667 & 0.0435 & 0.05 \end{pmatrix}$$

$$= \begin{pmatrix} 0.3688 & 0.3409 & 0.1827 & 0.0566 & 0.0481 \end{pmatrix}$$

3.5.3 Normalized Processing

Because

Table 6 Assignment Table of Evaluation Grade

Grade	Assignment
A(excellent)	[90, 100]
B(good)	[80, 89]
C(fair)	[70, 79]
D(pass)	[60, 69]
E(fail)	[0, 59]

$$0.3688 + 0.3409 + 0.1827 + 0.0566 + 0.0481 = 0.9971 \neq$$

$1\overline{G}$ must be carried out normalized processing, the processing result is:

$$\overline{G} = (0.3699\ 0.3419\ 0.1832\ 0.0568\ 0.0482)$$

3.6 The Processing of Evaluation Results

In this paper we use the weighted average method to process evaluation result. When we using the weighted average method, first of all we make quantification on evaluation set, namely when each level of the evaluation set assigned to value, we usually adopt expert scoring method. Here the five level of evaluation set assignment shown in Table 6.

To measure by using the average value of each interval length in every level agv, so the evaluation rating quantification matrix, thus the score obtained in the teaching evaluation system is:

$$S = \bar{G} \cdot V = (0.3699\ 0.3419\ 0.1832\ 0.0568\ 0.0482) \cdot (95\ 85\ 75\ 65\ 30)^T = 83.0800$$

Therefore, the score of this teachers teaching evaluation is 83.08 points, evaluation level is good.

4 Conclusion

This paper studies the GA-FCE new model to calculate the weight of each factor in AHP by using the standard genetic algorithm, and apply it to the analysis of higher vocational teaching evaluation. In the study we established evaluation index system that including professional ethics, theoretical teaching ability and practice teaching ability. We use actual data to take for instance analysis on this index system by using FCE method which is based on GA, and we achieve the desired effect. However, teaching evaluation is a dynamic process, it has the characteristic of instability and so on, therefore, for dynamic evolution of teaching evaluation, it needs to do more in-depth.

Acknowledgments Thanks to My tutor Prof Yu-bin Zhong and his Keynote Teaching Research Project of Guangzhou University.

References

1. Wang, Q.: The application of analytic hierarchy process in the comprehensive evaluation of university teachers.Xian: Acad. J. Xian Eng. Inst, 337–341(1998)
2. Zhong, Y., Mi, H., Chen, H.: Design and optimization algorithm of VCS under security protection.IEEE of International Conference on Web Information Systems and Mining, pp. 552–556(2009)
3. Xiaoping, W., Liming. The genetic algorithm theory, application and software implementation. Xian: Xi 'an Jiao Tong University Press, pp. 1–66 (2002)
4. Zhong, Y.: The FHSE Model of software system for Synthetic evaluating engerprising. J. Guangzhou Univ. **4**(4), 316–320 (2005)
5. Hu, Z.: The design and implementation of teachers teaching quality evaluation system which is based on the fuzzy comprehensive evaluation. Chongqing: The Engineering Degree Dissertation of Chongqing University, pp. 13–20(2006)
6. Zhong, Y.: The structures and expressions of power group. Fuzzy Inf. Eng 2(2) (2010)
7. Qin, S.: The principle and application of comprehensive evaluation. Electronics Industry Press, Beijing (2003)
8. Zhong, Y.: The thought of fuzzy mathematics modeling is infiltrated in MSMT. Adv. Intell Soft Comput **2**(62), 233–242 (2009)
9. Zhong, Y.: The design of a controller in Fuzzy PETRINET. Fuzzy Optim. Decis. Making, 7(4) (2008)
10. Chen B.: The development of higher vocational teaching quality evaluation and assessment system. Sichuan: Southwest University of Finance and Economics on Press, pp. 1–72(2009)
11. Zhong, Y.: The optimization for location for large commoditys regional distribution center, pp. 969–979. Advance in Soft, Computer (2007).

Theory and Practice of Cooperative Learning in Mathematical Modeling Teaching

Yu-bin Zhong, Yi-ming Lei, Xu-bin Wang, Lei Yang and Gen-hong Lin

Abstract Against the problems such as that teamwork of mathematical modeling is not strong enough, based on the characteristics and inherent laws of mathematical modeling, combined with the characteristics of the students, this paper presents several strategies. To the basic theory of cooperative learning as a guide, starting from the connotation of cooperative learning of mathematical modeling, analyzing the main factors that impacting the cooperative learning of mathematical modeling, this paper makes several teaching strategies of the cooperative learning of mathematical modeling for the research about the teaching strategies that the cooperative learning improves the modeling interest in learning, academic performance and learning ability. Empirical students have shown that the implementation of cooperative learning has a positive impact on improving student's modeling interest in learning, academic performance and learning ability and so on. In this way, we not only expand the theory of cooperative learning from high school to college, but also achieve the combine of theory and practice. It provides approach to improve student's comprehensive ability.

Keywords Mathematical modeling · Membership · Fuzzy clustering · Cooperative learning · Teaching mode.

1 Introduction

Under the guidance of teachers, Cooperative learning is group activities as the main body, peer cooperation and mutual assistance, team competition as the main forms. It has played an active role in the teaching of various disciplines to make the learning

Y. Zhong (✉) · Y. Lei · X. Wang · L. Yang · G. Lin
School of Mathematics and Information Sciences, Guangzhou University,
Guangzhou 510006, People's of Republic China
e-mail: Zhong_yb@163.com

B.-Y. Cao and H. Nasseri (eds.), *Fuzzy Information & Engineering and Operations Research & Management*, Advances in Intelligent Systems and Computing 211,
DOI: 10.1007/978-3-642-38667-1_52,

process a "re-creation"process under the guidance of the teacher [1, 2]. There are many relevant studies. As education critic Ellis and Fouts have asserted in the book, "co-teaching, if not the largest education reform, then, it is at least one of the largest" [3, 4]. There are few studies of Cooperative learning in the practice and teaching of mathematical modeling. To address this issue, starting from the concept, pattern and the theoretical basis of cooperative learning, this paper points out the purpose and significance of cooperative learning, explores the characteristics of cooperative learning, reveals the content of cooperative learning in the Mathematical Modeling course, explores influencing factors of cooperation learning way. From five aspects, the group choreography, timing, the choice and build of the problem, the promotion and protection of the cooperation and exchange, the role of teachers, this paper explores cooperative learning of undergraduate mathematical courses, respectively, to propose three pronged strategy of cooperative learning in improving college students interest, in mathematical modeling results and in the ability. So,we Introduce Mathematical Modeling Teaching Model (MMTM).

2 Principles and Steps of MMTM

2.1 Principles of MMTM

Cooperative learning is usually based on the study group as the basic form, systematically uses the interaction between the dynamic factors in teaching to promote students' active learning, uses group performance as evaluation criteria, team members together achieving the goal of teaching activities [5–7] . Higher mathematical modeling cooperative learning not only has inherent characteristics of cooperative learning, but also gives a new connotations.

2.1.1 Connotation of Cooperative Learning

In general, the focus of cooperative learning is interactive collaboration between student peers.The students cooperation is the proper meaning of cooperative learning. The cooperative learning of mathematical modeling is not only built on the basis of students cooperation, and has more for a wide range of areas of cooperation. First, the teachers cooperation require the teachers' behavior is no longer the individual behavior, but mutual respect between teachers, discussion, the organic integration, jointly improvement. This is also determined by mathematical modeling of the institutions of higher learning and teaching requirements of mathematical modeling. For now, in institutions of higher learning, mathematical modeling plays as a basic course aiming at specialized courses, which indicates that mathematical modeling courses and professional courses have the same educational object, the similar educational goal, the related educational content. This determines the possibility and necessity of cooperation between the mathematical modeling teachers, subjects and disciplines. Mathematical modeling teachers must participate in the students' application ability analysis, the development of applied learning projects, the integration of mathematical modeling course, the overall instructional design, and strive to do interdisciplinary communication and to improve students' application ability. Second, the teacher-student cooperation. Dialyses a new view of the teachers and students, and a new teaching philosophy, which means that equal dialogue and the initiative to participate in and build their own in the teaching of mathematical modeling is not only a teaching and learning activities, but also a educational context and spiritual atmosphere of a good spirit of worship of technology and focus on humanistic care filling between teachers and students. Teachers and students experience the meaning and significance of the value of mathematical modeling in the application of activities.

2.1.2 Cooperative Learning Penetrating the Modern Vocation

Faced with the challenges of science, technology, economic, social development, vocational education is undergoing a process of conversion from traditional mechanical Pedagogy to the modern evolutionary pedagogy. In mathematical modeling class cooperative learning is not only a simple transplantation of cooperative learning theory in the past, but also the emphasis on penetration of mathematical modeling from the reality of the teaching of vocational education mathematical modeling. It brings about the organic integration of the modern vocational learning theory and cooperative learning theory, constitutes a new type of vocational education strategiesin teaching and learning [8–10].

2.1.3 Professional Scene of Cooperative Learning

Cooperative learning is a teaching group activities as the main activities, which is often manifested in the mutual support among fellow. To a certain extent Mathematical modeling class cooperative learning is to expand the function of the professional learning team, combining mathematical modeling cognitive learning process with professional action, combining individual action and learning process of students with action space requested by outside. By simulation or the real work study group to improve the students' role capability, in simulated or real professional environment, integrate the cooperative learning of many elements to the teaching elements, and ultimately achieve the goals of mathematical modeling teaching.

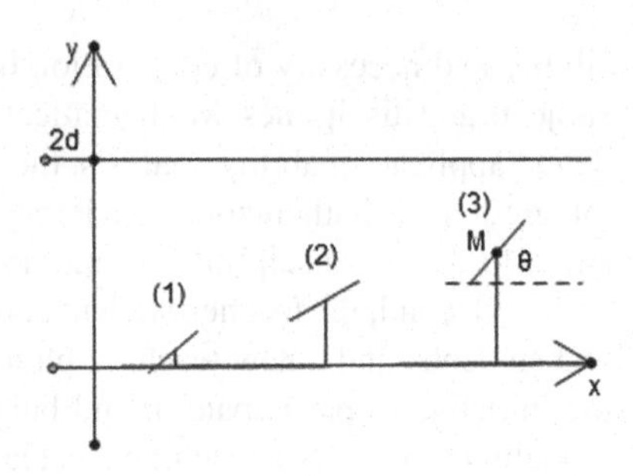

Fig. 1 Buffon random cast needle test

2.2 Cooperative Learning Teaching Model Steps

2.2.1 Combined with Professional, Clever Introduction

With practice, from the perspective of mathematical modeling applications and professional applications, we introduce the contents to be learned. For example, Buffon random cast needle test. assuming that there are numerous parallel lines (pairwise distance 2d) in the plane, Q: what is the probability of a needle with either a straight line intersecting? Let students draw on white paper enough invest needle, and measured with a ruler and a protractor, record each needle every random test results, and finally each group gives a lab report. If $l sin\theta > x$ (Set the length of the needle 2l),needle and parallel lines intersect (such as Fig. 1 (1)), otherwise do not intersect (such as (2)) (Fig. 2).

2.2.2 Explore Independently with the Problem

Taking advantage of students' interest and concerns during the introduction, we can guide students to self-learn the related content, while decomposing the major and difficult matter to design a self-thinking title, for students in the process self-thinking, to solve the shortcomings of teachers fast experiencing and fast abstracting instead of students' own. Students themselves complete the learning process from the initial introduction of the concept and expanding to establishing, making the mathematical

Fig. 2 point random

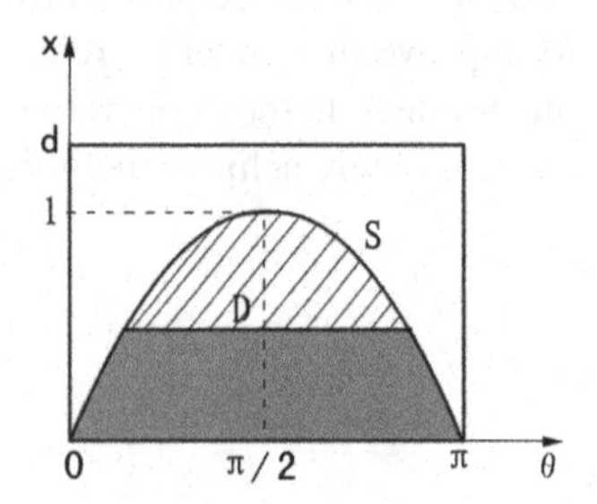

modeling process has flesh and blood, with rich connotation [11, 12]. Questions are originally raised by the teachers, after a period of time, by teachers and students co-sponsored, entirely by students, and that noting on their name and the name of the group after title can further stimulate the enthusiasm of students.

2.2.3 Cooperation and Discussion aiming at Questions

Mathematical modeling teaching will not be able to lead a thinking activity without questions. The practice shows that the set of one or two questions of moderate difficulty is significant to stimulate students' desire for knowledge, and fully mobilize the enthusiasm of the students' learning, lead students' to participate in class thinking [13]. Discussion and cooperation aiming at the problem not only can find the deficiencies of understanding and errors, and can also work together to construct a new understanding.

For example, have finished the learning of the cross-sectional area formula of prism $\sqrt{S_m} = \frac{\sqrt{S_u}+\sqrt{S_d}}{2}$, we can guide students to think about if the cross-section is not in the middle section, what is the relationship between $\sqrt{S_s}$, $\sqrt{S_u}$, $\sqrt{S_d}$. We can guide the students to observe the structure of the above formula. It's not difficult to find that this formula is very similar to the mid-point coordinate formula in the analytic geometry, while midpoint coordinates formula is the special case of the given fixed ratio coordinate formula, so we are interested whether $\sqrt{S_s}$, $\sqrt{S_u}$, $\sqrt{S_d}$ have the similar formula. Assume that the height of the prism is divided into two sections from top to bottom by the cross-section. Three situations (1) 1: 2, (2) 3: 2, (3) 5: 3, let different study groups analyse the cross-section area formula and cooperate to discuss one result of three. It's easy to get the following results:

$$(1)\sqrt{S_s} = \frac{\sqrt{S_u} + \frac{1}{2}\sqrt{S_d}}{1+\frac{1}{2}} (2)\sqrt{S_s} = \frac{\sqrt{S_u} + \frac{3}{2}\sqrt{S_d}}{1+\frac{3}{2}} (3)\sqrt{S_s} = \frac{\sqrt{S_u} + \frac{5}{3}\sqrt{S_d}}{1+\frac{5}{3}}.$$

The students were surprised to find that this formula are very similar to the fixed ratio coordinate formula, which prove the conclusion: if the height of the prism is divided into two sections from top to bottom by the cross-section and the ratio is , then we have $\sqrt{S_s} = \frac{\sqrt{S_u}+\lambda\sqrt{S_d}}{1+\lambda}$. Under the guidance of teachers, students found and proved a general formula by their own. Thus by the students' "re-creation "process, their desire to discover was enhanced and the learning of students are in the state of active cooperation, experiencing the joy of discovery. The enthusiasm to learn mathematical modeling is greatly enhanced.

In cooperative learning, the chairperson of cluster report the learning of his or her group, and assess course content, group members can add. Argument are allowed if groups have different views. Team members who are responsible for the interpretation can present their own thinking, monitoring and adjustment process by explaining their understanding and reasoning process to rivals. It not only avails operator to organize their own thinking and cognitive monitoringis more clearly, more effectively, and also makes students can more clearly observe his thinking and monitoring process. Thus students can evaluate and learn from others' effective part, and then through direct feedback so that students can recognize their own inadequacies, and under the guidance and help of rivals to do adjustment and supplement timely, making the cognitive structure completely.

2.2.4 Focus on Igration, Variations Innovation

More simply, the migration is to be able to use the things learned in new situations. The one drawback of the traditional teachings: students in classroom hear more but do less and have less time to solve the problem with the knowledge learned, can not be flexibility in the use of the knowledge learned. Fundamentally learning not just means to know a part of knowledge, but to use, enable it to function in the new scenario. To learn but not use are equal to not learn. Learn but not able to use are not very different from not learn. After students have obtained preliminary concepts, skills, and did some basic exercises, exchanged and discussed with partners, deepen the understanding of knowledge, got new ideas and methods, the part consolidates or expands the contents learned through practice and training. Then timely we shall guide the students to summarize the general conclusions of the new knowledge, new ideas, and organize into the knowledge system of mathematical modeling, and translate knowledge into the ability to cultivate students' good habits, study habits, innovation and practical ability. For example, we have state equation when learning queuing theory

$$\begin{cases} \frac{dP_n(t)}{dt} = -\lambda P_{n-1}(t) + \lambda P_n(t), \\ P_n(0) = 0 (n \geq 1), \end{cases}$$

$P_n(t)$ indicates possibility of state that the amount of customers is n at moment t.

Generally we have $P_n(t) = \frac{(\lambda t)^n}{n!} e^{-\lambda t}, (n = 0, 1, \ldots, t > 0)$ follow Poisson distribution.

Thus it's available to get variants of single server queuing model:

Variant : M/M/1 (infinite source, customers arrive independently, the law of reaching follows Poisson distribution: single desk, unlimited length of queue, first come first served, service time are independent and follow negative exponential distribution), there are queuing state equation:

$$\begin{cases} \dfrac{dP_0(t)}{dt} = -\lambda P_1(t) + \lambda P_n(t), \\ \dfrac{dP_n(t)}{dt} = -\lambda P_{n-1}(t) + \mu P_{n-1}(t) - (\lambda + \mu) P_n(t) (n > 1), \end{cases}$$

λ is the argument of Poisson distribution and μ is the one of negative exponential distribution.

3 An Empirical Study of Cooperative Learning

3.1 Experimental Research Object

The selection of our research objects applied a parallel selection to choose the lateral school classes, a longitudinal selection to choose two different examinations, using two sets of comparison. In 2010 Class A1 was the common class and Class A2 the experimental class. We applied cooperative learning mode to do the teaching of mathematical modeling of experimental class ,while common class the traditional teaching model. Each class and each period were given the same progress of teaching and the teaching contents.

3.2 Survey Method, Survey Results and Analysis

In order to seek the breakthrough in the research, and how to have a more targeted research, before the research of mathematical modeling cooperative learning model, I conducted a questionnaire survey on students' interest in learning and learning conditions of mathematical contest, and applied percentage to indicate the degree of membership.

The survey issued 120 questionnaires, 115 statistical, completed in March 2010, before the starting of experimental research. The survey of students' interest in learning mathematical modeling mainly applied the form of questionnaire.

Statistical methods: questionnaire as self-report questionnaire, of mainly 16 evaluation questions, sets the scores of a variety of answers for each question, cumulates total scores of students in the 16 issues, as the characteristic of students' interests and attitudes of studying mathematical modeling. In these 16 questions, 10 "affirmative" narrative, 6 "negation" narrative, 5 alternative answers to each question for each answer describing in a positive or negative degree of measuring students' learning interest and attitude of learning mathematical modeling, A- agree very much (2 points); B - agree (1 point); C - can not tell clear (0 points); D- do not agree with (-1 point); E - strongly disagree (-2 points).

The statistical results: Fig. 3 data show that: the experimental classes and common classes were 42.5 % and 40.2 % of the students showing low interest in learning the theme of the Mathematical Contest in Modeling, experimental classes and common classes of no more than a third of the students showing the interest.Therefore, the students' weak knowledge base have a direct impact on students' interest, giving out obstacles in education and teaching (Fig. 4).

First of all, we speculated that students' achievement follow a normal distribution, in order to prove this, we applied Jarque-Bera nonparametric normality test to students' achievement:

Calculate the test statistic: $JB = \frac{n}{6}(s^2 + \frac{(k-3)^2}{4})$ s is sample unbiased variance. k is sample kurtosis. Have known the critical value of JB statistic is CV = 5.0663, when

experimental class				common class			
options	selection numbers	membership	scores	options	selection numbers	membership	scores
A	48	42%	96	A	45	40.20%	90
B	29	25.70%	29	B	30	26.80%	30
C	21	18.60%	0	C	22	19.60%	0
D	8	7.10%	-8	D	7	6.30%	-7
E	7	6.20%	-14	E	8	7.10%	-16

Fig. 3 questionnaire results

class	number	mean	standard deviation
experimental class	56	0.5307	0.1776
common class	56	0.5150	0.1695

Fig. 4 pretest data

the statistics is greater than the critical value to reject the null hypothesis, indicating that the sample is not normal otherwise accept the null hypothesis. Calculate students' midterm performance of Class A1 to obtain that JB statistic is $1.4264 < CV$. Corresponding significant probability p = 0.3729 is much larger than the significant level of 0.05, indicating that the A1 class students' achievement follows a normal distribution. The calculation of A2 classes have $JB = 0.7016 < CV$, and $p = 0.50 > 0.05$, indicating that the A2 class results follow a normal distribution, too. With application of this method we tested that the results of samples of all classes meet the normal conditions.

In order to make sure the beginning of the experimental classes and common classes have similar base, we have done testing of questionnaire, and done the analysis of the difference of two normal populations: Under the situation that mean and variance are both unknown,I appied F-test to the data of two samples: $H_0 : \sigma_1^2 = \sigma_2^2$, $H_1 : \sigma_1^2 \neq \sigma_2^2$. Introduce statistics:$F = \frac{s_1^2}{s_2^2}$, s_1^2 and s_2^2 are unbiased variance of samples.

From results, we can conclude that $F = 1.0979 < F_0.05(60, 60) = 1.65 F_0.05$ (56, 56) , corresponding significance probability p=0.7315 so we accept the assumption H_0 . We can accept that the corresponding overall variance of two samples have no significant difference.

After the variance homogeneity test, we test the difference of the two samples' mathematical expectation. Have known that t test is always used to determine the differences of means when variances are equal. We set assumptions: $H_0 : \mu_1 = \mu_2$, $H_1 : \mu_1 \neq \mu_2, \sigma_1^2 = \sigma_2^2$. Introduce t statistic: $t = \frac{\overline{x_1} - \overline{x_2}}{\sqrt{\frac{s_1^2}{n_1} + \frac{s_2^2}{n_2}}}$, $\overline{x}$ is mean value. s^2 is unbiased variance. n is sample size.

After calculation we can conclude that t = 0.4791 is less than the critical value, given significant probability $p = 0.6328 > 0.05$ accepts null hypothesis H_0 . We can consider that the mathematical expectations of the two samples did not differ much.

By the above two results we can assume that there are no significant difference in the initial mathematical basis of the common classes and the experimental classes .Standard deviation and the mean of the two classes are as follows:

4 Analysis of Results

4.1 Composition and Selection of the Measured Objects

The objects of this study are two classes' students, the selection of study applied parallel to select the lateral classes, longitudinal to choose two examinations which are at different times, using two sets for comparison. Therefore the measured objects are selected from the vertical angle randomly, with its representative.

4.2 Analysis of Students' Achievement Data

4.2.1 Common and Experimental Classes Transverse Statistics

Results description and data analysis: before the experimental study, in the distribution of the scores section of the common classes and the experimental classes, almost equal. But after the experimental study, in experimental classes the number of students that have academic sense of accomplishment has increased, students with poor academic accomplishment significantly reduced. Therefore the mathematical modeling group cooperative learning model has improved students' interest, stimulated students' motivation to learn.

4.2.2 Longitudinal Statistics of Common and Experimental Classes

Results' description and data analysis: from the view of longitudinal analysis, the academic performance of the common classes and the experimental classes both have made progress. But in the experimental class the students that have good sense of academics made significant progress, while the number of students that have poor sense of academics are decreased. Therefore, mathematical modeling group cooperative learning model can better stimulate students, especially the students with poor academic accomplishment, and in the process of group cooperative learning,can improve

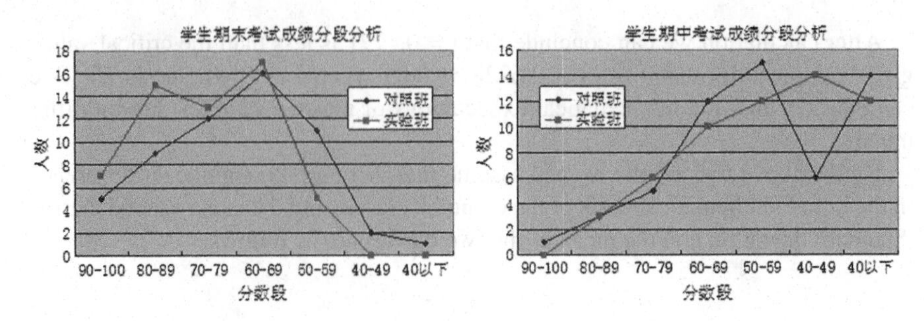

Fig. 5 horizontal analysis

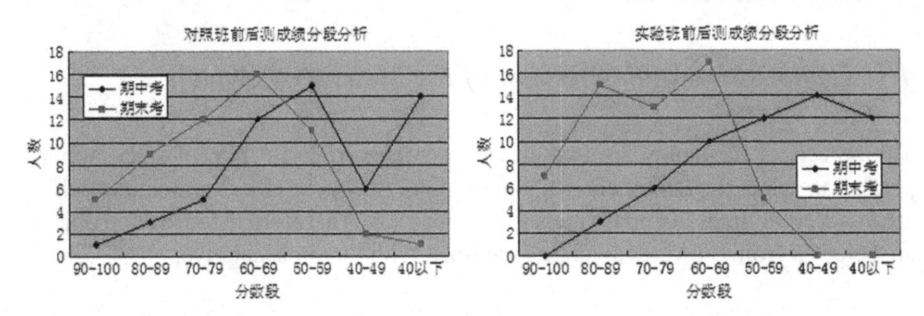

Fig. 6 longitudinal analysis

the learning and exchanges of students, being helpful for students' academically positive impact (Figs. 5 and 6).

4.2.3 Post-test Statistics of Students' Learning Status

Results' description and data analysis: First we applied difference test to pretest and posttest data on the experimental class so that we can conclude that the performance of experimental class have improved rather than the results of random error resulting increase. Then we applied difference test to two classes' post-test data so that we can conclude that there are obvious differences of teaching effectiveness between the experimental classes and common classes. Having applied F-test to pretest and post-test data of experimental class, we can get results $F = 1.9014, p = 0.04995 < 0.05$. There are obvious difference between two classes' variance. Because the mean of post-test data is significantly greater than the mean of the pretest data, we can conclude that the results of the experimental class have been significantly improved. Have applied F-test to post-test data of the two classes, we got $F = 1.8366, p = 0.03873 < 0.05$,rejecting the assumption that the two normal population have same variance, which indicated that the results of the two classes have shown a significant difference. From the aspect of mean and standard deviation, we can obtain that the standard deviation of common class is

experimental class				common class			
choice	selection numbers	percentage	scores	choice	selection numbers	percentage	scores
A	9	8%	18	A	36	31.90%	72
B	11	9.70%	11	B	27	23.90%	27
C	17	15.00%	0	C	25	22.10%	0
D	34	30.10%	34	D	13	11.50%	13
E	42	37.20%	84	E	12	10.60%	24

Fig. 7 questionnaire results

experimental class (pretest)				Experimental class (protest)			
choice	selection numbers	percentage	scores	choice	selection numbers	percentage	scores
A	48	42.00%	96	A	9	8%	18
B	29	25.70%	29	B	11	9.70%	11
C	21	18.60%	0	C	17	15.00%	0
D	8	7.10%	-8	D	34	30.10%	34
E	7	6.20%	-14	E	42	37.20%	84

Fig. 8 experimental class

13.4545, the one of experimental class is 11.7323. From the view of statistical, a set of data, the greater the standard deviation is, the greater the degree of data dispersion is, the more serious students' achievement polarization are. On the contrary, the smaller the standard deviation is, the smaller the degree of data scatter is, the more aggregation the students' achievement is, the less obvious the differentiation is. Therefore, experimental data indicated that the application of mathematical modeling group cooperative learning model is helpful for students' overall results stabilized. The size of the analysis's differences: by the comparison of the common classes and experimental classes' data, before the experimental study, the differences of the two classes' data Z value is 1.7178, while after the experimental study, the difference of the data Z value is 8.3764. The bigger the differences Z value is, the greater the differences of the two sets of data are. Therefore, experimental data indicated that with the help of the teaching of mathematical modeling group cooperative learning mode, the differences between students have decreased, the overall results have progressed significantly.

4.2.4 Post-test Statistics of Students' Interest

In the affective domain, among the aims' evaluation activities of mathematical modeling education , what have been studied more are about students' mathematical modeling learning interest and attitude. This is because students' mathematical modeling learning interest and attitude have more direct and significant impact on students' improving the ability of mathematical modeling. Thus as one of the situation variable in the various elements of the mathematical modeling, it's more and more important for the theory of mathematical modeling teaching. Generally used for the evaluation of students' mathematical modeling learning interest and attitude is questionnaire design, preparing a variety of issues on the students' behavior, to enable students to respond to these questions, then calculate the proportion of students' a variety of responses, do the analysis and get the conclusion. Results' description and analysis: The data show that: there are still 31.9 % of the common class's students that have little change of interest in learning mathematical modeling, while the proportion of the experimental class's students that have interest in learning the mathematical modeling is 37.2 %. At the same time, by conducting a longitudinal comparison of the experimental class itself, the proportion of students that have no interest in learning mathematical modeling has reduced from 42.5 % to 8 %, while the proportion of students that have interest in mathematical modeling study has rose from 6.2 % to 37.2 %. Therefore, the experimental data indicate that cooperative learning can enhance students' interest in learning to some extent, mobilize students' learning motivation and initiative to improve the quality of learning efficiency and effect.

5 Conclusion

The above empirical research indicated that in the institutions of higher the application of cooperative learning strategies in mathematical modeling teaching are effective, which will help improve students' interest in learning mathematical modeling, improve students' mathematical modeling performance, train students' ability of learning math modeling.So we can achieve the goal of improving students' comprehensive ability.

Acknowledgments This research was supported by the Higher Education Teaching Reform Project of Guangdong Province.

References

1. Wang T.: The concept and implementation of cooperative learning. China Personnel Press (2002)
2. Holtek force: Humble interpretation of the rise of cooperative fever in American primary and middle schools. Education Review **8**(3) (1990)
3. Zhong, Yubin: The FHSE model of software system for synthetic evaluating enterprising. J. Guangzhou Univ. **4**(4), 316–320 (2005)
4. Holtek forceZheng Shuzhen: The design of cooperative learning. Zhejiang Education Press (2006)
5. Holtek force: What kind of teaching tasks suitable for cooperative learning. People's Education **55**(5) (2004)

6. Slavin,R.E., Wang T.: Cooperative learning research: International Perspective. Shandong Educ. Res. **8**(1) (1994)
7. Sharon, S: The theory of cooperative learning. Shandong Educ. Res. 10(5) (1996)
8. Luan Q.: Some suggestions about cooperative learning in middle school mathematics teaching. Math. Teach. Learn. junior middle sch. **14**(3) (2005)
9. Hong-she, G.: Cooperative learning of senior high school mathematical modeling teaching. Henan Educ. **27**(8) (2004)
10. Guo, H.: The theory hypothesis and practice operation mode of team cooperative learning. J. Chin Educ. **19**(5) (1998)
11. Zhang B.: Theory of cooperative learning and its impact on students behavior and attitudes. Educ. Theor. Pract. **19**(9) (1999)
12. Qi Zeng: The basic elements of cooperative learning. J. Subject Educ. (5)(2000)
13. Shi, K., Yao, T., Yu, : To cultivate chemistry creative talents needs new ideas. China, Higher Educ. 39(17) (2003)
14. Zhong, Y: The structures and expressions of Power Group. Fuzzy Inf. Eng. **2**(2) (2010)
15. Zhong, Y: The thought of fuzzy mathematics Modeling Is Infiltrated in MSMT. Adv. Intell. Soft Comput. **2**(62), 233–242 (2009)

A Discrete-Time $Geo/G/1$ Retrial Queue with Preemptive Resume, Bernoulli Feedback and General Retrial Times

Cai-Min Wei, Yi-Yan Qin and Lu-Xiao He

Abstract In this work, we consider a discrete-time $Geo/G/1$ retrial queue with preemptive resume, Bernoulli feedback and general retrial times. We analyze the Markov chain underlying the considered queueing system and derive the generating functions of the system state, the orbit size and the system size distribution. Using probability generating function technique, some interesting and important performance measures are obtained. We also investigate the stochastic decomposition property and present some numerical examples.

Keywords Discrete-time retrial queue · Preemptive resume · Bernoulli feedback · General retrial times · Stochastic decomposition

1 Introduction

Queueing models are widely and successfully used as mathematical models of computer communications and manufacturing settings. Retrial phenomenon occurs when a new arriving customer who can't join the system decides to retry again after a random time. Retrial queues are queueing systems which arise naturally in computer systems and telecommunication networks and are characterized by the feature that arriving customers who find all servers busy leave the server temporarily and join a group of unsatisfied customers (which called orbit) in order to repeat their attempt again after a random time period.

C. Wei (✉) · L. He
Department of Mathematics, Shantou University, Shantou 515063, People's Republic of China
e-mail: cmwei@stu.edu.cn

L. He
e-mail: 10lxhe@stu.edu.cn

Y. Qin
College of Business, Guangxi University for Nationalities, Nanning 530006, People's Republic of China
e-mail: qinyiyan2002@sina.com

B.-Y. Cao and H. Nasseri (eds.), *Fuzzy Information & Engineering and Operations Research & Management*, Advances in Intelligent Systems and Computing 211, DOI: 10.1007/978-3-642-38667-1_53, © Springer-Verlag Berlin Heidelberg 2014

Although many continuous-time queueing models with preferred or feedback have been studied extensively in the past years [1–8], their discrete-time counterparts received less attention in the literature. However, in practice, discrete-time queues are more appropriate than their continuous-time counterparts for modelling computer and telecommunication systems in which time is slotted. For more detailed discussion and applications of discrete-time queues can be found in the books by Hunter [9], Takagi [10] and Woodward [11].

Queueing models with preemptive resume phenomenon is characterized by the fact that arriving customers have the priority to interrupt the customer in service to commence his own service with LCFS preemptive resume discipline. It occurs in many situations in our real life such as dealing with weather forecast information and treating to emergency patient in hospital. Liu and Wu [12] considered an $MAP/G/1$ G-queue with possible preemptive resume service discipline and multiple vacations. A $Geo^X/G/1$ queue with preemptive resume priority have been studied by Lee [13]. Liu and Wu [14] investigated a discrete-time $Geo/G/1$ retrial queue with preemptive resume and collisions.

Recently, there has been an increasing interest in the analysis of retrial queueing systems with feedback. Feedback is present for example in after-sales service and telecommunication systems where the messages with errors at the destination are sent again. This has been proved to be very useful and appropriate to model some situations where all the customers demand the main service and only some of them demand an extra service in day-to-day life. Hellerstein, Diao, Parekh, and Tilbury considered a good research of feedback control in the book [15] which investigates a practical treatment of the design and application of feedback control of computing system. Atencia and Moreno [16] studied a discrete-time $Geo^X//G_H/1$ retrial queue with Bernoulli feedback.

This work is an extension of the retrial queueing theory on preemptive resume and feedback into discrete-time retrial queues. We study a discrete-time $Geo/G/1$ retrial queue with preemptive resume, Bernoulli feedback and general retrial times. The system under study in this chapter, apart from its applications described before has theoretical interest because there is no same work in the previous.

The remainder of the chapter is organized as follows. In Sect. 2, we give the mathematical description and Markov chain, and derive the probability generating functions of the system state, the orbit size and the system size distribution. We investigate the stochastic decomposition law in Sect. 3. In Sect. 4, we present some numerical results to illustrate the impact of the collisions and impatience on the performance of the system. Finally, we give a conclusion in Sect. 5.

2 Model Description and Markov Chain

We consider a discrete-time retrial queue where the time axis is segmented into slots of equal length and all queueing activities occur around the slot boundaries, and may occur at the same time. For mathematical clarity, let the time axis be marked by $0, 1, \cdots, m, \cdots$. We suppose that the departures occur in the interval (m^-, m), and the arrivals and the retrials occur in the interval (m, m^+). we consider the model for early arrival system (EAS) policy. For more details see on the EAS discipline and related concepts can be found in Hunter [9].

New customers arrive the system according to a geometrical arrival process with probability p. We suppose that there is no waiting space in front of the server, and therefore, if an arriving customer finds the server idle, he commences his service immediately. Otherwise, if the server is busy at the arrival epoch, the arriving customer either interrupts the customer in service to commence his own service with probability α or leaves the service area and enters the orbit by himself with probability $\bar{\alpha} = 1 - \alpha$. The interrupted customer enters into the orbit. The service of the interrupted customer resumes from the beginning. Because we have interest in only distribution of the number of customers in the system and will not discuss the waiting and sojourn time distributions, we have no need to illuminate what place the interrupted customer will be in the orbit.

After service completion, the customer decides either to join the retrial group again for another service with probability θ or leaves the system with complementary probability $\bar{\theta} = 1 - \theta$.

When the server is idle and both an external arrival and a retrial occur at the same time, the external customer is given higher priority over the returning customers. Customers in the orbit are assumed to form a FCFS queue. It is only the customer at the head of the queue who makes retrials.

Service times are governed by probability distribution $\{s_i\}_{i=1}^{\infty}$ with generating function $S_1(x) = \sum_{i=1}^{\infty} s_i x^i$ and the n-th factorial moment β_n. Retrial times are assumed to follow a general distribution. Successive interretrial times follow a general distribution $\{r_i\}_{i=0}^{\infty}$ with generating function $R(x) = \sum_{i=0}^{\infty} r_i x^i$.

The interarrival times, the service times and the retrial times are assumed to be mutually independent. We will denote $\bar{p} = 1 - p$, $\bar{r_0} = 1 - r_0$. In order to avoid trivial cases, it is also supposed $0 < p < 1, 0 \leq \alpha \leq 1, 0 \leq \theta < 1$. At time m^+, the system can be described by the Markov process

$$X_m = (C_m, \xi_{0,m}, \xi_{1,m}, N_m),$$

where C_m denotes the state of the server 0 or 1 according to whether the server is idle or busy and N_m the number of repeated customers in the orbit. When $C_m = 0$ and $N_m > 0$, $\xi_{0,m}$ represents the remaining retrial time. When $C_m = 1$, $\xi_{1,m}$ represents the remaining service time of the customer currently being served.

It can be shown that $\{X_m, m \geq 1\}$ is the Markov chain of our queueing system, whose state space is

$$\{(0, 0); (0, i, k) : i \geq 1, k \geq 1; (1, i, k) : i \geq 1, k \geq 0\}.$$

Our object is to find the stationary distribution of the Markov chain $\{X_m; m \geq 1\}$ as follows:

$$\pi_{0,0}=\lim_{m\to\infty}P\{C_m=0,N_m=0\},\pi_{j,i,k}=\lim_{m\to\infty}P\{C_m=j,\xi_{j,m}=i,N_m=k\}.$$

The evolution of the chain is governed by the one-step transition probabilities

$$p_{yy'}=P[Y_{m+1}=y'|Y_m=y].$$

When $k=0$, we obtain

$$P_{(0,0)(0,0)}=\bar{p},\, P_{(1,1,0)(0,0)}=\bar{p}\bar{\theta}.$$

When $i\geq 1, k\geq 1$, we obtain

$$P_{(0,i+1,k)(0,i,k)}=\bar{p},\quad P_{(1,1,k-1)(0,i,k)}=\bar{p}\theta r_i,\quad P_{(1,1,k)(0,i,k)}=\bar{p}\bar{\theta}r_i.$$

When $i\geq 1, k\geq 0$, we obtain

$$P_{(0,0)(1,i,k)}=\delta_{0,k}ps_i,\quad P_{(0,1,k+1)(1,i,k)}=\bar{p}s_i,\quad P_{(0,j,k)(1,i,k)}=(1-\delta_{0,k})ps_i,\ j\geq 1,$$
$$P_{(1,1,k-1)(1,i,k)}=(1-\delta_{0,k})p\theta s_i,\quad P_{(1,1,k)(1,i,k)}=p\bar{\theta}s_i+\bar{p}\theta r_0 s_i,$$
$$P_{(1,1,k+1)(1,i,k)}=\bar{p}\bar{\theta}r_0 s_i,\quad P_{(1,i+1,k-1)(1,i,k)}=(1-\delta_{0,k})p\bar{\alpha},$$
$$P_{(1,i+1,k)(1,i,k)}=\bar{p},\quad P_{(1,j,k-1)(1,i,k)}=(1-\delta_{0,k})p\alpha s_i,\ j\geq 2.$$

The Kolmogorov equations for the stationary distribution of the system are:

$$\pi_{0,0}=\bar{p}\pi_{0,0}+\bar{p}\bar{\theta}\pi_{1,1,0} \tag{1}$$

$$\pi_{0,i,k}=\bar{p}\pi_{0,i+1,k}+\bar{p}\theta r_i\pi_{1,1,k-1}+\bar{p}\bar{\theta}r_i\pi_{1,1,k},\quad i\geq 1,\ k\geq 1, \tag{2}$$

$$\begin{aligned}\pi_{1,i,k}&=\delta_{0,k}ps_i\pi_{0,0}+\bar{p}s_i\pi_{0,1,k}\\&\quad+(1-\delta_{0,k})ps_i\sum_{j=1}^{\infty}\pi_{0,j,k}+(1-\delta_{0,k})p\theta s_i\pi_{1,1,k-1}\\&\quad+(p\bar{\theta}s_i+\bar{p}\theta r_0 s_i)\pi_{1,1,k}+\bar{p}\bar{\theta}r_0 s_i\pi_{1,1,k+1}\\&\quad+(1-\delta_{0,k})p\bar{\alpha}\pi_{1,i+1,k-1}+\bar{p}\pi_{1,i+1,k}\\&\quad+(1-\delta_{0,k})p\alpha s_i\sum_{j=2}^{\infty}\pi_{1,j,k-1},\quad i\geq 1,k\geq 0,\end{aligned} \tag{3}$$

where $\delta_{i,j}$ is the Kronecker's symbol, and the normalizing condition is

$$\pi_{0,0}+\sum_{i=1}^{\infty}\sum_{k=1}^{\infty}\pi_{0,i,k}+\sum_{i=1}^{\infty}\sum_{k=0}^{\infty}\pi_{1,i,k}=1.$$

To solve Eqs. (1)–(3), we introduce the following probability generating functions and auxiliary probability generating functions

$$\varphi_j(x,z) = \sum_{i=1}^{\infty}\sum_{k=b}^{\infty} \pi_{j,i,k} x^i z^k,$$
$$\varphi_{j,i}(z) = \sum_{k=b}^{\infty} \pi_{j,i,k} z^k, \ j=0, b=1, i \geq 1; \ j=1, b=0, i \geq 1.$$

Multiplying Eqs. (2), (3) by z^k, summing over k and using the boundary condition (1), we get

$$\varphi_{0,i}(z) = \bar{p}\varphi_{0,i+1}(z) + (\bar{\theta}+\theta z)\bar{p}r_i\varphi_{1,1}(z) - pr_i\pi_{0,0}, \tag{4}$$

$$\begin{aligned}\varphi_{1,i}(z) =& \frac{\bar{p}}{z}s_i\varphi_{0,1}(z) + [\frac{(\bar{p}r_0+pz)(\bar{\theta}+\theta z)}{z} - p\alpha z]s_i\varphi_{1,1}(z) + (\bar{p}+p\bar{\alpha}z)\varphi_{1,i+1}(z)\\ &+ ps_i\varphi_0(1,z) + p\alpha s_i z\varphi_1(1,z) + \frac{z-r_0}{z}ps_i\pi_{0,0}, \quad i \geq 1.\end{aligned} \tag{5}$$

Multiplying Eqs. (4), (5) by x^i and summing over i, we get

$$\frac{x-\bar{p}}{x}\varphi_0(x,z) = \bar{p}(\bar{\theta}+\theta z)(R(x)-r_0)\varphi_{1,1}(z) - \bar{p}\varphi_{0,1}(z) - p(R(x)-r_0)\pi_{0,0}, \tag{6}$$

$$\begin{aligned}\frac{x-\beta(z)}{x}\varphi_1(x,z) =& \left[\frac{(\bar{p}r_0+pz)(\bar{\theta}+\theta z)-p\alpha z^2}{z}S(x) - \beta(z)\right]\varphi_{1,1}(z)\\ &+ pS(x)\varphi_0(1,z)\\ &+ \frac{\bar{p}}{z}S(x)\varphi_{0,1}(z) + \frac{\bar{p}}{z}S(x)\varphi_{0,1}(z) + p\alpha zS(x)\varphi_1(1,z)\\ &+ \frac{z-r_0}{z}pS(x)\pi_{0,0},\end{aligned} \tag{7}$$

where

$$\beta(z) = \bar{p} + p\bar{\alpha}z.$$

To solve for $\varphi_0(1,z)$ and $\varphi_1(1,z)$, we put $x=1$ in Eqs. (6) and (7), we obtain

$$p\varphi_0(1,z) = \bar{p}(\bar{\theta}+\theta z)\bar{r_0}\varphi_{1,1}(z) - \bar{p}\varphi_{0,1}(z) - p\bar{r_0}\pi_{0,0}, \tag{8}$$
$$pz(1-z)\varphi_1(1,z) = \bar{p}(1-z)\varphi_{0,1}(z) + B(z)\varphi_{1,1}(z) - pr_0(1-z)\pi_{0,0}, \tag{9}$$

where

$$B(z) = [z + \bar{p}r_0(1 - z)](\bar{\theta} + \theta z) - \bar{p}z - pz^2.$$

Substituting the above equation into (6), we obtain

$$\frac{x - \beta(z)}{x}\varphi_1(x, z) = \{\frac{[z + \bar{p}r_0(1 - z)](\bar{\theta} + \theta z)}{z}S(x) + \alpha[\bar{p}r_0(\bar{\theta} + \theta z) - \theta z]S(x) - \beta(z)\}\varphi_{1,1}(z) + \frac{1 - \bar{\alpha}z}{z}\bar{p}S(x)\varphi_{0,1}(z) - \frac{1 - \bar{\alpha}z}{z}pr_0S(x)\pi_{0,0}. \tag{10}$$

Setting $x = \bar{p}$ in (6) and $x = \beta(z)$ in (10), we obtain

$$p(R(\bar{p}) - r_0)\pi_{0,0} = \bar{p}(\bar{\theta} + \theta z)(R(\bar{p}) - r_0)\varphi_{1,1}(z) + \bar{p}\varphi_{0,1}(z), \tag{11}$$

$$\frac{1 - \bar{\alpha}z}{z}pr_0S(\beta(z))\pi_{0,0} = \frac{1 - \bar{\alpha}z}{z}\bar{p}S(\beta(z))\varphi_{0,1}(z) + \{\frac{[z + \bar{p}r_0(1 - z)](\bar{\theta} + \theta z)}{z}S(\beta(z)) + \alpha[\bar{p}r_0(\bar{\theta} + \theta z) - \theta z]S(\beta(z)) - \beta(z)\}\varphi_{1,1}(z). \tag{12}$$

To solve for $\varphi_{0,1}(z)$ and $\varphi_{1,1}(z)$ by (11) and (12), we get

$$\varphi_{0,1}(z) = \frac{pz(R(\bar{p}) - r_0)[\beta(z) - (\bar{\theta} + \theta z - \alpha\theta z)S(\beta(z))]}{\Omega(z)} \cdot \frac{\pi_{0,0}}{\bar{p}}, \tag{13}$$

$$\varphi_{1,1}(z) = \frac{pR(\bar{p})(1 - \bar{\alpha}z)S(\beta(z)}{\Omega(z)}\pi_{0,0}, \tag{14}$$

where

$$\Omega(z) = \{[(1 - \bar{\alpha}z)\bar{p}R(\bar{p}) + z](\bar{\theta} + \theta z) - \alpha\theta z^2\}S(\beta(z)) - \beta(z)z.$$

In order to show that the above auxiliary probability generating functions are defined for $z \in [0, 1]$ and in $z = 1$ can be extended by continuity we give the following lemma.

Lemma 1 *For* $0 \leq z < 1$, *if*

$$[1 - \bar{\alpha}\bar{p}R(\bar{p}) + \theta(\alpha\bar{p}R(\bar{p}) + 1) - 2\alpha\theta]S(\beta(1)) + [\alpha\bar{p}R(\bar{p}) + 1 - \alpha\theta]p\bar{\alpha}S'(\beta(1)) < \bar{p} + p\bar{\alpha}(1 + z),$$

then $\Omega(z) > 0$.

Proof. Let us define the following functions

$$f(z) = [(1-\bar{\alpha}z)\bar{p}R(\bar{p}) + z](\bar{\theta} + \theta z) - \alpha\theta z^2\}S(\beta(z)), \quad g(z) = \beta(z)z.$$

In order to study the slope of the tangent of $f(z)$ and $g(z)$ we calculate:

$$\begin{aligned} f^{'}(1) &= [1-\bar{\alpha}\bar{p}R(\bar{p}) + \theta(\alpha\bar{p}R(\bar{p})+1) - 2\alpha\theta]S(\beta(1)) \\ &\quad + [\alpha\bar{p}R(\bar{p}) + 1 - \alpha\theta]p\bar{\alpha}S'(\beta(1)), \\ g^{'}(1) &= \bar{p} + 2p\bar{\alpha}. \end{aligned}$$

Due to the fact that $f(z)$ and $g(z)$ are convex functions, we observe that $f'(1) < g'(1)$, we have $f(z) > g(z)$ in $0 \le z < 1$.

Applying L'Hopitals' rule, we obtain

Lemma 2 *The following limits exist if* $1 - p\alpha < [\alpha\bar{p}R(\bar{p}) + 1 - \alpha\theta]S(1-p\alpha)$,

$$\begin{aligned} \lim_{z\to 1}\varphi_{0,1}(z) &= \frac{p(R(\bar{p}) - r_0)[1 - p\alpha - (1-\alpha\theta)S(1-p\alpha)]}{[\alpha\bar{p}R(\bar{p}) + 1 - \alpha\theta]S(1-p\alpha) + p\alpha - 1}, \\ \lim_{z\to 1}\varphi_{1,1}(z) &= \frac{p\alpha R(\bar{p})S(1-p\alpha)}{[\alpha\bar{p}R(\bar{p}) + 1 - \alpha\theta]S(1-p\alpha) + p\alpha - 1}\pi_{0,0}. \end{aligned}$$

We summarize the above results in the following theorem.

Theorem 1. *The probability generating functions of the stationary distribution of the chain are given by*

$$\begin{aligned} \varphi_0(x,z) &= \frac{R(x) - R(\bar{p})}{x - \bar{p}} \times \frac{pxz[\beta(z) - (\bar{\theta} + \theta z - \alpha\theta z)S(\beta(z))]}{\Omega(z)}\pi_{0,0}, \\ \varphi_1(x,z) &= \frac{S(x) - S(\beta(z))}{x - \beta(z)} \times \frac{pxR(\bar{p})(1-\bar{\alpha}z)\beta(z)}{\Omega(z)}\pi_{0,0}, \end{aligned}$$

where

$$\pi_{0,0} = \frac{(\bar{p}\alpha R(\bar{p}) + 1 - \alpha\theta)S(1-p\alpha) - (1-p\alpha)}{\alpha\bar{\theta}R(\bar{p})S(1-p\alpha)}.$$

Proof. In order to complete the proof we only have to substitute (13) and (14) into (6) and (10) and derive the probability generating functions $\varphi_0(x,z)$ and $\varphi_1(x,z)$ in its explicit form enunciated in the theorem.

Using the normalization condition, we find the unknown constant

$$\pi_{0,0} = \frac{(\bar{p}\alpha R(\bar{p}) + 1 - \alpha\theta)S(1-p\alpha) - (1-p\alpha)}{\alpha\bar{\theta}R(\bar{p})S(1-p\alpha)}.$$

This completes the proof.

In the following corollary, we calculate some important marginal probability generating functions and present their explicit expressions.

Corollary 1.

(1) The marginal generating function of the number of customers in the orbit when the server is idle is given by

$$\pi_{0,0}+\varphi_0(1,z)=\frac{R(\bar{p})\{[\bar{p}(1-\bar{\alpha}z)(\bar{\theta}+\theta z)+(\bar{\theta}+\theta z-\alpha\theta z)z]S(\beta(z))-z\beta(z)\}}{\Omega(z)}\pi_{0,0}.$$

(2) The marginal probability generating function of the number of customers in the orbit when the server is busy is given by

$$\varphi_1(1,z)=\frac{[1-S(\beta(z))]R(\bar{p})\beta(z)}{\Omega(z)}\pi_{0,0}.$$

(3) The probability generating function of the number of customers in the orbit (i.e. of the variable N) is given by

$$\begin{aligned}N(z)&=\pi_{0,0}+\varphi_0(1,z)+\varphi_1(1,z)\\&=\frac{R(\bar{p})\{\bar{p}(1-\bar{\alpha}z)(\bar{\theta}+\theta z)+(\bar{\theta}+\theta z-\alpha\theta z)z\,]\,S(\beta(z))+[1-z-S(\beta(z))]\beta(z)\}}{\Omega(z)}\pi_{0,0}.\end{aligned}$$

(4) The probability generating function of the number of customers in the system (of the variable L) is given by

$$\begin{aligned}L(z)&=\pi_{0,0}+\varphi_0(1,z)+z\varphi_1(1,z)\\&=\frac{R(\bar{p})\{\bar{p}(1-\bar{\alpha}z)(\bar{\theta}+\theta z)+(\bar{\theta}+\theta z-\alpha\theta z)z\,]\,S(\beta(z))-S(\beta(z))\beta(z)z\}}{\Omega(z)}\pi_{0,0}.\end{aligned}$$

The following corollary presents some important performance measures.

Corollary 2.

(1) The system is idle with probability

$$\pi_{0,0}=\frac{(\bar{p}\alpha R(\bar{p})+1-\alpha\theta)S(1-p\alpha)-(1-p\alpha)}{\alpha\bar{\theta}R(\bar{p})S(1-p\alpha)}.$$

(2) The system is occupied with probability

$$\varphi_0(1,1)+\varphi_1(1,1)=\frac{[(\bar{\theta}-\bar{p})\alpha R(\bar{p})+\alpha\theta-1]S(1-p\alpha)+(1-p\alpha)}{\alpha\bar{\theta}R(\bar{p})S(1-p\alpha)}.$$

(3) The server is idle with probability

$$\pi_{0,0}+\varphi_0(1,1)=\frac{(\bar{p}\alpha+1-\alpha\theta)S(1-p\alpha)-(1-p\alpha)}{\alpha\bar{\theta}S(1-p\alpha)}.$$

(4) The server is busy with probability

$$\varphi_1(1,1)=\frac{(1-p\alpha)[1-S(1-p\alpha)]}{\alpha\bar{\theta}S(1-p\alpha)}.$$

(5) The mean orbit size is given by

$$E[N]=N'(z)\mid_{z=1}.$$

(6) The mean system size is given by

$$E[L]=L'(z)\mid_{z=1}.$$

(7) The mean time a customer spends in the system (including the service time) is given by

$$W_S=\frac{E[L]}{P}.$$

Remark 1 (A Special Cases)

(1) When $\alpha=1,\theta=0$, $L(z)$ reduces to

$$L(z)=\frac{(\bar{P}+PZ)\{[1+\bar{P}R(\bar{p})]S(\bar{p})-\bar{p}\}}{[z+\bar{p}R(\bar{p})]S(\bar{p})-\bar{p}z},$$

this is the probability generating function of the number of customers in the model *Geo*/*G*/1 with LCFS preemptive resume discipline and general retrial times which according to the generating function of Remark 3.1 in Liu and Wu [14].

(2) When $r_0=1$ (i.e., $R(\bar{p})=1$), $L(z)$ reduces to

$$L(z)=\frac{\bar{p}(1-\bar{\alpha}z)(\bar{\theta}+\theta z)+(\bar{\theta}+\theta z-\alpha\theta z)z]S(\beta(z))-S(\beta(z))\beta(z)z}{\{[(1-\bar{\alpha}z)\bar{p}+z](\bar{\theta}+\theta z)-\alpha\theta z^2\}S(\beta(z))-\beta(z)z}\times\frac{(\bar{p}\alpha+1-\alpha\theta)S(1-p\alpha)-(1-p\alpha)}{\alpha\bar{\theta}S(1-p\alpha)},$$

which is the probability generating function for the number of customers in the standard $Geo/G/1/\infty$ queueing system with preemptive resume and feedback. Obviously, in this cases, the customer at the head of the orbit immediately get his service whenever the server is idle.

3 Stochastic Decomposition

The stochastic decomposition law for queueing systems was first given in Fuhrmann and Cooper [17]. This section we will present the stochastic decompositions of the system size distribution due to the fact that $L(z)$ can be written in the following form:

$$L(z) = L'(z) \times L''(z),$$

where

$$L'(z) = \frac{\bar{p}(1-\bar{\alpha}z)(\bar{\theta}+\theta z) + (\bar{\theta}+\theta z - \alpha\theta z)z]S(\beta(z)) - S(\beta(z))\beta(z)z}{[\bar{p}(1-\bar{\alpha}z)(\bar{\theta}+\theta z) + (\bar{\theta}+\theta z - \alpha\theta z)z]S(\beta(z)) - z\beta(z)\}}$$
$$\times \frac{(\bar{p}\alpha + 1 - \alpha\theta)S(1-p\alpha) - (1-p\alpha)}{\alpha\bar{\theta}S(1-p\alpha)},$$
$$L''(z) = \frac{[\bar{p}(1-\bar{\alpha}z)(\bar{\theta}+\theta z) + (\bar{\theta}+\theta z - \alpha\theta z)z]S(\beta(z)) - z\beta(z)}{\Omega(z)}$$
$$\times \frac{(\bar{p}\alpha R(\bar{p}) + 1 - \alpha\theta)S(1-p\alpha) - (1-p\alpha)}{\bar{p}\alpha + 1 - \alpha\theta)S(1-p\alpha) - (1-p\alpha)} = \frac{\pi_{0,0} + \varphi_0(1,z)}{\pi_{0,0} + \varphi_0(1,1)}.$$

We note that $L'(z)$ is the probability generating function of the number of customers in the standard $Geo/G/1$ queue with preemptive resume and feedback, as we obtained in Remark 1(2). This result can be summarized in the following theorem.

Theorem 2. *The total number of customers in the system (L) can be decomposed as the sum of two independent random variables, one of which is the number of customers in the $Geo/G/1$ queue system with preemptive resume and feedback (L') and the other is the number of repeated customers given that the server is idle (L''), i.e., $L = L' + L''$.*

4 Numerical Results

In this section, we build up a numerical study to investigate some numerical examples of the performance measures obtained in Sect. 3 in relation with the most specific parameters of our model. We consider two performance measures: the free probability $\pi_{0,0}$, the the probability $\varphi_1(1,1)$ that the server is busy. We assume that the arrival rate $p = 0.1$. Moreover, for the purpose of a numerical illustration, we assume that the service time distribution function is geometric with parameter 0.8, i.e., $S(x) = \frac{4x}{5-x}$, and the retrial times follow a geometric distribution with generating function $R(x) = \frac{r_0}{1-(1-r_0)x}$. We assume that the parametric values are chosen under the stability condition in all the below cases. We have presented three curves which correspond to $r_0 = 0.1, 0.4$ and 0.8 in Fig.1. In Fig. 1, we study the influence of the values of θ on the free probability $\pi_{0,0}$ for different $r_0 = 0.1, 0.4$ and 0.8. As expected, the free probability decreases with increasing the value of θ, obviously,

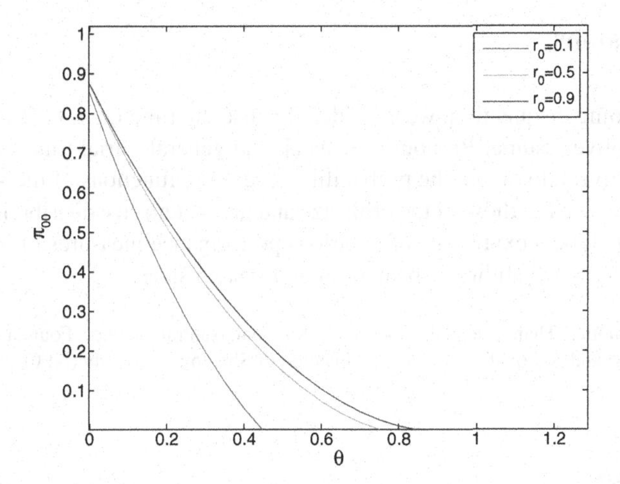

Fig. 1 The free probability $\pi_{0,0}$ versus θ for different r_0

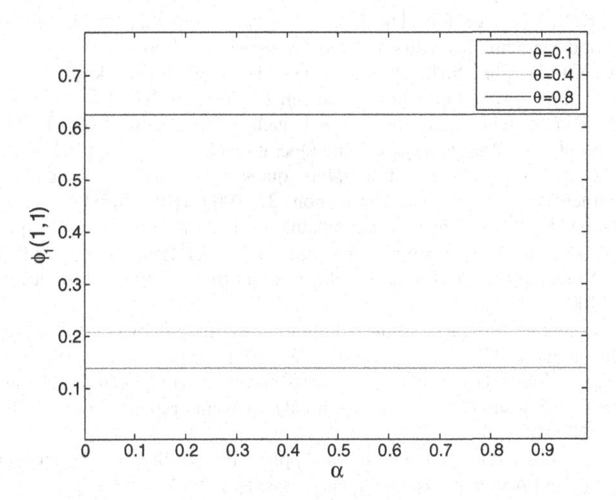

Fig. 2 The busy probability $\varphi_1(1, 1)$ versus α for different θ

we obtain that the probability that the system is free decreases. With another view, the parameter α is wiped off by simplifying the expression of $\pi_{0,0}$, that is to say, the parameter α does not affect the free probability.

In Fig. 2, we plot the busy probability $\varphi_1(1, 1)$ versus α for different $\theta = 0.1, 0.4$ and 0.8. As intuition tells us, the probability $\varphi_1(1, 1)$ (that the server is busy) is increasing as function of θ, this is due to the fact that the server will be more congested while the feedback probability increases. With another view, the parameter α is wiped off by simplifying the expression of $\varphi_1(1, 1)$, that is to say, the parameter α does not affect the busy probability. The same with α, the parameter r_0 does not affect the busy probability.

5 Conclusions

In the foregoing disquisition, we considered a discrete time $Geo/G/1$ retrial queue with preemptive resume, Bernoulli feedback and general retrial times. The system has been analyzed to obtain the probability generating functions of the system state distribution as well as those of the orbit size and the system size distributions. Hence, we obtain analytical expressions for various performance measures of interest such as idle and busy probabilities, mean orbit and system sizes.

Acknowledgments Thanks to the support by National Natural Science Foundation of China (No.71062008) and National Foundation Cultivation of Shantou University (NFC12002).

References

1. Choi, B.D., Kim, Y.C.: Lee Y.W: The $M/M/c$ retrial queue with geometric loss and feedback. Computers and Mathematics with Applications **36**, 41–52 (1998)
2. Choi, B.D.: Kim B, Choi S.H.: On the $M/G/1$ Bernoulli feedback queue with multi-class customers. Computers and Operations Research **27**, 269–286 (2000)
3. Choi, B.D.: Kim B, Choi S.H.: An $M/G/1$ queue with multiple types of feedback, gated vacations and FCFS policy. Computers and Operations Research **30**, 1289–1309 (2003)
4. Choi, D.I., Kim, T.S.: Analysis of a two-phase queueing system with vacations and Bernoulli feedback. Stochastic Analysis and Applications **21**, 1009–1019 (2003)
5. Krishna Kumar B, Vijayakumar A, Arivudainambi D.: An $M/G/1$ retrial queueing system with two-phase service and preemptive resume. Ann. Oper. Res., 113:61–79 (2002).
6. Drekic, S.: A preemptive resume queue with an expiry time for retained service. Perform. Eval. **54**, 59–74 (2003)
7. Artalejo, J.R., Dudin, A.N., Klimenok, V.I.: Stationary analysis of a retrial queue with preemptive repeated attempts. Oper. Res. Lett. **28**, 173–180 (2001)
8. Liu, Z., Wu, J., Yang, G.: An $M/G/1$ retrial G-queue with preemptive resume and feedback under N-policy subject to the server breakdowns and repairs. Comput. Math. Appl. **58**, 1792–1807 (2009)
9. Hunter J. J.: Mathematical Techniques of Applied Probability (vol.2), Discrete-time Models:Techniques and Applications. Academic Press, New York, (1983).
10. Takagi H.: Queueing analysis: A foundation of performance evaluation. Discrete-time systems(Vol.3). North-Holland: Amsterdam, (1993).
11. Woodward, M.E.: Communication and computer networks: Modelling with discrete-time queues. IEEE Computer Society Press, Los Alamitos, California (1994)
12. Liu, Z., Wu, J.: An $MAP/G/1$ G-queues with preemptive resume and multiple vacations. Appl. Math. Model. **33**, 1739–1748 (2009)
13. Lee, Y.: Discrete-time $Geo^X/G/1$ queue with preemptive resume priority. Math. Comput. Model. **34**, 243–250 (2001)
14. Liu, Z., Wu, J.: A discrete-time $Geo/G/1$ retrial queue with preemptive resume and collisions. Appl. Math. Model. **35**, 837–847 (2011)
15. Hellerstein J, & Diao Y, & Parekh S, & Tilbury D. M.: Feedback control of computing systems. Wiley-IEEE, (2004).
16. Atencia, I., Moreno, P.: Discrete-time $Geo^X/G_H/1$ retrial queue with Bernoulli feedback. Computers and Mathematics with Applications **47**, 1273–1294 (2004)
17. Fuhrmann, S.W., Cooper, R.B.: Stochastic decompositions in $M/G/1$ queue with generalized vacations. Operations Research **33**, 1117–1129 (1985)

Stability Analysis of T-S Fuzzy Systems with Knowledge on Membership Function Overlap

Zhi-hong Miao

Abstract A new approach for reducing the conservativeness in stability analysis and design of continuous-time T-S fuzzy control systems is proposed in this paper. Based on a fuzzy Lyapunov function together with knowledge on membership function overlap, previous stability and stabilization conditions are relaxed by the proposed approach. Both the stability and the stabilization conditions are written as linear matrix inequality (LMI) problems. Two examples are given to illustrate the effectiveness of the proposed approach.

Keywords Fuzzy Lyapunov function · Parallel distributed compensation (PDC) · T-S fuzzy model · Linear matrix inequality (LMI) · Nonlinear system

1 Introduction

Recently, Takagi-Sugeno (T-S) fuzzy control system [1] has been receiving increasing attention. The main reason is that this particular form can not only provides a general framework to represent the nonlinear system, but also offers an effective platform to facilitate the stability analysis and controller synthesis. Many beneficial results have been obtained in this research field in the past decade [2–4].

Generality, stability and stabilization of the T-S fuzzy system are usually investigated by using the direct Lyapunov method and stability conditions derived can be given in terms of linear matrix inequalities (LMIs). But this approach requires to find a common positive definite matrix solution for all local linear systems, which leads that the designed conditions are more conservative. The more the number of

Z. Miao (✉)
Department of Fire Protection Engineering, The Chinese People's Armed Police Force Academy, Hebei 065000 , Langfang, China
e-mail: miaozhh@21cn.com

B.-Y. Cao and H. Nasseri (eds.), *Fuzzy Information & Engineering and Operations Research & Management*, Advances in Intelligent Systems and Computing 211,
DOI: 10.1007/978-3-642-38667-1_54, © Springer-Verlag Berlin Heidelberg 2014

the fuzzy rules, the less the possibility of finding a common positive definite matrix solution. Certainly, many results have been obtained for improving these designed conditions [4–10].

Currently, an effective way to reduce the conservativeness has been proposed by using the fuzzy Lyapunov function [3] or piecewise Lyapunov functions instead of a common quadratic Lyapunov function. However, stabilization conditions for fuzzy Lyapunov functions and piecewise Lyapunov functions are in terms of bilinear matrix inequalities (BMIs) in general. Although, BMI conditions can be converted into LMI conditions by the way of the well-known completing square technique, in general such a conversion leads to conservative results. Another problem arising up in fuzzy Lyapunov approach is that the conditions of stability depend on the time derivative of the membership functions. To handle the difficulty, some strategies are proposed, for instance, by taking into account upper bounds the time derivative of the membership functions, then the stability conditions can be expressed as in terms of LMIs. Nevertheless, this approach does not take advantage of the knowledge on membership functions overlap. In [10] the stability condition is considered from the knowledge on membership functions overlap, but not using fuzzy Lyapunov function.

In this paper, we further extend the works in [8, 9] and [10], an improved approach which employs a fuzzy Lyapunov function is proposed to investigated the fuzzy control system. In this approach, taking advantage of the information of the derivatives and the overlap properties of membership functions, more relaxed the stability analysis and design conditions are achieved.

2 Takagi-Sugeno Fuzzy Model and Preliminaries

Consider the T-S fuzzy model described by the following rules:

$$\begin{aligned} &\mathbf{R}_i : \text{IF } z_1(t) \text{ is } F_1^i \text{ and } \cdots \text{ and } z_s(t) \text{ is } F_s^i \text{ THEN} \\ &\dot{\mathbf{x}}(t) = A_i\mathbf{x}(t) + B_i u(t), \ (i = 1, 2, \cdots, r) \end{aligned} \tag{1}$$

where $F_1^i, F_2^i, \cdots, F_s^i$ are fuzzy sets, $\mathbf{x}(t) = [x_1(t), x_2(t), \cdots, x_n(t)]^T$ is the state vector, $\mathbf{z}(t) = [z_1(t), z_2(t), \cdots, z_s(t)]^T \in R^s$ is the premise vector, $u(t) \in R^m$ is the control input vector, A_i and B_i the system matrices.

By using a singleton fuzzifier, product fuzzy inference and center-average defuzzifier, the system dynamics are described by

$$\dot{\mathbf{x}}(t) = \sum_{i=1}^{r} h_i(\mathbf{z}(t))[A_i\mathbf{x}(t) + B_i u(t)], \tag{2}$$

with $h_i(\mathbf{z}(t)) = \frac{w_i(\mathbf{z}(t))}{\sum_{i=1}^{r} w_i(\mathbf{z}(t))}$, $w_i(\mathbf{z}(t)) = \prod_{j=1}^{s} F_j^i(z_j(t))$. where $h_i(\mathbf{z}(t))$ are the normalized membership functions and the grade of membership of the premise variables in the respective fuzzy sets F_j^i are given as $F_j^i(z_j(t))$. The normalized membership functions satisfy the following properties

$$0 \leq h_i \leq 1, \sum_{i=1}^{r} h_i = 1, \sum_{i=1}^{r} \dot{h}_i = 0. \tag{3}$$

Based on the parallel distributed compensation (PDC) [2] method, the fuzzy state feedback controller as following is considered in this paper.

$$u(t) = -\sum_{i=1}^{r} h_i(\mathbf{z}(t))K_i\mathbf{x}(t) \tag{4}$$

where $K_i \in R^{m\times n}$ are the local linear state feedback gains. Substituting (4) into (2) yields

$$\begin{aligned}\dot{\mathbf{x}}(t) &= \sum_{i=1}^{r}\sum_{j=1}^{r} h_i(\mathbf{z}(t))h_j(\mathbf{z}(t))(A_i - B_iK_j)\mathbf{x}(t) = \sum_{i=1}^{r}\sum_{j=1}^{r} h_i(\mathbf{z}(t))h_j(\mathbf{z}(t))G_{ij}\mathbf{x}(t)\\ &= \left[\sum_{i=1}^{r} h_i^2 G_{ii} + 2\sum_{i=1}^{r}\sum_{i<j} h_ih_j\left(\frac{G_{ij}+G_{ji}}{2}\right)\right]\mathbf{x}(t)\end{aligned} \tag{5}$$

where $G_{ij} = A_i - B_iK_j$. when $u(t) \equiv 0$, then it becomes

$$\dot{\mathbf{x}}(t) = \sum_{i=1}^{r} h_i(\mathbf{z}(t))A_i\mathbf{x}(t), \tag{6}$$

In order to reduce the conservativeness of stability analysis and design, in [3], fuzzy Lyapunov functions was proposed as a more general alternative to the use of a common quadratic Lyapunov function, which can utilize some information of membership functions. The following Lemma gives some results

Lemma 1. *(Theorem 2 in [3]) Suppose that* $|\dot{h}_i| \leq \phi_i, (i = 1, 2, \cdots, r-1)$. *The T-S fuzzy system (6) is asymptotically stable if there exist symmetric matrices* $P_i > 0$, *such that*

$$\begin{aligned}&P_i \geq P_r, (i = 1, \cdots, r-1),\\ &\sum_{i=1}^{r} \phi_i(P_i - P_r) + \frac{1}{2}(A_i^T P_j + P_jA_i + A_j^T P_i + P_iA_j) < 0, (1 \leq i \leq j \leq r).\end{aligned}$$

Recently, Mozelli et al. proposed a less conservative result in [8], which improves above result.

Lemma 2. *(Theorem 6 in [8]) Suppose that* $|\dot{h}_i| \leq \phi_i, (i = 1, 2, \cdots, r-1)$. *The T-S fuzzy system (6)is asymptotically stable if there exist symmetric matrices* P_i, *such that*

$$P_i > 0, \ P_i + X \geq 0, (i = 1, \cdots, r)$$
$$\tilde{P}_\phi + \frac{1}{2}(A_i^T P_j + P_j A_i + A_j^T P_i + P_i A_j) < 0, (1 \leq i \leq j \leq r),$$

where $\tilde{P}_\phi = \sum_{k=1}^{r} \phi_k (P_k + X)$.

For a close-loop control system, to move forward a single step, the following Lemma can be obtained.

Lemma 3. *(Theorem 6 in [9]) Suppose that* $|\dot{h}_i| \leq \phi_i, (i = 1, 2, \cdots, r-1)$. *The T-S fuzzy system (6)is asymptotically stable if there exist symmetric matrices* Q_i, Z *and matrices* N, *such that*

(1) $Q_i > 0, (i = 1, \cdots, r)$, (2) $Q_i + Z \geq 0, (i = 1, \cdots, r)$
(3) $\Omega_{ii} < 0, (i = 1, \cdots, r)$, (4) $\Omega_{ij} + \Omega_{ji} < 0, (i < j)$

with

$$\Omega_{ij} = \begin{bmatrix} \Gamma_\phi - (A_i N + N^T A_i^T - B_i W_j - W_j^T B_i^T) & Q_i - \mu(N^T A_i^T - S_j B_i^T) + N \\ Q_i - \mu(A_i N - B_i W_j) + N^T & \mu(N + N^T) \end{bmatrix}.$$

where $\Gamma_\phi = \sum_{k=1}^{r} \phi_k (Q_k + Z)$. *in this case, the fuzzy controller gain matrices can be given by* $K_i = W_i N^{-1}$.

The following Lemma set up a less conservative stability condition by taking into account the knowledge of membership function overlap, but not using fuzzy Lyapunov function.

Lemma 4. *(Theorem 3 in [10]) Assume that* $0 \leq h_i h_j \leq \beta_{ij}$ $(i \leq j)$, *the T-S fuzzy system (6) is asymptotically stable if there exist symmetric matrices* P, R_{ij}, $(i \leq j)$, *and matrices* $X_{ij} = X_{ji}^T$, *such that*

$$P > 0, \ R_{ij} \geq 0, \ (i \leq j)$$
$$(G_{ii}^T P + P G_{ii}) - R_{ii} + \Lambda < X_{ii}, (i = 1, \cdots, r)$$
$$(G_{ij}^T P + P G_{ij}) + (G_{ji}^T P + P G_{ji}) - R_{ij} + 2\Lambda < X_{ij} + X_{ji}, (i < j)$$
$$\begin{bmatrix} X_{11}, \cdots X_{1r} \\ \vdots \ \ \ddots \ \ \vdots \\ X_{r1}, \cdots X_{rr} \end{bmatrix} < 0.$$

where $\Lambda = \sum_{k=1}^{r} \sum_{k \leq l \leq r} \beta_{kl} R_{kl}$.

3 Main Results

In this section, some sufficient conditions on stability and stabilization of fuzzy system are presented.

Firstly, for reducing the number of LMIs, we introduce matrices as follows

$$U_{ij} = \begin{cases} \beta_{ii} E - E_{ii}, & i = j \\ \beta_{ij} E - \frac{1}{2} E_{ij}, & i \neq j \end{cases} \tag{7}$$

here, $E \in R^{r \times r}$, and each element of E is 1. $E_{ij} \in R^{r \times r}$, and the i-th row and the j-th column element of E_{ij} is 1, other elements are 0, i.e.

$$E = \begin{bmatrix} 1 & \cdots & 1 \\ \vdots & \ddots & \vdots \\ 1 & \cdots & 1 \end{bmatrix}_{r \times r}, \quad E_{ij} = \begin{bmatrix} 0 & \cdots & 0 \\ & & 1 \\ & & \\ & 1 & \\ 0 & \cdots & 0 \end{bmatrix}_{r \times r},$$

If the normalized membership functions satisfy: $0 \leq h_k h_l \leq \beta_{kl} \leq 1$, then when $k \neq l$, we have

$$\begin{aligned} \beta_{kl} - h_k h_l &= \beta_{kl}(\sum_{i=1}^{r}\sum_{j=1}^{r} h_i h_j) - h_k h_l = \beta_{kl}(\sum_{i=1}^{r} h_i^2) + 2\beta_{kl}\sum_{i<j} h_i h_j - h_k h_l \\ &= \beta_{kl}\sum_{i=1}^{r} h_i^2 + 2\beta_{kl}\sum_{i \neq k, j \neq l}^{i<j} h_i h_j + 2(\beta_{kl} - \frac{1}{2}) h_k h_l \geq 0. \end{aligned}$$

With noting: $\bar{h} = [h_1, h_2, \cdots, h_r]^T$, above inequalities can be expressed as follows

$$\bar{h}^T U_{kl} \bar{h} \geq 0. \tag{8}$$

When $k = l$, that share same matrix inequalities.

3.1 Stability Analysis

Firstly, consider stability problem of the unforced T-S fuzzy system

$$\dot{\mathbf{x}}(t) = \sum_{i=1}^{r} h_i(\mathbf{z}(t)) A_i \mathbf{x}(t), \tag{9}$$

Theorem 1. *Assume that* $|\dot{h}_i| \leq \phi_i$, $0 \leq h_i h_j \leq \beta_{ij}$, $(i = 1, 2, \cdots, r, i \leq j)$, *the T-S fuzzy system (6) is asymptotically stable if there exist symmetric matrices* P_i, X, *and matrices* $Y_{ij} = Y_{ji}^T$, *scalars* $\eta_{ij} > 0$, $(i \leq j)$, *such that*

(1) $P_i > 0,\ P_i + X \geq 0, (i = 1, \cdots, r)$
(2) $\tilde{P}_\phi + (A_i^T P_i + P_i A_i) < Y_{ii}, (i = 1, \cdots, r)$
(3) $\tilde{P}_\phi + \frac{1}{2}(A_i^T P_j + P_j A_i + A_j^T P_i + P_i A_j) < Y_{ij} + Y_{ji}, (1 \leq i \leq j \leq r)$.
(4) $\begin{bmatrix} Y_{11}, & \cdots & Y_{1r} \\ \vdots & \ddots & \vdots \\ Y_{r1}, & \cdots & Y_{rr} \end{bmatrix} + \sum_{1 \leq i \leq j \leq r} \eta_{ij} \tilde{U}_{ij} < 0.$

where $\tilde{P}_\phi = \sum_{k=1}^r \phi_k (P_k + X)$, $\tilde{U}_{ij} = U_{ij} \otimes I_n$ *(*$\otimes$ *is Kronecker product,* I_n *is an identity matrix of order* n*).*

Proof: Consider a candidate fuzzy Lyapunov function

$$V(t) = \mathbf{x}^T(t) \sum_{i=1}^{r} h_i P_i \mathbf{x}(t).$$

The time derivative of (10) along the trajectories of (9) is

$$\dot{V}(t) = \mathbf{x}^T(t) \sum_{i=1}^{r} \dot{h}_i P_i \mathbf{x}(t) + \frac{1}{2} \sum_{i=1}^{r} \sum_{j=1}^{r} h_i h_j (A_i^T P_j + P_j A_i + A_j^T P_i + P_i A_j).$$

Note that $\sum_{i=1}^r \dot{h}_i X = 0$, we have

$$\begin{aligned}
\dot{V}(t) &\leq \mathbf{x}^T \sum_{i=1}^{r} \phi_i (P_i + X)\mathbf{x} + \frac{1}{2}\mathbf{x}^T \sum_{i=1}^{r} \sum_{j=1}^{r} h_i h_j (A_i^T P_j + P_j A_i + A_j^T P_i + P_i A_j)\mathbf{x} \\
&= \mathbf{x}^T \sum_{i=1}^{r} h_i^2 [\tilde{P}_\phi + A_i^T P_i + P_i A_i]\mathbf{x} \\
&\quad + 2\mathbf{x}^T \sum_{i=1}^{r} \sum_{i<j}^{r} h_i h_j [\tilde{P}_\phi + \frac{1}{2}(A_i^T P_j + P_j A_i + A_j^T P_i + P_i A_j)]\mathbf{x} \\
&\quad < \mathbf{x}^T \sum_{i=1}^{r} h_i^2 Y_{ii} \mathbf{x} + 2\mathbf{x}^T \sum_{i=1}^{r} \sum_{i<j}^{r} h_i h_j (Y_{ij} + Y_{ji})\mathbf{x}, (\forall \mathbf{x} \neq \mathbf{0})
\end{aligned}$$

Set $H = \bar{h} \otimes I_n$, above inequality can be rewritten as

$$\dot{V}(t) < \mathbf{x}^T H^T \begin{bmatrix} Y_{11}, & \cdots & Y_{1r} \\ \vdots & \ddots & \vdots \\ Y_{r1}, & \cdots & Y_{rr} \end{bmatrix} H\mathbf{x} = \mathbf{x}^T H^T \left\{ \begin{bmatrix} Y_{11}, & \cdots & Y_{1r} \\ \vdots & \ddots & \vdots \\ Y_{r1}, & \cdots & Y_{rr} \end{bmatrix} + \sum_{i \le j} \mu_{ij} \tilde{U}_{ij} \right\} H\mathbf{x}$$
$$- \mathbf{x}^T \sum_{i=1} \sum_{i \le j} \mu_{ij} H^T \tilde{U}_{ij} H\mathbf{x} \ \ (\forall \mathbf{x} \neq \mathbf{0})$$

From the condition (4) in this theorem, we have

$$\dot{V}(t) < -\mathbf{x}^T \sum_{i=1} \sum_{i \le j} \mu_{ij} H^T \tilde{U}_{ij} H\mathbf{x}, \ (\forall \mathbf{x} \neq \mathbf{0}).$$

From the properties of Kronecker product, we obtain

$$\begin{aligned} H^T \tilde{U}_{ij} H &= (\bar{h}^T \otimes I_n)(U_{ij} \otimes I_n)(\bar{h} \otimes I_n) = [(\bar{h}^T U_{ij}) \otimes I_n](\bar{h} \otimes I_n) \\ &= (\bar{h}^T U_{ij} \bar{h}) \otimes I_n = (\bar{h}^T U_{ij} \bar{h}) I_n. \end{aligned}$$

Therefore

$$\dot{V}(t) < -\sum_{i=1} \sum_{i \le j} \mu_{ij} \bar{h}^T U_{ij} \bar{h} \mathbf{x}^T \mathbf{x}. (\forall \mathbf{x} \neq \mathbf{0})$$

Note that the property (8), above inequality implies that fuzzy system is asymptotically stable. This completes the proof.

The following example illustrating the effectiveness of the proposed approach.

Example 1. Consider the same T-S fuzzy system give in [8]:

$$A_1 = \begin{bmatrix} -5 & -4 \\ -1 & a \end{bmatrix}, A_2 = \begin{bmatrix} -4 & -4 \\ \frac{3b-2}{5} & \frac{3a-4}{5} \end{bmatrix}, A_3 = \begin{bmatrix} -3 & -4 \\ \frac{2b-3}{5} & \frac{2a-6}{5} \end{bmatrix}, A_4 = \begin{bmatrix} -2 & -4 \\ b & -2 \end{bmatrix}.$$

where $a \in [-20, 0]$ and $b \in [0, 600]$. The membership function of the fuzzy sets F_j^i are

$$\alpha_i(x_i) = \begin{cases} (1 - \sin(x_i))/2, & |x_i| \le \pi/2 \\ 0, & x_i > \pi/2, \\ 1, & x_i < -\pi/2. \end{cases}, \ \beta_i(x_i) = 1 - \alpha_i(x_i).$$

The normalized membership functions are

$$h_1 = \alpha_1(x_1)\alpha_2(x_2), h_2 = \alpha_1(x_1)\beta_2(x_2), h_3 = \beta_1(x_1)\alpha_2(x_2), h_4 = \beta_1(x_1)\beta_2(x_2).$$

After some calculating, we have

$$\beta_{ii} = 1, (i = 1, 2, 3, 4), \beta_{12} = \beta_{13} = \beta_{24} = \beta_{34} = 0.25, \beta_{14} = \beta_{23} = 0.0624.$$

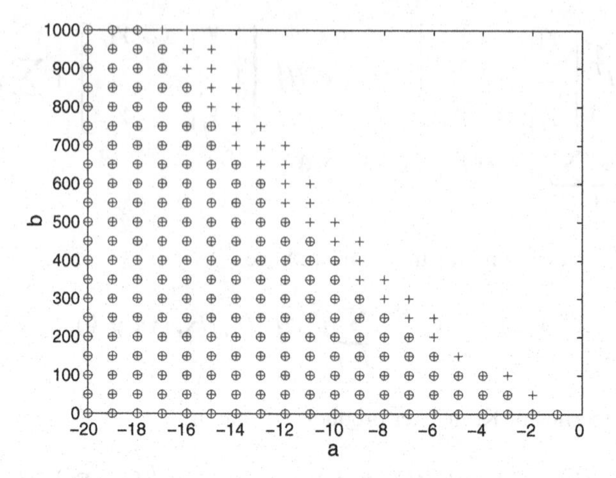

Fig. 1 The regions of feasible solution for Lemma 2(⊕) and Theorem 1(⊕, +)

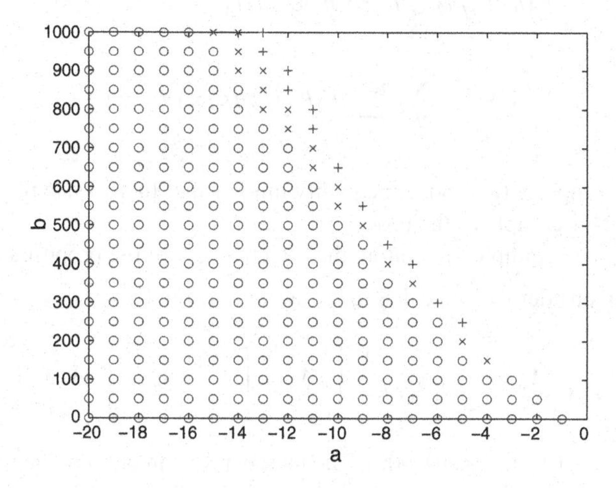

Fig. 2 The regions of feasible solution for Theorem 1((1):∘; (2): ∘, ×;(3):∘, ×, +)

The stability of this fuzzy system is checked by using Lemma 2(in) and the proposed approach in Theorem 1, respectively, with assuming $\phi_k = 0.85$. Using Matlab LMI toolbox, the feasible solutions are obtained. Figure 1 shows that the proposed approach yields a larger stable region than Lemma 2. It implies that conservativeness of stability condition is reduced by the proposed approach.

In next experiment, we change the upper bound of the normalized membership functions, Fig. 2 shows the feasible regions for Theorem 1 with three sets of the upper bound, i.e. (1) $\beta_{14} = \beta_{23} = 0.0624$, (2) $\beta_{14} = \beta_{23} = 0.02$; (3) $\beta_{14} = \beta_{23} = 0.002$.

From the Fig. 2, we know that if the fire degree of the overlap between the difference membership functions is lower, then conservativeness of the stability condition will be less. On the other hand, there are only $r(r+1)/2$ scalars μ_{ij} which represent the knowledge of membership functions overlap in Theorem 1.

Moveover, the conditions (2), (3) and (4) in Theorem 1 can can be rewritten as

(2a) $\tilde{P}_\phi + (A_i^T P_i + P_i A_i) + \Lambda < Y_{ii}, (i = 1, \cdots, r)$

(3a) $\tilde{P}_\phi + \frac{1}{2}(A_i^T P_j + P_j A_i + A_j^T P_i + P_i A_j) + 2\Lambda < Y_{ij} + Y_{ji}, (1 \leq i \leq j \leq r)$.

(4a) $\begin{bmatrix} Y_{11}, & \cdots & Y_{1r} \\ \vdots & \ddots & \vdots \\ Y_{r1}, & \cdots & Y_{rr} \end{bmatrix} < 0.$

where $\Lambda = \sum_{k=1}^r \sum_{k \leq l \leq r} \beta_{kl} R_{kl}$.

3.2 State Feedback Design

In this section, we will discuss the design condition of the PDC fuzzy controller for the close-loop system (5).

Theorem 2. *Assume that* $|\dot{h}_i| \leq \phi_i$, $0 \leq h_i h_j \leq \beta_{ij}$, $(i = 1, 2, \cdots, r, i \leq j)$, $\mu > 0$ *is a given scalar, the T-S fuzzy system (5) is asymptotically stable if there exist symmetric matrices* Q_i, Z, *and matrices* $Y_{ij} = Y_{ji}^T$, N, W_j, *and scalars* $\mu_{ij} > 0$, $(i \leq j)$

(1) $Q_i > 0,\ Q_i + Z \geq 0, (i = 1, \cdots, r)$

(2) $\Omega_{ii} \leq Y_{ii}, (i = 1, \cdots, r)$

(3) $\Omega_{ij} + \Omega_{ji} \leq Y_{ij} + Y_{ji}, (i < j)$,

(4) $\tilde{Y} + \sum_{1 \leq i \leq j \leq r} \mu_{ij} \tilde{U}_{ij} < 0$.

with

$$\Omega_{ij} = \begin{bmatrix} \Gamma_\phi - (A_i N + N^T A_i^T - B_i W_j - W_j^T B_i^T), & Q_i + N - \eta(N^T A_i^T - W_j^T B_i^T) \\ Q_j + N^T - \eta(A_i N - B_i W_j) & \eta(N + N^T) \end{bmatrix},$$

$$\tilde{Y} = \begin{bmatrix} Y_{11} & \cdots & Y_{1r} \\ \vdots & \ddots & \vdots \\ Y_{r1} & \cdots & Y_{rr} \end{bmatrix}, \quad \Gamma_\phi = \sum_{k=1}^r \phi_k (Q_k + Z), \quad \tilde{U}_{ij} = U_{ij} \otimes I_{2n}.$$

In this case, the control gains are given by $K_i = W_i N^{-1} (i = 1, 2, \cdots, r)$.

Proof: Consider a candidate fuzzy Lyapunov function as follows

$$V = \mathbf{x}^T \sum_{i=1}^r h_i P_i \mathbf{x}. \tag{10}$$

Its time-derivative is

$$\dot{V} = \mathbf{x}^T \sum_{k=1}^{r} \dot{h}_k P_k \mathbf{x} + \mathbf{x}^T \sum_{i=1}^{r} h_i P_i \dot{\mathbf{x}} + \dot{\mathbf{x}}^T \sum_{j=1}^{r} h_j P_j \mathbf{x}$$

Let X be a symmetric matrix, from the conditions $|\dot{h}_i| \le \phi_i, \sum_{i=1}^{r} h_i = 1$ and $\sum_{i=1} \dot{h}_i X = 0$. Consider the following null product which is similar to the procedure in [9].

$$2(\mathbf{x}^T M + \dot{\mathbf{x}}^T \mu M)(\dot{\mathbf{x}} - \sum_{i=1}^{r} \sum_{j=1}^{r} h_i h_j G_{ij} \mathbf{x}) = 0.$$

where M is a slack matrix variable, and $\mu > 0$ is a scalar. Then, we have

$$\begin{aligned} \dot{V} &\le \mathbf{x}^T \sum_{k=1}^{r} \phi_k (P_k + X) \mathbf{x} + \mathbf{x}^T \sum_{i=1}^{r} h_i P_i \dot{\mathbf{x}} + \dot{\mathbf{x}}^T \sum_{j=1}^{r} h_j P_j \mathbf{x} \\ &+ 2(\mathbf{x}^T M + \dot{\mathbf{x}}^T \mu M)(\dot{\mathbf{x}} - \sum_{i=1}^{r} \sum_{j=1}^{r} h_i h_j G_{ij} \mathbf{x}) \end{aligned}$$

After some operation, we have

$$\begin{aligned} \dot{V} \le & \sum_{i=1}^{r} \sum_{j=1}^{r} h_i h_j [\mathbf{x}^T \sum_{k=1}^{r} \phi_k (P_k + X) \mathbf{x} \\ & - \mathbf{x}^T (M G_{ij} + G_{ij}^T M^T) \mathbf{x} + \mathbf{x}^T (P_i + M - \mu G_{ij}^T M^T) \dot{\mathbf{x}} \\ & + \dot{\mathbf{x}}^T (P_j + M^T - \mu M G_{ij}) \mathbf{x} + \dot{\mathbf{x}}^T \mu (M + M^T) \dot{\mathbf{x}}] \end{aligned} \tag{11}$$

$$= \sum_{i=1}^{r} \sum_{j=1}^{r} h_i h_j [\mathbf{x}^T, \dot{\mathbf{x}}^T] \begin{bmatrix} \tilde{P}_\phi - (M G_{ij} + G_{ij}^T M^T) & P_i + M - \mu G_{ij}^T M^T \\ P_j + M^T - \mu M G_{ij} & \mu (M + M^T) \end{bmatrix} \begin{bmatrix} \mathbf{x} \\ \dot{\mathbf{x}} \end{bmatrix} \tag{12}$$

By introducing a new vector $\mathbf{z} = \begin{bmatrix} M^T & \\ & M^T \end{bmatrix} \begin{bmatrix} \mathbf{x} \\ \dot{\mathbf{x}} \end{bmatrix}$, The expression (11) can be rewritten as

$$\dot{V} \le \sum_{i=1}^{r} \sum_{j=1}^{r} h_i h_j \mathbf{z}^T \tag{13}$$

$$\begin{bmatrix} M^{-1} \tilde{P}_\phi M^{-T} - (G_{ij} M^{-T} + M^{-1} G_{ij}^T), & M^{-1} P_i M^{-T} + M^{-T} - \mu M^{-1} G_{ij}^T \\ M^{-1} P_j M^{-T} + M^{-1} - \mu G_{ij} M^{-T} & \mu (M^{-T} + M^{-1}) \end{bmatrix} \mathbf{z}.$$

Let

$$Q_k = M^{-1}P_kM^{-T}, Z = M^{-1}XM^{-T}, W_j = K_jM^{-T}, N = M^{-T},$$

$$\Gamma_\phi = M^{-1}\tilde{P}_\phi M^{-T} = M^{-1}\sum_{k=1}^{r}\phi_k(P_k + X)M^{-T} = \sum_{k=1}^{r}\phi_k(Q_k + Z).$$

The expression (13) becomes

$$\dot{V} \le \sum_{i=1}^{r}\sum_{j=1}^{r} h_i h_j \mathbf{z}^T \tag{14}$$

$$\begin{bmatrix} \Gamma_\phi - (A_iN + N^TA_i - B_iW_j - W_j^TB_i^T), & Q_i + N - \eta(N^TA_i^T - W_j^TB_i^T) \\ Q_j + M - \eta(A_iN - B_iW_j) & \eta(N + N^T) \end{bmatrix}\mathbf{z}$$

$$= \mathbf{z}^T\sum_{i=1}^{r}\sum_{j=1}^{r} h_i h_j \Omega_{ij}\mathbf{z} = \mathbf{z}^T\sum_{i=1}^{r}[h_i^2\Omega_{ii} + \sum_{i=1}^{r}\sum_{i<j} h_i h_j(\Omega_{ij} + \Omega_{ji})]\mathbf{z}.$$

Applying the conditions of this theorem to above expression

$$\dot{V} < \mathbf{z}^T\sum_{i=1}^{r}[h_i^2Y_{ii} + \sum_{i=1}^{r}\sum_{i<j} h_i h_j(Y_{ij} + Y_{ji})]\mathbf{z}, \quad (\forall \mathbf{z} \ne \mathbf{0})$$

Let $H = \bar{h} \otimes I_{2n}$, above inequality can be rewritten as

$$\begin{aligned} \dot{V} &< \mathbf{z}^TH^T\tilde{Y}H\mathbf{z} = \mathbf{z}^TH^T[\tilde{Y} + \sum_{i=1}\sum_{i\le j}\mu_{ij}\tilde{U}_{ij}]H\mathbf{z} - \sum_{i=1}\sum_{i\le j}\mu_{ij}\mathbf{z}^TH^T\tilde{U}_{ij}H\mathbf{z} \\ &< -\sum_{i=1}\sum_{i\le j}\mu_{ij}\mathbf{z}^TH^T\tilde{U}_{ij}H\mathbf{z}.(\forall \mathbf{z} \ne \mathbf{0}) \end{aligned}$$

After applying the property (8), we have

$$\dot{V} < -\sum_{i=1}\sum_{i\le j}\mu_{ij}\mathbf{z}^TH^T\tilde{U}_{ij}H\mathbf{z} = -\sum_{i=1}\sum_{i\le j}\mu_{ij}\bar{h}^TU_{ij}\bar{h}\mathbf{z}^T\mathbf{z} < 0. \ (\forall \mathbf{z} \ne \mathbf{0})$$

Then the fuzzy system is asymptotically stable. This completes the proof.

The conditions of Theorem 2 are formed as linear matrix inequalities with matrix variables Q_i, W_i, N, Z, and Y_{ij}, scalars μ_{ij}, which can be efficiently solved by using convex optimization techniques.

4 Conclusion

For a class of continuous-time T-S fuzzy systems, the knowledge on membership functions overlap has been considered in stability analysis via a fuzzy Lyapunov function. The previous stability and stabilization conditions have been relaxed by the proposed approach. Both the stability and the stabilization conditions have been written as linear matrix inequality (LMI) problems.

References

1. Takagi, T., Sugeno, M.: Fuzzy identification of systems and its application to modeling and control. IEEE Trans. Sys. Man. Cybern. **15**(1), 116–132 (1985)
2. Wang, H.O., Tanaka, K., Griffin, M.: An approach to fuzzy control of nonlinear systems: stability and design issues. IEEE Trans on Fuzzy Sys **4**(1), 14–23 (1996)
3. Tanaka, K., Hori, T., Wang, H.O.: A multiple Lyapunov function approach to stabilization of fuzzy control systems. IEEE Trans. on Fuzzy Sys. **11**(4), 582–589 (2003)
4. Kim, E., Lee, H.: New approaches to relaxed quadratic stability condition of fuzzy control systems. IEEE Trans. Fuzzy Sys. **8**(5), 523–534 (2000)
5. Liu, X.D., Zhang, Q.L.: New approaches to H_∞ controller designs based on fuzzy observers for T-S fuzzy systems via LMI. Automatica **39**(9), 1571–1582 (2003)
6. Jonhansson, M., Rantzer, A., Arzen, K.E.: Piecewise quadratic stability of fuzzy systems. IEEE Trans. on Fuzzy Sys. **7**(6), 713–722 (1999)
7. Rhee, B.J., Won, S.: A new Lyapunov function approach for a Takagi-Sugeno fuzzy control system design. Fuzzy Sets and Sys. **157**(9), 1211–1228 (2006)
8. Mozelli, L.A., Palhares, R.M., Souza, F.O., Mendes, E.M.: Reducing conservativeness in recent stability conditions of TS fuzzy systems. Automatica **45**(6), 1580–1583 (2009)
9. Mozelli, L.A., Palhares, R.M., Avellar, G.S.C.: A systematic approach to improve multiple Lyapunov function stability and stabilization conditions for fuzzy systems. Information Sciences **179**(8), 1149–1162 (2009)
10. Sala, A., Arino, C.: Relaxed stability and performance conditions for Takagi-Sugeno fuzzy systems wit konowledge on membership function overlap. IEEE Trans. Syst. Man Cybern. B Cybern. **37**(3), 727–732 (2007)
11. Lee, D.H., Park, J.B., Joo, Y.H.: A New fuzzy Lyapunov function for relaxed stability condition of continuous-time Takagi-Sugeno fuzzy systems. IEEE Trans. on Fuzzy Sys. **19**(4), 785–791 (2011)

Application of T-S Fuzzy Model in Candidate-well Selection for Hydraulic Fracturing

Xie Xiang-jun and Yu Ting

Abstract Hydraulic fracturing (HF) is the key technology of increasing production and injection for low permeable reservoirs. The candidate-well selection for HF is essential to oil and gas wells stimulation potential evaluation, which is crucial to improve fracturing operation efficiency and reduce HF investment risk. The candidate-well selection model is a high dimension, nonlinear, strong coupling, multi-input single-output system. However, the conventional methods, such as production performance comparisons can not be easy to use for this nonlinear model. As a solution, the advanced methods such as T-S models in this paper can be effectively used in the candidate-well selection for HF. First, the subtractive clustering (SC) algorithm is employed to partition the fuzzy space of the given input–output data, which is adopted as the initial premise structure and parameters. Second, the clusters obtained on the first stage are used to initialize the fuzzy c-means (FCM) algorithm, which can obtain optimal cluster number and cluster centers. Third, the consequent parameters are identified by using the orthogonal least-squares (OLS) algorithm. Finally, the proposed approach is successfully applied to candidate-well selection for HF in Hechuan gas field in Sichuan basin, and validation results have demonstrated the effectiveness of the proposed method.

Keywords Hydraulic fracturing · T-S fuzzy model · Stimulation potential · Candidate-well selection.

1 Introduction

The formation of Hechuan in Sichuan basin is a typical gas reservoir with low permeability and low porosity. Since the geological condition is so complex, natural flow yields low production. However, such reservoirs are capable of producing at

X. Xiang (✉) · Y. Ting
School of Science, Southwest Petroleum University, Chengdu 610500, China
e-mail: xiangjunxie@126.com

B.-Y. Cao and H. Nasseri (eds.), *Fuzzy Information & Engineering and Operations Research & Management*, Advances in Intelligent Systems and Computing 211,
DOI: 10.1007/978-3-642-38667-1_55, © Springer-Verlag Berlin Heidelberg 2014

commercial rates with the help of HF technology. The candidate-well selection for HF is crucial. Therefore, it is evident that to adopt this technology, considerable efforts have to be made in candidate-well selection. The model of candidate-well selection is a high dimension, nonlinear, strong coupling, multi-input single-output system. However, the conventional methods, including production performance comparisons, pattern recognition technology, production type curve matching [1–3], can not be easy to use for this nonlinear model. On the other hand, advanced methods such as T-S fuzzy systems have been proved to be useful in complex nonlinear system [4], especially, showing excellent ability in describing complicated dynamic of nonlinear behaviors of a process [5]. Therefore, T-S fuzzy model may be a good choice to describe such systems.

In identification of fuzzy models, in order to obtain model structure and parameter estimation, one of the most popular approaches is the widely used FCM algorithm. However, the FCM algorithm could not guarantee unique clustering result because initial cluster number and fuzzy c-partition matrix are chosen randomly. To solve the problem, an initialization method for the FCM algorithm based on the SC algorithm is proposed. In the paper, the OLS algorithm is proposed for the identification of the consequent parameters.

The paper is organized as follows. In Section 2, T-S fuzzy model is proposed. Section 3 figures obtainment of the initial clustering number and cluster centers of the FCM algorithm through SC algorithm. Section 4 describes the identification of the premise structure and parameter using FCM algorithm. Section 5 represents identification of the consequent parameter using OLS algorithm. Next, we deal with the problem of candidate-well selection for HF in Hechuan region.

2 Fuzzy Model

T-S fuzzy model, proposed in [6], is described by several fuzzy If-Then rules which locally represent linear relations of the input and output. Multi-input single-output (MISO) model is of the following form:

$$\begin{aligned} R_k : &\text{ If } x_1 \text{ is } A_{k1} \text{ and } x_2 \text{ is } A_{k2} \text{ and } \cdots \text{ and } x_m \text{ is } A_{km} \\ &\text{Then } y_k = b_{k0} + b_{k1}x_1 + b_{k2}x_2 + \cdots + b_{km}x_m \end{aligned} \tag{1}$$

where $k = 1, 2, 3, \cdots, c$, c is the number of fuzzy rules, $x_i (i = 1, 2, \cdots, m)$ is the i-th input variable, m is the dimension of input variables, A_{ki} is a fuzzy set, premise identification is to obtain the membership function of the fuzzy set, y_k is the i-th output, b_{ki} is the consequent parameters.

Given an input $(x_1^*, x_2^*, \cdots, x_m^*)$, the final output is inferred by a weighted mean defuzzification as follows:

$$y = \sum_{k=1}^{c} (\tau_k \cdot y_k) = \sum_{k=1}^{c} (\tau_k \cdot (b_{k0} + b_{k1}x_1^* + \cdots + b_{km}x_m^*)) \tag{2}$$

where the weight τ_k is the contribution degree function of the k-th rule of the premise for the output of the overall model.

3 FCM Algorithm

The fuzzy c-means (FCM) algorithm [7] is one of the most popular fuzzy clustering algorithms and it has widely been used in object segmentation [8], data preprocessing [9] and pattern recognition [10], Which is applied in premise structure and parameter identification in this paper.

Let $X=\{x_1,x_2,\cdots,x_n\} \subset R^m$, where n is the number of data points and m is the dimension of each vector x_k. Suppose c is the cluster number, the FCM algorithm is an iterative optimization that minimizes the following objective function:

$$J(U,V) = \sum_{j=1}^{c}\sum_{h=1}^{n} (u_{jh})^P (d_{jh})^2 \tag{3}$$

where:

P: the weighting exponent (typically p=2)

V: the cluster center vertor, $V = \{v_1, v_2, \cdots, v_c\}$

U : the fuzzy membership functions matrix, $U = (u_{jh})_{c\times n}$

u_{jh} : the value of the membership function of the h-th data point belonging to the j-th cluster center, $u_{jh}$is defined by

$$u_{jh} = \frac{1}{\sum_{t=1}^{c} (\frac{\|v_j - x_h\|}{\|v_t - x_h\|})^{\frac{2}{p-1}}}, j = 1, 2, \cdots, c \tag{4}$$

which is constrained with the following:

$$u_{jh} \in [0,1], \forall j, h \quad j = 1, 2, \cdots, c, h = 1, 2, \cdots, n \tag{5}$$

$$\sum_{j=1}^{c} u_{jh} = 1, \forall h \quad h = 1, 2, \cdots, n \tag{6}$$

$$0 < \sum_{h=1}^{n} u_{jh} < n, \forall j \quad j = 1, 2, \cdots, c \tag{7}$$

$d_{hj} : d_{hj} = \left\| x_h - v_j^l \right\|$, where denotes the distance from$x_h$to the cluster centre $v^l{}_j$ for the l-th iteration.

The cluster centers and fuzzy membership functions matrix are updated through an iterative process using Eqs. (8) and (9), respectively.

$$v_j^l = \sum_{h=1}^{n} (u_{jh}^l)^p x_k / \sum_{h=1}^{n} (u_{jh}^l)^p, j = 1, 2, \cdots, c \tag{8}$$

$$u_{jh}^{l+1} = 1 / \sum_{t=1}^{c} (\frac{d_{hj}}{d_{ht}})^{\frac{2}{p-1}}, j = 1, 2, \cdots, c \tag{9}$$

4 SC Algorithm

However, the FCM algorithm is sensitive to the initial value, which could not guarantee the best clustering result because initial cluster number and fuzzy c-partition matrix are chosen randomly. To solve the problem, an initialization method for the FCM algorithm based on the subtractive clustering (SC) algorithm is proposed [11]. SC algorithm, proposed in the literature [12], is one of the automated data-driven based methods for constructing the primary fuzzy models and has the benefit of avoiding the explosion of the rule base. In this paper, SC algorithm is used to find initial premise structure and parameters.

SC algorithm is an improved version of the mountain clustering method [13]. Consider a group of n data points$(\hat{x}_1, \hat{x}_2, \cdots, \hat{x}_n)$, where $\hat{x}_i$ is a vector of the feature space. SC algorithm labels each data point as a potential cluster center, the potential function of each data point can be assigned as follows:

$$D(i) = \sum_{j=1}^{m} \exp\left(-\frac{4||\hat{x}(i) - \hat{x}(j)||^2}{\gamma_\alpha^2}\right), (i = 1, 2, \cdots, n) \tag{10}$$

where$\gamma_\alpha$is a positive constant, which defines the neighborhood radius of each cluster center $\hat{x}(i)$. The density of surrounding data points is high, and these points have also high potential values.

Potential for each data point is calculated by the using of this potential function (10), and the one with the highest potential is selected as the first cluster center, after the k-th cluster center is determined, Let$\hat{x}^*(k)$be the k-th cluster center and$D^*(k)$be its potential. Next, the potential of remaining data point is revised by Eqs. (11), which quashes the potential for surrounding cluster centre points to be chosen as the next cluster.

$$D(i) = D(i) - D^*(k) \exp\left(-\frac{4||\hat{x}(i) - \hat{x}^*(k)||^2}{\gamma_\beta^2}\right) \tag{11}$$

The new cluster center is chosen as the point having the highest potential. Where$\gamma_\beta$is a positive constant. Typically$\gamma_\beta = 1.5\gamma_\alpha$, which is usually set to avoid obtaining closely spaced cluster centers.

This iteration procedure continues until stopping criteria is reached.

5 OLS Algorithm

Given an input $(x_1^*, x_2^*, \cdots, x_m^*)$, the final output of the model is calculated as follows:

$$\begin{aligned} y &= \sum_{k=1}^{c} \tau_k y_k = \sum_{k=1}^{c} \tau_k (b_{k0} + b_{k1}x_1 + b_{k2}x_2 + \cdots + b_{km}x_m) \\ &= [\tau_1, \tau_1 x_1, \cdots, \tau_1 x_m, \cdots, \tau_c, \tau_c x_1, \cdots, \tau_c x_m] \times [b_{10}, b_{11}, \cdots, b_{1m}, \cdots, b_{c0}, b_{c1}, \cdots, b_{cm}] \end{aligned} \tag{12}$$

Substituting n inputs into (12) yields matrix form (13).

$$Y = WB \tag{13}$$

where B is the consequent parameter.

The OLS algorithm is proposed for the identification of the consequent parameter B. This method can transform column vectors of matrix U into orthogonal columns u_i, which could be completed through Gram-Schmidt orthogonalization [14] as follows:

The matrix W is decomposed into

$$W = UR \tag{14}$$

where$U$ is a matrix with orthogonal columns u_i, and R is an upper triangular matrix with unity diagonal elements. Substituting (14) into (13) yields

$$Y = Ug \tag{15}$$

where g=RB, R is an invertible matrix.

The consequent parameter can be calculated by the using of the matrix relation

$$B = R^{-1}g \tag{16}$$

6 Application in Candidate-well Selection

The proposed identification approach is applied to candidate-well selection for HF in Hechuan region. In this paper, four attributes are given to forecast the wells deliverability after fracturing in Hechuan region: x_1: reservoir thickness, x_2: porosity,

Table 1 Clustering result of the FCM algorithm initialized by SC algorithm

Center cluster	x_1	x_2	x_3	x_4
1	12.4990	8.9873	50.6060	4.6733
2	16.2270	6.5983	39.8570	3.8323
3	19.6670	9.2518	64.5250	4.1126
4	11.2000	9.7002	85.3740	4.9502
5	29.0550	8.9350	53.3180	4.5605
6	24.5270	6.9376	36.4940	3.6664
7	8.7713	7.4699	31.6870	3.8774

x_3: gas saturation, x_4: structural feature parameters. The data set contains input-output data pairs of 45 fractured wells of Hechuan region, the number of data used for identification and prediction are 35 fractured wells and remaining 10 fractured wells, respectively. Table 1 show the center cluster by using a combination of FCM algorithm an SA.

Eq.(17) shows the identification result, the fuzzy model consists of seven If-Then rules.

$$\begin{aligned}
&R_1: \text{ If } x_1 \text{ is } 12.499 \text{ and } x_2 \text{ is } 8.9873 \text{ and } x_3 \text{ is } 50.606 \text{ and } x_4 \text{ is } 4.6733 \\
&\quad \text{Then } y_1 = 118.84 + 1.7828x_1 - 16.753x_2 - 0.4661x_3 + 2.7655x_4 \\
&R_2: \text{ if } x_1 \text{ is } 16.227 \text{ and } x_2 \text{ is } 6.5983 \text{ and } x_3\, is\, 39.857 \text{ and } x_4 \text{ is } 3.8323 \\
&\quad \text{Then } y_2 = -15.159 - 3.7744x_1 + 37.589x_2 - 3.4136x_3 - 0.83067x_4 \\
&R_3: \text{ if } x_1 \text{ is } 19.667 \text{ and } x_2 \text{ is } 9.2518 \text{ and } x_3\, is\, 64.525 \text{ and } x_4 \text{ is } 4.1126 \\
&\quad \text{Then } y_3 = 324.93 - 0.36133x_1 - 12.463x_2 - 2.63946x_3 - 7.6535x_4 \\
&R_4: \text{ if } x_1 \text{ is } 11.2 \text{ and } x_2 \text{ is } 9.7002 \text{ and } x_3 \text{ is } 85.374\, and\, x_4 \text{ is } 4.9502 \\
&\quad \text{Then } y_4 = 229.87 + 2.5752x_1 + 6.6395x_2 - 3.3085x_3 + 6.3284x_4 \\
&R_5: \text{ if } x_1 \text{ is } 29.055 \text{ and } x_2 \text{ is } 8.935 \text{ and } x_3 \text{ is } 53.318 \text{ and } x_4 \text{ is } 4.5605 \\
&\quad \text{Then } y_5 = -114.07 - 4.3485x_1 + 41.419x_2 - 4.574x_3 + 16.823x_4 \\
&R_6: \text{ if } x_1 \text{ is } 24.527 \text{ and } x_2 \text{ is } 6.9376 \text{ and } x_3 \text{ is } 36.494 \text{ and } x_4 \text{ is } 3.6664 \\
&\quad \text{Then } y_6 = -140.48 + 11.915x_1 - 70.034x_2 + 8.2184x_3 - 10.153x_4 \\
&R_7: \text{ if } x_1 \text{ is } 8.7713 \text{ and } x_2 \text{ is } 7.4699 \text{ and } x_3 \text{ is } 31.687 \text{ and } x_4 \text{ is } 3.8774 \\
&\quad \text{Then } y_7 = 54.793 - 1.0307x_1 + 10.691x_2 - 3.0819x_3 - 0.1965x_4
\end{aligned} \tag{17}$$

Where, contribution degree function of the k-th rule of the premise for the output of the overall model is defined as follows:

$$\tau^h(i) = \exp\left(-\frac{4\|\hat{x}(i) - x*(h)\|^2}{\gamma_\alpha^2}\right), h = 1, 2, \cdots, 7 \tag{18}$$

where, $x*(h)$is the$h$-th cluster center of fuzzy subspace.

Table 2 Comparison of forecasting production and actual production

Well name	y, $10^4m^3/d$	y^*, $10^4m^3/d$	Error, %
H001-15-x1	4.8182	4.9000	-1.67
H001-3-x2	3.3604	3.4100	-1.45
H124	8.0654	8.2300	-2.00
H4	0.6651	0.5400	23.16
H6	2.0906	1.9200	8.89
H1	4.8006	3.8500	24.69
H101	0.8178	0.6900	18.52
H7	1.5157	1.3900	9.04
H121	0.4869	0.7100	-31.41
H001-9-x1	0.9546	1.3200	-27.68

Table 2 shows the forecasting production y, actual production y^* and error values of 10 fractured gas wells in Hechuan region based on the proposed model.

The forecasting production agrees with the actual production in engineering accepting error, except H121 well, proving that the model in this paper is reasonable and reliable.

If the postfrac flow of the gas well is less than $0.6\times10^4m^3/d$, it is considered as low producing well and fracturing becomes unnecessary. Considering the model of precision, the low producing well is modified as $0.8\times10^4m^3/d$. The H4 well was suggested unfracturing since postfrac production forecast based on the model in this paper is less than $0.8\times10^4m^3/d$. However, the wells were still fracturing and the postfrac production was consistent with the forecast results.

7 Conclusion

In this paper, we have proposed T-S fuzzy model for candidate-well selection for HF in Hechuan region, which not only can conveniently realize the yield potential evaluation of the candidate well, but also can accurately forecast production after fracturing. It provides the theoretic basis for candidate-well selection in Hechuan area and also can extends to other oil fields.

References

1. Reeves, S. R., Bastian, P. A., Spivey, J. P., Flumerfelt, R. W., Mohaghegh, S. and Koperna, G. J.: Benchmarking of Restimulation Candidate Selection Techniques in Layered, Tight Gas Sand Formations Using Reservoir Stimulation. Paper SPE 63096 presented at the SPE Annual Technical Conference and Exhibition, 1–4 October, Dallas, Texas, USA (2000).

2. Reeves, S.R., Hill, D.G., Hopkins, C.W., Conway, M.W., Tiner, R.L., Mohaghegh, S.: Restimulation Technology for Tight Gas Sand Wells. Paper SPE 57492 presented at the SPE Annual Technical Conference and Exhibition. 3–6 October, Houston, Texas, USA (1999).
3. Zoveidavianpoor, M., Samsuri, A., Shadizadeh, S.R.: A review on conventional candidate-well selection for hydraulic fracturing in oil and gas wells. Int. J. Eng. Technol. **2**(1), 51–60 (2012).
4. Hellendoom, H., Driankov, D., (eds.): Fuzzy model identification: Selected approaches. Springer, Berlin (1997).
5. Xu, C.W., Lu, Y.Z.: Fuzzy model identification and self-learning for dynamic systems. IEEE Trans. Syst., Man, Cybern, 17(4), pp. 683–689 (1987).
6. Takagi, T., Sugeno, M.: Fuzzy identification of systems and its application to modeling and control. IEEE Trans. Syst., Man, Cybern., Part B Cybern. **15**(1), 116–132 (1985).
7. Bezdek, J.: Pattern recognition with fuzzy objective function algorithm. Plenum Press, New York (1981).
8. Chen, W.J., Giger, M.L., Bick, U.: A fuzzy c-means (FCM)-based approach for computerized segmentation of breast lesions in dynamic contrastenhanced MRI images. Acad. Radiol, 13 (1), pp. 63–72 (2006).
9. Shi, Y., Mizumoto, M., Yubazki, N., Otani, M.: Improvement of fuzzy rules generation based on fuzzy c-means clustering algorithm. J. Japan Soc. Fuzzy Theor. Syst. (inpress, in Japanese).
10. Chang, W.-Y., Yang, H.-T.: Application of fuzzy C-means clustering approach to partial discharge pattern recognition of Cast-Resin current transformers. Properties and applications of Dielectric Materials, 8th International Conference on Date of Conference, 372–375 (2006).
11. Yang, Q., Zhang, D., Tian, F.: An initialization method for fuzzy c-means algorithm using subtractive clustering. International conference on Intelligent Networks and Intelligent Systems, pp. 393–396 (2010).
12. Chiu, S.L.: Fuzzy model identification based on cluster estimation. J. Intell. Fuzzy Syst, 2(3), pp. 267–278 (1994).
13. Yager, R.R., Filev, D.P.: Approximate clustering via the mountain method. IEEE Trans. Syst., Man, Cybern 24(8), pp. 1279–1284 (1994).
14. Golub, G.H., Van Loan, C.F.: Matrix computations, 2nd ed. Johns Hopkins University Press, Baltimore (1989).